Linqiu Cao
Carrier-bound Immobilized Enzymes

Further Titles of Interest

J.-L. Reymond (Ed.)
Enzyme Assays
High-throughput Screening,
Genetic Selection and Fingerprinting

2006. ISBN 3-527-31095-9

A. Liese, K. Seelbach, C. Wandrey
Industrial Biotransformations

2nd Edition
2005. ISBN 3-527-31001-0

W. Aehle (Ed.)
Enzymes in Industry
Production and Applications

2nd Edition
2004. ISBN 3-527-29592-5

A. S. Bommarius, B. R. Riebel
Biocatalysis

2005. ISBN 3-527-31001-0

K. Drauz, H. Waldmann (Eds.)
Enzyme Catalysis in Organic Synthesis

2nd Edition
2002. ISBN 3-527-29949-1

Linqiu Cao

Carrier-bound Immobilized Enzymes

Principles, Applications and Design

WILEY-VCH Verlag GmbH & Co. KGaA

Author

Dr. Linqiu Cao
Aart van der Leeuwlaan 197
2624 PP Delft
The Netherlands

linqiu.cao@dmv-international.com

■ This book was carefully produced. Nevertheless, author and publisher do not warrant the information contained therein to be free of errors. Readers are advised to keep in mind that statements, data, illustrations, procedural details or other items may inadvertently be inaccurate.

Library of Congress Card No. applied for

British Library Cataloguing-in-Publication Data
A catalogue record for this book is available from the British Library

Bibliographic information published by Die Deutsche Bibliothek
Die Deutsche Bibliothek lists this publication in the Deutsche Nationalbibliografie; detailed bibliographic data is available in the Internet at http://dnd.ddb.de.

© 2005 WILEY-VCH Verlag GmbH & Co. KGaA, Weinheim
All rights reserved (including those of translation into other languages). No part of this book may be reproduced in any form – by photoprinting, microfilm, or any other means – nor transmitted or translated into a machine language without written permission from the publishers. Registered names, trademarks, etc. used in this book, even when not specifically marked as such, are not to be considered unprotected by law.

Printed in the Federal Republic of Germany
Printed on acid-free paper

Cover Design   SCHULZ, Grafik-Design, Fußgönheim
Typesetting   Fotosatz Detzner, Speyer
Printing   Strauss GmbH, Mörlenbach
Bookbinding   Litges & Dopf Buchbinderei, Heppenheim

ISBN-13: 978-3-527-31232-0
ISBN-10: 3-527-31232-3

Foreword

Enzymes are the biocatalysts of the living cell. Their excellent performance in cellular metabolism is due to their intrinsic catalytic properties. In addition, their activity may be further enhanced by functioning in distinct cellular compartments, e.g., within or attached to cellular membranes, or as multifunctional enzyme complexes such as a cellulose-degrading cellulosome.

Enzyme technologists have reinvented such natural forms of "enzyme immobilization". They have looked for their own bionic solutions to arrive at immobilized biocatalysts which can be used in analytical devices such as a glucose biosensor or in an industrial plant producing, e. g., chiral amines from racemic precursor material. These initial steps towards a benign "green technology" are presently gaining momentum. In fact, the emerging concept of "white biotechnology" (sustainable chemical processes built on renewable resources and biocatalysts, carried out in "biorefineries") builds not only on fermentation using metabolically engineered microorganisms, but as much on enzymes improved by protein engineering techniques which are used, attached to carrier material, as heterogeneous catalysts in an enzyme reactor. Quite often, the skill to stabilize and re-use an enzyme catalyst through immobilization has proven one of the key steps to render an enzymatic process economically viable.

With this book on "Carrier-bound immobilized enzymes", Linqiu Cao provides a comprehensive survey of this important field, covering both the history and the present state of immobilization procedures used in enzyme technology. After a short introduction to 100 years of enzyme immobilization, he discusses in great detail not only the methods by which enzymes can be adsorbed, covalently bound or entrapped, but also the laws governing their behaviour in these artificial environments. In the concluding chapter of his book, he also adds an authoritative survey of most recent developments such as enzyme immobilization using genetically engineered attachment points, artificial tags or enzymes whose properties have been changed through reversible binding to synthetic polymers. He thus provides both the industrial enzymologist and the researcher in an academic environment with a well-structured, easily accessible choice of options and protocols to solve their individual needs.

My compliments go to the author for the thorough collection and structuring of a plethora of data, and my wishes to the readers for continuing success in their research on and application of immobilized biocatalysts.

Stuttgart, May 2005 *Rolf D. Schmid*

Contents

Foreword V

1 Introduction: Immobilized Enzymes: Past, Present and Prospects *1*
1.1 Introduction *1*
1.2 The Past *4*
1.2.1 The Early Days (1916–1940s) *5*
1.2.2 The Underdeveloped Phase (1950s) *5*
1.2.3 The Developing Phase (1960s) *7*
1.2.4 The Developed Phase (1970s) *9*
1.2.5 The Post-developed Phase (1980s) *14*
1.2.6 Rational Design of Immobilized Enzymes (1990s–date) *16*
1.3 Immobilized Enzymes: Implications from the Past *20*
1.3.1 Methods of Immobilization *20*
1.3.2 Diversity versus Versatility *21*
1.3.3 Complimentary versus Alternative *23*
1.3.4 Modification versus Immobilization *25*
1.3.4.1 Enhanced Stability *25*
1.3.4.2 Enhanced Activity *26*
1.3.4.3 Improved Selectivity *29*
1.4 Prospective and Future Development *34*
1.4.1 The Room for Further Development *34*
1.4.2 An Integration Approach *36*
1.5 References *37*

2 Adsorption-based Immobilization *53*
2.1 Introduction *53*
2.2 Classification of Adsorption *54*
2.3 Principles Involved in Absorptive Enzyme Immobilization *55*
2.3.1 Monolayer Principle *56*
2.3.2 Stabilization Principle *57*
2.3.3 Enzyme Distribution *60*
2.4 Requirement of the Carriers *61*
2.4.1 Physical Requirements *61*

2.4.1.1 Pore-size and Available Surface *61*
2.4.1.2 Internal Structure *63*
2.4.1.3 Density of Binding Functionality *63*
2.4.1.4 Particle Size *64*
2.4.2 Chemical Nature of the Carriers *65*
2.4.2.1 Nature of Binding Functionality *65*
2.4.2.2 The Role of the Spacer *66*
2.4.2.3 The Nature of the Backbone *66*
2.5 Factors Which Dictate Enzyme Catalytic Performance *67*
2.5.1 Activity *67*
2.5.1.1 Diffusion-controlled Activity *67*
2.5.1.2 Conformation-controlled Activity *68*
2.5.1.3 Substrate-controlled Activity *70*
2.5.1.4 Loading-controlled Activity *70*
2.5.1.5 Medium-dependent Activity *72*
2.5.1.6 Microenvironment-dependent Activity *73*
2.5.1.7 Carrier Nature-dependent Activity *74*
2.5.1.8 Enzyme Nature-dependent Activity *75*
2.5.1.9 Additive-dependent Activity *75*
2.5.1.10 Hydrophilicity-dependent Activity *76*
2.5.1.11 Orientation-determined Activity *78*
2.5.1.12 Binding Nature-controlled Enzyme Activity *78*
2.5.1.13 Binding Density-controlled Enzyme Activity *80*
2.5.1.14 Reactor-dependent Activity *81*
2.5.1.15 Pore-size-dependent Activity *82*
2.5.1.16 Water-activity-dependent Activity *82*
2.5.2 Stability *83*
2.5.2.1 Conformation-controlled Stability *84*
2.5.2.2 Confinement-controlled Stability *85*
2.5.2.3 Enzyme Loading-dependent Stability *85*
2.5.2.4 Diffusion-controlled Stability *86*
2.5.2.5 Cross-linking-dependent Stability *86*
2.5.2.6 Carrier Nature-controlled Stability *86*
2.5.2.7 Aquaphilicity-controlled Stability *87*
2.5.2.8 Medium-controlled Stability *88*
2.5.2.9 Temperature-dependent Stability *88*
2.5.2.10 Microenvironment-controlled Stability *89*
2.5.2.11 Binding Nature-controlled Enzyme Stability *90*
2.5.2.12 Binding Density-controlled Enzyme Stability *91*
2.5.2.13 Additive-dependent Stability *91*
2.5.2.14 Enzyme Orientation-dependent Stability *91*
2.5.2.15 Enzyme-dependent Stability *91*
2.5.3 Selectivity *92*
2.5.3.1 Conformation-controlled Selectivity *93*
2.5.3.2 Diffusion-controlled Selectivity *94*

2.5.3.3   Binding Functionality-controlled Selectivity   *94*
2.5.3.4   Additive-controlled Selectivity   *95*
2.5.3.5   Orientation-controlled Selectivity   *96*
2.5.3.6   Medium-controlled Enantioselectivity   *96*
2.5.3.7   Water Activity-controlled Enantioselectivity   *97*
2.6   Preparation of Immobilized Enzymes by Adsorption   *97*
2.6.1   Conventional Adsorption   *97*
2.6.1.1   Non-specific Physical Adsorption   *99*
2.6.1.2   Ionic Adsorption   *100*
2.6.1.3   Hydrophobic Adsorption   *108*
2.6.1.4   Biospecific Adsorption   *113*
2.6.1.5   Affinity Adsorption   *116*
2.6.2   Unconventional Adsorption   *121*
2.6.2.1   Immobilization via Reversible Denaturation   *123*
2.6.2.2   Pseudo-covalent Immobilization   *123*
2.6.2.3   Mediated Adsorption   *124*
2.6.3   Adsorption-based Double Immobilization   *128*
2.6.3.1   Modification–Adsorption   *128*
2.6.3.2   Adsorption and Entrapment   *131*
2.6.3.3   Adsorption–Cross-linking   *134*
2.6.3.4   Adsorption–Covalent Attachment   *142*
2.7   References   *145*

3   Covalent Enzyme Immobilization   *169*
3.1   Introduction   *169*
3.2   Physical Nature of Carriers   *171*
3.2.1   The Surface of the Carriers   *172*
3.2.1.1   Internal and External Surface   *173*
3.2.1.2   Accessible Surface/Efficient Surface   *173*
3.2.1.3   A Theoretical Simulation   *175*
3.2.2   Density of Binding Sites   *176*
3.2.3   Pore Related Properties   *177*
3.2.3.1   The Porosity   *177*
3.2.3.2   Pore Size and Distribution   *178*
3.2.4   Particle Size   *180*
3.2.5   Shape of the Carriers   *182*
3.3   Chemical Nature of Carriers   *183*
3.3.1   Carrier-bound Active Groups (CAG)   *185*
3.3.2   Carrier-bound Inert Groups (CIG)   *187*
3.3.3   Spacer-Arm   *188*
3.4   Enzyme: Amino Acid Residues for Covalent Binding   *190*
3.4.1   Reactivity of Amino Acid Residues (AAR)   *191*
3.4.2   Position of Active Amino Acids   *192*
3.5   Factors Affecting Enzyme Performance   *193*
3.5.1   Activity Retention   *194*

3.5.1.1 Pore-size-dependent Activity *194*
3.5.1.2 CAG-controlled Activity *196*
3.5.1.3 CIG-controlled Retention of Activity *200*
3.5.1.4 Spacer-controlled Activity *205*
3.5.1.5 Enzyme Orientation-controlled Activity *210*
3.5.1.6 Binding Density-controlled Activity *210*
3.5.1.7 Diffusion-controlled Enzyme Activity *211*
3.5.1.8 Reactive Amino Acid Residues (RAAR)-controlled Activity *212*
3.5.1.9 Loading-dependent Activity *213*
3.5.1.10 Other Factors Controlling Activity *213*
3.5.2 Stability of Immobilized Enzymes *214*
3.5.2.1 Multipoint Attachment/binding density *215*
3.5.2.2 CAG-controlled Stability *217*
3.5.2.3 CIG-controlled Enzyme Stability *221*
3.5.2.4 Spacer-dependent Stability *224*
3.5.2.5 Molecular Confinement-controlled Stability *225*
3.5.2.6 Microenvironment-controlled Stability *227*
3.5.3 Selectivity of Immobilized Enzymes *228*
3.5.3.1 CAG-controlled Selectivity *228*
3.5.3.2 CIG-controlled Selectivity *230*
3.5.3.3 Spacer-controlled Enzyme Selectivity *232*
3.5.3.4 Diffusion-controlled Selectivity *233*
3.5.3.5 Aquaphilicity-controlled Selectivity *233*
3.5.3.6 Conformation-controlled Enantioselectivity *233*
3.5.3.7 Selectivity and Particle Size *234*
3.6 Preparation of Active Carriers *235*
3.6.1 Synthetic Active Carriers *237*
3.6.1.1 Polymers Bearing Acyl Azide *237*
3.6.1.2 Polymers Bearing Anhydrides *238*
3.6.1.3 Polymers Bearing Halogen Atoms *240*
3.6.1.4 Oxirane Functional Polymers *241*
3.6.1.5 Isocyanate/Thioisocyante Functional Polymers *244*
3.6.1.6 Polycarbonate *245*
3.6.1.7 Activated Carbonyl Polymers *247*
3.6.1.8 Polyphenolic Polymers *248*
3.6.1.9 Polymeric Carriers Bearing Aldehyde Groups *249*
3.6.1.10 Polymers Bearing Activated Ester *251*
3.6.1.11 Polymers Bearing Active Azalactone *252*
3.6.2 Inactive Pre-carriers *253*
3.6.2.1 Hydroxyl Functionality *253*
3.6.2.2 Polyacrylamide *254*
3.6.2.3 Insoluble Polyacrylic Acid or Derivatives *255*
3.6.2.4 Polymers Bearing Nitrile Groups *257*
3.6.2.5 Semi-synthetic Polysaccharides *258*
3.6.2.6 Synthetic Polypeptide *259*

3.6.2.7   Polymers Bearing Amino Groups   *260*
3.6.3   Interconversion of Inert Carriers   *260*
3.6.3.1   Polymers Containing Hydroxyl Groups   *261*
3.6.3.2   Activation of Separate Hydroxyl Groups   *268*
3.6.3.3   Polymers Containing Carboxylic or Ester Groups   *272*
3.6.3.4   Polymers Containing Amino Groups   *277*
3.6.3.5   Polymers Containing Amide Groups   *283*
3.6.3.6   Polymers Containing Nitrile Groups   *286*
3.6.3.7   Polymers Bearing Isonitrile Functional Groups   *287*
3.6.4   Interconversion of Active Functionality   *288*
3.6.4.1   Converting Epoxy Groups   *289*
3.6.4.2   Converting Anhydride to New Functionality   *290*
3.6.4.3   Aldehyde   *292*
3.7   References   *293*

4   Enzyme Entrapment   *317*
4.1   Introduction   *317*
4.2   Definition of Entrapment   *319*
4.3   Requirement of the Carriers   *321*
4.3.1   Physical Requirements   *321*
4.3.1.1   Pore Size   *321*
4.3.1.2   Porosity   *322*
4.3.1.3   Geometry   *322*
4.3.1.4   Particle Size   *323*
4.3.2   Chemical Requirements   *323*
4.3.2.1   Nature of the Active Functionality   *323*
4.3.2.2   Aquaphilicity of the Carriers   *324*
4.4   Effect of Entrapment   *324*
4.4.1   Activity of the Entrapped Enzyme   *324*
4.4.1.1   Loading-dependent Activity   *325*
4.4.1.2   Matrix-dependent Activity   *325*
4.4.1.3   Diffusion-controlled Enzyme Activity   *326*
4.4.1.4   Conformation-controlled Enzyme Activity   *327*
4.4.1.5   Additives-controlled Enzyme Activity   *328*
4.4.2   Stability   *328*
4.4.2.1   Confinement-determined Stability   *329*
4.4.2.2   Matrix-nature-dependent Stability   *329*
4.4.2.3   Enzyme-dependent Stability   *330*
4.4.2.4   Enzyme Structure-dependent Stability   *331*
4.4.3   Selectivity   *331*
4.4.3.1   Microenvironment-dependent Selectivity   *331*
4.4.3.2   Conformation-dependent Selectivity   *332*
4.4.3.3   Carrier Nature-dependent Selectivity   *333*
4.5   Preparation of Various Entrapped Enzymes   *333*
4.5.1   Conventional Entrapment Process   *334*

4.5.1.1	Formation of Entrapment Matrix by Chemical Cross-linking	334
4.5.1.2	Physical Entrapment	342
4.5.1.3	Covalent Entrapment	351
4.5.1.4	Sol–Gel Process	356
4.5.2	Non-conventional Entrapment	362
4.5.2.1	Post-loading Entrapment (PLE)	362
4.5.2.2	Entrapment-based Double-immobilization Technique	364
4.5.2.3	Modification and Entrapment	368
4.5.2.4	Supported Entrapment	371
4.6	References	379

5	**Enzyme Encapsulation**	**397**
5.1	Introduction	397
5.1.1	General Considerations	398
5.1.2	An Historical Overview	398
5.1.3	Pros and Cons of Micro-encapsulation	399
5.2	Classification of Encapsulation	399
5.2.1	Conventional Encapsulation	399
5.2.1.1	Encapsulation by an Interfacial Process	400
5.2.1.2	Encapsulation by Phase Inversion	401
5.2.2	Non-conventional Encapsulation	401
5.2.2.1	Encapsulation in Liquid Membrane	401
5.2.2.2	Encapsulation–Reticulation	401
5.2.3	Double Immobilization Based on Encapsulation	402
5.2.3.1	Encapsulation–Cross-linking	402
5.2.3.2	Encapsulation and Coating	402
5.2.3.3	Entrapment and Coating	403
5.2.3.4	Immobilization and Encapsulation	403
5.2.4	Post-loading Encapsulation	404
5.2.4.1	Encapsulation in Non-swellable Microcapsules	405
5.2.4.2	Encapsulation in Soft Microcapsules	405
5.3	Effect of Encapsulation	406
5.3.1	Activity of the Encapsulated Enzymes	406
5.3.2	Stability of the Encapsulated Enzymes	407
5.3.3	Enantioselectivity	407
5.4	Processes for Preparation of Encapsulated Enzymes	407
5.4.1	Interfacial Processes	407
5.4.1.1	Interfacial Cross-linking/Polymerization	408
5.4.1.2	Interfacial Physical Gelation	411
5.4.2	Surfactant-related Hollow Microsphere	412
5.4.2.1	Microemulsion-based Encapsulated Enzymes	412
5.4.2.2	Polymeric Micelles/Liposomes	416
5.4.2.3	Liposome capsules	416
5.4.2.4	Colloidal Liquid Aphrons (CLA)	421
5.4.3	Phase Inversion	423

5.4.3.1	Coacervation	423
5.4.3.2	The Double-emulsion Method	423
5.4.3.3	Modified Double-emulsion Methods	426
5.4.3.4	Other Methods	429
5.4.4	Pre-designed Capsules for Post-loading Encapsulation	429
5.4.4.1	Introduction	429
5.4.4.2	Theoretical Considerations	430
5.4.5	Non-conventional Encapsulation Processes	430
5.4.5.1	Encapsulation/Coating	430
5.4.5.2	Encapsulation/Cross-linking	432
5.4.5.3	Immobilization and Encapsulation	434
5.4.5.4	Immobilization and Encagement	434
5.5	References	438

6 Unconventional Enzyme Immobilization   449

6.1	Introduction	449
6.2	Coating-based Enzyme Immobilization	450
6.2.1	Monolayer Enzymes	451
6.2.2	Phase-inversion Coating	451
6.2.3	Multiple Enzyme Coating by Physical Adsorption	452
6.2.4	Mediated Formation of Multiple Enzyme Layers	452
6.2.5	Affinity-ligand-mediated Formation of Enzyme Coatings	455
6.2.6	Coating of Soluble Enzyme–Polymer Complexes	456
6.2.7	Enzymatically Gelified Multienzyme Layer	456
6.2.8	Sol–Gel Coating and Covalent Attachment	457
6.2.9	Electrochemical Deposition	457
6.2.10	Enzyme Coating by Use of Small Pore-size Carriers	458
6.3	Site-specific Immobilization	458
6.3.1	Site-specific Immobilization via Biospecific Ligand–Enzyme Interaction	463
6.3.2	Introduction of Chemical Tags	464
6.3.2.1	Oxidation of the Sugar Moiety of Enzymes	466
6.3.2.2	Introduction of a Cofactor into the Enzyme	466
6.3.2.3	Orientation and Covalent Binding	466
6.3.3	Immobilized Ligand (Substrate Analogue) Enzyme-binding	468
6.3.3.1	Immobilized Substrate or Substrate Analogues	470
6.3.3.2	Immobilized Non-substrate Ligand	470
6.3.4	Genetically Engineered Tags	473
6.3.4.1	Non-covalent Oriented Enzyme Immobilization	473
6.3.4.2	Covalent Orientation in Enzyme Immobilization	477
6.4	Immobilization in Organic Solvents	477
6.4.1	Covalent Attachment in Organic Solvents	478
6.4.2	Entrapment of Enzyme in Organic Solvent	479
6.4.3	Immobilization of Organic-soluble Enzyme Derivatives	480
6.4.4	Adsorption of Enzyme on to the Carrier in Organic Solvents	480

6.5	Imprinting–Immobilization *481*	
6.5.1	Imprinting-Multipoint Attachment *481*	
6.5.2	Imprinting–Cross-linking *482*	
6.5.3	Entrapment–Imprinting *484*	
6.5.4	Crystallization and Cross-linking *484*	
6.5.5	Aggregation and Cross-linking *485*	
6.5.6	Intra-molecular Cross-linking – Imprinting *485*	
6.5.7	Post-immobilization Imprinting *486*	
6.5.8	Lyophilization Imprinting *486*	
6.6	Stabilization–Immobilization *487*	
6.6.1	Stabilization by Ligand Binding *488*	
6.6.2	Stabilization by Addition of Stabilizer as the Excipient of the Conformation *490*	
6.6.3	Stabilization by Pre-immobilization Modification *490*	
6.6.3.1	Stabilization by Pre-immobilization Modification with Soluble Polymer *490*	
6.6.3.2	Stabilization by Pre-immobilization Chemical Modification *492*	
6.6.3.3	Chemical Cross-linking/Covalent Immobilization *493*	
6.7	Modification-based Enzyme Immobilization *493*	
6.7.1	Immobilization then Modification *494*	
6.7.2	Modification then Polymerization *494*	
6.7.3	Pre-immobilization Improvement Techniques (PIT) *496*	
6.7.3.1	Introduction of Extra Charge *497*	
6.7.3.2	Alteration of Enzyme Hydrophobicity *498*	
6.7.3.3	Formation of Polymer–Enzyme Conjugate *498*	
6.7.3.4	Introduction of Active Functionality for Covalent Binding *503*	
6.7.3.5	Introduction of Mediators *505*	
6.7.3.6	Interconversion of Amino Acid Residues (AAR) *505*	
6.7.3.7	Cross-linking/Immobilization *506*	
6.8	Post-Immobilization Techniques *506*	
6.8.1	Introduction *506*	
6.8.2	Classification of Post-treatments *507*	
6.8.3	Physical Methods *507*	
6.8.3.1	Increasing the pH *509*	
6.8.3.2	Solvent Washing *510*	
6.8.3.3	Lyophilization/Drying/Addition of Additives *510*	
6.8.3.4	pH Imprinting *511*	
6.8.3.5	Physical Entrapment *512*	
6.8.3.6	Thermal Activation *512*	
6.8.3.7	Activation by Denaturants *513*	
6.8.3.8	Post-immobilization by Physical Coating *513*	
6.8.3.9	Rehydration/water Activity Adjustment *514*	
6.8.3.10	Sonication *515*	
6.8.3.11	Acid or Alkaline Treatment *515*	
6.8.4	Chemical Methods *515*	

6.8.4.1 Consecutive Cross-linking of the Immobilized Enzymes *516*
6.8.4.2 Consecutive Chemical Modification of Immobilized Enzymes *518*
6.8.4.3 Conversion *518*
6.8.4.4 Consecutive Modification of the Carrier *519*
6.8.4.5 Chemical Coating *521*
6.8.5 Outlook *522*
6.9 Reversibly Soluble Immobilized Enzymes *522*
6.9.1 pH-responsive Smart Polymer *522*
6.9.2 Temperature-sensitive Smart Polymers *526*
6.9.3 Solvent-sensitive Enzyme–Polymer Conjugates *526*
6.9.4 Reversibly Soluble Immobilized Enzyme Based on Ionic Strength-sensitive Polymers *528*
6.9.5 Reversibly Soluble Immobilized Enzyme Based on Light-sensitive Polymers *528*
6.10 References *531*

Subject Index *551*

1
Introduction: Immobilized Enzymes: Past, Present and Prospects

1.1
Introduction

Since the second half of the last century, numerous efforts have been devoted to the development of insoluble immobilized enzymes for a variety of applications [2]; these applications can clearly benefit from use of the immobilized enzymes rather than the soluble counterparts, for instance as reusable heterogeneous biocatalysts, with the aim of reducing production costs by efficient recycling and control of the process [3, 4], as stable and reusable devices for analytical and medical applications [5–11], as selective adsorbents for purification of proteins and enzymes [12], as fundamental tools for solid-phase protein chemistry [13, 14] and as effective microdevices for controlled release of protein drugs [15] (Scheme 1.1).

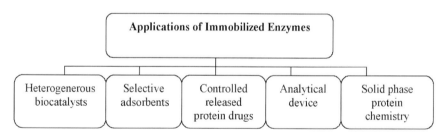

Scheme 1.1 Range of application of immobilized enzymes.

However, whatever the nature of an immobilized enzyme and no matter how it is prepared, any immobilized enzyme, by definition, must comprise two essential functions, namely the non-catalytic functions (NCF) that are designed to aid separation (e.g. isolation of catalysts from the application environment, reuse of the catalysts and control of the process) and the catalytic functions (CF) that are designed to convert the target compounds (or substrates) within the time and space desired (Scheme 1.2).

NCF are strongly connected with the physical and chemical nature of the non-catalytic part of the immobilized enzymes, especially the geometric properties, e.g. the shape, size, thickness, and length of the selected carrier, whereas the CF are linked to the catalytic properties, for example activity, selectivity, and stability, pH

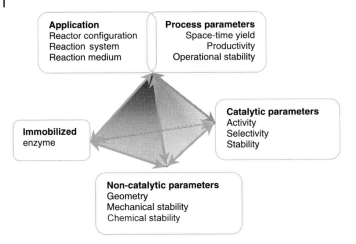

Scheme 1.2 Relationship between NCF and CF of an immobilized enzyme and its applications.

and temperature profiles. General criteria for selection of these two properties for robust immobilized enzymes as catalysts are proposed in Table 1.1 [16].

In practice, catalytic functions are designed in line with the desired activity, selectivity, substrate specificity, productivity and space–time yield, with the aim of achieving fewer side reactions, high tolerance of structural variation of the substrates, high productivity, high space–time yield, and high durability of the catalyst. On the other hand, the selection criteria for non-catalytic functions, especially geometric properties, are largely dependent on the design of reactor configurations (e.g. batch, stir-tank, column and plug-flow), the types of reaction medium (aqueous, organic solvent, or two-phase system), the reaction systems (slurry, liquid-to-liquid, liquid-to-solid, or solid-to-solid), and the process conditions (pH, temperature, pressure). The objectives when designing the non-catalytic properties are mainly to achieve easy separation of the immobilized enzymes from the reaction mixtures, broad reactor considerations (i.e. flexibility of reactor design), broad applicability in different reaction media and reaction systems, and facilitating process development, down-stream processing and, particularly, control of the process.

It is usually the peculiarities of these two essential elements, i.e. the non-catalytic functions and the catalytic functions that dictate the scope of the final application of the immobilized enzymes obtained. Conversely, the peculiarities of each application also dictate the design and selection of the two essential elements. In general, the NCF and CF of an immobilized enzyme are the two sides of a coin which are the basis of the scope of the final application, as illustrated in Scheme 1.2.

It is, therefore, hardly surprising that the main task of enzyme immobilization is to select a suitable immobilization method (carriers, conditions, and enzymes) to design an immobilized biocatalyst which can meet not only the catalytic needs (expressed as productivity, space–time yield, stability and selectivity) but also the non-catalytic needs (e.g. separation, control, down-streaming process) of a given appli-

Table 1.1 Criteria for robust immobilized enzymes (from Ref. [16])

Parameter	Requirement	Benefits
Non-catalytic function	Suitable particle size and shape	Aid separation, easy control of the reaction
	Suitable mechanical properties	Flexibility of reactor design
	Low water regain capability	Easy removal of water
	High stability in a variety of organic solvents	No change of pore radius and thus fewer diffusion constraints
Catalytic function	High volume activity (U g^{-1})	High productivity and space–time yield
	High selectivity	Fewer side reactions, easier downstream processing and separation of products, and less pollution
	Broad substrate specificity	Tolerance of structural variation of the substrates
	Stability in organic solvents	Shift of reaction equilibrium with the use of organic solvents
	Thermostability	Short reaction time by increasing temperature
	Operational stability	Cost-effective and lower cost-contribution for the product
	Conformational stability	Modulation of enzyme properties
Immobilized enzyme	Recyclability	Low cost-contribution of catalyst
	Broad applicability	Tolerance of process variation
	Reproducibility	Guarantee product quality
	Easy and quick design	Early insight into process development and avoidance of learning process
E and E consideration	Lower volume	Lower cost for the solid handling
	Easy disposal	Less environmental concern? Easy biodegradability?
	Rational design	Avoidance of laborious screening
	Safety for use	Meeting safety regulations
IPR	Innovative	Protection of IPR
	Attractive	Licensing
	Competitive	Strengthening marketing position

E and E: Economical and Ecological; IPR: Intellectual Property

cation. As a result, an immobilized enzyme can be labelled "robust" when its catalytic and the non-catalytic functions both meet the requirements of a specific application. Consequently, it is envisaged there are two possibilities in the development of a biocatalytic process – design of a process around an available immobilized enzyme and the design of an immobilized enzyme around a process.

The first possibility is obviously less desirable, because a ready-made immobilized enzyme (either commercially available or made in-house) is a specific immobilized enzyme only and is thus not necessarily the optimum catalyst for the desired processes, as exemplified by the fact that many types of carrier-bound immobilized penicillin G acylase which are regarded as robust immobilized catalysts for the production of 6-APA are not necessarily good catalysts for the kinetically controlled synthesis of semi-synthetic β-lactam antibiotics [17, 18]. This is largely ascribed to the fact that changing the process conditions often provokes a change of enzyme performance.

By contrast, the diversity of the processes (as reflected by different substrates, reaction types, reactor configurations, down-streaming processes) necessarily requires the design of specific immobilized enzymes which can match process requirements. Thus, it is hardly surprising that design of the immobilized enzyme around a process will dominate the future development of immobilized enzymes.

Although it is becoming increasingly appreciated that the availability of a robust immobilized enzyme in the early stage of process development will definitively enable early insight into process development and save costs not only in process development but also in production, the lack of guidelines to selection of the method of immobilization and the performance to be expected of an immobilized enzyme for a specific application seriously hampers application of a rational approach to the design of such robust immobilized enzymes [19].

In this regard, we attempt to analyse important developments in the history of enzyme immobilization and thus to provide readers with a fundamental basis for understanding and designing robust immobilized enzymes.

1.2
The Past

Although the chronological development of enzyme-immobilization techniques has been discussed intensively for several decades [20–22], it is still worth going back to several historical phases which were important milestones in the history of enzyme immobilization, to appreciate that the roots of enzyme-immobilization techniques are the basis of future development.

For the purpose of discussion, the development enzyme immobilization is classified according to five criteria:

- the number of methods developed,
- the number of materials used for enzyme immobilization,
- the number of binding types established,
- the degree of understanding of the factors influencing the performance of the immobilized enzymes, and
- the number of processes using immobilized enzymes.

Accordingly, the history of bio-immobilization can be divided into several phases:

- the early days (1916–1940s),
- the underdeveloped phase (1950s),
- the developing phase (1960s),
- the developed phase (1970s),
- the post-developed phase (1980s), and
- the rational design phase (1990s–present).

Although there might be some overlap in respect of the time and continuity of development, this classification reflects major developments in enzyme-immobilization techniques. Following this order, we briefly discuss what has been achieved in the last 90 years.

1.2.1
The Early Days (1916–1940s)

Although in 1916, Nelson and Griffin rediscovered that artificial carrier-bound invertase on $Al(OH)_3$ and charcoal was still catalytically active [1], the potential of bio-immobilization as a method of obtaining useful and reusable immobilized biocatalysts was unfortunately not recognized in the succeeding 40 years. This simple fortuitous discovery has, however, been widely recognized as the cornerstone of the various enzyme-immobilization techniques currently available, because in the last half century it actually stimulated much interest and effort in exploration of insolubilized active enzymes for various studies and industrial applications that can be better met with immobilized rather than free enzymes.

In these early days, bio-immobilization techniques were mainly used to prepare adsorbents for isolation of proteins by immunologists, via adsorption on simple inorganic carriers such as glass [23], alumina [24] or hydrophobic compound-coated glass [25].

Along with these prototypes of pseudo-immobilized enzymes (immobilized by reversible non-covalent physical adsorption), few irreversible immobilized enzymes prepared by covalent attachment were also reported in the literature at that time [26].

1.2.2
The Underdeveloped Phase (1950s)

Although in 1950s the method of enzyme immobilization was still dominated by physical methods, i.e. non-specific physical adsorption of enzymes or proteins on solid carriers, for example α-amylase adsorbed on activated carbon, bentonite or clay [27], AMP deaminase on silica [28], and chymotrypsin on kaolinite [29], the method of adsorption was gradually switched from simple physical adsorption to specific ionic adsorption, for instance, chymotrypsin on phosphocellulose [29], catalase on the ionic resin DEAE–cellulose [30, 31], DNase on cellulose [32, 33], lipase and catalase on styrenepolyaminostyrene (Amberlite XE-97) [34], and ribonuclease on the anionic exchanger Dowex-2 and the cationic exchanger Dowex-50 [35].

Along with physical methods of enzyme immobilization, however (e.g. non-specific adsorption, or ionic adsorption), other important methods of enzyme immobilization, for example covalent immobilization, were further investigated. Examples of enzymes were lipases and other enzymes or antibodies covalently bound to polyaminostyrene [34, 36–38], diazotized cellulose [7], poly(acrylic acid) chloride [40, 41], diazotized polyaminostyrene [36, 41, 42], and polyisocyanate [34, 38]. Unfortunately, those early-developed carriers were found to be less suitable for covalent enzyme immobilization, because of poor retention of activity (2–20 % of the native activity), probably attributable to the highly hydrophobic nature of the carriers used at that time [38–45] or the unsuitable active functionality such as diazonium salt, which often affords an immobilized enzyme with lower retention of activity [45].

Apart from the physical adsorption and covalent immobilization used in this period, it was demonstrated for the first time by Dickey that some enzymes such as AMP deaminase entrapped in the sol–gel inorganic matrix formed by silicic acid-derived glasses retained reasonable biological activity [28]. Unfortunately, the importance of this finding was not recognized in the succeeding 40 years [46–48].

In addition to the use of natural polymers, derivatives such as CM-cellulose [30] and DEAE-cellulose [31], and inorganic materials such as carbon [35], glass, kaolinite [39], and clays as carriers for enzyme immobilization, a few synthetic polymers, for example aminopolystyrene and polyisocyanate, prepared directly by poly-

Table 1.2 Survey of enzyme-immobilization techniques in the 1950s

Carriers	Activation or coupling methods	Techniques developed and important observations
Natural polymers and derivatives	Acylazide	Physical adsorption
Cellulose	Diazotium salt	Ionic adsorption [35]
DEAE-cellulose	Polyacrylic acid chloride	Covalent [36]
	Isocyanate [38]	Entrapment in sol-gel glass [28]
		Modification-adsorptive immobilization
Synthetic polymers		pH optimum shifting [35]
Amberlite		
Diaion		
Dowex		
Polystyrene		
Other polyacrylic polymers and derivatives		
Inorganic carriers		
Carbon		
Silica		
Kaolinite		
Clay		

merization of active monomers for covalent enzyme immobilization [37], and synthetic ionic adsorbents such as Amberlite XE-97 [34], Dowex-2, and Dowex 50 [35] for non-covalent enzyme immobilization by ionic adsorption [35, 37, 38, 41] were also added to the family of carriers used for enzyme immobilization (Table 1.2).

1.2.3
The Developing Phase (1960s)

Although different covalent methods of enzyme immobilization were the main focus of bio-immobilization at this time, the long-established non-covalent enzyme immobilization, i.e. adsorption [50] and entrapment [51–53] were further developed, as is reflected in the publications of the time (Ref. [54] and references cited therein). In addition, encapsulation of enzymes in semi-permeable spherical membranes (also called "artificial cells") was first proposed by Chang [55]. Enzyme entrapment techniques were also further extended by the use of synthetic polymeric gels such as PVA (polyvinyl alcohol) [56] or PAAm (polyacrylamide gel) [51] or the use of natural polymer derivatives such as nitrocellulose or starch [56] or silicon elastomers for the sol–gel process [57, 58]. Other techniques of enzyme immobilization, for example adsorptive cross-linking of enzymes on films and membranes [62], or beads for the formation of enzyme envelopes [62], were also developed.

Apart from the development of carrier-bound immobilized enzymes, it was also demonstrated that insoluble carrier-free immobilized enzymes could be prepared by cross-linking of crystalline enzymes [63] or dissolved enzymes [64], by use of a bifunctional cross-linker such as glutaraldehyde. Although the potential of cross-linking of enzyme crystals was not recognized at that time, intensive studies were devoted to preparation of these carrier-free immobilized enzymes, especially CLE (cross-linked dissolved enzymes), as immobilized enzymes. More than twenty enzymes of different classes were either directly cross-linked to form a variety of CLE or first adsorbed on inert supports, such as membranes, and subsequently cross-linked to form supported CLE (Ref. [54] and references cited therein). In the late 1960s, however, research emphasis switched mainly to carrier-bound immobilized enzymes; at this time a wide range of carriers was specifically developed for enzyme immobilization and several important organic reactions for binding enzymes to carriers were established, as is shown in Table 1.3.

From the middle to the end of the 1960s the scope of bio-immobilization was greatly extended owing to the use of more hydrophilic insoluble carriers with defined geometric properties, for example cross-linked dextran, agarose, and cellulose beads (Table 1.3) and particularly as a result of the use of new methods of activation, for example cyanogen bromide [65] and triazine for polysaccharide [66], isothiocyanate for coupling amino groups [67], and Woodward reagents [69] for activation of carboxyl groups. Furthermore, the preparation of synthetic carriers bearing active functionality such as polyanhydride [79] or polyisothiocyanate [67], etc., which could bind enzyme directly (Table 1.3), enabled relatively simple preparation of immobilized enzymes.

Table 1.3 Survey of enzyme-immobilization techniques in the 1960s

Carriers	Activation or coupling methods	Techniques developed and important observations
Synthetic polymers Poly(AAc-MAAn) PAAm PVA [56] Nylon Polystyrene [59, 60]	Cyanogen bromide [65] Triazine method [66] Glutaraldehyde for crosslinking and coupling [63] Woodward reagents [69] Anhydride	Entrapment of whole cells in synthetic gel [53] Encapsulation in artificial cell [57] Adsorption-cross-linking [62] Active site titration [71, 72] First industrial process with immobilized enzymes [52]
Natural polymers and derivatives DEAE-cellulose Sephadex Sepharose Starch *Semi-synthetic carriers* Collodion Nitrocellulose Epoxy ring-grafted natural polymer *Inorganic carriers* Carbon Clay Silica gel Hydroxyapatite Kaolinite	Isothiocyanate [67] Activation of carboxyl groups [68] Activation of hydroxyl group with monohalo-acetyl halides [70]	Modification-covalent immobilization [73, 171] Cross-linked enzyme (CLE) [63] and cross-linked enzyme crystals (CLEC) [63] Micro-environmental effect [76] Immobilization or post-treatment by denaturant [77] Binding mode was related to the enzyme stability [78] Importance of binding chemistry in terms of activity retention was appreciated [78]

The enzymes studied changed, moreover – from a few classic enzymes such as invertase, trypsin, urease and pepsin to a broad range of enzymes such as galactosidase, amyloglucosidase, urease [78], subtilisin, chymotrypsin [69], lactate dehydrogenase [81], apyrase [83], amino acylase [82], amino acid oxidase [86], catalase, peroxidase [84], hexokinase [85], cholinesterase [91], α-amylase [87], ATPase and adolase, alkaline phosphatase [88], penicillin G acylase [89], β-galactosidase [90], deoxyribonuclease [91], urate oxidase, and cholinesterase, etc., which were expected to have great application potential in chemical, pharmaceutical, and medical industrial sectors.

At the same time it was increasingly appreciated that the physical and chemical nature of the carriers, especially the microenvironment, for example their hydrophilic or hydrophobic nature, the charges on the carriers, and the binding chemistry also strongly dictated the catalytic characteristics of the enzyme, for example activity [76, 79, 92, 93], retention of activity [79, 94] and stability [87].

With increasing awareness that besides functioning as supports, i.e. as scaffolds for the enzyme molecules, the carriers could be used practically as the modifiers of enzyme properties, many carriers of different physical or chemical nature, different hydrophilicity or hydrophobicity, or different shape or size (for example beads, sheet, film, membrane [95] or capsules [55]) were developed to provide carriers with sufficient diversity. This was reflected by the shift of the carriers from a few classics, for example cellulose and its derivatives [44], inorganic carriers [86, 88, 97] and polystyrene and derivatives [37, 59, 60], to a broad variety ranging from naturally occurring materials such as agarose, Sephadex [83], Sepharose [65], glass [97], kaolinite, clay, DEAE-Sephadex, DEAE-cellulose [50], to synthetic carriers such as polyacrylamide [51], ethylene maleic acid copolymer [94], a co-polymer of methylacrylic acid and methylacrylic acid-*m*-fluoroanilide [96], nylon [98, 99], PVA-based carriers for covalent binding or entrapment [56], and a variety of synthetic ion-exchange resins such as Amberlite [100], Diaion and Dowex [101], which have defined chemical and physical properties.

It is also worthy of note that introduction of active-site titration has made it possible to assess the availability of the active site and how this immobilization was affected by incorrect orientation, by deactivation or by diffusion constraints [71, 72]. Meanwhile, the first example of resolution of a racemic compound catalysed by carrier-bound immobilized enzymes was also demonstrated and the first enzyme electrode appeared [6]. Glazer et al. demonstrated that introduction of extra functional groups to the enzyme before immobilization was an efficient means of controlling the mode of binding between the enzyme and the carrier [73, 75]. This technology also has other benefits, for example enzyme inactivation resulting from direct coupling of the enzyme to the resin might be avoided. This concept was later developed as modification–immobilization techniques, with the objective of improving the enzyme, e.g. by enhancement of its stability, activity and selectivity, before immobilization [103].

Remarkably, it was found that not only the soluble enzyme but also the enzyme crystals can be entrapped in a gel matrix with reasonable retention of activity [74].

By the end of 1960s the first industrial application of an immobilized enzyme (ionically bound l-amino acid acylase) for production of l-amino acids from racemic amino acid derivatives had been developed by a Japanese company [50]; this not only exemplified the practical (or industrial) value of immobilized enzymes but also inspired several new research interests; this was subsequently reflected by steadily increasing interest, by an explosive increase in publications on enzyme immobilization, and by the number of new immobilization techniques [174, 175].

1.2.4
The Developed Phase (1970s)

In the 1970s, enzyme immobilization continued to flourish into a maturing phase, although the methods used in this period were still labelled as "less rational". The methods developed in previous phases had been widely extended to several enzymes which were expected to have great industrial potential, for example α-amylase, acylase, penicillin G acylase, and invertase, etc. Achievements in this period have been the subject of several reviews [174, 175, 178].

Although the methods used for enzyme immobilization were not beyond the scope of the four basic methods already previously developed, namely covalent, adsorption, entrapment and encapsulation, many new method subgroups, for example affinity binding and coordination binding [105], and many novel variations of enzyme immobilization were developed (Tables 1.4 and 1.5).

The objective of the sophisticated immobilization techniques developed in 1970s was, primarily, improvement of the performance of the immobilized enzymes which could not be achieved by conventional methods of immobilization. For instance, enzymes can be entrapped in gel-matrix by copolymerization of an enzyme modified with double bonds in the presence of the monomers, leading to the formation of "plastic enzymes" with improved stability [110]. Entrapment of enzyme in the gel matrix can be followed by cross-linking, to reinforce the beads and to

Table 1.4 Survey of enzyme-immobilization techniques in the 1970s

Carriers	Activation or coupling methods	Important techniques developed
Active synthetic carriers Halogen Epoxy ring Aldehyde Anhydride Acylazide Carbonate Isocyanate	Ugi reaction [146, 147] Alkylation with epoxide [107] Aldehyde activation Carbonyldiimidazole for hydroxyl groups [109] Oxidization of glucosylated enzymes [120] Benzoquinone [121]	Reversibly covalent coupling and intra-molecular cross-linking [112] Affinity immobilization [115] Coordination immobilization [116] Oriented enzyme immobilization [115] Introduction of spacer [117] Complimentary multipoint attachment [118] Hydrophilicity–hydrophobicity balance
Functionalized prepolymers (for entrapment) PVA-SbQ PEG-DMA PEG-CA ENTP	Carbonate [122] Imidoester [123] Divinylsulphone Glutaraldehyde for polyacrylamide [124]	of the carrier [125] Enzyme immobilization in organic solvents Enzyme entrapment by reactive prepolymers [126, 127] Immobilization of enzymes to soluble supports [128] Reversibly soluble enzymes [128]
Inorganic carriers for covalent coupling Silica [118]		Modification and immobilization [103] Adsorption-covalent binding [104]
Natural polymers and derivatives Gelatin Alginate Agarose Collagen		

Table 1.5 Important technologies developed in the 1970s

Method	Remarks	Ref.
Covalent immobilization via spacer	With the aim of modulating retention of enzyme activity	114
Affinity immobilization	Combines mild immobilization conditions and reversibility of binding	115
Oriented enzyme immobilization	With the aim of enhancing activity retention compared with random immobilization	115
Coordination immobilization	Combines immobilization and regeneration of the carrier	116
Enzyme immobilization in organic solvents	With the aim of exploring other binding chemistry that works exclusively in the absence of water	117
Complimentary multi-point attachment	With the aim of enhancing the enzyme stability	118
Immobilization of enzymes to soluble supports	With the aim of acting on sparingly soluble substrates	128
Modification and immobilization	Combines the techniques of chemical modification with immobilization techniques	129
Reversibly soluble enzymes	Combines the advantages of soluble enzymes and immobilized enzymes	135
Stabilization and immobilization	Combines the techniques of enzyme stabilization with enzyme-immobilization techniques	136
Entrapment by wet spinning technique	High enzyme loading can be obtained	137, 138
Covalent entrapment	Enzyme entrapment and covalent binding of the enzyme molecules to the matrix occurred concomitantly	139

avoid leakage [111]; immobilization of the enzymes (either covalent or by affinity adsorption) via a suitable spacer can improve the enzyme activity [114, 115].

More importantly, inspired by the observation that chemical modification of enzymes often improves their characteristics, for example activity and stability, modified enzymes with improved properties, for example enhanced stability, have been further immobilized by a variety of suitable immobilization methods, for example adsorption on the cationic exchanger by introduction of carboxylic ions to the enzymes by succination [129] or entrapment in a polymeric matrix [130].

Another important discovery in the 1970s was that enzyme immobilization does not necessarily have to be performed in aqueous media – covalent coupling of an enzyme to a solid carrier or entrapment of an enzyme in a gel matrix can be performed in organic solvents [131, 132]; such methods have much attractive potential, for example modulation of enzyme conformation or extending the coupling chemistry beyond the scope of aqueous media. Unfortunately, this technology was not well developed at the time.

As with the carriers used in 1970s, different polymers with designed characteristics, for example tailored-made hydrophobicity or hydrophilicity, particle size and binding functionality, became available for bioimmobilization. By the end of the 1970s, several new synthetic or natural functionalized polymers with pre-designed chemical and physical nature, particularly natural polymer-based carriers bearing reactive functional groups such as aldehyde, cyclic carbonate, anhydride and acylazide, and synthetic polyacrylic polymers bearing different active functionality such as oxirane ring, aldehyde, anhydrides and carbonate [133], were specifically developed or designed for covalent enzyme immobilization [134].

Among these, synthetic polymers with epoxy groups [140, 141] and derivatives of natural polymers [142], which have defined chemical or physical nature and can be directly used to bind enzymes under mild conditions, attracted much attention [104–143]. An inter-conversion technique which was actually proposed by Manneck at the beginning of 1960s [37] was also widely used to convert the built-in active or inactive functionality into other suitable binding functionality for covalent immobilization [143–145].

More importantly, many new chemical reactions were identified and established for covalent coupling of enzymes to carriers; these included:

- the Ugi reaction [146, 147],
- acylation with an imidoester [149],
- carbohydrate coupling [150],
- use of N-hydroxysuccinimide esters for activation of carboxyl groups [151],
- coupling and concomitant purification via thio–disulphide interchange [152],
- oxirane coupling [153],
- the benzoquinone method [154], and
- reversible covalent coupling [112].

Remarkably, increasing attention was also directed toward the preparation of immobilized enzymes with designed geometric properties, for example beads [113], foam [155] or fibres [143], to suit various applications and reactor configurations.

During this period much deep insight was gained into the effect on the performance of the immobilized enzymes of factors such as the microenvironment effect of the carrier [155], the effect of the spacer or arm [158, 160], different modes of binding (chemistry, position and number) [170], enzyme loading [87, 167], changes in the conformation of the enzyme, diffusion constraints [161, 163], orientation of the enzyme [164], and the protective effect of substrate or inhibitor during immobilization, namely prevention of deactivation of enzyme from owing to modification of the active site [166] (Table 1.6).

Consequently, many new strategies were developed to improve the performance of the immobilized enzymes, for example the archetype of site-specific enzyme immobilization on the micelle [164], the stabilization–immobilization strategy [170], intramolecular crosslinking [118] and complimentary multipoint attachment [185]. Some of these achievements have been summarized in books and reviews [174, 159, 259].

Table 1.6 Important factors influencing performance of immobilized enzymes discovered in the 1970s

Factors	Implication or application	Ref.
Hydrophobic partition effect	Enhancement of reaction rate of hydrophobic substrate	125, 157
Microenvironment effect of the carrier	Hydrophilic nature often stabilizes the enzyme, whereas hydrophobic nature often destabilizes the protein	127, 156
Multipoint attachment effect	Enhancement of enzyme thermal stability	–
Spacer or arm various types of immobilized enzyme	With the aim of avoiding deactivation of the enzyme by incompatible interaction with protein-carrier or mitigating the steric hindrance	158–160
Diffusion constraints	Enzyme activity might decrease and stability increases	161, 162
Orientation of the enzyme	Site-specific enzyme immobilization techniques featured	163
Presence of substrates or inhibitor	Higher activity retention	164
Conformational changes or protection	Protection of the enzyme from conformational change during enzyme immobilization process leading to high activity retention	164
Physical post-treatments	Improvement of enzyme performance	165
Enzyme loading	Higher enzyme loading is essential to avoid lower enzyme activity expression	87, 166
Different binding mode	Activity and stability can be affected	167
Enzyme modification	Suitable chemical modification often leads to the improvement of enzyme stability	118
Enzyme modification/immobilization	Formation of active enzyme, which can be covalently bound to the inert carrier, can control the binding mode such as number of bonds formed between the carrier and enzyme, thus improving the activity retention	168
Physical structure of the carrier such as pore size	Activity retention was often pore-size-dependent	169
Stabilization-immobilization	Enzyme can, moreover, be stabilized before binding of enzyme to carrier	170
Physical nature of the carrier	Carriers with large pore size mitigates diffusion limitation, leading to higher activity retention	171
Hydrophilic-hydrophobic balance of the carrier	A delicate balance of hydrophilic and hydrophobic character of the selected carrier is essential for the activity and stability	172

Because of these in-depth investigations, the potential of enzyme-immobilization techniques in commercial processes has been completely recognized and many other commercial processes with use of enzymes have been under development, for example use of immobilized penicillin G acylase for production of 6-APA – the key intermediate in the synthesis of semi-synthetic β-lactam antibiotics – or the use of immobilized glucose isomerase for production of fructose syrup from glucose [119]. Other proposed applications of immobilized enzymes include controlled-release protein drugs and biomedical application as biosensors or artificial organs [176].

By the end of 1970s enzyme-immobilization techniques had matured to such extent that every enzyme could be immobilized by selecting a suitable method of immobilization (entrapment, encapsulation, covalent attachment, adsorption and combi-methods) or a suitable carrier (organic or inorganic, natural or synthetic, porous or non-porous, film, beads, foam, capsules or disks) and immobilization conditions (aqueous, organic solvents, pH, temperature, etc.). It was also increasingly appreciated that the main problem in enzyme immobilization was not immobilization of the enzymes on the carriers but how to obtain the performance desired for a given application by selecting a suitable immobilization approach from the numerous methods available.

1.2.5
The Post-developed Phase (1980s)

In this period, which spans the beginning to the end of the 1980s, incentives to design robust immobilized enzymes originated from the following potential of the immobilized enzymes:

- Immobilized enzymes might meet the increasing demand by manufacturers of pharmaceuticals and agrochemicals for enantiomerically pure compounds, because of their greater selectivity and specificity.
- Biocatalytic processes might meet increasingly strict environmental regulations, because of their mild reaction conditions and lower energy consumption.
- Biocatalytic processes can provide short-cuts compared with conventional chemical processes, because protecting chemistry can be abandoned, as was demonstrated in the production of 6-APA (6-aminopenicillanic acid, a core intermediate for semi-synthetic penicillins) [174].

It is worth mentioning that another important incentive in the search for robust immobilized enzymes in the 1980s and 1990s was the re-discovery that many enzymes are catalytically active and stable in organic solvents [177–179], thus enabling many reactions which cannot be performed in aqueous media.

Enzyme stability and activity are, however, usually lower in organic solvents, because of distortion of enzyme structure by the organic solvents used [177]. Consequently, much effort was devoted to elucidation of the effects of carriers or immobilization techniques on the catalytic behaviour of the immobilized enzymes obtained and of the effects of organic solvents on the enzyme activity, selectivity and stability under non-aqueous conditions. For example, encaging of enzymes in

Table 1.7 Survey of enzyme-immobilization techniques developed in the 1980s

Carriers	Activation methods or coupling reactions	Important techniques and findings
Synthetic microporous carriers Reactive carriers of pre-designed shape and size and active binding functionalities, e.g. Eupergit C and azalactone, were commercialized	Azalactone Tosylation (hydroxyl groups) [212] Chloroformate [213] 2-Fluro-1-methylpyridinium toluene-4-sulphonate for hydroxyl group [214] Carbonochloridate for activation of hydroxyl groups [215]	Encagement (double encagement) [180, 181] CLEC might be stable biocatalysts in organic solvents [188] Introduction of aquaphilicity [190] Dynamic immobilization technique [216] Deposition technique [201] Covalent multilayer immobilized enzyme [187]
Other types of carrier More than 100 other types of polymeric carrier have been made commercially available [143]		Post-loading entrapment [217] Organosoluble polymer–enzyme complex [191–194] Organosoluble lipid-coated enzyme [195] Introduction of genetically engineered tags [198] Introduction of orientation groups to the carrier Carrier-bound multipoint attachment [182–185] Stabilization and/or immobilization [218] Covalent immobilization of enzyme in organic solvents to design active enzyme in organic solvent [199] Imprinting of enzyme by entrapment [200] Covalent binding of enzyme to carriers might freeze the enzyme conformation induced by the effectors [202] Stabilization of immobilized enzyme by the presence of inhibitors [203]

symplex [180] or sandwich complexes [181] could drastically enhance enzyme stability under non-natural conditions; strengthening multipoint attachment to the carrier [182–185], instead of the complimentary multipoint attachment originally proposed by Martinek et al. [186], could stabilize the overall enzyme scaffold by trapping the hot area that is crucial for stabilization of the enzymes [187]; immobilizing an enzyme by a combination of covalent L–B–L(layer-by-layer) techniques and cross-linking could dramatically enhance enzyme loading and enzyme stability [188]; cross-linking of crystalline enzymes can be used to create stable biocatalysts for biotransformation, especially in organic solvents, because of their high stability in these solvents (see Table 1.7 for details) [189].

Besides efforts to prepare stable immobilized enzymes, several methods and concepts were developed to make the enzymes more active in organic solvents. First, the concept of the water activity of the reaction medium was proposed [190], and reliable comparison of different catalytic processes in low-water media became possible. The introduction of the concept "aquaphilicity" for the carrier enables quantitative measurement and screening of the desired carriers for immobilization of enzymes intended for use in organic solvents with regard to the close relationship between the carrier and the activity and selectivity of the immobilized enzymes [191]. The finding that physical and/or chemical modification of enzymes (for example chemical modification with activated PEG [192–195, 247] or lipid pairs [197], encapsulating the enzyme in micelles [196], and coating the enzyme surface with lipids) renders them soluble in organic solvents enabled the mass transfer limitation associated with the use of lyophilized enzyme powders to be surmounted [197]. Thus, higher activity can be obtained with these techniques.

Along with the cell-free enzymes, whole-cell associated biocatalysts, in the presence or absence of support materials, have also been successfully used in organic solvents [208–211].

Many techniques developed in the 1970s were further implemented to improve enzyme performance [272]. For instance, site-specific enzyme immobilization via a variety of genetically engineered tags attracted much attention [198] because of better retention of activity. Covalent attachment of the enzyme to the carrier in organic media was found to be unique in that the immobilized enzyme obtained is active in organic solvents whereas the immobilized enzyme prepared in aqueous medium is completely inactive, even though their hydrolytic activity was almost identical [199]. This inspired much research interest in preparing active and stable immobilized enzymes in organic solvents in 1990s, as will be discussed below.

1.2.6
Rational Design of Immobilized Enzymes (1990s–date)

Since the 1990s (Table 1.8) there has been an important transition in the development of immobilized enzymes. Approaches used for the design of immobilized enzymes have become increasingly more rational; this is reflected in the use of more integrated and sophisticated immobilization techniques to solve problems that cannot be easily solved by previously developed single immobilization approaches.

In this phase, the major focus of enzyme immobilization was on the development of robust enzymes that are not only active but also stable and selective in organic solvents. Although in the period from the 1970s to the 1980s it was recognized that many enzymes are active and stable in organic solvents under appropriate conditions, the enzymes used are usually less active or stable in organic solvents than in conventional aqueous media [177]. For this reason development of more robust immobilized enzymes which can work under hostile conditions, especially in non-aqueous media came to the forefront of many research interests in this period [219–231].

Table 1.8 Survey of important enzyme-immobilization techniques developed from the 1990s until the present

Strategies	Improvement	Remark	Ref.
Formation of plastic immobilized enzymes in organic solvent	Stability and activity in organic solvents	Stable in organic solvents	240, 241
Introduction of tentacle carriers	Substantial enhancement of enzyme loading	High enzyme loading and less diffusion limitation	242
Chemical post-immobilization techniques	Improvement of enzyme stability and activity	Such as pH-imprinting, consecutive modification, solvent washing and increasing pH, addition of additives	223, 243
Stabilization–immobilization strategy	Enzyme stability in organic solvents, followed by entrapment techniques	The enzyme is stabilized first, followed by another suitable immobilization strategy	245, 246
Engineering the microenvironment	Improvement of enzyme stability and activity	Improvement of enzyme stability in organic solvent	248
Strengthening the multipoint attachment	Improvement of stability	Increasing the number of bonds enhanced the enzyme stability	249
Site-directed enzyme immobilization	Improvement of enzyme activity and stability	Orientation of enzyme on the carrier surface improves activity retention	250
Imprinting-immobilization strategies	Improvement of enzyme selectivity by conformer selectors	Alter the enzyme selectivity by sol–gel techniques and/or cross-linking techniques	251
Improved sol–gel entrapment	Improvement of activity and selectivity	Selection of suitable monomers and conformer selector is essential	47, 48
Cross-linked enzyme aggregates	Improvement of enzyme activity and selectivity	Selectivity relative to CLEC was improved	16
Entrapped CLEA	Improvement of the mechanical stability and tailor-made particle size	CLEA can be prepared in a pre-designed hollow microsphere	252
Non-covalent L-B-L immobilization	Substantial enhancement of enzyme loading	Less diffusion limitation	254
Enzyme deposition techniques	Improvement of enzyme dispersion state in organic solvent	Monolayer principle	201

Among these, many efforts were devoted to the development of cross-linked enzyme crystals (CLEC) suitable for biotransformations in non-aqueous media or in organic–water mixtures, because of the greater stability of the enzymes under hostile conditions [232–234]. Remarkably, it has been noticed that the performance of the CLEC obtained is highly dependent on the predetermined conformation of the enzyme molecules in the crystal lattice. Thus, selection of a highly active enzyme conformation by varying the crystallization conditions becomes crucial for the creation of highly active, stable and selective CLEC.

Because the process of protein crystallization is homogenization of enzyme conformation, and the enzyme conformation in the crystal lattice is predetermined by the crystallization conditions, each type of cross-linked enzyme crystal of the same enzyme might represent only a specific immobilized enzyme whose conformation is homogeneous and fixed by the cross-linking. Although it is possible to crystallize an enzyme in different conformations, and thus to modulate its properties, this technology is obviously laborious and limited compared with carrier-bound methods.

Broad analysis of the performance of CLEC and comparison with conventional carrier-bound immobilized enzymes is still lacking. Cross-linked enzymes, especially CLEC, have occasionally been compared with lyophilized enzyme powder, which was proved to be not only less active but also less selective [235]. A few studies have also shown that the turnover frequency of cross-linked enzymes in organic solvents is generally lower than that of carrier-bound immobilized enzymes based on the same protein mass and the same reaction conditions, suggesting that rigidification or confinement of the enzyme molecules in the compact crystal lattice or diffusion limitation might be major factors responsible for the lower activity [236].

Increasing efforts have also been devoted to developing novel strategies for improving the performance of immobilized enzymes, for example their activity [237], selectivity [207, 251] and stability [239], by combining different immobilization techniques [236]. For example, biocatalytic plastics, which are prepared by polymerizing lipid-coated organic solvent-soluble enzymes bearing attached polymerizable double bonds, were found to be highly stable and active in organic solvents [240, 241] and polymerization of enzyme derivatives bearing unsaturated bonds in the presence of ligands or substrates proved to be an effective means of creation of insoluble enzymes with improved selectivity [207]. The use of dendrimeric (or tentacle) carriers has been found advantageous in that enzyme loading can be dramatically enhanced by at least one order of magnitude, with high retention of enzyme activity and stability [242]. Furthermore, sol–gel techniques were only recognized as an interesting immobilization technique in the 1990s – 40 years after the first use of these techniques for enzyme immobilization [46–48].

Apart from improvement by immobilization strategies, in the 1990s it was also increasingly appreciated that attachment of enzymes to the selected carriers is not the whole story of enzyme immobilization, because the performance of the immobilized enzymes can be substantially improved by use of various post-immobilization techniques; for example, strengthening the multipoint attachment often enhances enzyme stability [249] and consecutive treatment of immobilized enzymes by chemical or physical modification, or other activation and stabilization techniques, can dramatically improve enzyme performance [243], as will be discussed elsewhere in this book.

More important, the molecular imprinting techniques originally proposed in the 1970s, has been further developed and extended to several other, related areas [244]. The objective is to improve enzyme performance. For example, pH imprinting of the immobilized enzymes, including enzyme powders, enables maximum activity of the enzyme in anhydrous organic solvents [246], significant activity or selectivity improvement can be achieved by simply lyophilizing the enzymes with the ligands or transition-state analogues or by polymerizing the enzyme–ligand complex under more anhydrous conditions or in aqueous medium.

Remarkably, it has been found that even the stability of the immobilized enzymes can be imprinted. For instance, the temperature optimum of epoxy hydrolase immobilized on DEAE-cellulose was dramatically shifted from 35 to 45 °C if non-ionic detergent Triton X-100 was added during enzyme binding to the carrier [344], suggesting that the stable enzyme conformation induced by the additive was frozen on the carrier. Similarly, it was found that stability of *Candida rugosa* lipase, which was covalently immobilized on silanized controlled-pore silica (CPS) previously activated with glutaraldehyde in the presence of PEG-1500, was increased fivefold compared with the immobilized enzyme without addition of PEG-1500 [209]. Thus, in those cases, the stability or, more precisely, the enzyme conformation induced by the effectors (or additives) was imprinted.

Also worthy of note is that in the last few years of the 1990s it was discovered that not only enzyme crystals but also physical enzyme aggregates could be cross-linked to form catalytically active insoluble immobilized enzymes, nowadays known as CLEA [252, 255, 257]. This discovery might theoretically and/or practically open another possibility for design of robust, highly active, stable and selective immobilized enzymes [16].

As with cross-linked crystalline enzymes, however, the factors which hamper their industrial application lie not in their catalytic properties but in their non-catalytic part – they are usually small and their mechanical stability is usually very poor. This causes difficulties when they are applied to heterogeneous reaction systems, e.g. "solid-to-solid" reaction systems [256], in which large (>100 μm) immobilized enzymes are often chosen to facilitate separation by use of a sieve-plate reactor, as in the kinetically controlled enzymatic synthesis of β-lactam antibiotics [17].

It has recently been found that industrially robust CLEA, with greater activity both in organic solvents and in aqueous media, can be prepared by use of new cross-linking technology [252]. The use of preformed soft hollow microsphere has, moreover, enabled the preparation of CLEA with greater mechanical stability and tailor-made size, and which are thus, in principle, applicable to any reaction system, reactor configuration and reaction medium [253].

By combining the advantages of carrier-bound and carrier-free immobilized enzymes, CLEA with tailor-made properties with regard to both non-catalytic and catalytic function can be designed at will – an attractive proposition for industrial applications.

In general, the techniques currently used for creation of robust immobilized enzymes, which meet both catalytic requirements (desired activity, selectivity, and stability) and non-catalytic requirements (desired geometric properties such as shape, size and length) expected for a given process, are all characterized in that an combined method are used to solve problems that are unsolvable by the straightforward method.

1.3
Immobilized Enzymes: Implications from the Past

Having discussed the historical development of immobilized enzymes in the past 90 years, we are interested in the status of immobilization techniques. In this section, we briefly summarize what has been achieved, what more we can achieve, and what will be achieved in the near future.

1.3.1
Methods of Immobilization

More than 5000 publications, including patents, have been published on enzyme-immobilization techniques [259–263]. Several hundred enzymes have been immobilized in different forms and approximately a dozen immobilized enzymes, for example amino acylase, penicillin G acylase, many lipases, proteases, nitrilase, amylase, invertase, etc., have been increasingly used as indispensable catalysts in several industrial processes.

Although the basic methods of enzyme immobilization can be categorized into a few different methods only, for example adsorption, covalent bonding, entrapment, encapsulation, and cross-linking [264], hundreds of variations, based on combinations of these original methods, have been developed [265]. Correspondingly, many carriers of different physical and chemical nature or different occurrence have been designed for a variety of bio-immobilizations and bio-separations [143, 262, 263]. Rational combination of these enzyme-immobilization techniques with a great number of polymeric supports and feasible coupling chemistries leaves virtually no enzyme without a feasible immobilization route [266].

It has recently been increasingly demonstrated that rational combination of methods can often solve a problem that cannot be solved by an individual method. For instance, the physical entrapment of enzymes in a gel matrix often has drawbacks such as easy leakage, serious diffusion constraints, and lower stability than that for other immobilized enzymes. These drawbacks can, however, be easily solved by rational combination of different methods. For instance, higher stability can be achieved by means of the so-called pre-immobilization stabilization strategy [266] or post-immobilization strategy [267].

In the former case, the enzyme, for instance, can be first crosslinked to form stabilized enzyme preparations e.g. CLEA. Subsequent entrapment endows the CLEA with a suitable particle size and high mechanical stability [266]. Stabilization can be also achieved by chemical modification [267]. For instance, chemical modification of the soluble enzyme with a hydrophilic polymer often stabilizes the enzyme because of the introduction of a favourable hydrophilic microenvironment. Thus, the subsequent entrapment of the stabilized enzyme often leads to the formation of more stable enzyme, compared with the entrapped native enzymes [238, 267, 270].

In the later case, the entrapped enzyme can be further crosslinked, with the aim of enhancing the stability or avoidance of enzyme leakage. For instance, β-amylase from *Bacillus megaterium* immobilized in BSA gel matrix and subsequently covalently crosslinked was fourteen time more thermally stable than the native enzyme [269].

Because of these possibilities, a rational combination of the available methods will definitely facilitate the design of robust immobilized enzymes that can suit various applications. Consequently, use of immobilized enzymes is now becoming commonplace in many fields, for example chemical, medical, pharmaceutical and analytical applications [271–273], with the aim of enabling processes in continuous mode, control of the processes, overcoming cost constraints, and solving problems that were previously approached mainly by chemical means [274] or which could not easily be solved by chemical methods.

There is, nevertheless, still a significant lack of systematic analysis of the methods available. Most enzyme immobilization has been performed without any knowledge of structural information, and the relationship between the performance of the immobilized enzyme and the method selected for immobilization has, so far, rarely been defined or identified. Thus, a central task in the future development of immobilization techniques is probably not to develop new methods of immobilization but to establish guidelines linking the method selected with the performance expected.

1.3.2
Diversity versus Versatility

Despite our increasingly understanding of enzyme-immobilization techniques, and the numerous possible means of obtaining robust immobilized enzymes, development of a robust immobilized biocatalyst which can meet the requirements of modern biocatalytic processes – mild reaction conditions, high activity, high selectivity, high operational stability, high productivity, and low cost [275] – still relies on laborious trial-and-error experimental approaches [276]. Consequently, a crucial question is whether it is possible to develop a generic method or to establish generic guidelines for enzyme immobilization. Obviously, the answer to this question lies both in the reality of different immobilization techniques and the peculiarity of each individual application. The establishment of the guidelines necessitates systematic analysis of the methods available and the experimental information that has been obtained in the past. The poor comparability of many experimental results (obtained by different groups and people) seriously hampers the establishment of such universally applicable guidelines, however.

On the other hand, the peculiarities of applications, for example the types of reaction (hydrolytic reaction or reverse reaction), reaction medium (aqueous or organic solvents), reaction system (solid-to-solid, liquid-to-solid, liquid-to-liquid), reactor configuration (stir-tank, plug-flow), economic viability (cost contribution of the immobilized enzyme, space-time yield and productivity) and the intrinsic characteristics of the enzymes selected might differ from case to case. Thus differences between the peculiarities of each application also require specific solution of each individual application.

It must, therefore, be expected that choice of the method of immobilization is mainly dictated by the specific conditions and requirements of each application, which should selectively employ the positive attributives of the method selected. In this sense, the diversity of enzyme-immobilization techniques could be a powerful asset in the design of robust immobilized enzymes, because changes in the peculiarities of the applications often require design of new immobilized enzymes which fit the new applications.

It has, for instance, been demonstrated that differently carrier-bound immobilized penicillin G acylase (PGA) is not only suitable for catalysis of the hydrolysis of penicillin G for production of 6-APA – the nucleus of semi-synthetic β-lactam antibiotics (amoxicillin or ampicillin) [277] – it has also recently been increasingly applied to the synthesis of semi-synthetic β-lactam antibiotics (amoxicillin or ampicillin). In the hydrolysis many types of carrier-bound immobilized penicillin G acylase can be used, for example PGA immobilized on Eupergit C (PcA) or PGA immobilized on polyacrylamide, whereas for the synthesis of semi-synthetic antibiotics such as ampicillin, cephalotin, and cephalexin only few carrier-bound immobilized penicillin G acylases, for example gelatin-bead-bound or agarose bead-bound proved advantageous in terms of the high ratio of synthesized antibiotic to hydrolytic product [17, 249, 277].

Another example is the development of immobilized amino acid acylase for use in the production of chiral amino acids [278]. Among a number of preparations obtained by different methods of immobilization, several promising products, for example the enzyme ionically bound to DEAE-Sephadex, covalently bound to iodoacetyl cellulose, or entrapped in PAAm polyacrylamide gel matrix, were screened for further evaluation. Because of the possibility of regenerating the carrier, the stability of the immobilized enzyme, the ease of immobilization, and the cost of the immobilized enzyme, only DEAE-Sephadex was selected for the final process – resolution of racemic amino acid esters. Remarkably, although amino acid acylase immobilized on DEAE-Sephadex was the first enzyme used in commercial processes, the same immobilized enzyme was recently found to be inactive for resolution of racemic amines or alcohols in organic solvents [279].

The importance of diversity in enzyme immobilization techniques has recently been beautifully demonstrated by screening of carriers for immobilization of glycolate oxidase. Twenty-one different carriers were screened, ranging from natural polysaccharide-based carriers such as CNBr-activated agarose sepharose or epoxy activated agarose or sepharose, ionic exchange CH-sepharose, hydrophobic adsorbents such as phenyl sepharose, to synthetic organic carriers such as epoxy carriers such as Eupergit C, Eupergit C250 L, azalactone carrier such as Emphaze, ionic exchangers Bio-Rex 70, hydrophobic adsorbent such as Amberlite XAD 4, XAD 8, to inorganic carriers such as silanized CPG glass bead derivatives and silanized celite derivatives. The coupling mode covers three types, namely physical adsorption, ionic binding and covalent binding [280].

Remarkably, it was found that among the polysaccharide-based carriers immobilized enzyme with higher activity and retention of activity was obtained with CNBr-activated agarose Sepharose. In contrast, epoxy activated agarose or Sepharose usually afforded lower activity. Remarkably, comparable activity was obtained with synthetic epoxy carrier, i.e. Eupergit C [280]. This example strongly suggests that the performance of a carrier-bound immobilized enzyme is dictated by the physical and chemical nature of the carrier (e.g. chemical composition, binding chemistry, hydrophilicity, pore size and etc.) and that a good carrier or a suitable binding chemistry for an enzyme is not necessary the right one for other enzymes or other applications.

Because of the diversity of carrier nature in terms of the source (synthetic/natural, organic/inorganic), structure (porous/nonporous), the diversity in coupling chemistry, the nature of the interaction (physical, specific adsorption, covalent), designing or screening of a specific immobilized enzyme that suits a specific application becomes possible.

Thus, there is no doubt that changing the type of reaction (hydrolysis or condensation), the reaction medium (aqueous solutions or organic solvents), the reaction system (heterogeneous or homogeneous), the reaction conditions, or even the substrates might lead to a change in the criteria used to assess the robustness of the immobilized enzymes. On the other hand, the diversity of carrier nature (physical and chemical), the binding chemistry, and different immobilization methods provide us an indispensable tool for the design of robust immobilized enzymes.

1.3.3
Complimentary versus Alternative

Enzymes belong to the category of natural catalyst which includes DNA, RNA and catalytic antibodies. A unique function of enzymes is that all the reactions they catalyse, can be performed sequentially, selectively and precisely under mild physiological reaction conditions. This unique feature makes enzymes very attractive for synthetic chemistry, which is usually based on use of hazardous reactants and reaction conditions.

There is, however, no doubt that many are not ideal catalysts for industrial applications, for example in the manufacture of fine chemicals [281, 282] and pharmaceuticals and their intermediates [283], in which the enzymes are usually exposed to non-natural conditions such as high substrate concentrations, high pH, high temperature and the use of deleterious organic solvents. Accordingly, for most industrial applications, they must be modified either by genetic engineering or by chemical modification, with the objective of improving their selectivity, activity and durability under the process conditions. They must, furthermore, be used in the immobilized forms, to reduce production cost by facilitating downstream processing such as recycling and separation [284].

In the last decade, although it has been increasingly appreciated that genetic engineering is a powerful tool for improvement of enzyme performance, enzyme immobilization is the only technique, which can combine immobilization of an enzyme with improvement of enzyme performance, for example stability, selectivity and activity [285]. Thus, immobilization-improvement strategies might be very attractive for enzymes designed to be used in the immobilized form anyway (Scheme 1.3). In this sense, it is also increasingly recognized that rational immobilization of enzymes by combining immobilization and genetic engineering might be an alternative and complimentary technique for protein engineering.

Many examples have excitingly demonstrated that even for genetically engineered enzymes performance can be further improved by immobilization techniques and many examples have revealed that enzyme-immobilization techniques are indeed an indispensable complimentary tool in enzyme engineering, due to its potential for:

- combination of immobilization and improvement,
- modulation of enzyme performance by selecting appropriate method of immobilization, and
- combination of different immobilization methods.

As shown in Scheme 1.3, improvement by immobilization is obviously straightforward compared with genetic engineering. Furthermore, improvement by immobilization does not normally obviate immobilization of the exact structure.

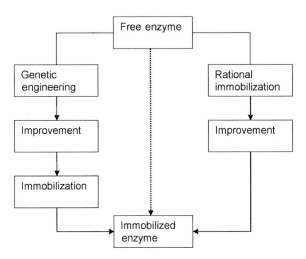

Scheme 1.3 Comparison of improvement of an enzyme by genetic engineering and by immobilization.

As more information becomes available about the relationship between the performance of the immobilized enzyme and the method selected, design of more robust immobilized enzymes at will, via the use of different immobilization techniques, might be a reality in the near future.

It is currently possible to draw the conclusion that immobilized enzymes might perform better than the native enzymes (improved stability, activity and stability) if the method is correctly selected. This will be discussed in Section 1.3.4 [285–287].

1.3.4
Modification versus Immobilization

As already noted, the problem of enzyme immobilization is not how to immobilize the enzyme but how to achieve the desired performance for a given application by selecting an appropriate means of immobilization. Thus, it is also important to distinguish the two concepts – modification and immobilization – before we enter discussion.

Although enzyme immobilization and improvement of enzyme performance by immobilization share the same principles, the emphasis is different. The former is mainly associated with efforts to find suitable immobilization methods for enzymes that must be immobilized for certain applications. Thus, the immobilization technique developed is mainly intended to retain the major catalytic functions of the native enzymes. In contrast, improvement-by-immobilization is focused mainly on utilization of available immobilization techniques to alter (or improve) enzyme performance, to suit the desired application. Thus, the native enzyme might be not suitable for a desired process, because of its poor performance such as lower activity, or stability or selectivity. Consequently, the technique to be developed should improve the performance of the enzyme besides immobilizing it. Because the success of the latter largely depends on knowledge acquired from experimental information from the former application, we recognize it is essential to provide detailed analysis of the results so far obtained from improvement, by immobilization, of three catalytic characteristics, i.e. activity, stability and selectivity, of the enzymes.

1.3.4.1 **Enhanced Stability**
Of these three important characteristics of enzymes stabilization by immobilization has been studied since the 1970s, when immobilized enzymes became increasingly used in industrial processes, in which the cost-contribution of the immobilized enzyme is often the indicator of process viability [285]. Since then, many useful strategies have been developed for stabilization of enzymes by immobilization, for example cross-linking, multipoint attachment and engineering of the microenvironment, confining the enzyme molecules, etc.

The stability of a native enzyme (i.e. a non-immobilized or modified enzyme) is principally determined by its intrinsic structure whereas the stability of an immobilized enzyme is highly dependent on many factors, including:

- the properties of its interaction with the carrier,
- the binding position and the number of the bonds,
- the freedom of the conformation change in the matrix,
- the microenvironment in which the enzyme molecule is located,
- the chemical and physical structure of the carrier,
- the properties of the spacer (for example, charged or neutral, hydrophilic or hydrophobic, size, length) linking the enzyme molecules to the carrier, and
- the conditions under which the enzyme molecules were immobilized.

Whatever the reason, the enhanced stability resulting from immobilization can often be ascribed to the intrinsic features of individual immobilization processes, for example:

- molecular confinement (which occurs often in the entrapment process, particularly the sol–gel process) [289];
- favourable microenvironment – achieved by selecting appropriate carriers [256] or engineering the microenvironment by post-immobilization techniques [258];
- chemical modification effect in covalent bonding (such as formation of an extra hydrogen bond as a result of chemical modification in the covalent immobilization process) [290]; and
- rigidification of conformation as a result of multipoint attachment [184].

It might, nevertheless, also be true that many stabilization factors can be integrated into one immobilization process, as in the stabilization and immobilization procedures [291] and three-dimensional immobilization (by cross-linking crystalline enzymes or enzyme aggregates) [16]. Also, it is very difficult to judge which method can give the most stable enzymes, because even the same method (let us say covalent immobilization) might lead to immobilized enzymes of different stability, depending on the carrier selected, the immobilization conditions (e.g. enzyme loading, pH, temperature, ionic strength, additives) [292] or subsequent treatment.

It will, however, never be found that an enzyme cannot be stabilized. Thus, stabilization by immobilization can be always achieved by selecting a suitable immobilization method. One can confidently state that stabilization by immobilization is currently no longer an exception, because of our increasing understanding of the immobilization processes. Remarkably, it has been found that even thermophilic enzymes or extremophilic enzymes [293] can be further stabilized by immobilization [239, 294–296, 230], suggesting that stabilization of enzyme by immobilization can be additive.

1.3.4.2 Enhanced Activity

Observation of the enhancement of enzyme activity by immobilization can be dated back to the early 1960s, when Goldstein et al. noted that for trypsin immobilized on a charged carrier the K_m for charged substrates could be reduced by a factor of fourteen [79], because of the so-called microenvironment effect.

It has been found that many types of enzyme immobilized by different immobilization techniques have higher activity than the native enzymes. For instance, epoxy hydrolase adsorbed on DEAE-cellulose by ionic bonding was more than twice as active as the native enzyme [344], lipase–lipid complex entrapped in n-vinyl-2-pyrrolidone gel matrix was 50-fold more active than the native enzyme [298]. Activation by immobilization is, however, often regarded as an extra benefit rather than a rational goal of enzyme immobilization.

Activity retention by carrier-bound immobilized enzymes is usually approximately 50 %. At high enzyme loading, especially, diffusion limitation might occur as a result of the unequal distribution of the enzyme within a porous carrier, leading to a reduction of apparent activity [299]. The conditions for high activity retention are often marginal, thus often requiring laborious screening of immobilization conditions such as enzyme loading, pH, carrier and binding chemistry [292].

Next to the microenvironment effect mentioned above, it has been demonstrated that immobilized enzymes can be more active than the native enzymes, when the inhibiting effect of the substrate was reduced. For example, immobilization of invertase from *Candida utilis* on porous cellulose beads led to reduced substrate inhibition and increased activity [301]. A positive partition effect (enrichment of substrates in the proximity of the enzymes) might also enhance enzyme activity as was observed for kinetically controlled synthesis of ampicillin with penicillin G acylase immobilized on a positively charged carrier [302] or horse liver alcohol dehydrogenase immobilized on poly(methylacrylate-co-acrylamide) matrix [124, 157].

Greater retention of enzyme activity can occasionally be achieved, especially for allosteric enzymes such as some lipases which have lids covering the active centre; conformational change increases the accessibility of the active centre [303–306]. In other instances of improvement of the molecular accessibility of enzymes by immobilization, enhancement of enzyme activity relative to that of the native enzyme powder can be achieved when the enzymes are intended for use in anhydrous organic solvents, for example lipase PS immobilized in sol–gel [236] or protease covalently bonded to silica, because of increasing dispersion of the enzyme molecules and conformation induction.

In general, the activity of the immobilized enzymes can be enhanced by at least ten different effects involved in enzyme immobilization:

- microenvironment effect,
- partition effect,
- diffusion effect (reducing the pH),
- conformational change,
- flexibility of conformational change,
- molecular orientation,
- water partition (especially in organic solvent),
- conformation flexibility,
- conformation induction, and
- binding mode.

For conformation-controlled activity it was found that the enzyme activity (U mg^{-1} protein immobilized) was strongly dependent on the nature of the carriers used. For instance, the activity of lipase PS (*Pseudomonas cepacia*) immobilized on Toyonite, Celite, glass and Amberlite was highly dependent on the nature of the carrier. The highest activity for transesterification in organic solvent was obtained with Toyonite (37.2 µmol min^{-1} mg^{-1}) and the lowest activity was obtained with Amberlite (0.4 µmol min^{-1} mg^{-1}) [307]; the difference in activity is approximately two orders of magnitude!

Enhancement of activity in organic solvents after immobilization by sol–gel processes was clearly demonstrated for lipases, the activity of which in organic solvents, relative to that of the native enzyme powders, can be increased at least fivefold by use of conformer selectors such as surfactants or crown ethers, suggesting that the presence of conformer selectors induced an active conformation, which is, however, frozen by the corresponding immobilization process [47].

Molecular orientation-controlled activity of enzymes was observed early in 1972 [163]. In connection with this observation and enzyme immobilization in organic solvents, it was also demonstrated that the lipase from *Mucor risopus* immobilized in organic solvent was more active in transesterification in organic solvent whereas the lipase immobilized in aqueous medium had almost no activity in organic solvents. The author suggested that the position of binding of the enzymes to the carrier in organic solvents is different from that when immobilization is performed in aqueous medium [199]. Many other types of immobilized enzyme, which can be categorized as immobilized enzymes with orderly oriented enzyme molecules generally have higher activity or stability relative to the counterpart (randomly immobilized enzymes), because of favourable accessibility or avoidance of the modification of the active site [308].

The effect of conformation flexibility is often in contradiction with enzyme activity, i.e. reduction of enzyme conformation flexibility often reduces the enzyme activity. This was confirmed initially by the observation that immobilization of an enzyme on a carrier via a suitable spacer often resulted in better retention of activity than if the enzyme was immobilized without a spacer [309–312].

In contrast, higher activity has been achieved by increasing conformational flexibility. For example, amino acid acylase immobilized ionically on DEAE-cellulose has high activity after post-treatment with a denaturant, which could possibly enhance the enzyme conformational flexibility [352]. For enzymes acting in low-water media, especially, enzyme conformational flexibility is much less than in aqueous media. Thus, if the water content of enzyme preparations is kept to a minimum that enables the enzyme to have the highest conformational flexibility, maximum activity may be achieved in organic solvent because of higher conformational flexibility. For quantitative control of water hydration level, water activity was developed [190].

Nevertheless, water activity is not the whole story of enzyme activity in organic solvents. The fact that dehydration history largely dictates enzyme activity rather than the water activity suggests that some dehydration processes might reversibly deactivate the enzyme [313]. Consequently, it is concluded that the water-activity concept is only valid when the enzyme preparation is not reversibly deactivated by the process for dehydration.

The effect of binding mode on the enzyme activity can be reflected by three factors – the number of bonds formed between the carrier and the enzyme molecules, the position of the bonds and the nature of the bonds. It is easily conceivable that the greater the number of bonds formed between the enzyme and the carrier, the lower the enzyme activity, as demonstrated by immobilization of β-galactosidase *E. coli* and *K. lactis* on thiolsulphinate-agarose and glutaraldehyde-agarose [313]. The greater retention of activity with thiolsulphinate-agarose can be largely ascribed to the fewer bonds formed between the enzyme and the carrier – thiolsulphinate-agarose [314]. Indeed, these two enzymes are much richer in the lysine residues than the cysteine residues. Thus, more bonds can be formed with glutaraldehyde-agarose, resulting in less retention of activity [314].

Interestingly, a recent example showed that α-amylase immobilized on thionyl chloride ($SOCl_2$) activated poly(Me methacrylate-acrylic acid) microspheres has 67.5 % retention of activity whereas 80.4 % was achieved with carbodiimide (CDI)-activated poly(Me methacrylate-acrylic acid) microspheres. Irrespective of whether the enzyme is immobilized on the same carrier with the same binding nature. It was, moreover, found that the former is twice as stable after storage for 1 month. On the other hand the free enzyme lost its activity completely in 20 days. Apparently, this difference can be solely ascribed to the difference in the position of the bonds formed [315].

1.3.4.3 Improved Selectivity

The selectivity of enzymes is nowadays becoming a powerful asset of enzyme-mediated asymmetric synthesis, because of the increasing need of the pharmaceutical industry for optically pure intermediates [312].

In general, the selectivity of enzymes includes [317]:
- substrate selectivity – the ability to distinguish and act on a subset of compounds within a larger group of chemically related compounds;
- stereoselectivity – the ability to act on a single enantiomer or diastereomer exclusively;
- regioselectivity – the ability to act exclusively on one location in a molecule;
- functional group selectivity – the ability to act on one functional group selectively in the presence of other equally reactive or more reactive functional groups, for example the selective acylation of amino alcohols [318].

Although a dramatic change of enzyme selectivity by genetic engineering has been beautifully demonstrated [225], there are also numerous attractive examples in which enzyme selectivity has been changed by a variety of immobilization techniques, for example covalent bonding, entrapment, and simple adsorption, as discussed in the following section (for details see Table 1.9). In several extreme instances it has been demonstrated that a non-selective enzyme such as chloroperoxidase was transformed into a stereoselective enzyme after immobilization [319]; the *S*-selective lipase has also been converted to *R*-selective CR lipase by covalent immobilization [320].

Table 1.9 Alteration of selectivity by immobilization

Method of immobilization	Selectivity	Remark	Enzyme	Ref.
Adsorption	Regioselectivity	Non-selective chloroperoxidase was transferred to stereoselective enzyme after immobilization	Chloroperoxidase	318
Adsorption, covalent	Enantio-selectivity	Selectivity was improved but dependent on the reaction conditions	Lipases	319 340
Covalent CPG	Product map		Subtilisin	324
Covalent	Product map	Resulted in different product map	Proteases	326
Covalent immobilization on glyoxal agarose	Different selectivity?	Resulted in different selectivity	Urokinase	327
Covalent?	Product map	Immobilized on phenol-formaldehyde resulted in different product composition	α-Amylase	328
Gelatin-entrapped, and surface-bound	Action pattern	The action pattern depended on how the enzymes were immobilized, namely entrapped or linked to the surface	Glucoamylase	329
Adsorption on DEAE-cellulose	Substrate selectivity	Resulted in changed substrate selectivity relative to the native enzyme	Dextransucrease	330
Adsorption	Enantio-selectivity	The selectivity of the enzyme is reduced	Lipase CAL-B	331
Covalent on agarose	Enantio-selectivity	Selectivity is dependent on the carrier and binding chemistry	Lipase CRL	332 333
Covalent on silica	Enantio-selectivity	The enantioselectivity was enhanced 7-fold relative to the free enzyme with trichlorotriazine as activating agent	CRL	332

Table 1.9 Continued

Method of immobilization	Selectivity	Remark	Enzyme	Ref.
Adsorption on celite	Enantio-selectivity	Simple adsorption enhanced the stability of *Candida rugosa* lipase against acetaldehyde, and the selectivity	Lipase CRL	334
Entrapment	Enantio-selectivity	Enhanced selectivity compared with the native enzymes	RML	335
Sol–gel entrapment	Enantio-selectivity	Chiral template influenced enzyme selectivity	Lipase	252
Entrapment in Ca-alginate gel beads	Enantio-selectivity	Increased threefold the enantioselectivity of pegylated PCL in Ca-alginate gel beads	Pegylated PCL	336
Sol–gel entrapment	Enantio-selectivity	Immobilization can trap different enzyme conformation	Fructose-1,6-bisphosphatase	337
Double immobilization technique	Substrate selectivity	(Adsorption on solid carrier, followed by entrapment in alginate beads) resulted in different product spectrum	2-Mannosidase	338
Covalent immobilization	Reaction selectivity	The ratio of condensation to hydrolysis in the kinetically controlled synthesis of β-lactam antibiotics depended on the immobilization methods	Penicillin G acylase	339
Covalent	Enantio-selectivity	The enantioselectivity of lipase MML immobilized on oxirane carrier is dependent on newly introduced functionalities	*Mucor miehei* lipase	340
Ionic binding	Enantio-selectivity	The selectivity is mainly dictated by the nature of pending charged groups	Alkylsulphatase	350

In general, the selectivity that can be influenced by the immobilization techniques can be classified into the following categories, according to the source of the effect:

1. Carrier-controlled selectivity
   a) pore size-controlled selectivity
   b) diffusion-controlled selectivity

2. Conformation-controlled selectivity
   a) microenvironment-controlled selectivity
   b) active centre-controlled selectivity

The effect of steric hindrance on enzyme selectivity, for example the product map, was observed in 1970s [324]. For example, the product pattern of CPG (controlled pore glass)-immobilized subtilsin-catalysed digestion of proteins can be affected by the pore size of the carrier used [325]. Similarly, immobilized ATP deaminase, β-galactosidase [325] and proteases also have different product maps [328, 329]. Urokinase covalently immobilized on glyoxal agarose has different selectivity [326]. α-Amylase immobilized on silica [329] or covalently bound to CNBr-activated carboxymethylcellulose [324] afforded products of composition different from that of the native enzyme. This was largely attributed to the fact that the size of the pores where the enzyme molecules are located determines the accessibility of the substrates, depending on their size.

Diffusion-controlled enantioselectivity was reported recently after a study of the enantioselectivity of lipase CAL-B in the transesterification in organic solvents [332]. For the first time it was reported that diffusion can reduce the enantioselectivity of enzymes. A relevant example worth mentioning is that simple adsorption of lipase CRL on Celite not only enhanced the stability of *Candida rugosa* lipase against acetaldehyde but also enhanced the enantioselectivity up to threefold [331]. It is possible that improvement of enzyme dispersion enhanced the enantioselectivity of the immobilized enzymes relative to the enzyme powders.

The important implication of this discovery is that in diffusion-controlled enantioselectivity reduced enantioselectivity is always accompanied by reduced reaction rate [332]. When screening an enzyme for resolution of racemic compounds it is essential to ensure that the enzyme preparation selected has no diffusion constraints. Otherwise the real potential of the enzyme might be overlooked [332].

Not only can immobilization change the selectivity (product map or enantioselectivity), the presence of diffusion constraints can also affect the selectivity between two reactions that might occur in parallel in the same reaction system. One example is the kinetically controlled synthesis of peptides or β-lactam antibiotics in which one of the reactants, for example an amino acid ester (or generally called active acyl donor), can be integrated into the desired product (S) or hydrolysed into the unwanted amino acid (H) [17]. Thus the S/H ratio was regarded as a criterion of the viability of the corresponding process [255].

As with conformation-controlled selectivity, there are often difficulties distinguishing microenvironment effect from conformation change. For instance, entrapment of RML in cellulose acetate–TiO_2 gel fibre improved selectivity in the hydrolysis of 1,2-diacetoxypropane, compared with that of native enzymes [336], and the enantioselectivity of pegylated PCL was increased threefold by entrapment in Ca-alginate gel beads [341]. In such cases the lipases might adopt a conformation different from that in the native enzymes owing to interaction between the carrier and the enzyme (change of the enzyme conformation) or to the micro-environmental effect (pH gradient).

The micro-environmental effect has, however, been clearly demonstrated for 1,2-α-mannosidase, for which a double immobilization technique, adsorption on china clay or cellulose DE-52, followed by entrapment in alginate beads, was used; the product spectrum obtained depended on the carrier used for adsorption before entrapment in sodium alginate [339]. Similarly, the substrate selectivity of dextransucrease adsorbed on DEAE-cellulose was different from that of the native enzyme [329].

Most strikingly, it has recently been found that the enantioselectivity of CRL immobilized on silica activated with 2,4,6-trichloro-1,3,5-triazine was approximately seven times higher than that of the soluble enzyme whereas CLR immobilized on agarose activated with tosylate was only four times more selective than the native enzyme [333], implying that chemical modification of the enzyme by active carriers can also affect enzyme selectivity.

Similarly, it was recently found that enzyme activity and selectivity can be also influenced by the nature of the pendant binding functionality. For example, the enantioselectivity of alkylsulphatase immobilized on anionic exchangers such as DEAE-Sephadex, TEAE-cellulose, and Ecetola-cellulose differed substantially, depending on the pendant ionic groups. Immobilization of alkylsulphatase on Ecetola-cellulose enhanced the selectivity severalfold in the hydrolysis of *sec*-alkyl sulphates. Because TEAE-cellulose and Ecetola cellulose differ mainly in the spacer, the selectivity of the immobilized enzyme is mainly dictated by the side chain and the spacer of the binding functionality. Enhancement of the selectivity might be because the charged groups might be able to approach certain negatively charged domains or sites (e.g. the active sites) [351]

Conformation-controlled selectivity was also recently observed for so-called molecular impinging techniques, which are based on the hypothesis that the conformation induced by a ligand can be frozen by physical or chemical means such as lyophilization or cross-linking or molecular confinement. One possible explanation is that the population of some enzyme conformers is enhanced by the conformer selectors used and, consequently, enzyme selectivity toward some substrates can be improved, as is exemplified by the so-called molecular imprinting techniques (MIT) [251].

When improving the selectivity of enzymes by immobilization it is essential to pay attention to medium engineering, because microenvironment–controlled selectivity is not only related to the carrier selected but also to the medium used. Immobilization of the enzyme often results in a change in the optimum pH or temperature. Thus, enzyme characteristics such as activity and selectivity, which are closely related to the pH and temperature, might be correspondingly changed. The optimum pH for selectivity expression might also be different from that of the native enzyme; this was shown by a recent study of catalysis of the resolution of (*R,S*)-

mandelic acid methyl ester by immobilized CRL [320]. In this process the extent of selectivity enhancement was strongly related to the pH of the medium used.

In general, improvement of enzyme enantioselectivity by immobilization might be attractive, because of its simplicity and universal applicability and because it usually obviates the need for detailed structural information.

As discussed above, enzyme immobilization can be regarded as a modification process. It is hardly surprising that the performance of the immobilized enzyme depends on the modification (e.g. the immobilization conditions), the nature of the modifier (i.e. the selected carriers) and the nature of the enzymes (source, purity and strain) to be modified.

With regard to the similarity of enzyme immobilization and chemical modification [321], many methods and principles which are widely used for chemical modification of enzymes to enhance enzyme functionality can also be used to improve the performance of the carrier-bound immobilized enzymes. For instance, the stabilization of enzymes by chemical modification can usually be achieved by two major approaches – rigidification of enzyme scaffold with the use of a bifunctional crosslinker and engineering the microenvironment by introduction of new functional groups which favour of hydrophobic interaction (by hydrophobization of the enzyme surface) or hydrophilization of the enzyme surface (because of mitigation of unfavourable hydrophobic interaction) or formation of new salt bridges or hydrogen bonds (because of the introduction of polar groups) [322]. Similarly, these two principles have been also increasingly applied to improve the enzyme performance for instance the stability, selectivity and activity [323].

1.4
Prospective and Future Development

1.4.1
The Room for Further Development

Although the best method of immobilization might differ from enzyme to enzyme, from application to application and from carrier to carrier, depending on the peculiarities of each specific application, criteria for assessing the robustness of the immobilized enzymes remain the same – industrial immobilized enzymes must be highly active (high activity in a unit of volume, $U\ g^{-1}$ or mL^{-1}), highly selective (to reduce side reactions), highly stable (to reduce cost by effective reuse), cost-effective (low cost contribution thus economically attractive), safe to use (to meet safety regulations) and innovative (for recognition as intellectual property).

As with the volume activity (U enzyme g^{-1} carrier used), most enzymes bound to carriers with particle sizes above 100 μm (minimum size requirements for a carrier-bound immobilized enzyme [284]) have a loading (or payload) ranging from 0.001 to 0.1. The volume ratio of catalyst to reactor is usually in the range 10–20 %. Thus the productivity of most immobilized enzymes is still much lower than in chemical processes, mainly because of the small number of active sites per kg of

1.4 Prospective and Future Development

biocatalyst (low volume activity) [342]. For currently available porous carriers, moreover, activity retention at maximum enzyme loading is often below 50 %, because of diffusion constraints [299].

Although development of carrier-free enzymes such as CLEA [16] or CLEC [233] can eliminate the use of the extra non-catalytic mass-carrier, the intrinsic drawbacks associated with the carrier-free immobilized enzymes, for example narrow reactor configuration (because of the small sizes), laborious screening of conditions for aggregation, crystallization and cross-linking, can hardly make them the first choice for the bioprocess engineers.

Because the carrier not only functions as a scaffold for the enzyme molecules but also strongly modifies the enzyme characteristics, it is conceivable that abandoning the carrier might simultaneously reject a powerful means of modulating enzyme properties (both non-catalytic and catalytic function) which would easily be obtained by use of appropriate carriers, binding chemistry and immobilization methods.

As a result, it is to be expected that the focus in bio-immobilization should be the development of a new method of enzyme immobilization that combines the advantages of carrier-free and carrier-bound methods. In other words, the new method of enzyme immobilization should be able to provide high enzyme loading (close to that of carrier-free enzymes), high retention of activity, and broad reactor configurations. No currently available method can meet these criteria. Thus, the development of carriers with a predetermined chemical and physical nature, especially suitable geometric properties and binding chemistry, which can bind (or hold) enzyme directly under mild conditions and thus can be used in different reactor configurations, will continue to be the major focus of future developments.

As regards the stability of the immobilized enzymes, it is known that any type of immobilization method (entrapment, encapsulation or covalent entrapment or adsorption [335]) has the potential to stabilize the enzymes relative to the native enzymes or that an immobilized enzyme can be better-stabilized than others immobilized by different methods. For example, lipase from *Candida rugosa* entrapped in alginate gel was found to be more stable than the covalently bound enzyme on Eupergit C or the enzyme encapsulated in a sol–gel matrix [347], and immobilized glucoamylase entrapped in polyacrylamide gels was more stable than that covalently bound to SP-Sephadex C-50 [346]. Another striking example is that pronase and chymotrypsin covalently attached to PDMS film are less stable than the entrapped enzymes [349].

It is, therefore, appreciated that each enzyme-immobilization technique is unique and thus the possibility of improving enzyme performance such as activity, selectivity and stability, and pH optimum, is limited. For example, although multipoint attachment can improve enzyme stability, the extent of this stabilization might be limited because only a part of the protein surface is rigidified. Often, adsorption of enzyme on carriers cannot be used to improve, significantly, enzyme performance such as stability, compared with covalent enzyme immobilization. For instance, covalently immobilized limonoid glycosyltransferase is much more stable than its non-covalent adsorbed counterpart [350]. On the other hand, entrapment of the enzyme in hard sol–gel matrix can often be used to stabilize the overall molecule in a spatially restricted three-dimensional matrix.

Thus, the matrix-entrapped enzymes are, occasionally, even more stable than the covalently immobilized enzymes [349]. In contrast, encapsulation of enzymes in semi-permeable capsules often has less effect on enzyme stability, because neither the microenvironment of the enzyme nor the structure of the enzyme molecules is significantly modified. Thus, it is not surprising that combination of a variety of immobilization techniques will increasingly be used to solve problems which cannot be solved by any single immobilization technique.

With regard to improvement of enzyme selectivity, although, as noted above, there are many exciting examples of immobilized enzymes for which selectivity, e.g. reaction selectivity, substrate selectivity, stereoselectivity or chemical selectivity, can be affected by the immobilization procedure [339, 343, 344], perhaps combined with reaction medium engineering [341], improvement of enzyme selectivity by immobilization is still, fundamentally, a new endeavour, lacking guidelines that can be used to guide practical experiments. Nevertheless, as with increasing understanding of the relationship between enzyme selectivity and the structural changes resulting from genetic engineering or other chemical modification, increasing interest in improvement of enzyme selectivity by immobilization can be expected in the near future.

1.4.2
An Integration Approach

As noted above, a vast number of methods of immobilization are currently available. Thus, the major problem in enzyme immobilization is not how to immobilize enzymes, but how to design the performance of the immobilized enzyme at will. Unfortunately, the approaches currently used to design robust industrial immobilized enzymes are, without exception, labelled as "irrational", because they often result from screening of several immobilized enzymes and are not designed. As a result, many industrial processes might be operating under suboptimum conditions because of a lack of robust immobilized enzymes.

Another difficulty in rational design is that the comparability of different methods of immobilization is often very poor, mainly owing to inconsistency in the enzymes used (for example source, purity, contamination), the immobilization conditions (time, pH, additives, ionic strength), the assay (substrate, concentration, temperatures), the preconditioning of the carrier and the post-treatment of catalysts. In addition, many data and results reported in the literature are often incomplete, so many conclusions or explanations are not only obscure but also controversial and misleading.

We therefore surmise it might be more realistic to use a Lego approach. In other words, if the enzyme-immobilization method (or approach) can be generally divided into several essential steps (or components), individual optimization of these by use of a rational design might lead to the more rational creation of a robust immobilized enzyme. Analysis of all the methods of immobilization currently avail-

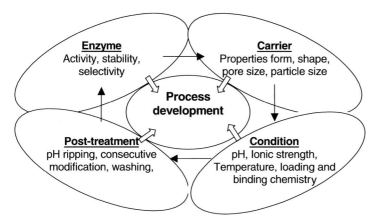

Scheme 1.4 Illustration of general procedures for enzyme immobilization.

able has led to the proposal of a rational general approach to enzyme immobilization based on three stages, selection of enzymes, selection of carriers, and selection of conditions and post-treatments, as shown in Scheme 1.4.

Although many books and reviews dealing with enzyme immobilization appeared in the second half of the last century, the subject still lacks systematic analysis of a general approach to enzyme immobilization, because the books available merely report the feasibility or list the different immobilization techniques [265]. In this context, the author of this book will try to delineate the basic principles governing the individual approaches used to design robust enzymes (Scheme 1.4) and to provide a rational basis for future development of immobilized enzymes.

1.5
References

1 Nelson JM, Griffin EG: Adsorption of invertase. *J Am Chem Soc* 1916, 38:1109–1115
2 Silman IH, Katchalski E: Water-insoluble derivatives of enzymes, antigens and antibodies. *Ann Rev Biochem* 1966, 35:837–877
3 Vandamme EJ: Peptide antibiotic production through immobilized biocatalyst technology. *Enzyme Microb Technol* 1983, 5:403–415
4 Schulze B, Wubbolts MG: Biocatalysis for industrial production of fine chemicals. *Curr Opin Biotechnol* 1999, 10:609–615
5 Stetter H: *Enzymatische Analyse*, Verlag Chemie, Weinheim, 1951
6 Clark LC Jr, Lyons C: Electrode systems for continuous monitoring in cardiovascular surgery. *Ann NY Acad Sci* 1962, *102*:29–45
7 Campbell DH, Luescher FL, Lerman LS: Immologic adsorbents. I. Isolation of antibody by means of a cellulose-protein antigen. *Proc Natl Acad Sci USA* 1951, 37:575–578
8 Watanabe S, Shimizu Y, Teramatsu T, Murachi T, Hino T: Application of immobilized enzymes for biomaterials used in surgery. *Methods Enzymol* 1988, 137:545–551
9 Chang TMS: *Medical application of immobilised enzymes and proteins*, Vols I and II, Plenum Press, New York, 1977

10 Klein MD, Langer R: Immobilised enzymes in clinical medicines: an emerging approaches to new drug therapies. *Trends Biotechnol* 1986, 4:179–186

11 Kircka LJ, Thorpe GHG: Immobilised enzymes in analysis. *Trends Biotechnol* 1986, 4:253–258

12 Dunlap BR: *Immobilised chemicals and affinity chromatography*, Plenum Press, New York, 1974

13 Bickerstaff GF: Application of immobilised enzymes to fundamental studies on enzyme structure and function. In: Wiseman A. (Ed.) *Topics in enzyme and fermentation biotechnology*. Ellis Horwood, Chichester, UK, 1984, pp:162–201

14 Martinek K, Mozhaev VV: Immobilization of enzymes: an approach to fundamental studies in biochemistry. *Adv Enzymol* 1985, 57:179–249

15 Cristallini CLL, Cascone MG, Polacco G, Lupinacci D, Barbani N: Enzyme-based bioartificial polymeric materials: the α-amylase–poly(vinyl alcohol) system. *Polym Int* 1997, 44:510–516

16 Cao LQ, van Langen L, Sheldon RA: Immobilised enzymes: carrier-bound or carrier-free? *Curr Opin Biotechnol* 2003, 14:387–394

17 Bruggink A, Roos EC, de Vroom E: Penicillin acylase in the industrial production of β-lactam antibiotics. *Org Process Res Dev* 1998, 2:128–133

18 Parmar A, Kumar H, Marwaha SS, Kennedy JF: Advances in enzymatic transformation of penicillins to 6-aminopenicillanic acid (6-APA). *Biotechnol Adv* 2000, 18:289–301

19 Van Roon J, Beeftink R, Schroen K, Tramper H: Assessment of intraparticle biocatalytic distribution as a tool in rational formulation. *Curr Opin Biotechnol* 2002, 13:398–405

20 Trevan MD: *Immobilised enzymes*. Wiley, London, 1980

21 Hartmeier W: *Immobilised Biocatalysts*, Springer, 1988

22 Gerbsch N, Buchholz R: New processes and actual trends in biotechnology. *FEMS Microbiol Rev* 1995, 16:259–269

23 Harkins WD, Fourt L, Fourt PC: Immunochemistry of catalase II. Activity in multilayers. *J Biol Chem* 1940, 132:111–118

24 Gale EF, Epps MR: Studies on bacterial amino-acid decarboxylases 1. 1-(+)-lysine decarboxylase. *Biochem J* 1944, 38:232–242

25 Langmuir I, Schaefer VJ: Activities of urease and monolayers. *J Am Chem Soc* 1938, 60:1351–1360

26 Micheel F, Evers J: Synthesis of cellulose-bound proteins. *Macromol Chem* 1949. 3:200–209

27 Stone I: Method of making dextrose using starch glucogenase. US patent 2,717,852, 1955

28 Dickey FH: Specific adsorption. *J Phys Chem* 1955, 59:695–707

29 Mclaren AD: Concerning the pH dependence of enzyme reactions on cells, particulates and in solution. *Science* 1957, 125:697–697

30 Metz MA, Schuster R: Isolation and proteolysis enzymes from solution as dry stable derivatives of cellulosic ion exchangers. *J Am Chem Soc* 1959, 81:4024–4028

31 Mitz MA: New soluble active derivatives of an enzyme as a model for study of cellular metabolism. *Science* 1956, 123:1076–1077

32 Fletcher GL, Okada S: Protection of deoxyribonuclease from ionizing radiation by adsorbents. *Nature* 1955, 176:882–883

33 Fletcher GL, Okada S: Effect of adsorbing materials on radiation inactivation of deoxyribonuclease I. *Radiation Res* 1959, 11:291–298

34 Brandenberg H: Methods for linking enzymes to insoluble carriers. *Angew Chem* 1955, 67:661–661

35 Barnet LB, Bull HB: The optimum pH of adsorbed ribonuclease. *Biochim Biophys Acta* 1959, 36:244–246

36 Grubhofer N, Schleith L: Coupling of proteins on diazotized polyaminostyrene. *Hoppe-Seylers Z Physiol Chem* 1954, 297:108–112

37 Manecke G, Singer S: Chemical transformation of polystyrene. *Makromol Chem* 1960, 37:119–142

38 Brandenberg H: Methods for linking enzymes to insoluble carriers. *Angew Chem* 1955, 67:661–661

39 McLaren AD, Estermann EF: The adsorption and reactions of enzymes and proteins after filtration on (and through) a microcellulose membrane. *Biochem Biophys Res Commun* 1956, 61:158–173

40 Isliker HC: Purification of antibodies by means of antigens linked to ion-exchange resin. *Ann NY Acad Sci* 1953, 3:225–238

41 Grubhofer N, Schleith L: Modified ion-exchange resins as specific adsorbent. *Naturwissenschaften* 1953, 40:508–508

42 Manecke G, Gillert KE: Serologich Spezifische Adsorbentien. *Naturwissenschaften* 1955, 42:212–213

43 Gyenes L, Rose B, Sehon AH: Isolation of antibodies on antigen–polystyrene conjugates. *Nature* 1958, 181:1465–1466

44 Mitz MA, Summaria LJ: Synthesis of biologically active cellulose derivatives of enzymes. *Nature* 1961, 189:576–560

45 Newirth TL, Diegelman MA, Pye EK, Kallen RG. Multiple immobilized enzyme reactors. Determination of pyruvate and phosphoenolpyruvate concentrations using immobilized lactate dehydrogenase and pyruvate kinase. *Biotechnol Bioeng* 1973, 15:1089–1100

46 Braun S, Rappoport S, Zusman R, Avnir D, Ottolenghi M: Biochemically active sol–gel glasses for the trapping of enzymes. *Mater Lett* 1990, 10:1–5

47 Reetz MT, Wenkel R, Avnir D: Entrapment of lipases in hydrophobic sol–gel materials: efficient heterogeneous biocatalysts in aqueous medium. *Synthesis* 2000, 6:781–783

48 Reetz MT, Tielman P, Wiesenhoefer W, Koenen W, Zonta A: Second generation sol–gel encapsulated lipases: robust heterogeneous biocatalyst. *Adv Synth Catal* 2003, 345:717–728

49 Johnson P, Whateley TL: On the use of polymerizing silica gel systems for the immobilization of trypsin. *J Colloid Interf Sci* 1971, 37: 557–563

50 Tosa T, Mori T, Fuse N, Chibata I: Studies on continuous enzyme reactions Part V Kinetic and industrial application of aminoacylase column for continuous optical resolution of acyl-dl-amino acids. *Biotechnol Bioeng* 1967, 9:603–615

51 Bernfeld P, Wan J: Antigens and enzymes made insoluble by entrapping them into lattice of synthetic polymers. *Science* 1963, 142:678–679

52 Hicks GP, Updike SJ: The preparation and characterisation of lyophilised polyacrylamide enzyme gels for chemical analysis. *Anal Chem* 1966, 38:726–729

53 Mosbach K, Mosbach R: Entrapment of enzymes and microorganisms in synthetic cross-linked polymers and their applications in column techniques. *Acta Chem Scand* 1966, 20:2807–2810

54 Zarborsky OR: *Immobilised Enzymes*. CRC Press, Cleveland, 1973

55 Chang TMS: Semipermeable microcapsules. *Science* 1964, 146:524-525

56 Leuschner F: shaped structures for biological processes. German Patent 1,227,855, 1966

57 Brown HD, Patel AB, Chattopadhyay SK: Enzyme entrapment within hydrophobic and hydrophilic matrices. *J Biomed Mater Res* 1968, 2:231–235

58 Pennington SN, Brown HD, Patel AB, Chattopadhyay SK: Silastic entrapment of glucose oxidase-peroxidase and acetylcholine esterase. J Biomed Mater Res 1968, 2:443–446

59 Manecke G, Foerster HJ: Reactive polystyrene as carriers of enzymes and proteins. *Makromol Chem* 1969, 91:136–154

60 Kent LH, Slade JHR: Immunochemically-active cross-linked polystyrene preparations. *Biochem J* 1960, 77:12–17

61 Goldman R, Goldstain L, Katshaski E: Water-insoluble enzyme derivatives and artificial enzyme membranes. In: Stark GR (Ed.) *Biochemical aspects of reactions on solid supports*, Academic Press, New York, 1971

62 Haynes R, Walsh KA: Enzyme envelopes on colloidal particles. *Biochem Biophys Res Commun* 1969, 36:235–242

63 Quiocho FA, Richards FM: Inter-molecular cross-linking of a protein in the crystalline state: carboxypeptidase A.

Proc Natl Acad Sci USA 1964, 52:833–839

64 Quiocho FA, Richards FM: Enzyme behavior of carboxypeptidase-A in the solid state. *Biochemistry* 1966, 5:4062–4076

65 Axen R, Porath J, Ernback S: Chemical coupling of peptides and proteins to poly-saccharides by means of cyanogen halides. *Nature* 1967, 214:1302–1304

66 Kay G, Crook EM: Coupling of enzymes to cellulose using chloro-*s*-triazines. *Nature* 1967, 216:514–515

67 Manecke G, Gunzel G: Polymere Isothiocyanate zur Darstellung hochwirksamer Enzymharze. *Naturwissenschaften* 1967, 54:531–533

68 Weliky N, Brown FS, Dale EC: Carrier-bound proteins: Properties of peroxidase bound to insoluble carboxymethylcellulose particles. *Arch Biochem Biophys* 1969, 131:1–8

69 Patel RP, Lopiekes DV, Brown SR, Price S: Derivatives of proteins II. Coupling of α-chymotrypsin to carboxyl containing polymers by use of *N*-ethyl-5-phenylisoxazolium-3′-sulfonate. *Biopolymers* 1967, 5:577–582

70 Jagendorf AT, patchornik A and Sela M: Use of antibody bound to modified cellulose as an immunospecific adsorbent of antigen, *Biochim Biophys Acta* 1963, 78:516–528

71 Chase T Jr, Shaw E: *p*-Nitrophenyl-*p*′-quqnidinobezoate HCl – A new site titrant for trypsin. *Arch Biochem Biophys Res Commun* 1967, 29:508–514

72 Haynes R, Walsh KA: Enzyme envelopes on colloidal particles. *Biochem Biophys Res Commun* 1969, 36:235–242

73 Glazer AN, Bar-Eli A, Katchalski E: Preparation and characterization of polytyrosyl trypsin. *J Biol Chem* 1962, 237:1832–1838

74 Bernfeld P, Bieber RE, Watson DM: Kinetics of water-insoluble phosphoglycerate mutase. Biochim Biophys Acta - Enzymol 1969, 191:570–578

75 Katchalski E, Weizman N, Levin Y, Blumberg S: preparation of water-insoluble enzyme derivatives, U.S. patent, 3706633, 1965

76 Caviezel O: Inactivation of streptokinase by polyaminostyrene. *Schweiz Ned Wochenstr* 1965, 94:1194

77 Tosa T, Mori T, Chibata I: Studies on continuous enzyme reactions Part VI. Enzymatic properties of DEAE-Sephadex–aminoacylase complex. *Agric Biol Chem* 1969, 33:1053–1059

78 Barkers SA, Somers PJ: Preparation and stability of exo-amylolytic enzymes chemically coupled to microcrystalline cellulose. *Carbohydr Res* 1969, 9:257–263

79 Goldstein L, Levin Y, Pecht M, Katchalski E: A water-insoluble polyanionic derivatives of trypsin, effect of the polyelectrolyte carrier on the kinetic behaviour of the bound trypsin. *Biochemistry* 1964, 3:1914–1919

80 Riesel E, Katchalski E: Preparation and properties of water-insoluble derivatives of urease. *J Biol Chem* 1964, 23:1521–1524

81 Wilson RJH, Kay G, Lilly MD: The preparation and kinetics of lactate dehydrogenase attached to water insoluble particles of sheets. *Biochem J* 1968, 108:845–853

82 Kirimura J, Yoshida J: Water-insoluble copolymer of acylase with an amino acid anhydride. US Patent 3,234,356, 1966

83 Brown HD, Patel AB, Chattopadhyay SK, Pennington SN: Support matrix for apyrase. *Enzymologia* 1968, 35:233–238

84 Weliky N, Brown FS, Dale EC: Carrier-bound proteins: properties of peroxidase bound to insoluble carboxymethylcellulose particles. *Arch Biochem Biophys* 1969, 131:1–8

85 Bohnensack R, Augustin W, Hofmann E: Chemical coupling of hexokinase from yeast on Sephadex. *Experientia* 1969, 25:348

86 Weetall HH, Baum G: Preparation and characterisation of insolubilised L-amino acid oxidase. *Biotechnol Bioeng* 1970, 12:399–407

87 Barker SA, Somers PJ, Epton R: Preparation and properties of α-amylase chemically coupled to microcrystalline cellulose. *Carbohydr Res* 1968, 8:491–497

88 Weetall HH: Alkaline phosphatase insolubilised by covalent linkage to porous glass. *Nature* 1969, 223:959–960

89 Self DA, Kay G, Lilly MD, Dunnill P: Conversation of benzylpenicillin to 6-aminopenicillanic acid using an insoluble derivative of penicillin amidase. *Biotechnol Bioeng* 1969, 11:337–348

90 Sharp AK, Kay G, Lilly MD: The kinetics of β-galactosidase attached to porous cellulose sheets. *Biotechnol Bioeng* 1969, 11:363–382

91 Axen R, Heilbronn E, Winter A: Preparation and properties of cholinesterase, covalently bound to Sepharose. *Biochim Biophys Acta* 1969, 191:478–481

92 Crook EM: Enzymes on solid matrixes. In: Sols A (Ed.) Metab Regul Enzyme Action, Fed Eur Biochem Soc, Meet, 6th, 1969, 1970:297–308

93 Katchalski-Katzir E, Silman I, Goldman R: Effect of the microenvironment on the mode of action of immobilized enzymes. *Adv Enzymol* 1971, 34:445–536

94 Levin Y, Pecht M, Goldstain L, Katchalski E: A water-insoluble polyanionic derivative of trypsin. *Biochemistry* 1964, 3:1905–1913

95 Whittam R, Edwards BA, Wheeler KP: An approach to the study of enzyme action in artificial membrane. *Biochem J* 1968, 107:3p-4p

96 Manecke G, Singer S: Umsetzungen an Copolymerisaten des Methacryl-siiurefluoranilids. *Makromol Chem* 1960, 39:13–25

97 Weetall HH, Hersh LS: Urease covalently coupled to porous glass. *Biochim Biophys Acta* 1969, 185:464–465

98 Sundaram PV, Hornby WE: Preparation and properties of urease chemically attached to nylon tube. *FEBS Lett* 1970, 10:325–327

99 Inman DJ, Hornby WE: The immobilization of enzymes on nylon structures and their use in automated analyses. *Biochem J* 1972, 129:255–62

100 Tosa T, Moris T, Fuse N, Chibata I: Studies on continuous enzyme reactions I. Screening of carriers for preparation of water-insoluble aminoacylase. *Enzymologia* 1966, 3:214–224

101 Bachler MJ, Strandberg GW, Smiley KL: Starch conversion by immobilised glucoamylase. *Biotechnol Bioeng* 1970, 12:85–92

102 Manecke G, Gunzel G: Verwendung eines nitrierten Copolymerisates aus Methacrylsäure und Methacrylsäure-*m*-fluoranilid zur Darstellung von Enzymharzen sowie zu Racematspaltungs- und Gerbungsversuchen, *Makromol Chem* 1962, 51:199–216

103 Zaborsky OR: Alteration of enzymic properties prior to immobilization. *Biotechnol Bioeng Symp* 1972, 3:211–217

104 Torchilin VP, Tishchenko EG, Smirnov VN: Covalent immobilization of enzymes on ionogenic carriers. Effect of electrostatic complex formation prior to immobilization. *J Solid-Phase Biochem* 1977, 2:19–29

105 Ahmad A, Bishayee S, Bachhawat BK: Novel method for immobilization of chicken brain arylsulfatase A using concanavalin A. *Biochem Biophys Res Commun* 1973, 53:730–736

106 Vretblad P, Axen R: Covalent fixation of pepsin to agrose derivatives. *FEBS Lett* 1971, 254–256

107 Vretblad P: Immobilisation of ligands for biospecific affinity chromatography via their hydroxyl groups. The cyclohexa-amylose-α-amylase system. *FEBS Lett* 1974, 47:86–89

108 Goldstein L: New polyamine carrier for the immobilization of proteins. Water-insoluble derivatives of pepsin and trypsin. *Biochim Biophys Acta* 1973, 327:132–137.

109 Bethell GS, Ayres J, Hancock WS, Hearn MTW: a novel method of activation of crosslinked agarose with 1,1-carbonyldiimidazole which gives a matrix for affinity chromatography devoid of additional charged groups. *J Biol Chem* 1979, 254:2572–2574

110 Beaucamp K, Bergmeyer H, Ulrich B, Heinz K, Jaworek D, Nelboeck-Hochstetter M: Carrier-bound proteins. Ger Offen, DE 71-2128743, 1972

111 Gondo S, Koya H: Solubilized collagen fibril as a supporting material for enzyme immobilization. *Biotechnol Bioeng* 1978, 20:2007–2010

112 Royer GP, Ikeda S, Aso K: Crosslinking of reversibly Immobilised enzymes. *FEBS Lett* 1977, 80:89–94

113 Nillson H, Mosbach R, Mosbach K: Use of polymerisation of acrylic monomers for immobilisation of enzymes. *Biochim Biophys Acta* 1972, 268:253

114 Taylor JB, Swaisgood HE: Kinetic study on the effect of coupling distance between insoluble trypsin and its carrier matrix. *Biochim Biophys Acta* 1972, 284:268–277

115 Hipwell MC, Harvey MJ, Dean PDG: Affinity chromatography on an homologous series of immobilized N6-amega-aminoalkyl AMP. Effect of ligand-matrix spacer length on ligand–enzyme interaction. *FEBS Lett* 1974, 42:355–359

116 Epton R, Mclaren JV, Thomas TH: Poly(N-acrylolyl-4-and-5-amino salicylic acid) Part III Uses as their titanium complexes for the insolubilisation of enzymes. *Carbohydr Res* 1973, 27:11–20

117 Brown E, Racois A, Gueniffey H: Preparation and properties of urease derivatives insoluble in water. *Tetrahedron Lett* 1970, 25:2139–2142

118 Martinek K, Klibanov AM, Coldmacher VS, Berezin LV: The principles of enzyme stabilisation. *Biochim Biophys Acta* 1977, 485:1–12

119 Messing RA, Filbert AM: Immobilized glucose isomerase for the continuous conversation of glucose to fructose. *J Agric Food Chem* 1975, 23:920–923

120 Royer GP, Liberatore FA, Green GM: Immobilization of enzymes on aldehydic matrixes by reductive alkylation. *Biochem Biophys Res Commun* 1975, 64:478–484

121 Brandt J, Andersson LO, Porath J: Covalent attachment of proteins to polysaccharide carriers by means of benzoquinone. *Biochim Biophys Acta* 1975, 386:196–202

122 Kennedy JF, Barker SA, Rosevear A: Preparation of a water-insoluble trans-2,3-cyclic carbonate derivative of macroporous cellulose and its use as a matrix for enzyme immobilisation. *J Chem Soc* 1973,20:2293–2299

123 Zaborsky OR: Immobilization of enzymes with imido ester-containing polymers. In: Olson AC, 120Cooney CL (Eds) *Immobilized enzymes food microbial processes*. Publisher: Plenum, New York 1975:187–203

124 Johansson, AC, Mosbach K: Acrylic copolymers as matrixes for the immobilization of enzymes. I. Covalent binding or entrapping of various enzymes to bead-formed acrylic copolymers *Biochim Biophys Acta* 1974, 370:339–347.

125 Kipper H, Egorov Kh-R, Kivisilla K: Characteristics of polymer-modified inorganic carriers. *Tr Tallin Politekh Inst* 1979, 465:33–39

126 Miyairi S: An enzyme-polymer film prepared with the use of poly(vinyl alcohol) bearing photosensitive aromatic azido groups. *Biochim Biophys Acta* 1979, 571:374–377

127 Klug JH: Poly(ureaurethane) foams containing immobilized active. US Patent 3,905,923, 1975

128 Wykes JR, Dunnill P, Lilly MD: Immobilisation of α-amylase by attachment to soluble support materials. *Biochim Biophys Acta* 1971, 250:522–529

129 Kamogashira T, Mihara S, Tamaoka H, Doi T: 6-Aminopenicillanic acid by penicillin hydrolysis with immobilized enzyme. *Jpn Kokai Tokkyo Koho* 1972, JP 47028187

130 Hueper F: Water soluble, polymeric substrate covalently bound penicillin acylase for preparing 6-aminopenicillanic acid. DE 2,312,824, 1974

131 Monsan P et al.: Nouvelle Methode de Preparation d'Enzymes Fixes sur des Supports Mineraux, *C R Acad Sci Paris T* 1971, 273:33–36

132 Bartling GJ, Brown HD, Chattopadhyay SK: Synthesis of matrix-supported enzyme in non-aqueous conditions. *Nature (London)* 1973, 243:342–344

133 Mosbach K: Matrix bound enzymes, Part 1. The use of different acrylic copolymers as matrix. *Acta Chem Scand* 1970, 24:2084–2092

134 Goldstein L, Manecke G: The chemistry of enzyme immobilization. In Wingard LB Jr, Katchalski-Katzir E, Goldstein L (Eds) *Applied Biochemistry and Bioengineering*, Academic Press, New York, 1976, p. 23

135 Charles M, Coughlin RW, Hasselberger FX: Soluble–insoluble enzyme catalysts. *Biotechnol Bioeng* 1974, 16:1553–1556

136 Hixson HF: Water-soluble enzyme–polymer grafts. Thermal stabilization of glucose oxidase. *Biotechnol Bioeng* 1973, 15:1011–1016

137 Dinelli D: Fiber-entrapped enzymes. *Process Biochem* 1972, 7:9–12

138 Marconi W, Bartoli F, Cecere F, Galli G, Morisi F: Synthesis of penicillins and cephalosporins by penicillin acylase entrapped in fibers. *Agric Biol Chem* 1975, 39:277–279

139 Adalsteinsson O, Lamotte A, Baddour RF, Colton CK, Pollak A, Whitesides GM: Preparation and magnetic filtration of polyacrylamide gels containing covalently immobilized proteins and a ferrofluid. *J Mol Catal* 1979, 6:199–225.

140 Kraemer KL, Pennerwise H, Plainer H, Reisner W, Sproessler BG: Enzymes covalently bound to acrylic gel beads. I. Interaction of hydrophilic anionic gel beads with biomacromolecules. *J Polym Sci Symp* 1974, 47:77–89

141 Kramer DM, Lehmann K, Pennewiss M, Plainer H: Photo-beads and oxirane beads as solid supports for catalysis and bio-specific adsorption. In: Peeters H (Ed.). *Protides of Biological Fluids*, 23th Colloquium. Pergamon Press, Oxford, 1975:505–511

142 Cuatrecasas P, Parikh I: Adsorbents for affinity chromatography. Use of *N*-hydroxysuccinimide esters of agarose. *Biochemistry* 1972, 11:2291–2299

143 White CA, Kennedy JF: Popular matrices for enzyme and other immobilizations. *Enzyme Microb Technol* 1980, 2:82–90

144 Goldstein L: New polyamine carrier for the immobilization of proteins. Water-insoluble derivatives of pepsin and trypsin. *Biochim Biophys Acta* 1973, 327:132–137

145 Manecke G, Korenzecher R: Reactive copolymers of *N*-vinyl-2-pyrrolidone for the immobilization of enzymes. *Makromol Chem* 1977, 178:1729–38

146 Axen R, Vretblad P, Porath J: The use of isocyanides for the attachment of biologically active substances to polymers. *Acta Chem Stand* 1971, 25:1129–1132

147 Vretblad P, Axen R: Use of isocyanides for the immobilization of biological molecules. *Acta Chem Scand* 1973, 27:2769–2780

148 Shaw E: Chemical modification by active site –directed reagents. In: Boyer PD (Ed.) *The enzymes*, Academic Press, New York, 1970, pp. 91–146

149 Zarbosky OR: Immobilisation of enzymes with imidoester-containing polymers. In: Olson AE, Cooney CL (Eds) *Immobilised enzymes in food and microbial processes*. Plenum Press, New York, 1974

150 Marshall JJ: Manipulation of the properties of enzyme by covalent attachment of carbohydrate. *Trends Biochem Adv Enzymol* 1978, 57:179–249

151 Carcases P, Parikh I: Adsorbents for affinity chromatography, Use of *N*-hydroxysuccinimide ester of agarose. *Biochemistry* 1972, 11:2291–2299

152 Carlsson J, Axen R, Brocklehurst K, Crook EM: Immobilisation of urease by thio–disulfide interchange with concomitant purification. *Eur J Biochem* 1974, 44:189–194

153 Porath J, Sundberg I: High capacity chemisorbents for protein immobilisation. *Nature (London) New Biology* 1972, 238, 261–262

154 Brandt J, Andersson LO, Porath J: Covalent attachment of proteins to polysaccharide carriers by means of benzoquinone. *Biochim Biophys Acta* 1975, 386:196–202

155 Hartdegen FJ, Swann WE: Polyurethane foams containing bound immobilized proteins. *Ger Offen* 1976, DE 2,612,138

156 Gabel D, Porath J: Molecular properties of immobilized proteins. *Biochem J* 1972, 127:13–14

157 Johansson AC, Mosbach K: Acrylic copolymers as matrixes for the immobilization of 150 enzymes. II. Effect of a hydrophobic microenvironment on enzyme reactions studied with alcohol dehydrogenase immobilized to different acrylic copolymers. *Biochim Biophys Acta* 1974, 370:348–353

158 Cuatrecasas P: Protein purification by affinity chromatography-derivatizations of agarose and polyacrylamide beads. *J Biol Chem* 1970, 12:3059–3065

159 Drobnik J, Saudek V, Svec F, Kalal J, Vojtisek V, Barta M: Enzyme immobilization techniques on poly(glycidyl methacrylate-co-ethylene dimethacrylate) carrier with penicillin amidase as model. *Biotechnol Bioeng* 1979, 21:1317–1332

160 Manecke G, Polakowaski D: Some carriers for the immobilisation of enzymes based on copolymers of derivatized poly(vinyl alcohol) and on copolymers of methacrylates with different spacer lengths. *J Chromatogr* 1981, 21:13–24

161 Ollis DF: Diffusion influence in denaturable insolubilised enzyme catalysts. *Biotechnol. Bioeng* 1972, 14:871–884

162 Engasser JM, Horvath C: Inhibition of bound enzymes. 1. Antienergistic interaction of chemical and diffusion. *Biochemistry* 1974, 13:3845–3849

163 Bernath FR, Vieth WR: Lysozyme activity in the presence of nonionic detergent micelles. *Biotechnol Bioeng* 1972, 14:737–752

164 Miwa N, Ohtomo K: Enzyme immobilization in the presence of substrates and inhibitors. *Jpn. Kokai Tokkyo Koho* 1975, JP 56,045,591

165 Kennedy JF, Barker SA, Rosevear A: Use of a poly(allyl carbonate) for the preparation of active, water insoluble derivatives of enzymes. *J Chem Soc Perkin Trans* 1972, 1:2568–2573

166 Koch-Schmidt AC, Mosbach K: Studies on conformation of soluble and immobilized enzymes using differential scanning calorimetry. II. Specific activity and thermal stability of enzymes bound weakly and strongly to Sepharose CL 4B. *Biochemistry* 1977, 16:2105–2109

167 Barker SA, Somers PJ: Cross-linked polyacrylamide derivatives as water-insoluble carriers of amylolytic enzymes. *Carbohydr Res* 1970, 14:287–296

168 Katchalski E, Goldstein L, Levin Y, Blumberg S: Water-insoluble enzyme derivatives, U.S. patent 1972, US 3706633

169 Messing RA: Potential applications of molecular inclusion to beer processing. *Brewers Digest* 1971, 46:60–63

170 Zaborsky OR: Alteration of enzymic properties prior to immobilization. *Biotechnol Bioeng Symp* 1972, 3:211–217

171 Wierzbicki LE, Edwards VH, Kosikowski FV: Immobilization of microbial lactases by covalent attachment to porous glass. *J Food Sci* 1974, 38:1070–1073.

172 Sugiura M, Isobe M: Studies on the mechanism of lipase reaction. III. Adsorption of Chromobacterium lipase on hydrophobic glass beads. *Chem Pharm Bull* 1976, 24:72–78.

173 Zaborsky OR: *Immobilized Enzymes*. CRC Press, Cleveland, 1973

174 Mosbach K: Immobilized Enzymes. *Methods Enzymol* 1976, 44

175 Buchholz K, Kasche V: *Biokatalysatoren und Enzym-technologie*. VCH, 1997:7–11

176 Mosbach K: Immobilised enzymes. *FEBS Lett* 1976, 62:E80-E95

177 Klibanov AM: Why are enzymes less active in organic solvents than in water? *Trends Biotechnol* 1997, 15:97–101

178 Schmitke JL, Stern LJ, Klibanov AM: The crystal structure of subtilisin Carlsberg in anhydrous dioxane and its comparison with those in water and acetonitrile. *Proc Natl Acad Sci USA* 1997, 94:4250–4255

179 Zaks A, Kllbanov AM: Enzyme-catalyzed processes in organic solvents. *Proc Natl Acad Sci USA* 1985, 82:3192–3196

180 Osada Y, Iino Y, Numajiri Y: Preparation and behavior of enzymes immobilized by polymer–polymer complexes. *Chem Lett* 1982, 4:559–562

181 Tor R, Dror Y, Freeman A: Enzyme stabilization by bilayer "encagement". *Enzyme Microb Technol* 1989, 11:306–312

182 Mozhaev VV, Melik-Nubarov, NS Sergeeva, MV, Sikrnis V, Martinek K: Strategy for stabilising enzymes I, Increasing stability of enzymes via their multipoint interaction with a support. *Biocatalysis* 1990, 3:179–187

183 Blanco RM, Calvete JJ, Guisan JM: Immobilisation-stabilisation of enzymes: variables that control the intensity of the trypsin (amine) agarose (aldehyde) multi-point covalent

attachment. *Enzyme Microb Technol* 1988, 11:353–359

184 Guisan JM: Aldehyde gels as activated support for immobilisation-stabilisation of enzymes. *Enzyme Microb Technol* 1988, 10:357–382

185 Martinek K, Klibanov AM, Goldmacher VS, Berezin IV: The principles of enzyme stabilization: 1. Increase in thermostability of enzymes covalently bound to a complimentary surface of a polymer support in a multipoint fashion. *Biochim Biophys Acta* 1977, 485:1–12

186 Burteau N, Burton S, Crichton RR: Stabilization and immobilization of penicillin amidase. *FEBS Lett* 1989, 258:185–189

187 Ho GH, Liao CC: Multi-layer immobilized enzyme compositions, US Patent 4,506,015 A, 1985

188 Lee KM, Blaghen M, Samama JP, Biellmann JF: Cross-linked crystalline horse liver alcohol dehydrogenase as a redox catalyst: activity and stability towards organic solvent. *Bioorg Chem* 1986, 14:202–210

189 Hahn-Hagerdal B: Water activity a possible external regulator in biotechnical processes. *Enzyme Microb Technol* 1986, 8:322–327

190 Reslow M, Adlercreutz P, Mattiason B: On the importance of the support material for bioorganic synthesis: Influence of water partition between solvent, enzyme and solid support in water poor reaction media. *Eur J Biochem* 1992, 172:573–578

191 Takahashi KNH, Yoshimoto T, Saito Y, Inada Y: A chemical modification to make horseradish peroxidase soluble and active in benzene. *Biochim Biophys Res Commun* 1984, 121:261–265

192 Inada Y, Nisklmnra II, Takakashi K, Yoshimoto, Saha AR, Saito Y: Ester synthesis catalyzed by polyethylene glycol-modified lipase in benzene. *Biochem Biophys Res Commun* 1984, 122:845–850

193 Takabaskl K, Ajlma A, Yoshimoto T, Inada Y: Polyethylene glycol-modified catalase exhibits unexpectedly high activity in benzene. *Biochem Biophys Res Commun* 1984, 125:761–766

194 Takahashi K, Nishiiura H, Yoshiioto T, Okada M, Ajima A, Matsashima A, Tamaura Y, Saito Y, Inada Y: Polyethylene glycol-modified enzymes trap water on their surface and exert enzymic activity in organic solvents. *Biotechnol Lett* 1984, 6:765–770

195 Okahata Y, Fujimoto Y, Ijiro K: Lipase–lipid complex as a resolution catalyst of racemic alcohols in organic solvents. *Tetrahedron Lett* 1988, 29:5133–5134

196 Martinek K, Levashov AV, Myachko N, Khmebdtski YL, Rerezin IV: Micellar enzymology. *Eur J Biochem* 1986, 155:453–468

197 Okahata Y, Hatano A, Ijiro K: Enhancing enantioselectivity of a lipid-coated lipase via imprinting method for esterification in organic solvents, *Tetrahedron Asymmetry* 1995, 6:1311–1322

198 Persson M, Bulow L, Mosbach K: Purification and site-specific immobilization of genetically engineered glucose dehydrogenase on thiopropyl-Sepharose, *FEBS Lett* 1990, 270:41–44

199 Stark MB, Kolmberg K: Covalent immobilisation of lipase in organic solvents. *Biotechnol Bioeng* 1989, 34:942–950

200 Glad M, Norrloew O, Sellergren B, Siegbahn N, Mosbach K: Use of silane monomers for molecular imprinting and enzyme entrapment in polysiloxane-coated porous silica. *J Chromatogr* 1985, 347:11–23

201 Wehtje E, Adlercreutz P, Mattiasson B:Improved activity retention of enzymes deposited on solid supports. *Biotechnol Bioeng* 1993, 41:171–178

202 Barbotin JN, Breuil M:Immobilization of glutamate dehydrogenase into proteic films. Stability and kinetic modulation by effectors. *Biochim Biophys Acta* 1978, 525 :18–27

203 Blanco RM.; Guisan JM: Protecting effect of competitive inhibitors during very intense insolubilized enzyme-activated support multipoint attachments: trypsin (amine)-agarose (aldehyde) system. *Enzyme Microb Technol* 1988, 10: 227–232

204 Ohya Y, Miyaoka J, Ouchi T: Recruitment of enzyme activity in albumin by molecular imprinting. *Macromol Rapid Commun* 1996, 17:871–874

205 Slade CJ, Vulfson EN: Induction of catalytic activity in proteins by lyophilization in the presence of a transition state analogue. *Biotechnol Bioeng* 1998, 57:211–215

206 Rich JO, Dordick JS: Controlling subtilisin activity and selectivity in organic media by imprinting with nucleophilic substrates. *J Am Chem Soc* 1997, 119:3245–3252

207 Peissker F, Fischer L: Cross-linking of imprinted proteases to maintain tailor-made substrate selectivity in aqueous solutions. *Bioorg Med Chem* 1999, 7:2231–2237

208 Nakashima T, Fokada H, Kyotant S, Morikawa H: Culture conditions for intracellular lipase production by *Rhizopus chinensis* and its immobilization within biomass support particles. *J Ferment Technol* 1988, 66:441–448

209 Soares CMF, de Castro HF, Santana MHA, Zanin GM: Selection of stabilizing additive for lipase immobilization on controlled pore silica by factorial design. *Appl biochem Biotechnol* 2001, 91–93: 703–18

210 Fukui S, Tanaka A, Iida T: Immobilisation of biocatalysts for bioprocesses in organic solvent media. In: Lane C, Tramper J, Lilly MD (Eds) *Biocatalysis in organic media*, Elsevier, Amsterdam, 1987, pp 21–44

211 Niiolova P, Ward OP: Whole cell biocatalysis in nonconventional media. *J Ind Microbiol* 1993, 12:76–86

212 Nilsson N, Mosbach K: *p*-Toluenesulfonyl chloride as an activating agent of agarose for the preparation of immobilized affinity ligands and proteins. *Eur J Biochem* 1980, 112:397–402

213 Miron T, Wilchek M: Activation of trisacryl gels with chloroformates and their use for affinity chromatography and protein immobilization. *Appl Biochem Biotechnol* 1985, 11:445–456

214 Ngo TT: facile activation of Sepharose hydroxyl groups by 2-fluro-1-methylpyridinium toluene-4-sulfonate: preparation of affinity and covalent chromatographic matrix. *Bio/Technology* 1986, 4:134–137

215 Buettner W, Becker M, Rupprich Ch, Boeden H.-F, Henkelin P Loth F and Dautzenberg H:A novel carbonochloridate for activation of supports containing hydroxyl groups. *Biotechnol. Bioeng* 1989, 33:26–31

216 Scardi V, Cantarella M, Gianfreda L, Palescandolo R, Alfani F, Greco G Jr: Enzyme immobilization by means of ultrafiltration techniques. *Biochimie* 1980, 62:635–643

217 Milstein O, Nicklas B, Huettermann A: Oxidation of aromatic compounds in organic solvents with laccase from *Trametes versicolor*. *Appl Microbiol Biotechnol* 1989, 31:70–74

218 Burteau N, Burton S, Crichton RR: Stabilization and immobilization of penicillin amidase. *FEBS Lett* 1989, 258:185–189

219 Zaks A, Klibanov AM: Enzyme catalysis in organic medium at 100°C. *Science* 1984, 224:1249–1251

220 Laane C, Tramper JH, Lilly MD: Biocatalysis in organic media, Elsevier, Amsterdam, 1987

221 Zaks A, Klibanov AM: Enzymatic catalysis in nonaqueous solvents. *J Biol Chem* 1988, 263:3194–3201

222 Inada Y, Takahashi K, Yoshimoto T, Ajii A, Matsushima A, Saito Y: Application of polyethylene glycolmodified enzymes in biotechnological processes: organic solvent-soluble enzymes. *Trends Biotechnol* 1986, 4:190–194

223 Blanco RM, Guisan JM: Additional stabilisation of PGA-agarose derivatives by chemical modification with aldehydes. *Enzyme Microb Technol* 1992, 14:489–495

224 Xu K, Griebenow K, Klibanov AM: Correlation between catalytic activity and secondary structure of subtilisin dissolved in organic solvents. Biotechnol Bioeng 1997, 56:485–491

225 Reetz MT, Zonta A, Schimossek K, Liebeton K, Jaeger K.-E: Creation of

enantioselective biocatalysts for organic chemistry by *in vitro* evolution. *Angew Chem Int Ed* 1997,36:2830–2832

226 Dabulis K, Klibanov AM: Dramatic enhancement of enzymatic activity in organic solvents by lyoprotectants. *Biotechnol Bioeng* 1993, 41:566–571

227 Vidal MW, Barletta G: Improved enzyme activity and enantioselectivity in organic solvents by methyl-β-cyclodextrin. *J Am Chem Soc* 1999, 121: 8157–8163

228 Dordic JS: Principles and Applications of Non-aqueous Enzymology. In: Blanch HW, Clark DS (Eds) *Applied biocatalysis*, Marcel Dekker, New York, 1991, pp. 1–52

229 Khmelnitsky YL, Rich JO: Biocatalysis in nonaqueous solvents, *Curr Opin Chem Biol* 1999, 3:47–53

230 Kawakami K, Abe T, Yoshida T: Silicon-immobilised biocatalyst effective for bioconversions in non-aqueous media. *Enzyme Microb Technol* 1999, 14:371–375

231 Kise H, Hayakawa A: Immobilisation of protease to porous chitosan beads and their catalysis for ester and peptide synthesis in organic solvents. *Enzyme Microb Technol* 1991, 13:584–588

232 St Clair NL, Navia MA: Cross-linked enzyme crystals as robust biocatalysts. *J Am Chem Soc* 1992, 114:7314–7316

233 Lalonde J: Practical catalysis with enzyme crystals. *Chemtech* 1997, 27:38–45

234 Margolin AL: Novel crystalline catalysts. *Trends Biotechnol* 1996, 14:223–230

235 Colombo G, Ottolina G, Carrea G, Bernardi A, Scolastico C: Application of structure-based thermodynamic calculations to the rationalization of the enantioselectivity of subtilisin in organic solvents. *Tetrahedron Asymmetry* 1998, 9:1205–1214

236 Secundo F, Spadaro S, Carrea G, Overbeek PLA: Optimisation of *Pseudomonas cepacia* lipase preparations for catalysis in organic solvents. *Biotechnol Bioeng* 1999, 62:554–561

237 Partridge J, Halling PJ, Moore BD: Practical route to high activity enzyme preparations for synthesis in organic media. *J Chem Soc Chem Commun* 1998, 7:841–842

238 Bille V, Plainchamp D, Lavielle S, Chassaing G, Remacle J:Effect of the microenvironment on the kinetic properties of immobilized enzymes. *Eur J Biochem* 1989, 180: 41–7

239 Fernandex-Lafuente R, Cowan DA, Wood ANP: Hyperstabilization of a thermophilic esterase by multipoint covalent attachment. *Enzyme Microb Technol* 1995, 17:366–372

240 Wang P, Sergeeva MV, Lim L, Dordick JS: Biocatalytic plastics as active and stable materials for biotransformations. *Nat Biotechnol* 1997, 15:789–793

241 Dordick JS, Novick SJ, Sergeeva MV: Biocatalytic plastics. *Chem Ind (London)* 1998, 17–20

242 Matoba S, Tsuneda S, Saito K, Sugo T: Highly efficient enzyme recovery using a porous membrane with immobilized tentacle polymer chains. *Bio/Technology* 1995, 13:795–797

243 Rocha JMS, Gil MH, Garcia FAP: Effects of additives on the activity of a covalently immobilised lipase in organic media. *J Biotechnol* 1998, 66:61–67

244 Costantino HR, Griebenow K, Langer R, Klibanov AM: On the pH memory of lyophilized compounds containing protein functional groups. *Biotechnol Bioeng* 1997, 53:345–348

245 Yang Z, Mesiano AJ, Venkatasubramanian S, Gross SH, Harris JM, Russell AJ: Activity and stability of enzymes incorporated into acrylic polymers. *J Am Chem Soc* 1995, 117:4843–4850

246 Yang, Z. et al.: Synthesis of protein-containing polymers in organic solvents. *Biotechnol Bioeng* 1995, 45:10–17

247 Delgado C, Francis, GE, Fisher D: The uses and properties of PEG-linked proteins. *Crit Rev Ther Drug Carrier Syst* 1992, 9:249–304

248 Abian O, Mateo C, Fernandez-Lorente G, Palomo JM, Fernandez-Lafuente R, Guisan JM: Stabilization of immobilized enzymes against water-soluble organic cosolvents and generation of hyper-hydrophilic microenvironments surrounding enzyme molecules. *Biocatal Biotransform* 2001, 19:489–503

249 Guisan JM, Alvaro G, Fernandez-Lafuente R, Rosell CM, Garcia JL, Tagliani A: Stabilization of heterodimeric enzyme by multipoint covalent immobilization: penicillin G acylase from *Kluyvera citrophila*. *Biotechnol Bioeng* 1993, 42:455–64

250 Butterfield DA, Bhattacharyya D, Daunert S, Bachas L: Catalytic biofunctional membranes containing site-specifically immobilized enzyme arrays: a review. *J Membr Sci* 2001, 18:29–37

251 Furukawa S, Ono T, Ijima H, Kawakami K: Effect of imprinting sol–gel immobilized lipase with chiral template substrates in esterification of (R)-(+)- and (S)-(–)-glycidol. *J Mol Catal B: Enzymatic* 2002, 17:23–28

252 Cao L, Elzinga J: Cross-linked enzyme aggregates and crosslinker agents therefore, WO 03/066,850, 2003

253 Hilal N, Nigmatullin R, Alpatova A: Immobilization of cross-linked lipase aggregates within microporous polymeric membranes. *J Membr Sci* 2004, 238:131–141

254 Gemeiner P, Docolomansky P, Vikartovska A, Stefuca V: Amplification of flow-255 microcalorimetry signal by means of multiple bioaffinity layering of lectin and glycoenzyme. *Biotechnol Appl Biochem* 1998, 28:155–161

255 Cao L, van Rantwijk F, Sheldon RA: Cross-linked enzyme aggregates: a simple and effective method for the immobilization of penicillin acylase. *Org Lett* 2000, 2:1361–1364

256 Cao L, Fischer A, Bornscheuer UT, Schmid RD: Lipase-catalyzed solid phase preparation of sugar fatty acid esters. *Biocatal Biotransform* 1997, 14:269–283

257 Cao L, van Rantwijk F, Sheldon RA: Cross-linked enzyme aggregates: a simple and effective method for the immobilization of penicillin acylase. *Org Lett* 2000, 2:1361–1364

258 Ovsejevi K, Brena B, Batista-Viera F, Carlsson J: Immobilization of β-galactosidase on thiolsulfonate-agarose. *Enzyme Microb Technol* 1995, 17:151–156

259 Messing RA: Carriers. In *Immobilized enzymes for industrial reactors*. Messing RA (Ed.) Academic Press, New York, pp.63–77, 1975

260 Bickerstaff GF: Immobilization of enzymes and cells, Humana Press, Totowa, NJ, 1997

261 Katchalski-Katzir E, Kraemer DM: Eupergit C, a carrier for immobilization of enzymes of industrial potential. *J Mol Catal B Enzym* 2000, 10:157–176

262 Taylor RF: *Protein immobilization: fundamentals and applications*. Marcel Dekker, New York

263 Gemeiner P: Materials for enzyme engineering. In: Gemeiner P (Ed) *Enzyme engineering*, Ellis Horwood, New York, 1992, pp.13–119

264 Mosbach K: Immobilised Enzymes. *TIBS* 1980, 5:1–3

265 Katzbauer B, Narodoslawsky M, Moser A: Classification system for immobilisation techniques. *Bioprocess Eng* 1995, 12:173–179

266 Akgol S, Kacar Y, Denizli A, Arica MY: Hydrolysis of sucrose by invertase immobilized onto novel magnetic polyvinylalcohol microspheres. *Food Chem* 2001, 74:281–288

267 Wilson L, Illanes A, Pessela BC, Abian O, Fernandez-Lafuente R, Guisan JM: Encapsulation of crosslinked penicillin G acylase aggregates in lentkas: evaluation of a novel biocatalyst in organic media. *Biotechnol Bioeng* 2004, 86:558–562

268 Andreopoulos FM, Roberts MJ, Bentley MD, Harris JM, Beckman EJ, Russell AJ: Photo-immobilization of organophosphorus hydrolase within a PEG-based hydrogel. *Biotechnol Bioeng* 1999, 65:579–588

269 Ray R, Jana SC, Nanda G: Biochemical approaches of increasing thermostability of β-amylase from *Bacillus megaterium* B, *FEBS Lett* 1994, 356:30–32

270 Mohapatra SC, Hsu JT: Immobilization of α-chymotrypsin for use in batch and continuous reactors. *J Chem Technol Biotechnol* 2000, 75:519–525

271 Weetall HH, Suzuki S (Eds) *Immobilised enzyme technology*, Plenum Press, New York, 1975

272 Mosbach K: Immobilised enzymes and cells. *Methods Enzymol B* 1987, 135

273 Mosbach K: Immobilised enzymes and cells, Part D, *Methods Enzymol*, 137, 1988
274 Woodley JM: Immobilised biocatalysts. In Smith K (Ed.) *Solid supports and catalysts in organic synthesis*, Ellis Horwood, 1992
275 Cheetham PSJ: What makes a good biocatalyst? *J Biotechnol* 1998, 66:3–10
276 Van Roon J, Beeftink R, Schroen K, Tramper H: Assessment of intraparticle biocatalytic distribution as a tool in rational formulation. *Curr Opin Biotechnol* 2002, 13:398–405
277 Terreni M, Pagani G, Ubiali D, Fernandez-Lafuente R, Mateo C, Guisan JM: Modulation of penicillin acylase properties via immobilization techniques: one-pot chemoenzymatic synthesis of cephamandole from cephalosporin C. *Bioorg Med Chem Lett* 2001, 11:2429–2432
278 Chibata I, Tosa T, Sato T, Mori T, Matuo Y: In: Terui G (Ed.) *Fermentation technology today*, 1972, pp 383–389
279 Bakker M: Immobilisation of metalloenzymes and their application in non-natural conversions. PhD Thesis, Technical University Delft, The Netherlands, 2000.
280 Seip JE, Faber JE, Gavagen DL, Anton, R de Cosimo: Glyoxylic acid production using immobilised glycolate oxidase and catalase, *Bioorg Med Chem* 1994, 2:371–378
281 Liese A, Filho MV: Production of fine chemicals using biocatalysis. *Curr Opin Biotechnol* 1999, 10:595–603
282 Schulze B, Wubbolts MG: Biocatalysis for industrial production of fine chemicals. *Curr Opin Biotechnol* 1999, 10:609–615
283 Bommarius AS, Schwarm M, Drauz K: Biocatalysis to amino acid based chiral pharmaceuticals – examples and perspectives. *J Mol Cat B: Enzym* 1998, 5:1–11
284 Tischer W, Kasche V: Immobilized enzymes: crystals or carriers? *Trends Biotechnol* 1999, 17:326–335
285 Clark DS: Can immobilisation be exploited to modify enzyme activity? *Trends Biotechnol* 1994, 12:439–443
286 Cabral JMS, Kennedy JF: Immobilisation techniques for altering thermal stability of enzymes. In: Gupta MN (Ed.) *Thermostability of enzymes*, Springer, Berlin, pp 163–179
287 Rocchietti S, Urrutia ASV, Pregnolato M, Tagliani A, Guisan JM, Fernandez-Lafuente R, Terreni M: Influence of the enzyme derivative preparation and substrate structure on the enantioselectivity of penicillin G acylase. *Enzyme Microb Technol* 2002, 31:88–93
288 Klibanov AM: Enzyme stabilization by immobilization. *Anal Biochem* 1979, 93:1–25
289 Bismuto E, Martelli PL, De Maio A, Mita DG, Irace G, Casadio R: Effect of molecular confinement on internal enzyme dynamics: Frequency domain fluorimetry and molecular dynamics simulation studies. *Biopolymers* 2002, 67:85–95
290 Arica MY, Yavuz H, Denizli A: Immobilization of glucoamylase on the plain and on the spacer arm-attached poly(HEMA-EGDMA) microspheres. *J Appl Polym Sci* 2001, 81:2702–2710
291 Moreno JM, Fagain CO: Stabilisation of alanine aminotransferase by consecutive modification and immobilisation. *Biotechnol Lett* 1996, 18:51–56
292 Taylor RF: A comparison of various commercially available liquid chromatographic supports for immobilisation of enzymes and immunoglobulins. *Anal Chim Acta* 1985, 172:241–248
293 Sellek GA, Chaudhuri JB: Biocatalysis in organic media using enzymes from extremophiles. *Enzyme Microb Technol* 1999, 25:471–482
294 Lee DC, Lee SG, Kim HS: Production of d-*p*-hydroxyphenylglycine from d,l-5-(4-hydroxyphenyl) hydantoin using immobilized thermostable D-hydantoinase from *Bacillus stearothermophilus* SD-1. *Enzyme Microb Technol* 1996, 18:35–40
295 Fruhwirth GO, Paar A, Gudelj M, Cavaco-Paulo A, Robra KH, Guebitz GM: An immobilized catalase peroxidase from the alkalothermophilic Bacillus SF for the treatment of textile-bleaching effluents. *Appl Microb Biotechnol* 2000, 60:313–319

296 Piller K, Daniel RM, Petach HH: Properties and stabilization of an extracellular β-glucosidase from the extremely thermophilic archaebacteria Thermococcus strain AN1: enzyme activity at 130 °C. *Biochim Biophys Acta* 1996 1292:197–205

297 Fernandez-Lafuente R, Guisan JM, Ali S, Cowan D: Immobilization of functionally unstable catechol-2,3-dioxygenase greatly improves operational stability. *Enzyme Microb Technol* 2000, 26:568–573

298 Goto M, Hatanakaa C, Goto M: Immobilization of surfactant-lipase complexes and their high heat resistance in organic media. *Biocheml Eng J* 2005, in press

299 Janssen MHA, van Langen LM, Pereira SRM, van Rantwijk F, Sheldon RA: Evaluation of the performance of immobilised penicillin G acylase using active-site titration. *Biotechnol Bioeng* 78:425–432

300 Simpson HD, Haufler UR, Daniel RM: An extremely thermostable xylanase from the thermophilic eubacterium Thermotoga. *Biochem J* 1991, 277:413–417

301 Dickensheet PA, Chen LF, Tsao GT: Characterization of yeast invertase immobilised on porous cellulose beads. *Biotechnol Boeng* 1977, 19:365–375

302 Kasche V, Haufler U, Riechmann L: Kinetically controlled semisynthesis of β-lactam antibiotics and peptides. *Ann NY Acad Sci* 1984, 434:99–105

303 Hsu AF, Foglia TA, Shen S: Immobilization of *Pseudomonas cepacia* lipase in a phyllosilicate sol–gel matrix. Effectiveness as a biocatalyst. *Biotechnol Appl Biochem* 2000, 31:179–183

304 Ruckenstein E, Wang X: Lipase immobilized on hydrophobic porous polymer supports prepared by concentrated emulsion polymerization and their activity in the hydrolysis of triacylglycerides. *Biotechnol Bioeng* 1993, 42:821–828

305 Bastida A, Sabuquillo P, Armisen P, Fernandez-Lafuente R, Huguet J, Guisan JM: A single step purification, immobilization, and hyperactivation of lipases via interfacial adsorption on strongly hydrophobic supports. *Biotechnol Bioeng* 1998, 58:486–493

306 Horiuti Y, Imamura S: Stimulation of Chromobacterium lipase activity and prevention of its adsorption to palmitoyl cellulose by hydrophobic binding of fatty acids. *J Biochem* 83: 1381–1385

307 Kanori M, Hori T, Yamashita Y, Hirose Y, Naoshima Y: A new inorganic ceramic support, Toyonite and the reactivity and enantioselectivity of the immobilised lipase. *J Mol Catal B Enzymatic* 2000, 9:269–274

308 Turkova J: Oriented immobilization of biologically active proteins as a tool for revealing protein interactions and function. *J Chromatogr B: Biomed Sci Appl* 1999, 722:11–31.

309 Hayashi T, Ikada Y: Protease immobilization onto polyacrolein microspheres. *Biotechnol Bioeng* 1990, 35:518–24

310 Rexova-Benkova L, Mrackova-Dobrotova M: Effect of immobilisation of *Aspergillus niger* extracellular endo-d-galacturonanase on kinetics and action pattern. *Carbohydr Res* 1981, 98:115–122

311 Itoyama K, Tanibe H, Hayashi T, Ikada Y: Spacer effects on enzymatic activity of papain immobilized onto porous chitosan beads. *Biomaterials* 1994, 15:107–112

312 Itoyama K, Tokura S, Hayashi T: Lipoprotein lipase immobilization onto porous chitosan beads. *Biotechnol Prog* 1994, 10:225–9

313 Cassell S, Halling J: Effect of thermodynamic water activity on thermolysin-catalysed peptide synthesis in organic two phase synthesis. *Enzyme Microb Technol* 1988, 10:486–491

314 Giacomini C, Irazoqui G, Batista-Viera F, Brena BM: Influence of the immobilization chemistry on the properties of immobilized β-galactosidases. *J Mol Catal B: Enzymatic* 2001,11:597–606

315 Aksoy S, Tumturk H, Hasirci N: Stability of α-amylase immobilized on poly(methyl methacrylate-acrylic acid) microspheres. J Biotechnol1998, 60: 37–46

316 Klibanov AM: Asymmetric transformations catalysed by enzymes in organic solvents. *Acc Chem Res* 1990, 23: 114-120

317 Rozzell JD: Commercial scale biocatalysis: myths and realities. *Bioorg Med Chem* 1999, 7:2253–2261

318 Lundell K, Kanerva LT: Enantiomers of ring-substituted 2-amino-1-phenyl-ethanols by *Pseudomonas cepacia* lipase. *Tetrahedron Asymmetry* 1995, 6:2281–2286

319 Aoun S, Baboulene M: Regioselectivity bromohydroxylation of alkenes catalysed by chloroperoxidase: advantages of the immobilisation of enzyme on taic. *J Mol Catal B Enzym*atic 1998, 4:101–109

320 Palomo JM, Fernandez-Lorente G, Mateo C, Ortiz C, Fernandez-Lafuente R, Guisan JM: Modulation of the enantioselectivity of lipases via controlled immobilization and medium engineering: hydrolytic resolution of mandelic acid esters. *Enzym Microb Technol* 2002, 31:775–783

321 DeSantis G, Jones JB: Chemical modification of enzymes for enhanced functionality. *Curr Opin Biotechnol* 1999, 10:324–330.

322 Mozheav VV Melik-Nubarov, NS, Sksnis V, Martinek K: Strategy for stabilising enzymes. Part two: Increasing enzyme stability by selective chemical modification. *Biocatalysis* 1990, 3:181–196

323 Cao L: Immobilised enzymes: science or art? *Curr Opin Chem Biol* 2005, 9:219–226 (and the references cited therein)

324 Linko Y, Saarinen RL, Linko M: Starch conversion by soluble and immobilised α-amylase. *Biotechnol Bioeng* 1975, 17:153–165

325 Nishio T, Hayashi R: Digestion of protein substrates by subtilisin: Immobilization changes the pattern of products. *Arch Biochem Biophys* 1984, 229:304–311

326 Hernaiz, MJ, Crout DHG: A highly selective synthesis of N-acetyllactosamine catalyzed by immobilised β-galactosidase from *Bacillus circulans*. *J Mol Catal B Enzym* 2000, 10:403–408

327 Wolodko WT, Kay CM: Rabbit cardiac myosin. 1. proteolytic fragmentation with insoluble papain. *Can J Biochem* 1975, 53:175–188

328 Suh CW, Choi GS, Lee EK: Enzymic cleavage of fusion protein using immobilized urokinase covalently conjugated to glyoxyl-agarose. *Biotechnol Appl Biochem* 2003, 37:149–155

329 Boundy JA, Smiley KL, Swanson CL, Hofreiter BT: Exoenzymic activity of α-amylase immobilized on a phenol-formaldehyde resin. *Carbohydr Res* 1976, 48:239–244

330 Kennedy JF, Cabral JMS, Kalogerakis B: Comparison of action patterns of gelatin-entrapped and surface-bound glucoamylase on an α-amylase degraded starch substrate: a critical examination of reversion products. *Enzyme Microb Technol* 1985, 7:22–8.

331 Ogino S: Formation of the fructose–rich polymer by water-insoluble dextran-sucrease and presence of glycogen value-lowering factor. *Agric Biol Chem* 1970, 34:1268

332 Rotticci D, Norin T, Hult K: Mass transport limitations reduce the effective stereospecificity in enzyme-catalyzed kinetic resolution. *Org Lett* 2000, 2:1373–1376

333 Sanchez EM, Bello JF, Roig MG, Burguillo FJ, Moreno JM, Sinisterra JV: Kinetic and enantioselective behaviour of the lipase from *Candida cylindracea*. A comparative study between the soluble enzyme and the enzyme immobilised on agarose and silica gels. *Enzyme Microb Technol* 1996, 18:468–476

334 Moreno JM, Samoza A, de Campo C, Liama EF, JV. Sinisterra : Organic reactions catalyzed by immobilized lipases. Part I. Hydrolysis of 2-aryl propionic and 2-aryl butyric esters with immobilized Candida cylindraceu lipase. *J Mol Catal A: Chemical* 1995, 95:179–192

335 Kaga H, Siegmund B, Neufellner E, Faber K, Paltauf F: Stabilization of Candida lipase against acetaldehyde by adsorption onto Celite. *Biotechnol Tech* 8:369–374

336 Ikeda Y, Kurokawa Y: Hydrolysis of 1,2-diacetoxypropane by immobilized lipase on cellulose acetate-TiO_2 gel fiber derived from the sol–gel method. *J Sol–Gel Sci Technol* 21:221–226

337 Mohapatra SC, Hsu JT: Optimizing lipase activity, enantioselectivity, and stability with medium engineering and immobilization for β-blocker synthesis. *Biotechnol Bioeng* 1999, 64:213–220

338 McIninch JK, Kantrowitz ER: Use of silicate sol–gels to trap the R and T quaternary conformational states of pig kidney fructose-1,6-bisphosphatase. *Biochim Biophys Acta* 2001, 1547:320–328

339 Suwasono S, Rastall RA: Synthesis of oligosaccharides using immobilised 1,2-mannosides from *Apergillus phoenicis*, Immobilisation-dependent modulation of product spectrum. *Biotechnol Lett* 1998, 20:15–17

340 Rosell CM, Fernandez-Lafuente R, Guisan JM: Modification of enzyme properties by the use of inhibitors during their stabilization by multipoint covalent attachment. *Biocatal Biotransform* 1995, 12:67–76

341 Palomo JM, Munoz G, Fernandez-Lorente G, Mateo C, Fuentes M, Guisan JM, Fernand Lafuente R: Modulation of *Mucor miehei* lipase properties via directed immobilization on different hetero-functional epoxy resins. Hydrolytic resolution of (R,S)-2-butyroyl-2-phenylacetic acid. *J Mol Catal B: Enzymatic* 2003, 21:201–210

342 Straathof AJJ, Panke S, Schmid A: The production of fine chemicals by biotransformations. *Curr Opin Biotechnol* 2002, 13:548–556

343 Partridge J, Halling PJ, Moore BD: Practical route to high activity enzyme preparations for synthesis in organic media. *Chem Commun* 1998, 7: 841–842

344 Ursini A, Maragni P, Bismara C, Tamburini B: Enzymatic method of preparation of optically active *trans*-2-amino cyclohexanol derivatives. *Synth Commun* 1999, 29:1369–1377

345 Kroutil W, Orru RVA, Faber K: Stabilization of Nocardia EH1 epoxide hydrolase by immobilization. *Biotech Lett* 1998, 20:373–377

346 Moriyama S, Noda A, Nakanishi K, Matsuno R, Kamikubo T: Thermal stability of immobilized glucoamylase entrapped in polyacrylamide gels and bound to SP-Sephadex C-50. *Agric Biol Chem* 1980, 44:2047–2054

347 Matsumoto M, Ohashi K: Effect of immobilization on thermostability of lipase from *Candida rugosa*. *Biochem Eng J* 2003, 14:75–77

348 Wasserman BP, Burke D, Jacobson BS: Immobilization of glucoamylase from *Aspergillus niger* on polyethylenimine-coated nonporous glass beads. *Enzyme Microb Technol* 1982, 4:107–109

349 Kim Y, Dordick J, Clark D: Siloxane-based biocatalytic films and paints for use as reactive coatings. *Biotechnol Bioeng* 2001, 72:475–482

350 Karim MR, Hashinaga F: Preparation and properties of immobilised pumnelo, limonoid glycosyltransferase. *Process Biochem* 2002, 38:809–814

351 Pogorevc M, Strauss UT, Riermeier T, Faber K: Selectivity enhancement in enantioselectivity hydrolysis of sec-alkyl sulfates by an alkylsulfatase from Rhodococus rubber DSM 44541. *Tetrahedron Asymmetry* 2002, 13:1443–1447

352 Tosa T, Mori T and Chibata I: Activation of water-insoluble aminoacylase by protein denaturing agents. *Enzymologia* 1971, 40:49–53

2
Adsorption-based Immobilization

2.1
Introduction

Adsorption-based enzyme immobilization was among the first enzyme immobilization methods. One of the first immobilized enzymes prepared by adsorption was reported by Nillson and Griffin in 1916 [1], when it was shown that invertase physically adsorbed by charcoal was still catalytically active. The first industrially used immobilized enzyme was prepared by adsorption of amino acid acylase on DEAE-cellulose [2].

In the last few decades, adsorptive enzyme immobilization techniques have been intensively studied, because of the intrinsic advantages of adsorptive enzyme-immobilization methods:

- reversibility, which enables not only the purification of proteins but also the re-use of the carriers;
- simplicity, which enables enzyme immobilization under mild conditions;
- possible high retention of activity because there is no chemical modification [3], in contrast with covalent enzyme immobilization [4].

In general, however, the immobilized enzymes prepared by adsorption tend to leak from the carriers, owing to the relatively weak interaction between the enzyme and the carrier, which can be destroyed by desorption forces such as high ionic strength, pH, etc. Occasionally the enzymes can be strongly adsorbed on suitable carriers [5] or the adsorption is stable enough under the application conditions.

Consequently, a number of variations have been developed in recent decades to solve this intrinsic drawback. Examples are adsorption–cross-linking; modification–adsorption; selective adsorption–covalent attachment; and adsorption–coating, etc.

In this chapter both conventional adsorption-based enzyme-immobilization methods and various new variations developed in the last few decades are presented. Each section starts from a brief introduction and a summary of each method including the advantages and disadvantages, preparation, and requirements of the corresponding carriers, followed by an illustrative table which presents the reader with comments about the method and selected references.

2.2
Classification of Adsorption

As noted above, adsorption of enzymes by carriers without chemical modification of the enzyme was one of the earliest developed methods of enzyme immobilization. This type of enzyme immobilization covers many subclasses, however.

Although there are many classifications of enzyme immobilization via non-covalent bonding, for example non-specific physical adsorption [118, 119], ionic bonding [7, 10], coordination (metal chelating) [11], and affinity adsorption [12], non-covalent carrier-bound enzyme immobilization can, methodologically, be generally classified into the following categories (Scheme 2.1):

- non-specific physical adsorption, in which the enzyme is adsorbed via non-specific forces such as van der Waals forces, hydrogen bonds, and hydrophilic interaction;
- bio-specific adsorption which normally uses immobilized ligands for adsorption of the enzymes, and is thus bio-specific;
- affinity adsorption to either immobilized dyes or immobilized metals [13, 14];
- electrostatic interaction (also ionic binding), which is based on the charge–charge interaction between the carrier and the enzymes [7–10]; and
- hydrophobic interaction, which is based on interaction of hydrophobic regions of the enzyme and carrier.

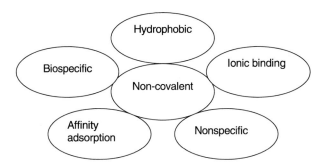

Scheme 2.1 Classification of attachment of enzymes to carriers by non-covalent adsorption.

Basically, any type of carrier (synthetic, naturally occurring, or insoluble organic or inorganic) can be used to adsorb the enzymes via non-specific binding (physical adsorption). For adsorption other than by non-specific physical adsorption, the corresponding functionality, for instance charged groups for ionic adsorption, hydrophobic tails for hydrophobic interaction, immobilized ligands for affinity adsorption, etc. must be introduced to the carriers.

As shown in Scheme 2.2, non-covalent enzyme immobilization relies largely on the presence of a specific binding ligand, which must first be introduced to the carrier, with the aim of mitigating non-specific adsorption and thus intensifying spe-

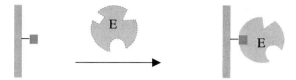

Scheme 2.2 Non-covalent immobilization of an enzyme on a carrier.

cific adsorption. For instance, porous zirconia-based particles with nominal pore diameters of 160 and 1000 Å modified with two different affinity ligands (iminodiacetic acid-Cu(II) and Con A (concanavalin A) were used to immobilize enzymes, leading to the finding that only one type of interaction was present and non-specific interactions with the support surface played an insignificant role [11], suggesting that the presence of specific binding functionality suppressed the non-specific interaction and that the immobilized binding functionality can thus conceal non-specific binding sites on the carriers from binding enzyme.

Compared with non-specific adsorption, one characteristic of specific adsorption is often that the enzyme molecules are, to a large extent, oriented on the surface (Scheme 2.3). This is believed to occur mainly because the carrier-bound functionality is often able to orient the enzyme molecules in a direction allowed by the nature of the binding and the spatial complementary effect, especially for biospecific adsorption.

Consequently, it is envisaged that selection of binding functionality is an important aspect of creating robust immobilized enzymes, for example immobilization of alkaline phosphatase on phospholipid Langmuir films [15].

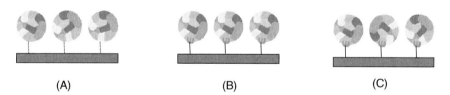

Scheme 2.3 Enzymes adsorbed on the surface: (A) Non-specific adsorption (random mode); (B) bio-specific adsorption (often site directed); (C) hydrophobic or ionic adsorption (hydrophobic pitch-directed adsorption).

2.3
Principles Involved in Absorptive Enzyme Immobilization

During the process of enzyme immobilization by non-covalent adsorption on a carrier a significant change of the microenvironment of the enzyme molecules can occur. The highly solvated enzyme molecules surrounded by the solvent (H_2O) and other solutes will be desolvated. Thus, the native forces that maintain the solvated structure (or native structure) can be disturbed. As a result, to enhance their interaction the enzyme molecules must adapt their conformation, undoubtedly leading to a change of the enzyme stability, selectivity and activity. Thus, it is not surprising that the principles involved in non-covalent enzyme immobilization should be minimization of changes such as loss of activity and stability and maximization of activity and stability.

Here, two principles are often encountered in non-covalent enzyme immobilization – the monolayer principle and the stability principle; these will be discussed in the following sections.

2.3.1
Monolayer Principle

Early in the 1970s it was observed that activity retention and enzyme stability were largely dependent on enzyme loading [16]. The mechanism, however, was unclear until the 1980s, when it was found that the simple deposition of enzyme on a solid carrier might constitute a very simple method for immobilization of an enzyme to be used in non-aqueous media [17].

It has, however, also been observed that immobilization via an adsorption–deposition strategy often distorts the conformation of the enzyme, leading to loss of the enzyme activity [18, 19], even though there was no chemical modification of the enzyme molecules. The fact that activity retention is closely related to the thermostability of the enzyme also suggests that the conformation of the less stable enzyme might undergo significant change [19]. It was, on the other hand, found that the change of conformation of enzyme adsorbed on certain type of carrier was strongly related to enzyme loading [20], suggesting that enzyme molecules tend to maximize contacts with the carrier at lower enzyme loading whereas contact with the carrier is minimized at higher enzyme loading [21]. Several studies led to the discovery that the minimum enzyme loading required to achieve the highest retention of activity was equivalent to the formation of a monolayer (2–3 mg protein m^{-2}) [17, 20, 22].

It was, furthermore, found that addition of additives such bovine serum albumin, gelatin, or casein can stabilize the enzyme (Scheme 2.4), when the loading of the enzyme is below monolayer coverage [22].

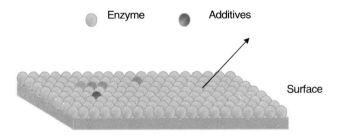

Scheme 2.4 Illustration of the monolayer principle for enzyme immobilization.

This requirement of minimum monolayer coverage can be ascribed to the potential of protein molecules to maximize contact with the carrier surface by deforming or unfolding, thus resulting in loss of the activity, because of conformation changes, when coverage of the carrier surface by the protein is below the monolayer. This argument is also confirmed by a recent study which revealed that the structural change was also loading-dependent for the protein molecules adsorbed on liquid/solid interfaces [23].

The requirement of monolayer coverage is, however, enzyme and carrier-dependent. For many "soft enzymes" it is necessary to meet the monolayer requirement [24, 25] whereas for some" hard" enzymes such as *Candida antarctica* lipase (CAL-B) and *Candida rugosa* lipase (CRL) or other enzymes such as trypsin [26] or dextransucrase [27, 28] this requirement is often not necessary, because the enzyme is also very active even at lower surface coverage.

It is worthwhile pointing out that although this monolayer concept was derived from the adsorption experiment it applies also to covalent enzyme immobilization. It was, for instance, also found not only that activity retention is closely related to the loading and the diffusion constraints [23], but also that the stability of the immobilized enzyme is determined by the loading, for example covalent binding of enzyme to porous carriers [29].

2.3.2
Stabilization Principle

As noted above, immobilization might stabilize or destabilize the enzyme [30] whereas sometimes enzyme stability is not affected by immobilization, or one method of immobilization is better than others with regard to stabilization. Thus, a central question arises: which factors, in addition to enzyme loading, determine enzyme stability? The answer to this question lies to a large extent in discovering what can be changed by interaction of the enzyme with the carrier.

Many examples have demonstrated that the physical interaction between the enzyme and carrier might change several enzyme characteristics, for example enzyme conformation, pH optimum, specific activity, or enzyme selectivity. These change of enzyme characteristics are closely related to changes of enzyme structure, i.e. they are caused by the change of enzyme structure during or after enzyme immobilization.

For example, alteration of the pH optimum of egg white lysozyme on ultra-fine silica particles from 4.5 to 7.0 suggested that the net charge in the neighbourhood of the enzyme might be different from that of the native enzyme, leading to the change of enzyme conformation and loss of the enzyme activity [31]. Similarly, immobilization of porcine trypsin, horseradish peroxidase, and bovine catalase on negatively charged ultra-fine silica particles (average diameter 20 nm) led to the discovery that the enzymes adsorbed at pH around and above their isoelectric points (pI) had high activity. In contrast, enzymes adsorbed at pH below their pI had significantly diminished activity and large CD spectral changes were observed on adsorption. These results suggested that the negatively charged zone of the corresponding enzyme might be involved in the activity of the enzyme [29].

Studies on the interaction of proteins with polystyrene carriers prepared from different co-monomers revealed that the ease of dissociation of the adsorbed proteins from the surfaces followed the order PS/PHEA > PS/PAA > PS/PMAA > PS (Scheme 2.5).

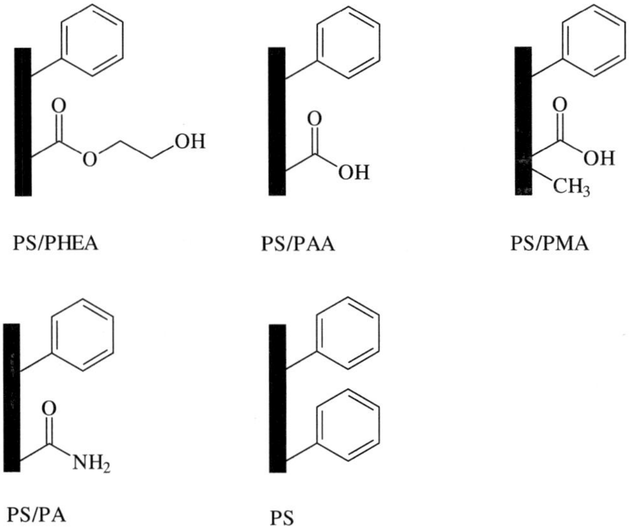

Scheme 2.5 Effect of the nature of the carrier on the stability of the enzyme.

Remarkably, this order also reflected the order of activity retention when the proteins were immobilized on the same carriers, suggesting that interaction of the protein might involve a structural change. The fact that hydrophobic interaction is often favoured by the hydrophobic interface suggested that proteins undergo conformational changes when immobilized on the carriers [32].

Immobilization of subtilisin on solid surfaces led to the conclusion that denaturation of this enzyme was several times greater than for a hydrophilic enzyme, suggesting that on this hydrophobic surface the enzyme is often labile [33]. Interestingly, changing particular lysine residues into phenylalanine sometimes improved enzyme stability, suggesting that the enzyme might be preferentially adsorbed on the hydrophobic surface via its specific hydrophobic regions.

These examples clearly showed that any immobilization process might distort the native structure of the enzyme in three possible directions or degrees, namely:

- reinforcement of the structure of the enzyme,
- loss of native enzyme structure, or
- no appreciable change [34].

The extent of the change of enzyme structure did, however, largely depend on the methods used. As shown in Scheme 2.6, the possible conformations of the immobilized enzymes (E (M)) can be classified into three groups – E (N) (native enzyme

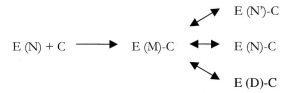

Scheme 2.6 Schematic illustration of conformational changes after immobilization. E (N) represents the native enzyme conformation and E (M) the modified enzyme conformation; the stabilized and destabilized forms are represented by E (N′) and E (D), respectively.

conformation), E (N′) (stabilized native enzyme conformation), and E (D) (destabilized enzyme conformation).

For the destabilized enzyme this can be plausibly ascribed to unmatched carrier–enzyme interaction [23]. For example, glucoamylase immobilized on cationic silica was less thermally stable than the native enzyme, whereas the same enzyme immobilized on anionic resin (SP-Sephadex C-50) was also less stable than the native enzyme [35], suggesting that the charge–charge interaction of the enzyme with the carrier might induce a conformation that is less thermostable than that of the native enzyme.

Further explanation is forthcoming on the basis that immobilization of the enzyme on to the carrier can be simply regarded as a physical and chemical reaction between enzyme (E) and carrier (C), leading to the formation of immobilized enzyme species (EC) which should be of lower energy but not necessarily more stable or active, as confirmed by the study of Sandwick and Schray, who found that conformational change is endothermic and that increased tendency toward enzyme inactivation by the lower-energy surface was observed [23].

If the conformation of the enzyme in the EC conjugate resembles the conformation of native enzymes, the enzyme will be stabilized by the carrier used. Otherwise, the enzyme will be destabilized (Scheme 2.7).

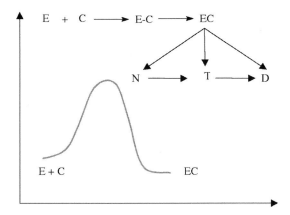

Scheme 2.7 Enzyme stabilization by the carrier.

Indeed, this was validated by study of the structure, stability, and activity of enzymes such as α-chymotrypsin and cutinase adsorbed on the carriers of different hydrophobicity [20], which led to the discovery that either the denatured enzyme (destabilized) or the stabilized enzyme is stabilized by the selected carrier. In other words, destabilization is the result of stabilizing the denatured conformation by the carrier used [20], for example, when the hydrophobic carrier such as Perflex (a fluorocarbon-based carrier) is used for immobilization of many enzymes [36]. By contrast, immobilization of CRL on hydrophilic carriers led to the formation of immobilized enzymes with similar activity and stability to the native enzyme, suggesting that the conformation of lipase was not disturbed by the carrier used [37].

For hydrophilic enzymes it is, therefore, often necessary to select hydrophilic carriers whereas for hydrophobic enzymes it is necessary to select hydrophobic carriers, for activity retention and enzyme stability. This is easy to understand – the native state of the hydrophilic enzymes is often hydrophilic and the denatured state is mostly hydrophobic, because of unfolding, which often leads to exposure of the hydrophobic domains. In contrast, the hydrophobic enzyme probably already has a hydrophobic native state and thus can be stabilized or activated by the hydrophobic carriers, as is observed for many lipases.

2.3.3
Enzyme Distribution

Immobilization of an enzyme in a porous carrier usually follows the moving front theory (or shrinking core theory) – the enzyme molecules first occupy the outer shell of the carrier and move slowly to the interior of the carrier when the nearest part of the carrier is occupied. This theory is also confirmed by studying enzymes adsorbed in carriers by physical adsorption, as exemplified by immobilization of β-galactosidase on porous supports [38]. Remarkably, it was found that most of the enzyme molecules were unable to move into half of the beads after incubation for 10 h [38]. It was also found that the distribution of the enzyme molecules on the carrier has a significant effect on enzyme activity expression, enzyme kinetics, and enzyme stability as exemplified by many studies.

Control of enzyme distribution can be achieved in several ways, for example kinetically controlled enzyme distribution as proposed by Borchert and Buchholz [39]. In general, three types of enzyme distribution on carriers can be distinguished – convex (obtained as a result of incomplete adsorption), uniform (obtained as a result of complete adsorption), and concave (obtained by backwashing a pellet of uniform profile with an appropriate solvent) [39].

Among the three types of distribution, activity expression is highest with convex distribution and lower retention of activity was obtained with concave distribution. The former is apparently suitable for enzymatic reactions that suffer from diffusion constraint, resulting in increase of the reaction selectivity [76], whereas the latter is suitable for the reactions that suffer from inhibition kinetics [40].

2.4 Requirement of the Carriers

As for any other type of carrier-bound immobilized enzyme, the performance of the adsorptive immobilized enzyme is also largely determined by the physical and chemical nature of the carriers. Although the nature of the carriers might differ greatly from each other, for enzyme binding to the surface of carriers there are certainly some general requirements of the physical and chemical nature of the carrier.

For the purpose of this discussion, carrier nature is grouped into two categories, physical and chemical. The former covers surface area, pore size, and binding density whereas the chemical nature of the carrier necessarily includes the nature of backbone of the carrier (chemical composition), binding functionality that takes part directly in the binding with the enzyme molecules, and the spacer that links the binding site to the carrier backbone.

The critical components of a carrier-bound immobilized enzyme are illustrated in Scheme 2.8.

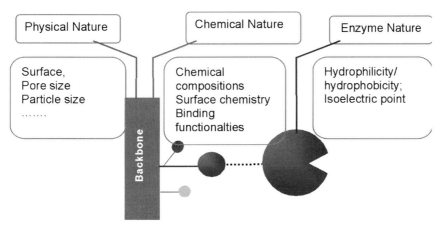

Scheme 2.8 Illustration of carrier-adsorptive enzyme immobilization.

2.4.1 Physical Requirements

As already discussed, the selected carrier functions as a chemical and physical modifier of the enzyme to be immobilized. The physical nature of the selected carrier must therefore meet some requirements.

2.4.1.1 Pore-size and Available Surface

As with the porous carriers, it is often possible to achieve high enzyme loading. This is largely because high surface area is associated with porous carriers. High surface area is, however, not a prerequisite for obtaining higher enzyme loading.

Early in the mid-1970s it was observed that not all porous carriers are applicable for enzyme immobilization, because of the pore size limitation [41]. In contrast, there were certain requirements of the internal geometry of the carriers if high enzyme loading and activity retention were to be achieved. This is understandable. First, the enzyme should be able to enter the pore, to be adsorbed in the interior of the carriers. Conceivably, the pore size of the selected carrier should meet three requirements.

- First, pore size should be at least of the same order as enzyme size.
- Second, enzyme-conformation mobility should be not significantly reduced compared with that of the free enzyme. For catalysis to occur the enzyme should have similar freedom to change its conformation as might occur in the free enzyme, thus necessitating the presence of larger pore size than the size of the enzyme molecules.
- Last, diffusion constraints should be mitigated to retain high apparent activity.

In all cases, a suitable pore size is required. In practice it is necessary to compromise the specific activity and the volume activity by selecting an appropriate pore size and available surface.

By studying the adsorption of lipase on porous inorganic carriers, some useful insights have been obtained. It was found that activity retention depended on pore size in the range below 100 nm. Above this value, however, the activity is independent of the pore size, suggesting that diffusion constraints are negligible when the pore size is above 100 nm and there is no substrate concentration gradient [42]. Similarly, immobilization of CRL on different hydrophobic carriers of different pore size (divided into two groups, >100 nm and <100 nm) also led to the discovery that pore size >100 nm results in increased accessibility of the pores, thus increasing enzyme loading and retention of activity [43].

In an ideal situation, the enzyme molecules might be uniformly spread over the available surface. In other words, monolayer of enzymes can be formed on the wall of the pores, as shown in Scheme 2.9. In fact, this has been confirmed by many recent studies as mentioned above [52, 82].

Although this pore-size requirement was deduced from adsorption experiments, it is obvious that this guideline should also apply to other types of enzyme immobilization, when porous carriers are used. In other words, the pore size of the por-

 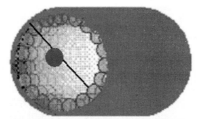

Scheme 2.9 Schematic illustration of enzyme molecules adsorbed on the wall of the pores

ous carriers used for any type of enzyme immobilization should exceed a certain value, to mitigate diffusion constraints and maximize enzyme activity [44].

In fact, the carriers often used for covalent enzyme immobilization, e.g. the commercially available Eupergit C (pore size 10–20 nm) have pore sizes far below 100 nm. Several studies have revealed that retention of activity by enzymes immobilized on Eupergit C is significantly lower than when immobilized on Eupergit C 250 L, which has a pore size of approximately 100 nm. Consequently, the activity of enzymes immobilized on carriers with pore size below 100 nm is also lower and strongly related to enzyme loading and pore size [45].

In general, activity expression with porous carriers is below 50% particularly at high enzyme loading, because of the presence of diffusion limitation, as confirmed by active site titration experiments [46]. Thus, when considering activity retention by an enzyme immobilized on a porous carrier it is absolutely necessary to specify the enzyme loading, because the distribution of the enzyme molecules in porous carriers is often loading-dependent, as discussed above.

Future development of porous carriers for enzyme immobilization should be directed toward the preparation of porous carriers, which not only have bigger pore size but also uniform pore-size distribution.

2.4.1.2 Internal Structure

The internal structure of a carrier refers to pore volume, porosity, tortuosity, pore size, pore size distribution, and pore mode (interconnected or closed).

The number of enzyme molecules immobilized or adsorbed is obviously related to pore size, when the specific area is kept relatively constant. For example, immobilization of urease on macroporous carriers with a specific area (500–600 $cm^2\,g^{-1}$) led to the discovery that enzyme activity increased with the pore diameter and that activity also increased with increasing specific area when the pore size was >18 nm [196].

Studies of lipase immobilized on different types of ion exchanger, i.e. Duolite ES562, Duolite ES568 and Spherosil DEA (148 nm) have revealed that the highest activity expression was obtained with Duolite ES562, which has a large proportion of narrow pores, suggesting that immobilization occurred mainly in the external pores, thus being less diffusion-controlled. In contrast, for diffusion-controlled activity, higher retention of activity was obtained with Spherosil DEA and Duolite ES568, because of the larger pore size of the former [17].

As regards the nature of the microenvironment, the activity, optimum pH, and temperature ranges of immobilized urease are larger on polar (acrylonitrile) than on non-polar (styrene) resins [196].

2.4.1.3 Density of Binding Functionality

Like covalent enzyme immobilization, it is also essential to control the density of the binding functionality in the non-covalent binding, with regard to activity retention, pH optimum, and enzyme stability. For instance, when urease was immobilized on a copolymer of 2-(N,N-dimethylamino)ethyl methacrylate and methacrylic acid with the acid in the range of 24–68 mol% the immobilized enzyme had increasing activity with increasing acid content; enzyme activity was, however, inhibited if acidity was above 84 mol% [47]. In contrast, increasing the itaconic acid concentration in poly(N-VP-co-IA) hydrogel led to enhanced enzyme loading, activity, and storage stability of the immobilized α-amylase [48].

The effect of binding density on performance could be explained as follows: first, the density of binding functionality affects the microenvironment of the carrier; second, the conformational flexibility of the enzymes is obviously related to number of the links to the enzymes; and third, loading is, naturally, dependent on the number of binding sites.

This was clearly observed by immobilizing α-chymotrypsin on the hydrogel-coated polymer film support obtained by radiation graft copolymerization of 2-hydroxyethyl methacrylate (HEMA) and methacrylic acid (MAAc) to silicone rubber films. It was shown by active-site titration with diisopropylfluoro phosphate (DFP) and hydrolysis of the model substrate, N-acetyl-l-tyrosine ethyl ester (ATEE) that increasing the MAAc content of the hydrogel resulted not only in increased enzyme loading but also increased fraction active towards DFP, obviously because of the larger number of binding sites. With increasing number of binding sites, however, it was found there were greater shifts of the pH optimum to higher pH for the ATEE assay and that the specific activity for ATEE fell sharply with increasing MAAc content of the hydrogel. This is obviously ascribed to the microenvironmental effect – at high MAAc content the microenvironment of the enzyme was changed [49].

Increasing the binding density also causes an increase in non-specific interaction, as revealed by studying the effect of concentration of immobilized inhibitor on the biospecific chromatography of pepsins [50].

2.4.1.4 Particle Size

The particle size of the carrier selected for enzyme immobilization often dictates reactor configuration. In general, large particle size is required for the packed-column reactor, to reduce pressure drop. For heterogeneous reactions especially, in which a sieve-plate reactor separates the immobilized enzymes from the reaction mixture, it is often necessary to use immobilized enzymes on carriers of large particle size; one example is for the semi-synthetic β-lactam antibiotics, which might crystallize from the reaction mixture [51].

However, increasing the pore size obviously increases diffusion limitation, leading to a decrease of apparent enzyme activity, for example immobilization of lipase on Poly(methyl methacrylate) (PMMA) [52].

Thus, it is envisaged that reducing the particle size will definitely increase the expression of enzyme activity. For instance, the activity of glucose isomerase from *Actinoplanes missouriensis* immobilized on porous DEAE-cellulose beads by simple adsorption was less influenced by the particle size when the bead size was ca. 35 mesh [53].

Intra-particle diffusion can, however, be overcome for large particles by locating the enzyme molecules on the surface of large particles only. One example is the immobilization of glucoamylase on Amberlite CG-50 Types I, II, and III as carriers. Even with Amberlite CG-50 Type I, the largest carrier used, immobilized glucoamylase with high specific activity could be prepared by locating the enzyme molecules on the outer surface of the carrier, suggesting that the major factors causing a drop in activity in this system was participation of the intra-particle diffusion process [54].

Practically, the influence of the particle size on enzyme activity can be determined by grinding. If significant increase in the activity, when reducing the particle sizes by grinding them into small ones, can be obtained, it will definitively suggest the presence of a severe internal mass limitation [471].

2.4.2
Chemical Nature of the Carriers

As pointed by Messing, there are, in principle, no inert carriers. Consequently, closer contact and subsequent interaction will disturb the native forces that maintain the native structure of the enzymes, leading to a new conformation, depending on the surface chemistry of the carriers used [20]. Thus, judicious choice of the carrier is with desired chemical nature is very important.

The chemical nature of carriers selected for non-covalent enzyme immobilization depends both on the chemical composition of the backbone and the surface chemistry, for example pendant functionality that dictates the type of adsorption – namely ionic adsorption, coordination, bio-affinity, or hydrophobic. Other functionality not directly involved in the binding might, moreover, also be critical in terms of enzyme activity, stability, and selectivity. In general, these not only determine the orientation of the enzyme on the carrier surface but also dictate largely how the enzyme interacts with the carrier (interaction via hydrophilic or hydrophobic regions or negatively or positively charged zones).

2.4.2.1 Nature of Binding Functionality
The nature of the binding functionality encompasses the properties of the functionality that participate in the binding. Binding functionality can be described by:

- charge properties,
- polarity of the functionality,
- hydrophobicity,
- size.

As for covalent enzyme immobilization, the different nature of the ionic groups of the carriers might lead to different enzyme retention of activity [9, 55], suggesting that enzyme orientation or conformation on the surface of the carrier might depend on the nature of the binding functionality of the carrier.

It was, for instance, found that the adsorption capacity of trypsin on a copolymer of styrene and 2-hydroxyethyl methacrylate was largely dependent on the ratio of the comonomer, namely 2-hydroxyethyl methacrylate. Increasing the ratio of 2-hydroxyethyl methacrylate led to increased hydrophilicity, thus reducing the loading [56].

It has also been found that not only enzyme loading but also enzyme performance such as activity [57], stability [24] and selectivity [58] could be affected by slight differences between the chemical nature of the carrier [20]. These facts exemplify the importance of designing suitable carriers for the desirable immobilization.

Changing one type of binding functionality to another can often change the performance of the resulting immobilized enzyme. For example, immobilized amino acid acylase was approximately 40 times more stable on tannin cellulose than on immobilized on DEAE-cellulose [59].

Recently, influence of the nature of the non-covalent binding functionality on the activity of the immobilized enzyme in organic solvent has been convincingly demonstrated by the fact that the specific activity of PSL (*Pseudomonas cepacia* lipase) immobilized on TN-M (Toyonite 200-M) is twice active as that immobilized on TM-A (Toyonite 200-A) for the transesterification of racemic 1-phenylethanol with vinylacetate. The fact that two ceramics-based carrier differ from each other only in the pendant binding functionality (hydrophobic methyaryloyloxy group for TN-M and hydrophilic amino group for TN-A), suggests that their activity is dictated only by their individual binding functionality [472].

2.4.2.2 The Role of the Spacer

Early in the 1970s it was discovered that direct adsorption of some proteins on some carriers failed, owing to incompatibility of the carriers and the protein to be adsorbed. To enhance the interaction of the protein to be adsorbed with the carrier used, a suitable spacer must be introduced. This observation was soon applied to immobilization of enzymes on carriers via non-covalent adsorption [9, 60]. Today, the use of spacers has been also extended to covalent immobilization of enzymes [61].

The presence of a suitable spacer, which can be characterized by their nature (hydrophilic/hydrophobic, negatively/positively charged/neutral, linear/branched, molecular size and shape), often improves some of the properties of the immobilized enzymes, for example enhanced enzyme stability, because of the more favourable environment created by the presence of a suitable spacer, enhanced retention of activity, because of the improved flexible conformational flexibility (which is usually a prerequisite for higher activity) and higher selectivity [62].

2.4.2.3 The Nature of the Backbone

It is often found that the activity of the immobilized enzyme is very dependent on the nature of the carrier used. For example, it was found that *Rhizopus niveus* lipase immobilized on the hydrophilic Celite, Spherosil, had more hydrolytic activity than when immobilized on Duolite XAD 761 whereas the highest synthetic activity was obtained by use of Duolite XAD 761. This result suggested that enzyme conformation was different on the different carriers. It is most likely that the enzyme is in a conformation on the hydrophobic carrier that is active either in organic solvent or in aqueous medium, whereas the conformation of the enzyme on the hydrophilic carrier is most probably active only in aqueous medium. This behaviour is often observed for lipases [63].

As with the nature of the carrier, it is conceivable that the hydrophilic environment normally favours higher enzyme stability. Thus, immobilized lipase is more stable on PEI-coated silica than on silanized carriers, which are generally more hydrophobic than PEI-coated silica [354].

2.5
Factors Which Dictate Enzyme Catalytic Performance

The catalytic performance of an immobilized enzyme denotes the enzyme activity, stability and selectivity. As discussed above, the performance of immobilized enzymes is determined by the enzyme, the carrier, and the immobilization conditions.

Here, the important factors, which influence the catalytic performance of a noncovalently immobilized enzyme denotes the enzyme activity, stability and selectivity, are summarized and analyzed.

2.5.1
Activity

Although the activity of the immobilized enzyme can be determined by the nature of the carrier, nature of the enzyme, the immobilization conditions, etc., it is often very difficult to find a clear boundary between these factors. Often, the final characteristics of the immobilized enzyme (IME) are a result of synergetic interaction of these factors, although one may often play a more important role than the others.

2.5.1.1 Diffusion-controlled Activity

If activity is diffusion-controlled, enzyme activity is often lower if porous carriers are used, because of mass-transfer limitations. The activity of the immobilized enzymes may, however, be determined not only by the nature of the carrier but also by the distribution of the enzyme molecules within the porous carrier and by substrate concentrations. Thus, high retention of activity can be achieved by designing a tentacle carrier with a more open structure [64], or by using carriers with larger pores [354]. Immobilization of papain on porous silica with pores in the range 30–90 nm led to the finding that pore size should be 5–10 times the molecular size of the enzyme regarding the high retention of activity [354].

It is also often found that activity retention for a certain enzyme and carrier is often loading-dependent [23, 39, 65–68]. For example, the volume activity of glucoamylase immobilized on silanized porous glass is constant in the range 180–300 mg g^{-1} carrier. The specific activity (U mg^{-1} protein), however, decreases as a result of diffusion limitation [69]. This phenomenon is ascribed to the different distribution of the enzyme molecules. Thus, kinetically controlled enzyme immobilization [70, 71], for example immobilization of β-glucosidase on the macroporous resin cation exchanger Amberlite DP-1 [72] or immobilization of lysozyme on polystyrene ion-exchange beads [73], can be used to control enzyme distribution. At low enzyme loading, the enzyme molecules occupy only the outer layer of the porous particles. Thus, the extent of the diffusion is less if the enzyme molecules are located on the surface only. As loading increases the enzyme molecules must penetrate the interior of the carrier, thus leading to the lower activity expression.

Consequently, penetration of the enzyme into carriers can be controlled by incubation time [74]. Immobilization of enzyme on a small-pore carrier, for example Sephadex C-25-type cation exchangers, might avoid this problem, because penetration of enzyme molecules into the core of the matrix can be avoided [74]. This model was validated by studying the adsorption of β-galactosidase on Duolite ion-exchange resins, porous spherical supports [74]. Active site titration experiments revealed that diffusion limitation can lead to 60–70% loss of activity [76].

In addition to these factors it is found that the activity of an immobilized enzyme also depends on substrate concentration. This means that at higher substrate concentrations the enzyme might have the same activity as the native enzyme whereas enzyme activity might be lower at lower substrate concentrations; one example is the immobilization of alkaline phosphatase on activated alumina particles [63].

It might be worthwhile pointing out that diffusion limitation can also be positively used to increase enzyme activity. For example, lipase from *Pseudomonas fluorescens* immobilized on macroporous anion-exchange resin by adsorption and cross-linking strategies was less susceptible to substrate or product-inhibition effects [73], owing to the positive effect of diffusion limitation. Similarly, less substrate inhibition was also observed for milk xanthine oxidase immobilized by covalent attachment on CNBr-activated Sepharose 4B and by adsorption on n-octylamine-substituted Sepharose 4B, because of substrate diffusion limitation in the pores of the carrier beads (internal diffusion limitation) [77]. In these examples the presence of diffusion limitation reduces the concentration of the substrate or product, thus increasing enzyme activity under the conditions used for the reaction.

2.5.1.2 Conformation-controlled Activity

Changes of enzyme activity as a result of immobilization can generally be ascribed to effects such as diffusion limitation or conformation change. For conformation-controlled activity it has been found that enzyme activity (U mg^{-1} protein immobilized) was highly dependent on the nature of the carriers used. For example, papain immobilized on polysulphone membrane retained only 12% of the native activity whereas 25% of native activity was retained when the enzyme was immobilized on hydroxylethylcellulose-coated polyethersulphone, suggesting that the change of conformation is mainly related to the nature of the carrier, as revealed by ESR study [23, 65].

It has also recently been demonstrated that lipase from *Rhizopus oryzae* immobilized on hydrophobic carriers was an order of magnitude more active than that immobilized on hydrophilic carriers [78]. Similarly, studies of *Rhizopus niveus* lipase immobilized on Duolite XAD 761, Celite Hyflo-Supercel, Spherosil XOA 200, and silica revealed that the highest synthetic activity was obtained with Duolite XAD 761 whereas the highest hydrolytic activity was obtained by use of Celite Hyflo-Supercel. Also, immobilization on hydrophilic carriers usually enhanced hydrolytic activity compared with that of the free enzyme. More remarkably, the initial synthetic activity can be increased 12.2-fold [79].

It is most probable that lipase immobilized on the hydrophobic carrier adopted a different conformation (active conformation) than that immobilized on the hydrophilic carrier – a typical phenomena of interfacially active enzymes. In other words, the conformation of the lipase immobilized on hydrophilic carrier is an active conformation either in aqueous medium or in organic solvents whereas the conformation of lipase immobilized on the hydrophilic carrier is most active in aqueous medium, because of the interfacial activation, and less active in organic solvents, because the lipase cannot freely adopt an active conformation in organic solvents. This hypothesis has been confirmed by covalently immobilizing lipase in organic solvent instead of in aqueous medium [80].

This conformation difference ultimately alters the substrate-specificity, thus the enzyme might have different activity against different substrates. Conformation-controlled activity can be probed only by use of different substrates. For example, alcohol dehydrogenase from horse liver immobilized on octyl-Sepharose CL-4B, a hydrophobic analogue of Sepharose, has very low affinity for cyclohexanol compared with the same enzyme immobilized covalently on CNBr-Sepharose. In contrast, both immobilized enzymes have almost the same activity for ethanol, irrespective of the method of immobilization. EPR spectroscopy revealed that immobilization of alcohol dehydrogenase on the hydrophobic carrier significantly narrowed the size of the active centre [84]. Accordingly, loss of activity as a result of conformation change during immobilization can be minimized by adding substrate or inhibitors during binding – to prevent the conformation change during immobilization; this is exemplified in many examples [71].

The extent to which activity is conformation-controlled is usually easy to judge on the basis of activity retention. For instance, several enzymes immobilized on PCTFE retained only 4% activity, suggesting the loss of activity is mainly a result of loss of conformation [81]. Similarly, the low retention of activity (~40%) of *endo-d*-galacturonanase adsorbed on porous PET with payload of 0.00525 is most probably ascribed to a conformation change induced by the hydrophobic carrier [82]. Occasionally, however, the conformation change might be essential for allosteric enzymes. Thus, high retention of activity (more than 100%) can be obtained with some enzymes, e.g. lipases, because of the activation effect of immobilization [83].

2.5.1.3 Substrate-controlled Activity

Many examples have demonstrated that the substrate specificity of immobilized enzymes might be significantly different from that of the free enzymes as a result of the immobilization. This can be explained as the result of a conformation change (change of enzyme geometry) or the presence of diffusion constraints.

In the former instance, enzyme activity depends largely on the size of the substrates [18]. Because of steric hindrance, the large substrates diffuse slowly into the pores of the carrier, leading to low activity. As with conformation change, the rate of reaction of the immobilized enzymes with the native enzyme can be increased or decreased, depending on the substrates [73].For example, HLAH immobilized on octyl-Sepharose CL-4B had different activity against ethanol and cyclohexanol, because of a the change in the geometry of the active site [84]. Similarly, the lipase from *Candida cylindracea* immobilized on hydrophobic supports gave the highest esterification rates for the secondary alcohol (R,S)-1-phenylethanol whereas for the primary alcohol heptanol the highest reaction rate was obtained with the free enzyme powder [85].

It is most probable that the geometry of the active centre is different in the two cases. The former might have a very open active centre (open conformation) whereas the latter possibly has a narrower active centre. Similarly, it was found that *Candida cylindracea* lipase immobilized on hydrophobic carriers had higher rate of reaction in the esterification of the secondary alcohol (R,S)-1-phenylethanol than same enzyme immobilized on hydrophilic carriers. When the primary alcohol heptanol was used, however, the reaction rate was highest for the free lipase [85]. Lipase CAL-B immobilized on octyl-silica also had strong substrate-dependent activity; the activity measured with p-NPP was only 10–20% that measured with tributyrin [87].

2.5.1.4 Loading-controlled Activity

As noted above, the activity of immobilized enzymes is often loading-dependent. Typically, the activity (U g^{-1}) of the immobilized enzyme increases with increasing enzyme loading then often reaches a plateau.

Study of immobilization of *Canadida rugosa* lipase on porous chitosan beads by a non-specific interaction revealed that the volumetric activity of the immobilized enzyme varies from 48 to 480 U g^{-1} of support) when the loading varied from 0.1 g to 1 g crude lipase g^{-1} chitosan. The highest retention of activity, however, was obtained when the loading was 120 U lipase g^{-1} carrier. A further increase in the enzyme loading had no effect on retention of activity, suggesting that formation of multilayer of enzyme might be the major reason for the lower retention of activity [86].

Similarly, the activity of CAL-B immobilized on octyl-silica was also found to depend on enzyme loading (Scheme 2.10) [87].

In general, factors that are closely related to the enzyme loading are diffusion limitation (or enzyme distribution) and conformation changes. Diffusion constraints occurring as a result of penetration of the enzyme molecules into the interior of the beads might reduce the activity [73] whereas the enzyme often undergoes conformational changes to adapt itself to the carrier surface [23], leading to activity

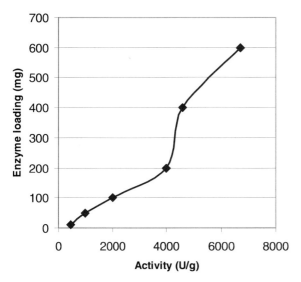

Scheme 2.10 Illustration of loading–activity relationship for immobilized CAL-B (adapted from Ref. [87]).

loss. Therefore, enzyme loading up to the amount sufficient to form a monolayer is often required to avoid conformation change, for example the immobilization of exopolygalacturonase on poly(ethylene terephthalate) with 51.8% activity and a payload of 0.0912 [82].

Thus activity retention and enzyme loading can be used as indicators of conformation change or diffusion constraints. For example, immobilization of lactate dehydrogenase on porous Al_2O_3 [88] led to only 20% retention of activity at lower enzyme coverage on the carrier surface, suggesting that the enzyme might be deformed at lower enzyme coverage [88]. Similarly, only 42.4% retention of activity was obtained when *endo-d*-galacturonanase was immobilized on porous poly(ethylene terephthalate) with a enzyme/carrier ratio of 5.25 mg g^{-1}. Thus, activity retention is mainly determined by conformation change.

In contrast, a strong activity–loading relationship was observed for lipase PPL immobilized on Amberlite IRA 938. When loading was more than 0.15 g g^{-1} carrier, interesterification activity remained unchanged, because of diffusion limitation [89]. Immobilization of xanthine dehydrogenase by adsorption on a Sepharose derivative prepared by reaction of *n*-octylamine with CNBr-activated Sepharose 4B led to the discovery that retention of activity decreased, as protein loading was increased. Obviously, diffusion constraints play an important role in activity expression [24].

Immobilization of amyloglucosidase from *Aspergillus niger* by adsorption on to a hexyl-Sepharose containing 0.51 mol hexyl group mol^{-1} galactose resulted in full retention of activity at lower enzyme coverage and 68% at the maximum loading (17 mg g^{-1} wet carrier). That immobilized enzymes obtained from the same enzyme coverage but with different concentrations of salts in the immobilization medium had comparable activity and operational stability implied that the activity of the immobilized enzyme is mainly determined by diffusion constraints. The conformation change, however, contributed less to activity retention but much more to the stability [24].

The effect of the microenvironment on enzyme activity can be clearly shown by exclusion of diffusion limitation. For instance, for sweet potato α-amylase [90] immobilized by adsorption on to an agarose gel, activity and stability were dependent on enzyme-loading (with saturation at an enzyme content of 35 mg mL^{-1} packed gel). The stability increased with increasing enzyme loading whereas retention of activity was inversely related to enzyme loading, being 50% at low enzyme loading. If the contribution of diffusion limitation to enzyme activity is negligible at lower enzyme loading, loss of activity can apparently be ascribed to the conformation change.

2.5.1.5 Medium-dependent Activity

In recent decades, especially since the 1980s, it has very often been observed that the activity of the immobilized enzyme is largely determined by the conditions of immobilization, for example pH, temperature, ionic strength, and the presence of additives. Among the various properties of the medium, pH, ionic strength, and temperature, etc., are of critical importance. It is often found that variation of the pH leads to a change in the activity of the immobilized enzyme relative to that of the native enzyme [91], although the pH optimum of the immobilized enzyme might not change.

For instance, it was recently found that the activity of lipase from *Nigella sativa* seeds immobilized by adsorption on to Celite 535 from phosphate buffer solutions of different pH from 5.0–8.0 at 25 °C varied, depending on the pH of the adsorption medium, with pH 6 being the optimum in terms of activity [92]. This suggests that the conformation of the enzyme at the moment of immobilization is very important. This is understandable, because a change in the conformation of the enzyme often becomes much more difficult after immobilization, owing to rigidification of the enzyme molecule as a result of multipoint attachment.

Along with pH, ionic strength, and temperature, it is also found that addition of additives such as crown ether, sugar, or surfactants can affect activity expression. For instance, the specific activity of chlorophyllase immobilized on silica gel was increased by 50% when a suitable amount of lipid was added to the reaction mixtures [93].

The nature of the immobilization medium can affect not only enzyme conformation but also the nature of the carrier. For instance, the pH of the medium determines the net charge on the enzyme, owing to the isoelectric point, but can also affect the charge on the carrier if this is affected by pH. Thus adsorption of chloroperoxidase (CPO) on hydrophobic and chemically neutral talc at pH below the IP led to hydrophobic adsorption (up to 8 mg CPO g^{-1} talc) but strong inhibition of activity expression whereas adsorption at pH above the IP (pH 6) led to higher activity expression (~61% for halogenation and up to 72% for oxidation). Excellent enzymic activity (80–126%) was obtained, however, irrespective of the pH of adsorption, when the hydrophilic and slightly acidic calcined talc, type CLST was used, although the loading was lower (2.5 mg CPO g^{-1} talc) [94].

2.5.1.6 Microenvironment-dependent Activity

The term "microenvironment" denotes the nature of the carrier in the proximity of the enzyme molecules. Microenvironment-controlled activity was first demonstrated by Goldstein and Katschalski in the early 1960s, when it was found that the pH profile of the enzymes as compared with the native was shifted and K_m was changed depending on the charge properties of the carrier, the substrate, and the ionic strength [95].

For example, immobilization of trypsin on a negatively charged polymer led to decrease of K_m of more than one order of magnitude with a positively charged substrate [95]. The deviation of the pH optimum and the kinetic behaviour relative to those of the free enzyme is obviously because of the different levels of protons around the enzyme and the bulk medium. Similarly, it is found that high retention of activity (more than 100%) can be achieved by adsorbing some enzymes, for example lipases, on to the hydrophobic carriers, because of the interfacial activation effect [57].

However, the microenvironment effect is not limited to changes of pH optimum – it can also provoke changes of enzyme conformation. Thus, direct interaction of the charge of the carriers with the enzyme might cause a change of enzyme conformation or orientation of the enzyme, as a result of disturbance of the forces involved in maintaining the native structure. For example, retention of activity of *Candida cylindracea* lipase immobilized on the copolymer of methyl acrylate and divinylbenzene (PMA–DVB) to which the positively charged functionality was introduced by aminolysis with a diaminoalkane was severalfold higher than for the parent carrier [96]. That the pH optimum of the lipase *Candida cylindracea* on AP-MA–DVA was not shifted compared with the free lipase and the lipase immobilized on PMA–DVB suggested that the increased activity on aminolated carriers was exclusively ascribed to a change of enzyme conformation or molecular orientation because of charge–charge interaction (the enzyme at pH 7 is negatively charged and the carrier is positively charged).

The effect of the microenvironment on enzyme activity can be clearly shown by exclusion of diffusion limitation. The effect of microenvironment on enzyme activity was recently beautifully demonstrated by immobilization of organophosphorus hydrolase (OPH) on different meso-porous silicas bearing different binding functionality, for example negatively charged FMS (HCOOH–CH$_2$–CH$_2$), positively charged (NH$_2$), and UMS (native negatively charged surface). It was found that UMS afforded a protein loading of 3.1% (w/w) only and a specific activity of 935 units mg^{-1} of adsorbed protein; a higher protein loading of 4.7% (w/w) and significantly enhanced specific activity of 4182 units mg^{-1} was obtained with the negatively charged FMS as carrier. Unexpectedly, the specific activity with 2% NH$_2$-

FMS was as high as 2691 units mg^{-1} with a protein loading of 3.8% (w/w) whereas 20% NH$_2$-FMS afforded almost zero retention of activity and 0.25% loading. Remarkably the difference between the former and the latter is only 2% replacement of the surface COOH groups [97]. This result clearly suggested that the microenvironment of the binding functionality is of vital importance.

The nature of the microenvironment is also very dependent on binding density. Study of the purification of guanidinobenzoatase by affinity chromatography with different ligand concentrations (prepared by controlled immobilization of a ligand on agarose) revealed that a low concentration of ligand (2 µmol mL^{-1}, corresponding to modification of only 5% of active groups in the commercial resin) favours specific adsorption. In contrast, at higher ligand concentration (10 µmol mL^{-1} of gel or higher) the matrix has much lower specific binding activity for the targeting protein, suggesting that the nature of the microenvironment was strongly related to the concentration of the binding functionality [98].

The effect of the nature of the carriers, for example the microenvironment, on enzyme activity is interestingly demonstrated by the immobilization of trypsin on functionalized SBA-15. It was found that loading and activity retention depends on the procedures used to prepare the carriers, namely in-situ or post-synthesis, suggesting that the microenvironment of the carriers (around the CAG) is different [99].

Immobilization of *Mucor miehei* lipase on Indion 850, a styrene–divinylbenzene polymer resin, led to an immobilized enzyme with 1.4 times higher activity than its solution counterpart and enhanced thermal stability [100].

2.5.1.7 Carrier Nature-dependent Activity

It is often found that the activity of an immobilized enzyme is highly dependent on the nature of the carrier used. For example, it was found that for *Rhizopus niveus* lipase immobilized on hydrophilic Celite, use of Spherosil resulted in higher hydrolytic activity than Duolite XAD 761 whereas the highest synthetic activity was obtained with Duolite XAD 761. This result suggests that the enzyme conformation was different on the different carrier. It is most probable that the enzyme on the hydrophobic carrier is in a conformation that is active either in organic solvent or aqueous medium whereas the conformation of the enzyme on hydrophilic carrier is most probably only active in aqueous medium. This behaviour is often observed for lipases [79].

Immobilization of penicillin G acylase on large porous silica materials prepared by condensation of TEOS (tetraethylorthosilicate) and its derivatives led to the finding that not only the volume activity but also the specific activity of the immobilized penicillin acylase depends on the nature of the pedant groups [101]. High specific activity was obtained with VTES as carriers (Scheme 2.11).

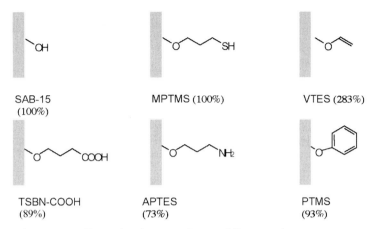

Scheme 2.11 Different silica derivatives bearing different pendant groups for immobilization of penicillin G acylase.

2.5.1.8 Enzyme Nature-dependent Activity

Activity retention is also highly dependent on the nature of the enzyme, for example the distribution of its charged or hydrophobic regions. Thus, it is hardly surprising that the enzyme activity retention on some carriers is enzyme-dependent (a good carrier for one enzyme is not necessarily a good carrier for another), as was demonstrated by the work of Bower et al., who found that activity retention of T4 lysozyme on colloidal silica was very dependent on the net charge and on charge location [102].

Because it is known that a good carrier for an enzyme or for an application is not necessarily a good one for another enzyme or application, it is not surprising that activity retention is also enzyme-dependent in biospecific immobilization [103]. Thus, it is not surprising that retention of activity might be different for enzymes immobilized on the same carrier, even when they are immobilized on a carrier bearing immobilized biospecific ligand [103]. Different retention of activity (82, 60, 16, 80%, respectively) was observed for invertase, glucose amylase, cellulase, and glucose oxidase immobilized on CCL-lectin-seralose [103], suggesting that the activity is enzyme-dependent.

2.5.1.9 Additive-dependent Activity

Because it is known that the performance of an immobilized enzyme is highly dependent on the properties of the medium, it is not surprising that the activity of the immobilized enzyme also depends on any additives which can affect the interaction of the enzyme with the carrier and the conformation of the enzyme. Immobilization of chlorophyllase on silica by physical adsorption led to the finding that the presence of optimized amounts of selected membrane lipids enhance the specific activity of the immobilized chlorophyllase by approximately 50%, suggesting that the presence of the additives might alter the conformation of the enzyme, making it more active [93].

Additives might also alter the microenvironment, thus enhancing enzyme activity. For instance, immobilization of horse liver alcohol dehydrogenase (HLAD) on porous glass in the presence of the bovine serum albumin enhanced the activity (from 1.6% to 9.8% of the native activity) [104]. Use of glycophase porous glass enhanced the activity to 43% of native HLAD activity [104]. More recently, it was found that co-precipitation of *Pseudomonas fluorescens* lipase (PFL) with some hydrophobic compounds enhanced the catalytic activity in the reaction of (R,S)-1-1phenylethanol acetation with vinyl acetate in *tert*-butyl methyl ether. Remarkably, the presence of hexadecane-1,2-diol during the immobilization of this enzyme (PFL) on celite 545 via physical adsorption or on Eupergit C 250 L via covalent binding was able to increase the esterification activity several fold relative to the native enzyme [476], suggesting that the induced active conformation was frozen on the carrier and was independent on the nature of the immobilization (physical adsorption or covalent binding).

Thus, the factors that can disturb hydrophobic interaction can affect the binding of the enzyme to the carriers. For example, the presence of the substrates can improve enzyme stability [24], use of the inhibitor or activator both affect the adsorption behaviour of the enzymes, because of a conformation change, and adsorption of phosphorylase b on butyl agarose involves cooperative and multivalent interactions and is accompanied by metastable states [105].

2.5.1.10 Hydrophilicity-dependent Activity

Since the 1970s it has been widely appreciated that a delicate hydrophilic–hydrophobic balance of the carrier is very important for activity expression. Some enzymes prefer a hydrophilic carrier whereas others have preference for hydrophilic carriers. For instance, immobilization of *Rhizopus niveus* lipase on different types of carrier such as inorganic Celite or silica and organic carriers such as Duolite resulted in immobilized enzyme with different ratios of synthetic activity to hydrolytic activity. The highest S/H ratio was obtained with the Duolite and lowest ratio was obtained with Celite Hyflo-Supercel [79].

The effect of the hydrophilicity of the carrier on enzyme activity was beautifully demonstrated by immobilizing PPL lipase on cellulose beads of different hydrophobicity [24]. In general, highly hydrophobic carriers enhanced lipase activity more than less hydrophobic carriers [107].

Similar behaviour has been observed for many other enzymes. For instance, high loss of activity was observed when trypsin was immobilized on PS microsphere whereas higher activity (almost 100%) was obtained when 10% (molar ratio) 2-hydroxyethyl methacrylate was added to obtain a PS/HEMA composite [56]. Similarly, trypsin immobilized on poly(2-hydroxyethyl methacrylate)/polystyrene (PHEMA/PS) composite microspheres had 65–100% of the activity of free trypsin when the HEMA content was 5 mol% [108].

This observation is readily explained by the concept of aquaphilicity, which is, by definition, the partition coefficient of water between a carrier and a hydrophobic organic solvent, e.g. propyl ether [109]. According to the original concept, aquaphilicity can be measured either by dispersing the dried carrier in a wet hydrophobic organic solvent (propyl ether or ether saturated with water) or by dispersing the wet carrier (or enzyme preparation) in a dry hydrophobic organic solvent.

An interesting experiment has been performed by Marron-Brignone and coworkers, who immobilized firefly luciferase on Langmuir films of different hydrophilicity and hydrophobicity formed by attaching the same lipids in two different orientations (either with polar head oriented on the surface or vice versa), as shown in Scheme 2.12. It was found that close contact between the protein and the film surface provoked a change in its kinetic behaviour, irrespective of the nature of the surface [110].

The development of this concept was actually spurred by non-aqueous enzymology. At the end of the 1980s it was found that the activity of carrier-bound immobilized enzymes such as (alpha-chymotrypsin and horse liver alcohol dehydrogenase in organic solvents was closely related to the amount of water available in the vicinity of the enzyme molecules [109]. It is, therefore, extremely important and beneficial to know the capacity of a carrier to strip water or to maintain a certain level of water within it. To this end, the concept of aquaphilicity was introduced for quantitative measurement of the water-adsorbing capacity of carriers in organic solvents. It was initially found that the rate of a reaction catalysed by carrier-bound immobilized enzymes, especially in apolar organic solvents, decreases as the aquaphilicity of the carrier increases, suggesting that the polar carrier strips the water from the enzymes and that there is a good correlation between enzyme activity and aquaphilicity [111].

However, before this concept was introduced, a similar term – the water-retaining capacity or hydrophilicity–hydrophobicity balance [112] was occasionally applied [111]. Remarkably, although this concept (water-retaining capacity) was initially used to express the relationship between carrier characteristics and enzyme

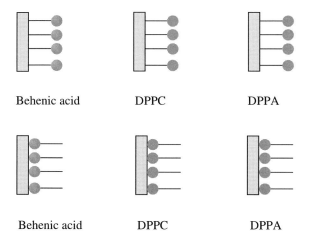

Scheme 2.12 Immobilization of firefly on lipidic Langmuir–Blodgett films whose surfaces were either hydrophilic or hydrophobic, depending on the orientation of lipid molecules.

activity in organic solvents, it was recently found that several enzyme characteristics, e.g. activity retention, enzyme conformation, enzyme selectivity, and enzyme stability, are highly dependent on the aquaphilicity of the carrier. It has, moreover, also been discovered that the concept of aquaphilicity is not only limited to non-aqueous biocatalysis. Indeed, it is also valid for use of immobilized enzymes in aqueous medium, as discovered by Cao and co-workers [113, 114].

2.5.1.11 Orientation-determined Activity

Because the interaction between an enzyme and a carrier often involves participation of specific binding, the orientation of the enzyme on the carrier surface might differ depending on the nature of the interaction, for example the position of the amino acid residues that are involved in the interaction, on the nature of the microenvironment, the nature of the binding site, and the spatial arrangement of the carrier-bound binding site and enzyme-binding amino acid residues (complementary).

As a consequence, enzyme molecules adsorbed on the surface of carriers with specific binding functionality might be preferentially orientated. Thus, expression of activity differs from enzyme to enzyme and from carrier to carrier.

As pointed out elsewhere in this book, the performance of a carrier-bound immobilized enzyme is largely determined by the mode of the interaction between the enzyme molecule and the carrier, for example number of interaction sites, enzyme orientation, etc. Often, high activity expression can be achieved with biospecific interaction, when the essential amino acid residues are not involved in the binding [115]. The effect of molecular orientation on enzyme activity is, however, not only limited to the accessibility of the active centre – an inappropriate interaction might lead to the complete loss of activity, because of a change of conformation, even without any covalent modification of the amino acid residues.

Studies have revealed that the orientation of a physically adsorbed enzyme can also affect its activity. An interesting example is the immobilization of chymotrypsin on montmorillonite [116]. The activity of the immobilized enzyme was found to be pH-dependent. Remarkably, loss of enzyme activity in a particular pH range was mainly attributed to incorrect orientation of the essential catalytic cavity on the negatively charged surface. Consequently, changing the pH to the extent that these essential groups lose their charges led to a change of enzyme orientation, and activity was recovered.

One intrinsic advantage of enzyme immobilization by biospecific interaction is that specific sites (such as active sites or glycosylated moieties) interact with immobilized ligand. Thus, greater retention of activity can be achieved [115–117] by avoiding binding of the enzyme via its active centre [118].

2.5.1.12 Binding Nature-controlled Enzyme Activity

As with covalent enzyme immobilization, differences between the nature of the binding groups of the carriers might lead to different retention of enzyme activity [9, 55], suggesting that enzyme orientation or the conformation of the enzymes on the surface of the carriers might be different and might depend on the nature of the binding functionality of the carriers.

Retention of activity of enzymes adsorbed on carriers via ionic interaction is also largely dependent on the nature of the enzyme itself. Thus, the activity of invertase and glucose isomerase immobilized on polystyrene-based ion-exchange fibres (Ionex) differed by at least 50–100% [119].

As with covalent enzyme immobilization, the nature of the binding functionality is also very important in non-covalent enzyme immobilization. Thus, variation of the type of the binding functionality, or use of the same type of functionality but of different size or hydrophilicity, might lead to a change in activity retention.

The effect of the binding functionality of the carrier on the performance of immobilized proteins was beautifully demonstrated by immobilizing haemoglobin on CM-Sephadex C-50 by ionic adsorption or covalent attachment. Adsorption resulted in reduction of affinity for oxygen and haem–haem interaction at low loading whereas higher loading of the gel resulted in increased cooperativity, although oxygen affinity remained unchanged. In contrast, covalent binding of haemoglobin resulted in an increased oxygen affinity, especially at lower enzyme loadings. This result suggested that the function of the protein is influenced by the nature of the binding functionality [120].

The effect of binding functionality on activity retention was well exemplified by amino acylase immobilized on polysaccharide-based adsorbents. It was found that amino acylase immobilized on tannin aminohexyl cellulose was 40 times more thermally stable than DEAE-Sephadex-aminoacylase and had 5 times higher volume activity than DEAE-Sephadex [121]. This is not an isolated example; many other examples have also shown that activity retention is often higher for enzyme adsorbed on an affinity adsorbent [59].

Interestingly, it has been found that variation of the pendant charged groups can also have a large influence on activity retention and immobilization efficiency in ionic binding. Ichijo et al. studied the effect of different ionic pendant groups on the binding of invertase on functionalized PVA (Scheme 2.13). The best result was obtained with dimethyl aminated PVA [55].

Similarly, it was found that the activity and stability of immobilized mandelate racemase from *Pseudomonas putida* is largely affected by the binding functionality attached to the cellulose backbone. Among the four cellulose-based ionic exchange

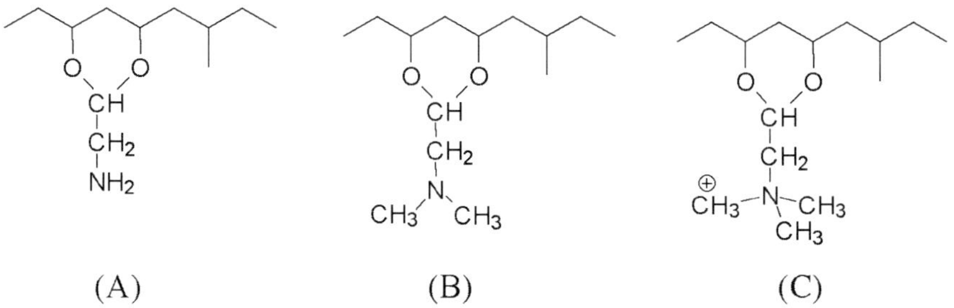

Scheme 2.13 Illustration of the structure of functionalized PVA with different binding groups.

resins, i.e. CM cellulose, AE (Aminoethyl)-cellulose, DEAE cellulose and TEAE cellulose, DEAE and TEAE cellulose gave the best results, regarding the high activation degree (250% for TEAE cellulose and 270% for DEAE cellulose) and high stability after lyophilisation [486].

Occasionally, lack of covalent modification can lead to greater retention of activity compared with covalent enzyme immobilization, for example adsorption of penicillin G acylase on MCM 41 [140] or immobilization of yeast mitochondrial alcohol dehydrogenase by adsorption on silanized silica bearing benzaldehyde compared with covalent immobilization via a silane/glutaraldehyde covalent technique [74].

Apart from these factors which affect enzyme activity, it has been found that retention of activity and enzyme loading also depend on the properties of the binding functionality [296]. For example, it has been found that the loading of lysozyme or catalase immobilized on poly(HEMA-co-MMA)-MS-CB/Poly(HEMA-co-MMA)-MS-CB-F3GA/or Cu(II) depends on the presence of transition metal ions. Ten times higher loading can be achieved in the presence of metal ions [296, 297]. The loading of the enzyme on an affinity adsorbent is five times higher than on an ionic adsorbent [98].

2.5.1.13 Binding Density-controlled Enzyme Activity

In the ionic binding of enzymes to ion-exchange resins, the binding density refers to the density of the ionic functionality that can participate in the binding. Like covalent immobilization of enzyme to the active carriers, the binding density was found to be crucial to the enzyme activity. For instance, the amount of α-amylase immobilized on poly(N-vinyl 2-pyrrolidone/itaconic acid) (P(VP/IA)) hydrogels depends on the concentration of IA. Increasing IA led to an increase in the adsorption capacity of amylase from 2.30 to 3.40 mg α-amylase g^{-1} dry gel and the activity was increased from 49.9% to 77.4%, suggesting that the specific activity of the enzyme also increased with increasing IA concentration in the gel [48]. Compared with the native enzyme, K_m increased by 150–200% and V_{max} was increased to 100–130% that of the native enzyme. Remarkably, enzyme stability was also significantly increased.

Similarly, it was also found that V_{max} for amylase immobilized on poly(acrylamide/maleic acid) [P(AAm/MA)] hydrogels also depends on the MA content of the gels. The storage stability of the immobilized enzyme is, moreover, also dependent on the MA content. Although the activity of the free enzyme was completely lost after 20 days, adsorbed enzyme retained 47–59% of its original activity after 20 days, depending on the amount of MA in the hydrogels [122].

As noted above, the binding density is directly related to the microenvironmental effect. However, not only the volume activity but also the specific activity is related to binding density. The volume activity is related to the number of the binding sites available for binding the enzyme, thus the activity is related to the binding density, whereas the specific activity is determined by the microenvironment. It is, therefore, necessary to distinguish between two types of activity, although both are all related to the binding density.

Study of the immobilization of invertase on two types of gel, for example PAAm and poly(acrylamide/maleic acid) (P(AAm/MA)) has revealed that enzyme volume activity and specific activity increase with increasing concentration of MA in the system [123], suggesting that the binding density affects both the microenvironment (pH gradient) and the binding capacity. More interestingly, it was found that the storage stability of the immobilized enzyme was increased two to threefold.

The effect of binding density on the activity of the immobilized enzyme was demonstrated by Caldewell et al. early in the mid-1970s, when they found that α-amylase immobilized on hexyl Sepharose 6B had maximum activity when the hexyl to galactose ratio was 0.51 in the range from 0.02 to 0.70 mol hexyl side chain mol^{-1} galactose residue [124]. The activity depends on the density of the binding functionality, suggesting that a critical concentration of the hydrophobic moieties is essential for maximum loading [125].

Like other types of immobilized enzyme, the density of the binding functionality also affects enzyme activity and the amount of enzyme immobilized. For biospecific immobilization, however, the effect of binding density is mainly ascribed to a change in the microenvironment. In other words, an increase in the binding functionality might lead to the increase in multiple non-specific interactions between the isolated enzyme and specific sorbents. Also, increasing the binding functionality might not lead to an increase in the activity of immobilized enzymes, because of steric hindrance. For instance, 99% of 0.85 µmol immobilized inhibitors participated in the formation of the specific complex between pepsin and the adsorbent, whereas 26% of 4.5 µmol immobilized inhibitor participated in the binding [50].

Studies have revealed that the density of the affinity binding site also affects the adsorption behaviour of protein on the carriers. For example, guanidinobenzoatase can be specifically immobilized on CM-agarose gels containing 5% immobilized agmatine whereas CM-agarose gels containing 10% immobilized agmatine functions as a very good anionic-exchange support able to adsorb most plasma proteins non-specifically, because of steric hindrances promoted by the interaction between each individual immobilized ligand and the corresponding binding pocket in the target protein [98]. Thus, in practice, it is essential to control the binding density, especially to promote specific binding; the specific binding site is obtained by controlled modification of other active binding sites.

2.5.1.14 Reactor-dependent Activity

The apparent activity of the immobilized enzyme might be reactor-dependent. For example, glucose isomerase from mycelium from a Streptomyces species immobilized on DEAE-cellulose was slightly less active in shallow-packed bed reactors than in stirred batch reactors. The lower velocity was attributed principally to channelling [34, 126].

2.5.1.15 Pore-size-dependent Activity

Immobilization of lipase *Mucor miehei* on several types of carrier such as Spherosil, Duolite S562, and Duolite 568 led to the finding that the activity of the immobilized lipase on Spherosil of mean pore size 148 nm was loading-dependent [17], clearly suggesting that diffusion limitation occurred with this carrier also, whereas use of S 562, of narrow pore size, minimized the diffusion limitation by excluding the enzyme from penetrating deeply into the pores of the carriers, thus no internal diffusion limitation was observed.

In contrast, study of esterification activity of lipase *Mucor miehei* (MML) immobilized on hydrophobic controlled-pore glass of different pore size clearly revealed that the esterification activity also depends on the pore size of the hydrophobic silica. When the pore size of CPG is below 100 nm, esterification activity increased with increasing the pore size, whereas esterification activity was less influenced by pore size when the pore size was above 100 nm, strongly suggested that diffusion limitation became less when the pore size was above certain limit [127]. Indeed, the fact that the esterification activity of the immobilized MML decreased with increasing particle size also suggested that diffusion limitation has significant effect on activity expression [127].

2.5.1.16 Water-activity-dependent Activity

In recent decades it has been increasingly appreciated that the catalytic activity or selectivity of an enzyme preparations (either enzyme powders or immobilized enzymes) in a non-aqueous medium is largely dictated by the water activity in the catalytic system (including water, enzyme preparation, solvent and the substrates). In a non-equilibrated reaction system, the water might be redistributed among the solvent and the enzyme preparation or carrier). Thus, it is not surprising that the enzyme activity might differ from carrier to carrier and vary with the water activity, as revealed in a number of studies [128–130].

For instance, lipase *Rhizopus niveus* immobilized on Duolite XAD 761 by physical adsorption had the maximum activity for a water activity in the range of 0.32–0.52. Remarkably, immobilization of lipase *Rhizopus niveus* on Duolite XAD led to 12.2-fold enhancement in the synthetic activity compared with the free enzyme [130].

Interestingly, it was found that the water activity-dependent activity also depended on enzyme loading. For instance, a lipase immobilized on Accurel EP-100 with high protein loading (40 mg g^{-1}) had a specific activity of 6.1 µmol min^{-1} mg^{-1} protein at $a_w = 0.11$, whereas the same immobilized enzyme with low protein loading (4 mg g^{-1}) had the highest specific activity at $a_w = 0.75$. This might reflect that the enzyme conformation is different at different enzyme loading [131].

It was recently found that the activity of lipase PS immobilized on carriers such as Toyonite, Celite, glass and Amberlite was strongly dependent on the nature of the carriers. The highest activity for transesterification in organic solvent was obtained with Toyonite (37.2 µmol min^{-1} mg^{-1}) whereas the lowest activity was obtained with Amberlite (0.4 µmol min^{-1} mg^{-1}) [132].

Because the hydrophilicity of the surface of Toyonite coated with methacryloyloxy groups is quite comparable with that of Amberlite XAD-7, the different enzyme activity might be ascribed to the different amount of water in the carriers. If the water content in the Toyonite is higher than that in Amberlite XAD-7 it is highly probably that the water activity of the whole reaction system with Toyonite-bound lipase is higher, leading to an undoubtedly higher reaction rate [132].

2.5.2
Stability

Enzyme stability involves several categories, e.g. thermal stability, operational stability, and stability under extremes such as extreme pH, heating, and organic solvents. Depending on the method of immobilization, the extent of stabilization might be different in different categories. In other words, an immobilized enzyme which is thermally stabilized is not necessarily stabilized against extremes such as solvents.

As already noted, the stability of enzymes immobilized by adsorption depends on the properties of the enzyme, the carrier (chemical and physical nature), binding functionality, the microenvironment of the enzyme, and the temperature measured. Bearing in mind that immobilized enzymes are always the stabilized species (either stabilized native species or stabilized denatured species), the stability of the enzyme immobilized on the carrier via adsorption can go in three directions:

- The stability of the immobilized enzyme is unchanged. For example, when partially purified glucose isomerase from *Bacillus coagulans* was immobilized by adsorption on DEAE-cellulose its stability was similar to that of the free enzyme [34]. Similarly, immobilizing tyrosinase from potato (*Solanum tuberosum*) on chitin led to 95% retention of activity but no dramatic changes in the thermal stability of the enzyme [133]. *Endo-d-galacturonanase* separated from a commercial pectinase preparation was irreversibly adsorbed by porous beads of polyethylene terephthalate without significant alteration in the thermal stability [5].

- The stability of the immobilized enzyme is enhanced. For example, *Candida rugosa* lipase adsorbed on poly(methyl methacrylate) (PMMA) was up to three times more stable in buffer than the free lipase [52]. The thermostability of trehalose phosphorylase at 30 °C was increased 35-fold by adsorption on anion-exchange resin [46].

- The stability of the enzyme is reduced. For example, glutamate dehydrogenase adsorbed on palmityl-substituted Sepharose 4B by hydrophobic interactions was significantly less stable upon adsorption, whereas the stability of trypsin was unaltered [26]. Similarly, adsorption of aldehyde oxidase on *n*-hexyl and *n*-octylamine-substituted Sepharose 4B and DEAE-Sepharose 6B resulted in the best retention of enzyme activity but the stability relative to that of the native enzyme was low [121].

In general, the factors that dictate the stability of the adsorbed enzymes can be generally classified into several categories:

- microenvironment-controlled,
- binding nature-controlled,
- conformationally controlled,
- confinement-dependent stability [134],
- temperature-dependent stability,
- orientation-controlled stability,
- enzyme loading-dependent stability,
- diffusion-controlled stability,
- carrier-nature-dependent stability.

In this section, each of these factors will be discussed. Although the stability of the immobilized enzyme is often affected by many factors, one of the factors is often dominant. Thus, it might be extremely instructive to discuss these factors separately.

2.5.2.1 Conformation-controlled Stability

With conformation-controlled stability, the stability is often lower than for the native enzyme when the enzyme and carrier are not compatible [135]. For example, immobilization of alkaline phosphatase on Na sepiolite or other organic–inorganic carriers [136] by adsorption led to reduced enzyme stability, although the surface of the carrier was completely covered by enzyme molecules [137]. Thus, the reduced enzyme stability might be ascribed to conformation change, which led to the formation of an unstable enzyme conformation [138], often because of hydrophobic interaction.

Similarly, yeast alcohol dehydrogenase immobilized on polyethylene terephthalate was not only less active (1.2% of the specific activity of the soluble enzyme) but also less stable than the native enzyme, suggesting that the activity loss was mainly ascribed to a change of conformation on the hydrophobic carriers [138]. In contrast, stability can be enhanced when the carrier and enzyme are compatible [137].

Interestingly, it has been shown that immobilization of enzymes on hydrophobic carriers often not only results in a change of enzyme conformation but actually requires a change of conformation to enhance the interaction. In such circumstances the conformation change often reduces the stability of the enzyme [26], because the conformation of the immobilized enzyme resembles that of the denatured enzyme.

The conformation of the enzyme exposed to the conditions under which its stability is measured dictates enzyme stability. Consequently, it is conceivable that an immobilized enzyme with a rigid conformation should be more stable than that with flexible enzyme conformation.

On the other hand, interaction of the enzyme with the carrier under the same conditions might also play an important role. In other words, the immobilization might induce an enzyme conformation that is less stable than that of the native enzyme. For example, amyloglucosidase adsorbed on hexyl-substituted Sepharose 6B is less stable than the free enzyme but enhancement of the stability was obtained in the presence of the substrate [24], suggesting that immobilization induced the formation of a less stable conformation whereas the presence of the substrates tightens the enzyme conformation. In this example enzyme stability is controlled by the conformation in the medium to which the enzyme molecules are exposed.

However, in another example interaction of the enzyme with the carrier might induce an enzyme conformation that is stable or unstable relative to the native enzyme. For instance, yeast alcohol dehydrogenase adsorbed on polyethylene terephthalate not only had a much lower specific activity (1.2 % of the specific activity of the soluble enzyme) but was also less stable than the native enzyme [24], suggesting that loss of the enzyme activity is mainly ascribed to alteration of the conformation of the enzyme to one which is a less stable.

Conformation controlled enzyme stability is often observed for immobilization of enzymes on hydrophobic carriers. The reason is that the enzyme molecules often undergo conformational changes during adsorption; one example is the immobilization of thromboxane synthase on Ph-Sepharose beads by adsorption [138].

2.5.2.2 Confinement-controlled Stability

When molecular confinement is predominant the enzymes can be tightly adsorbed and confined on or in inorganic carriers such as MCM 41 [139, 140]. Often, stability (e.g. thermostability, operational stability, or storage stability) can be improved by simple adsorption of enzymes or their derivatives on inorganic carriers [141, 142, 144]. However, enzyme activity and stability depends on the surface characteristics and size-matching between pore size and the molecular diameters, as has been found for horseradish peroxidase (HRP) [143]. This effect can be explained as the result of molecular confinement: The enzyme molecules that match the pore size of the carrier are probably confined in the pores, thus being less flexible than enzyme that has more flexibility in a big pore.

As with confinement-dependent stability, immobilization of enzymes in pores of similar dimension to the enzyme can lead to increased enzyme stability, for example adsorption of enzyme on porous zeolite [134]. Obviously, enzyme mobility is reduced, resulting in an increase of unfolding enthalpy.

2.5.2.3 Enzyme Loading-dependent Stability

As for loading-dependent stability, the stability of the immobilized enzyme usually increases with enzyme loading. This is largely because of the deformation of the enzyme which occurs on the surface of the carrier at low enzyme loading. This was exemplified by adsorption of alcohol dehydrogenase on polyaminomethylstyrene (PAMS) [158].

Loading-dependent stability has very often been observed with carrier-bound immobilized enzymes (either covalent or non-covalent). Interpretation of the result can be based on the monolayer principles discussed above [107, 158]. It has, for example, been observed that increasing immobilization time reduced the enzyme activity of Lipase CR on PMMA; enhanced stability relative to the native enzyme was obtained [52]. Here increasing the immobilization time might increase enzyme loading.

Investigation of the thermal stability of the protease α-chymotrypsin has led to the discovery that the thermal stability of protease α-chymotrypsin adsorbed on a polystyrene surface was loading-dependent. This phenomenon can be explained as a result of enzyme deformation or conformation change [135].

2.5.2.4 Diffusion-controlled Stability
Diffusion controlled enzyme stability was first observed by Ollis at the beginning of the 1970s [145]. This effect has, however, received less attention in recent decades.

2.5.2.5 Cross-linking-dependent Stability
The function of cross-linking has been clearly confirmed by studying a similar immobilization process – cross-linking of covalently immobilized enzymes. For example, glutaraldehyde-aided cross-linking of pectinlyase covalently immobilized on several carriers, e.g. Eupergit C, CDI-activated hydrolysed XAD-7, and glutaraldehyde-activated Nylon 6, was found to improve enzyme stability compared with the original, corresponding, carrier-bound enzyme [146]. Similarly, cross-linking, with glutaraldehyde or 3,5-difluoronitrobenzene, of immobilized porcine pancreas lipase on PAAm beads (via carbodiimide coupling) also improved the thermal stability [147].

As well as enhancing thermostability, immobilization of enzymes on carriers by adsorption then cross-linking often improved operational stability by mitigation of enzyme leakage. For instance, immobilization of a cephalosporin acetylesterase by adsorption on bentonite, followed by cross-linking with glutaraldehyde or bisdimethyladipimidate rendered the immobilized enzyme more operationally stable [150]. Remarkably, it was found that the degree of stabilization was also largely dependent on the nature of the cross-linker, as revealed by cross-linking of α-l-rhamnopyranosidase with a number of cross-linkers such as glutaraldehyde, diexpoxyoctane, suberimidate [473].

2.5.2.6 Carrier Nature-controlled Stability
It is conceivable that a carrier with a hydrophilic environment will normally favour higher enzyme stability. Thus, the stability of lipase immobilized on PEI-coated silica is more stable than that on silanized carriers, which are generally more hydrophobic than PEI-coated silica (Scheme 2.14) [354].

It has, however, been found that two types of cellulase immobilized on Aminosilica-1 were less stable than the free forms at temperatures higher than 50 °C, suggesting that the deactivation of enzyme might involve a two-step mechanism [151]. Moreover, immobilized glucoamylase bound to SP-Sephadex by ionic interaction was less stable than free glucoamylase [152].

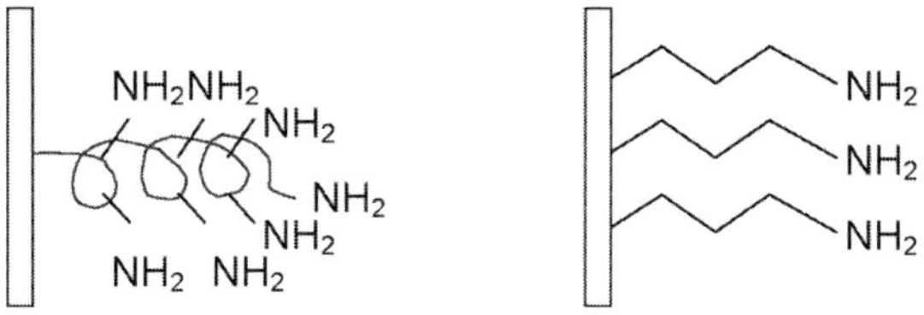

Scheme 2.14 Silanized silica and PEI-coated silica.

Interestingly, it has been found that immobilization of glucoamylase (GA) on PEI-coated non-porous glass by adsorption and cross-linking led to the formation of immobilized GA with 80% of the enzyme's total activity provided (i.e. the activity retention is 80%) but less thermal stability than the soluble enzyme [441]. On the basis of this result it was possible to draw the conclusion that among the different factors, the nature of the carrier is much more important than cross-linking processes. Although cross-linking usually enhanced enzyme stability [146, 153–157], this stability was often determined by the adsorption step.

In other words, if the first step leads to an unstable immobilized enzyme, because the enzyme is incompatible with the carrier, the cross-linking step cannot increase stability. For instance, glucoamylase immobilized on PEI-coated non-porous glass beads was less stable than the free enzyme, suggesting that the stability of the immobilized enzyme was mainly determined by the first-step, i.e. adsorption of glucoamylase on the charged glass beads might disturb the native charge–charge interaction, leading to a less stable enzyme conformation [146].

With regard to the nature of the binding functionality, it is worth mentioning that use of affinity adsorbents might be advantageous because enzyme stability and activity might be simultaneously enhanced [301] compared with, other nonspecific, adsorption.

2.5.2.7 Aquaphilicity-controlled Stability

In the last decade, much effort has been devoted to the relationship of enzyme activity or selectivity with the aquaphilicity of the carrier used. However, less is known about the effect of carrier's aquaphilicity on the enzyme stability.

For certain enzymes such as a-chymotrypsin and cutinase, it was found by Nord et al. that the ratio of the native-like (N) state to more perturbed (stabilised or destabilised) (P) states generally increases with decreasing hydrophobicity of the sorbent surface, thus suggesting that enzyme structure is much more disturbed and destabilized by the hydrophobic carrier than by the hydrophilic carrier [494].

However, this conclusion is not valid for the lipolytic lipase, which has much higher stability upon immobilisation on hydrophobic PVC than on hydrophilic agarose [60]. Thus, Thus, the stability of an enzyme adsorbed on a carrier is not only dependent on the enzyme but also on the nature of the carrier.

Remarkably, hydrophobic interaction of an enzyme with a carrier often provokes a change of enzyme conformation, thus activity can be increased for hydrophobic enzymes such as lipases [160, 161]. In contrast, enzyme stability is often reduced, when the enzymes are known as "soft enzymes" [24, 26]. For hard enzymes such as trypsin [26] and dextransucrase [27] the stability are often not changed, or can be increased [162], whereas the stability of the soft enzymes such as amylase can be significantly reduced compared with the native enzyme [90].

2.5.2.8 Medium-controlled Stability

It is well known that the enzyme stability is closely related to enzyme conformation. Thus, it is not surprising that media that can affect enzyme conformation and the properties of the enzyme interaction can also influence enzyme stability [158].

Thus, modulation of pH value, ionic strength, type of buffers, and addition of additives such as sugars, surfactants or other conformer selectors such as crown ethers during the immobilization is often able to affect the stability of carrier-bound immobilized enzymes.

Remarkably, it was found that the stability of lipoxygenase, when immobilised on Eupergit C in phosphate buffer of pH 7.5 and low ionic strength (0.05 M), displayed twentyfold stability relative to that immobilised in a medium with high ionic strength [492].

It is conceivable that pH value might exert profound influence on the protonation of ionogenic groups of the enzymes or the exchanger resins. As a result, the types of interaction (hydrophobic or hydrophilic) between the enzyme and the carrier might be pH-dependent [236]. Thus, the enzyme orientation, conformation and the binding strength between the enzyme and the carrier might be strong pH-dependent as well.

Change of the medium from aqueous to organic solvent tremendously enhanced the operational stability of lipase *Candida rugosa* immobilized on styrene-divinylbenzene (STY-DVB) copolymer via physical adsorption. The enzyme immobilised in heptan (with 7–10 % water) lost less than 25 % of the initial activity after 12 cycles, while the one immobilised in aqueous medium completely lost the activity after 3 cycles [493], thus justifying the exploitation of medium engineering for modulation of enzyme performance.

2.5.2.9 Temperature-dependent Stability

Many examples have revealed that the relative stability of an immobilized enzyme (compared with the free enzyme) can be temperature-dependent. The stability of an immobilized enzyme is usually greater than that of the free enzyme below a certain temperature whereas the stability is lower than that of the free enzyme above that temperature. For example, cellulases from Robillarda sp. Y-20 and *Trichoderma reesei* immobilized on aminosilica-1 by ion exchange were less stable than the free forms at temperatures higher than 50 °C [151].

These facts suggested that thermal deactivation of the immobilized enzymes involves two-step mechanisms. Below a certain temperature the enzyme is more stable than the free enzyme. With increasing temperature, however, the presence of the carrier might accelerate the deactivation.

In some cases, the immobilized enzyme displayed higher stability relative to the soluble enzyme in all range of temperatures tested. However, the relative thermal stabilization degree varies, depending on the temperature, under which the thermal stability is measured [301].

2.5.2.10 Microenvironment-controlled Stability

In general, enzyme stability is a consequence of interaction of the enzyme molecules with the surroundings. Thus, it is not surprising that enzyme stability is strongly determined by the nature of the microenvironment.

Studies have revealed that the thermal stability of immobilized enzymes depends both on the nature of the carrier and the temperature used. Below a certain temperature the immobilized enzyme might be more stable than the free enzyme whereas above that temperature it is less stable. For example, cellulases immobilized on aminosilica-1 by physical adsorption were less stable than the free forms at temperatures higher than 50 °C [151], suggesting that at higher temperature the transition from the native state to the denatured state was accelerated by the carrier, resulting in two-step deactivation. Similarly, it was found that invertase adsorbed on AM and AL was more stable to thermal deactivation than the free enzyme whereas invertase adsorbed on M was much less stable (M is a montmorillonite, AL is a noncrystalline Al hydroxide and AM is montmorillonite partially coated with OH-Al ions [163]). This can be ascribed to hydrophilic surface of AL and AM.

Glucoamylase adsorbed ionically on Duolite A7 is slightly less stable than the covalently immobilized enzymes on s-triazinyl Duolite A7 [164, 165]; similarly, the same enzyme immobilized on cationic non-porous glass beads [441] or SP-Sephadex (sulphopropyl Sephadex) by ionic interaction [35] was also less stable than the soluble enzymes.

There are, in contrast, many examples which exemplify that the stability of the immobilized enzyme, for example its thermal stability, storage stability, and operational stability can be significantly improved, compared with the native enzyme, on immobilization by adsorption on different carriers. For example, simple adsorption of glucose oxidase on activated carbon increased its operational stability against H_2O_2 [166]; similarly, immobilization of *Candida rugosa* lipase on Celite might enhance its stability against deactivation by one of the toxic products, acetaldehyde, during transesterification with vinyl acetate as irreversible acyl donor [322]. It might be worth pointing out, however, that the nature of the carrier is much more important with regard to the stabilization effect. For example, immobilization of the same enzyme on hydrophobic PVA-based carrier bearing a long hydrophobic tail was superior to immobilization on Celite in terms of enzyme stability (operational and temperature) and activity [167].

2.5.2.11 Binding Nature-controlled Enzyme Stability

The nature of the binding namely the binding position, number, and enzyme orientation dictates the specificity of the binding. Thus, variation of binding functionality on the same carrier is the best way to study this effect, because of the avoidance of other effects, particularly diffusion limitation.

Recently, immobilization of aldehyde oxidase on different carriers showed that n-hexyl and n-octylamine-substituted Sepharose 4B and DEAE-Sepharose 6B resulted in the best activity retention (Scheme 2.15). The storage stability of the immobilized enzyme on n-octylamine-substituted Sepharose 4B at pH 7.8 and 9.0 was, however, much lower than that of the free enzyme [121]. This result suggested that stability and activity are governed by different factors [121].

In binding nature-controlled stability it is remarkable that among the various adsorption methods enzymes immobilized on a bio-specific affinity adsorbent can often be more stable than on other adsorbents. For instance, amino acid acylase immobilized on tannin-bound chitosan or cellulose was highly thermostable (40 times more thermally stable than DEAE-Sephadex-aminoacylase and with five times higher volume activity than DEAE-Sephadex) [59] suggesting that the specific non-covalent interaction between the enzyme and tannin stabilizes the enzyme [168]. Similarly, enhancement of thermostability was also observed for GOD immobilized on a dye-based affinity p-HEMA membranes–CB-Fe(III), which was prepared by covalent immobilization of the dye Cibacron blue, followed by chelation of Fe(III) ion [169]. Apparently, the properties of the interaction, i.e. position of binding, enzyme orientation, and number of binding sites, differ from enzyme to enzyme and from carrier to carrier.

Although there are few examples of comparison of the stability of enzymes immobilized by affinity adsorption, the stability of immobilized enzymes is usually enhanced by affinity adsorption [297, 300, 301] compared with other types of non-covalent binding. For example, amino acid acylase immobilized on tannin cellulose was approximately 40 times more stable the enzyme immobilized on DEAE-cellulose [59].

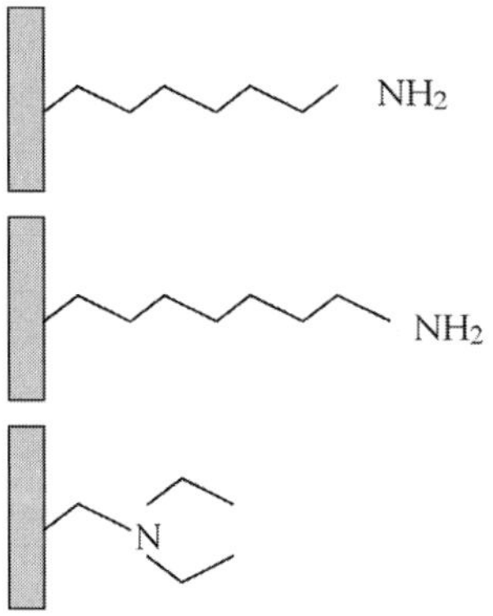

Scheme 2.15 Illustration of different Sepharose derivatives

2.5.2.12 Binding Density-controlled Enzyme Stability

There are many examples which demonstrate that enzyme stability is related to the concentration of binding sites on the carrier. This is easy to understand – increasing the number of attachments of the enzyme to the carrier will reduce enzyme mobility thus increasing enzyme stability.

When immobilizing an enzyme by non-covalent adsorption, increasing the binding density will increase enzyme stability because of multivalent non-specific interaction, in the same way as for covalent enzyme immobilization. For example, α-amylase immobilized on poly(acrylamide/maleic acid) hydrogels had greater storage stability than the free enzyme (depending on the concentration of maleic acid in the gels) [122].

2.5.2.13 Additive-dependent Stability

Factors that disturb the hydrophobic interaction can affect the binding of the enzyme to the carriers. For example, the presence of other substrates can improve the enzyme stability [24] and the use of the inhibitors or activators can affect the adsorption behaviour of the enzymes, because of conformational changes, the adsorption of phosphorylase b on butyl agarose involves cooperative and multivalent interactions and is accompanied by metastable states [105].

2.5.2.14 Enzyme Orientation-dependent Stability

In the same way as for enzyme activity, enzyme stability can be also affected by the orientation of enzyme molecules on the solid/liquid interface. It was recently demonstrated that the stability, after immobilization, of site-specific variants of subtilisin BPN' depended on the orientation of the enzyme on the carrier [33]. The improved surface stability of subtilisin was attributed to the change of the mode of interaction of the enzyme with the carrier [33].

2.5.2.15 Enzyme-dependent Stability

The extent of stabilization also depends on the nature of the enzyme used [170]. Thus, a method of enzyme immobilisation, which is better-suited for an enzyme, is not necessarily the best for another enzyme as mentioned in previously in Chapter 1.

However, the fact that the nature of the enzyme to be immobilised can be altered before immobilisation by a number of pre-immobilization modification methods e.g. introduction of extra ionogenic groups to enzymes [385, 386], hydrophobization of enzyme [381, 383, 388], suggests that the same method can still be used, if the enzyme to be immobilised can be properly pre-modified and thus improved.

Remarkably it has been found that these pre-immobilization enzyme modification methods are not only effective in enhancement of enzyme binding to the carriers that were not suitable for the native enzymes but also in improvement of the enzyme stability, due to the increased interaction between the carrier and the enzyme [381, 385] or due to the pre-stabilisation effect by the modification. A striking example is that mannan-modified penicillin G acylase (neoglycoenzyme) immobilized on Con A-cellulose beads displayed even higher thermal stability than a covalently bound penicillin G acylase [271].

2.5.3
Selectivity

In recent decades, the need for pure enantiomers by the agricultural and pharmaceutical industries has grown steadily. Because of high selectivity and mild reaction conditions, the use of enzymes to obtain these pure enantiomers has matured to a technology enabling many industrial processes with high selectivity and economic viability.

Since the 1980s much insight into application of pure enzymatic methods has been gained by use of the following approaches:
- medium engineering – by alteration of the reaction medium, for example organic solvents, the selectivity can be improved,
- substrate engineering – the structure of the substrate can be adjusted to affect reaction selectivity,
- process engineering,
- enzyme engineering.

Significant progress and impressive proof of principle have been achieved in improving enzyme selectivity by genetic manipulation of the enzyme molecules [171]. It has also been shown that the selectivity of an enzyme can be affected by the immobilization method selected.

As noted above, attachment of enzymes to carriers by non-covalent adsorption can disturb the conformation of an enzyme; it is, therefore, conceivable that adsorption of enzymes on a carrier might alter any selectivity that is closely related to the geometry of the immobilized enzymes.

Usually, the selectivity of an enzyme molecule can be classified in the following categories:
- Substrate selectivity – the ability to distinguish and act on a subset of compounds within a larger group of chemically related compounds.
- Stereoselectivity – the ability to act on a single enantiomer or diastereomer selectively.
- Regioselectivity – the ability to act on one location in a molecule selectively.
- Functional group selectivity – the ability to act on one functional group selectively in the presence of other equally reactive or more reactive functional groups, for example selective acylation of amino alcohols.
- Reaction selectivity – the ability to distinguish two types of reaction that might occur concomitantly and competitively in the same reaction system, e.g. kinetically controlled synthesis of peptides.

It has been reported in numerous publications that these types of selectivity can be also affected or altered by immobilization. Here, some representative examples will be summarized, according to the factors that effect enzyme selectivity during immobilization.

2.5.3.1 Conformation-controlled Selectivity

As the term implies, conformation-controlled selectivity is selectivity determined by enzyme conformation. Thus, conformation-controlled selectivity could be either substrate selectivity or enantioselectivity. In both, the enzyme adsorbed on the carrier might adopt a different conformation, thus the geometry of the active centre might be changed, leading to different selectivity. In this case, the enzyme conformation is "frozen" by the selected method of enzyme immobilization.

For example, simple adsorption on Celite enhanced the stability of *Candida rugosa* lipase against acetaldehyde [322] and the enantioselectivity was enhanced up to threefold. Similarly, immobilization of *Rhizopus oryzae* lipase (ROL) by adsorption on carriers of different nature led to the discovery that the enantioselectivity largely depends on the hydrophilicity of the carrier: hydrophobic carriers were usually less selective than hydrophilic carriers, indicating that the conformation of the enzyme on hydrophobic carriers is much looser than on hydrophilic carriers [78]. Remarkably, we found that immobilized enzymes which are very active are often not very selective. For instance, lipase ROL immobilized on Sepabeads by ionic adsorption followed by coating dextran sulphate (DS) was 25 times more enantioselective than the same enzyme immobilized on octyl Sepharose; the rate of reaction was, however, one order of magnitude lower than for immobilization on hydrophobic carriers. For interfacially active enzymes such as lipases, it is evident that the selectivity is often conformation-controlled due to the fact that the lid opening (due to conformation change) is often a prerequisite for higher activity and selectivity for ether the carrier-bound immobilized enzyme [161] or CLEC of lipases [185].

Occasionally, the fact that some lipases have preferential substrates, not only for high activity [67] but also for higher selectivity [186], strongly suggests the selectivity is conformation-controlled.

Immobilization of horse liver alcohol dehydrogenase (HLAD) by hydrophobic adsorption on octyl-Sepharose 4B [84] also resulted in a change of substrate selectivity. It was found that HLAD immobilized on octyl-Sepharose was less selective for hydrophobic alcohols, because of a change in the geometry of the active centre [84].

Similarly, it was found that the rate of reaction of *Candida cylindracea* lipase in esterification of the secondary alcohol (R,S)-1-phenylethanol was higher when immobilized on hydrophobic carriers than on hydrophilic carriers. When the primary alcohol heptanol was used the reaction rate was highest for the free lipase [85].

Interestingly, it has recently been found that the enantioselectivity of *Pseudomonas fluorescens* lipase (PFL) adsorbed on hydrophobic carriers (i.e. octyl agarose or decaoctyl sepabeads) for hydrolytic resolution of fully soluble (R,S)-2-hydroxy-4-phenylbutanoic acid ethyl ester was at least one order of magnitude greater than that of the native enzyme [172].

Remarkably, the activity of an immobilized enzyme is also clearly related to the enantioselectivity and hydrophobicity of the carrier. In general, the greater the hydrophobicity, the higher the activity or selectivity, suggesting that the hydrophobic carrier induced an active and selective enzyme conformation [172]. Similarly, it has been found that the enantioselectivity of adsorbed CRL also highly dependent on the pH of the medium used, implying that engineering the medium is also a crucial step in evaluation of the selectivity of immobilized enzymes [4742].

This is easy to understand: the conformation of an enzyme (either free or immobilized) is strongly related to the nature of the medium such as pH, ionic strength, properties of the solvent. On the other hand, change of the enzyme conformation by immobilization will definitively provoke the change of the optimum conditions for maximum selectivity. Thus, the conditions under which the native enzyme might have the optimum selectivity might not apply to the immobilized enzymes.

2.5.3.2 Diffusion-controlled Selectivity

In diffusion-controlled selectivity, several types of selectivity can be distinguished – diffusion-controlled reaction enantioselectivity, diffusion-controlled product selectivity, and diffusion-controlled reaction selectivity.

Enantioselectivity can be reduced by the presence of diffusion limitation. For instance, lipases adsorbed on carriers might have lower enantioselectivity than the free counterpart [173]. It is sometimes difficult to distinguish this effect (diffusion limitation) on enantioselectivity. Crushing the immobilized enzymes into small particles might be means of determining whether the selectivity of immobilized enzymes is affected by diffusion limitation [173].

Diffusion-controlled selectivity was recently observed in a study of the enantioselectivity of CAL-B (*Candida antarctica* lipase B) in the transesterification in organic solvents [173]. For the first time it was reported that the presence of the diffusion can reduce the enantioselectivity of enzymes. In fact, a similar observation was made by Moreno and his co-workers in 1995, who found that the selectivity of CRL immobilized on Al_2O_3 depended not only on substrate concentration but also on the rate of stirring [174]. Interestingly, it was found that the diffusion-controlled enantioselectivity was also affected by the enzyme loading [475]. Although the volume activity [u/g] of *Candida rugosa* lipase (CRL) slightly increased with increasing enzyme loading, the enantioselectivity of immobilized CRL decreased eightfold correspondingly, suggesting that a careful control of enzyme loading is critical in designing enantioselective immobilised enzymes.

The presence of the diffusion limitation might also effect enzyme selectivity by altering the action pattern of the immobilized enzyme. For example, Aspergillus endopolygalacturonase (EC 3.2.1.15) immobilized by adsorption on porous powder poly(ethylene terephthalate) had a pattern of action dependent on the external diffusion effect [139].

2.5.3.3 Binding Functionality-controlled Selectivity

It was recently demonstrated that immobilization of alkylsulphatase on anion exchangers such as DEAE-Sephadex, TEAE-cellulose, and Ecetola-cellulose can affect

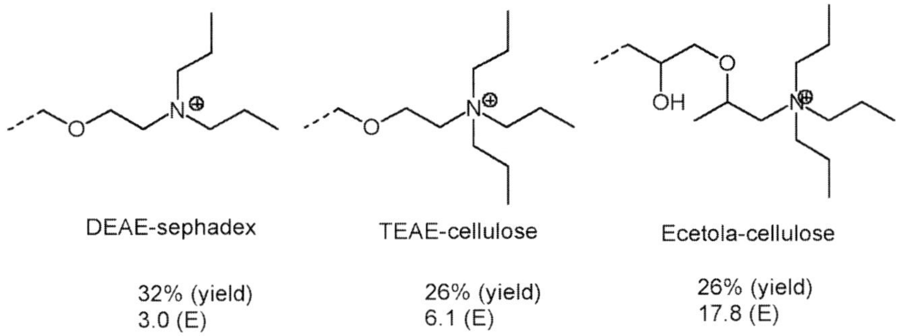

Scheme 2.16 Effect of the carrier on the E value for hydrolysis of alkylsulphates.

	DEAE-sephadex	TEAE-cellulose	Ecetola-cellulose
Yield	32%	26%	26%
E	3.0	6.1	17.8

enzyme selectivity in the hydrolysis of *sec*-alkyl sulphates by alkylsulphatase [175]. The best result was obtained with Ecetola-cellulose as carrier (Scheme 2.16).

Because TEAE-cellulose and Ecetola cellulose differ mainly in the spacer, the selectivity of the immobilized enzyme is mainly determined by the side chain and the spacer. The selectivity might be enhanced because the charged groups might be able to approach certain negatively charged domains or sites (e.g. the active centre), thus affecting the geometry of the active centre. Indeed, addition of 0.2% cetyl trimethylammonium bromide significantly enhanced the selectivity compared with that of the enzyme adsorbed on Ecetola cellulose.

More interestingly, it was found that the selectivity of this enzyme can be increased more than 200-fold in the presence of 5 mm $FeCl_3$, suggesting that interaction of Fe^{3+} with amino acid residues close to the active centre might tighten the conformation of the enzyme, thus enhancing selectivity. This was also reflected by higher enzyme selectivity but lower reactivity.

2.5.3.4 Additive-controlled Selectivity

It has been widely observed that the presence of some additives might significantly alter enzyme activity, stability, and selectivity. Whereas polyols and amino acids are often used to enhance enzyme activity and stability in aqueous media [176, 177], other additives such as crown ethers, surfactants and the conformer selectors are often used to enhance the enzyme activity, stability and selectivity in non-aqueous media [178–181].

In contrast to the use of additives for the improvement of the performance of the free enzymes (such as the shelf life) or the lyophilized enzyme powder (for use in non-aqueous media), it was recently demonstrated that the selectivity of an immobilized enzyme can be improved by addition of additives to aqueous medium [182]. Partial purification and immobilization of crude *Candida rugosa* lipase OF have recently been combined by simple adsorption of the enzyme on SP-Sephadex C-50 – a cation-exchange resin. More interestingly, it was found that the enzyme immobilized at pH 3.5 was more enantioselective than the native enzyme in the hydrolysis of the 2-chloroethyl ester of racemic ketoprofen [182].

Remarkably, it was found that immobilization improved the enantioselectivity from 1.4 (free enzyme) to 6.1 (immobilized) in the absence of Tween 80 whereas enantioselectivity was increased 13 and 50-fold for the free and immobilized enzymes, respectively, in the presence of the surfactant [182]

2.5.3.5 Orientation-controlled Selectivity

The molecular orientation of enzyme molecules immobilized on a carrier by non-covalent adsorption is largely determined by the nature of the surface, the binding functionality of the carriers, and the nature of the immobilization medium.

That enzyme activity can be affected by molecular orientation suggests also that the molecular orientation of an enzyme on a carrier might also affect its selectivity. In fact, several interesting examples have shown that non-covalently immobilized enzymes might have better activity and selectivity than the native enzymes or other types of immobilized enzyme, for example covalently immobilized enzymes [477, 478].

Recently, influence of molecular orientation on the selectivity of non-covalently immobilized enzyme has been indirectly exemplified by immobilizing *A. niger* epoxide hydrolase (ANEH) onto Eupergit C, which is partially derivatized with ethylene diamine (EDA). The fact that the E-values of EDA-Eupergit C-immobilized ANEH were more than twofold enhanced relative to ANEH immobilised on original Eupergit C suggests that the enhanced selectivity is mainly due to the favourable molecular orientation on EDA-Eupergit C carrier [478].

2.5.3.6 Medium-controlled Enantioselectivity

Lipase PFL immobilized on octyl agarose/decaoctyl Sepabeads is more selective than the covalently immobilized enzyme or the native enzyme [172]. Similarly, the selectivity of lipase *Mucor miehei* (MML) immobilized on Eupergit C derivatives by adsorption and cross-linking was found to depend on the nature of the binding functionality [183].

For CAL-B immobilized on octyl agarose it has been found that selectivity in the hydrolysis of mandelic acid methyl ester is highly dependent on the pH and the temperature of the medium. High enantioselectivity was usually obtained at low pH and temperature [184]. This might be because the conformation of the enzyme was significantly affected at pH close to its pI [31]. A tighter conformation might be obtained at low pH, because exposure of the hydrophobic domains of CAL-B led to higher selectivity.

A similar observation was also made with CRL [184]. Interestingly, it was found that the effect of pH on the enantioselectivity of the immobilized enzyme depended on the method of immobilization. Among three methods investigated, i.e. covalent with glutaraldehyde as coupling agent and non-covalent adsorption on either a hydrophobic carrier (octyl-agarose) or PEI-coated resin, it was found that enhancement of the enantioselectivity (ratio of E at pH 5.0 to that at pH 7.0) in the hydrolysis of the same compound were 1.33, 2, and 10, respectively.

2.5.3.7 Water Activity-controlled Enantioselectivity

The influence of water activity applies only to the non-aqueous media, where the water activity (a_w) is far below 1. Water activity is strongly related to the enzyme conformation flexibility, geometry of the active centre and the diffusion constraints in organic medium. Thus, water activity has been found to be one of most important determinants for the enzyme catalytic functions e.g. activity, selectivity and stability in non-aqueous media [479, 480]. However, the effect of water activity on the selectivity of the immobilised enzymes differ from enzyme to enzyme, from support to support, from solvent to solvent and from substrate to substrate [487–489].

For instance, the selectivity of Novozyme 435 (CALB)-catalyzed kinetically controlled resolution of 3-bromo-1-phenoxy-2-propanol (esterification) with irreversible/reversible donor increases with increasing water activity in all organic solvents tested, while for 1-phenoxy-2 pentanol and 3-chloro-1-phenoxy-2-propanol the enantioselectivity decreases by increasing water activity [481]. Similarly, more than twofold increase in the enantioselectivity of Novozyme 435 (CALB)-catalyzed esterification of ibuprofen at lower water activity was observed [490].

In some cases, the influence of water activity on the enzyme enantioselectivity is not evident. For instance, water activity has almost no influence on the selectivity of lipase PS and lipolytic lipase for the transesterification between the (±) sulcatol and vinyl acetate in organic solvents when water activity varies from 0.1 to 0.53 [482]. Similarly, the enantioselectivity of lipase *Pseudomonas* sp immobilised on Duolite A 568 was only slightly influenced by water activity in the resolution of racemic 2-methyl-1-pentanol in a fixed bed reactor [483]; Also, the enantioselectivity of cutinase, which was immobilised on inorganic carriers i.e. zeolite NaA and NA Y and on Accurel PA-6 is not influenced by water activity in the reaction between vinyl butyrate and 2-phenyl-1-propanol in acetonitrile, as the water activity increased from 0.2 to 0.7 [485]. These examples might suggest that the structure of the immobilised enzyme might be rigid on the carrier and thus less influenced by the water activity.

Interestingly, it was found that the fatty acid selectivity of lipases i.e. lipase from *Pseudomonas cepacia, Rhizomucor miehei, Candida antaractica* B, *Candida rugosa* lipase immobilised on celite and synthetic resins was affected by the immobilisation and water activity in an unpredicted manner [484].

2.6 Preparation of Immobilized Enzymes by Adsorption

2.6.1 Conventional Adsorption

Conventional adsorptive enzyme immobilization denotes immobilization via noncovalent binding to ready-made carriers (e.g. with defined chemical and physical nature). Thus, the forces involved in the adsorption can be non-specific physical adsorption [188, 189] or specific adsorption mechanisms such as hydrophobic adsorption [3], ionic adsorption [48], affinitive, coordination and biospecific adsorption [12, 170].

Consequently, it is necessary to design different carriers not only with appropriate physical nature such as pore size and particle size but also specific binding functionality which enables the enzyme to participate in the desired binding. Thus, the binding functionality of the carrier basically determines the type of interaction between it and the enzyme.

As already discussed, the dominant types of interaction involved in adsorptive enzyme immobilization (other types of interaction might be also present, but negligible) can be categorized into the following types.

- Ionic binding (interaction of the immobilized ionogenic groups e.g. negatively or positively charged groups with those of the enzyme molecules).
- Biospecific interaction (interaction of immobilized ligands such as antibodies, substrates/substrate analogues, or inhibitors with the complementary binding site of the targeting protein molecules).
- Non-specific (non-specific groups on the surface, thus hydrophobic, hydrophilic, and electrostatic interaction might coexist).
- Hydrophobic (interaction of the carrier-bound hydrophobic functionalities with the hydrophobic regions of the protein molecules).
- Coordination (interaction of immobilized transition metal ions such as Cu^{2+} or Ni^{2+} with amino acid residues such as -SH, -COOH, -histidyl, etc., on the enzyme surface).
- Affinity binding (interaction of carrier-bound immobilized ligands such as such as immobilized dyes or analogues with the enzymes).

As shown in Scheme 2.17, except for non-specific binding, other types of binding usually require the presence of the corresponding bonding functionality tethered to the surface of the carriers. In contrast, for non-specific binding the carriers used usually lack the long tethering groups or the tethering groups are very short and

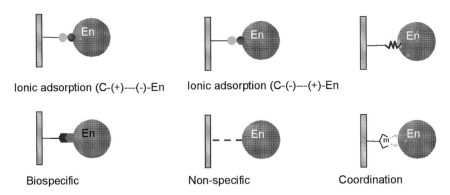

Scheme 2.17 Illustration of different means of adsorptive immobilization of enzymes.

the adsorption nature is usually determined by the overall nature of the carrier not by the tethering groups.

On the other hand, derivatization of a non-specific carrier often leads to formation of a specific adsorbent. For example, silica gel can be used as non-specific adsorbent for enzyme immobilization (mainly by hydrophilic interaction) whereas silica gel bearing long alkyl chains can be used as specific adsorbent for enzyme adsorption by hydrophobic interaction. Under these conditions the hydrophilic nature of the backbone is shielded [190]. Introduction of specific binding groups can often enhance the adsorption capacity. For example, the specific adsorption capacity of enzymes on poly(HMA-co-MMA) microspheres was increased by almost an order of magnitude by linking the affinity dyes to the carrier [296].

Another means of distinguishing between non-specific physical and specific interaction is that the force involved in the non-specific interaction is usually very weak as compared with specific adsorption.

2.6.1.1 Non-specific Physical Adsorption

Enzyme immobilization by physical adsorption usually denotes adsorption of the enzyme by the carrier as a result of the action of several non-specific forces. It is, therefore, difficult to identify which of hydrophobic, hydrophilic, and ionic adsorption is involved in non-specific adsorptive enzyme immobilization. Although some carriers lack the specific functionality responsible for specific adsorption, it is still possible to categorize their interactions as hydrophobic, hydrophilic and ionic.

Like other types of carrier-bound immobilized enzyme, the performance of immobilized enzymes, for example their stability, activity, selectivity, temperature optimum, or pH optimum or profile, are determined by the physicochemical nature of the carrier, for example the nature of the surface (charged or neutral), hydrophobic or hydrophilic balance [108, 191], pore size [82], and surface area [192, 193], the method of immobilization and conditions such as pH, temperature, ionic strength, immobilization time [52], presence of additives [370], enzyme loading, and the properties of the enzymes to be immobilized, for example the size of the enzyme [88], whether it is charged or neutral, hydrophilic or hydrophobic surface [191].

Although the performance of an enzyme immobilized by simple adsorption is determined by many factors, it is possible to alter the properties of the enzymes such as:

- enhanced/decreased stability [20, 135],
- pH optimum shifting or altered pH profile [91],
- enhanced/reduced activity [83],
- higher retention of activity than other immobilization methods [188],
- irreversible adsorption [5],
- alteration of selectivity [58].

Although there is no covalent bond formation between the enzyme and the carrier, the immobilized enzyme might lose some or all of its activity, for one or more of the following reasons:

- diffusion constraints, determined by the structure of the carrier [134] and enzyme loading [73];
- conformational change, which results from interaction of the enzyme and the carrier [18, 23, 187];
- confinement effect, which reduces molecular mobility and thus enzyme activity [189].

Similarly, the stability can be also increased or reduced, for the following reasons:

- conformational change (tighter conformation usually leads to greater stability whereas a looser conformation usually results in lower stability);
- alteration of microenvironment (more hydrophilic environment leads to more stability or vice versa);
- confinement (enzyme molecules are more stable in more confined environment because of restricted mobility);
- diffusion-controlled stability.

In general, three classification methods can be used to define the carriers that can be used for non-specific adsorption as follows:

- geometry (bead [9, 13, 66, 187], film, membrane, amorphous, fibrous [218–220], foam [450, 455];
- structure (microporous [83, 92], mesoporous [139, 140, 207–210], macroporous [146, 197–200], non-porous [212], gel, tentacle carrier [245–247, 252, 253];
- occurrence (synthetic, natural, organic/inorganic, semi-synthetic, composite).

Practically, however, it might be useful to combine the three classification methods to define a carrier.

Although many non-specific carriers can be used for immobilization of enzymes, they must fulfil the requirements of a robustly immobilized enzyme – i.e. high activity, high stability and selectivity for a specific application. The characteristics of each type of carrier are listed in Table 2.1. Examples of enzymes immobilized via non-specific adsorption are listed in Table 2.2.

2.6.1.2 Ionic Adsorption

In ionic adsorption binding of the enzyme to the carrier is a result of interaction of the charged groups of the carrier with charged amino acid residues on the enzymes, for example the amino groups of the lysine and the carboxyl groups of the glutamic acid or aspartic acid.

In general, the carriers that can be used for non-specific adsorption can be all modified, with the aim of introducing ionogenic groups. In practice, the carriers that can be used for adsorption of an enzyme by charge-charge interaction can be classified into three groups:

- Synthetic carriers that can be used directly as adsorbents such as copolymer gels of acrylamide/maleic acid or itaconic acid [12] or other polymers prepared by condensation of phenol and formaldehyde. Duolite can be used directly as weak cation exchanger for adsorption of protein [231].

- Derivatives of synthetic polymers. Some synthetic inert polymers such as polymers of poly(styrene-co-divinylbenzene) can be derivatized to form ionic adsorbents. These include some Dowex resins such as Dowex SBR-P, Amberlite IRA-904 [146], Amberlite IRC 50 [422], Amberlite CG-50, Diaion CR-20 [232], derivatives of polyacrylamide, Duolite A-7, Duolite S-761, porous PhOH-H_2CO resin (Doulite ES 762 [231]), derivatives of synthetic polymers such as PMMA-PEI [244], etc.

- Derivatives of cross-linked polysaccharides such as DEAE-cellulose, QAE-cellulose, SP-cellulose, DEAE-Sephadex, QAE–Sephadex, CM-Sephadex, DEAE-dextran, CH-Sepharose 4B (-NH(CH_2)+COOH), AH-Sepharose 4B (-NH-(CH_2)$_6$-NH_2), Q-Sepharose (CH_2-N+(CH_3)$_3$, S-Sepharose (-CH_2-SO_3-), CM-Sepharose (OCH_2COOH).

The binding functionality can be basically classified into cationic and anionic, as illustrated in Scheme 2.18. This ionic binding functionality can be prepared in situ during polymerization [12] or post-synthesis by derivatizing ready-made polymers such as functionalized poly(styrene–divinylbenzene), e.g. Dowex, and cross-linked polysaccharide carriers [34, 146, 238].

Because of differences between binding functionality in terms of size, spacer and microenvironment, it is not surprising that the use of different binding functionality might affect the performance, e.g. activity, stability or selectivity, of the immobilized enzymes.

In the early 1980s it was observed that in ionic adsorption the nature of the charged groups (spacer, polarity or volume) was crucially important with regard to retention of activity [55]. Occasionally the wrong binding functionality led to complete loss of enzyme activity. For instance, immobilization of LGTase (limonoid glucosyltransferase) on Amberlite 40 or Dowex 1X4 (Cl-) resulted in complete loss of activity whereas immobilization of the same enzyme on carriers (Toyopearl, Cellulose, Sepharose) bearing the DEAE moiety retained 20–50% activity [233]. In general, this effect (designing appropriate binding functionality) was not further studied for creation of robust immobilized enzymes based on the ionic binding.

In the same way as for covalent enzyme immobilization (as will be discussed later in this book) the performance of the immobilized enzymes, for example their activity and stability, on ionic carriers can also be affected by the hydrophobicity/hydrophilicity balance [158, 236], the density of exchanger capacity [68], enzyme loading [241], etc.

Retention of enzyme activity [specific activity of the immobilized enzyme (U mg^{-1} protein bound)/specific activity of the native enzyme (U mg^{-1} protein)] varies over a wide range (from several percent to more than 100%), depending on the carrier selected [99, 354], binding functionalities [55] and the immobilization conditions such as pH, temperature, ionic strength or additives [255]. Stabilization factors resulting from immobilization of an enzyme vary from highly destabilized to stabilization by two orders of magnitude [59] relative to the native enzyme.

Table 2.1 Classification of carriers for physical adsorption

Classification	Subclass	Remark	Ref.
Geometry	Bead	Beads with defined size and various functional groups and internal structures are widely used as adsorbents for various separation procedures/enzyme immobilization	9, 13, 66
	Film	Often, the films used for enzyme immobilization are supported. In other words, the films are often a polymer matrix coating upon which the enzyme can be immobilized via normal enzyme immobilization techniques	293, 298, 379
	Membrane	Which can combine the catalytic action and separation in one embodiment	301, 252, 246
	Irregular	Not frequently used	
	Fibrous	Fibre-based carriers have the advantages of high volume activity, presumably because of the high surface available for the enzyme immobilization	218–220
	Foam	Mainly used for enzyme entrapment. In special cases pre-designed foam can be used for enzyme adsorption	450, 455
Structure	Microporous	Microporous carriers are carriers which generally have pores in the range 0.1–10 μm	83, 206
	Mesoporous	Mesoporous materials have pore size in the range 3–10 nm. The pore sizes are, therefore, usually in the same range as the enzymes to be immobilized	139, 140, 143, 207–210
	Macroporous	Macroporous carriers refer to the materials with permanent pores in the range 8–1000 nm and specific areas in the range of 25–100 m^2 g^{-1}	197–200
	Non-porous	Less diffusion limitation might be expected, but enzyme loading might be lower because the available specific surface area is 2–3 order of magnitude less than for porous carriers	221
	Gel	In a strict sense gel materials are generally regarded as non-porous materials, because they do not have permanent pores and the pores only exist when they swell in a solvent	327–330
	Tentacle carrier	Preparation of tentacle carriers can be realized by grafting, coating, and covalent attachment	245–247, 252, 253
Occurrence	Synthetic carriers	These are playing an increasing role in bioseparation and bioimmobilization because of their defined composition and structure	192–195
	Natural carrier	Soils, clay, Celite, bones, etc., which can be directly used for enzyme immobilization	79, 201, 216
	Inorganic	Zeolite, alumina, clay, silica, and molecular sieves are often used for enzyme immobilization. Regeneration of the carrier is often possible	116, 141, 213, 215
	Semi-synthetic	These are semi-synthetic carriers which are prepared by modification of naturally occurring polymers	
	Composite	Composites, belonging to the category of the hybrid materials, which are prepared by combination of more than two types of different material with different properties, sources, or structures	222–225

Table 2.2 Adsorption of enzyme via non-specific interaction

Polymer	Nature of adsorption	Comments	Enzyme	Ref.
PET	Hydrophobic	Retention of activity was higher than when covalently immobilized on CNBr-activated Sepharose	Pectin esterases	188
NaY Zeolite/Accurel PA6	Hydrophobic	The payload is 0.025; activity retention is 85%. IME is more stable than on Accurel PA6 (no activity loss was observed after 45 days). Enzyme is desorbed more easily from Accurel PA6 than from NaY zeolite	Cutinase	189
Poly(chlorotrifluoroethylene) (PCTFE)	Hydrophobic	Immobilized LDH had only 0.2 mg g^{-1} payload and 4% activity relative to the native enzyme, suggesting that this carrier was not desirable	LDH, URA, GDH	190
Talc (hydrophobic support)	Hydrophobic	Adsorption was favoured by the hydrophobicity of the support; modification of hydrophobic–hydrophilic balance produced IME with high enzyme activity	Proteases, lipases and peroxidases	191
Porous PMA	Hydrophobic	The pore structure and the functional groups are the basic factors affecting the performance of the immobilized enzyme	Aminoacylase	192
Polystyrene latex/Accurel EP400	Hydrophobic	Retention of enzyme activity is lower than when adsorbed on EP 400	Lipase	193
PET (Sorsilen) Poly(ethylene terephthalate)	Hydrophobic	There was no enzyme leakage when the immobilized enzyme was used in a CH$_3$Cl–water two-phase system for esterification of acetyl-l-tryptophan with EtOH	Chymotrypsin	194
Hydrophobic Accurel	Hydrophobic	The activity of free and modified lipase was increased ninefold by adsorption on EP 100. Compared with the native enzyme, the adsorbed enzyme has lower activity	PEGylated CAL-B, PEG-OE-CALB, ON-CAL-B	195

Table 2.2 Continued

Polymer	Nature of adsorption	Comments	Enzyme	Ref.
Poly(ST-co-DVB) Poly(ST-co-AN)	Hydrophobic	Polar carrier is favourable for activity and optimum pH and temperature ranges	Urease	196
Macroporous copolymer support	Hydrophobic	Payload 0.0154 with 62% retention of activity	CRL	197
Copolymer of methyl acrylate and divinyl benzene	Hydrophobic	Introduction of PCG led to a severalfold increase in activity	CRL	198
Amberlite XAD	Hydrophobic	The diffusion limitation of IME was related to the penetration depth of the enzymes in beads and can be controlled by manipulating conditions during immobilization	Trypsin	199
Silica (30–90 nm)	Hydrophobic	For efficient molecular inclusion of enzymes the pore size should be 5–10 times the molecules size of the enzyme	Papain	200
Fibrous clay, sepiolite	Hydrophobic	Adsorption was promoted at high ionic strength	Urease	201
Hydrophobic HDPE (high-density polyethylene)	Hydrophobic	The immobilized lipase maintained its activity over broader temperature (25–55 °C) and pH (4–8) ranges than soluble lipases	CRL and ANL	202
Porous powder poly(ethylene terephthalate)	Hydrophobic	For endopolygalacturonase bound to the support by hydrophobic interactions, external diffusion effects were regarded as the factors governing enzyme action	Aspergillus endopolygalacturonase	203
Hydrophobic zeolite type Y	Hydrophobic	Payload is 0.0082 and activity retention is 33%	Lipase from *Candida cylindracea*	204
Hexadecyltrimethyl-ammonium (HDTMA)-smectite	Hydrophobic	The immobilized enzyme has the same activity as the free enzyme. The payload is 0.1 (irreversibly bound) and the thermostability is reduced after immobilization	Urease	205

2.6 Preparation of Immobilized Enzymes by Adsorption

Table 2.2 Continued

Polymer	Nature of adsorption	Comments	Enzyme	Ref.
Controlled-pore glass (CPG)/ A microporous ceramic	Hydrophilic	Better than conventional silica glass carriers, with regard to lower leakage of enzyme; the carrier is more stable than CPG	β-Glucosidase	206
MCM-41 (4 nm)	Hydrophilic	Enzyme loading is inversely proportional to the size of the enzyme molecules. The payload for all the enzymes tested was below 0.01	Cytochrome C Papain Trypsin	207
FSM-16, MCM-41	Hydrophilic	Pore size should exceed the molecular size of the enzymes	HRP, subtilisin	208
MCM-41, MCM-48, Nb-TMS1	Hydrophilic	Adsorption was dependent on molecular sieve component The IME cannot be removed at pH < 9. Silanization prevented significant leaching at high pH	Cytochrome C	209
M41S silica	Hydrophilic	The payload was temperature-dependent: 472 mg g^{-1} at 0°C, 362 mg g^{-1} at 25°C and 119 mg g^{-1} at 40°C. Indeed, activity retention was approximately 47% and 78%	Lipase	210
Vermiculite	Hydrophilic	Retention of activity was greater than with the covalent method based on the same carrier	Neutral protease	211
Non-porous silica	Hydrophilic	Support provides a suitable model for studying the effect of EMTL on the reaction rate in the absence of IDL	Penicillin acylase	212
Mesoporous silica matrixes	Adsorption	Enzyme properties depend on template content and on sol–gel matrix pore dimensions	HRP	213
MCM-41	Adsorption	Adsorption results in higher activity than covalent bonding	Penicillin G acylase	214
Silica gel containing CaSO$_4$ (Silica G)	Adsorption	Silica gel is a more effective carrier than alumina or CPG	Prostaglandin synthetase	215

Table 2.2 Continued

Polymer	Nature of adsorption	Comments	Enzyme	Ref.
Celite	Adsorption	The aquaphilicity of the carrier is more important, in terms of enzyme activity in organic solvents	α-Chymotrypsin HLADH	216
Na Y Zeolite	Adsorption	Enzyme thermostability depends on the temperature and on the properties of the carriers used (NaY, NaA, alumina, etc.)	Recombinant cutinase	217
Cellulose acetate and titanium isopropoxide	Coordination	The fibre is harder than alginate gel; it acts as an anion at low pH and as a cation at high pH	β-Galactosidase; α-Chymotrypsin	218
Transparent cellulose acetate – a metal alkoxide gel	Coordination	Coordination bonding between hydroxyl groups on the pyranose ring and the metal	Enzymes	219
TiO_2-coated cellulose microfibres	Coordination	The activity of the immobilized enzyme is reduced, presumably because of its interaction with the oxide surface	Horseradish peroxidase (HRP)	220
PTFE membrane with polyurethane coating	Coulomb force	The IME can be recovered by applying a reversed electric potential	AGD	221
Polystyrene seed particles copolymer	Adsorption	Thermosensitive polymer; adsorbed on the composite particle at a temperature above LCST	Trypsin	222
Cross-linked starch entrapped in alginate	Adsorption	Fewer mass-transfer constraints than for beads with un-cross-linked starch	α-Amylase	223
ZrO_2-coated glass carriers	Coordination	A metal whose oxide is more water-durable than is the carrier	Glucose isomerase and glucoamylase	224
Pumice–metal ($TiCl_4$) complex (coating)	Coordination	50.7% of the original enzyme activity was obtained	α-Amylase	225

* Payload is defined as ratio of the mass of the immobilized enzyme to mass of carrier used

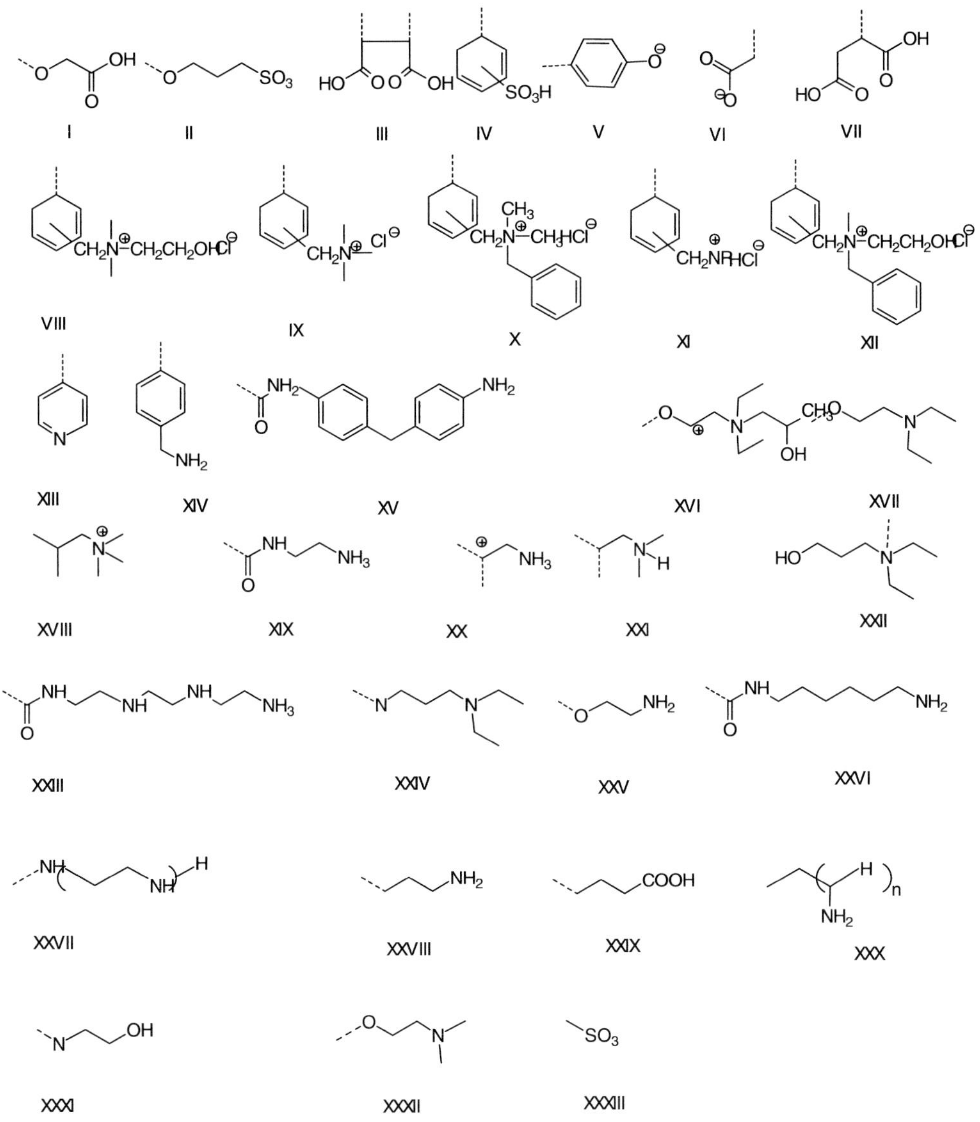

Scheme 2.18 Charged groups of ion exchangers commonly used for enzyme immobilization.

It is worth remarking that the hydrophilicity of the backbones of the carriers used for enzyme immobilization varies greatly. The carriers can be classified into the following groups:

- very hydrophilic natural polymer derivatives such as AH-Sepharose 4B (NH$_2$) [238], aminoethylcellulose [240], DEAE-cellulose [34], DEAE-Cellulofine (ionic binding) [254], DEAE-cellulose [256], agarose [243];

- very hydrophilic synthetic gels such as poly(vinyl ether)-monoethanolamine) [239], aminoethylcellulose [240] poly(acrylamide/maleic acid) hydrogels [12], functionalized PVA [55];
- inorganic silica (60 A)-PEI, amino-alkylated silica [351];
- hydrophobic carriers such as Amberlite DP-1 [59], Dowex 1-X2 [59], Duolite A7 (a weakly basic anion exchanger) [234], Duolite A 568 [67].

However, the actual hydrophilicity of the carrier depends on:

- the density of the binding functionality as exemplified in Ref. [47],
- the nature of the binding groups,
- the nature of the secondary groups,
- the density of secondary functionality,
- the properties of the spacer.

The hydrophobicity of ionic carriers is occasionally pH-dependent so changing the pH can affect the protonation of ionogenic groups such as pyridine groups. At a particular pH pyridine groups are protonated and a carrier bearing pyridine groups can function as anion exchanger, whereas hydrophobic interactions will predominate when the pyridine groups are in the electronically neutral state [236].

In addition, the hydrophobicity of carriers bearing weak acidic or basic groups are also pH-dependent. Thus, it is not surprising that a hydrophobic backbone bearing weak carboxymethyl groups will become increasingly hydrophobic as the pH of the medium is reduced below the pK_a values of the carboxyl groups.

Hydrophobicity differences affect not only enzyme loading but also enzyme performance.

2.6.1.3 Hydrophobic Adsorption

Hydrophobic adsorption is based on interactions between hydrophobic moieties on the carrier with hydrophobic regions or domains on the enzymes. Thus, the surface chemistry of the carriers used for hydrophobic adsorption is dominated by the pendant hydrophobic groups, as illustrated in Scheme 2.19. Without exception, these pendent groups are all hydrophobic. The lengths of the hydrophobic groups can also be modulated.

Other types of adsorption on non-specific hydrophobic carriers such as XAD-7 are also mainly governed by hydrophobic interactions. These are, however, categorized into the non-specific physical adsorption.

As shown in Table 2.4 carriers for hydrophobic adsorption of enzymes are usually prepared by derivatization of ready-made carriers, for example natural polymers such as cellulose [107], agarose beads [161], Sepharose [24, 107, 125, 158, 258], etc.. Alternatively, inorganic carriers such as silica [162, 268] or synthetic carriers such as Eupergit C, Sepabeads [267] or poly(vinyl alcohol) beads [266] can be also used to prepare hydrophobic adsorbents for enzyme immobilization.

Hydrophobic functionality is usually covalently bound to ready-made carriers. Occasionally, however, the hydrophobic compound can simply be coated on the carrier [269].

2.6 Preparation of Immobilized Enzymes by Adsorption

Table 2.3 Enzyme immobilized by ionic adsorption on insoluble synthetic carrier

Polymer	Method	Comments	Enzyme	Ref.
Duolite A7 (a weakly basic anion exchanger)	XXI?	Similar results were obtained by physical adsorption and by a covalent method with same type of carrier. However the covalently immobilized enzyme is 50% more stable	Glucoamylase	234
Sulphonated PVDF film	-SO$_3$- (XXXIII)	Enzyme activity is proportional to ionic-exchanger capacity (IEC)	Urokinase	235
Poly(VP-co-EGDMA and/or TEGDMA)	XIII	Cross-linker properties and cross-linking density influence structure such as pore size and degree of swelling. High loading was obtained with very hydrophilic carriers	Urease	236
PET fibres-g-MA-co-AAm	Adsorption?	The graft yield strongly affected enzyme activity and stability; immobilized urease was more stable toward acidic and alkaline pH, high temperature, and storage conditions	Urease	237
AH-Sepharose 4B (-NH$_2$)	XXVI	Covalent immobilization deactivated enzyme whereas physical adsorption resulted in high activity (U mL^{-1} gel)	Phenol hydroxylase	238
P(vinylether-monoethanolamine)	XXV	High enzyme activity was obtained and adsorption was irreversible	Catalase from *Penicillium vitale*	239
Aminoethylcellulose	XXVII	High retention of activity and carrier regeneration; it is, however, of limited applicability	Esterase from *Bacillus subtilis*	240
PEI-coated non-porous glass microbeads	XXVII	Enzyme activity approaches the maximum at 50% saturation	Glucose oxidase/catalase	241
Aminoalkylsilane-alumina (As-alumina)	XXVIII	The activity retention is 13.5–50%	*Bacillus subtilis* α-amylase	242
(Agarose, silica, polymeric resins)-PEI	XXVII	~100% retention of activity was obtained, with rapid immobilization and enhancement of stability of IME	CRL, galactosidase, DAAO)	243
Macroporous Me acrylate-divinylbenzene copolymers	XXX	High operational stability was obtained. Activity retention was more than 90% after continuous use for one month	Aminoacylase	244
Ozonized PET fibres-g-PAA	I	Adsorbed trypsin was inhibited more easily than the covalently immobilized enzyme	Trypsin	245
Hollow-fibre membrane-g-GMA-DEA	XIX	A novel "tentacle-type" porous membrane with 50-fold greater enzyme loading than the native membrane	Urease	246

Table 2.3 Continued

Polymer	Method	Comments	Enzyme	Ref.
Cotton cloth-enzyme-PEI aggregates	XXI	(250 mg g^{-1} support) with approximately 90–95 % efficiency	β-Galactosidase	247
Dowex 66 (weak anionic)	?	Pretreatment of the carriers with polar organic solvent substantially enhanced enzyme loading	*Pseudomonas fluorescens* lipase	248
A weakly basic anion exchange resin: Duolite A7	XXI?	Enzyme properties of the immobilized aspartase were found to be fit for industrial purposes	Aspartase	249
Cross-linked anion exchange resin – Indion 48		This method is inexpensive, does not require cross-linking agents, and results in firm binding of the enzyme to the resin	d-Xylose isomerase	250
Anion-exchange PHFM-g-GMA-diethylamine	XXXI	Loading of 38 and 110 mg citase g^{-1}, respectively, was obtained (ca. 7–9 layers)	Citase	251
Porous hollow-fibre membrane-g-GMA opened with EA	XXXI	Negligible mass-transfer resistance of the starch to the α-amylase due to convective flow	α-Amylase	252
Poly(AN-co-MMA-co-SVS) membranes-g-AMPSA (or DMAEM)	XXXII	Glucose oxidase was immobilized on modified acrylonitrile copolymer membranes with DMAEM and AMPSA and had high relative activity	Glucose oxidase	253
DEAE-Cellulofine (ionic binding)	XVII	Among various methods such as covalent and entrapment, ionic binding proved to be the best method	Bromoperoxidase of *C. pilulifera*	254
DEAE-cellulose	XVII	Addition of Triton X-100 during the immobilization enhanced the enzyme stability	Nocardia EH1 epoxide hydrolase	255
DEAE-cellulose	XVII	The properties of IME were similar to those of the free enzyme. Immobilization led to lower K_m (app) and higher V_m (app)	Epoxide hydrolase	256
Non-porous polystyrene/poly(sodium styrene sulphonate) (PS/PNaSS) microspheres	O-CH$_2$CH$_2$N(CH$_3$)$_2$ (XXXII), -SO$_3$- (XXXIII)	PS/PNaSS microspheres enable very simple, mild, and time-saving enzyme immobilization	Amyloglucosidase	257

SVS: sodium vinylsulphonate: DMAEM: 2-dimethylaminoethyl methacrylate; AMPSA: 2-acrylamido-2-methylpropanesulphonic acid; GMA: glycidyl methacrylate; PP: polypropylene

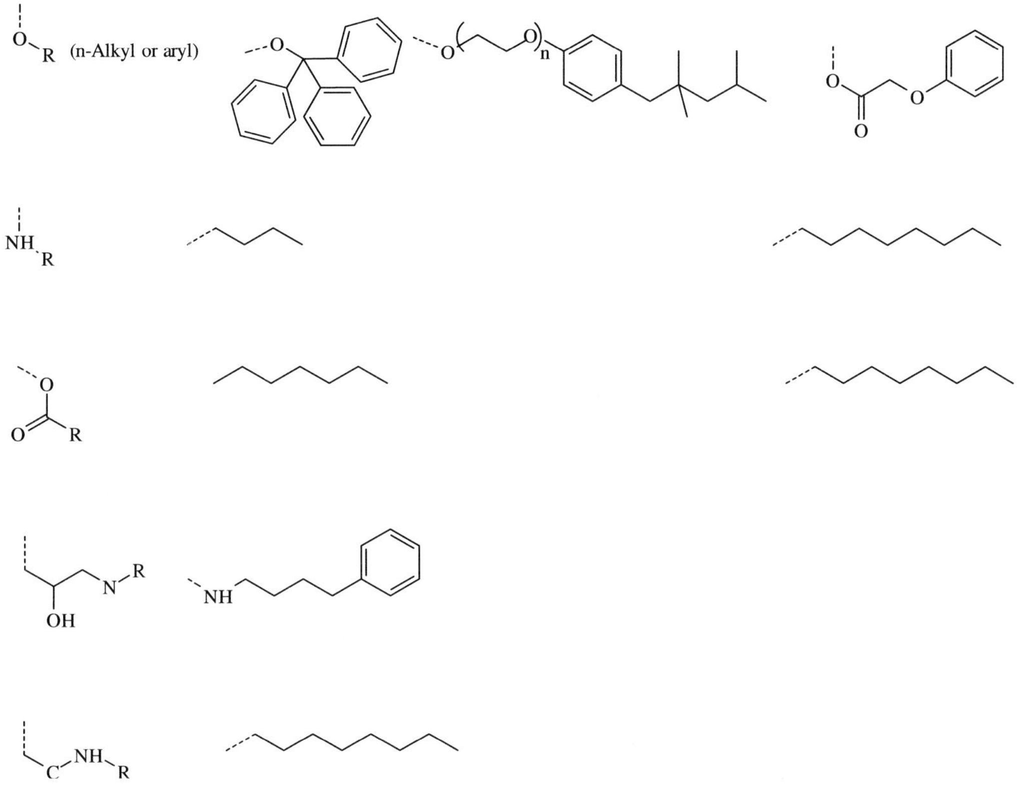

Scheme 2.19 Hydrophobic functionality for binding enzymes.

The adsorption capacity of the carriers depends on many factors:

- carrier structure,
- density of binding functionality [125],
- properties of the binding functionality [55],
- presence of activators or inhibitors [258],
- nature of the protein,
- the medium, in which the interaction occurs.

Retention of activity by enzymes adsorbed by hydrophobic interactions usually depends on:

- enzyme loading [24, 158],
- hydrophobicity of the carrier; more than 100 % activity is possible [160, 161],
- size of the substrate [84],
- length of the hydrophobic tails [467],
- diffusion constraints,
- binding density.

Table 2.4 Immobilized enzyme via hydrophobic adsorption on insoluble synthetic carrier

Polymeric carrier	Hydrophobic tail	Comments	Enzyme	Ref.
Sepharose gels	Butyl, hexyl, and octyl	Nearly completely adsorbed by the last two gels. Inhibitor and activator both influenced the adsorption behaviour of the enzymes, because of the conformational change	*Escherichia coli* phosphoenolpyruvate carboxylase	258
Sepharose 4B	N-Butylamine (A-C$_4$) or n-octyl-amine (A-C$_8$)	Acidic enzymes can be virtually irreversibly immobilized on the carrier with A-C8	Xanthine oxidase, LDH DNase I, AKP, and urease	259
N-alkyl or aryl amino-agar beads	n-Alkyl or aryl amino	The payload is 0.020–0.035 with 80% retention of activity and increased thermostability	MPS	260
Triton X-100-substituted Sepharose 4B	Triton X-100	Specific interactions involved the hydrophobic region of Triton X-100 and apolar regions on the protein surface	BSA, Hb, GLD, Cyt C, pepsin	261
Palmityl-substituted Sepharose 4B	Palmityl	Immobilization conditions for individual enzymes are determined by the properties of the enzyme itself	Lysozyme, trypsin, α-CT, etc.	262
Remazol blue (RB)-cellulose beads	Remazol blue (RB)	Instead of the expected bio-specificity in the interaction of LDH with RB-cellulose beads, the predominant interaction is hydrophobic	LDH	263
Cellulose esters of both alkyl and aryl carboxylic acids	Alkyl and aryl carboxylic acids	The bound enzymes retained their catalytic activity almost completely	Ten of the enzymes tested	264
Adipic acid dihydrazido-Sepharose 4B	Sialic acids and sialyl-glycoconjugates	Immobilization is via its CO$_2$H group, C-7 to C-9 side chain, or its NH$_2$ function as d-neuraminic acid-β-me glycoside or 2-deoxy-2,3-didehydro-neuraminic acid; hydrophobic interaction is involved	*Clostridium perfringens* sialidase	265
Different esters of cross-linked poly(vinyl alcohol) (PVA)	Alkyl carboxylic acids	Lipase CRL showed greater affinity for carbon chain length ranging from 8 to 12. The result was better than with Celite 545	Lipase of *Candida rugosa*	266
Phenylbutylamine-Eupergit/Eupergit 250L	Phenylbutylamine	The pore size of the support had a strong effect on the activity but did not influence stability	*Staphylococcus carnosus* lipase	267
Hexadecyl silica	Hexadecyl	Simply heating hexadecanol and silica gel led to the formation of the hydrophobic carrier	Three enzymes?	268
Triton X-100 and triton X-405, adsorbed on silica gel	Triton X-100	Under restricted conditions the immobilized enzyme was very stable	Lysozyme and α-chymotrypsin	269
Agarose (4 XL or 6 XL)	Hydrophobic tails such as octyl	Immobilization and purification can be combined into one step	Penicillin acylase	270

MPS: Microsome prostaglandin synthetase

The stability of the immobilized enzymes can be increased [162], decreased [24] or unaltered [26] by hydrophobic adsorption.

Substantial enhancement of substrate selectivity on hydrophobic interaction, compared with either the native enzyme or the covalently bound immobilized enzyme was observed when lipase *Pseudomonas florences* (PFL) was simply adsorbed on a hydrophobic carrier [172], demonstrating that enzyme immobilization techniques can be complementary to genetic approaches for influencing enzyme performance.

2.6.1.4 Biospecific Adsorption

As discussed above, biospecific adsorption is based principally on intrinsic interaction between the specific functional groups from enzyme and carrier.

This technology was initially designed for separation and purification of enzymes and other proteins, because of the reversibility and selectivity of the adsorption. Remarkably, enzymes immobilized on bioaffinity adsorbents such as immobilized Con A often had high retention of activity, probably because of the orderly oriented immobilization (site-specific immobilization) [12, 273, 440]. Stability, for example during storage and operation, and thermal stability, was also often enhanced, probably because of complementary interaction of the enzyme with an immobilized bioligand, e.g. immobilized Con A or tannin [12], or immobilized dyes [300, 301].

Immobilized substrates, substrate analogues, or inhibitors are often used to purify enzymes; enzymes immobilized on carrier-bound substrates, substrate analogues or inhibitors are rarely used as biocatalysts. One reason is, obviously, the weak interaction of the enzyme with the carriers. Also, occupation of the active centre by the substrate might inhibit the activity.

Accordingly, biospecific enzyme immobilization can be classified into three groups, antibody–antigen based bioaffinity, for example with an immobilized antibody, group-specific bioaffinity interactions, for example that with immobilized Con A (specifically for glycosylated enzymes), and ligand–acceptor interactions, such as those with an immobilized substrate (or inhibitor) [284].

Occasionally designed ligands can be also used to immobilize enzyme selectively. For instance, thermostable lactase from *A. oryzae* has been immobilized on *p*-(*N*-acetyl-L-tyrosine azo) benzamidoethyl-CL-Sepharose 4B with a dramatic increase in the apparent thermal stability of the lactase [285].

Interestingly, pseudo-affinity immobilization can be achieved by complexation of an enzyme with a ligand that is, coincidently, also the ligand of another immobilized enzyme. In this way, immobilized carbonic anhydrase can be used to adsorb conjugates of an aryl sulphonamide with hexokinase, lysozyme, and glucose 6-phosphate dehydrogenase [374].

This technology was initially designed for separation and purification of enzymes, because of the high reversibility and high selectivity of adsorption.

In 1910 it was reported that α-amylase could be selectively adsorbed on insoluble starch and, possibly in the 1960s, this technology was proposed as an efficient tool

Table 2.5 Bioaffinity adsorption

Polymeric carrier	Ligand/binding site	Comments	Enzyme	Ref.
Triazine bead celluloses MT-100	Concanavalin A/mannan	The specific activity and storage and operational stability of the penicillin G acylase immobilized on Con A-cellulose is even better than that covalently immobilized on Eupergit C	Synthetic glycolated penicillin G acylase	271
Con A-Sepharose	Con A/sugar moiety	Compared with the native enzyme, there is no significant change in the enzyme kinetics, suggesting that the active site is far from the binding site	Laccase (p-diphenol: oxygen oxidoreductase	272
Chitosan	Tannin	Highest retention of activity, reduced pH optimum, and enhanced optimum temperature	*Aspergillus niger* NRC 107 xylanase and (R)-xylosidase	273
Sepharose	Anti-CHT-IgG I	Biospecific immobilization offered great advantages over the covalent immobilized counterpart, because of the oriented immobilization	CHT	274
Cellulose beads	CCBMPB	90.12% retention of activity; reversibility of the binding > lowering pH	HRP	275
Sepharose	Monoclonal anti-β-galactosidase	The antibody immobilized rates of adsorption and desorption of the enzyme from the immobilized antibody and the specific activity of immunosorbents were affected	β-Galactosidase	276
Sepharose	Monoclonal antibody pH 8	The IME retained kinetic properties similar to those of the native enzyme and responded to activators and inhibitor in the same way as the native enzyme	Tryptophan hydroxylase	277
Porous glass	Biotin	Neither refolding of the fusion protein nor interaction of the streptavidin domain with immobilized biotin altered the structure of the substrate binding site	Streptavidin-β-galactosidase	278
Chitosan	Tannin	The optimum temperature was reduced, but thermostability increased	Chitinase and (nahase)	279
Zirconia surfaces, organosilanes	Con A	Only one type of interaction is present, with non-specific interactions with the support surface playing an insignificant role	Horseradish peroxidase	280
Polystyrene and silica beads	Protein A and a monoclonal antibody	K_{cat}/K_m was more than seven times that of the RIE and storage stability was superior	Subtilisin	281
Glass beads (porous or non-porous)	Biotin	The activity of IME was four times that of FE	CRL	282

2.6 Preparation of Immobilized Enzymes by Adsorption

Table 2.5 Continued

Polymeric carrier	Ligand/binding site	Comments	Enzyme	Ref.
Poly(2-HEMA-co-EDMA)	Anti-chymotrypsin antibodies	Immunosorbent adsorbed 166.7 μg chymotrypsin g^{-1} dry carrier with 100% retention of activity	Chymotrypsin	283
Sub-micron ferrite	Soybean trypsin inhibitor	Selective recovery of trypsin was obtained by adsorption with immobilized inhibitor	Trypsin	284
Sepharose	N6-(6-aminohexyl)-5'-AMP	Linear temperature gradients were successfully employed for elution of enzymes from affinity adsorbents	Glycerokinase and yeast alcohol dehydrogenase	286
Amine-derivatized agarose	Triazine scaffold bis-substituted with 5-aminoindan	The ligand, a triazine scaffold bis-substituted with 5-aminoindan, which mimics natural protein–carbohydrate interactions, can adsorb glycosylated enzyme	Glucose oxidase	287
Agarose-AA	Aminohexyl	Glycine enzyme can be desorbed with 0.5 M NaCl and 0.4 M phosphate or 0.14 M HAD; enhanced enzyme activity was obtained	Prephenate dehydratase	288
CNBr-activated-Sepharose 4B	Aminolevulinate	Ionic strength influenced the binding behaviour	Bovine liver I dehydratase	289
Sepharose	Divinylsulphone-activated methicillin	Carrier bearing substrate analogues were used to demonstrate reversible adsorption of enzyme	Penicillinase	290
Affinity-ligand-modified liposomes	p-Aminobenz-amidine (PAB) as the affinity ligand for trypsin	Recovery yield from the crude mixture was 68%; trypsin purity was 98%	Trypsin	291
Sepharose	Histamine/carboxyl-histidyl	The dissociation constant depends on the nature of the ligands, suggesting that the nature of interaction involved is different	Catechol-2,3-dioxygenase	292
PP-g-GMA-AAs	Amino acid residues	Payload increased with increasing density of amino acid groups in the order: l-Phe > d,l-Phe > d,l-try > l-cys but was pH-dependent: loading at pH 7.4 was higher than at pH 9.0	Urokinase	293
Sepharose CL-4B	Analogue of a Phe-Arg dipeptide	110-Fold purification was obtained in one step	Pancreatic kallikrein	294

CCBMPB: catechol[2-(diethylamino)carbonyl-4-bromomethyl phenylboronate for separation of enzymes. For enzyme immobilization, affinity adsorption can be dated back to the mid-1970s, when Ahmad found that chicken brain arylsulphatase A immobilized on Sepharose-Con A carriers was still catalytically active [12].

2.6.1.5 Affinity Adsorption

Along with the above mentioned bio-affinity adsorption-based enzyme immobilization, for example adsorption based on immobilized antibody, immobilized Con A (lectin), immobilized substrate and substrate analogues, immobilized inhibitors or cofactors, and other affinity-based adsorption such as immobilized metal affinity (IMA) [295], immobilized dyes, etc., have also recently been investigated for immobilization of enzymes.

Affinity based immobilization with the use of immobilized dyes The use of carrier-bound immobilized dyes for the reversible protein immobilization with the aim of purifying the proteins can be dated to the beginning of the 1980s, when Lowe and coworkers used immobilized procion dyes for purification of inosinate dehydrogenase [13]. The use of this concept for immobilization of enzymes for biocatalytic purposes has, however, been pursued only since the beginning of the 1990s. The potential of immobilized dyes as adsorbents for proteins was, however, observed in 1960s.

The dyes frequently used as immobilized ligands for affinity enzyme immobilization are given in Scheme 2.20. These dyes used as ligands for affinity enzyme immobilization fall in the same categories as dye-affinity for enzyme purification. They include triazine dyes e.g. Cibacron blue F3G-A [296, 297, 302, 303, 305], Procion red HE-3B, Pricon green, Pricon yellow [298, 299], Pricon brown [301], Congo red [306] etc. The dye ligands can be covalently bound to a porous carrier or a non-porous carrier from natural source such as Agarose beads or synthetic sources such as hydrophilic polyacrylic carriers.

It is worth mentioning that some transition metal ions, for example as Fe(III), Ni(II), and Cu(II) can be chelated to the immobilized dyes, leading to the formation of new adsorbents which have affinity behaviour analogous to that of the mixed ligands. In these cases, the binding capacity of the affinity adsorbent is usually enhanced compared with the immobilized dyes without the transition metal ions [296, 297].

Affinity-based immobilization with the use of chelated transition metals Because protein purification based on different chromatographic processes is a special case of enzyme immobilization, use of immobilized metal affinity based adsorption for enzyme (or protein) immobilization can be dated back to 1975, when Porath and coworkers introduced a new chromatographic process for protein purification with the name "metal chelate affinity chromatography" (MCAC) [14]. Soon after, this concept was extended by the same research group for enzyme immobilization with the aim of regenerating the carrier and increasing the stability of the enzyme [325, 326].

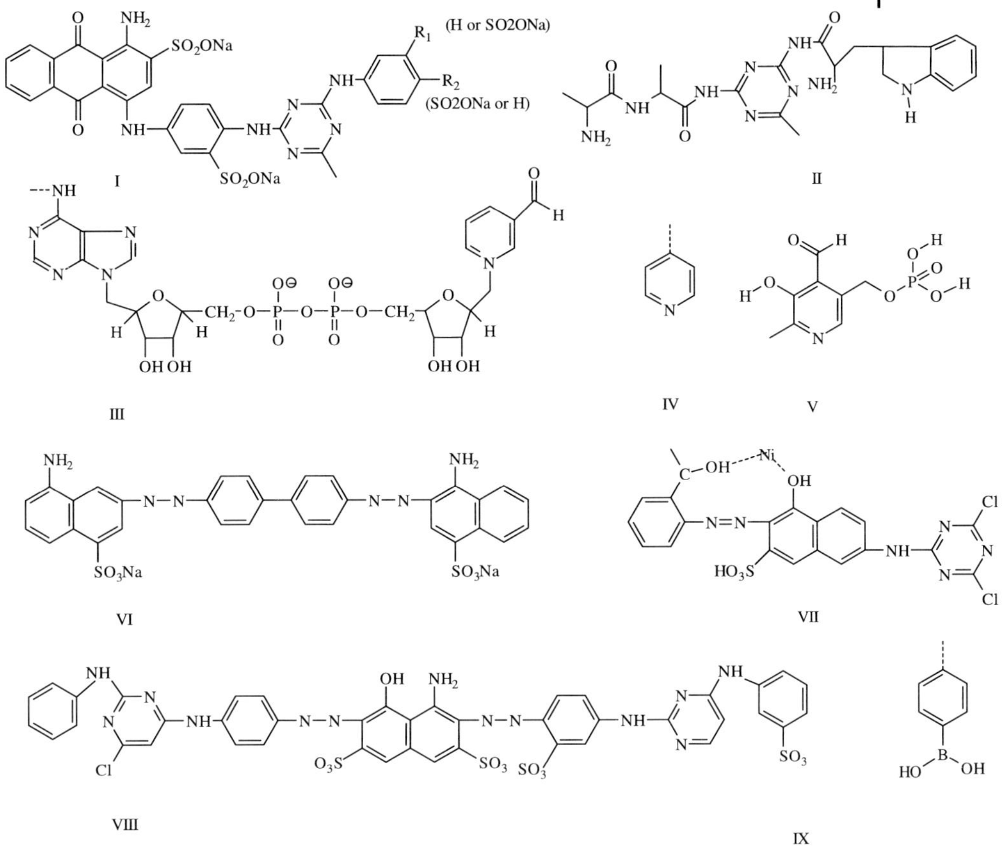

Scheme 2.20 Dye-based ligands frequently used for adsorptive affinity enzyme immobilization.

In fact, use of carrier bound immobilized metal ions for enzyme immobilization was probably pioneered by Epton et al., who used poly(N-acrylolyl-4-and-5-aminosalicylic acid)–titanium complexes for insolubilization of enzymes [170]. Although the potential of immobilized metal affinity enzyme-immobilization methods, for example the possibility of regenerating the carrier and thus reducing production costs and disposal of the carriers, have been recognized [336, 337], in practice this method has not received great attention in industrial sectors where immobilized enzymes might be needed.

In general, binding of an enzyme to an immobilized metal adsorbent can be achieved by interaction of specific binding sites on the enzymes (e.g. genetically introduced tags [330] or chemically introduced tags) with the immobilized metals and non-specific binding of the amino acid residues, for example exposed histidine(s), phosphorylated side chains, carboxylic amino acids, cysteine, tryptophan, phenylalanine, and tyrosine.

Table 2.6 Affinitive adsorption for enzyme immobilization: (affinitive dyes)

Polymeric carrier	Dye (ligand)	Comments	Enzyme	Ref.
Poly(HEMA-co-MMA)-MS-CB-F3GA/or Cu(II)	I	Payload on poly(HEMA-MMA) microspheres was 0.0036, whereas on Cibacron blue F3G-A or Cu(II) the payload was increased to 0.248 or 0.319, respectively	Lysozyme	296
(pHEMA)-cb/(pHEMA)-CB-Fe(III)	I	30% increase in enzyme loading was achieved with Fe(III), compared with poly(HEMA)-CB. Slightly higher K_m relative to the native enzyme; increased stability with immobilization, especially in the presence of Fe(III)	Catalase	297
Procion green H-E4BD-attached pHEMA	VIII	The enzyme could be repeatedly adsorbed and desorbed from the dye-attached pHEMA film without any significant loss in adsorption capacity	Lysozyme	298
Non-porous monodisperse silicas/triazine dyes	I	Procion red HE3B, Procion red MX5B, and Cibacron blue F3G-A	LDH, and MDH, ARD	299
Poly(HEMA) membranes/CB-F3G-A	I	The storage stability of the enzyme was found to increase upon immobilization	β-Galactosidase	300
HMF (polyamide)-procion brown Mx-5 BR-Ni(II)	VII	The payload is 0.078, activity retention is 37%, and affinity immobilization enhanced the thermostability 3–10 times, depending on the temperature	Urease	301
Supported alkanethiol SAM-CB-F3G-A	I	NAD^+-binding pocket is not involved in the binding of the dyes	LDH	302
Agarose-Cb	I	Affinity adsorbent	Sulphurtransferase	303
Activated silica/reactive triazine dyes	I	LDH, hexokinase, AKP, carboxypeptidase, tryptophanyl-tRNA synthetase		304
Cibacron blue F3G-A	I	Procion red H-8BN, Procion yellow H-A and Cibacron blue F3G-A	Carboxypeptidase G2, hexokinase	305
Poly(HEMA-co-MMA)-Congo red/or CR-Fe^{3+}	VI	The payloads are 0.126 and 0.165 for P-CR-GOD and P-CR-Fe^{3+}, respectively	GOD	306
Sepharose CL-6B	Reactive blue 2	Each enzyme molecule was retained by a single immobilized dye molecule at the binding site for NADH	Rabbit muscle lactate dehydrogenase	307
Nitrocellulose membrane	Cibacron blue F3G-A		Thermostable α-amylase	308

Minimum requirements for binding to immobilized metal ions were proposed by Sulkowski in 1989. He pointed out that the binding behaviour of the immobilized metals with the amino acid residues depends not only on their degree of exposure but also on their numbers and separation distance [322].

As shown in Scheme 2.21, for enzyme immobilization, chelating ligands usually fall into the same range of the IMAC (Immobilized metal affinity chromatography). They can be classified into the following categories:

- carboxymethylated amines such as tetraethylene pentamine (TEPA) or carboxymethylated aspartic acid (CM-ASP), iminodiacetic acid (IDA), nitrilotriacetic acid (NTA);
- carboxymethylated aspartic acid (CM-Asp), tris-carboxymethyl ethylene diamine (TED);
- not chemically related to carboxymethylated amines such as dye-resistant yellow 2KT, O-phosphoserine (OPS) and 8-hydroxyquinoline (8-HQ);
- multidentate amino compounds such as EDA, ethylenediamine, DET, diethylenetriamine, or TET, triethylenetetramine [331, 332];
- others such as amino acids, peptides, salicylaldehyde, 8-hydroxylquinoline (8-HQ).

The metals ions that can be used for immobilized metal ion affinity immobilization are Cu(II), Ni(II), Co(II), Co(III),Fe(II), Fe(III), Zn(II), Ca(II), and Al(III), etc. The following combinations are often encountered:

- salicylaldehyde-Cu(II),
- 8-hydroxyquinoline-Al(III)/Fe(II)/Yb(III)/Ca(II) [318],
- iminodiacetic acid (IDA)-Cu(II)/Zn(II)/Ni(II)/Co(II),
- dipicolylamine (DPA)-Zn(II)/Ni(II) [309],
- orthophosphoric (OPS) Fe(III)/Al(III) [313],
- N-(2-pyridylmethylaminoacetate)-Cu(II) [318],
- 2,6-diaminomethylpyridine-Cu(II) [319],
- nitrilotriacetic acid (NTA)-Ni(II),
- carboxymethylated aspartic acid (CM-Asp)-Ca(II)/Co(II),
- N,N,N'-tris(carboxymethylethylenediamine) (TED)-Cu(II)/Zn(II),
- EDDA (ethylenediamine-N,N'-diacetic acid)/Fe(II)/Cu(II)/Ni(II)/Zn(II) [309],
- cystine-Fe(III),
- tris(2-aminoethyl)amine (TREN)-Cu(II) [314],
- imidazole-Cu(II) [315],
- 1,4,7-triazacyclononane (TACN)-Cu(II)/Ni(II)/Zn(II) [316].

In general, enzymes immobilized on adsorbents bearing immobilized metal ions are often more active, because of the mild immobilization conditions. With porous carriers, however, the activity might be reduced by diffusion constraints [307].

The backbone of the adsorbent used for affinity enzyme immobilization generally follows the same principles as for other types of enzyme immobilization. They must be physically and chemically stable during processing and have to fulfil special requirements proposed by Ueda et al. [317] – ease of derivatization, low non-

120 2 Adsorption-based Immobilization

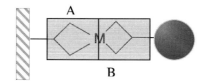

A =

I, II, III, IV, V, VI, VII, VIII, IX, X, XI, XII

IDA — XIII
NIA — XIV
XV
CM-Asp — XVI
TED — XVII
XVIII
XIX
XX

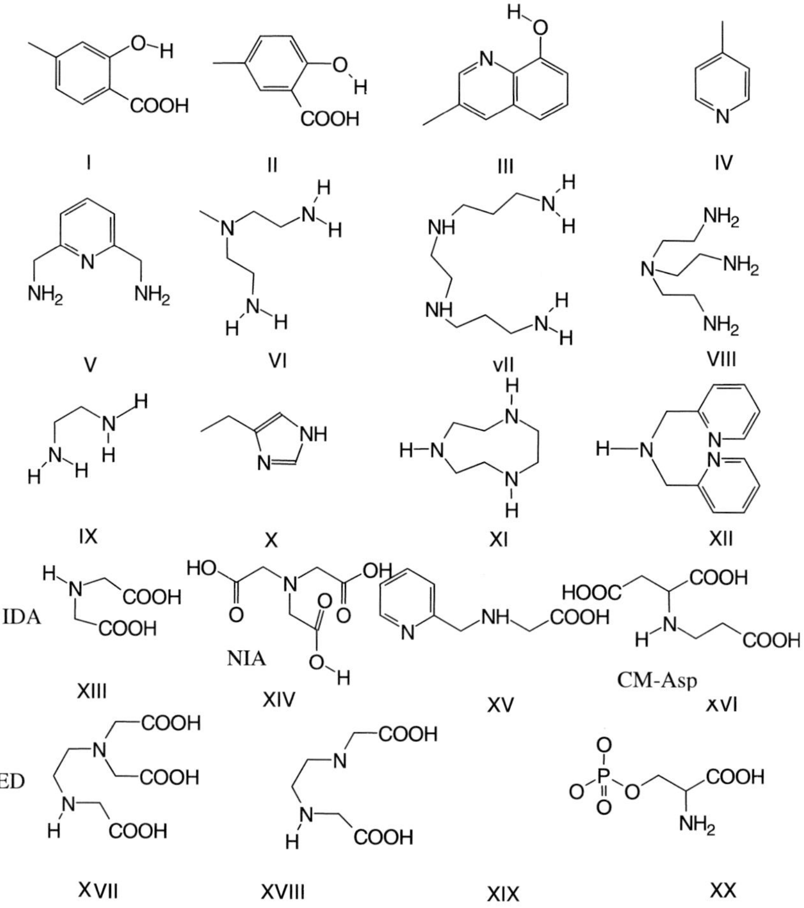

B = Histidine, Cysteine, Aspartic acid, Glutamic acid, Lysine, Argine Tryptophan, Tyrosine, Phenylalaine, N-Terminus

M = Cu(II), Ni(II), Co(II), Co(III), Fe(II), Fe(III), Zn(II), Ca(II),

Scheme 2.21 Immobilization of enzymes by coordination bonding.

specific adsorption; and the possibility of regeneration of carriers without degradation of the matrix.

The effect of the support on catalase separation using IMAC was recently beautifully demonstrated by Hidalgo et al., who found that two similar epoxy carriers, Eupergit C 250 and Sepabeads FP-EP3, derivatized with IDA ligand and coupled with Cu^{2+} differ greatly in their binding capacity toward catalase from *Thermus thermophilus*. The former has almost no binding capacity for catalase whereas the latter can bind 80% of the catalase in the original extract [320].

The effect of ligands on binding efficiency has been demonstrated by Meszarosova et al., who introduced EDT (XVII), quinoline-8-ol (III), and *N*-(2-pyridine methyl)glycine) (XV) to a hydrophilic methacrylate polymer and incorporated metal ions for separation of horseradish-peroxidase-specific IgG1. It was found that the immobilized metal adsorbent prepared with quinolin-8-ol (III) gave the best results (selectivity 10 times higher than with the other two), probably because of different steric arrangements of the chelating ligands [320]. The structure of the ligands, for example the length of the spacer, might also have had a large effect on the elution behaviour of the proteins, as demonstrated by Anspach and Birger [323, 327].

Compared with other types of enzyme immobilization, immobilization based on metal affinity can combine both enzyme immobilization and purification [338]. Enzyme loading can also be higher than is achieved with other enzyme-immobilization methods [324]. For one or more of the following reasons, however, this method is hardly the first choice for enzyme immobilization:

- enzyme leakage and low recyclability of the immobilized enzyme [333, 338],
- carrier recyclability [307],
- complicated chemistry involved in the preparation of the immobilized ligands,
- higher cost of ligand development,
- genetic or chemical modification of enzyme is sometimes necessary [330–334].

However, as for other types of adsorptive enzyme immobilization method, some of these intrinsic drawbacks might be overcome by combining different immobilization techniques. For example, combining adsorption with cross-linking can be used to reduce enzyme leakage in non-covalent enzyme immobilization [338].

It was recently found that immobilized metal affinity adsorbents can be prepared simply by coating silica with chitosan [339]. Not only were non-specific interactions reduced but the binding capacity was increased. More interestingly, the HAP (hydroxyapatite) can be used as an immobilized metal affinity adsorbent simply by loading it with the desired metals [340].

2.6.2
Unconventional Adsorption

Unconventional adsorption is adsorption of the enzyme by any method other than simple mixing with the carrier, as in the methods described above. The adsorbents obtained are usually found to be complicated.

Table 2.7 Metal affinity adsorption

Polymer	Ligand	Comments	Enzyme	Ref.
Sepharose 4B, silica	IDA/Cu(II)	Regeneration of carrier is possible. Introduction of suitable spacer between the ligand and the carrier is essential for the adsorption	PGA	327
Sepharose	Activated by Cu^{2+} ions	Activity retention is around 80%. The carrier can be repeatedly used without significant reduction of the capacity of adsorption	Papain	328
A chelating gel	Fe^{3+} ions	The capacities were approximately 5 mg protein mL^{-1} adsorbent	Phosphorylase and LDH	329
Agarose resin	Ni^{2+}	The IME can withstand repeated buffer changes without substantial activity loss	His6-tags-RNA	330
Polystyrene supports	EDA/DET/TET chelated with Co(II), Ni(II), Cu(II), Fe(III), or Co(III)	IME has slightly low K_m and V_{max} reduced by a factor of 2. High thermal stability was obtained by oxidation of Co(II) to Co(III) of IME; high stability was also found for Fe(III) bearing IME	Penicillin amidohydrolase	331
Polystyrene or silocrome supports	Co(II)-EDA, Cu(II)-EDA	Co(II)-EDA and Cu(II)-EDA-polystyrene and all metal complexes of DEA-polystyrene had the highest activity. Enzyme immobilized on Co(II) and Cu(II) complexes of EDA-polystyrene has the highest activity retention	Penicillin amidohydrolase	332
Poly(vinyl alcohol)	Glutamic acid resin/Cu(II)	The optimum pH for IME was 5.59 and 6.64, and K_m was 1.70×10^{-2} mol L^{-1}, which are similar to those for free PPO. The IME can be recycled five times with 20% retention of activity	Polyphenol oxidase	333
Porous silica matrixes	Alkylamine derivatives of Ti(IV)	The K_m of IME was higher, because of diffusion resistance and the pH optimum was slightly lower than for the free enzyme. The optimum temperature was reduced to 60 °C after immobilization	Glucoamylase	334
An agarose gel	α-Amino acid residue-Cu(II)	The method is useful not only for immobilization but also for cross-linking of proteins. It is characterized as being site-specific	Salicylaldehyde modified chymotrypsin	335
Sepharose 4B	Epibromohydrin-IDA/Cu^{2+}, Zn^{2+}	Sepharose 4B-Epi-IDA-Cu^{2+} was the best carrier with regard to activity retention. The technique is advantageous in that costs and disposal problems can be reduced because of carrier re-use	Laccase	337
Silica	IDA/Cu(II)	Enzyme leakage can be avoided by post-immobilization cross-linking with glutaraldehyde	Poly(his) tagged hydantoinase	338

EDA, ethylenediamine, DET, diethylenetriamine, or TET, triethylenetetramine, LDH: lactate dehydrogenase

2.6 Preparation of Immobilized Enzymes by Adsorption

2.6.2.1 Immobilization via Reversible Denaturation

Often it is found that the native enzyme cannot be adsorbed on the carrier because of mismatching hydrophobicity [381]. If so, modification of the enzyme molecules with the objective of the improving the hydrophobicity has proved to be an efficient method [381].

Along with this method, it was also found that some hydrophilic enzymes can be also immobilized on hydrophobic carriers by so-called reversible denaturation, which covers two different methods, namely heat-induced denaturation and pH-induced denaturation. The former involves the use of heat to partially unfold the protein, leading to exposure of the hydrophobic regions. The protein can then be adsorbed on the hydrophobic carrier without the need for chemical methods; an example is the immobilization of carbonic anhydrase on palmityl-substituted Sepharose 4B [328]. Similarly, limited heating of lactose-phlorizin hydrolase, as a representative membrane enzyme, was found also to enhance binding, owing to better availability of the hydrophobic sites [341].

In the second method reducing the pH leads to the exposure of the hydrophobic cluster to the bulk medium. As a result, the reversibly denatured protein can be immobilized on to the hydrophobic carrier; an example is the immobilization of urease on a hydrophobic carrier.

2.6.2.2 Pseudo-covalent Immobilization

It is often found that attempts to immobilize a hydrophilic enzyme on a hydrophobic carrier are unsuccessful. To solve this problem an amphiphile, which contains both hydrophilic and hydrophobic moieties, can be adsorbed on the hydrophobic surface by hydrophobic interaction. The remaining hydrophilic moieties can be used to bind the enzyme, depending on the nature of the hydrophilic moieties. The principle is illustrated in Scheme 2.22. With this method, the activity of alkaline phosphatase (AP) immobilized on a hydrophobic surface coated with polyvinylbenzyl lactonoylamide (PVLA), which was oxidized after coating, was increased more than sixfold compared with immobilization by simple adsorption [290].

Similarly, enzymes can also be immobilized on hydrophobic carriers by coating the carrier with a compound bearing a hydrophobic tail but active functionality, followed by covalent binding of the enzyme, as demonstrated by Wu and Means, who

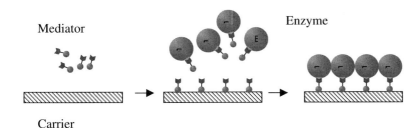

Scheme 2.22 Enzyme immobilized on parent polymeric beads with a mediator.

immobilized trypsin by reaction with NaBH$_3$CN and dodecylaldehyde-coated octyl-Sepharose [388]. The enzyme can be quantitatively immobilized. The same procedure can also be used to immobilize *Escherichia coli* asparaginase and yeast alcohol dehydrogenase in high yield. At the beginning of the 1980s a similar approach was used to adsorb modified trypsin, i.e. linoleoyl trypsin, on the surface of a liposome [334].

Another technique, which has been widely used in recent decades, is based on coating of a carrier with PEI (polyethyleneimine) then activation of the remaining amino groups with glutaraldehyde [346].

It is worth pointing out that the presence of the extra polymer as mediator could alter the tortuosity of the matrix, as in the immobilization of invertase on porous alumina using adsorbed PEI which was subsequently activated with glutaraldehyde. Consequently, it is conceivable that the amount of PEI and the molecular weight should be optimized in respect of enzyme activity and diffusion limitation [347].

By applying this method, α-chymotrypsin and pancreatin have been immobilized with high stability toward the high-molecular-weight substrate casein [348]. Similarly, macroporous beads of synthetic polymers such as methacrylate–di(methacryloyloxymethyl)naphthalene copolymer coated with keratin or polyamide-6 were used for immobilization of glucoamylase and peroxidase with good yields (7.62–10.5 mg protein g^{-1} carrier) [349]. Penicillin G acylase has been covalently immobilized on aluminium oxide coated with PEI with high (63%) retention of activity [351].

Remarkably, several enzymes have been immobilized on PMG Polymethylglutamate-coated glass with high retention of activity (>90%) and high payload [353].

Small molecules can be used for the same purpose as the polymers. For example, lactase was adsorbed on granular carbon with improved binding and enzyme activity when *p*-aminophenol and 1-phenol-2-amino-4-sulphonic acid were adsorbed by the carbon followed by activation with glutaraldehyde [356]. The use of small molecules for non-covalent coating was pioneered by Grey and Livingstone in 1974, when they coated cellulose with *m*-diaminobenzene, followed by polymerization and activation. The enzyme can be covalently bound to the carrier via the non-covalently bound active polymer [358, 359].

2.6.2.3 Mediated Adsorption

Occasionally adsorption of an enzyme by a carrier fails because of incompatible interactions between them, especially when a hydrophobic enzyme is being adsorbed on a hydrophilic carrier. In such circumstances a mediator with amphiphilic properties can be used to enhance the interaction. Use of this methodology can be dated back to the mid-1970s, when Katchalski et al. immobilized amyloglucosidase with 80–100% retention of activity on an alumina carrier treated with procion brown MX5BR (a monoazo dichlorotriazine dye) [372].

By use of similar methodology, adsorption of catalase and horseradish peroxidase (HRP) on silica beads with 80-nm pores can be enhanced by three orders of

2.6 Preparation of Immobilized Enzymes by Adsorption

Table 2.8 Non-conventional adsorption by pseudo-covalent immobilization

Carrier	Method	Remark	Enzyme	Ref.
Hydrophobic surface	Carrier surface was coated with PVLA	This method is very similar to the modification-adsorption strategy. However, the enzyme can be regarded as covalently immobilized	Alkaline phosphatase (AP)	345
Octyl-Sepharose	Dodecylaldehyde was coated to the carrier	Immobilization of trypsin by reaction with $NaBH_3CN$ and dodecyl-aldehyde-coated octyl-Sepharose led to quantitative immobilization	Trypsin	338
Acrylate-based polymer beads-PEI-GAH	PEI coating was activated with glutaraldehyde	10% recovered as active enzyme; the presence of glycine after the coupling is essential; activity is reduced by long coupling	acetate kinase	346
Alumina	PEI coating	The tortuosity was increased from 2.0 for plain alumina to 2.75 for invertase immobilized on alumina, because of the presence of the polymer and the enzyme	Invertase	347
Silica	PEI/PAA coating	The IME has good activity retention, rigidity and durability	Trypsin	348
Macroporous MA-di (MAOM)NAP copolymer beads	Keratin or polyamide coating	A loading of 7.62–10.5 mg protein g^{-1} carrier was obtained	GA and Pox	349
Macroporous polymer beads	PEI coating	43% retention of activity; enhanced thermostability, less change in pH optimum; increased temperature optimum; reduced K_m	PGA	350
Aluminium oxide	PEI coating	Increase in activity retention 63% and decrease in adsorption of the enzyme 40%	PGA	351
Glass beads coated with poly(MAA-co-vinyl Me ester)	Covalent	Shell-structured immobilized enzymes supported by a fluid-impervious spherical core	Trypsin	352
PMG coated on glass beads		>90% retention of activity; payload 0.02–0.3	GOD, POD, TP CT, URA, AA	353
Silica gel	PEI coating	Lipase covalently bound to PEI-coated silica had higher activity and stability than the lipase covalently bound to the silanized carrier	Bacillus strarothermophillus lipase	354

Table 2.8 Continued

Carrier	Method	Remark	Enzyme	Ref.
Porous p-trimethylamine-polystyrene (TMPS) beads	Molecular deposition	Hexamethylenebis(trimethylammonium)iodide as mediator	Glucose isomerase	355
PCMS beads	Coated with PEI, activated by glutaraldehyde	Payload: 0.019 with 80% activity; less diffusion limitation; five times higher stability than free enzyme	Invertase	356
Granular C	–	Granular C was coated with p-aminophenol and 1-phenol-2-amino-4-sulphonic acid followed by activation with GAH	Laccase	357
Cellulose	Diazotized m-diaminobenzene	Diazotized m-diaminobenzene was polymerized in situ on cellulose	Enzymes	358, 359
Glass beads	Coated with PMG and APG	High loading, high retention of activity, markedly improved heat stability	Urease	360
Porous glass	PEI coated	Enhanced stabilization against denaturants such as urea and heating	Pronase	361
Metal powders	Coated with copolymer: ABA-FAL	Enzymes can be easily separated by applying a magnetic field	β-Glucosidase	362, 363
CP silicas	Coated copolymers of HAMA and ADMA activated by ECH	–	α-Chymotrypsin	364
Silochrome or silica gel	Coated polymers such as PhOH-HCHO or epoxy resins	Increased hydrophilicity of the carrier surface; no significant change in the porosity; enhanced enzyme binding stability	α-Chymotrypsin and pancreatin	365

PEF-g-DMAEMA-HEM: Polyester fibre-g-dimethylaminoethyl methacrylate-hydroxyethyl methacrylate copolymer; MBSG: Micro bead silica gel/(MB-800)/poly[(N-acetylimino)ethylene] and poly[(N-nonanoylimino)ethylene ABA-FAL: p-amino benzoic acid-formaldehyde

2.6 Preparation of Immobilized Enzymes by Adsorption

Table 2.9 Non-conventional adsorption by mediators

Carrier	Method	Remark	Enzyme	Ref.
Silica gel	Amphiphilic block copolymer consisting of PPNI/PAIE	The amount of catalase by adsorbed by silica-gel in aqueous solution at pH 7.0 was significantly increased	Catalase	370
Silica gel	Amphiphilic block copolymer consisting of PPNI/PAIE	The presence of polymers such as poly[(N-pentanoylimino)ethylene] and poly[(N-acetylimino)ethylene] enhanced the enzyme loading	Catalase	371
Alumina carrier	Procion brown MX5BR-azo dichlorotriazine dye	80–100% retention of activity was achieved	Amyloglucosidase	372
BSA-coated alumina	BSA	The presence of BSA enhanced enzyme activity compared with that without BSA	Alcohol dehydrogenase or aminoacylase	373
Carrier bearing immobilized carbonic anhydrase	Aryl sulphonamide	–	Hexokinase, lysozyme, and glucose 6-phosphate dehydrogenase	374
Octadecyl-Si40, octyl-Sepharose and butyl-fractogel	Ca^{2+}	The effective concentration of $CaCl_2$ required is inversely proportional to the hydrophobicity of the carrier: less $CaCl_2$ is required with highly hydrophobic carrier	Phospholipase	375
Silica gel	Surfactant	Adsorption of the enzyme was found to be affected by the pH and ionic strength of the buffer	Chitinase	376
A hydrophobic carrier	Pluronic-F108 bearing nitrilotriacetic acid (NTA) chelating group	More than 95% retention of activity was obtained	Histidine-tagged firefly luciferase (FFL)	377
Montmorillonite	Coated with hydroxylamine	Enzyme was immobilized the clay surfaces	Tyrosinase	378
Glass beads involving 5-diazosalicylic acid bonded to a titanium(IV) oxide film	Adsorption	Increased K_m for starch, reduced K_m for maltose; pH optimum shifted to acid pH	Glucoamylase	379

PPNI: poly[(N-pentanoylimino)ethylene]; PAIE: poly[(N-acetylimino)ethylene]

magnitude and tenfold, respectively, in the presence of [(N-acetylimino)ethylene] and poly[(N-nonanoylimino)ethylene] copolymer, respectively [370].

Alcohol dehydrogenase or aminoacylase adsorbed on BSA-coated alumina had enhanced enzyme activity [373]. By use of aryl sulphonamide several enzymes, for example hexokinase, lysozyme, and glucose 6-phosphate dehydrogenase, can be strongly adsorbed on a carrier bearing immobilized carbonic anhydrase [374].

Occasionally the presence of mediators alters the conformation of an enzyme, increasing enzyme adsorption by some carriers. For instance, the presence of calcium was found to be essential for adsorption of phospholipase D on hydrophobic carriers such as octadecyl-Si40, octyl-Sepharose and butyl-Fractogel. Interestingly, it was found that the effective concentration of $CaCl_2$ is inversely proportional to the hydrophobicity of the carrier – i.e. less $CaCl_2$ is required with highly hydrophobic carriers [269, 375].

Surfactants can be also adsorbed on carriers as mediators and the enzyme can be adsorbed on the surface of the immobilized surfactant monolayer. Adsorption of the product depends on the nature and orientation of the surfactant on the carrier. With non-ionic surfactants and silica gel adsorption of the enzyme was found to be affected by pH and the ionic strength of the buffer [376].

Interestingly, modified surfactants such as Pluronic F108 bearing nitrilotriacetic acid (NTA) chelating groups can be used as a mediators which, when adsorbed on hydrophobic carriers, can be used for specific adsorption of histidine-tagged firefly luciferase (FFL) with more than 95 % activity [377].

2.6.3
Adsorption-based Double Immobilization

Adsorption-based double immobilization denotes methods which combine adsorption-based enzyme immobilization with other related enzyme-immobilization techniques. The aim is to immobilized enzymes which cannot be immobilized by normal adsorption methods.

2.6.3.1 Modification–Adsorption

Very often it is not possible to adsorb some enzymes or proteins on certain types of carrier [25] or the adsorption capacity of the carrier is very low [296]. To surmount this problem, two strategies have been developed:

- modification of the protein surface with the objective of enhancing adsorption on non-specific carriers [381];
- modification of the carrier surface with the aim of introducing new specific functionality [296].

Because of the introduction of new chemical functionality, interaction of the enzyme with the carrier can be enhanced. The interaction force between the enzyme and the carrier might be electrostatic, hydrophilic or hydrophobic, as illustrated in Scheme 2.23.

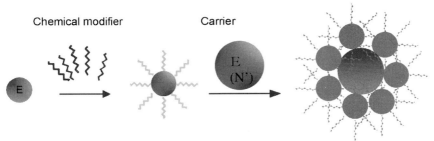

Scheme 2.23 Modification–adsorption methods.

The concept that modification of the enzyme surface with the desired functional groups can enhance interaction of the protein with the corresponding surface was probably first patented at the beginning of the 1970s, following the first industrial application of ionically adsorbed enzymes [382]. The method is generally applicable to two different types of carrier, ion-exchange resins such as DEAE-cellulose and hydrophobic carriers such as EP 100 [383].

It has been found that the introduction of extra charges such as succination of enzymes can enhance the adsorption of enzymes on some ion-exchange resins [1] and the modification of the enzyme with some compounds also led to an increase in the efficiency of enzyme binding [380].

Use of aromatic compounds such as methyl benzimidate hydrochloride to make the enzyme surface hydrophobic enhanced adsorption on hydrophobic carriers such as Amberlite XAD 7 [381]. Modification of enzymes such as trypsin, yeast alcohol dehydrogenase, and *Escherichia coli* asparaginase with hydrophobic imidoesters led to almost quantitative adsorption of protein on XAD-7 polymer beads [381].

Modification of the enzyme surface has another unexpected effect, namely stabilization of the enzyme [195] or mitigation of the deleterious effect of the carriers [36].

As with immobilization by different types of adsorption, enzyme activity can be determined by many factors such as the nature of the carriers, properties of the enzymes, and the immobilization conditions as summarized above. The activity of enzymes immobilized in this manner also depend on the following factors:

- extent of modification,
- the properties of the enzyme,
- the properties of the modifiers,
- the nature of the carriers.

It has been demonstrated that increasing the extent of modification of α-chymotrypsin with stearic acid *N*-hydroxysuccinimide ester led to an increase of adsorption of the enzyme on or in liposomes, suggesting that the enzyme might undergo conformational changes, when the extent of modification is increased [389]. Similarly, introduction of excess negative charges by succination might also lead to loss

Table 2.10 Modification-adsorption immobilization

Polymer	Method	Comments	Enzyme	Ref.
Amberlite XAD-7/hydrophobic imidoester	Adsorption	Hydrophobic imidoester-modified enzymes can be adsorbed on the carrier with high retention of activity	Trypsin, alcohol dehydrogenase, asparaginase	381
Polymer beads/PEG (1900) activated p-NPCF*	Adsorption	Enzymes are PEGylated with monomethoxy PEG (1900) activated with p-nitrophenyl chloroformate	Lipase from Candida rugosa	383
A vinylpyridine copolymer/succinic anhydride	Adsorption	25% retention of activity was obtained; the optimum temperature for enzyme action was shifted from 50 to 55°C	Cyclodextrin glycosyltransferase	385
Amberlite IRA 900/succinic anhydride	Adsorption	Optimum temperature of CGTase immobilized on Amberlite IRA 900 was shifted from 55 to 50°C	Cyclodextrin glucanotransferase	386
DEAE-cellulose/succinic anhydride	Adsorption	Although adsorption of native lysozyme is driven by electrostatic attraction to the surface, adsorption of succinylated lysozyme is controlled by hydrophobic interactions	Hen egg white lysozyme	387
Dodecylaldehyde-coated octyl-Sepharose/reductive alkylated trypsin	Adsorption	Modification and adsorption was performed in one-pot process; NaBH$_3$CN reduced the Schiff base resulting in formation of modified enzyme adsorbed on the hydrophobic carrier, the binding was very strong	Trypsin	388
Liposome/hydrophobilized with stearic acid N-hydroxysuccinimide ester	Adsorption	Modification of the enzyme enhanced interaction of the enzyme with the liposome	α-chymotrypsin	389
Concanavalin A-Sepharose	Adsorption	The thermal stability of the immobilized I preparations was increased compared with the solid enzyme if it was treated with glutaraldehyde before adsorption on concanavalin	β-Glucosidase	390
Alumina through their phosphate function	Adsorption	Converted them into semi-synthetic phosphoproteins by pyridoxal 5′-phosphate; high retention of activity	Several proteases	391
	Adsorption	After the modification, HPR can be easily adsorbed on the polystyrene latexes with retention of activity	Oxidized horseradish peroxidase (HRP)	392
Biotinylated aminopropyl glass	Adsorption	Interaction between enzyme and ligand led to a decrease of enzyme activity. However, the activity is substrate-size dependent, suggesting that diffusion constraints had a strong influence on enzyme activity	Biotinylated transglutaminase	393
Divalent metal ion-iminodiacetic acid-agarose	Adsorption	The artificial tag was introduced by reacting the oxidized enzyme with histidine	Histidine modified glycolated horseradish peroxidase	394

*p-nitrophenyl chloroformate

of the activity, because of conformational changes [385]. Thus, in practice it is necessary to control the amount of modification.

Modification of lipase with PEG often enhances the activity of the enzyme compared with the unmodified enzyme immobilized on the same carrier [195]. Modification of the enzyme with a suitable modifier can mitigate incompatible interactions between carriers and enzyme, as a result of shielding effects [36], leading to greater retention of activity relative to the native enzyme. The stability of enzymes immobilized in this way is usually determined by the modifiers [195].

2.6.3.2 Adsorption and Entrapment

Although physical adsorption is the one of the easiest means of immobilization of enzymes, an intrinsic drawback is leakage of the immobilized enzyme from the carrier surface under some reaction conditions. To prevent enzyme leakage, the adsorbed enzyme preparation can be further entrapped in a gel matrix, as illustrated in Scheme 2.24.

By use of this method penicillin V acylase from actinomycetes was immobilized by a two-step method, in which the enzyme was first adsorbed on kieselguhr and then entrapped in the polyacrylamide gel [399].

Similarly, glucoamylase [404] from *Aspergillus niger* was first adsorbed on gelatinized corn starch and then entrapped in alginate fibre, leading to the formation of an immobilized enzyme which was stable and active for 21 days without any decrease in the activity under the reaction conditions. Similarly, glucose oxidase adsorbed on manganese dioxide particles has been entrapped in calcium alginate gel beads [405].

Partially purified acylase from *Xanthomonas compestris* has been adsorbed on bentonite then entrapped in alginate beads [408]. With this method, the reusability of the immobilized enzyme for the synthesis of cephalexin was significantly increased.

Similarly, the enzymes horseradish peroxidase (HRP) modified with histidine and choline oxidase (Chox) have been adsorbed on IDA- Sepharose and DEAE-

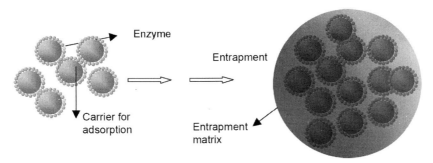

Scheme 2.24 Adsorption and entrapment.

Table 2.11 Physical adsorption and entrapment (PA/ENT)

Carrier for adsorption/ matrix for entrapment	Methods	Comments	Enzyme	Ref.
Silica/PGAAm	Adsorption/ entrapment	The first example of double immobilization (adsorption–entrapment)	Whole cells	396
Drierite/poly(2-HEMA)*	PA/ENT	Considerable enhancement of enzyme activity and durability in batch enzyme reactions	α-Amylase	397
Celite/(ENTP-4000)	PA/ENT	Addition of Celite increased the activity sixfold compared with that without Celite	A very pure lipase from *R. arrhizus*	398
Kieselguhr/PAAm Silanized	PA/ENT	Shift of pH optimum 1 unit upwards	Penicillin V acylase	399
γ-Alumina/alginate	PA/ENT	Entrapment of adsorbed glucoamylase on γ-alumina into Al alginate	Glucoamylase	400
γ-Alumina/alginate	PA/ENT	Entrapment of adsorbed glucoamylase on γ-alumina into Al alginate	Glucoamylase	401
60–120 mesh γ-alumina/ Ca arginate	PA/ENT	The shape of the immobilized enzyme is important with regard to diffusion limitation and enzyme activity	Glucose oxidase	402
China clay/sodium alginate beads	PA/ENT	Selectivity of the adsorbed/entrapped enzyme was changed	1,2-mannosidase	403
Cellulose DE-52/sodium alginate beads	PA/ENT	Selectivity of the adsorbed/entrapped enzyme was changed	1,2-mannosidase	403
Gelatinized corn starch/ alginate fibre	PA/ENT	The optimum conditions were not affected by immobilization and the optimum pH and temperature for free and immobilized enzyme were 4.3 and 60 °C, respectively	Glucoamylase	404
Manganese dioxide particles/ calcium alginate gel beads	PA/ENT	Reduction of product inhibition	Glucose oxidase	405
Alumina beads/poly-HEMA (or poly-HPMA)	Adsorption/ entrapment	The hydrolysis yield varied with the irradiation dose used for pretreatment, the properties of polymer matrix, and the cross-linking conditions of the enzyme	Cross-linked cellulase	406

Table 2.11 Continued

Carrier for adsorption/ matrix for entrapment	Methods	Comments	Enzyme	Ref.
Copolymers of MMA TMP/MA/AN/polyamines with dialdehydes	PA/ENT	The IME was stable for continuous use >6 times, whereas the immobilized I preparation without the top polymer layer lost most of its activity after 3 uses	α-amylase	407
Bentonite/alginate beads	PA/ENT	Immobilization of this enzyme, by initial adsorption on bentonite followed by entrapment in alginate beads, helped to stabilize the enzyme and rendered the biocatalyst reusable	Penicillin acylase	408
IDA-Sepharose/PVA-SbQ photopolymer	MAC/ entrapment	Enzyme loading is high and the stability of the enzyme was enhanced	Histidine modified HRP	409
DEAE-Sepharose/PVA-SbQ photopolymer	Ionic adsorption/ entrapment	Enzyme loading is high and the stability of the enzyme was enhanced	Choline oxidase	409
FSM-16/hybrid silica gel	Adsorption/ entrapment	FSM of big pore size promoted activity enhancement by thermal treatment	HRP	410
Concanavalin A-Sepharose/ Ca alginate gel spheres	Affinity adsorption/ entrapment	Only 35% retention of activity was obtained, when assayed with 10 mm cellobiose. However, twice the activity can be obtained with 100 mm cellobiose	d-Glucosidase	411
Macroporous silica-adsorbed enzyme entrapped in pectin gel	Adsorption– entrapment	Enables more agitation without causing shearing	Glucoamylase	412

* Poly(2-hydroxyethyl methacrylate)

Sepharose by ion exchange or affinity adsorption then entrapped in PVA-SbQ photopolymer for the preparation of a biosensor [416–421].

Immobilized protease prepared by adsorption on Celite then entrapment in hybrid silica gel via a sol–gel process is eight times more active in organic solvent in the transesterification of vinyl n-butyrate with 3-methyl-1-butanol (isoamyl alcohol) than the enzyme on Celite [395].

Because adsorption and entrapment are hybrid methods of enzyme immobilization between the adsorption and entrapment, the factors that govern the activity of the adsorbed enzyme and entrapped enzyme are valid here also.

Thus, the activity of adsorption–entrapment-based immobilized enzymes depends on:

- the interaction of the enzyme with the first carrier used for adsorption [398],
- diffusion constraints [401, 411],
- substrate concentration [411].

One characteristic of this technique is that the immobilized enzyme is usually more stable than the enzyme without entrapment. The stabilization usually stems from effects such as:

- confinement effect [397],
- favourable environment [408, 409],
- confinement and favourable environment [407].

Remarkably, the selectivity of the immobilized enzymes can be changed by the carriers used for initial adsorption. One example is 1,2-mannosidase from *A. phoenicis*, which is first immobilized on china clay or cellulose DE-52 and then entrapped in sodium alginate beads. It was found that when the enzymes were immobilized on china clay and DE-52 the predominant product was mannose-(1-6)-man with lesser amounts of mannose-(1-2) and mannose-(1-3). When the immobilized enzyme was entrapped in alginate beads man-(1-2) was the predominant linkage [403]. This suggests that the selectivity is determined by the interaction of the enzyme with the china clay or DE-cellulose.

However, the selectivity of an enzyme immobilized by adsorption–entrapment can be also determined by the entrapment matrix in the presence of a conformer selector, because of the confinement effect of the sol–gel matrix [395].

2.6.3.3 Adsorption–Cross-linking

An interesting method of enzyme immobilization that was intensively investigated in the 1960s and 1970s is the combination of adsorption and cross-linking [418], as illustrated in Scheme 2.25.

The carriers used for this purpose can be ion exchangers, hydrophobic adsorbents and biospecific adsorbents such as Con A-bound Sepharose. Affinity adsorption of the enzyme on the carrier surface, then cross-linking with glutaraldehyde, often furnishes an enzyme with very good retention of activity and thermostability [440] and greater operational stability because of mitigation of enzyme leakage [150].

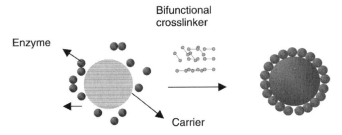

Scheme 2.25 Cross-linked enzyme on the solid surface. Enzyme molecules are first adsorbed on the carrier surface to form an enzyme layer, followed by cross-linking with a bifunctional cross-linker.

Using this concept, penicillin G acylase has been immobilized on different carriers such as fine granular gelatin by adsorption and cross-linking with glutaraldehyde [148, 149] on chitosan-based support followed by cross-linking with glutaraldehyde [150, 414, 415, 424], on derivatized cross-linked agarose, phenoxyacetylcellulose, oxirane-polyacrylamide resin, polyacrylamide foam, cross-linked polymethyl methacrylate/glycol dimethacrylate copolymer, cross-linked polystyrene, and porous glass [451].

That the stability of immobilized penicillin G acylase at 60°C was more than ten times that of the free enzyme suggested that the crosslinking was able to stabilize the enzyme [149]. Interestingly, for penicillin G acylase, there is a clear relationship between the aquaphilicity of the carrier and activity retention. Usually, high retention of activity (50–91%) was obtained with the hydrophilic polymers. In contrast, lower retention of activity was observed with more hydrophobic carriers [451].

Advantages of this method might be that a number of non-specific carriers could be used for the purposes of enzyme immobilization. Additionally, the cross-linker can be added before, during, or after the adsorption. With penicillin G acylase up to 170 U g^{-1} could be achieved with phenyl Sepharose and oxirane-polyacrylamide. In contrast, lower retention of activity (only 20–30% of the original activity) was obtained by using the same method for immobilization of penicillin G acylase on fine (180–320 μm) granular gelatin followed by cross-linking of the penicillin G acylase with glutaraldehyde [148]. This immobilized penicillin G acylase has high operational stability in the hydrolysis of penicillin G (10%) in a column reactor, however. These results indicated unambiguously that fair comparison of each method is quite difficult if the source of the enzyme and immobilization conditions are not the same.

Enzyme molecules can be cross-linked in the presence of insoluble carriers. On the basis of this method, penicillin G acylase has been immobilized on several carriers, for example as DEAE-cellulose, cellulose, CM-cellulose, ion exchangers, acryl amide gel, polyamides, silica gel, and Al_2O_3 [451]. This kind of immobilized enzyme might be a mixture of several components, for example cross-linked enzyme (CLE), cross-linked enzyme on the carrier, cross-linked enzyme–carrier conjugates. It cannot, therefore, be regarded as an homogeneous preparation.

Chitin, chitosan and the derivatized chitosan diethylaminoethyl chitosan (DE-chitosan) were tested for immobilization of l-rhamnopyranosidase with different bifunctional agents (glutaraldehyde, diepoxyoctane, suberimidate and carbodiimide) [421]. The presence of rhamnose and N-hydroxysuccinimide (NHS) during cross-linking enhanced the action of carbodiimide and so increased the immobilization yield and activity.

The enzyme can first be adsorbed on an anion-exchange group-containing porous membrane in hollow-fibre form, followed by cross-linking [413]. This method was recently used to immobilize ascorbic acid oxidase (AsOM) on a hollow fibrous membrane. The membrane was first charged with enzyme solution, then the enzyme was cross-linked in the pores, resulting in the formation of immobilized enzyme with high payload (0.13, i.e. 130 mg g^{-1} fibre).

Similarly, multilayers of enzyme can be formed on anion-exchange porous hollow-fibre membranes prepared by graft polymerization of an epoxy group-containing monomer and then reaction with diethylamine. The enzyme is first adsorbed on to the membrane and gelified by transglutaminase. In this way, a multilayer enzyme is formed with payload of 0.038 to 0.11 [251].

Another interesting strategy for adsorption–cross-linking is adsorption of enzymes on polyelectrolytes, for example on PEI-coated carriers. It has been found that immobilization of several enzymes such as glucose oxidase and catalase resulted in higher specific activity (U g^{-1} carrier) than the enzyme covalently immobilized on the aminopropyl-activated glass microbeads; this could be ascribed to the "tentacle" effect [433]. The polyelectrolyte is adsorbed on a carrier surface which bears pendant tentacle-like polymeric chains. The enzyme is adsorbed on the tentacles as a multilayer, whereas covalently bound immobilized enzymes can form monolayers only (Scheme 2.26).

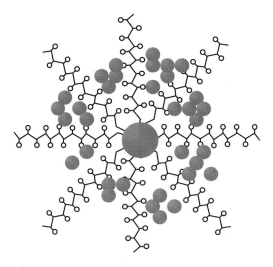

Scheme 2.26 Adsorption and cross-linking of enzymes by use of tentacle carriers.

Activity retention is determined by two steps, adsorption and cross-linking. Thus, it is expected that the following factors will dominate activity retention:

- diffusion limitation [154],
- pretreatment of the carriers,
- addition of additives (substrate or others) [421, 422],
- nature of the cross-linker [157, 421],
- the properties of the enzymes [103],
- the nature of the carriers,
- the conditions used for immobilization.

The approximate extent of activity loss as a result of cross-linking and diffusion limitation can be estimated. For instance, N-carbamyl-d-amino acid amido hydrolase immobilized on Duolite A-568 (anionic) and Chitopearl 3003 led to 60% activity loss by desorption and 28% activity loss by cross-linking, suggesting that diffusion limitation in the first step might be the prevailing reason for loss of activity [157].

For carrier-dependent enzyme activity the activity is determined mainly by the carrier and not by cross-linking. For instance, immobilization of lactase on granular carbon by adsorption and cross-linking with glutaraldehyde resulted in activity <10% that of the free enzyme, despite the higher adsorption capacity of this carrier for laccase. The low retention of activity was probably a consequence of incompatibility of enzyme and carrier [465].

As with the chemical nature of the carrier, the nature of binding of the carrier is of crucial importance to activity retention. Usually higher retention of activity is obtained by use of carriers bearing biospecific ligands such as con A as binding sites [440]. This is largely because the binding conditions are mild and the binding site is probably far from the active centre.

The microenvironment of the carriers also has a significant effect on activity retention. This is reflected by the observation that retention of activity of GOD and CAT immobilized on PEI-coated glass microbeads is higher than for the same enzymes immobilized on aminopropyl-activated glass microbeads [433].

Pretreatment of the carriers with solvents can enhance the enzyme loading and activity retention, especially when using hydrophobic carriers [434].

When diffusion limitation occurs, the occurrence of mass-transfer limitation might reduce enzyme activity [442] or increase activity because of elimination of the product inhibition [154]. On the other hand, immobilization of enzyme on a non-porous carrier might exclude the presence of the diffusion limitation. Thus the loss of the activity might be mainly because of cross-linking [441].

Use of additives can enhance activity – adsorption of cholesterol oxidase on Amberlite IRC-50 and subsequent cross-linking with GAH by addition of Triton X-100 resulted in 4.5-fold enhancement of activity [422]. Addition of substrate or analogues can prevent the activity loss which arises as a result of cross-linking [421].

Extra cross-linking also has a significant effect on retention of enzyme activity [434]. Occasionally enzymes are deactivated by cross-linking [103] and some cross-linkers can poison enzymes; selection of the cross-linker should thus avoid involvement of the amino acid residues that are essential for the enzyme function [157 421].

The resistance of enzyme against cross-linking with glutaraldehyde might differ from enzyme to enzyme. For instance, immobilization of lactase on macroporous ion-exchange carriers achieved a payload of 0.1 but activity retention was only 10%, suggesting that cross-linking led to more than 50% activity loss (assuming diffusion limitation led to 40% activity loss).

The stability of the enzyme immobilized by adsorption–cross-linking is affected by:

- the nature of the carrier,
- the nature of the binding functionality,
- the properties of the enzyme,
- cross-linking.

Among the different factors, the nature of the carriers is much more important than cross-linking processes. Although cross-linking usually enhances enzyme stability [154–157], the stability of enzymes immobilized in this way is often determined by the adsorption step. In other words, if the first step leads to the formation of unstable immobilized enzymes, because of incompatible interaction of the enzyme with the carrier, the second step of cross-linking cannot increase the stability. For instance, glucoamylase immobilized on PEI-coated non-porous glass beads was less stable than the free enzyme; this suggested that the stability of the immobilized enzyme was determined by the first step, adsorption, i.e. adsorption of glucoamylase disturbs the native charge–charge interaction, leading to a change to a less stable enzyme conformation [146].

As for the nature of the binding functionality, it is worth mentioning that use of affinity adsorbents might be advantageous in that enzyme stability and activity might be simultaneously enhanced [271], compared to other non-specific adsorption technique, as demonstrated in a recent study [271].

The function of cross-linking has been clearly confirmed by a similar immobilization process – cross-linking of covalently immobilized enzymes. For example, glutaraldehyde-aided cross-linking of covalently immobilized pectinlyase on several carriers, i.e. Eupergit C, CDI-activated hydrolysed XAD-7, and glutaraldehyde-activated Nylon 6, was found to improve the stability of the enzyme compared with the original carrier-bound enzymes [158]. Similarly, cross-linking with glutaraldehyde or 3,5-difluoronitrobenzene of porcine pancreas lipase immobilized on PAAm beads (by carbodiimide coupling) also improved the thermal stability [419].

Apart from the enhanced thermostability which results from cross-linking, immobilization of enzymes on carriers by adsorption then cross-linking often improves operational stability by mitigation of enzyme leakage. For example, immobilization of a cephalosporin acetylesterase by adsorption on bentonite then cross-linking with glutaraldehyde or bisdimethyladipimidate resulted in higher operational stability of the immobilized enzyme [150].

A change in selectivity as a result of adsorption and subsequent cross-linking has been observed for several enzymes, for example as β-glucanase and glucoamylase. It has been found that β-glucanase immobilized on phenol-formaldehyde resin (Duolite) had a different pattern of action than the crude enzyme [423]. Similarly,

2.6 Preparation of Immobilized Enzymes by Adsorption | 139

Table 2.12 Immobilization of enzyme by adsorption–cross-linking

Polymer	Method	Comments	Enzyme	Ref.
DE-chitosan	Glutaraldehyde, diepoxyoctane, suberimidate and carbodiimide	Activity retention depends on the cross-linker; addition of substrates enhanced activity retention	β-l-Rhamnopyranosidase	421
Phenol–HCHO resin	Adsorption–cross-linking	High operational stability (>500 h, with more than 90% conversion)	Lactase	422
Amberlite IRC-50	Adsorption–cross-linking	Addition of 0.15% Triton X-100 increased enzyme activity 4.5-fold	Cholesterol oxidase	423
Phenol–formaldehyde resin (Duolite)	Adsorption–cross-linking	IME had an action pattern different from that of the crude enzyme	β-Glucanase	424
Weak cation-exchange resin: poly(MA-co-EG DMA-co-AAc)	Adsorption and cross-linking	Enhanced optimum temperature and slightly reduced pH optimum; six times higher K_m	PGA	425
Macroporous anion-exchange resin	Adsorption–cross-linking	The microenvironment created around the lipase by immobilization eliminates product inhibition. The stability of lipase against chemical denaturation was enhanced	*Pseudomonas fluorescens* lipase	426
PBA-eupergit and PBA eupergit 250L	Adsorption–cross-linking	The carrier is modified carrier; no desorption of the enzyme occurred	Lipase	427
Agarose beads/NAD analogue	Cofactor binding site	The subsequent cross-linking with glutaraldehyde enhances the binding of the enzyme	LDH	428
Stainless-steel beads	Adsorption–cross-linking	Similar K_m to the native enzyme was obtained	Pancreatic lipase	429
Chitosan	Adsorption–cross-linking	The immobilized laccase had a lower specific activity and a lower substrate affinity than the free enzyme but enhanced stability to various conditions such as temperature, pH, and storage time	Laccase	430
Trimaleylchitosan	Adsorption–cross-linking	50% of the enzyme is active; enhanced stability; higher K_m and smaller V_{max}	Endo-polygalacturonase	431
Silica gel	Adsorption–cross-linking	Immobilization by adsorption and subsequent cross-linking afforded 100% yield	Laccase	432
γ-Alumina	Cross-linking with 0.1% glutaraldehyde	Immobilization did not alter enzyme performance	Endo-pectin lyase	432

140 | 2 Adsorption-based Immobilization

Table 2.12 Continued

Polymer	Method	Comments	Enzyme	Ref.
PEI-coated glass microbeads	Adsorption–cross-linking	78–87% retention of activity was obtained; higher retention of activity than covalently bound enzyme via aminopropyl-activated glass microbeads	GOD, CAT	433
Microporous polypropylene	Adsorption–cross-linking	Pretreatment of the carriers with organic solvents increased activity retention; cross-linking diminished activity but increased stability	CRL	434
Microporous polypropylene	Adsorption–cross-linking	–	α-Amylase and amyloglucosidase	435
Phenol–formaldehyde-based anion-exchange resin/glutaraldehyde	Adsorption–cross-linking	Payload: 0.103; about 10% retention of activity, compared with specific activity	Lactase	436
XAD-7	Adsorption–cross-linking	Cross-linking with GAH increased the thermostability of the enzyme	CRL	437
Duolite A-568 and Chitopearl 3003	Adsorption–cross-linking	The enhanced stability with glutaraldehyde crosslinking justified the use of this methodology	N-Carbamyl-d-amino acid amidohydrolase	438
Silica gel	Adsorption–cross-linking	Monolayer adsorption	Trypsin	439
Sepharose-con A	Adsorption–cross-linking	96% retention of activity and enhanced operational stability	Glucoamylase, peroxidase, glucose oxidase, and carboxypeptidase Y	440
Cationic non-porous glass beads	Adsorption–cross-linking	Over 80% retention of activity	Glucoamylase	441
Hollow fibres	Adsorption–cross-linking	K_m was 200 times different because of diffusion constraints	l-Asparaginase	442
XAD-7	Adsorption–cross-linking	The enzyme can be used in water-immiscible organic solvent and plug-flow-type reactor	Thermolysin	443
Modified Sephadex G-100 as ionic resins	Adsorption–cross-linking	High payload and activity retention and enhanced storage stability	β-Galactosidase	444
DEAE-cellulose	Adsorption–cross-linking	80% retention of activity; increased shelf-life, thermostability, and stability in a variety of organic solvents	d-Hydantoinase	445
Anion-exchange resin: HFM-GMA-DEA	Adsorption–cross-linking	Pores are rimmed by enzyme; payload 0.11 (0.055); cross-linked with transglutaminase	Citase	446

2.6 Preparation of Immobilized Enzymes by Adsorption | 141

Table 2.12 Continued

Polymer	Method	Comments	Enzyme	Ref.
HFM-GMA-DEA	Adsorption–cross-linking	Subsequent enzyme cross-linking with transglutaminase, Payload varies from 0.003–0.11, which is equivalent to 0.3–9.8 monolayers	Citase	447
HFM-GMA-DEA	Adsorption–cross-linking	Cross-linked with GAH; payload is 0.13, which is equivalent to 12 monolayers	AsOM	448
Porous hollow-fibre-g-anion-exchange polymer chain	Adsorption–cross-linking	Payload is 1.2, which is ascribed to the high density of the binding functionality; the adsorbed enzyme is cross-linked with transglutaminase	Urease	449
Porous polyurethane	Adsorption–cross-linking	Cross-linking enhanced the stability of the immobilized enzyme. The specific activity of the IM is higher than the free enzyme, suggesting that the enzyme molecules are activated	Lipase from *Candida rugosa*	450
Porous carriers	Adsorption–cross-linking	Many non-specific carriers can be used for enzyme immobilization	Penicillin G acylase	451
Porous ceramic monolith	Adsorption–intermolecular cross-linking	The immobilized enzyme has enhanced thermostability The activity of the IM was diffusion-controlled	β-Galactosidase from *Aspergillus niger*	452
Con A-bead cellulose matrix	Adsorption–intermolecular cross-linking	Compared with other types of immobilized enzyme, the performance of the enzyme immobilized on Con A cellulose beads was similar to that of the free enzyme, regarding the activity retention and stability and inactivation mechanism	Yeast invertase	453
DEAE-cellulose	Adsorption–cross-linking	The cross-linking prevented leakage of the enzyme in >0.2 m methanol	AOD (alcohol oxidase)	454
Polyurethane (PU) sheet impregnated with 5% (w/v) gelatin	Adsorption and cross-linking	–	Urokinase	455
Sephadex G-100 modified with chloroacetic acid and CEAH	Adsorption and cross-linking	Anionic and cationic carriers enhanced storage stability; broadened pH range	β-Galactosidase	456
Con A-bead cellulose matrix	Adsorption and cross-linking	Biospecific adsorption gives the immobilized enzyme its native properties and it can thus be used as a model of the free enzyme	Yeast invertase	457

immobilization of glucoamylase on polyethyleneimine-coated non-porous glass beads eliminated the problem of product reversion. The immobilized enzyme was, however, less stable than the soluble enzyme (not immobilized) [438].

The pattern of product formation by polyphenol oxidase immobilized on Nylon and PES membranes was found to depend on the matrix used [420].

2.6.3.4 Adsorption–Covalent Attachment

Covalent immobilization of an enzyme on a carrier is also influenced by the physical nature of the reactants. In general, covalent enzyme immobilization on a carrier can occur via two steps, adsorption and covalent binding. In the first step, the enzymes must approach the proximity of the carrier as a result of physical interactions during which, where the chemically inactive/active functional groups are located. In the second step, the spatially closely arranged reactive groups can react with each other, leading to formation of the covalent bond between the enzyme and the carrier (Scheme 2.27).

The importance of the physical interaction of the enzyme with the carrier before covalent binding was observed at the end of the 1970s, when Torchilin et al., when coupling chymotrypsin to insoluble and water-soluble copolymers of acrylamide and acrylic acid, using a water-soluble carbodiimide as activating agent, showed that formation of electrostatic complexes between the carrier and the enzyme before immobilization was essential for both high retention of activity and high thermostability [458]. This study implied that the first step, interaction of the enzyme with the carrier (for example orientation) governed the second step, covalent binding (e.g. orientation, binding numbers) [459].

The role of the first adsorption step in covalent immobilization of enzymes was beautifully demonstrated in 1981, when synthetic carriers bearing the same backbone but different leaving groups were used for immobilization of chymotrypsin [460]. The carriers were prepared from four different monomers bearing different activated leaving groups. If interaction of the carrier with the enzyme has no ef-

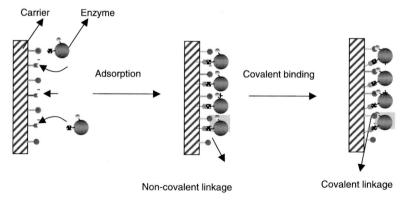

Scheme 2.27 Adsorption and covalent binding.

fect on the immobilized enzymes obtained, the performance of all the immobilized enzymes should be similar.

In fact, significantly different activity retention was observed. Because the linkage of the enzyme with the carrier and the nature of the carrier were same in all the resulting immobilized enzyme preparations, the differences were ascribed to differences in the first step of adsorption, which resulted in different modes of covalent binding (orientation, steric hindrance). In extensive work at the beginning of the 1990s the strategy of varying the nature of the leaving groups was used to modulate the orientation of the enzymes on the carriers [461].

In line with the same principle, synthetic carriers such as Eupergit C have recently been modified with the objective of introducing extra pendant functional groups and enhancing physical interaction before to covalent binding [183, 463]. Although the concentration of the newly introduced functionality is lower (10–20%), the pattern of enzyme immobilization (orientation, binding site) was altered, depending on the nature of the new functionality [463].

Indeed, covalent enzyme immobilization on synthetic carriers such as epoxy carriers or other relatively hydrophobic carriers, for example copolymers of 2-hydroxyethyl methacrylate with ethylene dimethacrylate bearing epoxy groups, was often affected by the properties of the salts and their concentration [462]. High salt concentration often promotes the efficiency of binding, because of exposure of buried amino acid residues.

Under these conditions enzyme activity is determined by two processes, the adsorption process and the covalent binding process, and thus by the factors that govern activity retention in the individual processes. In general, diffusion constraints, chemical nature, inert functionality and active binding functionality, and immobilization conditions (e.g. pH, ionic strength, etc.) will effect enzyme activity.

For example, immobilization of α-chymotrypsin on synthetic copolymers of acrylamide and acrylic acid led to the finding that formation of an ionic complex before covalent binding is essential for high retention of activity and higher stability, suggesting that the formation of the complex prevents random covalent binding in the next step. Thus, higher retention can be obtained by avoiding covalent binding.

Occasionally, although the non-covalent interaction of the carriers with the enzyme does not result in loss of activity, the orientation of the enzyme molecules on the carrier surface might be different. Thus, the binding mode e.g. molecule orientation and number of bonds might be different, resulting in significant differences in activity retention; for example immobilization of trypsin on carriers bearing different leaving groups [460].

The nature of the carrier also affects the binding and the conditions for binding. Immobilization on hydrophobic carriers of enzymes such as penicillin G acylase [462], aminoacylase, proteases, amidohydrolase, and carboxypeptidase A often requires use of a buffer of higher ionic strength. Obviously, the interaction between the enzyme and the carrier is quite weak (because of incompatible aquaphilicity) and the high ionic strength promotes interaction of the enzyme with the carrier because of exposure of the buried hydrophobic amino acid residues [462].

Table 2.13 Adsorption and covalent adsorption

Polymer	Method	Comments	Enzyme	Ref.
Synthetic copolymers of acrylamide and acrylic acid	Covalent	Formation of the ionic complex is essential for high retention of activity and greater stability	Chymotrypsin	458
Soluble carboxylic carrier (PAAC-PAAm)	CDI coupling method	pH and ionic strength affected the coupling process. Increasing the ionic strength reduces the reaction rate	Fibrinolysin	459
PMND, PMBS, PMBB, PMBT	Covalent	Despite the same backbone and the same type of binding mode, the immobilized enzyme differs greatly in activity retention, depending on the leaving groups	Trypsin	460
Activated agarose	Covalent	Different leaving groups lead to the formation of immobilized enzymes with different orientation, suggesting that adsorption is the dominating step	Enzymes	461
Separon HEMA containing epoxy groups	Covalent	At specific concentration, the salting-out ions cause a protein-matrix hydrophobic interaction which is a prerequisite for the covalent bond formation	Aminoacylase, proteases, amidohydrolase, carboxypeptidase A	462
Eupergit C derivatives	Coordination-covalent	Coordination	Penicillin G acylase	463
p-Aminobenzamidohexyl Sepharose 4B/pyridoxal 5′-phosphate	CBS (covalently)	Immobilization of the enzyme via the immobilized cofactor significantly improved activity retention	Tryptophanase	464

In the same way as for conventional covalent enzyme immobilization, enzyme stability is governed by:

- the nature of the carrier,
- the nature of CAF and CIF,
- immobilization conditions,
- binding mode.

The fact that formation of an ionic complex between the PAAm bearing carboxyl functionality and enzymes bearing NH_2 functionality is essential for higher stability suggests that the binding mode (i.e. position and number of linkages) affects enzyme stability [458].

2.7
References

1. Nelson JM, Griffin EG: Adsorption of Invertase. *J Am Chem Soc* 1916, 38:1109–1916
2. Tosa T, Mori T, Fuse N, Chibata I: Studies on continuous enzyme reactions Part V Kinetics and industrial application of aminoacylase column for continuous optical resolution of acyl-*dl* amino acids. *Biotechnol Bioeng* 1967, 9:603–615
3. Sharma S, Yamazaki H: Preparation of hydrophobic cotton cloth. *Biotechnol Lett* 1984, 6:301–306
4. Albayrak N, Yang ST: Immobilisation of *Aspergillus oryzae* β-galactosidase on tosylated cotton cloth. *Enzyme Microb Technol* 2002, 31:371–383
5. Rexova-Benkova L, Omelkova J, Kubanek V: *endo-d*-Galacturonanase immobilised by adsorption on porous poly(ethylene terephthalate). *Collect Czech Chem Commun* 1982, 47:2716–2723
6. Barnet LB, Bull HB: The optimum pH of adsorbed ribonuclease. *Biochim Biophys Acta* 1959, 36:244–246
7. Maxim S, Flondor A, Carpov A: Ionic binding of biologically active proteins on cross-linked acrylic macromolecular supports. *Biotechnol Bioeng* 1987, 30:593–597
8. Maxim S, Flondor A, Pasa R, Popa M: Immobilised pectolytic enzymes on acrylic supports. *Acta Biotechnol* 1992, 12:497–507
9. Maxim S, Flondor A: New crosslinked functionalized acrylic copolymers for biotechnological applications. *Rev Roum Chim* 1989, 34:1389–1395
10. Maxim S, Flondor A, Carpov A, Rugina V, Cojocaru D, Bontas I, Topala N: Functionalized crosslinked acrylic crosslinkings as supports for biologically active proteins. *Biotechnol Bioeng* 1986, 28:294–296
11. Moriya K, Tanizawa K, Kanaoka Y: Immobilised chymotrypsin by means of Schiff base copper(II) chelate. *Biochem Biophys Res Commun* 1989, 162:408–414
12. Ahmad A, Bishayee S, Bachhawat B: Novel method for immobilisation of chicken brain arylsulfatase A using concanavalin A. *Biochem Biophys Res Commun* 1973, 53:730–736
13. Lowe CR, Hans M, Spibey N, Drabble WT: The purification of inosine 5′-monophosphate dehydrogenase from *Escherichia coli* by affinity chromatography on immobilised procion dyes. *Anal Biochem* 1980, 104:23–28
14. Porath J, Carlsson J, Olsson I, Belfrage G: Metal chelate affinity chromatography, a new approach to protein fractionation. *Nature* 1975, 258:598–599
15. Caseli L, Furriel RPM, de Andrade JF, Leone FA, Zaniquelli MED: Surface density as a significant parameter for the enzymatic activity of two forms of

alkaline phosphatase immobilised on phospholipid Langmuir films. *J Colloid Interface Sci* 2004, 275:123–130

16 Barker SA, Somers PJ, Epton R: Preparation and properties of α-amylase chemically coupled to microcrystalline cellulose. *Carbohydr Res* 1968, 8:491–497

17 Ison AP, Macrae AR, Smith CG, Bosley J: Mass transfer effects in solvent-free fat interesterification reactions: influences on catalyst design. *Biotechnol Bioeng* 1994, 43:122–130

18 Kondo A, Urabe T: Relationships between molecular states (conformation and orientation) and activities of α-amylase adsorbed on ultrafine silica particles. *Appl. Microb Biotechnol* 1995, 43:801–807

19 Bower CK, Xu Q, McGuire J: Activity losses among T4 lysozyme variants after adsorption to colloidal silica. *Biotechnol Bioeng* 1989, 58:658–662

20 Zoungrana T, Findenegg GH, Norde W: Structure, stability and activity of adsorbed enzymes. *J Colloid Interface Sci* 1997, 190:437–448

21 Bosley JA, Peilow AD: Immobilisation of lipases on porous polypropylene: reduction in esterification efficiency at low loadings. *J Am Oil Chem Soc* 1997, 74:107–111

22 Wehtje E, Adlercreutz P, Mattiasson B: Improved activity retention of enzymes deposited on solid supports. *Biotechnol Bioeng* 1993, 41:171–178

23 Sandwick RK, Schray KJ: Conformational states of enzymes bound to surfaces. *J Colloid Interface Sci* 1988, 121:1–12

24 Caldwell KD, Axen R, Bergwall M, Porath J: Immobilisation of enzymes based on hydrophobic interaction. II. Preparation and properties of an amyloglucosidase adsorbate. *Biotechnol Bioeng* 1976, 18:1589–1604

25 Azari F, Nemat-Gorgani M: Reversible denaturation of carbonic anhydrase provides a method for its adsorptive immobilisation. *Biotechnol Bioeng* 1999, 62:193–199

26 Nemat-Gorgani M, Karimian K: Enzyme immobilisation on palmityl-Sepharose. *Biotechnol Bioeng* 1983, 25:2617–2629

27 Chang, HN, Ghim YS, Cho YR, Landis DA, Reilly PJ: Immobilisation of *Leuconostoc mesenteroides* dextransucrase to porous phenoxyacetyl cellulose beads. *Biotechnol Bioeng* 1981, 23:2647–2653

28 Melander W, Horvath C: Hydrophobic interactions in purification and utilization of enzymes. *Enzyme Eng* 1978, 4:355–363

29 Kondo A, Murakami F, Kawagoe M, Higashitani K: Kinetic and circular dichroism studies of enzymes adsorbed on ultrafine silica particles. *Appl Microb Biotechnol* 1993, 39:726–731

30 Gupta MN: Thermostabilisation of proteins. *Biotechnol Appl Biochem* 1991, 14:1–11

31 Matsuyama H, Yamamoto T, Furuyoshi S, Teramoto M, Kondo A: Spectral changes of lysozyme adsorbed on ultrafine silica particles. *Biosci Biotechnol Biochem* 1993, 57:992–993

32 Bale MD, Danielson SJ, Daiss JL, Goppert KE, Sutton RC: Influence of copolymer composition on protein adsorption and structural rearrangements at the polymer surface. *J Colloid Interface Sci* 1989, 132:176–187

33 Brode PF 3rd, Erwin CR, Rauch DS, Lucas DS, Rubingh DN: Enzyme behaviour at surfaces. Site-specific variants of subtilisin BPN′ with enhanced surface stability. *J Biol Chem* 1994, 269:23538–23543

34 Huitron C, Limon-Lason J: Immobilisation of glucose isomerase to ion-exchange materials. *Biotechnol Bioeng* 1978, 20:1377–1391

35 Moriyama S, Noda A, Nakanishi K, Matsuno R, Kamikubo T: Thermal stability of immobilised glucoamylase entrapped in polyacrylamide gels and bound to SP-Sephadex C-50. *Agric Biol Chem* 1980, 44:2047–2054

36 Boivin P, Kobos RK, Papa SL, Scouten WH: Immobilisation of perfluoroalkylated enzymes in a biologically active state on to Perflex support. *Biotechnol Appl Biochem* 1991, 14:155–169

37 Guit RPM, Kloosterman M, Meindersma GW, Mayer M, Meijer EM: Lipase kinetics: hydrolysis of triacetin by lipase from *Candida cylindracea* in a

hollow-fibre membrane reactor. *Biotechnol Bioeng* 1991, 38:727–732

38 Pedersen H, Furler L, Venkatasubramanian K, Prenosil J, Stuker E: Enzyme adsorption in porous supports: local thermodynamic equilibrium model. *Biotechnol Bioeng* 1985, 27:961–971

39 Borchert A, Buchholz K: Improved biocatalyst effectiveness by controlled immobilisation of enzymes. *Biotechnol Bioeng* 1984, 26:727–736

40 Juang HD, Weng HS: Performance of biocatalysts with nonuniformly distributed immobilised enzyme. *Biotechnol Bioeng* 1984, 26:623–626

41 Messing RA: Adsorption and inorganic bridge formation. *Methods Enzymol* 1976, 44:148–169

42 Bosley J: Turning lipases into industrial biocatalysts. *Biochem Soc Trans* 1997, 2:174–178

43 Al-Duri B, Robinson E, MacNerlan S, Bailie P: Hydrolysis of edible oils by lipase immobilised on hydrophobic supports: effect of internal support structure, *J Am Oil Chem Soc* 1995, 72:1351–1395

44 Artemova AA, Voroshilova OI, Nikitin YS, Khokhlova TD: Macroporous silica in chromatography and immobilisation of biopolymers. *Adv Colloid Interface Sci* 1986, 25:235–248

45 Hernaiz MJ, Crout DHG: Immobilisation/stabilisation on Eupergit C of the β-galactosidase from *B. circulans* and an α-galactosidase from *Aspergillus oryzae*. *Enzyme Microb Technol* 2000, 27:26–32

46 Klimacek M, Eis C, Nidetzky B: Continuous production of α-trehalose by immobilised fungal trehalose phosphorylase. *Biotechnol Tech* 1999, 13:243–248

47 Dupommier A, Merle-Aubry L, Merle Y, Selegny E: Enzyme-synthetic polyampholyte systems. *Makromol Chem* 1986, 187:211–221

48 Tumturk HR, Caykara T, Kantoglu O, Guven O: Adsorption of α-amylase on to poly(N-vinyl 2-pyrrolidone/itaconic acid) hydrogels. *Nuclear Instr Method Phys Res, Section B: Beam Interact Mater Atom* 1999, 151:238–241

49 Venkataraman S, Horbett TA, Hoffman AS: The reactivity of α-chymotrypsin immobilised on radiation-grafted hydrogel surfaces. *J Biomed Mater Res* 1977, 11:111–123

50 Turkova J, Blaha K, Adamova K: Effect of concentration of immobilised inhibitor on the biospecific chromatography of pepsins. *J Chromatogr* 1982, 236:375–383

51 Bruggink A, Roos EC, de Vroom E: Penicillin acylase in the industrial production of β-lactam antibiotics. *Org Process Res Dev* 1998, 2:128–133

52 Wan Yunus WMZ, Shalleh AB, Ismail A, Ampon K, Razak CAN, Basri M: Poly(methyl methacrylate) as a matrix for immobilisation of lipase. *Appl Biochem Biotechnol* 1992, 36:97–105

53 Chen LF, Gong CS, Tsao GT: Immobilised glucose isomerase on DEAE cellulose beads. *Starch/Staerke* 1981, 33:58–63

54 Miyamoto K, Fujii T, Tamaoki N, Okazaki M, Miura Y: Intraparticle diffusion in the reaction catalysed by immobilised glucoamylase. *Hakko Kogaku Zasshi (Jpn)* 1973, 51:566–574

55 Ichijo H, Suehiro T, Yamauchi A, Ogawa S: Fibrous support for immobilisation of enzymes. *J Appl Polym Sci* 1982, 27:1665–1674

56 Kamei S, Okubo M, Matsuda T, Matsumoto T: Hydroxyethyl methacrylate copolymer microspheres and its enzymic activity. *Colloid Polym Sci* 1986, 264:743–747

57 Ruckenstein E, Wang X: Lipase immobilised on hydrophobic porous polymer supports prepared by concentrated emulsion polymerization and their activity in the hydrolysis of triacylglycerides. *Biotechnol Bioeng* 1993, 42:821–828

58 Kaga H; Siegmund B; Neufellner E; Faber K; Paltauf F: Stabilization of Candida lipase against acetaldehyde by adsorption onto Celite. *Biotechnol Tech* 1994, 8:369–74.

59 Watanabe T, Mori T, Tosa T, Chibata I: Immobilisation of aminoacylase by adsorption to tannin immobilised on aminohexyl cellulose. *Biotechnol Bioeng* 1979, 21:477–486

60 Shaw JF, Chang R, Wang FF, Wang YJ: Lipolytic activity of a lipase immobilised on six selected supporting materials. *Biotechnol Bioeng* 1990, 35:132–137

61 Taylor JB, Swaisgood HE: Kinetic study on the effect of coupling distance between insoluble trypsin and its carrier matrix. *Biochim Biophys Acta* 1972, 284:268–277

62 Cao L: Immobilised enzymes: science or art? *Curr Opin in Chem Biol* 2005, in press

63 Koga J, Yamaguchi K, Gondo S: Immobilisation of alkaline phosphatase on activated alumina particles. *Biotechnol Bioeng* 1984, 26:100–103

64 Matoba S, Tsuneda S, Saito K, Sugo T: Highly efficient enzyme recovery using a porous membrane with immobilised tentacle polymer chains. *Bio/Technology* 1995, 13:795–797

65 Ganapathi S, Butterfield DA, Bhattacharyya D: Flat-sheet and hollow fibre membrane bioreactors: a study of the kinetics and active site conformational changes of immobilised papain including sorption studies of reaction constituents. *J Chem Technol Biotechnol* 1995, 64:157–164

66 Buchholz K: Non-uniform enzyme distribution inside carriers. *Biotechnol Lett* 1979, 1:451–456

67 Indlekofer M, Brotz F, Bauer A, Reuss M: Stereoselective bioconversions in continuously operated fixed bed reactors: modeling and process optimization. *Biotechnol Bioeng* 1996, 52:459–471

68 Ling D, Wu G, Wang C, Wang F, Song G: The preparation and characterisation of an immobilised l-glutamic decarboxylase and its application for determination of l-glutamic acid. *Enzyme Microb Technol* 2000, 27:516–521

69 Miller E, Sugier H: Immobilisation of glucoamylase on porous glass. *Acta Biotechnol* 1988, 8:503–508

70 Kminkova M, Proskova A, Kucera J: Immobilisation of mold β-galactosidase. *Collect Czech Chem Commun* 1988, 53:3214–3219

71 Duggal SK, Buchholz K: Effects of immobilisation on intrinsic kinetics and selectivity of trypsin. *Eur J Appl Microb Biotechnol* 1982, 16:81–87

72 Venardos D, Klei HE, Sundstrom DW: Conversion of cellobiose to glucose using immobilised β-glucosidase reactors. *Enzyme Microb Technol* 1980, 2:112–116

73 Crapisi A, Lante A, Pasini G, Spettoli P: Enhanced microbial cell lysis by the use of lysozyme immobilised on different carriers. *Process Biochem* 1993, 28:17–21

74 Pedersen H, Furler L, Venkatasubramanian K, Prenosil J, Stuker E: Enzyme adsorption in porous supports: local thermodynamic equilibrium model. *Biotechnol Bioeng* 1985, 27:961–971

75 Brougham MJ, Johnson DB: Studies on the stability of soluble and immobilised alcohol dehydrogenase from yeast mitochondria. *Enzyme Eng* 1980, 5:431–434

76 Janssen MHA, van Langen LM, Pereira SRM, van Rantwijk F, Sheldon RA: Evaluation of the performance of immobilised penicillin G acylase using active-site titration. *Biotechnol Bioeng* 2002, 78:425–432

77 Tramper J, Mueller F, Van der Plas HC: Immobilised xanthine oxidase: kinetics, (in)stability, and stabilisation by coimmobilisation with superoxide dismutase and catalase. *Biotechnol Bioeng* 1978, 20:1507–1522

78 Palomo JM, Segura RL, Fernandez-Lorente G, Guisan JM, Fernandez-Lafuente R: Enzymatic resolution of (±)-glycidyl butyrate in aqueous media. Strong modulation of the properties of the lipase from *Rhizopus oryzae* via immobilisation techniques. *Tetrahedron Asymmetry* 2004, 15:1157–1161

79 Tweddell RJ, Kermasha S, Combes D, Marty A: Immobilisation of lipase from *Rhizopus niveus*: a way to enhance its synthetic activity in organic solvent. *Biocatal Biotrans* 1999, 16:411–426

80 Stark MB, Kolmberg K: Covalent immobilisation of lipase in organic solvents. *Biotechnol Bioeng* 1989, 34:942–950

81 Danielson ND, Bossu TM, Kruempelman M: Immobilisation of enzymes on poly(chlorotrifluoroethylene) particles packed in small columns. *Anal Lett* 1982, 15:1289–1300

82 Heinrichova K, Zliechovcova D: Poly(ethylene terephthalate) immobilised exo-*d*-galacturonanase. *Collect Czech Chem Commun* 1986, 51:723–730

83 Ruckenstein E, Wang X: A novel support for the immobilisation of lipase and the effects of the details of its preparation on the hydrolysis of triacylglycerides. *Biotechnol Tech* 1993, 7:117–122

84 Skerker PS, Clark DS: Catalytic properties and active-site structural features of immobilised horse liver alcohol dehydrogenase. *Biotechnol Bioeng* 1988, 32:148–158

85 Norin M, Boutelje J, Holmberg E, Hult K: Lipase immobilised by adsorption. Effect of support hydrophobicity on the reaction rate of ester synthesis in cyclohexane. *Appl Microb Biotechnol* 1988, 28:527–530

86 Pereira EB, De Castro HF, Moraes FF, Zanin GM : Kinetic studies of lipase from *Candida rugosa*:A Comparative Study between free and immobilized enzyme onto porous chitosan beads. *Appl Biochem Biotechnol* 2001 Vol. 91–93, 739–748

87 Blanco RM, Terreros P, Fernandez-Perez M, Otero C, Diaz-Gonzalez G: Functionalization of mesoporous silica for lipase immobilisation. Characterisation of the support and the catalysts. *J Mol Catal B: Enzyme* 2004, 30:83–93

88 Kovalenko GA, Shitova NB, Sokolovskii VD: Immobilisation of oxyreductases on inorganic supports based on alumina. Immobilisation of lactate dehydrogenase on alumina by adsorption. *Biotechnol Bioeng* 1981, 23:1721–1734

89 Aydemir T, Telefoncu A: The interesterification of fats by immobilised stereospecific lipase. *NATO ASI Series, Series E: Appl Sci* 1992, 210:505–506

90 Caldwell KD, Axen R, Bergwall M, Porath J: Immobilisation of enzymes based on hydrophobic interaction. I. Preparation and properties of a α-amylase adsorbate. *Biotechnol Bioeng* 1976, 18:1573–1588

91 Yoshida M, Kimura T, Ogata M, Nakakuki T: Immobilisation of the exo-maltotetrahydrolase and some properties of the enzyme. *Denpun Kagaku* 1988, 35:245–252

92 Akova A, Ustun G: Activity and adsorption of lipase from *Nigella sativa* seeds on Celite at different pH values. *Biotechnol Lett* 2000, 22:355–359

93 Gaffar R, Kermasha S, Bisakowski B: Biocatalysis of immobilised chlorophyllase in a ternary micellar system. *J Biotechnol* 1999, 75:45–55

94 Aoun S, Chebli C, Baboulene M: Noncovalent immobilisation of chloroperoxidase on to talc: catalytic properties of a new biocatalyst. *Enzyme Microb Technol* 1998, 23:380–385

95 Goldstein L, Levin Y, Pecht M, Katchalski E: A water-insoluble polyanionic derivatives of trypsin, effect of the polyelectrolyte carrier on the kinetic behaviour of the bound trypsin. *Biochemistry* 1964, 3:1914–1919

96 Xu H, Li M, He B: Immobilisation of *Candida cyclindracea* lipase on methyl acrylate–divinyl benzene copolymer and its derivatives. *Enzyme Microb Technol* 1995, 17:194–199

97 Lei CH, Shin YS, Liu J, Ackerman EJ: Entrapping enzyme in a functionalized nanoporous support. *J Am Chem Soc* 2002, 124:11 242–11 243

98 Murza A, Fernandez-Lafuente R, Guisan JM: Essential role of the concentration of immobilised ligands in affinity chromatography: purification of guanidinobenzoatase on an ionized ligand. *J Chromatogr B Biomed Sci Appl* 2000, 740:211–218

99 Yiu HHP, Wright PA, Botting NP: Enzyme immobilisation using SBA-15 mesoporous molecular sieves with functionalized surfaces. *J Mol Catal B: Enzym* 2001, 15:81–92

100 Gandhi NN, Vijayalakshmi V, Sawant SB, Joshi JB: Immobilisation of *Mucor miehei* lipase on ion exchange resins. *Chem Eng J (Lausanne)* 1996, 61:149–156

101 Chong SM, Zhao XS: Design of large-pore mesoporous materials for immobilisation of penicillin G acylase biocatalyst, *Catal Today* 2004, 93/95: 293–299

102 Bower CK, Sananikone S, Bothwell MK, McGuire J: Activity losses among T4 lysozyme charge variants after adsorption to colloidal silica. *Biotechnol Bioeng* 1999, 64:373–376

103 Ahmad S, Anwar A, Saleemuddin M: Immobilisation and stabilisation of invertase on Cajanus Cajan Lectin support, *Bioresour Technol* 2001, 79:121–127

104 Deetz JS, Rozzell JD: Enzymic catalysis by alcohol dehydrogenases in organic solvents. *Ann NY Acad Sci* 1988, 542:230–234

105 Muller HM, Schuber F: Properties of native calf spleen NAD+ glycohydrolase immobilised by hydrophobic interaction. *Biochem Int* 1988, 17:953–963

106 Caldwell KD, Axen R, Bergwall M, Porath J: Immobilisation of enzymes based on hydrophobic interaction. II. Preparation and properties of an amyloglucosidase adsorbate. *Biotechnol Bioeng* 1976, 18:1589–1604

107 Kery V, Haplova J, Tihlarik K, Schmidt S: Factors influencing the activity and thermostability of immobilised porcine pancreatic lipase: *J Chem Technol Biotechnol* 1990, 48:201–207

108 Kamei S, Okubo M, Matsumoto T: Adsorption of trypsin on to poly(2-hydroxyethyl methacrylate)/polystyrene composite microspheres and its enzymatic activity. *J Appl Polym Sci* 1987, 34:1439–1446

109 Reslow M, Adlercreutz P, Mattiason B: On the importance of the support material for bioorganic synthesis. Influence of water partition between solvent, enzyme and solid support in water poor reaction media. *Eur J Biochem* 1988, 172:573–578

110 Marron-Brignone L, Morelis RM, Coulet PR: Hydrophilicity or hydrophobicity of the surface on the enzyme kinetic behaviour. *Langmuir* 1996, 12:5674–5680

111 Bryjak J, Noworyta A, Trochimczuk A: Immobilisation of penicillin-acylase on acrylic carriers. *Bioprocess Eng* 1989, 4:159–162

112 Sokolovskii VD, Kovalenko GA: Immobilisation of oxidoreductases on inorganic supports based on alumina: the role of mutual correspondence of enzyme-support hydrophobic– hydrophilic characters. *Biotechnol Bioeng* 1988, 32:916–919

113 Cao L, 2001, unpublished data

114 Sheldon RA, van Langen LM, Cao L, Janssen MHA: Biocatalysts and biocatalysis in the synthesis of β-lactam antibiotics. In: Bruggink A (Ed) *Synthesis of β-lactam antibiotics.* Kluwer Academic, 2001

115 Haiech J, Predeleanu R, Watterson DM, Ladant D, Bellalou J, Ullmann A, Barzu O: Affinity-based chromatography utilizing genetically engineered proteins. Interaction of Bordetella pertussis adenylate cyclase with calmodulin. *J Biol Chem* 1988, 263:4259–4262

116 Baron MH, Revault M, Servagent-Noinville S, Abadie J, Quiquampoix H: Chymotrypsin adsorption on montmorillonite: enzymatic activity and kinetic FTIR structural analysis. *J Colloid Interface Sci* 1999, 214:319–332

117 Kovba GV, Rubtsova MY, Egorov AM: Chemiluminescent biosensors with immobilised peroxidase. *J Biolumin Chemilumin* 1997, 12:33–36

118 Wallace EF, Lovenberg W: Carbohydrate moiety of dopamine β-hydroxylase. Interaction of the enzyme with concanavalin A. *Proc Natl Acad Sci USA* 1974, 71:3217–3220

119 Yoshioka T, Shimamura M: Studies of polystyrene-based ion-exchange fibre. IV. A novel fibre-form material for adsorption and immobilisation of biologically-active proteins. *Bull Chem Soc Jpn* 1986, 59:399–403

120 Lampe J, Pommerening K: Oxygen binding to carrier fixed haemoglobin. *Abhandl Akad Wissenschaft DDR* 1975: 187–189

121 Angelino SAGF, Mueller Franz, Van der Plas HC: The use of enzymes in organic synthesis. Part 14. Purification and immobilisation of rabbit liver aldehyde oxidase. *Biotechnol Bioeng* 1985, 27:447–455

122 Tumturk H, Caykara T, Sen M, Guven O: Adsorption of α-amylase on to poly(acrylamide/maleic acid) hydrogels. *Radiat Phys Chem* 1999, 55: 713–716

123 Arslan F, Tumturk H, Caykara T, Sen M, Guven O: The effect of gel composition on the adsorption of invertase on poly(acrylamide/maleic acid) hydrogels. *Food Chem* 2000, 70:33–38

124 Caldwell KD, Axen R, Bergwall M, Olsson I, Porath J: Immobilisation of enzymes based on hydrophobic interaction. III. Adsorbent substituent density and its impact on the immobilisation of α-amylase. *Biotechnol Bioeng* 1976, 18:1605–1614

125 Caldwell KD, Axen R, Porath J: Reversible immobilisation of enzymes to hydrophobic agarose gels. *Biotechnol Bioeng* 1976, 18:433–438

126 Lloyd NE, Khaleeluddin K: A kinetic comparison of Streptomyces glucose isomerase in free solution and adsorbed on DEAE-cellulose. *Cereal Chem* 1976, 53:270–282

127 Bosley, JA, Clayton JC: Blueprint for a lipase support: use of hydrophobic controlled-pore glasses as model systems. *Biotechnol Bioeng* 1994, 43:934–938

128 Adlercreutz P: On the importance of the support material for enzymatic synthesis in organic media. Support effects at controlled water activity. *Eur J Biochem* 1991, 199:609–614

129 Serralha FN, Lopes JM, Lemos F, Prazeres DMF, Aires-Barros MR, Cabral JMS, Ramoa Ribeiro F: Zeolites as supports for an enzymic alcoholysis reaction. *J Mol Catal B: Enzymatic* 1998, 4:303–311.

130 Tweddell RJ, Kermasha S, Combes D, Marty A: Immobilization of lipase from *Rhizopus niveus*: a way to enhance its synthetic activity in organic solvent. *Biocatal Biotrans* 1999, 16: 411–426

131 Persson M, Wehtje E, Adlercreutz P: Immobilisation of lipases by adsorption and deposition: high protein loading gives lower water activity optimum. *Biotechnol Lett* 2000, 22:1571–1575

132 Secundo F, Carrea G, Soregaroli C, Varinelli D, Morrone R: Activity of different *Candida antarctica* lipase B formulations in organic solvents. *Biotechnol Bioeng* 2001, 73:157–163

133 Batra R, Gupta MN: Non-covalent immobilisation of potato (*Solanum tuberosum*) polyphenol oxidase on chitin. *Biotechnol Appl Biochem* 1994, 19:209–215

134 Li B, Inagaki S, Miyazaki C, Takahashi H: Preparation of highly ordered mesoporous silica materials and application as immobilised enzyme supports in organic solvent. *Chem J Int* 2000, 2:48

135 Zoungrana T, Norde W: Thermal stability and enzymic activity of α-chymotrypsin adsorbed on polystyrene surfaces. *Colloid Surf B: Biointerf* 1997, 9:157–167

136 Carrasco MS, Rad JC, Gonzalez-Carcedo S: Immobilisation of alkaline phosphatase by sorption on Na-sepiolite. *Bioresour Technol* 1995, 51:175–181

137 Simon LM, Heinrichova K, Veszelka I, Szajani B: Effects of polyethylene terephthalate on yeast alcohol dehydrogenase. *Acta Biochim Biophys Hung* 1990, 25:1–7

138 Jung JY, Yun HS, Kim EK: Hydrolysis of olive oil by lipase, immobilised on hydrophobic support. *J Microb Biotechnol* 1997, 7:151–156

139 Rexova-Benkova L, Omelkova J, Veruovic B, Kubanek V: Mode of action of endopolygalacturonase immobilised by adsorption and by covalent binding via amino groups. *Biotechnol Bioeng* 1989, 34(1):79–85

140 He J, Li X, Evans DG, Duan X, Li C: A new support for the immobilisation of penicillin acylase. *J Mol Catal B: Enzym* 2000, 11:45–53

141 Nakajima N, Ishihara K, Sugimoto M, Nakahara T, Tsuji H: Further stabilisation of by chemical modification and immobilisation. *Biosci Biotechnol Biochem* 2002, 66:2739–2742

142 Solomon B, Levin Y: Adsorption of amyloglucosidase on inorganic carriers. *Biotechnol Bioeng* 1975, 17: 1323–1333

143 Takahashi H, Li B, Sasaki T, Miyazaki C, Kajino T, Inagaki S: Catalytic activity in organic solvents and stability of immobilised enzymes depend on the pore size and surface characteristics of mesoporous silica. *Chem Mater* 2000, 12:3301–3305

144 Pifferi PG, Spagna G, Busetto L: The immobilisation of endopolygalacturonase on γ-alumina. *J Mol Catal* 1987, 42:137–149

145 Ollis DF: Diffusion influence in denaturable insolubilised enzyme catalysts, *Biotechnol Bioeng* 1972, 14:871–884

146 Suekane M: Immobilisation of glucose isomerase. *Z Allg Mikrobiol* 1982, 22:565–576

147 Spagna G, Pifferi PG, Tramontini M: Immobilisation and stabilisation of pectinlyase on synthetic polymers for application in the beverage industry. *J Mol Catal A Chem* 1993, 101:99–105

148 Yu H, Liu G, Ye Q, Chen J, Yuan J, Zhang N: A novel method for enzyme immobilization-granular gelatin for adsorbing and crosslinking penicillin acylase. *Kangshengsu* (Chinese) 1988, 13:177–180

149 Lin G, Yu H, Ye Q, Yuan J, Chen J: Properties of penicillin acylase immobilized on granular gelatin. *Kangshengsu* 1988, 13:92–94

150 Abbott BJ, Fukuda DS: Preparation and properties of a cephalosporin acetylesterase adsorbed on to bentonite. *Antimicrob Agent Chemother* 1975, 8:282–288

151 Kitaoka M, Taniguchi H, Sasaki T: A simple method of cellulase immobilisation on a modified silica support. *J Ferm Bioeng* 1989, 67:182–185

152 Moriyama S, Noda A, Nakanishi K, Matsuno R, Kamikubo T: Thermal stability of immobilised glucoamylase entrapped in polyacrylamide gels and bound to SP-Sephadex C-50. *Agric Biol Chem* 1980, 44:2047–2054

153 Stanley WL, Olson AC: Chemistry of immobilising enzymes. *J Food Sci* 1974, 39:660–666

154 Kosugi Y, Suzuki H: Functional immobilisation of lipase eliminating lipolysis product inhibition. *Biotechnol Bioeng* 1992, 40:369–374

155 Miyanaga M, Ohmori M, Imamura K, Sakiyama T, Nakanishi K: Stability of immobilised thermolysin in organic solvents. *J Biosci Bioeng* 1999, 87(4):463–472

156 Basri M, Ampon K, Wan Yunus WMZ, Razak CAN, Salleh AB: Stability of hydrophobic lipase derivatives immobilised on organic polymer beads. *Appl Biochem Biotechnol* 1994, 48:173–183

157 Nanba H, Ikenaka Y, Yamada Y, Yajima K, Takano M, Ohkubo K, Hiraishi Y, Yamada K, Takahashi S: Immobilisation of N-carbamyl-d-amino acid amidohydrolase. *Biosci Biotechnol Biochem* 1998, 62:1839–1844

158 Schopp W, Grunow M: Immobilisation of yeast ADH by adsorption on to polyaminomethylstyrene. *Appl Microb Biotechnol* 1986, 24:271–276

159 Tramper J, Aelino SA, Muller F, van der Plas HC: Kinetics and stability of immobilised chicken liver xanthine dehydrogenase. *Biotechnol Bioeng* 1979, 21:1767–1786

160 Horiuti Y, Imamura S: Stimulation of Chromobacterium lipase activity and prevention of its adsorption to palmitoyl cellulose by hydrophobic binding of fatty acids. *J Biochem* 1978, 83:1381–1385

161 Bastida A, Sabuquillo P, Armisen P, Fernandez-Lafuente R, Huguet J, Guisan JM: A single step purification, immobilisation, and hyperactivation of lipases via interfacial adsorption on strongly hydrophobic supports. *Biotechnol Bioeng* 1998, 58:486–493

162 Melander W, Horvath C: Hydrophobic interactions in purification and utilization of enzymes. *Enzyme Eng* 1978, 4:355–363

163 Gianfreda L, Rao MA, Violante A: Invertase (β-fructosidase): effects of montmorillonite, aluminium hydroxide and Al(OH)x-montmorillonite complex on activity and kinetic properties. *Soil Biol Biochem* 1991, 23:581–587

164 Morikawa Y, Tezuka T, Teranishi M, Kimura K, Fujimoto Y, Samejima H: Dichloro-s-triazinyl resin as a carrier of immobilised enzymes. *Agric Biol Chem* 1976, 40:1137–1142

165 Yokote Y, Kimura K, Samejima H: Production of high fructose syrup by glucose isomerase immobilised on phenol–formaldehyde resin. *Staerke* 1975, 27:302–306

166 Cho YK, Bailey JE: Enzyme immobilisation on activated carbon: alleviation of enzyme deactivation by hydrogen peroxide. *Biotechnol Bioeng* 1977, 19:769–775

167 Carbone K, Casarci M: New immobilisation system for *Candida rugosa* lipase: characterisation and applications. *Prog Biotechnol* 1998, 15:553–558

168 Sakai K, Uchiyama T, Matahira Y, Nanjo F: Immobilisation of chitinolytic

enzymes and continuous production of N-acetylglucosamine with the immobilised enzymes. *J Ferm Bioeng* 1991, 72:168–172

169 Arica MY, Denizli A, Baran T, Hasirci V: Dye derived and metal incorporated affinity poly(2-hydroxyethyl methacrylate) membranes for use in enzyme immobilisation. *Polym Int* 1998, 46:345–352

170 Epton R, Mclaren JV, Thomas TH: Poly(N-acrylolyl-4-and-5-amino salicylic acid) Part III Uses as their titanium complexes for the insolubilisation of enzymes. *Carbohydr Res* 1973, 27:11–20

171 Reetz MT, Zonta A, Schimossek K, Liebeton K, Jaeger KE: Creation of enantioselective biocatalysts for organic chemistry by *in vitro* evolution. *Angew Chem Int Ed* 1997, 36:2830–2832

172 Fernandez-Lorente G, Terreni M, Mateo C, Bastida A, Fernandez-Lafuente R, Dalmases P, Huguet J, Guisan JM: Modulation of lipase properties in macro-aqueous systems by controlled enzyme immobilisation: enantioselective hydrolysis of a chiral ester by immobilised Pseudomonas lipase. *Enzyme Microb Technol* 2001, 28:389–396

173 Palomo JM, Fernandez-Lorente G, Mateo C, Ortiz C, Rotticci D, Norin T, Hult K: Mass transport limitations reduce the effective stereospecificity in enzyme-catalysed kinetic resolution. *Org Lett* 2000, 2:1373–1376

174 Moreno JCM, Samoza A, de Campo C, Llama EF, Sinisterra JCV: Organic reactions catalysed by immobilised lipases. Part I. Hydrolysis of 2-aryl propionic and 2-aryl butyric esters with immobilised *Candida cylindracea* lipase. *J Mol Catal A: Chem* 1995, 95:179–192

175 Pogorevc M, Strauss UT, Riermeier T, Faber K: Selectivity enhancement in enantioselectivity hydrolysis of *sec*-alkyl sulfates by an alkylsulfatase from Rhodococus rubber DSM 44541. *Tetrahedron Asymmetry* 2002, 13:1443–1447

176 Lee JC, Timasheff SN. 1981. The stabilization of proteins by sucrose. *J Biol Chem* 256:7193–7201.

177 Arakawa T, Timasheff SN. 1985. The stabilization of proteins by osmolytes. *Biophys J* 47:411–414.

178 Ye, W., Combes, D. and Monsan, P: Influence of additives on the thermostability of glucose oxidase. *Enzyme Microb Technol* 1988, 10:498–502

179 van Unen D, Engbersen J, Reinhoudt D: Large acceleration of α-chymotrypsin-catalyzed dipeptide formation by 18-Crown-6 in organic solvents. Biotechnol Bioeng 1998, 59:553–556.

180 Triantafyllou A, Wehtje E, Adlercreutz P, Mattiasson B: How do additives affect enzyme activity and stability in nonaqueous media. *Biotechnol Bioeng* 1997, 54:67–76

181 Van Unen DJ, Sakodinskaya IK, Engbersen JFJ, Reinhoudt DN: Crown ether activation of cross-linked substilisin Carslberg crystals in organic solvents. *J Chem. Soc, Perkin Trans I* 1998, 3341-3343.

182 Liu YY, Xu JH, Wu HY, Shen D: Integration of purification with immobilisation of *Candida rugosa* lipase for kinetic resolution of racemic ketoprofen. *J Biotechnol* 2004, 110: 209–217

183 Palomo JM, Munoz G, Fernandez-Lorente G, Mateo C, Fuentes M, Guisan JM, Fernandez-Lafuente R: Modulation of *Mucor miehei* lipase properties via directed immobilisation on different hetero-functional epoxy resins. Hydrolytic resolution of (R,S)-2-butyroyl-2-phenylacetic acid. *J Mol Catal B: Enzym* 2003, 21:201–210

184 Fernandez-Lorentea G, Fernandez-Lafuente R, Palomoa JM, Mateo C, Bastida A, Cocab J, Haramboureb T, Hernandez-Justiz O, Terrenic M, Guisan JM: Biocatalyst engineering exerts a dramatic effect on selectivity of hydrolysis catalysed by immobilised lipases in aqueous medium, *J Mol Catal B: Enzymatic* 2001, 11:649–645

185 Overbeeke PLA, Govardhan C, Khalaf N, Jongejan JA, Heijnen JJ: Influence of lid conformation on lipase enantioselectivity. *J Mol Catal B: Enzymatic* 2000, 10:385–393

186 Fernandez-Lafuente R, Armisen P, Sabuquillo P, Fernandez-Lorente G, Guisan JM: Immobilisation of lipases by selective adsorption on hydrophobic supports. *Chem Phys* 1998, 93:185–197

187 Norde W, Zoungrana T: Surface-induced changes in the structure and activity of enzymes physically immobilised at solid/liquid interfaces. *Biotechnol Appl Biochem* 1998, 28:133–143

188 Markovic O, Machova E: Immobilisation of pectin esterase from tomatoes and *Aspergillus foetidus* on various supports. *Collect Czech Chem Commun* 1985, 50:2021–2027

189 Goncalves APV, Lopes JM, Lemos F, Ribeiro FR, Prazeres DMF, Cabral JMS, Aires-Barros MR: Effect of the immobilisation support on the hydrolytic activity of a cutinase from *Fusarium solani pisi*. *Enzyme Microb Technol* 1997, 20:93–101

190 Nemat-Gorgani M, Karimian K: Use of hexadecyl Fractosil as a hydrophobic carrier for adsorptive immobilisation of proteins. *Biotechnol Bioeng* 1986, 28:1037–1043

191 Arseguel D, Baboulene M: Adsorption of various enzymes on to talcs; scope and limitations of this new immobilisation system. *Biocatal Biotrans* 1995, 12:267–279

192 He BL, Li MQ, Wang DB: Studies on aminoacylase from *Aspergillus oryzae* immobilised on polyacrylate copolymer. *React Polym* 1992, 17:341–346

193 Murray M, Rooney D, Van Neikerk M, Montenegro A, Weatherley LR: Immobilisation of lipase on to lipophilic polymer particles and application to oil hydrolysis. *Process Biochem* 1997, 32:479–486

194 Turkova J, Vanek T, Turkova R, Veruovic B, Kubanek V: Chymotrypsin adsorbed to a polyethylene terephthalate support (SORSILEN) as a suitable model for the investigation of enzymically catalysed organic syntheses in non-aqueous media. *Biotechnol Lett* 1982, 4:165–170

195 Koops BC, Papadimou E, Verheij HM, Slotboom AJ, Egmond MR: Activity and stability of chemically modified *Candida antarctica* lipase B adsorbed on solid supports. *Appl Microb Biotechnol* 1999, 52:791–796

196 Xu HD, Liu YX, Chen BN, Yang HF, Yao GR, Wang RX, Huang YP, Chen WS: Study of urease immobilised on macroporous resins as carriers. *Polym Prepri (Am Chem Soc, Div Polym Chem)* 1986, 27:466–467

197 Mojovic L, Knezevic Z, Popadic R, Jovanovic S: Immobilisation of lipase from *Candida rugosa* on a polymer support. *Appl Microb Biotechnol* 1998, 50:676–681

198 Xu H, Li M, He B: Immobilisation of *Candida cyclindracea* lipase on methyl acrylate–divinyl benzene copolymer and its derivatives. *Enzyme Microb Technol* 1995, 17:194–199

199 Ampon K: Distribution of an enzyme in porous polymer beads. *J Chem Technol Biotechnol* 1992, 55:185–190

200 Messing RA: Potential applications of molecular inclusion to beer processing. *Brewers Dig* 1971, 46:60–63

201 Garcia-Segura JM, Cid C, Martin de Llano J, Gavilanes JG: Sepiolite-supported urease. *Br Polym J* 1987, 19:517–522

202 Salis A, Sanjust E, Solinas V, Monduzzi M: Characterisation of Accurel MP1004 powder and its uses as support for lipase immobilisation. *J Mol Catal B Enzym*atic 2003, 24/25:75–82

203 Rexova-Benkova L, Omelkova J, Veruovic B, Kubanek V: Mode of action of endopolygalacturonase immobilised by adsorption and by covalent binding via amino groups. *Biotechnol Bioeng* 1989, 34:79–85

204 Knezevic Z, Mojovic L, Adnadjevic B: Immobilisation of lipase on a hydrophobic zeolite type Y. *J Serbian Chem Soc* 1998, 63:257–264

205 Boyd SA, Mortland MM: Urease activity on a clay–organic complex. *Soil Sci Soc Am J* 1985, 49:619–622

206 Suzuki T, Toriyama M, Hosono H, Abe Y: Application of a microporous glass-ceramics with a skeleton of calcium titanic phosphate ($CaTi_4(PO_4)_6$) to carriers for immobilisation of enzymes. *J Ferm Bioeng* 1991, 72: 384–391

207 Diaz JF, Balkus Jr KJ: Enzyme immobilisation in MCM-41 molecular sieve. *J Mol Catal B: Enzym* 1996, 2:115–126

208 Takahashi H, Li B, Sasaki T, Miyazaki C, Kajino T, Inagaki S: Immobilised enzymes in ordered mesoporous silica materials and improvement of their stability and catalytic activity in an organic solvent. *Micropor Mesopor Mater* 2001, 44/45:755–762

209 Gimon-Kinsel ME, Jimenez VL, Washmon L, Balkus J Jr: Mesoporous molecular sieve immobilised enzymes. *Stud Surf Sci Catal* 1998, 117:373–380

210 Macario A, Calabro V, Curcio S, De Paola M, Giordano G, Iorio G, Katovic A: Preparation of mesoporous materials as a support for the immobilisation of lipase. *Stud Surf Sci Catal* B 2002, 142:1561–1568

211 Chellapandian M, Sastry CA: Vermiculite as an economic support for immobilisation of neutral protease. *Bioprocess Eng* 1992, 8:27–31

212 Kheirolomoom A, Khorasheh F, Fazelinia H: Influence of external mass-transfer limitation on apparent kinetic parameters of penicillin G acylase immobilised on non-porous ultrafine silica particles. *J Biosci Bioeng* 2002, 93:125–129

213 Xu JG, Feng QW, Dong H, Wei Y: Stability of immobilised horseradish peroxidase in mesoporous silica sol–gel materials. *Book of Abstracts, 219th ACS National Meeting,* 2000

214 He J, Li X, Evans DG, Duan X, Li C: A new support for the immobilisation of penicillin acylase. *J Mol Catal B: Enzym* 2000, 11:45–53

215 Ghosh D, Mukherjee E, Dutta J: Stabilisation of prostaglandin synthetase by immobilisation of goat seminal microsomes on silica gel-G. *Indian J Biochem Biophys* 1990, 27:76–80

216 Adlercreutz P: On the importance of the support material for enzymatic synthesis in organic media. Support effects at controlled water activity. *Eur J Biochem* 1991, 199:609–614

217 Srrralha FN, Lopes JM, Aires-Barros MR, Prazeres DMF, Cabral JMS, Lemos F, Ribeiro FR: Stability of a recombinant cutinase immobilised on zeolites. *Enzyme Microb Technol* 2002, 31:29–34

218 Kurokawa Y, Suzuki K, Tamai Y: Adsorption and enzyme (β-galactosidase and α-chymotrypsin): immobilisation properties of gel fibre prepared by the gel formation of cellulose acetate and titanium iso-propoxide. *Biotechnol Bioeng* 1998, 59:651–656

219 Kurokawa Y, Ohta H, Okubo M, Takahashi M: Formation and use in enzyme immobilisation of cellulose acetate–metal alkoxide gels. *Carbohydr Polym* 1994, 23:1–4

220 Da Silva LRD, Gushikem Y, Kubota LT: Horseradish peroxidase enzyme immobilised on titanium(IV) oxide coated cellulose microfibers: study of the enzymic activity by flow injection system. *Colloid Surf B: Biointerf* 1996, 6:309–315

221 Furusaki S, Asai N: Enzyme immobilisation by the Coulomb force. *Biotechnol Bioeng* 1983, 25:2209–2219

222 Okubo M, Ahmad H: Adsorption of enzymes on to submicron-sized temperature-sensitive composite polymer particles and its activity. *Colloid Surf A: Physicochem Eng Aspect* 1999, 153:429–433

223 Somers WAC, Lojenga AK, Bonte A, Rozie HJ, Visser J, Rombouts FM, van't Riet K: Alginate–starch copolymers and immobilised starch as affinity adsorbents for α-amylase. *Biotechnol Appl Biochem* 1993, 18:9–24

224 Tomb WH, Weetall HH: Enzymes bound to carriers having a metal oxide surface layer. *US Pat 3,783,101,* 1974

225 Mutsugawa T, Fujishima T, Kuninaka A, Yoshino H: Immobilised enzyme composite. *Jpn Kokai Tokkyo Koho 52,114,090* 1977

226 Murray M, Rooney D, Van Neikerk M, Montenegro A, Weatherley LR: Immobilisation of lipase on to lipophilic polymer particles and application to oil hydrolysis. *Process Biochem* 1997, 32:479–486

227 Deetz JS, Rozzell JD: Enzymic catalysis by alcohol dehydrogenases in organic solvents. *Ann NY Acad Sci* 1988, 542:230–234

228 Taylor MJ, Cheryan M, Richardson T, Olson NF: Pepsin immobilised on

inorganic supports for the continuous coagulation of skim milk. *Biotechnol Bioeng* 1977, 19:683–700

229 Garwood GA, Mortland MM, Pinnavaia TJ: Immobilisation of glucose oxidase on montmorillonite clay: hydrophobic and ionic modes of binding. *J Mol Catal* 1983, 22:153–163

230 Ghosh BK, Mukherjea RN, Bhattacharya P: Nature of enzyme– matrix binding in trypsin immobilised on molecular sieve type 4A. *Indian J Chem Section A: Inorg Phys Theoret Anal* 1982, 21:111–113

231 Okos MR, Grulke EA, Syverson A: Hydrolysis of lactose in acid whey using β-galactosidase adsorbed to a phenol formaldehyde resin. *J Food Sci* 1978, 43:566–571

232 Kobayashi K, Kageyama B, Yagi S, Sonoyama T: A novel method of immobilizing penicillin acylase on basic anion exchange resin with hexamethylene diisocyanate and p-hydroxybenz-aldehyde. *J Ferment Bioeng* 1992, 74:410–12

233 Karim MR, Hashinaga C: Preparation and properties of immobilised pumnelo, limonoid glycosyltransferase. *Process Biochem* 2002, 38:809–814

234 Kimura K, Yokote Y, Fujita M, Samejima H: Production of high-fructose syrup using glucoamylase and glucose isomerase immobilised on phenol–formaldehyde resin. *Enzyme Eng* 1978, 3:531–536

235 Aoshima R, Kanda Y, Takada A, Yamashita A: Sulfonated poly(vinylidene fluoride) as a biomaterial: immobilisation of urokinase and biocompatibility. *J Biomed Mater Res* 1982, 16:289–299

236 Sugii A, Ogawa N, Matsumoto M: Preparation of new hydrophilic poly(vinylpyridine) beads and their application to immobilisation of urease. *J Appl Polym Sci* 1986, 32:4931–4938

237 Elcin YM, Sacak M: Methacrylic acid–acrylamide-γ-poly(ethylene-terephthalate) fibers for urea hydrolysis. *J Chem Technol Biotechnol* 1995, 63:174–180

238 Kjellen KG, Neujahr HY: Immobilisation of phenol hydroxylase. *Biotechnol Bioeng* 1979, 21:715–719

239 Bektenova GA, Kudaibergenov SE, Bekturov EA: Interaction of catalase with cationic hydrogels: influence of pH, kinetics of process and isotherms of adsorption. *Polym Adv Technol* 1999, 10:141–145

240 Konecny J: The immobilisation of a stable esterase by entrapment, covalent binding and adsorption. *Enzyme Eng* 1978, 3:11–18

241 Wasserman BP, Jacobson BS, Hultin HO: Explanation of anomalous binding kinetics with a high yield immobilised enzyme system. *Biochim Biophys Acta* 1981, 657:52–57

242 Abdel-Naby MA, Hashem AM, Esawy MA, Abdel-Fattah AF: Immobilisation of *Bacillus subtilis* α-amylase and characterisation of its enzymic properties. *Microbiol Res* 1999, 153:319–325

243 Mateo C, Abian O, Fernandez-Lafuente R, Guisan JM: Reversible enzyme immobilisation via a very strong and nondistorting ionic adsorption on support-polyethylenimine composites. *Biotechnol Bioeng* 2000, 68:98–105

244 Li MQ, He BL: Immobilisation of aminoacylase on modified methyl acrylate–divinylbenzene copolymers. *Chinese J React Polym* 1993, 2:34–41

245 Kulik EA, Kato K, Ivanchenko MI, Ikada Y: Trypsin immobilisation on to polymer surface through grafted layer and its reaction with inhibitors. *Biomaterials* 1993, 14:763–769

246 Matoba S, Tsuneda S, Saito K, Sugo T: Highly efficient enzyme recovery using a porous membrane with immobilised tentacle polymer chains. *Bio/Technol* 1995, 13:795–797

247 Albayrak N, Yang ST: Immobilisation of β-galactosidase on fibrous matrix by polyethyleneimine for production of galacto-oligosaccharides from lactose. *Biotechnol Prog* 2002, 18:240–251

248 Kogusi Y, Takahashi K, Lopez C: Large scale immobilisation of lipase from *Pseudomonas fluorescens* Biotype I and application for sardine oil hydrolysis. *J Am Oil Chem* 1995, 72:1281–1285

249 Yokote Y, Maeda S, Yabushita H Noguchi S, Kimura K, Samejima H: Production of l-aspartic acid by *E. coli*

aspartase immobilised on phenol–formaldehyde resin. *J Solid-Phase Biochem* 1978, 3:247–261
250 Pawar HS, Deshmukh DR: Immobilisation of d-xylose (d-glucose) isomerase from a Chainia species. *Prep Biochem* 1994, 24:143–150
251 Kawakita H, Sugita K, Saito K, Tamada M, Sugo T, Kawamoto H: Optimization of reaction conditions in production of cycloisomaltooligosaccharides using enzyme immobilised in multilayers on to pore surface of porous hollow-fibre membranes. *J Membr Sci* 2002, 205:175–182
252 Miura S, Kubota N, Kawakita H, Saito K, Sugita K, Watanabe K, Sugo T: High-throughput hydrolysis of starch during permeation across α-amylase-immobilised porous hollow-fibre membranes. *Radiat Phys Chem* 2002, 63:143–149
253 Godjevargova T, Dimov A: Grafting of acrylonitrile copolymer membranes with hydrophilic monomers for immobilisation of glucose oxidase. *J Appl Polym Sci* 1995, 57:487–491
254 Itoh N, Cheng LY, Izumi Y, Yamada H: Immobilised bromoperoxidase of *Corallina pilulifera* as a multifunctional halogenating biocatalyst. *J Biotechnol* 1987, 5:29–38
255 Kroutil W, Orru RVA, Faber K: Stabilization of Nocardia EH1 epoxide hydrolase by immobilization. *Biotech Lett* 1998, 20:373–377
256 Karboune S, Amourache L, Nellaiah H, Morisseau C, Baratti J: Immobilisation of the epoxide hydrolase from *Aspergillus niger*. *Biotechnol Lett* 2001, 23:1633–1639
257 Oh JT, Kim JH: Preparation and properties of immobilised amyloglucosidase on non-porous PS/PNaSS microspheres. *Enzyme Microb Technol* 2000, 27:356–361
258 Izui K, Fujita N, Katsuki H: Phosphoenolpyruvate carboxylase of *Escherichia coli*. Hydrophobic chromatography using specific elution with allosteric inhibitor. *J Biochem* 1982, 92:423–432
259 Hofstee BHJ, Otillio NF: Immobilisation of enzymes through noncovalent binding to substituted agaroses. *Biochem Biophysl Res Commun* 1973, 53:1137–1144

260 Ma L, Wang XT, You DL, Tang S, Huang ZL, Cheng YH: Immobilisation of prostaglandin synthetase by hydrophobic adsorption. *Appl Biochem Biotechnol* 1996, 56:223–233
261 Nemat-Gorgani M, Karimian K: Interaction of proteins with Triton X-100-substituted Sepharose 4B. *Biotechnol Bioeng* 1984, 26:565–572
262 Madry N, Zocher R, Grodzki K, Kleinkauf H: Selective synthesis of depsipeptides by the immobilised multienzyme enniatin synthetase. *Appl Microb Biotechnol* 1984, 20:83–86
263 Mislovicova D, Gemeiner P, Breier A: Study of porous cellulose beads as an affinity adsorbent via quantitative measurements of interactions of lactate dehydrogenase with immobilised anthraquinone dyes. *Enzyme Microb Technol* 1988, 10:568–573
264 Butler LG: Enzyme immobilisation by adsorption on hydrophobic derivatives of cellulose and other hydrophilic materials. *Arch Biochem Biophys* 1975, 171:645–650
265 Corfield AP, Clarice DAC, Wember M, Schauer R: The interaction of *Clostridium perfringens* sialidase with immobilised sialic acids and sialyl-glycoconjugates. *Glycoconjugat J* 1985, 2:45–60
266 Carbone K, Casarci M, Varrone M: Crosslinked poly(vinyl alcohol) supports for the immobilisation of a lipolytic enzyme. *J Appl Polym Sci* 1999, 74:1881–1889
267 Warmuth W, Wenzig E, Mersmann A: Selection of a support for immobilisation of a microbial lipase for the hydrolysis of triglycerides. *Bioprocess Eng* 1995, 12:87–93
268 Nemat-Gorgani M, Karimian K, Mohanazadeh F: Synthesis, characterisation, and properties of hexadecyl silica: a novel hydrophobic matrix for protein immobilisation. *J Am Chem Soc* 1985, 107:4756–4759
269 Kondo K, Matsumoto M, Maeda R: Kinetics of hydrolysis reaction of glycol chitin with a novel enzyme immobilised through nonionic surfactant adsorbed on silica gel. *Ad Chitin Sci* 1997, 2:220–222

270 Adikane HV, Singh RK, Thakar DM, Nene SN: Single-step purification and immobilisation of penicillin acylase using hydrophobic ligands. *Appl Biochem Biotechnol* 2001, 94:127–134

271 Mislovicová D, Masárová J, Vikartovská A, Gemeiner P, Michalková E: Biospecific immobilization of mannan–penicillin G acylase neoglycoenzyme on Concanavalin A-bead cellulose. *J Biotechnol* 2004, 110:11–19

272 Froehner SC, Eriksson KE: Properties of the glycoprotein laccase immobilised by two methods. *Acta Chem Scand Series B: Org Chem Biochem* 1975, 29:691–694

273 Abdel-Naby MA: Immobilisation of *Aspergillus Niger* NRC 107 xylanase and β-xylosidase, and properties of the immobilised enzymes. *Appl Biochem Biotechnol* 1993, 38:69–81

274 Fusek M, Turkova J, Stovickova J, Franek F: Polyclonal anti-chymotrypsin antibodies suitable for oriented immobilisation of chymotrypsin. *Biotechnol Lett* 1988, 10:85–90

275 Liu XC, Scouten WH: Studies on oriented and reversible immobilisation of glycoprotein using novel boronate affinity gel. *J Mol Recognit* 1996, 9:462–467

276 Fowell SL, Chase HA: A comparison of some activated matrixes for preparation of immunosorbents. *J Biotechnol* 1986, 4:355–368

277 Johansen PA, Jennings I, Cotton RGH, Kuhn DM: Immobilisation of tryptophan hydroxylase by immune adsorption: a method to study regulation of catalytic activity. *Brain Res Bull* 1992, 29:949–953

278 Huang XL, Walsh MK, Swaisgood HE: Simultaneous isolation and immobilisation of streptavidin-β-galactosidase: some kinetic characteristics of the immobilised enzyme and regeneration of bioreactors. *Enzyme Microb Technol* 1996, 19:378–383

279 Sakai K, Uchiyama T, Matahira Y, Nanjo F: Immobilisation of chitinolytic enzymes and continuous production of N-acetylglucosamine with the immobilised enzymes. *J Ferm Bioeng* 1991, 72:168–172

280 Wirth HJ, Hearn MTW: High-performance liquid chromatography of amino acids, peptides and proteins. Modified porous zirconia as sorbents in affinity chromatography. *J Chromatogr* 1993, 646:143–151

281 Wang JQ, Bhattacharyya D, Bachas LG: Improving the activity of immobilised subtilisin by site-directed attachment through a genetically engineered affinity tag. *Fresenius J Anal Chem* 2001, 369:280–285

282 Lee P, Swaisgood HE: Characterisation of a chemically conjugated lipase bioreactor. *J Agric Food Chem* 1997, 45:3350–3356

283 Bilkova Z, Mazurova J, Churacek J, Hora D, Turkova J: Oriented immobilisation of chymotrypsin by use of suitable antibodies coupled to a non-porous solid support. *J Chromatogr A* 1999, 852:141–149

284 Halling PJ, Dunnill P: Recovery of free enzymes from product liquors by bioaffinity adsorption: trypsin binding by immobilised soybean inhibitor. *Eur J Appl Microb Biotechnol* 1979, 6:195–205

285 Friend, BA, Shahani KM: Characterization and evaluation of Aspergillus oryzae lactase coupled to a regenerable support. *Biotechnol Bioeng* 1982, 24:329–45

286 Harvey MJ, Lowe CR, Dean PDG: Affinity chromatography on immobilised adenosine 5′-monophosphate. 5. Influence of temperature on the binding of dehydrogenases and kinases. *Eur J Biochem* 1974, 41:353–357

287 Palanisamy UD, Winzor DJ, Lowe CR: Synthesis and evaluation of affinity adsorbents for glycoproteins: an artificial. *J Chromatogr B, Biomed Sci Appl* 2000, 746:265–281

288 Campbell P, Glover GI: Activation of prephenate dehydratase by adsorption to agarose derivatives. *J Solid-Phase Biochem* 1979, 3:107–117

289 Stella AM, Batlle AM: Del C porphyrin biosynthesis – immobilised enzymes and ligands. V. Purification of aminolevulinate dehydratase from bovine liver by affinity chromatography. *Int J Biochem* 1977, 8:353–358

290 Coombe RG, George AM: An alternative coupling procedure for preparing

activated Sepharose for affinity chromatography of penicillinase. *Aust J Biol Sci* 1976, 29:305–316

291 Powers JD, Kilpatrick PK, Carbonell RG: Trypsin purification by affinity binding to small unilamellar liposomes. *Biotechnol Bioeng* 1990, 36:506–519

292 Haupt K, Vijayalakshmi MA: Interaction of catechol-2,3-dioxygenase of *Pseudomonas putida* with immobilised histidine and histamine. *J Chromatogr* 1993, 644:289–297

293 Choi SH, Lee KP, Lee JG: Adsorption behaviour of urokinase by polypropylene film modified with amino acids as affinity groups. *Microchem J* 2001, 68:205–213

294 Burton NP, Lowe CR: Design of novel affinity adsorbents for the purification of trypsin-like proteases. *J Mol Recognit* 1992, 5:55–68

295 Beitle RR, Ataai MM: Immobilised metal affinity chromatography and related techniques. *AIChE Symp Ser* 1992, 290:34–44

296 Denizli A, Yavuz H, Garipcan B, Arica MY: Non-porous monosize polymeric sorbents: dye and metal chelate affinity separation of lysozyme. *J Appl Polym Sci* 2000, 76:115–124

297 Arica MY, Denizli A, Salih B, Piskin E, Hasirci V: Catalase adsorption on to Cibacron blue F3G-A and Fe(III)-derivatized poly(hydroxyethyl methacrylate) membranes and application to a continuous system. *J Membr Sci* 1997, 129:65–76

298 Kacar Y, Yakup AM: Preparation of reversibly immobilised lysozyme on to procion green H-E4BD-attached poly(hydroxyethylmethacrylate) film for hydrolysis of bacterial cells. *Food Chem* 2001, 75:325–332

299 Anspach B, Unger KK, Davies J, Hearn MTW: Affinity chromatography with triazine dyes immobilised on to activated non-porous monodisperse silicas. *J Chromatogr* 1988, 457:195–204

300 Baran T, Arica MY, Denizli A, Hasirci V: Comparison of β-galactosidase immobilisation by entrapment in and adsorption on poly(2-hydroxyethyl-methacrylate) membranes. *Polym Int* 1997, 44:530–536

301 Akgol S, Gulay YY, Denizli BA, Arica MY: Reversible immobilisation of urease on to procion Brown MX-5 BR-Ni(II) attached polyamide hollow fibre membranes. *Process Biochem* 2002, 38:675–683

302 Schlereth DD: Preparation of gold surfaces with biospecific affinity for NAD(H)-dependent lactate dehydrogenase. *Sens Actuators B: Chem* 1997, 43:78–86

303 Hughes P, Lowe CR, Sherwood RF: Metal ion-promoted binding of proteins to immobilised triazine dye affinity adsorbents. *Biochim Biophys Acta* 1982, 700:90–100

304 Small DAP, Atkinson T, Lowe CR: High-performance liquid affinity chromatography of enzymes on silica-immobilised triazine dyes. *J Chromatogr* 1981, 216:175–190

305 Horowitz PM: Purification of thiosulfate sulfurtransferase by selective immobilisation on blue agarose. *Anal Biochem* 1978, 86:751–753

306 Yavuz H, Bayramoglu G, Kacar Y, Denizli A, Yakup AM: Congo Red attached monosize poly(HEMA-co-MMA) microspheres for use in reversible enzyme immobilisation. *Biochem Eng J* 2002, 10:1–8

307 Liu YC, Ledger R, Stellwagen E: Quantitative analysis of protein: immobilised dye interaction. *J Biol Chem* 1984, 259:3796–3799

308 Tanyolaç D, Yürüksoy BI, and Özdural AR: Immobilization of a thermostable α-amylase, Termamyl®, onto nitrocellulose membrane by Cibacron blue F3G-A dye binding. *Biochemical Engineering Journal* 1998, 2:179–186

309 Porath J, Hansen P: Cascade-mode multiaffinity chromatography: fractionation of human serum proteins. *J Chromatogr* 1991, 550:751–764

310 Bacolod MD, El RZ: High-performance metal chelate interaction chromatography of proteins with silica-bound ethylenediamine-N,N'-diacetic acid. *J Chromatogr* 1990, 512:237–247

311 Liu YU, Yu S: Application of a new chelating resin (IDAO) in immobilised metal ion affinity chromatography. *Fenxi Ceshi Tongbao* 1991, 10:16–20

312 Zachariou M, Hearn MTW: High-performance liquid chromatography of amino acids, peptides and proteins: CXXI. 8-hydroxyquinoline-metal chelate chromatographic support: an additional mode of selectivity in immobilised-metal affinity chromatography. *J Chromatogr* 1992, 599:171–177

313 Zachariou M, Traverso I, Hearn MT: High-performance liquid chromatography of amino acids, peptides and proteins: CXXXI. O-phosphoserine as a new chelating ligand for use with hard Lewis metal ions in the immobilised-metal affinity chromatography of proteins. *J Chromatogr.* 1993, 646:107–120

314 Boden V, Winzerling JJ, Vijayalakshmi M, Porath J: Rapid one-step purification of goat immunoglobulins by immobilised metal ion affinity chromatography. *J Immunol Methods* 1995, 181:225–232

315 Millot MC, Herve F, Sebille B: Retention behaviour of proteins on poly(vinylimidazole)–copper(II) complexes supported on silica: application to the fractionation of desialylated human α 1-acid glycoprotein variants. *J Chromatogr B: Biomed Appl* 1995, 664:55–57

316 Jiang W, Graham B, Spiccia L, Hearn MT: Protein selectivity with immobilised metal ion-tacn sorbents: chromatographic studies with human serum proteins and several other globular proteins. *Anal Biochem* 1998, 255:47–48

317 Ueda EKM, Goutb PW, Morganti L: Current and prospective applications of metal ion protein binding. *J Chromatogr A* 2003, 988:1–23

318 Chaouk H, Hearn MT: New ligand, N-(2-pyridylmethyl) aminoacetate, for use in the immobilised metal ion affinity chromatographic separation of proteins. *J Chromatogr A* 1999, 852:105–115

319 Chaouk H, Hearn MT: Examination of the protein binding behaviour of immobilised copper(II)–2,6-diaminomethylpyridine and its application in the immobilised metal ion affinity chromatographic separation of several human serum proteins. *J. Biochem Biophys Methods* 1999, 39:161–177

320 Hidalgo A, Betancor L, Gallego FL, Moreno R, Berenguer J, Fernandez-Lafuente R, Guisan JM: Purification of a catalase from *Thermus thermophilus* via IMAC chromatography: Effect of the support. *Biotechnol progr* 2004, 70:1578–1582

321 Meszarosova K, Tischchenko G, Bouchal K, Bleha M: Immobilised metal affinity adsorbents based on hydrophilic methacrylate polymers and their interaction with immunoglobulins. *React Funct Polym* 2003, 56:27–35

322 Sulkowski E: The saga of IMAC and MIT. *BioEssays* 1989, 10:170–175

323 Anspach FB: Silica-based metal chelate affinity sorbents. II. Adsorption and elution behaviour of proteins on iminodiacetic acid affinity sorbents prepared via different immobilisation techniques. *J Chromatogr A* 1994, 676:249–266

324 Kirstein D: Chelate mediated immobilisation of proteins. *Adv. Mol Cell Biol A* 1996, 15:247–256

325 Coulet P, Carlsson J, Porath J: Immobilisation of enzymes on metal-chelate regenerable carriers. *Biotechnol Bioeng* 1981, 23:663–668

326 Ljungquist C, Breitholtz A, Brink-Nilsson H, Moks T, Uhlen M, Nilsson B: Immobilisation and affinity purification of recombinant proteins using histidine peptide fusions. *Eur J Biochem* 1989, 186:563–569

327 Anspach FB, Altmann-Haase G: Immobilised-metal-chelate regenerable carriers: (I) adsorption and stability of penicillin G amidohydrolase from *Escherichia coli*. *Biotechnol Appl Biochem* 1994, 20:313–322

328 Afaq S, Iqbal J: Immobilisation and stabilisation of papain on chelating Sepharose: a metal chelate regenerable carrier. *EJB Electron J Biotechnol* 2001, 4:1–5

329 Chaga G, Andersson L, Ersson B, Porath J: Purification of two muscle enzymes by chromatography on immobilised ferric ions. *Biotechnol Appl Biochem* 1989, 11:424–431

330 Kashlev M, Martin E, Polyakov A, Severinov K, Nikiforov V, Goldfarb A: Histidine-tagged RNA polymerase: dissection of the transcription cycle using immobilised enzyme. *Gene* 1993, 130:9–14

331 Yamskov IA, Budanov MV, Davankov V: Coordination-ionic enzyme immobilisation. Effect of the metal and stationary ligand on the properties of immobilised penicillin amidohydrolase A. *Biokhimiya (Moscow)* 1981, 46:1603–1608

332 Yamskov IA, Budanov MV, Davankov VA: Coordination-ionic enzyme immobilisation. II. Penicillin amidohydrolase binding through cobalt, nickel and iron complexes. *Bioorg. Khim* 1980, 6:1404–1408

333 Lei FH, Huang ZY, Shi XH: Immobilisation of polyphenol oxidase by coordination with Cu(II) in poly(vinyl alcohol) grafted glutamic acid resin. *Yingyong Huaxue* 2001, 18:881–884

334 Cabral JMS, Kennedy JF, Novais JM: Investigation of the operational stabilities and kinetics of glucoamylase immobilised on alkylamine derivatives of titanium(IV)-activated porous inorganic supports. *Enzyme Microb Technol* 1982, 4:343–348

335 Moriya K, Tanizawa K, Kanaoka Y: Immobilised chymotrypsin by means of Schiff base copper(II) chelate. *Biochem Biophys Res Commun* 1989, 162:408–414

336 Piacquadio P, De Stefano G, Sammartino M, Sciancalepore V: Apple juice stabilisation by laccase (EC. 1.10.3.2) immobilised on metal-chelate regenerable carriers. *Industrie delle Bevande* 1998, 27:378–383

337 Servili M, De Stefano G, Piacquadio P, Sciancalepore V: A novel method for removing phenols from grape must. *Am J Enology Viticulture* 2000, 51:357–361

338 Hoa LF, Lia SY, Lin SC, Hsu WH: Integrated enzyme purification and immobilisation processes with immobilised metal affinity adsorbents. *Process Biochem* 2004, 39:1573–1581

339 Shi QH, Tai Y, Dong XY, Bai S, Sun Y: Chitosan-coated silica beads as immobilised metal affinity support for protein adsorption, *Biochem Eng J* 2003, 16:317–322

340 Suen RB, Lin SC, Hsu WH: Hydroxyapatite-based immobilised metal affinity adsorbents for protein purification. *J Chromatogr* 2004, 1048:31–39

341 Habibi-Rezaei M, Nemat-Gorgani M: Adsorptive immobilisation of intestinal brush border membrane on Triton X-100-substituted Sepharose 4B. *Appl Biochem Biotechnol* 2002, 97:79–90

342 Azari F, Hosseinkhani S, Nemat-Gorgani M: Use of reversible denaturation for adsorptive immobilisation of urease. *Appl Biochem Biotechnol* 2001, 94:265–277

343 Hosseinkhani S, Nemat-Gorgani M: Partial unfolding of carbonic anhydrase provides a method for its immobilisation on hydrophobic adsorbents and protects it against irreversible thermoinactivation. *Enzyme Microb Technol* 2003, 33:179–184

344 Goldmacher VS: Immobilisation of protein molecules on liposomes Achoraggregates artificially bound unsaturated hydrocarbon tails. *Biochem Pharmacol* 1983, 32:1207–1210

345 Suzuki N, Quesenberry MS, Wang JK, Lee RT, Kobayashi K, Lee YC: Efficient immobilisation of proteins by modification of plate surface with polystyrene derivatives. *Anal Biochem* 1997, 247:412–416

346 Solomon BA, Chen CC, Colton CK: Immobilisation of acetate kinase on functionalized solid-core polymeric beads. *Enzyme Eng* 1978, 4:105–108

347 Hu MC, Haering ER, Geankoplis CJ: Diffusion and adsorption phenomena in an immobilised enzyme reactor using adsorbed polymer for attachment of the enzyme in porous alumina particles. *Chem Eng Sci* 1985, 40:2241–2248

348 Royer GP, Green GM, Sinha BK: Rigid support for the immobilisation of enzymes. *J Macromol Sci Chem* 1976, A-10:289–307

349 Fiedurek J, Lobarzewski J, Wojcik A, Wolski T: Optimization of enzyme immobilisation on keratin- or polyamide-coated bead-shaped polymeric matrix. *Biotechnol Bioeng* 1986, 28:747–750

350 Bahulekar RV, Prabhune AA, SivaRaman H, Ponrathnam S: Immobilisation of penicillin G acylase on functionalized macroporous polymer beads. *Polymer* 1993, 34:163–166

351 Bahulekar RV, Ponrathnam S, Uphade BS, Ayyangar NR, Kumar KK, Shewale JG: Immobilisation of penicillin G acylase on to alumina: effect of hydrophilicity. *Biotechnol Tech* 1991, 5:401–404

352 Horvath C: Pellicular immobilised enzymes. *Biochim Biophys Acta* 1974, 358:164–177

353 Minamoto Y, Yugai Y: Preparation and properties of various enzymes covalently immobilised on polymethylglutamate. *Chem Pharm Bull* 1980, 28:2052–2058

354 Hwang S, Lee KT, Park JW, Min BR, Haam SJ, Ahn IS, Jung JK: Stabilisation of *Bacillus strarothermophillus* lipase immobilised on surface-modified silica gels. *Biochem Eng J* 2004, 17:85–90

355 Ge YB, Zhou H, Kong W, Tong Y, Wang SY, Li W: Immobilisation of glucose isomerase and its application in continuous production of high fructose syrup. *Appl Biochem Biotechnol* 1998, 69:203–215

356 Bahar T, Tuncel A: Immobilisation of invertase on to crosslinked poly(*p*-chloromethylstyrene) beads. *J Appl Polym Sci* 2002, 83:1268–1279

357 Liu CC, Lahoda EJ, Galasco RT, Wingard LB Jr: Immobilisation of lactase on carbon. *Biotechnol Bioeng* 1975, 17:1695–1696

358 Charles JG, Catherine ML: Properties of enzymes immobilised by the diazotized *m*-diaminobenzene. *Methods Biotechnol Bioeng* 1977, 19:349–364

359 Charles JG, Catherine ML, Jones CM, Baker SA: *BBA* 1974, 341:457

360 Yugari YY, Minamoto Y, Komiya K, Mitsugi K, Mimura N: Immobilisation of enzyme on PMG and its application to the urea monitoring apparatus. *Enzyme Eng* 1978, 4:223–225

361 Royer GP, Green GM: Immobilised pronase. *Papers presented at the meeting – Am Chem Soc, Div Org Coat Plastics Chem* 1971, 31:357–360

362 Chaplin MF, Kennedy JF: Magnetic, immobilised derivatives of enzymes. *Carbohydr Res* 1976, 50:267–274

363 Kennedy JF, Chaplin MF: Reactive matrixes. *US patent 4,070,246*, 1978

364 Zacharieva EI, Konstantinov CI: Coated macroporous carriers with oxirane groups and their reactivity. *Biomaterials* 1993, 14:953–957

365 Kipper H, Egorov KhR, Kivisilla K: Characteristics of polymer-modified inorganic carriers. *Tr. Tallin. Politekh. Inst.* 1979, 465:33–39

366 Bahulekar RV, Prabhune AA, SivaRaman H, Ponrathnam S: Immobilisation of penicillin G acylase on functionalized macroporous polymer beads. *Polymer* 1993, 34:163–166

367 Lee YW, Stanish I, Rastogi V, Cheng TC, Singh A: Sustained enzyme activity of organophosphorus hydrolase in polymer encased multilayer assemblies. *Langmuir* 2003, 19:1330–1336

368 Minamoto Y, Yugai Y: Preparation and properties of various enzymes covalently immobilised on polymethylglutamate. *Chem Pharm Bull* 1980, 28:2052–2058

369 Ge YB, Zhou H, Kong W, Tong Y, Wang SY, Li W: Immobilisation of glucose isomerase and its application in continuous production of high fructose syrup. *Appl Biochem Biotechnol* 1998, 69:203–215

370 Naka K, Kitano N, Iwaki K, Ohki A, Maeda S: Adsorption of amphiphilic blocks copolymer-catalase aggregate at silica-gel. *Chem Lett* 1995, 11:1037–1038

371 Naka K, Iwaki K, Kitano N, Ohki A, Maeda S: Adsorption of enzymes at silica-gel with amphiphilic blocks copolymers. *Polym J (Tokyo)* 1996, 28:911–915

372 Katchalski E, Levin Y, Solomon B: Insoluble enzymes. *US Patent 3873426*, 1975

373 Aizawa M, Coughlin RW, Charles M: Activation of alumina with bovine serum albumin for immobilising enzymes. *Biotechnol Bioeng* 1975, 17:1369–1372

374 Horowitz R, Whitesides GM: Generalized affinity chromatography: enzyme-sulfonamide conjugates can be isolated by adsorption on immobilised carbonic anhydrase. *J Am Chem Soc* 1978, 100:4632–4633

375 Lambrecht R, Ulbrich-Hofmann R: The adsorptive immobilisation of phospholipase D mediated by calcium ions. *Biotechnol Bioeng* 1993, 41:833–836

376 Kondo K, Matsumoto M, Maeda R: Kinetics of hydrolysis reaction of glycol chitin with a novel enzyme immobilised through nonionic surfactant adsorbed on silica gel. *Adv Chitin Sci* 1997, 2:220–227

377 Ho Chih-Hu, Limberis L, Stewart RJ: Metal-chelating Pluronics for immobilisation of histidine-tagged proteins at interfaces: firefly luciferase on the polystyrene beads. *Jiemian Kexue Huizhi* 1997, 20:95–106

378 Naidja A, Huang PM, Bollag JM: Activity of tyrosinase immobilised on hydroxyaluminium–montmorillonite complexes. *J Mol Catal A: Chem* 1997, 115:305–316

379 Kennedy JF, Chaplin MF: An investigation of the properties of glucoamylase immobilised on glass beads involving 5-diazosalicylic acid bonded to a titanium(IV) oxide film. *Enzyme Microb Technol* 1979, 1:197–200

380 Nys PS, Savitskaya EM, Shellenberg NN: Penicillin amidase from *Escherichia coli*. Comparative study on the stability of penicillin amidase immobilised by various methods. *Antibiotiki* (Moscow) 1981, 22:130–136

381 Ampon K, Means GE: Immobilisation of proteins on Organic polymer beads. *Biotechnol Bioeng* 1988, 32:689–697

382 Kamogashira T, Mihara S, Tamaoka H, Doi T: 6-Aminopenicillanic acid by penicillin hydrolysis with immobilised enzyme. *Jpn Kokai Tokkyo Koho* 47028187, 1972

383 Cao L, Bornscheuer UT, Schmid RD: Lipase-catalyzedsolid-phase synthesis of sugar esters.Influence of immobilization on productivity and stability of the enzyme. *J Mol Catal B: Enzymatic* 1999, 6:279–285

384 Basri M, Ampon K, Yunus WMW, Razak CNAR, Selleh B: Immobilisation of hydrophobic lipase derivatives on to organic polymer beads. *J Chem Technol Biotechnol* 1994, 59:37–44

385 Nakamura N, Horikoshi K: Production of Schardinger β-dextrin by soluble and immobilised cyclodextrin glycosyltransferase of an alkalophilic Bacillus sp. *Biotechnol Bioeng* 1977, 19:87–99

386 Lee SH, Shin HD, Lee YH: Evaluation of immobilisation methods for cyclodextrin glucanotransferase and characterisation of its enzymic properties. *J Microb Biotechnol* 1991, 1:54–62

387 Barroug A, Fastrez J, Lemaitre J, Rouxhet P: Adsorption of succinylated lysozyme on hydroxyapatite. *J Colloid Interface Sci* 1997, 189:37–42

388 Wu HL, Means GE: Immobilisation of proteins by reductive alkylation with hydrophobic aldehydes *Biotechnol Bioeng* 1981, 23:855–861

389 Kozlova NO, Bruskovskaya IB, Melik-Nubarov NS, Yaroslavov AA, Kabanov VA: Catalytic properties and conformation of hydrophobized chymotrypsin incorporated into a bilayer lipid membrane. *FEBS Lett* 1999, 461:141–144

390 Woodward J, Wohlpart DL: Properties of native and immobilised preparations of β-d-glucosidase from *Aspergillus niger*. *J Chem Technol Biotechnol* 1982, 32:547–552

391 Pugniere M, San JC, Coletti-Previero MA, Previero A: Immobilisation of enzymes on alumina by means of pyridoxal 5′-phosphate. *Biosci Rep* 1988, 8:263–269

392 Basinska T, Slomkowski S: Attachment of horseradish peroxidase (HRP) on to the poly(styrene/acrolein) latexes and on to their derivatives with amino groups on the surface; activity of immobilised enzyme. *Colloid Polym Sci* 1995, 273:431–438

393 Huang XL, Catignani GL, Swaisgood HE: Immobilisation of biotinylated transglutaminase by bioselective adsorption to immobilised avidin and characterisation of the immobilised activity. *J Agric Food Chem* 1995, 43:895–901

394 Chaga G:A general method for immobilization of glycoproteins on regenerable immobilized metal-ion carriers: application to glucose oxidase from Penicillium chrysogenum and. Biotechnol Appl Biochem 1994, 20:43–53

395 Furukawa S, Ono T, Ijima H, Kawakami K: Activation of protease by sol–gel entrapment into organically modified hybrid silicates. *Biotechnol Lett* 2002, 24:13–16

396 Atrat P, Groh H: Steroid transformation with immobilised microorganisms. VI. The reverse reaction of steroid-1-dehydrogenases from different micoorganisms in immobilised state. *Z Alg Microbiol* 1981, 21:3–6

397 Kumakura M, Yoshida M, Asano M, Kaetsu I: Immobilisation of enzymes by radiation-induced polymerization of glass-forming monomers. Double entrapping of enzymes in the presence of various additives. *J Solid-Phase Biochem* 1977, 2:279–288

398 Nordvi B, Holmsen H: Effect of polyhydroxy compounds on the activity of lipase from *Rhizopus arrhizus* in organic solvent. *Prog Biotechnol* 1992, 8:355–361

399 Chauhan S, Nichkawade A, Iyengar MRS, Chattoo BB: Chainia penicillin V acylase: strain characteristics, enzyme immobilisation, and kinetic studies. *Curr Microbiol* 37:186–190

400 Yamade K, Fukushima S: Continuous alcohol production from starchy materials with a novel immobilised cell/enzyme bioreactor. *J Ferm Bioeng* 1989, 67:97–101

401 Fukushima S, Yamade K: Continuous ethanol production from starch using immobilised enzyme and cells. *Baiotekunoroji Kenkyu Hokokusho* (in Jpn) 1986:53–66

402 Fukushima S: External diffusion of enzymes entrapped with calcium arginate in packed beds. In: Weetall HH, Suzuki S (Eds) *Immobilised Enzyme Technol., Proc. U.S.–Jpn. Coop. Program Semin.* 1975, pp 225–240

403 Suwasono S, Rastall RA: Synthesis of oligosaccharides using immobilised 1,2-mannosides from *Aspergillus phoenicis*, Immobilisation-dependent modulation of product spectrum. *Biotechnol Lett* 1998, 20:15–17

404 Tanriseven A, Uludag YB, Dogan S: A novel method for the immobilisation of glucoamylase to produce glucose from maltodextrin. *Enzyme Microb Technol* 2002, 30:406–409

405 Bao J, Furumoto K, Fukunaga K, Nakao K: A kinetic study on air oxidation of glucose catalysed by immobilised glucose oxidase for production of calcium gluconate. *Biochem Eng J* 2001, 8:91–102

406 Kumakura M: Formation of immobilised enzyme beads by radiation and chemical reaction. *ACH – Models in Chemistry* 2000, 137:427–437

407 Nitto Electric Industrial Co Ltd. Preparation of immobilised enzymes. *Jpn. Kokai Tokkyo Koho 82-154856*, 1984

408 Dharmarajan TS, Deshpande JV, Divekar K: Production of cephalexin through immobilised *Xanthomonas compestris* acylase. *Indian J Pharm Sci* 1994, 56:126–128

409 Tsafack VC, Marquette CA, Pizzolato F, Blum LJ: Chemiluminescent choline biosensor using histidine-modified peroxidase immobilised on metal-chelate substituted beads and choline oxidase immobilised on anion-exchanger beads co-entrapped in a photocrosslinkable polymer. *Biosens Bioelectron* 2000, 15:125–133

410 Li B, Takahashi H: New immobilisation method for enzyme stabilisation involving a mesoporous material and an organic/inorganic hybrid gel. *Biotechnol Lett* 2000, 22:1953–1958

411 Lee JM, Woodward J: Properties and application of immobilised β-d-glucosidase coentrapped with *Zymomonas mobilis* in calcium alginate. *Biotechnol Bioeng* 1983, 25:2441–2451

412 Goncalves LRB, Giordano RLC, Giordano RC: A bidisperse model to study the hydrolysis of maltose using glucoamylase immobilised in silica and wrapped in pectin gel. *Braz J Chem Eng* 1997, 14:333–339

413 Kawai T, Saito K, Sugita K, Sugo T, Misaki H: Immobilisation of ascorbic acid oxidase in multilayers on to porous hollow-fibre membrane. *J Membr Sci* 2001, 191:207–213

414 Braun J, LeChanu P, Le Goffic F: The immobilisation of penicillin G acylase on chitosan. *Biotechnol Bioeng* 1989, 33:242–246

415 Koilpillai L et al: Immobilisation of penicillin G acylase on methacrylate polymers. *J Chem Technol Biotechnol* 1990, 49:173–182

416 Tsafack VC, Marquette CA, Pizzolato F, Blum LJ: Chemiluminescent choline biosensor using histidine-modified peroxidase immobilised on metal-chelate substituted beads and choline oxidase immobilised on anion-exchanger beads co-entrapped in a photocrosslinkable polymer. *Biosens Bioelectron* 2000, 15:125–133

417 Leca B, Blum LJ: Luminol electrochemiluminescence with screen-printed electrodes for low-cost disposable oxidase-based optical sensors. *Analyst* 2000, 125:789–791

418 Haynes R, Walsh KA: Enzyme envelopes on colloidal particles. *Biochem Biophys Res Commun* 1969, 36:235–242

419 Bagi K, Simon LM, Szajani B: Immobilisation and characterisation of porcine pancreas lipase. *Enzyme Microb Technol* 1997, 20:531–535

420 Boshoff A, Edwards W, Leukes WD, Rose PD, Burton SG: Immobilisation of polyphenol oxidase on nylon and polyethersulfone membranes: Effect on product formation. *Desalination* 1998, 115:307–312

421 Spagna G, Barbagallo RN, Casarini D, Pifferi PG: A novel chitosan derivative to immobilise α-l-rhamnopyranosidase from *Aspergillus niger* for application in beverage technologies. *Enzyme Microb Technol* 2001, 28:427–438

422 Stanley WL, Olson AC: Chemistry of immobilising enzymes. *J Food Sci* 1974, 39:660–666

423 Svensson B, Ottesen M: Immobilisation of β-glucanase and studies on its degradation of barley β-glucan. *Carlsberg Res Commun* 1978, 43:5–14

424 Koilpillai L, Gadre RA, Bhatnagar S, Raman RC, Ponrathnam S, Kumar KK, Ambekar GR, Shewale J: Immobilisation of penicillin G acylase on methacrylate polymers. *J Chem Technol* 1990, 49:173–182

425 Kosugi Y, Suzuki H: Functional immobilisation of lipase eliminating lipolysis product inhibition. *Biotechnol Bioeng* 1992, 40:369–374

426 Warmuth W, Wenzig E, Mersmann A: Selection of a support for immobilisation of a microbial lipase for the hydrolysis of triglycerides. *Bioprocess Eng* 1995, 12:87–93

427 Mansson MO, Siegbahn N, Mosbach K: Site-to-site directed immobilisation of enzymes with bis-NAD analogues. *Proc Natl Acad Sci USA* 1983, 80:1487–1489

428 Lieberman RB, Ollis DF: Hydrolysis of particulate tributyrin in a fluidized lipase reactor. *Biotechnol Bioeng* 1975, 17:1401–1419

429 D'Annibale A, Stazi SR, Vinciguerra V, Di Mattia E, Sermanni GG: Characterisation of immobilised laccase from *Lentinula edodes* and its use in olive-mill wastewater treatment. *Process Biochem* 1999, 34:697–706

430 Pifferi PG, Tramontini M, Malacarne A: Immobilisation of endo-polygalacturonase from *Aspergillus niger* on various types of macromolecular supports. *Biotechnol Bioeng* 1989, 33:1258–1266

431 Lante A, Crapisi A, Pasini G, Zamorani A, Spettoli P: Characteristics of laccase immobilised on different supports for wine-making technology. *Adv Mol Cell Biol A* 1996, 15:229–236

432 Dinnella C, Lanzarini G, Stagni A, Palleschi C: Immobilisation of an endo-pectin lyase on γ-alumina: study of factors influencing the biocatalytic matrix stability. *J Chem Technol Biotechnol* 1994, 59:237–241

433 Wasserman BP, Hultin HO, Jacobson BS: High-yield method for immobilisation of enzymes. *Biotechnol Bioeng* 1980, 22:271–287

434 D'Souza SF, Melo JS: A method for the preparation of coimmobilisates by adhesion using polyethylenimine. *Enzyme Microb Technol* 1991, 13:508–511

435 Reiss M, Heibges A, Metzger J, Hartmeier W: Determination of BOD-values of starch-containing waste water by a BOD-biosensor. *Biosens Bioelectron* 1998, 13:1083–1090

436 Hirohara H, Yamamoto H, Kawano E, Nabeshima S, Mitsuda S, Nagase T: Immobilised lactase and its use. *Eur. Pat. Appl. 37667 A2*, 1981

437 Gray CJ, Narang JS, Barker SA: Immobilisation of lipase from *Candida cylindracea* and its use in the synthesis of menthol esters by transesterification. *Enzyme Microb Technol* 1990, 12:800–807

438 Nanba H, Ikenaka Y, Yamada Y, Yajima K, Takano M, Ohkubo K, Hiraishi Y, Yamada K, Takahashi S: Immobilisation of N-carbamyl-d-amino acid amidohydrolase. *Biosci Biotechnol Biochem* 1998, 62:1839–1844

439 Haynes R, Walsh KA: Enzyme envelopes on colloidal particles. *Biochem Biophys Res Commun* 1969, 36:235–242

440 Hsiao HY, Royer GP: Immobilisation of glycoenzymes through carbohydrate side chains. *Arch Biochem Biophys* 1979, 198:379–385

441 Wasserman BP, Burke D, Jacobson BS: Immobilisation of glucoamylase from *Aspergillus niger* on polyethylenimine-coated non-porous glass beads. *Enzyme Microb Technol* 1982, 4:107–109

442 Pietta PG, Agnellini D, Mazzola G, Vecchio G, Colombi S, Bianchi G: Immobilisation and characterisation of enzymes on hollow fibers for a possible use in the biomedical field. *Enzyme Eng* 1980, 5:235–237

443 Nakanishi K, Matsuno R: Continuous peptide synthesis in a water-immiscible organic solvent with an immobilised enzyme. *Ann NY Acad Sci* 1990, 613:652–655

444 Li XY, Meng YF, Zhao KH, Tu WX: Characteristics of immobilised β-galactosidase from *Cicer arietinum*. *ACH – Models Chem* 1997, 134:383–393

445 Rai R, Taneja V: Production of d-amino acids using immobilised d-hydantoinase from lentil, Lens esculenta, seeds. *Appl Microb Biotechnol* 1998, 50:658–662

446 Kawai T, Kawakita H, Sugita K, Saito K, Tamada M, Sugo T, Kawamoto H: Conversion of dextran to cycloisomalto-oligosaccharides using an enzyme-immobilised porous hollow-fibre membrane. *J Agric Food Chem* 2002, 50:1073–1076

447 Hidetaka K, Kazuyuki S, Kyoichi S, Masao T, Takanobu S, Hiroshi K: Production of cycloisomaltooligo-saccharides from dextran using enzyme immobilised in multilayers on to porous membranes. *Biotechnol Prog* 2002, 18:465–469

448 Kawai T, Saito K, Sugita K, Sugo T, Misaki H: Immobilisation of ascorbic acid oxidase in multilayers on to porous hollow-fibre membrane. *J Membr Sci* 2001, 191:207–213

449 Kobayshi S, Yonezu S, Kawakita H, Saito K, Sugita K, Tamada M, Sugo T, Lee W: Highly multilayered urease decomposes highly concentrated urea. *Biotechnol Prog* 2003, 19:396–399

450 Wang X, Ruckenstein E: Preparation of porous polyurethane particles and their use in enzyme immobilisation. *Biotechnol Prog* 1993, 9:661–665

451 Kraemer D, Plainer H, Sproessier B, Uhlig H, Schnee R:Method for immobilised proteins. US patent 4819439, 1989

452 Papayannakos N, Markas G, Kekos D: Studies on modeling and simulation of lactose hydrolysis by free and immobilised β-galactosidase from *Aspergillus niger*. *Chem Eng J* 1993, 52:1–12

453 Vrabel P, Polakovic M, Godo S, Bales V, Docolomansky P, Gemeiner P: Influence of immobilisation on the thermal inactivation of yeast invertase. *Enzyme Microb Technol* 1997, 21:196–202

454 Ostafe V, Chiriac A, Serban M: Immobilisation of alcohol oxidase on DEAE-cellulose. *Analele Universitatii de Vest din Timisoara, Seria Chimie* 1994, 3:65–72

455 Ryu GH, Park SY, Cho HI, Min BG: Enhancement of antithrombogenicity by urokinase immobilisation on the gelatin-adsorbed polyurethane surface. *Seoul J Med* 1989, 30:37–46

456 Li XY, Meng YF, Zhao KH, Tu WX: Characteristics of immobilised β-galactosidase from *Cicer arietinum*. *ACH – Models Chem* 1997, 134:383–393

457 Vrabel P, Polakovic M, Godo S, Bales V, Docolomansky P, Gemeiner P: Influence of immobilisation on the thermal inactivation of yeast invertase. *Enzyme Microb Technol* 1997, 21:196–202

458 Torchilin VP, Tishchenko EG, Smirnov VN: Covalent immobilisation of enzymes on ionogenic carriers. Effect of electrostatic complex formation prior to immobilisation. *J Solid-Phase Biochem* 1977, 2:19–29

459 Papisov MI, Maksimenko AV, Torchilin VP: Optimization of reaction conditions during enzyme immobilisation on soluble carboxyl-containing carriers. *Enzyme Microbial Technol* 1985, 7:11–16

460 Lu ZZ, Yang JJ, Leng LC, Feng XD, Li DC: Studies on biologically active *p*-(methacrylamido)benzoic acid esters. *Kexue Tongbao (Chinese Edition)* 1981, 23:1433–1435

461 Voivodov K, Chan WH, Scouten W: Chemical approaches to oriented protein immobilisation. *Makromol Chem Macromol Symp* 1993, 275–283

462 Smalla K, Turkova J, Coupek J, Hermann P: Influence of salts on the covalent immobilisation of proteins to modified copolymers of 2-hydroxyethyl methacrylate with ethylene dimethacrylate. *Biotechnol Appl Biochem* 1988, 10:21–31

463 Mateo C, Fernandez-Lorente G, Abian O, Fernandez-Lafuente R, Guisan JM: Multifunctional epoxy supports: a new tool to improve the covalent immobilisation of proteins. The promotion of physical adsorptions of proteins on the supports before their covalent linkage. *Biomacromolecules* 2000, 1:739–745

464 Fukui S, Ikeda S: Immobilised enzyme requiring pyridoxal-5′-phosphate as coenzyme. *Jpn. Kokai Tokkyo Koho* 50,042,086, 1975

465 Liu CC, Lahoda EJ, Galasco RT, Wingard LB Jr: Immobilisation of lactase on carbon. *Biotechnol Bioeng* 1975, 17:1695–1696

466 Jennissen HP: The binding and regulation of biologically active proteins on cellular interfaces: model studies of enzyme adsorption on hydrophobic binding site lattices and biomembranes. *Adv Enzyme Regul* 1981, 19:377–406

467 Hosseinkhani S, Szittner R, Nemat-Gorgani M, Meighen EA: Adsorptiive immobilisation of bacterial luciferases on alkyl-substituted Sepharose 4B. *Enzyme Microb Technol* 2003, 32:186–193

468 Santarelli X, Fitton V, Verdoni N, Cassagne C: Preparation, evaluation and application of new pseudo-affinity chromatographic supports for penicillin acylase purification. *J Chromatogr B Biomed Sci* 2000, 739:163–172

469 Iborra JL, Cortes E, Manjon A, Ferragut J, Llorca F: Affinity chromatography of frog epidermis dopa-oxidase. *J Solid-Phase Biochem* 1976, 1:91–100

470 Ikeda S, Hara H, Sugimoto S, Fukui S: Immobilised derivative of pyridoxal 5′-phosphate. Application to affinity chromatography of tryptophanase and tyrosine phenol-lyase. *FEBS Lett* 1975, 56:307–311

471 Weibel MK, Barrios R, Delooto R, Humphrey AE: Immobilised enzymes: pectinase covalently coupled to porous glass particles. *Biotechnol Bioeng* 1975, 17:85–98

472 Kamori M, Yamashita Y Naoshima Y: Enzyme immobilisation utilizing a porous ceramics support for chiral synthesis. *Chrality* 2002, 14:558–561

473 Spagna G, Barbagallo RN, Casarini D, Pifferi PG: A novel chitosan derivative to immobilize a-l-rhamnopyranosidas. *Enzyme Microb Technol* 2001, 28:427–438

474 Palomo JM, Fernandez-Lorente G, Mateo C, Oritz C, Fernandez-Lafuente R, Guisan JM: Modulation of the enantioselectivity of lipase via controlled modification and medium engineering: hydrolytic resolution of mandelic acid esters. *Enzyme Microb Technol* 2002, 31:775–738

475 Ujang Z, Husain WH, Seng MC, Rashid AHA: The kinetic resolution of 2-(4-chlorophenoxy)propionic acid using *Candida rugosa* lipase. *Process Biochem* 2003, 38:1483–1488

476 Gorokhova IV, Ivanov AE, Zubov VP, Co-precipitation of Pseudomonas fluoroscens lipase with hydrophobic compounds as an approach to its immobilization for catalysis in non-aqueous media. *Russian J Bioorg Chem* 2002, 28:38–43

477 Palomo JM, Oritz C, Fernandez-Lorente G, Fuentes M, Guisan JM, Fernandez-Lafuente R: Lipase-lipase interactions as a new tool to immobilize and modulate the lipase properties. *Enzyme Microb Technol* 2005, 36:447–454

478 Mateo C, Archelas A, Fernandez-Lafuente R, Guisan JM, Furstoss R:

Enzymatic transformation. Immobilized A niger epoxide hydrolase as a novel biocatalytic tool for repeated-batch hydrolytic kinetic resolution of epoxides. *Org Biomol Chem* 2003, 1:2739–2743

479 Valivety RH, Halling PJ, Peilow AD, Macrea AR:Relationship between water activity and catalytical activity of lipases in organic media. Effect of supports, loading and enzyme preparation. *Eur J Biochem* 1994, 222:461–466

480 Yng L, Dordick JS, Garde S: Hydration of enzyme in non-aqueous media is consistent with solvent dependence of its activity. *Biophys J* 2004, 87:812–821

481 Jacobsen EE, Anthonsen T: Water content influences the selectivity of CALB-catalyzed kinetic resolution of phenoxy methyl-substituted secondary alcohols alcohols. *Can J Chem/Rev Can Chim* 2002, 80:577–581

482 Bovara R, Carrea G, Ottolina G, Riva S: Water activity does not influence the antioselectivity of lipase PS and lipoprotein lipase in organic solvents. Biotechnol Lett 1993, 15:169–174

483 Indlekofer M, Brotz f, Bauer A, Reuss M: Stereoselective bioconversion in continuously operated fixed bed reactors: Modeling and process optimization. Biotechnol Bioeng 1996, 52:459–471

484 Lee C-H, Parkin KC: Effect of water activity and immobilization on fatty acid selectivity for esterification reaction mediated by lipases. Biotechnol Bioeng 2001, 75:219–227

485 Vidinha P, Harper N, Micaelo NM, Lourenco NMT, da Silva MDRG, Cabral JMS, Afonso CAM, Soares CM, Barreiros S: effect of immobilization support, water activity and enzyme ionizatrion stat on cutinase activity and enantioselectivity in organic media. Biotechnol Bioeng 2004, 85:442–449

486 Strauss UT, Kandelbauer A, Faber K: Stabilization and activity-enhancement of mandelate racemase from *Pseudomonas putida* ATCC 12336 by immobilization. *Biotechnol Lett* 2000, 22: 515–520

487 Wehtje E, Costes D, Adelercrutz P: Enantioselectivity of lipases: effect of water activity. *J Mole Catal B, Enzymatic* 1997, 3:221–230

488 Guieysse D Salagnaad C, Monsan P, Remaud-simeon M: Resolution of 2-brmo-O-tolyl-carboxylic acid by transesterification using lipases from *Rhizomucor miehei* and *Pseudomonas cepacia. Tetrahedron Asymmetry* 2001, 12:2473–2480.

489 Ducret A, Trani M, Lortie R: Lipase-catalyzed enantioselective esterification of ibuprofen in organic solvents under controlled water activity. Enzyme microb. Technol 1998, 22:212–216

490 Pierre P, Lortie R: Influence of water activity on the enantioselective esterification of (R, S)-Ibuprofen by *Candida antarctica* lipase B in solventless media. *Biotechnol Bioeng* 1999, 63:502–504

491 Orsat B, Drtina GJ, Williams MG, Kilbanov AM: effect of support material and enzyme pre-treatment on enantioselectivity of immobilized subtilisin in organic solvents. *Biotechnol Bioeng* 1994, 44:1265–1269

492 Chikere AC, Galunsky B, Schunemann V, Kasche V: Stability of immbilized soybean lipoxygenases: Stability of immobilized soybean lipoxygenases: influence of coupling conditions on the ionization state of the active site Fe. *Enzyme Microb Technol* 2001, 28:168–175

493 Oliveira PC de, Alves GM, de Castro HF: Immobilisation studies and catalytic properties of microbial lipase onto styrene±divinylbenzene copolymer. *Biochem Eng J* 2000, 5:63–71

494 Norde W, Zoungrana T: Surface-induced changes in the structure and activity of enzymes physically immobilized at solid/liquid interfaces. *Biotechnol Appl Biochem* 1998, 28:133–143

3
Covalent Enzyme Immobilization

3.1
Introduction

Covalently binding enzymes to a suitable carrier is the second method developed for enzyme immobilization. Since the 1950s covalent immobilization of enzyme has flourished and is now an important method of enzyme immobilization, because covalent bonds usually provide the strongest linkages between enzyme and carrier, compared with other types of enzyme immobilization method such as non-covalent adsorption-based enzyme immobilization [1]. Thus, leakage of enzyme from the matrix used is often minimized with covalently bound immobilized enzymes.

In general, covalent bonding of an enzyme to a carrier is based on chemical reaction between the active amino acid residues (AAR) located on the enzyme surface (A) and active functionalities that are attached to the carrier surface (B), or vice versa, as illustrated in Scheme 3.1. To achieve efficient linkage, the functionality of the carrier and/or the enzyme must be activated before immobilization. Often, carriers are activated before their use for binding enzymes. In a few cases where control of the mode of binding or number of bonds is necessary [2], the enzyme molecules should be activated before binding to the carrier [3].

Intuitively, the selected carrier functions only as a scaffold. It has, however, been increasingly appreciated that the selected carrier also strongly dictates the performance of the immobilized enzymes. In a strict sense, covalently immobilized

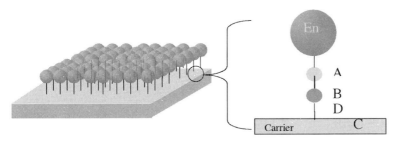

Scheme 3.1 Covalent immobilization of enzyme on the carrier: (A) active amino acid residue; (B) binding functionality of the carrier; (C) carrier; (D) spacer.

Carrier-bound Immobilized Enzymes: Principles, Application and Design. Linqiu Cao
Copyright © 2005 WILEY-VCH Verlag GmbH & Co. KGaA, Weinheim
ISBN: 3-527-31232-3

enzymes on carriers can be regarded as chemically/physically modified enzymes, whose physical and chemical nature is modified by the carrier used. Thus, it is not surprising that the physical and chemical nature of the selected carrier strongly dictates the performance, for example activity, selectivity, stability, and, particularly, the application, of the immobilized enzymes obtained.

In general, covalent binding of enzymes to carriers belongs to the category radical method of enzyme immobilization. This is reflected not only by the irreversibility of the binding, freezing of the enzyme conformation (because of the multipoint attachment), alteration of the chemical entity (because of the chemical modification) but also by the fact that enzyme performance, such as activity, selectivity and stability can be radically improved in comparison not only with other methods of enzyme immobilization but also with the native enzyme.

As shown in Scheme 3.1, a covalently immobilized enzyme can be regarded as a composite consisting of the components carrier, spacer, linkage and enzyme. Thus, the following properties are expected to affect the performance of a covalently carrier-bound immobilized enzyme:

- physical nature of the carrier (e.g. pore size, particle size, porosity, shape, etc.);
- the chemical nature of the carrier (chemical composition of the backbone; active functionality; other non-active functionality);
- the nature of the linkage or binding chemistry;
- the conformation of the enzyme at the moment it is immobilized or after immobilization;
- enzyme orientation;
- the nature and length of the spacer;
- the properties of the medium used for binding the enzymes;
- the number of bonds formed between the enzyme and the carrier;
- enzyme distribution on or within the carriers.

Remarkably, the medium used for enzyme immobilization is usually 99.9% aqueous. Thus, many reactions that are prone to hydrolysis or less feasible (active) in aqueous medium cannot be used to covalently bind enzymes to carriers. In a few cases, immobilization of enzyme in organic medium has been reported [4, 5]. The aim was to extend the scope of chemistry for binding enzymes to carriers. Occasionally, the control of activity and selectivity of the immobilized enzyme by immobilization in organic solvent is also possible [6]. Unfortunately, less attention has been paid to this method, although the discovery that enzymes can be dissolved in some organic solvents with high retention of activity might spur new interest in exploiting this potential in order to achieve tailor-made activity and selectivity [7].

For the purpose of this discussion, the carrier is defined as a composite, consisting of two different functional components, namely the physical component, that is strongly related to the physical nature of the carrier, e.g. pore size, porosity and pore-size distribution, particle size, morphology, mechanical stability, swelling factor, hydrophilicity, etc., and the chemical component of the carriers that are related to the chemical composition, surface chemistry, active functionality, and other tethering inert functionality, etc.

3.2
Physical Nature of Carriers

The physical nature of a carrier includes dimensional properties, which refer often to the surface area, shape, size, and internal structure (e.g. the pore size, the pore-size distribution, and tortuosity) of the carriers. In general, these properties can be quantitatively described.

Study of the effect of the physical properties on the performance of the immobilized enzymes can be traced back to the early 1970s when different types of carrier, for instance different occurrence, different structure, and different shape and size, became available for bio-immobilization. In this period there was increasing interest in the development of robust carriers which could provide higher loading of enzyme and higher retention of activity. Thus, for researchers or engineers it is obviously beneficial to know the requirements of the physical nature of the carrier, to enable design of a desired immobilized enzyme with high enzyme loading and higher retention of activity.

Consequently, it has been found that the performance of a number of carrier-bound immobilized enzymes depends not only on the chemical nature of the carriers but also very much on the physical nature of the carrier such as the shape, size, pore size, porosity, and mode of pore distribution [8]. Thus, variation of physical properties such as surface area, pore volume, and mean pore radius can influence not only the binding of enzyme and activity expression but also the performance of the immobilized enzymes such as specific activity (U mg^{-1} protein), volume activity (U g^{-1} carrier), selectivity (reaction selectivity, enantioselectivity, or product map) and even enzyme stability [9].

Thus, rational selection of these physical properties not only satisfies the requirements of the geometric properties needed for certain reactor configurations and process conditions but also addresses a number of problems e.g. activity, selectivity and stability of the immobilized enzymes, which are also closely related to the physical nature of the carriers such as the diffusion constraints, binding density, available surface for binding, porosity, and pore volume.

In general, the requirement of pore geometry is determined by the physical/chemical peculiarities of the enzyme, substrate(s) and product(s) and by the nature of reaction being catalysed, while geometric properties such as shape, size and length are largely dictated by the peculiarities of the applications (reactor configuration, reaction system and reaction medium).

Apart from the two geometry-related properties, other important dimension-related properties are the surface and density of binding functionality. The former is closely related to enzyme loading while the latter has a strong connection with enzyme stability or activity retention. Hence, it is expected that the criteria for selecting these physical properties might differ for enzyme to enzyme, from carrier to carrier, and from application to application [10]. However, there must be some general requirements.

As early as the 1960s it was recognized that the success of enzyme immobilization or bio-immobilization (e.g. pseudo-immobilization in the case of reversible separation) was largely dependent on the physical and chemical nature of the carriers, for instance the charge, particle size, pore size and hydrophilicity or aquaphilicity, chemical composition, pendant functionality, nature of the spacer and binding chemistry [11].

Accordingly, a great number of synthetic or natural carriers with tailor-made chemical and physical properties, for instance, with different shapes/sizes, porous/non-porous structures, different aquaphilicity and different binding functionalities were specifically designed for various bio-immobilization and bio-separation procedures.

However, for many new applications, it is still necessary to re-assess the physical/chemical nature of these available carriers before binding enzyme to carrier. This is mainly ascribed to the lack of the guidelines that link the make-up of the carriers to the performance of the immobilized enzymes obtained which are intended for a given application.

The important physical properties include pore size, porosity, pore volume, particle size, tortuosity, mode of the pore (such as closed or interconnected) and pore-size distribution, aquaphilicity, etc., which can be generally quantified or measured. In contrast, the chemical nature of the carriers includes the make-up of the backbone, spacer, pendant functionality, binding chemistry, aquaphilicity and chemical microenvironment [12].

Thanks to the diversity and broad availability of the carrier, it is always possible to find a feasible method of immobilization for the selected enzyme. On the other hand, the diversity in enzyme nature and application often complicates the establishment of such universally applicable rules for the selection of carriers and the method of enzyme immobilization. Thus, it is not surprising that there is no universal carrier for enzyme immobilization and a good carrier for one enzyme applied in a specific process might be not the proper carrier for another enzyme or the same enzyme applied in a different process [13], as exemplified by the design of immobilized penicillin G acylase for the enzymatic synthesis of semi-synthetic antibiotics [14].

Consequently, the change of application, reaction type (hydrolysis, or condensation), reaction medium (aqueous or organic solvents) or the enzyme and substrates often necessitates the design or screening of new carriers or immobilization method.

3.2.1
The Surface of the Carriers

The surface of a carrier might include external surface and internal surface (for a non-porous carrier the internal surface is zero whereas for a highly porous carrier, the external surface is negligible). For enzyme immobilization it is conceivable that high internal surface is highly desirable if high loading is required. However, the fact that the available surface (i.e. the surface available for enzyme occupation) dictates the maximal payload of carrier-bound immobilized enzymes, suggests that it might be useful (or informative) to classify the surface as accessible surface and inaccessible surface. The former might cover the external surface and the surface enzyme molecules can reach by diffusion whereas the latter refers to the surface the enzyme cannot reach because of physical barriers (small pore size or the closed or dead-ended pores).

3.2.1.1 Internal and External Surface
As noted above, the surface of a carrier can be classified into internal surface and external surface. The former is the total surface contributed by the pores, while the latter refers only to the outer surface (Scheme 3.2).

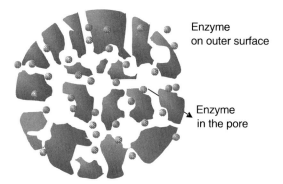

Scheme 3.2 Internal and external surfaces of a porous carrier.

The internal surface is generally about 100 to 1000 times the external surface for a macro-porous carrier, depending on the diameter of the particles. Thus, for such carriers the enzyme loading on the external surface can be neglected in comparison with the overall enzyme loading [15].

For enzyme immobilization, it is obviously advantageous to use carriers with high surface area with regard to enzyme loading. However, high surface area is not the only prerequisite for high enzyme loading, because the surface to be occupied should be accessible to the enzymes to be immobilized, as will be discussed in the following section.

3.2.1.2 Accessible Surface/Efficient Surface
The accessible surface area of a carrier is the surface area accessible to the enzymes to be immobilized. In other words, the enzymes can freely move to the accessible surface and interact with it, leading to binding of the enzyme to the carrier and full occupation of the accessible area [16].

In general, the loading (or payload, which is generally defined as g protein bound g^{-1} carrier used) and activity retention of a carrier-bound immobilized enzyme is to a large extent related to the pore size rather than the surface. For instance, in an early study of immobilization of papain in porous glass it was found that little or no protein was immobilized in the pores when glass pores approached the molecular dimensions of the protein. Thus, for efficient immobilization, the pores need to be 3–9 times larger than the size of the enzymes [18].

Early in 1970s it was also found that the loading of trypsin on amino silica by covalent bonding was strongly related to the pore size but not to the surface. This suggests that the pore size usually dictates the loading of the carrier (Scheme 3.3) [19].

Similarly, it was found that penicillin G acylase was immobilized on triacrylate carrier (surface area 245 m^2 g^{-1} and pore radius 3.62 nm) with a very low payload (0.01), approximately one tenth the payload on Eupergit C (~0.1) which has a similar surface area (180 m^2 g^{-1}) but much greater pore size (10–20 nm). Obviously, in the former example the enzyme can not enter the pores of the carrier (size of penicillin G acylase is about 5 nm) and the enzymes were apparently immobilized mainly on the external surface.

These facts suggest that accessible surface, rather than total surface area (which is dependent largely on the pore size) is much more important with regard to enzyme loading and the performance of the immobilized enzyme [17].

In practice it is possible to calculate the surface area needed for a certain number of enzyme molecules, if the size of the enzyme is known: the area occupied by 1 µmol enzyme in a monolayer is approximately 125 m^2, if the average diameter of a spherical enzyme is 4 nm. If the molecular weight is 25 kDa, the payload of the enzyme on 125 m^2 is 0.25. This value was very close to the real adsorption behaviour of trypsin on colloidal silica (100 mg protein per 230 mg colloidal silica with specific area 230 m^2 g^{-1}) [20]. Therefore, monolayer adsorption of the enzyme on the surface can reach approximately 2 mg enzyme m^{-2}. Although this value might vary from enzyme to enzyme, it is generally true for most enzymes.

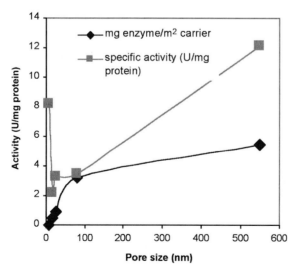

Scheme 3.3 Relationship between pore size and specific activity and enzyme loading (mg m^{-2}).

On the other hand, it has been found that 1.2 µmol penicillin G acylase (~100 mg) was immobilized on 1 g Eupergit C or Eupergit C 250 L. If the surface area needed for occupation of 200 mg protein (with the assumption that the enzyme preparation contains 50% impure protein) in both examples is equal, then 100 m² should be enough for penicillin G acylase monolayer formation. However, the specific surface area of Eupergit C is approximately 180 m² g^{-1} [21]. Thus, the efficiency of the surface area of Eupergit C for penicillin G acylase is around 50%. This lower efficiency of occupation might reflect either that the enzyme is not immobilized in the monolayer or that the pore sizes are not uniformly distributed, because of the limitation of the pore size or inaccessibility of the surface.

A similar observation was made when lipase PS was immobilized on EP100 (45 m² g^{-1} specific surface and particle void 0.75). The maximum loading capacity was approximately 40 mg lipase PS g^{-1} EP100. Thus, the efficiency of surface usage was also approximately 50%. However, maximum activity retention is 30% of maximum enzyme loading with 10% retention of activity, suggesting that the enzyme is suffered from serious diffusion limitation [22].

Immobilization of CAL-B (*Candida antarctica* lipase; size 6.92×5.05×8.67 nm) on octylsilica reached monolayer occupation at 200 mg g^{-1} carrier (254 m² g^{-1}). More or less 100 m² is enough for 200 mg lipase. The efficiency of surface utility is approximately 40%. Thus, 60% of the surface is probably not accessible to the enzymes, which was also confirmed by the fact that only 23% of the pore volume was occupied by the enzymes [23].

However, it has been also found that the loading of enzymes on the same carrier differs from enzyme to enzyme. For instance, the protein loading for β-galactosidase, albumin and γ-globulin on epoxy-activated acrylic beads was 40 mg/g, 140 mg/g and 120 mg/g, respectively, suggesting that the coverage of the inner surface of the carrier is protein size dependent, as illustrated in Scheme 3.4 [24].

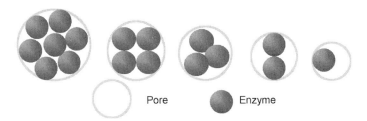

Scheme 3.4 Relationship between pore size and surface occupation efficiency.

3.2.1.3 A Theoretical Simulation

Because it is known that the pore size of carrier is not only related to the available surface but also to diffusion limitation and the mobility of the immobilized enzyme molecules, it should be possible to postulate theoretically the pore-size requirement for some enzymes, if their molecule size are known.

As shown in Scheme 3.4, the simulation of the enzyme on the surface of the pore suggests that the efficient area occupied by the immobilized enzyme decreases as pore size decreases. Thus, highly efficient surface occupation can only be obtained when the minimum pore size of the carrier is three times the main axis of the enzyme. The carrier should also have a very narrow pore-size distribution, for better retention of activity.

For instance, the commercially available carriers Eupergit C and Eupergit C 250L differ mainly in pore size. The pore size of Eupergit C 250 is more or less 10 times bigger than that of Eupergit C. Although similar enzyme payloads (~0.1) can be achieved with both, retention of activity of penicillin G acylase on Eupergit C 250 was almost twice that on Eupergit C. Similar result was obtained by immobilizing β-galactosidase from *B. circulans* and an α-galactosidase on Eupergit C [26]. These results suggested that pore size dictates not only the efficiency of the surface occupation but also the expression of activity as discussed below [27].

However, many recent studies have revealed that the pore size should be much bigger than the minimum requirement for the pore size. This can be possibly ascribed to the following requirements:

- the pore size should be big enough to let the enzyme enter (pore size > enzyme size);
- the pore size should be big enough that the enzyme has freedom of movement to change its conformation (pore size > three times the main axis of the enzyme);
- the pore size should be big enough that diffusion limitation can be mitigated.

Several studies which involved other enzymes or other types of immobilized enzymes have demonstrated that the minimum size for overcoming the diffusion constraints is usually about 8 times the size of the enzyme, as discussed below.

3.2.2
Density of Binding Sites

Binding density refers to the number of active binding functionalities per unit of area (let us say 10 nm^2) on the carrier surface, as shown in Scheme 3.5.

It is conceivable that the loading of enzyme and the number of bonds between the enzyme and the carrier is closely related to the binding density. Besides, the nature of the microenvironment of the surface is also largely dependent on the concentration of the binding density. Consequently, other characteristics of the immobilized enzymes such as activity retention, enzyme stability and selectivity can be also influenced by the concentration of the binding functionality i.e. the binding

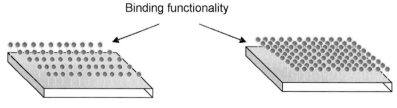

Scheme 3.5 Binding density on the carrier surface.

density. In particular, the enzyme stability can be strikingly influenced by the binding density, because the degree of cross-linking between the protein and the rigid carrier is largely dictated by these properties [27].

The use of cross-linked acrolein-2-hydroxyethyl methacrylate polymeric carrier for enzyme immobilization led to the discovery that the enzyme activity of horseradish peroxidase, immobilized on cross-linked acrolein–2-hydroxyethyl methacrylate polymer particles, depended on the acrolein content, reaching a maximum at 0.14 mol fraction and decreasing with further increase in acrolein content of the particle [28].

3.2.3
Pore Related Properties

Pore-related properties include pore size, pore-size distribution, pore volume, pore model and porosity. For enzyme immobilization, pore size and porosity are the most important properties, with respect to enzyme loading, activity retention and even enzyme stability.

3.2.3.1 The Porosity

The porosity is the ratio of volume within the particle to the external volume of the particle. For synthetic carriers, the porosity varies from 0–1.0 (when the porosity is zero, there are no pores in the carriers; conversely, the carrier is completely hollow when the porosity is close to 1). For most synthetic carriers the porosity is between 0.5 and 0.8.

Based on the observation that the payload of commercially available carriers such as Eupergit C 250 is approximately 0.1–0.2 (approx. 0.1–0.2 g enzyme can be loaded into 1 g carrier), it is possible to conclude that only 15–30% of the porosity of these carriers is occupied by enzyme molecules and that the other pores are not accessible to the enzyme (closed pore or the pore size is smaller than the enzyme size), as validated by the ratio of the enzyme-occupied surface to that of the overall surface (~25%) mentioned above.

For enzyme immobilization, porosity is a crucial factor. It influences not only the enzyme loading [29], but also the retention of activity of the enzyme [30–32] and the performance of the immobilized enzymes [33]. For synthetic carriers prepared by suspension polymerization, the porosity is affected by many factors, for example:

- pore-generating solvent [31],
- properties of the cross-linker [32],
- cross-linker concentration [30],
- process conditions [30].

Consequently, the retention of activity after immobilization (E = [Activity measured]/[Activity bound to carrier] or [specific activity of the immobilized enzyme]/[specific activity of the native enzyme on the basis of the weight of the protein]) and enzyme loading can be influenced by the porosity [31]. Moreover, the surface microenvironment of the carriers is also influenced by the porosity [29]. For example, increasing the porosity and the specific surface of a fibre-based biocatalyst reduced K_m and the energy of activation for hydrolysis of benzyl penicillin but increased V_{max} [33], indicating that porosity is closely related to diffusion constraints. Similarly, Ferreira et al. used different derivatized silicas with different porosities and surface areas. They found that the highest retention of activity was obtained with carriers of high porosity and surface area [34].

3.2.3.2 Pore Size and Distribution

Studies have revealed that most of the carriers used for enzyme immobilization have pore size distribution in a wide range. Consequently, not only the enzyme loading but also activity retention is highly dependent on the pore-size distribution [35]. To reach high enzyme loading and to reduce diffusion constraints, the pore-size distribution should exceed the size of the enzymes, as for macroporous carriers such as Duolite ES568 and Spherosil DEA (148 nm) [35]. However, this may provoke another problem namely diffusion constraints, when the enzyme can enter the depth of the carrier's particles, leading inevitably to lower retention of activity [35].

On the other hand, the use of the carriers with heterogeneous pore size distribution might provide another means of controlling the enzyme distribution. For instance, use of Duolite ES562, which has many narrow pores and few external pores [35], led to the preparation of immobilized enzyme without significant internal diffusion constraints.

As noted above, the use of Eupergit C 250 L instead of Eupergit C for immobilization of penicillin G acylase increased activity retention from 40% to 70% with comparable enzyme loading. Thus, the different activity retention was obviously ascribed to the big pores of Eupergit C 250 L [26]. Many other examples have also shown that increasing the pore size enhanced not only enzyme loading but also activity retention and specific activity.

However, a question arises here: what is the minimum pore size for which diffusion constraints can be generally overcome? Because each enzyme differs in their size, the minimum pore-size requirement for each enzyme might also be different. However, there must be some general requirements. It is known that diffusion of the substrate in the pores is dictated by the properties of the partition layer, which is approximately 20 nm, and above the 20 nm the diffusion constraint caused by the partition layer is generally neglected.

As shown in Scheme 3.6, the distance from the enzyme layer on the wall of the pore to the centre of the pore should be bigger than 20 nm, namely $r_p - d_e > 20$ nm. Consequently, $r_p > (20 + d_e)$ and the minimum pore size should be $2 r_p$, i.e. $(40 + 2 d_e)$. If the minimum enzyme size is 5 nm, the minimum pore size should be 50 nm. This value is very close to the value reported for the penicillin G acylase [36] and lipase on controlled-pore glass [37].

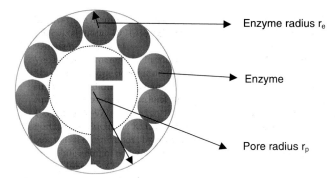

Scheme 3.6 Requirement of minimum pore size for elimination of diffusion constraints.

This is approximately 8–10 times of the size of penicillin G acylase. Similar pore-size requirements have also been reported for other enzymes such as glucose isomerase [38] and lipases [39] or trypsin or co-immobilized pectinesterase and *endo*-d-polygalacturonase pectinesterase [40]. Thus, diffusion limitation can be partially overcome by use of carriers with large pores.

As already noted, the minimum pore size is enzyme-dependent. For enzymes with a main radius 5 nm, the optimum pore size is usually 50 nm. However, for some enzymes with bigger radius, for example fructofuranosidase from *Aureobasidium* sp. covalently immobilized on alkylamine porous silica activated with glutaraldehyde, the optimum pore diameter of porous silica for immobilization of the enzyme was 91.7 nm [41]. However, for urease with Mw 490 KD pore size of 100–200 nm was required to achieve maximal binding capacity and activity retention [38].

Thus, it was concluded by Artyomova and coworkers that maximum capacity and, consequently, the maximum accessible surface for enzyme penetration can be obtained when the pore size of silicas used for enzyme immobilization exceeded by 5 to 10 times the protein globule sizes [38]

Apart from the efficiency of immobilization, the application and the process conditions can be also affected by the pore dimensions. Immobilization of penicillin G acylase on carrier with pore size > 15 nm led to precipitation of the products, if precipitation of products could occur during the reaction [42]. Interestingly, it was found that the activity of the same immobilized penicillin G acylase on different carriers was highly pore-size dependent. Moreover, the reaction selectivity was different for the hydrolytic reaction and the condensation reaction [8].

Loading of inulinase on porous silica with controlled surface and pore size has been demonstrated (Scheme 3.7). It was found that enzyme loading decreased with increasing pore size, because the surface available for enzyme occupation usually decreased with increasing pore size, as discussed above.

Interestingly, it was found that the specific activity (U mg^{-1} protein) of the immobilized enzyme first decreases as the pore size increases. When the pore size approaches 60 nm, the specific activity is lowest, possibly suggesting that the internal diffusion limitation prevails and the enzyme distribution in the particles

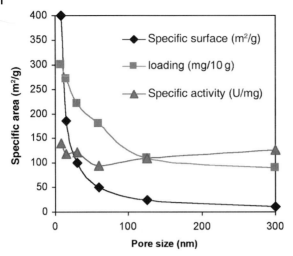

Scheme 3.7 Relationship between enzyme loading, specific surface, and pore size of CPG.

changes significantly. When the pore size is beyond 60 nm, the specific activity increases again, suggesting that internal diffusion limitation is becoming less [43]. This minimum pore-size requirement is consistent with our hypothesis described above.

3.2.4
Particle Size

The particle size is not only important in terms of enzyme loading and enzyme activity expression but is also important with regard to selection of reaction systems and reactor configuration, enzyme re-use, and even enzyme selectivity. In general, the particle size of the immobilized enzyme is an important non-catalytic property.

For nonporous carriers, the smaller the particle size the larger the surface available for enzyme immobilization. Thus, small particle size favours high enzyme loading [44].

However, at the same time, it might provoke another problem – separation of the immobilized enzymes from the reaction mixture, especially for heterogeneous reaction mixtures where the product or substrates might be insoluble and have the similar size. Besides, for column reactor, the small particle size might result in serious pressure drop in the column. Thus, in practice, the particle size should not be below 50 μm for either non-porous or porous carriers. Thus, it is not surprising that the commercially available carriers such as Eupergit C have particle sizes of approximately 100 μm [26].

With porous carriers the presence of the diffusion constraints often reduces the expression of the activity, because of the presence of the concentration gradient, as shown by immobilization of chymotrypsin, trypsin, and glucose oxidase immobi-

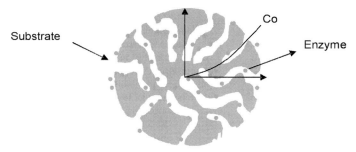

Scheme 3.8 Diffusion limitation in a porous carrier.

lized on the copolymer of N-ethyl-N-(2-methacroylethyl)-p-phenylenediamine and 2-hydroxyethylmethacrylate by an oxidative coupling reaction. It was found that the enzyme activity depended on the particle size of the carrier [45].

Thus, although the enzyme might not be deactivated it might be functionally inactive, as discussed above. In this case, the particle size should be controlled to ensure that the selected particle size conforms to the requirements of the desired process conditions, while activity expression is not compromised (Scheme 3.8).

Activity expression usually decreases as loading increases for porous carriers. This is because of penetration of the catalysts into the depth of the particles. Thus, for carriers of larger particle size, activity expression can still be controlled by enzyme loading, for example the immobilization of α-chymotrypsin on a glutaraldehyde-activated silicate support [43].

Control of the particles is not only important for the carrier-bound immobilized enzymes but also for carrier-free immobilized enzymes such as cross-linked enzyme crystals (CLEC) [46, 47]. To achieve high retention of activity (by reduction of diffusion constraints), it is also essential to control the size and the shape of the crystals [48].

For heterogeneous reaction systems in which the substrates or products might be solid, immobilized enzymes of big particle size are desired, to facilitate separation of the immobilized enzymes; one example is the solid-to-solid synthesis of semi-synthetic β-lactam antibiotics [14]. In these cases the spherical particles of the immobilized enzymes should be >100 μm, in order to be retained on conventional sieve plates, even in large enzyme reactors. By contrast, the immobilized enzymes, e.g. carrier-bound or carrier-free immobilized enzymes (CLEC or CLEA), which have particles less than 100 μm, other solutions should be found, for example, the preparation of magnetic enzymes [49] or the use of a double-immobilization technique, e.g. entrapment of the small pre-immobilized enzymes in a gel matrix of the desired size [50].

The types of reactor configuration that can be distinguished are batch, stirred-tank, column and plug-flow. Apparently, the use of immobilized enzymes of large particle size is advantageous in that the catalysts can be easily separated from the reaction medium for the purpose of recycling or control of the reaction progress.

3.2.5
Shape of the Carriers

The shape of the carrier can be classified into two types, i.e. irregular and regular shapes such as beads (A); fibres (B); hollow spheres (C), thin films (D), discs (E) [51], and membranes (F), as shown in Scheme 3.9.

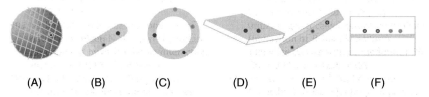

(A)   (B)   (C)   (D)   (E)   (F)

Scheme 3.9 Carriers of different shape.

Selection of the geometric properties for an immobilized enzyme is largely dependent on the peculiarity of certain applications. In general, the following must be taken into consideration:

- reactor configuration (column, stir-tank, plug-flow, etc.),
- flow characteristics,
- pressure,
- temperature,
- stirring mode,
- diffusion constraints,
- reaction systems,
- enzyme loading.

Selection of the geometric properties is often closely related to control of the process, downstreaming the process, and re-use of the catalysts. Obviously, the geometric requirement is important in heterogeneous reaction systems, where the substrates or products might exist mainly in the solid state. In this case, particles of the immobilized enzyme have to be significantly larger than that of these solid products or substrates [52], in order to facilitate separation of catalysts from the products or substrates.

However, external geometrical properties such as particles size, diameters and lengths of fibres, thickness of films, disk or the hollow sphere membranes, are often related to diffusion constraints. Thus, retention of enzyme activity by immobilizing the enzyme on the carrier must also be taken into account when selecting suitable carriers. For example, various fibrous carriers are advantageous in that they usually have a greater outer surface and the kinetics of the immobilized enzyme are usually good (e.g. greater retention of activity) [53].

Apart from these considerations, the method of immobilization and the ease of preparation of the carrier must be taken into consideration, because each type of geometry might involve different preparation and immobilization methods, thus necessitating the use of different types of material and preparation conditions. For instance, beaded immobilized enzymes can be prepared by adsorption, covalent binding, entrapment, encapsulation, or a combination of these methods e.g. adsorption–cross-linking and adsorption–encapsulation, which require not only different methods of immobilization but also different strategies to realize the geometric requirements.

In general, the particle size of the immobilized enzymes should match the reaction systems and the reactor configurations. The pore size of the immobilized enzymes should be selected to reduce mass transfer limitations and to obtain high retention of activity.

3.3
Chemical Nature of Carriers

The chemical nature of a carrier can be considered as a dimensionless property that generally refers to the chemical composition of the backbone, the surface chemistry (e.g. the side groups), and the nature of the spacer, active or inactive pendant functionality tethered to the surface, hydrophilicity or hydrophobicity, aquaphilicity, and the microenvironment in which the enzyme molecules are encapsulated.

As shown in Scheme 3.10, these dimensionless properties of a carrier can be divided into several essential components, namely the chemical nature of the backbone (CNB) and the surface-tethered functional groups which can be principally placed into two sub-groups, carrier-bound active groups (CAG) that take part directly in the binding and carrier–bound inert groups (CIG) that do not directly take part in the binding.

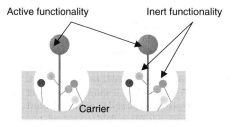

Scheme 3.10 The chemical nature of the carrier.

Although CAG will participate in the binding of the enzyme to the carrier, CIG do not form any covalent linkage with the enzyme but can also exert great influence on the performance of the immobilized enzymes such as activity retention, binding efficiency, stability and selectivity. Apart from these pendant groups, the spacers that link the CIG or CAG to the backbone of the carrier are also important factors influencing enzyme performance.

Immobilization of enzyme is usually regarded as a process by which the enzyme molecules are brought into close contact with the surface of the carrier (Scheme 3.11). Consequently, the close contact and the subsequent interaction will inevita-

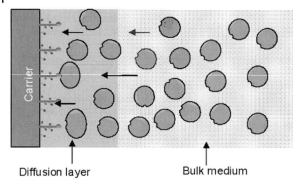

Scheme 3.11 Conformation change of enzyme approaching the carrier – effect of the chemical nature of the carrier on the enzyme molecules.

bly disturb the native forces that maintain the structure of the native enzymes, leading to modification of enzyme structure and function, depending on the chemical nature of the carriers and medium used [54].

Whatever the new conformation of the enzyme molecules on the solid surface, interestingly, it has been found that the enzyme usually adopts a conformation stabilized by interaction of the enzyme and the carrier [55]. If the new induced conformation resembles the structure of the native enzyme, the immobilized enzyme can be stabilized. Otherwise, it will be deactivated. Several other characteristics of the immobilized enzymes can also be affected by the nature of the surface, for example:

- stability,
- activity,
- selectivity,
- enzyme conformation,
- orientation,
- pH profile,
- temperature profiles.

It is, therefore, expected that the availability of the guidelines that link these enzyme characteristics to the chemical nature of the carriers to be designed or selected is of paramount importance for researchers and engineers, especially those working in the sphere of process development, who are often eager to obtain early insight into process development by the availability of a robust immobilized enzyme [56].

Often, the best carrier for a given enzyme differs from enzyme to enzyme and from application to application [14]. Thus, it is expected that many "ready-made" carriers are not necessarily the best-suited carriers for the enzymes to be immobilized.

In this context, it is surmised that engineering these pre-designed carriers by chemical modification [57], grafting [58, 59] or coating [60–62], or post-immobilization modification [63] might be a promising direction for relatively simple creation of robust carrier-bound immobilized enzymes.

3.3.1
Carrier-bound Active Groups (CAG)

By definition, the carrier-bound active groups (CAG) are the functionalities that take part in the covalent binding of proteins to the carrier, namely the X group illustrated in Scheme 3.12.

The frequently used CAG are given in Scheme 3.13. They cover, in general, the functionalities polyanhydride (A), polycarbonate (B), polyaldehyde (C), polyepoxide (D), polyacylazide (E), polyisocyanate (F), polycarboxylic acid phenyl ester (G), and polyazlactone (H), etc.

Two methods are frequently used for preparation of active functionality, preparation of the active carrier by direct polymerization of the active monomers in the presence of other suitable monomers and preparation of the active carriers by different activation techniques (interconversion techniques). However, different CIG might be inevitably introduced by the latter method.

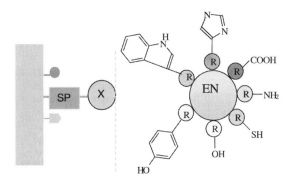

Scheme 3.12 Reactivity of CAG toward different amino acid residues on the enzyme (EN) surface.

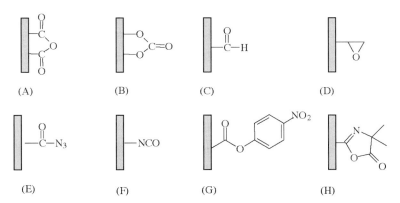

Scheme 3.13 Synthetic carriers bearing active functionality: (A) polyanhydride; (B) polycarbonate; (C) polyaldehyde; (D) polyepoxide; (E) polyacylazide; (F) polyisocyanate; (G) polycarboxylic acid phenyl ester; (H) polyazlactone.

Historically, numerous carrier-bound CAG, for example polymeric acyl azide [64], polymeric anhydride, polymeric isocyanine, polymeric carbonate, polymeric aldehydes, polyepoxy carriers, and polymeric triazine have been used to immobilize enzymes. In recent decades, however, interest in covalent coupling of enzymes to carriers has switched mainly to polymeric carriers bearing epoxy groups, carbonyl groups, or amino groups. Examples of the carriers currently used for enzyme immobilization are Eupergit C, Sepabeads (bearing intrinsic epoxy groups), oxidized polysaccharide (aldehyde groups), or other aminolated carriers that can be converted easily into a variety of new functionality ready for enzyme immobilization [65].

In general, designing functionality which can react with protein to form covalent bond is dictated mainly by the properties of the functionality (for example stability), reactivity, and the retention of the activity. Thus, the optimum reaction conditions for coupling protein to the carriers differ from one enzyme to another and from functionality to functionality.

According to the functionality of the enzymes involved in binding to the carriers and the functionality of the carriers to be used for binding enzyme molecules, it has been established that the coupling reaction can be placed in the nine groups, as illustrated in Scheme 3.14.

For more reviews of binding chemistry, readers are referred to Taylor [66].

It has been observed that each type of binding chemistry requires specific functionality between the carriers and the enzymes. It is also very difficult to predict which carrier-bound active binding group (CAG) is best-suited for a given enzyme, because other factors such as nature of the backbone also dictate the activity of the binding functionality and the activity of the resulting immobilized enzyme, and in particular the immobilization conditions, as has been demonstrated in many studies [113, 115, 332].

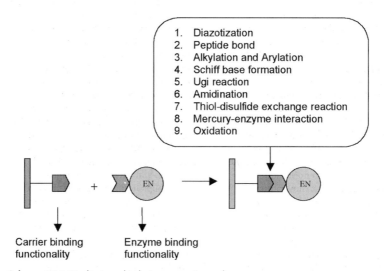

Scheme 3.14 Binding modes between carrier and enzyme.

One of early examples was provided by Bowers and Carr, who immobilized hexokinase on controlled pore glass with different surface chemistry. Although the binding chemistry is the same when immobilizing enzyme via glutaraldehyde coupling on the silanized carrier or aldehyde coupling on the oxidized glycophase, 30 fold higher activity was obtained in the case of the glycophase-coated carrier (which bears aldehyde groups) [67].

3.3.2
Carrier-bound Inert Groups (CIG)

By definition, CIG of a carrier are the chemically inert moieties that are directly linked to the backbone of a polymeric carrier via covalent bonds or linked to the spacer that connects the CAG to the backbone as linear or branched moieties, as illustrated in Scheme 3.15.

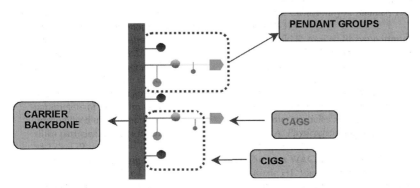

Scheme 3.15 Carrier-bound pendant groups.

However, CIG are occasionally also linked at the terminus of the CAG as leaving groups [75]. In this case, CIG will be cleared from the CAG after formation of covalent bonds. Although these leaving CIG have no effect after the immobilization of the enzyme, the performance of the immobilized enzymes can be affected by the CIG during the immobilization process.

As noted above, although CIG do not participate in any chemical reaction, it cannot be excluded that CIG might participate in the non-covalent binding event in a covalent enzyme immobilization process. Occasionally CIG might also influence the covalent binding. Thus, variation of the CIG might modify the microenvironment of the carriers, thus influencing the performance of the immobilized enzymes for example:

- selectivity [76],
- activity expression [75],
- enzyme binding [77],
- operational stability [78],
- thermostability [79, 80].

Recently, the influence of CIGs on the performance of covalently immobilized enzymes has been beautifully exemplified in several pioneering studies [117–122].

Compared with CAG, the influence of CIG has been not well documented. However, much useful information can be derived from non-covalent adsorption. Thus, the CIG can be classified into the following groups based on the classification of non-covalent enzyme immobilization:

- charged CIG, e.g. carriers bearing $-NH_2$ or $-COOH$ groups;
- hydrophobic CIG, which can be introduced by use of hydrophobic monomers or cross-linkers;
- hydrophilic CIG, which can be introduced by use of hydrophilic monomers;
- biospecific CIG: biospecific ligands which can specifically bind some enzymes can be introduced into the carriers.

Depending on the nature of CIG and the nature of immobilized enzymes, major enzyme performance such as stability, activity, and selectivity can all be influenced by CIG.

3.3.3
Spacer-Arm

In addition to the CAG or CIG, the spacers that link the CAG or CIG are of crucial importance. Usually, spacers can affect the binding of bio-molecules to the carriers and the performance of the resulting immobilized systems [81, 82].

Investigation on the effect of the spacer on enzyme performance can be dated back to beginning of the 1970s, when Taylor and co-workers immobilized trypsin on an insoluble polystyrene matrix mediated by peptide chains of different length [66]. It was found that K_{cat} was independent of the chain length or column flow rate but K_m was highly dependent on both properties. Design of appropriate spacers for a variety of bio-immobilization and bio-separations has since become the subject of intensive research in recent decades [83–86].

In fact, the study of spacer stems from the affinity chromatography in the same period, when the people found that the distance between the immobilized ligand (functional group) and carrier's backbone is crucial, with respect to the binding.

The importance of the proper distance was convincingly demonstrated in the study by Markus and Balbinder, who found that introduction of 8 spacer atoms into the immobilized anthranilate derivatives (from $-NH-ph-COOH$ to $-NH(CH_2)_6NHC(O)CH_2CH_2C(O)NH-ph-COOH$) enhanced the amount of enzyme bound to the carrier by more than 10 fold [87].

The necessity for a matched ligand can be explained as follows: The interaction of the ligands with the targeting molecules is often complementary in spatially very confined points within the ligand and the targeting molecules. Thus, the ligand and the binding site should be accessible to each other. Consequently, only introduction of an appropriate spacer can overcome the steric hindrance and enable the interaction, due to the enhanced flexibility of the ligand to approach the targeting molecules.

Although in covalent enzyme immobilization there is usually no difficulty for the active functional groups to approach some active amino acid residues to bind the enzyme, the fact that the binding mode (position, number of bonds) and the flexibility of the attached enzyme molecules will be influenced by the presence of the spacer, justifying the use of a correct spacer [87].

As previously discussed, the spacer is an essential part of the pendant groups of the carriers. Often, the spacer is treated together with the CAG (the carrier-bound active groups). In principle, any carrier-bound immobilized enzyme has a spacer.

Much evidence has shown that direct immobilization of an enzyme on some activated carriers (with only active functional groups) leads to lower immobilization efficiency, and an increase [88] or decrease [89, 94, 95] of enzyme stability, for the following reasons:

- steric hindrance [90, 91];
- environmental effects of the surface (hydrophobic surfaces often cause inactivation of the immobilized enzyme);
- properties of the matrices;
- unfavourable interaction of enzyme molecules with carriers [92];
- mismatched orientation of enzyme on the carriers such as involvement of active centre in the binding [468];
- non-compatible microenvironment;
- "over-dosed" multipoint attachment;
- influence of enzyme mobility or conformational flexibility [86, 93].

In contrast, introduction of a spacer often improves the performance of the immobilized enzyme, for example enzyme stability, retention of activity, and specific activity [93], because of the higher molecular mobility in the presence of spacers between the enzyme and the carrier, as revealed by ESR spectroscopy of the immobilized enzymes with or without spacers [93]. Enzyme stability against extremes such as high pH, temperature, and organic solvents can also be improved. These results suggest that the presence of an appropriate spacer may mitigate the influence of the solid surface and convey flexibility to the enzyme, and alteration of the enzyme microenvironment.

In general, a spacer can be characterized by the following properties:
- length and size,
- structure,
- shape (linear or globular),
- hydrophobicity/hydrophilicity,
- positively charged/negatively charged/neutral.

Thus, it is expected that the properties of spacers such as length [88, 91, 94], hydrophilicity/hydrophobicity, and charged/neutral character can exert a conspicuous influence on binding capability, retention of activity [95], stability [88] and catalytic performance. In general, hydrophobic spacers have a more profound effect on non-specific binding capacity, than hydrophilic spacers [96].

In principle, many compounds, varying from small molecules to macromolecules, can be used as spacers:

- various bifunctional compounds, for example bifunctional cross-linkers such as glutaraldehyde[116, 125, 149];
- linear polymers such as PEG-diamine [89] or PEO acid [97], dextran [98] and PEI [99];
- functionalize polysaccharides such as aldehyde dextran, amino dextran and proteins such as bovine serum albumin (BSA);
- polyether [91].

3.4
Enzyme: Amino Acid Residues for Covalent Binding

To achieve efficient linkage and high retention of activity, the selected enzyme and carrier should be in a condition that their individual functionality located on their surface can come close to each other and finally covalent bonds can be formed between them.

Two essential requirements must be met to achieve efficient covalent binding between the enzyme and the carrier. The reactivity of the functional groups of the selected carrier and the enzyme that are going to react with each other must match each other in the aqueous medium or organic solvent used for the reaction, to achieve enzyme stability and reactivity. Second, the functional groups should be sterically accessible to each other under the selected immobilization conditions before binding

It is well known that enzymes or proteins usually contain 20 ordinary l-amino acids, all of which have a free R residue [100]. However, only half of these can be used for covalent enzyme immobilization (Scheme 3.16), namely:

- amino groups of N-terminal amino acids (NAA) and (ε)-amino groups of lysine (Lys);
- γ and β carboxyl groups of glutamic acid (Glu) and aspartic acid (Asp) and C-terminal carboxyl groups;
- the guanidinyl group of arginine (Arg);
- the sulphydryl group of cysteine (Cys);
- the imidazolyl group of histidine (His);
- the thioether moiety of methionine (Met);
- the indolyl group of tryptophan (Trp);
- phenolic hydroxyl groups of tyrosine (Tyr).

Moreover, some carbohydrate moieties in the glycosylated enzymes can also be used for enzyme immobilization, after conversion of the vicinal hydroxyl groups to aldehyde groups by oxidation [101, 102]. More information is available elsewhere [2, 103,104].

As illustrated in Scheme 3.16, some reactive amino acid residues are frequently used in enzyme immobilization, whereas others are seldom used. The major rea-

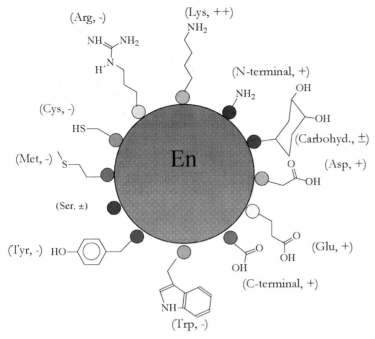

Scheme 3.16 Reactive amino acid residues: + frequently used in enzyme immobilization; − not used; ± not frequently used; ++ very frequently used.

son for this is that the reactivity of these reactive groups differ greatly. Thus, the conditions used to favour binding with the frequently used amino acids might be less hazardous than for the not frequently used amino acid residues.

Considering that amino acid residues are often involved in some critical function of the enzyme, i.e. catalysis, the conformation and function of the enzyme after binding might be changed or completed diminished. It is thus extremely informative to know which amino acid residues are located in the active centre or close to the active centre.

3.4.1
Reactivity of Amino Acid Residues (AAR)

It is known that the so-called amino acid residues (AAR) usually differ from each other in reactivity, because of their different chemical compositions and the microenvironment in the protein molecules [103, 104].

In general, the reactivity of the active amino acid side-chains of the protein is dictated by factors such as chemical nature, microenvironment; nature of the medium, for example pH and ionic strength, and the composition of the medium.

The chemical reactions, which occur mainly in aqueous media, can be classified into the following groups:

- peptide formation (Lys, Asp, Glu),
- diazotization (Arg, Cys, His, Lys, Tyr, Try),
- alkylation (Lys),
- arylation (Lys),
- Ugi reaction (Lys, Asp, Glu),
- thiodisulphide (Met),
- Schiff base formation (Lys, His, Cys, Tyr),
- amidation (Lys).

Thus, it is very difficult to find a specific amino acid-directed reagent, suggesting that heterogeneity of the immobilized enzymes, in terms of binding mode and performance, is often encountered in enzyme immobilization [105]. This is often because of the participation of many different amino acid residues in the same type of reaction or the location of many similar amino acids in different domains of the proteins, as discussed in the following section.

3.4.2
Position of Active Amino Acids

It is well known that a protein is composed of a hard hydrophobic core and a "soft skin", due to the preferential distribution of the amino acid residues in such a structure. In other words, the extent of exposure of each specific amino acid residue to the aqueous medium is not equal, depending on the nature of the amino acid and the position in the 3D structure.

In general, this structural feature of globular proteins is maintained by the tendency of hydrophobic amino residues to remain in the interior of the proteins and hydrophilic amino residues to remain at the surface where they can form hydrogen bonds with water molecules. Usually, the dynamic structure of a protein is the

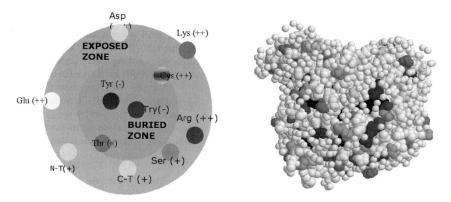

Scheme 3.17 Reactive enzyme residues and utility for covalent binding: + frequently used; ++ very frequently used; − not used; ± not frequently used. Amino acid residues on the surface are very exposed; amino acid residues in the cavity are hidden.

result of delicate balance between a number of forces such as hydrogen bonding, which can be influenced by the dielectric balance, ionic strength, pH and temperature.

The positions of the amino acid residues which are often used in covalent enzyme immobilization are schematically illustrated in Scheme 3.17. Some types of amino acid, for example the hydrophilic amino acids such as lysine, and aspartic and glutamic acids are preferentially distributed on the surface of the proteins in hydrophilic clusters, whereas some amino acids, for example Thr, Tyr, and Tyr, are often buried in the core. Occasionally, however, hydrophilic amino acids are also located in hydrophobic clusters.

However, the extent of exposure of some amino acid residues can be varied, depending on the properties of the medium. For instance, increasing the ionic strength or adding organic solvents often helps to expose hydrophobic amino acids.

According to Scheme 3.17, there are several important implications of this:
- any immobilized enzyme will be heterogeneous except after site-directed enzyme immobilization;
- immobilization via buried amino acid residues might need specific conditions. For instance, a high salt concentration favours binding to buried amino acids;
- enzyme properties can be possibly modulated by rational design of binding chemistry.

As a result, it is hardly surprising that selection of the binding chemistry and the immobilization conditions is of crucial importance in covalent enzyme immobilization as discussed in the following sections

3.5
Factors Affecting Enzyme Performance

For rational design of a robust immobilized enzyme, it is essential to know which factors can dictate enzyme performance. As shown in Scheme 3.18, a carrier-bound immobilized enzyme consists of the following essential components:

- the carrier (backbone),
- CIG (carrier-bound inert pendant groups),
- CAG (carrier-bound active groups) that participate in the binding,
- the spacer (SPA), that might be linear or branched,
- AAR (active amino acid residues) that participate in the binding,
- the enzyme (or conformation).

Thus, it is hardly surprising that variation of these properties might lead to the change of the enzyme structure or performance. However, the structure or conformation of a carrier-bound immobilized enzyme is also dictated by many factors such as the immobilization conditions (pH, temperature, additives, medium). Although the optimum for a individual condition may differ from enzyme to enzyme, from carrier to carrier, from reaction to reaction, and from application to ap-

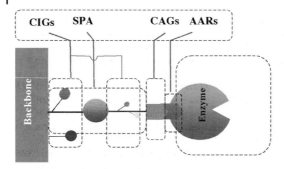

Scheme 3.18 The essential components of covalently immobilized enzymes: Ca, carrier; SPA, spacer arm; CABF, carrier-bound active functionality; CAIF, carrier-bound inert functionality; EBF, enzyme-bound active functionality.

plication, the general requirements for the carrier might be simplicity of the immobilization, availability of the carrier, stability of the binding under the application conditions, enzyme performance in the desired application, and requirements for the ideal process development.

3.5.1
Activity Retention

Activity retention can be affected either by the dimensional physical nature of the immobilized enzymes such as pore size, porosity, tortuosity, particle size, morphology, and specific surface, and by non-dimensional properties such as hydrophobicity, hydrophilicity, aquaphilicity, water-retention capacity and microenvironment, and the chemical nature of the carriers such as the nature of CAG, CIGS, and spacer. In addition, the immobilization conditions such as pH, ionic strength, temperature, additives, enzyme loading and immobilization time are also crucial.

3.5.1.1 Pore-size-dependent Activity

The activity of many enzymes, for example penicillin G acylase on a variety of porous carriers [107, 108], or lipases [109,110] immobilized on porous carriers are often highly dependent on the pore size [19, 109] and pore-size distribution of the carriers [19]. This phenomenon can be generally explained as a result of the non-uniform distribution of either the enzyme molecules or the substrates within the porous carrier. As revealed by active site titration, retention of activity is indeed turnover-related, suggesting that the enzyme molecule is active but the turnover frequency is lower than the native enzyme molecule [110].

On the other hand, it implies that immobilization of enzyme molecules in the internal domains of porous carriers is a slow diffusion process [111, 112, 114]. In other words, binding of the enzymes to the internal pores of the carrier also obey the mass transfer law and are thus governed by the diffusion constraints. Consequently, the enzyme molecules that can enter the pores of the carrier, are usually

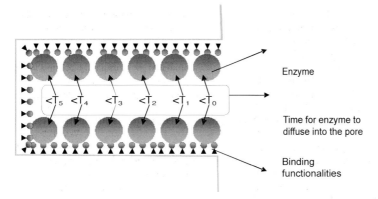

Scheme 3.19 Relationship between enzyme position in the pores and the time needed to diffuse in.

first bound to the binding sites close to the surface. When the binding sites located close to the surface have been occupied, the enzyme molecules will diffuse further into the depths of the pores, etc., as illustrated in Scheme 3.19.

Thus, it is not surprising that activity retention depends on enzyme loading, because at high enzyme loading the enzymes have to penetrate deep into the carriers and thus approach the bottom of the pores. Although the enzymes are not functionally damaged, as revealed by the active site titration experiments [110], they are almost inactive, because the concentration of the substrate in the depth of the pores is negligible.

As diffusion constraint is pore-size-dependent, use of a carrier with large pores should be able to mitigate the diffusion limitation, which will translate into high retention of activity. Indeed, the use of aminoalkylated silica of pore size from 7 to 550 nm for immobilization of trypsin led to the discovery that high specific activity (U mg^{-1} protein immobilized) was obtained by use of the silica with the highest pore size [19].

Remarkably, it was found that although the specific activity decreased as pore size decreased, when the pore size of the carrier approached the size of the enzyme molecules the specific activity increased again, possibly suggesting that the enzyme was mainly immobilized on the outer surface of the carrier when the pore size is small or close to the size of the enzyme (in this example the efficiency of coverage of the carrier surface with enzyme was only 1.3%, calculated in accordance with the fact that the payload was 11 mg g^{-1} silica with a specific surface area is 445 m^2 g^{-1} and that the surface needed to immobilize 11 mg trypsin is about 5 m^2). However, when the pore size increased, the enzyme molecules are able to diffuse into the pores and the activity is diffusion-controlled until the pore size becomes big enough to eliminate the diffusion constraints, as shown in Scheme 3.20 [19].

Similar results were also obtained by use of the Eupergit C and Eupergit C 250 for immobilization of penicillin G acylase. Retention of activity with the latter was twice that with the former, because of the 10 times bigger pore size [26, 178]. Stud-

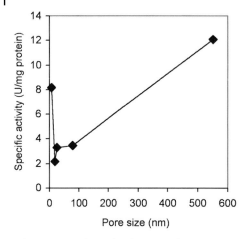

Scheme 3.20 Relationship between silica pore size and the specific activity of immobilized trypsin (data from [19]).

ies on subtilisin Carlsberg covalently immobilized on the same silica derivatives with different pore sizes revealed that activity retention on the large pore size silica was almost twice that for small pore material, irrespective of comparable enzyme loading. Moreover, immobilization of this enzyme on the same parent carrier with different spacers or binding chemistry might result in twentyfold enhancement of specific activity [34].

3.5.1.2 CAG-controlled Activity

Although the effect of different CAG in enzyme immobilization was observed early in the 1960s, study of the effect of different CAG on enzyme activity was possibly pioneered by Barker and Somers, who found that retention of activity of α-amylase on PAAm-based carrier was 16.1%, 6.1% and 9.5% when immobilized with the different active functionality PAAm-acylazide, PAAm-diazo and PAAm-isothiocyanato, respectively [112]. On the same carriers with different CAG, retention of activity by α-amylase was 0%, 1.5% and 0.8% (Scheme 3.21).

For carrier B (acylazide) it is obvious that the most exposed lysine residues can be bound to the carrier. Thus, the enzyme might be stabilized, because of multipoint attachment. In contrast, for carrier A (diazonium salt) the binding of the tyrosine residues might provoke a change of enzyme conformation, leading to loss of the enzyme activity. The hydrophobic backbone might also accelerate thermal deactivation (the hydrophobic microenvironment favours deactivation). In carrier C, because of the hydrophobic effect of the spacer, it is most likely that the lysine residues that participate in the binding are located in the hydrophobic clusters. Consequently, the binding number might be lowest among the three different CAG.

Thus, the CAG generally dictate the nature of the binding chemistry but the binding mode (position of EBF and thus enzyme orientation) is generally dictated

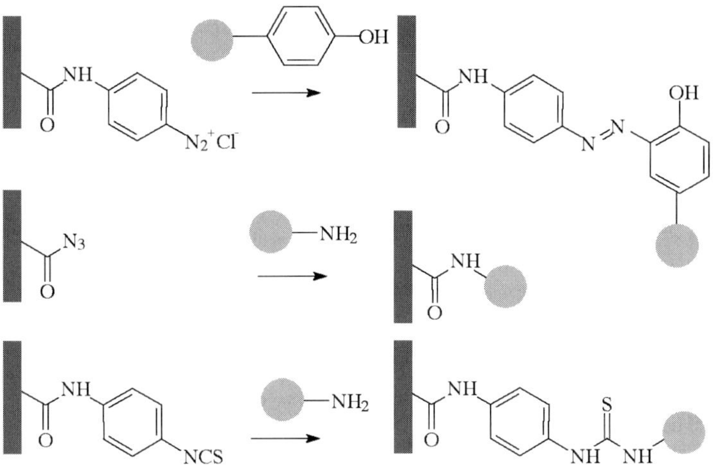

Scheme 3.21 Immobilization of α-amylase on differently functionalized PAAm.

by the backbone and the spacer. Thus, retention of activity also depends greatly on the CAG. In practice, it is essential to discover the best CAG for the enzyme to be immobilized. Although there are no rules to follow, the selected CAG should be able to bind the enzyme under very mild conditions. Moreover, the CAG should be group-specific.

Systematic research on the effect of binding functionality on retention of the activity was only conducted by Taylor in 1985, when he simultaneously used several types of carrier such as glass, agarose-based carrier, and synthetic polymeric carriers such as Eupergit C (epoxy synthetic carriers), which had different functionality such as NHS or CDI such as glass–CDI, glass–NHS or agarose–NHS, or CNBr-activated Sepharose, for immobilization of three different enzymes, i.e. alkaline phosphatase, glucose oxidase, and peroxidase at different pH [113]. The important conclusion was that the optimum immobilization conditions differ from enzyme to enzyme and from carrier to carrier, when the same type of CAG is selected.

Liu et al. studied the influence of different CAG on the performance of penicillin G acylase immobilized on poly(vinyl acetate-co-divinylbenzene) beads (Scheme 3.22). It was found that these CAG influence not only the efficiency of immobilization but also the operational stability of the immobilized enzymes. The highest specific activity (852 U g^{-1}, 57 mg protein g^{-1} wet carrier) was achieved with immobilized p-benzoquinone as CAG [114].

When diazotizing the carrier, the functional amino acid residues that participate in the coupling reaction between the carrier and the proteins are histidine and tyrosine groups, as indicated in Scheme 3.32. Binding of these groups might provoke the change of enzyme conformation thus resulting in the loss of large portion of the activity but often the stability of the enzyme is higher. In contrast, the CAG (aldehyde group) of the pendant functionalities in (A) are located far from the backbone. Because of the higher hydrophobicity of the spacer, binding via the aldehyde

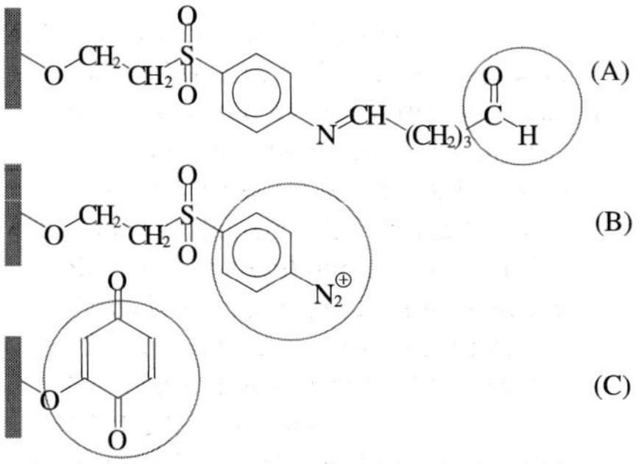

Scheme 3.22 Introduction of different pendant groups to poly(vinyl acetate-co-divinylbenzene) beads [114].

group might involve the lysine residues located in the hydrophobic clusters, leading to formation of much more active immobilized enzyme. The activity loss might be ascribed to modification of the lysine residues close to the active centre. The greater retention of activity with carrier (C) might be because only the highly exposed lysine residues are involved in the binding. Thus, it is conceivable that multipoint attachment is hardly possible in this instance because of steric hindrance. As a result, it is expected that immobilized penicillin G acylase is also less stabilized than other preparations (Scheme 3.23).

Similarly, it has been found that high retention of activity and stability of glycolate oxidase was obtained with CNBr-activated agarose or Sepharose, compared with epoxy activated agarose or Sepharose (Scheme 3.24). Although these two CAGs are located on the same backbone and both are specific for the amino functionality of the lysine residues, almost 10 times greater retention of activity was obtained with CNBr-activated carriers, suggesting that CAG can distinguish the positions of EBG (enzyme-bound groups). In other word, the lysine residues of glycolate oxidase that take part in the binding are different in these two cases, resulting in different binding mode i.e. orientation and number of binding.

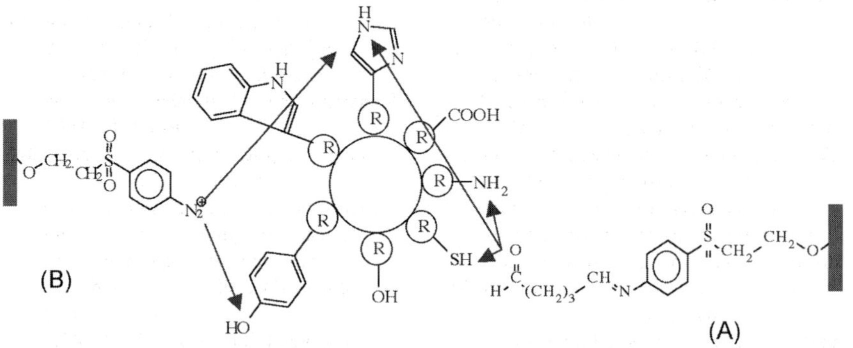

Scheme 3.23 Different possibilities of coupling in enzyme immobilization with different pendant groups.

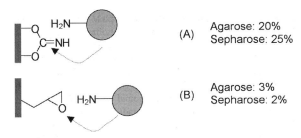

Scheme 3.24 Immobilization of glycolate oxidase on agarose and Sepharose with different CAG.

The different retention of activity also suggests that the lysine residues which participate in the binding might be different. For CNBr-activated carriers, only the highly exposed lysine residues can participate in the reaction, because of steric hindrance, whereas the lysine residues located in the active centre might participate in the reaction of epoxy Sepharose, leading to loss of a large amount of the activity. However, when the microenvironment of the epoxy carrier changes as a result of the use of a carrier with a more hydrophobic backbone, for example Eupergit C, retention of activity can be improved, as discussed previously [116], suggesting that the effect of CAG is largely dictated by the microenvironment. This is understandable, because the reactivity of the CAG is dictated by the microenvironment where they are accommodated.

Immobilization of β-lactamase on polystyrene derivatives with different pendant active functionality such as GAH, diazonium salt, and isocyanate groups (Scheme 3.25), led to the discovery that the highest specific activity and retention of activity were obtained with glutaraldehyde-activated PS derivatives, followed by diazotized carrier and carrier bearing isocyanate groups [116].

Because there are no diffusion constraints with the nonporous carrier, it seems most likely that the loss of activity is ascribed solely to enzyme deactivation, because of reaction of essential amino acid residues near the active centre. For carrier

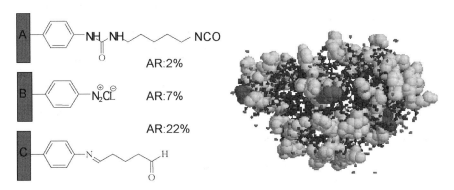

Scheme 3.25 Immobilization of β-lactamase on derivatized polystyrene.

C (aldehyde), the spacer is more hydrophobic. Thus, only the lysine residues of hydrophobic cluster might be involved in the binding and consequently multipoint attachment can be avoided. Often, the diazonium salts or isocyanate groups can react with a number of amino residues than can aldehyde functionality. Thus, the possibility of enzyme deactivation as a result of use of this functionality might be greater.

Thus, there is no doubt that the selection of the binding chemistry should be carefully considered together with the nature of the backbone and the nature of the enzyme [115]. Apart from the nature of the carrier and the enzyme, the immobilization conditions might play an important role as well. The influence of the coupling conditions such as pH on the activity retention has been systematically studied [113].

3.5.1.3 CIG-controlled Retention of Activity

As defined above, CIG (carrier-bound inert groups) do not participate in binding but may have profound effects on the binding and the performance of the resulting immobilized enzyme. An early observation that variation of the CIG of the synthetic active polymer bearing anhydride groups, by changing the cross-linkers of monomers, could also influence the hydrophobicity of the carrier and, further, retention of activity of the enzyme was made in 1970s [117]. Accordingly, since the 1970s research interest in the preparation of ready-made carriers for enzyme immobilization has switched to carriers with more hydrophilic CIG. For instance, the commercial polyacrylic carrier-Eupergit C is usually prepared in the presence of more hydrophilic monomer such as acrylamide, to improve the microenvironment of the binding site.

Recently, synthesis of a series of carriers with the same CAG but different CIG by varying the cross-linkers used in suspension polymerization [116] has enabled a precise study of the CIG effect. The use of these carriers for immobilization of penicillin G acylase led to the conclusion that enzyme activity and retention of activity were closely related to the properties of the cross-linkers, suggesting that the microenvironment of the carrier, especially the CIG are more crucial, with regard to retention of activity and enzyme-loading capacity [118].

Similarly, Mauz used a number of new hydrophilic monomers, for example vinyl or allyl ethers, (meth)acrylates of glycerol carbonate or vinylene carbonate and comonomers such as N-vinylpyrrolidone and cross-linkers such as methylenebizacrylamide or butanediol divinyl ether and N,N'-divinylalkyleneurea to prepare hydrophilic beaded polymer bearing active carbonate groups for enzyme immobilization [119]. Remarkably, these carriers can be used to bind large amounts of penicillin G acylase with very much higher payload (1.28) and retention of activity [119], suggesting that the presence of hydrophilic CIG is crucial for activity retention.

On the other hand, it was observed that for some enzymes such as lipase, hydrophobic CIG are more important than hydrophilic carriers, with regard to the specific activity of the enzyme. For example, immobilization of *Candida rugosa* lipase (CRL) on synthetic hydrophobic carrier poly(styrene-co-divinylbenzene) polymer

3.5 Factors Affecting Enzyme Performance

particles (specific surface area, pore volume, and pore diameter 165 m² g⁻¹, 1.63 mL g⁻¹, and 198 nm, respectively) resulted in a specific activity of 15.2 U mg⁻¹ (0.068 mg lipase/10 mg carrier) compared with 4.45–4.27 (0.058–0.144) with Sephadex G-100 or Eupergit C as a carrier. Obviously, the hydrophobic nature of the carriers favours activity expression of lipases, which often can be activated by the presence of a hydrophobic interface [120].

In other work the effect of CIG on the binding and activity expression of trypsin was studied by synthesizing a set of carriers bearing the same backbone but different leaving groups (Scheme 3.26). It is obvious that the microenvironment around the binding functionality is different, depending on the properties of the leaving groups (or the monomer used). Consequently, the orientation of the enzyme on the carrier might be also governed by the properties of the side-chain, as shown in Scheme 3.26 [121].

As expected, although the payloads of these polymers were quite high and similar for trypsin (147, 149, 138, and 142 mg g⁻¹ carrier, respectively), recovery of the activity was only 31, 25, 8, and 8%, respectively. The final binding nature (peptide bond) was the same for all the immobilized enzymes, which suggests that the nature of the leaving groups may be another factor which can also affect the activity of the immobilized enzyme obtained. It is most probable that the leaving groups can determine which lysine residue can participate the binding of the enzyme to the carrier. In other words, the microenvironment of the binding functionality governs both enzyme orientation and retention of activity, as shown in Scheme 3.26.

By studying the structure of the carrier (CIG of the leaving groups) and the enzyme, it was found that carriers (A) and (B) have neutral hydrophobic leaving groups. In contrast, carriers (C) and (D) bear positively charged leaving groups

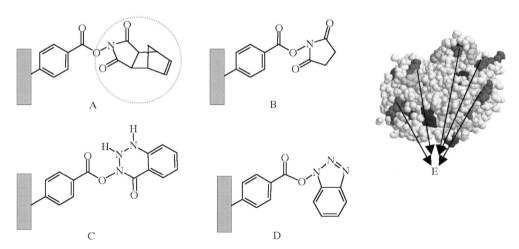

Scheme 3.26 Monomers used for preparation of activated polymeric carbonyl carriers:
(A) poly(N-p-methacryloxybenzoyloxy-5-norbornene-2,3-dicarboximide);
(B) poly(N-p-methacryloxybenzoyloxysuccinimide);
(C) poly(N-p-methacryloxybenzoyloxy-4-oxo-3,4-dihydro-1,2,3-benzotriazine);
(D) poly(N-p-methacryloxybenzoyloxybenzotriazole).

(Scheme 3.26). Moreover, the active centre of trypsin is negatively charged, thus, often accepting the positively charged substrates [122] Thus it is probable that lysine residues located close to the active centre could be bound to the carrier, because of the orientation effect of the negatively charged leaving groups (Scheme 3.26E). As a result, retention of activity with carriers C and D could be lower than with carriers A and B, because of to the high probability of modifying the essential amino acid residues involved in the catalysis.

In a similar example it was found that urate oxidase (uricase) immobilized on CM-HASCL-35 (cross-linked heavy amylose starch bearing COOH groups) by use of 1-ethyl-3-(3-dimethylaminopropyl)carbodiimide as a coupling agent resulted in 53% retention of activity whereas use of N-ethyl-5-phenylisoxazolium-3'-sulphonate (Woodward reagent K) resulted in high binding but totally inhibited enzyme activity [122]. The difference was obviously ascribed to the same effect as mentioned above.

Controlled introduction of other CIG by modification of CAG provides another efficient means of studying the effect of the CIG on the performance of the immobilized enzymes, because use of the same carrier but with different pendant functionality can easily exclude the influences of other factors such as variation of the nature of the carrier, diffusion constraints, and nature of the binding [106].

Investigation of the effect of controlled modification of the active carrier on enzyme activity can be dated back to the early 1970s, when it was found that optimum retention of activity of immobilized trypsin could be achieved with EMA samples of 50–70% anhydride content, which was obtained by controlled hydrolysis [123]. Obviously, the controlled hydrolysis led to modification of the microenvironment of the carrier, which can further affect the performance of the immobilized enzymes obtained. Analogous to this concept, the controlled aminolysis of PMA with diaminoalkane led to the formation of a new adsorbent, which could be used to immobilize lipase CRL (*Candida rugosa* lipase) with high retention of activity [124]. Apparently, introduction of charged functionality favoured more active orientation of the enzyme molecules.

Interconversion of CAG into new CAG is another means of studying the influence of pendant groups on retention of activity. Often, along with introduction of new CAG, new CIG can be introduced and therefore the charge, polarity and length of the pendant groups can be changed. However, it is very difficult to interpret the results obtained, because the interconversion techniques often also provoke changes of other conditions beside changes of CAG. For instance, using the interconversion technique, Drobnik [125] was able to improve the efficiency of immobilization of penicillin G acylase by introduction of spacers bearing diamino groups, followed by activation with glutaraldehyde or other activators, as shown in Scheme 3.27. Obviously, it is difficult to ascribe the greater retention of activity solely to the use of the spacer or to the change of the CIG, because the different binding chemistries might also contribute to the different retention of activity (Scheme 3.27D).

In a similar approach, penicillin G acylase was immobilized on 2,3-epoxypropyl methacrylate-based polymers by introduction of a spacer (HAD and lysine). Maxi-

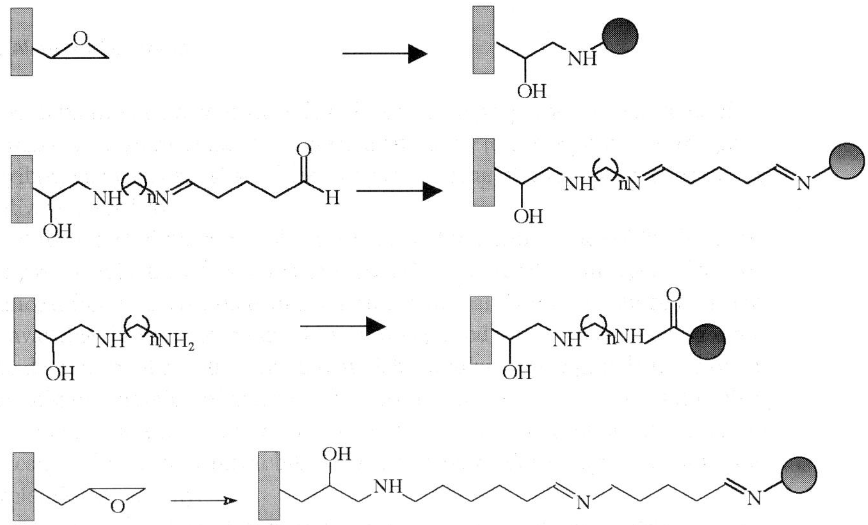

Scheme 3.27 Immobilization of enzymes on oxirane ring carriers using different activation methods.

mum retention of activity among the various approaches tested, approximately 20%, resulted from use of HAD as spacer [126]. The lower retention of activity was ascribed to the lower availability of the active binding site. However, in our opinion the lower retention of activity should be ascribed to the effect of modification of CAG located at the terminus of a hydrophobic pendant CIG.

Because penicillin G acylase has a very hydrophobic and negatively charged active centre [127], the lysine residues located closely to the active centre might possibly be modified, leading to the drastic conformation change and thus loss of enzyme activity [128]. That activity retention was increased to 43% by treating an epoxy polymeric carrier with PEI and consecutive activation by glutaraldehyde validated this hypothesis [61].

The effect on activity retention of the inert pendant groups (CIG) adjacent to the CAG, was beautifully demonstrated by preparing a copolymer of N-vinylpyrrolidone with allyl ester and glycidyl crotonate, as shown in Scheme 3.28.

The two water-soluble polymers differ from each other only slightly in the nature of the spacer (i.e. hydrophobicity of the spacer) that links the CAG to the backbone and the CIG bound to the backbone (H instead of –CH$_3$). The high retention of activity with carrier A might be because the microenvironment of the CAG of this type of polymer is much polar than that of polymer B. Consequently, the binding position with polymer A might be limited to the hydrophilic clusters of lysine residues, whereas the binding position with polymer B is mainly oriented to the lysine residues located in the hydrophobic clusters of α-chymotrypsin [129].

In fact, two lysine residues are accommodated in the hydrophobic clusters: one is located in the active centre and another is in the domain close to the active cen-

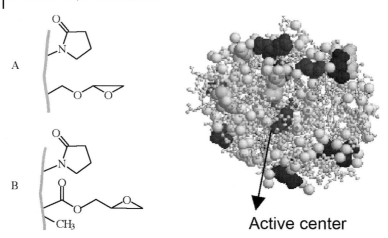

Scheme 3.28 Structures of the carriers and the enzymes immobilized (α-CT).

tre (Scheme 3.28). Thus, these two lysine residues which are essential for the function of the enzymes might be modified, leading to loss of activity during immobilization.

Apart from the positive partitioning effect, a negative partitioning effect was also observed with the CIG. For instance, for penicillin G acylase immobilized on *p*-aminoarylsilochrome anomalous dependence of activity on substrate concentration was observed; this was ascribed to specific sorption of substrate or phenylacetic acid on the support bearing the benzyl ring, leading to hindrance of transport of substrate or inhibition [130].

Similarly, it was found that the oxirane acrylic polymer cross-linked with divinylbenzene (Scheme 3.29) is also not suitable for immobilization of penicillin G acylase, because of an incompatible interaction of penicillin G acylase with the benzyl ring [131]. However, the same carriers might be very suitable for immobilization of other enzymes such as lipase [124] or amino acid oxidase [10]. Thus, the criteria for a good carrier might differ from enzyme to enzyme.

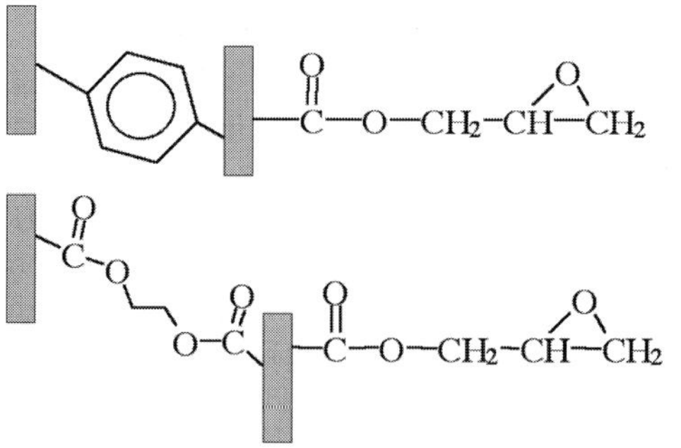

Scheme 3.29 Effect of chemical composition on the performance of immobilized enzymes.

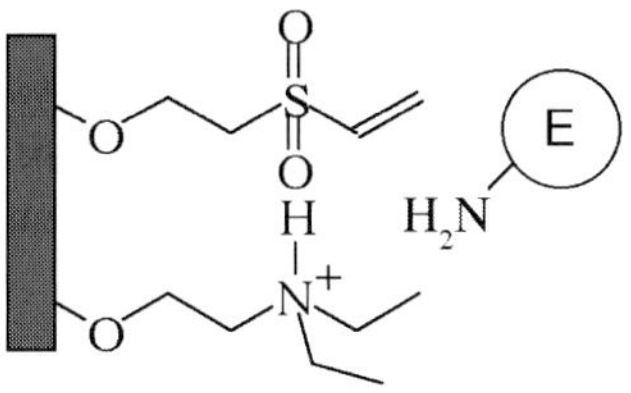

3.5 Factors Affecting Enzyme Performance

Scheme 3.30 Immobilization of laccase on DEAE-cellulose (–OH).

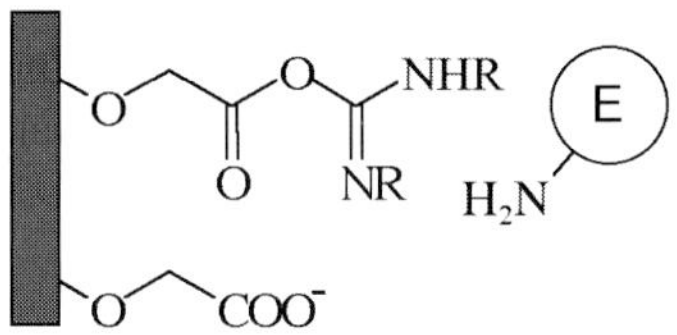

Scheme 3.31 Immobilization of laccase on CM-cellulose (–COOH).

The function of CIG in the immobilization of laccase on DEAE-cellulose (–OH) activated with DVS and on CM-cellulose (–COOH) activated with CDI has been clearly demonstrated (Schemes 3.30 and 3.31). It was found that carrier A is suitable for immobilization of *C. unicolour* laccase with regard to high retention of activity and storage stability, whereas carrier B does not enable binding of the enzyme. This result suggested that adsorption of this acidic enzyme is favoured by the presence of positively charged groups and that adsorption of the enzyme on the negatively charged enzyme is not favoured [132].

From the previous discussion, the presence of CIG in covalent enzyme immobilization should not be overlooked, because enzyme loading, retention of activity, stability, and selectivity can all be affected by the presence of the appropriate CIG. It is worthwhile pointing out that this observation also applies to other types of immobilized enzymes.

3.5.1.4 Spacer-controlled Activity

Interpretation of the effect of nature of the spacer on enzyme activity retention is often complicated, because the presence of the spacer often changes the microenvironment, orientation, and conformational flexibility of the enzymes.

However, if an appropriate spacer (length, hydrophilicity, branched, or linear) is selected, greater retention of activity is often achieved (Scheme 3.32). For higher molecular weight substrates, in particular, introduction of an appropriate spacer

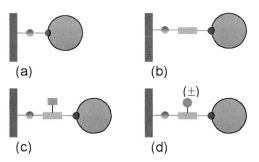

Scheme 3.32 Spacer-mediated carrier-bound immobilized enzymes:
(a) no spacer; (b) linear spacer; (c) branched spacer; (d) charged spacer.

often confers more flexibility on the enzyme, thus significantly increasing the activity of the enzyme, because of reduction of steric hindrance, compared with the immobilized enzyme without the spacer [125,133]. The presence of a suitable spacer often results in the enzyme having kinetic behaviour similar to that of the native enzyme, which might be attractive for immobilized enzymes suffering from diffusion constraints, for example immobilized penicillin G acylase used for kinetically controlled synthesis of semi-synthetic β-lactam antibiotics [14].

Rexova-Benkova and Mrackova-Dobrotova were among the first few people to study the effect of the length of the spacer on retention of activity and the kinetic behaviour of *Aspergillus niger* extracellular *endo*-d-galacturonanase immobilized on a polyacrylic polymer (Spheron) with the use of glycine, (-alanine, 4-aminobutanoic acid, gly–gly and 6-aminohexanoic acid as spacers [135]. It was found that activity retention increases with increasing spacer length, as shown in Scheme 3.33. Among these, the highest activity was obtained with 6-aminohexanoic acid. Remarkably, K_m values for all immobilized enzymes were quite similar, but V_{max} depended on the length of the spacer.

Similarly, papain immobilized on chitosan beads with spacers had almost the same activity toward small or large substrates as the native enzyme [136]. However, the papain immobilized directly on the surface of the chitosan beads without any spacer was more stable than the enzyme immobilized with a spacer. Obviously, the effect of the spacer on the activity could be explained in terms of the mobility of the immobilized papain molecule.

Similar results were also obtained with lipoprotein lipase (LPL) of *Pseudomonas fluorescens* immobilized on porous chitosan beads [137].

The conformational flexibility of thermolysin from *Bacillus thermoproteolyticus rokko* immobilized on an anionic polymer latex (AA-2, prepared by polymerization of styrene, divinylbenzene, and acrylic acid) or acrolein-containing latex (AL, obtained by copolymerization of styrene and acrolein) was reduced by immobilization [138]. In contrast, by introduction of a spacer between the latex and the enzyme, this effect was reduced, i.e. activity was higher with the spacer. Also, direct

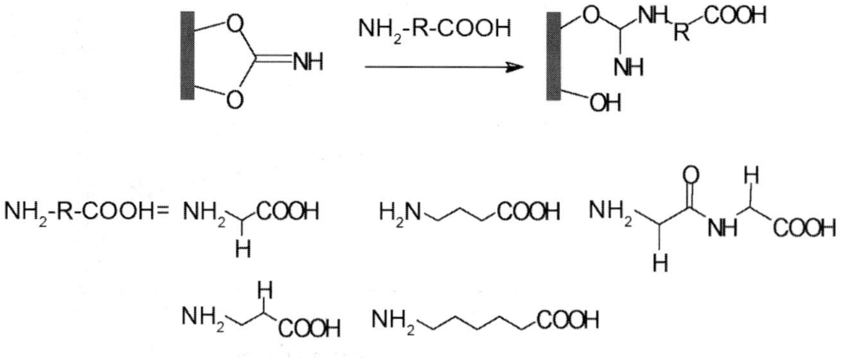

Scheme 3.33 Immobilization of endo-D-galacturonanase on the polyacrylic polymer Spheron by use of different spacers.

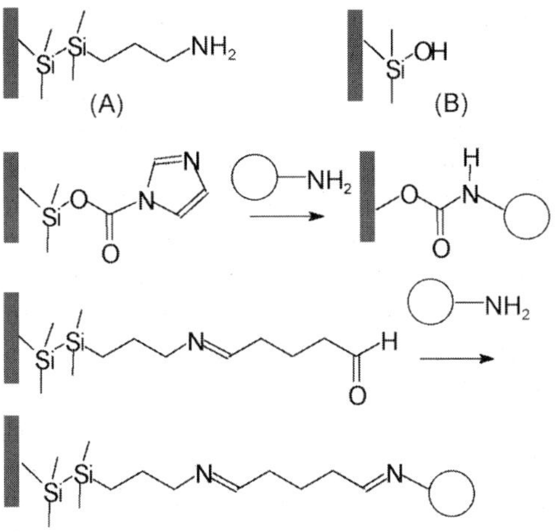

Scheme 3.34 Immobilization of subtilisin Carlsberg on silica.

immobilization of subtilisin Carlsberg on derivatized silica via CDI coupling (B) resulted in only ca. 1/20 of the activity of derivatives obtained by using spacers (A), as illustrated in Scheme 3.34, suggesting that spacer is able to render the enzyme more flexible [34].

Immobilization of glucosidase on poly(2-hydroxyethylmethacrylate-ethyleneglycol dimethacrylate) (poly(HEMA-EGDMA)) microspheres via a spacer arm resulted in an increase in the apparent activity of the immobilized enzyme compared with the enzyme immobilized on plain microspheres, depending on the coupling method (Scheme 3.35).

The highest retention of activity was obtained with CNBr coupling. Remarkably, the enzyme immobilized on CNBr-activated carrier (coupling) is almost twice as thermostable than the CDI-coupled enzyme [139], suggesting that the presence of

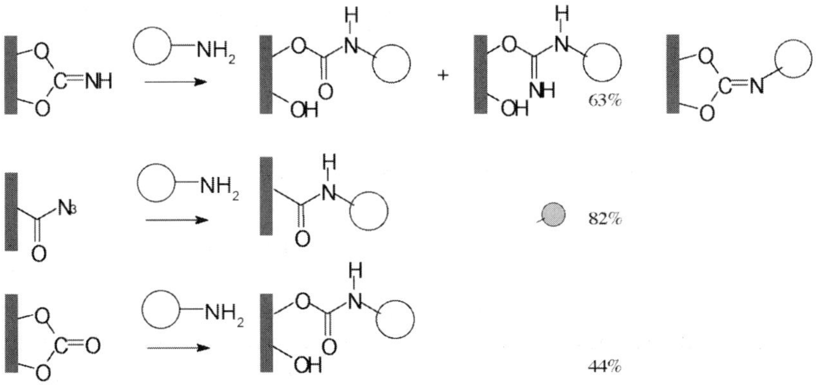

Scheme 3.35 Immobilization of glucosidase on poly(HEMA-EGDMA) microspheres with and without spacers.

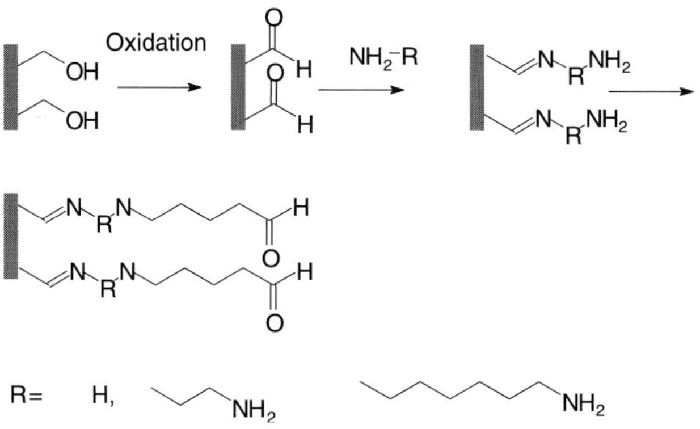

Scheme 3.36 Immobilization of glucosidase on PMMA, with spacers.

the spacer improved the retention of activity but that the thermostability is largely dependent on the CAG, which dictates the nature of binding mode such as position and number of bonds as discussed above.

There is, however, usually an optimum length for the spacer-arm. Thus, the residual activity can increase with increasing spacer length, as demonstrated by glucosidase immobilized on modified PMMA monosized microspheres by use of different spacers, for example ammonium, ethylenediamine, and hexamethylenediamine (HMDA). The highest payload and retention of activity was obtained with carriers having HMDA as spacer (Scheme 3.36).

This was presumably because of the increased flexibility of the enzyme conformation and mitigation of the non-specific influence of the carriers [140]. A follow-up study, however, revealed that for GOD immobilized on glutaraldehyde-activated poly(EGDMA/AAm) copolymer beads with HMDA as spacer an optimum HMDA content was required, because at high HMDA concentrations the HMDA might function as a cross-linker, i.e. two terminal amino groups will react with adjacent aldehyde groups on the surface [141]. A plausible explanation is that the risk that the active centre might be modified by CAG increases when the spacer grows to over certain length [94, 142]. The optimum length of the spacer might, however, differ from enzyme to enzyme. An alternative explanation is that the enzyme linked to the carrier via a different spacer might affect the nature of the binding of the amino acid side-chains involved in the covalent coupling of the enzyme to the carrier [143].

Trypsin was immobilized on PF (polyester fleece) grafted with a number of spacers such as aldehyde dextran, PEG diamine, aminodextran and albumin, as shown in Scheme 3.37. The immobilized enzymes obtained with the spacers were usually more active (10–30 times higher than the counterpart) and were also more thermostable, because of the favourable environment created by the hydrophilic spacer. Remarkably, enzyme loading was also increased by use of amino dextran or albumin spacers [89].

In a similar experiment the activity of β-glucosidase and trypsin immobilized on nylon via macromolecules such as chitosan and PEI was increased 1.5–2.0 times

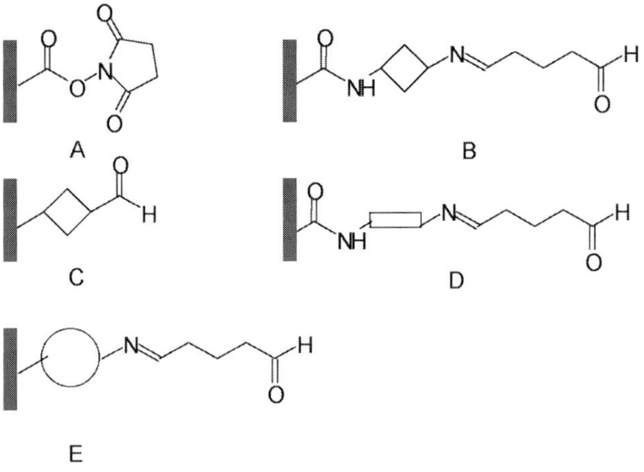

Scheme 3.37 Immobilization of trypsin on a hydrophobic carrier with different spacers: (A) no spacer; (B) aminodextran as spacer; (C) aldehyde dextran as spacer; (D) amino-PEG as spacer; (E) albumin as spacer.

compared with the counterpart without spacer (Scheme 3.38) [144]. It is obvious that apart from the spatial effect of the spacer, i.e. an increase in molecular mobility, introduction of spacers can also lead to increased enzyme activity, because of extra microenvironment effect of the spacer [145], which might also activate the enzymes, leading to increased activity relative to the native enzymes.

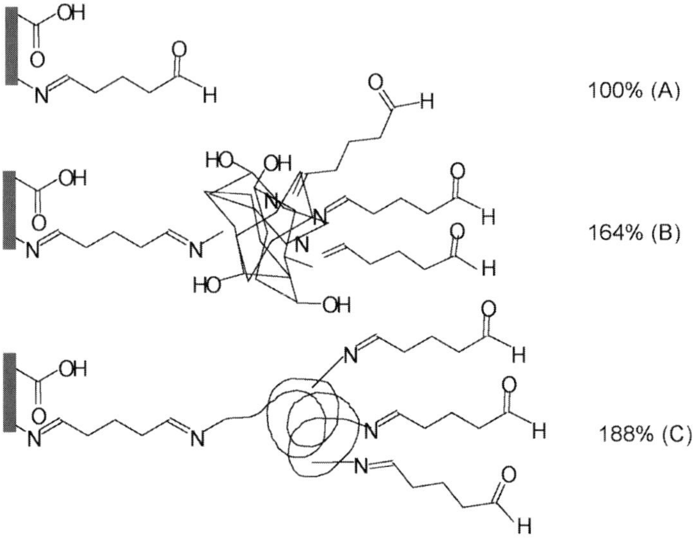

Scheme 3.38 Immobilization of enzyme via macromolecular spacers: (A) no spacer; (B) chitosan as spacer; (C) PEI as spacer.

Although it was shown that significant improvement can be obtained by introduction of a spacer, the change of CAG and the microenvironment caused by the introduction of spacer might complicate the interpretation of the obtained results. For instance, lipase immobilized on epoxy group-containing poly(GMA-HEMA-EGDMA) microspheres retained only 9% of activity of the native enzyme (in terms of specific activity), while introduction of 1,6-diaminohexane into this epoxy carrier, followed by activation with glutaraldehyde led to 5 times more activity as compared with the native epoxy carrier [146]. More strikingly, it was found that a lipase immobilized on BrCN-activated agarose or Sephadex carriers displayed very low activity, while higher activity retention was obtained by immobilizing this lipase on aminolated Sepharose 2B with 4,4'-methylenedianiline as spacer, as compared to the same material without spacer [147]. The author ascribed this higher activity to the decreased steric hindrance, due to the presence of longer spacer. However, the presence of hydrophobic spacer might also significantly alter the microenvironment, thus leading to the higher activity retention.

Obviously, the role of the spacer is more complicated. Generally, the introduction of spacer not only improves the conformational flexibility of the attached enzyme molecules, but may also alter the microenvironment and the binding mode.

3.5.1.5 Enzyme Orientation-controlled Activity

Enzyme orientation-controlled activity is largely ascribed to the fact that enzyme activity might be reduced if essential amino acid residues located close to the active site are involved in enzyme immobilization.

For instance, approximately 90% loss of activity was observed when CFTase (*Thermoanaerobacter* sp) was immobilized on Eupergit C [148]. Obviously, immobilization of this enzyme on Eupergit C with oxirane groups might involve the essential amino acids such as lysine residues, because the same enzyme immobilized on CNBr Sepharose and EAH-Sepharose also had very low activity. The same enzymes, but from different strain (*Bacillus macerans*), immobilized on aminated silica via glutaraldehyde activation retained 27% activity [149] and retention of activity was less than 5% when the same enzyme was immobilized on polyacrylamide-type support with carboxyl functional groups activated by H_2O-soluble carbodiimide, which enables the coupling of the lysine residue of the enzyme with the carboxylic groups of the carrier [150].

For silanized silica-NH_2 the pendant functional groups are more hydrophobic, thus, only the lysine close to the hydrophobic cluster can be modified. Indeed, higher retention of activity was observed when CFTase from a similar strain (*Paenibacillus macerans*) was immobilized on aminated PVA-activated with glutaraldehyde [151].

3.5.1.6 Binding Density-controlled Activity

Historically, study of the binding density can be dated back to the year 1970, when a group of scientists started to study the influence of the ligand density of the absorbent on the chromatographic behaviour of proteins. However, for enzyme immobilization, studies of binding density and its effect on characteristics such as enzyme activity and enzyme stability are mainly limited to stability, whereas activity retention is less well studied.

Usually, binding density affects not only the number of bonds formed between the enzyme molecules and the carriers and also the microenvironment. Thus, the number of the enzyme molecules immobilized per units of surface (10 nm^2) (volume activity), the enzyme mobility, stability, etc. can be affected by the binding density.

Thus, it has been found that increasing the binding density can lead to an increased enzyme loading [152]. It was, however, also observed that high binding density on the carrier surface often has a negative effect on the efficiency of the immobilized enzyme e.g. reduction of activity retention, presumably because of rigidification of the conformation of the enzyme, as demonstrated by immobilizing Ribonuclease A and α-chymotrypsin on CNBr activated-Sepharose CL 4B with different binding density [153].

Similarly, the activity of penicillin G acylase immobilized on a grafted polyacrylonitrile prepared by photo-initiated grafted copolymer of glycidyl methacrylate and 2-hydroxyethyl methacrylate was found to also be strongly dependent on epoxy group concentration. The activity increase was not, however, linearly related to the concentration of epoxy groups, indicating that high density of binding functionality might lead to an irreversible change of enzyme conformation (deactivation) [154].

Apart from the multipoint attachment effect (which usually increase the enzyme conformation rigidity and thus decreases the enzyme activity), it has been also found that the binding density might be also related to the microenvironment effect. For instance, it was found that the highest retention of activity of trypsin immobilized on ethylene–maleic anhydride copolymers (EMA) with well-defined and controlled anhydride content (achieved by controlled hydrolysis) could be achieved with EMA samples containing 50–70% anhydride [155]. This first example suggested that the nature of the microenvironment is also related to the binding density. A similar microenvironmental effect was observed with other carriers. For example, the activity of α-chymotrypsin (CT), immobilized on silicone rubber film-γ-HEMA-co-MAAc (activated by NHS) increased with increasing MAAc content. On the other hand, the specific activity for ATEE (negatively charged substrate) declined sharply as the content of the MMAc increased, because of repulsion by the ATEE. In contrast, the rate of hydrolysis of BAEE, a positively charged substrate, by immobilized CT at pH 11, is almost four times greater than that by free CT at its pH optimum [156]. Thus, the higher binding density not only rigidifies the enzyme conformation but also alters the microenvironment of the carrier, leading to a change of the enzyme kinetic properties.

3.5.1.7 Diffusion-controlled Enzyme Activity

In general, the presence of diffusion constraints leads to formation of concentration gradients. The concentration of substrates gradually decreases from the bulk medium to the centre of the particles containing the immobilized enzymes. Thus, the activity of the enzyme molecules in the interior of the particles will be different from those located on the surface, depending on the concentration gradient. As previously discussed (see Chapter 2), the enzyme activity might be increased or decreased in the case of diffusion-controlled activity [157]. In the majority of cases, the activity of the enzyme is decreased due to the presence of the concentration gradient.

In order to overcome the diffusion limitation and thus increase the enzyme activity, many strategies have been developed such as:

- the use of the carriers with large pore sizes [157],
- controlled distribution of enzyme molecules as mentioned above,
- use of nonporous carriers or carriers with small pores.

Many studies have shown that diffusion limitation can be overcome as the pore size exceeds certain limits [39]. The relationship between pore size of the carrier and activity retention has been clearly shown in several studies [26, 38–40].

In the case of controlled distribution of enzyme molecules, the enzyme molecules are mainly located on the surface and thus face almost the same substrate concentration, leading to higher activity retention. The use of nonporous carrier or carriers with small pores is based on the same principle, since the enzyme cannot penetrate the particles [158].

3.5.1.8 Reactive Amino Acid Residues (RAAR)-controlled Activity

In the early 1970s, concurrent with the establishment of many coupling chemistries for covalent enzyme immobilization, it had been appreciated that the activity of the immobilized enzyme is largely dictated by the nature of reactive amino acid residues (RAARs) of the enzyme that take part in the binding [112].

The effect of RAARs on the activity of the resulting immobilized enzymes can be divided into three types:

- Differences in enzyme orientation,
- Differences in the number of bonds formed,
- Involvement of essential amino groups in the binding.

In extreme cases, the enzymes might be completely deactivated, if the RAARs that are bound to the carriers are the amino acids residues essential for catalytic function. Thus, immobilization of indolyl-3-acetic acid oxidase on carriers bearing amino groups with the aid of a carbodiimide led to the formation of immobilized enzyme with higher activity retention than if coupled through the amino groups to the carrier [159]. This is mainly due to the fact that free carboxyl groups are involved in the catalysis of peroxidase and indole-3-acetic acid oxidase [159].

Similarly, immobilization of phenol hydroxylase on AH-Sepharose 4B and modified nylon nets via carbodiimide-aided condensation reaction yielded immobilized enzyme with satisfactory activity, while immobilization through lysine residues gave inactive or poorly active immobilized enzyme, suggesting that lysine is probably part of the active centre of the enzyme [160].

In some cases, RAARs are not involved in the active centre but instead located in the domains that that are essential for the enzyme function. Consequently, modification of these amino acid residues might also reduce the enzyme activity. For instance, it was found that immobilization of penicillin acylase on aminoalkylated carrier via carbodiimide-aided condensation of carrier-bound amino functionalities and the carboxylic groups of the enzyme resulted in higher activity retention, compared with the immobilization via the lysine residues of the enzyme with the epoxy carriers [125]. This could be ascribed to the fact that many highly exposed lysine residues are located in the proximity of the active center of penicillin G acylase. By contrast only a few highly exposed glutamic residues are located in this labile region [125].

3.5.1.9 Loading-dependent Activity

It has long been known that enzyme activity is often loading-dependent. As noted above, influence of the degree of loading on enzyme activity might either be related to the monolayer principle or to the diffusion limitation. In the former case, low activity may result when the loading of the enzyme is below the monolayer coverage, due to the deactivation of the enzyme molecules as discussed previously (see Chapter 2).

On the other hand, diffusion limitation has also to be taken into account, as the loading of enzyme into carrier increases. In this case, the enzyme molecules that are immobilized in the depth of the carrier display lower apparent activity than those located on the surface. A recent study has revealed that the turnover frequency of penicillin G acylase immobilized on Eupergit C 250 L displayed strong loading-dependence, due to the presence of diffusion limitation [277]. In general, because of diffusion limitation, the specific activity of the immobilized enzyme is usually 30–60% of that measured in the free enzyme, depending on enzyme loading [277].

This loading-activity relationship is quite common with many types of porous carriers and enzymes. For instance, the activity of *Aspergillus tarmarii* xylanase immobilized on Duolite A 147 activated with glutaraldehyde varies in the range of 30–60%, depending on the enzyme loading [161]. It was also found that the activity of a protease-Flavourzyme immobilized on Lewait R258 K (activated by glutaraldehyde) was strongly related to the degree of enzyme loading. An increase in the enzyme loading led to the increase of the enzyme activity (U/g carrier) but the specific activity (U/mg protein) was decreased. The author suggested that the enzyme is deactivated by the unreacted aldehyde groups. Apparently, the activity of the immobilized enzyme is mainly dictated by the diffusion limitation [162]

In practical terms, a compromise must be found between the activity expression (and retention) and the volume activity.

3.5.1.10 Other Factors Controlling Activity

Apart from the factors mentioned above, other factors might be also important, for example conformation, conformation-flexibility-controlled activity [163], and partitioning-dependent activity. Partitioning-dependent activity is usually ascribed to selective adsorption of the substrate (or product) in the proximity of the enzyme molecules, thus leading to increased (or reduced) activity [164].

3.5.2
Stability of Immobilized Enzymes

The stability of an immobilized enzyme is an important characteristic for the practical application, since greater stability results in longevity of the immobilized enzyme and large amount of recycling, which consequently translate into low cost contribution of the catalyst in the final production cost.

Thus, the search for or creation of stable catalysts has been an important target of enzyme immobilization since the early part of the second half of 20 century, when the immobilized enzymes were put to practical use. Consequently, since the 1970s much effort has been dedicated to the development of robust immobilized enzymes or the elucidation of the mechanism of stabilization of the enzymes or the factors that affect the stability of the immobilized enzymes [165].

Nowadays, it is recognized that the stability of covalently immobilized enzymes can be affected by many factors such as physical/chemical nature of the carrier, the microenvironment of the immobilized enzyme molecules, the binding mode, for example the nature of the linkage, the binding number and the position of the binding on the enzyme surface. Currently it is clear that the stability of covalently immobilized carrier-bound enzymes is dictated by many factors, for example:

- chemical/physical nature of the carrier selected,
- the binding density,
- the nature of the binding functionality,
- the nature of the linkage,
- the length of the spacer,
- the immobilization conditions,
- the binding site,
- the microenvironment.

Accordingly, many methods or strategies have been developed for improvement of the stability of the immobilized enzymes, for example:

- complementary multipoint attachment,
- carrier-bound multipoint attachment,
- introduction of extra polar groups,
- engineering the microenvironment,
- enhancement of the confinement effect,
- engineering the binding chemistry,
- immobilization–stabilization strategy,
- double immobilization strategies,
- pre-immobilization stabilization–immobilization strategy,
- post-immobilization strategies.

Among these strategies, various post-immobilization technologies have been increasingly used, presumably due to the fact that many factors influencing enzyme stability are now known. Besides, tailoring the stability after the enzyme has been immobilized has the added advantage that the enzyme is often already stabilized to some extent. Thus, further modifications can be carried out under harsh conditions under which the free enzyme is probably already deactivated.

3.5.2.1 Multipoint Attachment/binding density

In the period from the mid 1970s to the end of the 1980s it was found that the stability of covalently immobilized enzymes was closely related to the number of bonds formed between the enzyme and the carrier [166, 167]. This simple discovery is principally consistent with the concept that a reactive rigid or solid carrier can also be regarded as a chemical cross-linker, which can make the unfolding of the peptide chains much more difficult than for the native enzymes (see [166] and references cited therein).

Obviously, high binding density leads to an increase of the number of bonds between the enzyme and carrier, as shown in Scheme 3.39. Consequently, enhancement of multiple-point attachment of enzyme to carrier often leads to an increase of enzyme stability [166, 169, 170]. However, the activity is usually reversibly proportional to the number of bonds suggesting that this is due to the molecular rigidification. This is easy to understand, because too many attachments will inevitably result in rigidification of the conformation of the enzyme compared with enzymes with less binding attachments (Scheme 3.39).

Although carrier-bound multiple-point attachment (or cross-linking) for stabilization of the carrier-bound immobilized enzyme was recognized as early as the beginning of 1970s by studying the molecular properties of the immobilized enzymes and the native enzymes [171], further exploration of this concept for design of stable carrier-bound immobilized enzymes was proposed at the end of 1980s [65]. Along with carrier-bound multiple-point attachment, complementary multipoint attachment was one of the earliest methods for preparation of stable catalysts developed by a group of Russian scientists at the end of the 1970s [172], although the relevant immobilization method had already been devised at the beginning of 1970s [173].

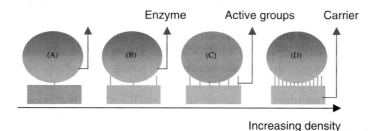

Scheme 3.39 Effect of the density of the active functionality on the immobilized enzyme; increasing binding density might result in multipoint attachment.

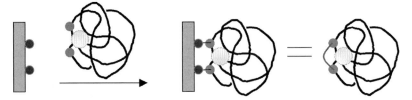

Scheme 3.40 The cross-linking functions of an active solid surface.

These technologies were based on the hypothesis that enzyme stability can be enhanced by tying more attachments to the enzyme surface, thus reducing the conformational flexibility of the enzyme, as illustrated in Scheme 3.40. Remarkably, the contact area between the enzyme and the carrier surface might be limited (usually 10–20% of the whole enzyme surface). However, significant stabilization has been observed for some enzymes. Unfortunately, no comparison of carrier-bound multipoint attachment and surface complementary stabilization has been reported.

It has been found that stabilization of enzymes by multipoint attachment to carriers depends on the number of bonds between the enzyme and the carrier [174]; the amino acid residues selected should be the richest and broadly distributed. It was, however, also indirectly confirmed that highly stable enzymes might be less active than less stable but highly active enzymes [175, 176].

Another scenario of the relationship between multipoint attachment and activity expression and the stability of the immobilized enzyme was demonstrated by the work of Bryjak and co-workers [175]. From the data they presented it was concluded that high retention of activity can be achieved by use of a carrier with high water-retention capacity (ratio of the weight of fully hydrated carrier to that of the dry carrier) whereas higher stability was obtained with the carriers with lower water-retention capacity.

Obviously, carriers with high water-retention capacity also have lower binding density. Consequently, it is expected that the immobilized enzyme was also less stabilized than by use of a carrier with lower water-retention capacity but high binding density [175]. Thus, for practical application, it might be desirable to find a compromise between activity retention and enzyme stability by selecting a carrier with suitable binding density.

Nevertheless, it has been found that the number of bonds between enzyme and carrier necessary to achieve maximum stabilization varies from enzyme to enzyme and from carrier to carrier [177]. On the other hand, it has been also revealed that immobilized enzymes which have the same activity might have different stability, depending on the number of bonds [170], suggesting that the performance of individual enzymes is governed by different principles.

In fact, the density of the active binding functionality of carriers such as Eupergit C 250 (600 µmol g^{-1}) is often in excess of enzyme loading (~1 µmol for an enzyme with Mw 80 KD [277]). Thus, the binding capacity is not efficiently used for binding the enzyme [178]. Thus, practically, it is often necessary to quench the active functionality after enzyme immobilization, to reduce the possibility of enzyme deactivation by subsequent slow reaction between the enzyme and carrier. However, positive use of this excess active functionality to increase enzyme stability by promoting further multipoint attachment or alteration of the enzyme microenvironment is another proposition [184].

The binding density of the carriers can be controlled, for instance, by controlling the ratio of the active monomers or by controlled activation (or modification). For instance, the controlled activation of agarose 10 B CL with glycidol followed by oxidization to an aldehyde group, can give a gel with aldehyde concentration ranging from 0.5 active aldehyde groups to 20 active groups per 10 nm^2 – approximately 10% of protein surface with Mw 34 kD [177].

Commercially available active carriers such as Eupergit C have a surface binding density of ~30 active groups per 10 nm^2, assuming the concentration of active density is 600 µmol g^{-1} and the specific surface area is 200 m^2 g^{-1}.

3.5.2.2 CAG-controlled Stability

While engineering the stability of enzymes by immobilization has been subject of much research interest in recent decades [179, 180, 183, 185], the effect of coupling chemistry on enzyme stability [145] has not received much attention.

The effect of CAG (chemically active binding groups) of the carriers on enzyme stability might be attributed to the difference between the binding mode (position, number of linkage, and nature), as validated by the site-specific immobilization of trypsin-like protease, for which the stability was strongly dependent on binding position [181]. In other words, CAG may dictate the number of bonds formed between the enzyme and the carrier, the orientation of the enzyme, and the nature of the binding.

The study of CAG effects on the enzyme stability can be traced back to the 1960s, when Bakker and Sommers used cellulose derivatives bearing different functionalities for the immobilization of β- and γ-amylase and found that the β-amylase immobilized via diazo-coupling on cellulose gave the most stable enzyme [186]. A similar effect of CAG on the enzyme stability was also found with α-amylase immobilized on cellulose or aerosol bearing different CAGs such as activated CDI or diisocynate-activated groups [182].

The effect of CAG on enzyme stability was possibly first studied by Wojcik and co-workers [58]. Enzymes such as peroxidase, glucoamylase and urease were immobilized on grafted silica gels by means of different CAG such as oxirane groups or aldehyde groups linked to the parent carrier via a diamine spacer (Scheme 3.41). It was found that better activity retention and high storage stability were obtained by Schiff-base formation rather than alkylation with oxirane groups. Again, this result can be interpreted in terms of the different reactivity of the active functionality toward the amino acid resides of the enzymes.

Apparently, the highly exposed lysine residues will take part in the binding. Because of the relatively high hydrophobicity of the CAG microenvironment, it is most likely that more homogeneous binding can be achieved. In contrast, with active oxirane groups heterogeneous binding was possibly achieved.

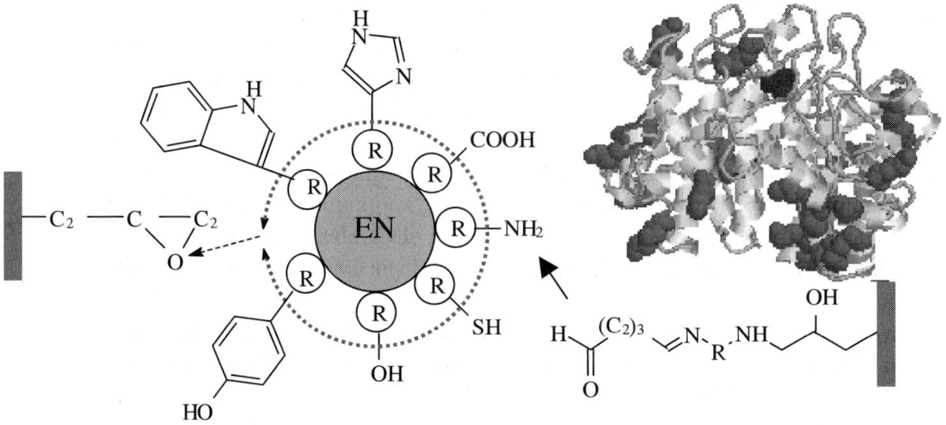

Scheme 3.41 Immobilization of an enzyme on a carrier containing different active functionality.

The effect of CAG on enzyme stability with regard to multipoint attachment has been demonstrated by immobilization of β-galactosidase on agarose with different pendant groups such as glutaraldehyde–agarose (Glut–agarose) and thiolsulphinate–agarose (TSI–agarose). It was discovered that TSI-gels furnished higher yields after immobilization, i.e. 60–85% compared with 36–40% with Glut-gels. However, the glut–agarose derivatives usually had better thermal and solvent stability than the TSI derivatives [183].

It is very likely that the stability of the enzyme is strongly related to the flexibility of the enzyme conformation. In other word, immobilization via the –SH groups of cysteine residues, which are not abundant amino acid residues, will result in the formation of single point attachment. In contrast, immobilization via abundant lysine residues will often result in multipoint attachment. Consequently, the conformation of the enzyme might be more rigid after multipoint attachment and the enzyme is expected to be much more stabilized. Indeed, as shown in Scheme 3.42, the distribution of cysteine and lysine residues of β-galactosidase indicates that cysteine residues are all separately located on the protein surface. Thus, it is very difficult to form multipoint attachment. Conversely, lysine residues are present on the protein surface in clusters and also very close to the interface of the subunits. As a result, the 3D conformation might be stabilized by attachment of lysine residues to a carrier bearing aldehyde functionality.

The highly flexible enzyme conformation of TSI-agarose-bound β-galactosidase was confirmed in an earlier study [184], which showed that approximately 150% activity relative to that of the native enzyme was obtained. It seems likely that activity retention is reversibly proportional to enzyme stability. This statement might be true for multipoint attachment, because too many attachments tied to the enzymes obviously stabilize the enzyme but reduce the conformational flexibility of the immobilized enzymes, which is usually a prerequisite for higher activity expression.

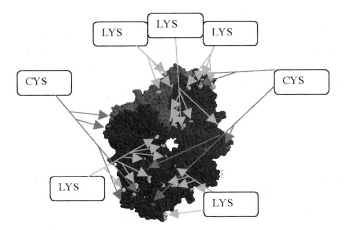

Scheme 3.42 Distribution of Cys and Lys on the surface of galactosidase.

In a similar experiment, β-glucosidase was immobilized on an agarose-based carrier by two different types of binding, Schiff base formation between the aldehyde agarose and the enzyme and peptide bond formation between the amino group of agarose and carboxylic groups of the enzyme. It was discovered that β-glucosidase immobilized by peptide-bond formation is approximately twice as stable as that bound by Schiff base formation and four times more stable than the free enzyme. This can be also ascribed to enhanced multipoint attachment as a result of peptide bond formation. In contrast, the lower stability obtained by Schiff base formation can be attributed to the small number of lysine residues distributed on the surface of β-glucosidase. Thus, the immobilized enzyme (A) must be more active than (B) because of its high molecular flexibility [68].

The effect of CAG nature on enzyme stability was beautifully exemplified by Arica and co-workers, who found that the thermal stability of glucoamylase attached to magnetic poly(methyl methacrylate) microspheres via a spacer depends on the linkage between the enzyme and the spacer. The one with an arginine linkage is more stable than that with the amide linkage, even though the rest of the immobilized enzyme is exactly the same [181].

The different stability suggested either that the nature of the link affects enzyme stability or that unreacted CAG ($-NH_2$) has a negative effect on the enzyme stability. Because the lysine residues are not as abundant as Asp or Glu, the effect of multipoint attachment can be ruled out. Most likely, the positively charged CAG ($-NH_2$) has a negative effect on enzyme stability; this has been confirmed by other experiments. As the active centre is negatively charged, binding of CAG (positively charged) to the glutamic acid located in this region possibly leads to the formation of less stable species, as illustrated in Scheme 3.43.

Similarly, in 1998 Yang and Chase have conducted an interesting study on the effect of the different carrier-bearing functionality on the stability of the immobilized enzymes [185]. PVA-coated perfluoro polymer, activated by SESA (*p*-β-sul-

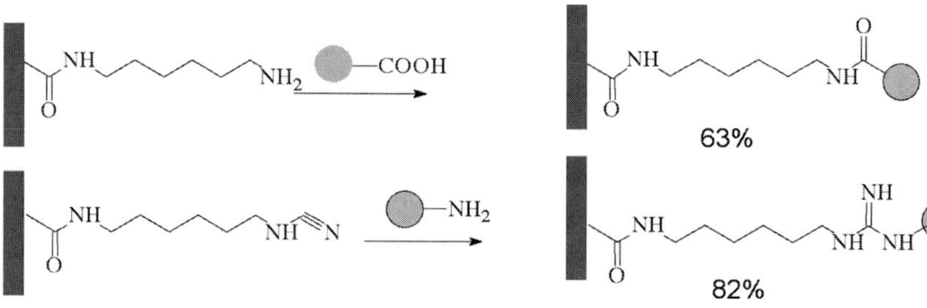

3 Covalent Enzyme Immobilization

Scheme 3.43 Immobilization of glucoamylase on PMMA microspheres via spacers.

phate-(ethylsulphonide)-aniline), triazine, CDI, and tresyl chloride was used for the immobilization of the α-amylase (Scheme 3.44).

It was found that the thermostability of the immobilized enzyme at 90 °C follows the order: CDI (A) >> tresyl chloride (B) = SESA (D) >> triazine (C). Studies of the three-dimensional structure and the distribution of amino acid residues revealed that α-amylase is a negatively charged enzyme and the charge of the active centre is also negative. It is very probable that binding with carrier A and carrier B (via CDA) is via the lysine residue that is far from the active centre, whereas binding via carriers C and D is via the lysine residue close to the active centre, because the binding functionality of this carrier is positively charged.

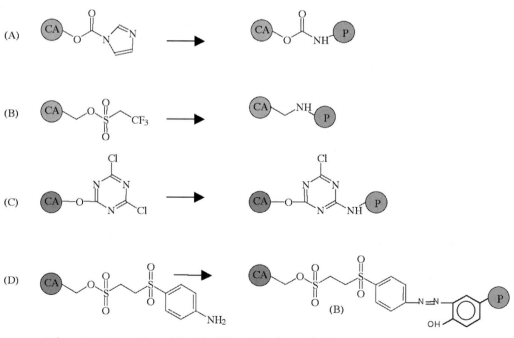

Scheme 3.44 Enzyme immobilized by different coupling methods.

Thus, the observed thermostability of the immobilized enzymes can be readily explained by the hypothesis that binding functionality might affect the binding mode, for example position and number of bonds. Thus, it is not surprising that the stability of the immobilized enzymes can be significantly improved by variation of the method of activation of the carriers.

3.5.2.3 CIG-controlled Enzyme Stability

While CAG dictates the binding mode (nature, number and position of the binding), CIG dictates largely the enzyme microenvironment. Thus, the effect of CIG on the stability of the immobilized enzymes obtained could be ascribed to differences between the microenvironments of the enzymes, when the CIG varies, as shown in Scheme 3.45.

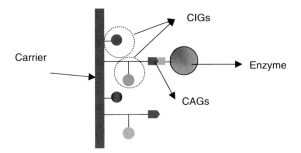

Scheme 3.45 Effect of CIG on the microenvironment of an enzyme immobilized on a carrier.

Consequently, it is to be expected that the presence of an appropriate CIG will create a microenvironment favourable for stabilization of enzymes. Moreover, during the immobilization process, the nature of the CIG may influence the orientation of enzyme molecules on the carrier surface, leading to the formation of immobilized enzyme of different stability, as confirmed by the site-specific immobilization of trypsin-like protease [187]. Also, introduction of an appropriate CIG might shield the enzyme from direct contact with the carrier; this might prevent the enzyme from undergoing a deleterious conformational change, especially on a hydrophobic surface [188].

The effect of the CIG on the enzyme stability was demonstrated beautifully by the work of Cardias et al. [79], who found that the thermostability of penicillin G acylase immobilized on the activated silica surface depended on the nature of the CIG, although they are immobilized on the same carrier via the same CAG, as shown in Scheme 3.46.

Compared with carrier S-2, S-1 is different mainly in the nature of the CIG. However, penicillin G acylase immobilized on S-2 is much more stable than that on S-2. It seems likely that the long hydrophobic spacer renders the carrier's surface more hydrophobic. Consequently, the binding mode (properties of the linkage, enzyme orientation) might be different. Besides, length of the pendant groups can also affect the mobility of the enzyme molecules.

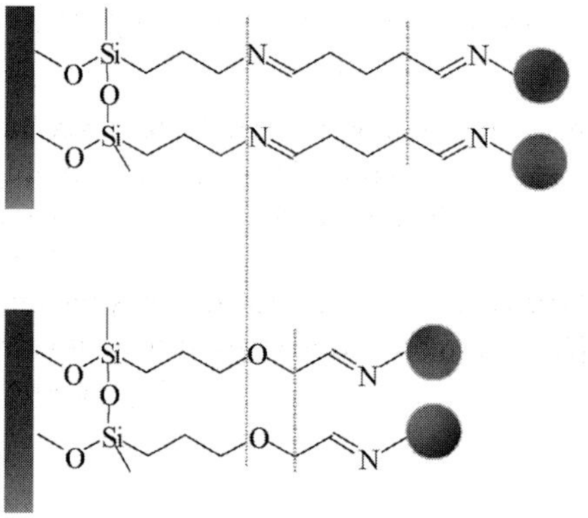

Scheme 3.46 Silica activated by different methods.

Because most of the lysine residues of penicillin G acylase are located in hydrophilic clusters, binding on S-2 (hydrophilic surface) might be multipoint whereas binding on S-1 (hydrophobic surface) might be by single-point attachment. Thus, enzyme immobilized on a hydrophobic carrier might be less stable than an enzyme immobilized on a hydrophilic carrier (Scheme 3.47).

Another example of the effect of CIG on enzyme stability was demonstrated by converting excess CAG into inert CIG via the so-called post-immobilization techniques. It was, for instance, found that the stability of β-galactosidase immobilized on the thiosulphonate Sepharose beaded gel was highly dependent on enzyme loading and also on the properties of the blocking agents used to neutralize excess active binding functionalities [184], as illustrated in Scheme 3.48.

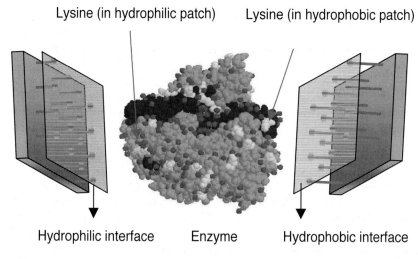

Scheme 3.47 Effect of pendant groups on the microenvironment of the carrier surface.

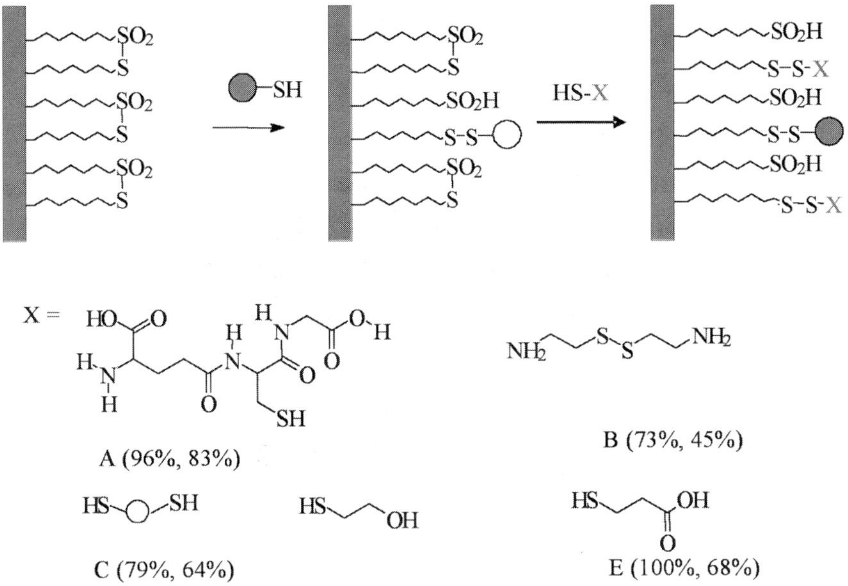

Scheme 3.48

X =

A (96%, 83%) — glutathione

B (73%, 45%) — NH₂-CH₂-CH₂-S-S-CH₂-CH₂-NH₂

C (79%, 64%) — HS—◯—SH

HS-CH₂-CH₂-OH

E (100%, 68%) — HS-CH₂-C(=O)-OH

Scheme 3.48 Immobilization of β-galactosidase on thiosulphonate Sepharose and blocking of the active groups with different quenching agents.

Among the blocking agents tested it was found that glutathione was best, with regard to high retention of activity after blocking (96 %) and high thermal stability (83 % retention of activity after heating at 50 °C for 30 min). In contrast, the non-blocked enzyme was completely deactivated.

As with the properties of the blocking agents, it seems likely that the negatively charged CIG prevents the enzyme from close contact with the carrier, thus protecting the enzyme from the deactivation caused by the interaction of the enzyme with the carrier.

Although at higher enzyme loading (0.114), the immobilized enzyme has almost the same stability as that after use of the blocking agent, the very low retention of activity (27 %) can hardly justify the practical use of this method for obtaining stable immobilized enzymes by more enzyme loading. In contrast, the fact that higher retention of activity (up to 80 % of the native activity) and higher enzyme stability can both be achieved by use of the blocking agent makes this strategy much more attractive for enzymes that are expensive in terms of price per unit.

In addition to covalent enzyme immobilization, the nature of the CIG also has an important effect on enzyme activity and stability in non-covalent immobilization, for example adsorptive immobilization on carriers and entrapment in a matrix.

Another striking example is the immobilization of aminoacylase on polysaccharide-based carriers such as aminohexyl cellulose with immobilized tannin and Sephadex with DEAE as the non-covalent binding functionality. The binding of amino acylase on cellulose with tannin afforded an immobilized enzyme 40 times more thermostable and with five times higher volume activity than that bound to DEAE–Sephadex [189].

As discussed above, the carriers selected can be regarded as chemical and physical modifiers of enzyme molecules, and in the same way as conventional small molecules such as amino acids, sugars (polyols), and osmolytes [190] are often used to stabilize enzymes or proteins, they are also applicable in this case [191]. Thus, it is inferred that design of appropriate CIG should follow the principles used to stabilize enzymes by chemical and physical modification of the enzymes by use of these small molecules [191, 192].

The effect of CIG as leaving groups on enzyme stability was recently demonstrated by immobilizing α-amylase on poly(methyl-methacrylate-acrylic acid) microspheres activated by thionyl chloride (TC) and carbodiimide (CDI). It was found that enzyme immobilized by the former method is twice as stable after storage for 1 month as compared with the latter method. On the other hand, the free enzyme lost its activity completely within 20 days [193].

One possible explanation might be that multipoint attachment is mediated by TC-activated carrier as compared to CDI-activated carrier, due to the fact that more amino groups of the lysine residues will take part in TC-binding, while less bonds will be formed with CDI as coupling agents because of steric hindrance.

3.5.2.4 Spacer-dependent Stability

As noted above, many examples demonstrated that the stability of a carrier-bound immobilized enzyme mediated by a spacer was lower than that without spacer [88, 180], presumably because enzyme molecules immobilized on carriers via a spacer usually have molecular mobility similar to that of the native enzyme. This was shown by the work of Yodoya et al., who immobilized Bromelain via peptide spacers of varying chain length ($(gly)_n$) as shown in Scheme 3.49 [133].

Obviously, as the length of the spacer increases, the molecular mobility approaches that of the free enzymes. By a similar approach it was also found that immobilization of several enzymes such as lipase and protease on polyacrolein with spacer led to immobilized enzymes that were less stable than those without a

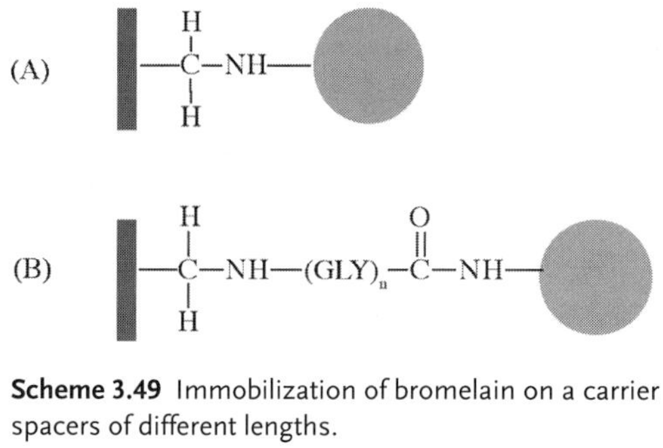

Scheme 3.49 Immobilization of bromelain on a carrier by use of peptide spacers of different lengths.

spacer [134]. It is, however, still very difficult to draw the conclusion that slightly lower stability than the counterpart (without spacer) can be attributed to the effect of the spacer, because the different stability sometimes also originates from the nature of the linkage and the spacer itself especially the nature of the CAG.

Conversely, there are also many examples which also demonstrate that the stability of immobilized enzyme mediated by a spacer can be increased, for instance, the stability of galactosyltransferase to stirring and its thermostability and storage stability were greatly enhanced by immobilization with the presence of spacer [194]. It seems likely that here the nature of the spacer might dictate the stability of the immobilized enzymes. For instance, a hydrophilic macromolecular spacer such as albumin can increase severalfold not only retention of activity but also the pH stability of trypsin on polyester fleece. Moreover, the storage stability was also prolonged [89] and the activity toward the macromolecular substrate was also increased [98].

Papain has been immobilized on polystyrene beads and PAAm polymers by use of spacers of different nature and length [195]. With long, flexible, hydrophilic polyethylene glycol as spacer, increased payload and activity retention were obtained. It was also found that hydrophilic polyacrylamide-based supports were more efficient for immobilization than hydrophobic polystyrene-based supports.

Similarly, horseradish peroxidase immobilized on derivatized nylon tube via a polyether as spacer was not only very stable but also had a similar K_m to the free enzyme [91]. Also, alkaline phosphatase and trypsin covalently bound to isothiocyanate-carrying hydrophilic and macroporous polymer resins via a hydrophilic spacer group between the carrier resins were more stable [196]. This increased stability could be ascribed to the artificially created hydrophilic environment around the enzymes by the use of hydrophilic spacers.

Thus, it is reasonable to believe that the favorable environment created by the presence of the hydrophilic spacer is largely responsible for the greater thermostability.

3.5.2.5 Molecular Confinement-controlled Stability

Molecular confinement effects on the stability of enzyme entrapped in the confined matrix have recently gained increasing attention, because many studies have showed that sol-gel entrapped enzymes display surprisingly higher thermostability as compared with other types of immobilized enzymes [197]. A study of the molecular state of the enzyme entrapped in the matrix revealed that the enhanced stability is mainly due to the molecular confinement, which is able to restrict the molecular movement, thus reducing the possibility of deactivation [198].

In the early 1970s immobilization of Taka-amylase on a variety of polysaccharide based carriers of different pore size, for example Sephadex G-25 (A), Sephadex G-200 (B), Sepharose 6B (C), and Sepharose 2B (D), led to the discovery that the stability followed the order, T6B > T2B = TG200 (amylase bound on Sephadex G-200) > TG25 (amylase bound on Sephadex G-25) = soluble amylase (not immobilized), suggesting that enzyme immobilized on the internal pores is more stable than that on the surface [199]. Obviously, confining enzyme molecules in a limited space reduces the conformational flexibility of the enzyme. Thus, the stability, especially the thermal stability, of the enzyme is enhanced.

Although this technique is currently often used for encapsulation or entrapment in the sol–gel system, use of confinement to enhance enzyme stability in conjunction with covalent immobilization is less appreciated. This is largely because the confinement effect is apparently in contradiction with the enzyme-attachment process, as demonstrated in a recent study which showed that increased pore size and surface area usually lead to the higher activity [118], whereas the reverse relationship is observed for the stability of immobilized enzymes, indicating that molecular confinement might stabilize the enzymes but might reduce the enzyme activity [118].Thus, the carrier used for attachment must contain large pores to enable the enzyme molecules to freely diffuse into the pores; on the other hand, the confinement effect requires the enzyme molecules to be closely packed.

The relationship between pore size and enzyme stability was also demonstrated recently by covalent immobilization of α-amylase on silica derivatives such as SBA15, silica-N, and silica-AC of pore size varying from 7.6 nm to 30 nm. Interestingly, it was found that the specific activity of the immobilized enzymes followed the order silica-AC > silica-N > SBA15, whereas operational stability was in the order SBA15 > silica-AC > silica-N [200]. Similar pore size requirements of stability and activity were also found when horseradish peroxidase (HRP) was immobilized the mesoporous silica materials FSM-16, MCM-41, and SBA-15 with pore diameter from 27 to 92 Å [201].

However, the effect of pore size on the enzyme stability is also closely related to the size of the enzyme in relation to the pore size of the carrier. Often only when both properties match is a significant stabilization effect obtained. For instance, immobilization of bacterial lipase from *Staphylococcus carnosus* (I) on porous carriers with pore sizes above the size of the enzymes led to the discovery that the pore size of the support had no effect on enzyme stability [19].

The greater stability of enzymes immobilized on carriers with small pores is presumably because of reduced molecular mobility, owing to confinement of the enzyme molecules in the small pores, as shown in Scheme 3.50.

It was interestingly demonstrated by Messing that the stabilized immobilized catalase (size 18.3 nm)-glucose oxidase (8.4 nm) multi-enzyme system on TiO_2 car-

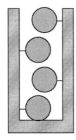
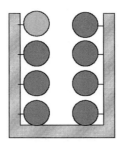

Scheme 3.50 Molecular mobility of enzyme immobilized in carriers of different pore size.

rier can be obtained only with carriers bearing pores more or less twice the main axis of both enzymes (pore size of TiO_2, 35 nm) [202, 203].

In general, to obtain improved stability by covalently immobilizing enzymes in ready-made carriers with small pores is less advantageous, due to the dilemma mentioned above, namely that small pore sizes often provoke a serious diffusion limitation, thus reducing the activity expression.

3.5.2.6 Microenvironment-controlled Stability

The microenvironment of the carrier is another important factor affecting the stability of immobilized enzymes. However, regarding the fact that microenvironment of the carrier is dictated by the chemical composition of the backbone, the nature of the CAG, CIG and the spacer [98], it is quite difficult to discuss the effect of the microenvironment separately.

Currently, engineering the microenvironment of the enzyme molecules is increasingly used to improve enzyme stability. In general, engineering the microenvironment of the immobilized enzyme falls into one of two strategies, namely pre-immobilization modification of the ready-made carrier and post-immobilization modification of the ready-made immobilized enzymes, e.g. by introducing suitable blocking agents [184].

In the first case, alteration of microenvironment can be easily achieved by quenching the remaining excess active functionalities using blocking agents [204], which can be classified into two groups: small molecules (e.g. amino acids or other amines), and macromolecules (e.g. bovine serum albumin, gelatine, polyethyleneimine (PEI) [205] and aminolated polyethylene glycol (PEG)). These blocking agents (also called quenching agents) can easily react with the active functionalities, such as epoxy rings or aldehyde groups of the commonly used carriers without the use of additional activating chemistry.

Often, a stabilizing hydrophilic microenvironment can be created by introduction of hydrophilic macromolecules into the proximity of the enzyme. For example, consecutive modification of penicillin V acylase from *Streptomyces lavendulae* immobilized on Eupergit C with bovine serum albumin led to the formation of a new biocatalyst not only with enhanced activity (1.5-fold), but also enhanced stability. Remarkably, this biocatalyst could be recycled for at least 50 consecutive batch reactions without loss of catalytic activity, strongly suggesting that introduction of a hydrophilic environment can lead to the stabilization of the enzyme environment [206]. In another study with small blocking agents it was found that the stabilizing effect was largely dependent on the nature of the quenching agents. For instance, among the quenching agents studied (L-lysine, L-glycine and ethanolamine), ethanolamine is the best for the immobilization of GL-7-ACA acylase on aldehyde silica, both in terms of activity retention and enzyme stability [207].

Hydrophilization of enzyme microenvironment can also be realized by introduction of a hydrophilic spacer. Enzyme stability can be improved over that of the enzyme immobilized on the parent carrier via the same coupling chemistry, as demonstrated by covalent immobilization of *Bacillus stearothermophilus* lipase (BSL) to the glutaraldehyde-activated PEI-coated silica. BSL immobilized in this way displayed improved stability and operational stability, as compared with the same lipase immobilized to glutaraldehyde-activated silanized carrier [208]. Most likely, the presence of PEI created a favourable microenvironment in which the enzyme is much more stable.

3.5.3
Selectivity of Immobilized Enzymes

Selectivity is a powerful property of enzymes. Because of physical and chemical modification of the properties of the native enzyme, covalent immobilization of enzyme on carriers inevitably changes enzyme selectivity e.g. enantioselectivity or reaction selectivity.

In covalent enzyme immobilization it has been found that the selectivity of the immobilized enzymes can be influenced by:

- diffusion constraints,
- aquaphilicity,
- carrier hydrophobicity,
- the microenvironment,
- the presence of conformer selectors.

3.5.3.1 CAG-controlled Selectivity

There is currently no example which demonstrates that enantioselectivity can be influenced by variation of binding CAG without also changing other properties such as spacer, carrier and other pendant groups (e.g. CIG), although several other examples indicate that changing the microenvironment of the CAG can lead to a change of the enzyme selectivity, suggesting that changing only the CAG, if possible, can also influence selectivity.

For instance, it was found very recently that not only activity retention but also the enantioselectivity of *Mucor miehei* lipase was very dependent on the environment of the binding moieties of carriers obtained by controlled modification of an epoxy carrier, as illustrated in Scheme 3.51.

Among various derivatization techniques it was found that type C gave the highest activity at pH 7, whereas type B gave the highest activity at pH 5. Among the various preparations, only type E was selective for the *R* enantiomer. Type C had high enantioselectivity toward the *S* enantiomer among other enzymes having *S* selectivity for the hydrolytic resolution of (*R*,*S*)-2-butyroyl-2-phenylacetic acid [76]. Obviously, the individual microenvironment around the binding sites influences the mode of binding (position of the binding site).

Because all the enzymes were immobilized on the same carriers with same enzyme loading (0.0017), the effect of diffusion limitation on enantioselectivity can be disregarded. Thus, it suggested that the highly active enzymes are also highly selective, suggesting that activity and selectivity are both closely related to molecular mobility and that high selectivity can never be obtained with enzymes that have lower activity.

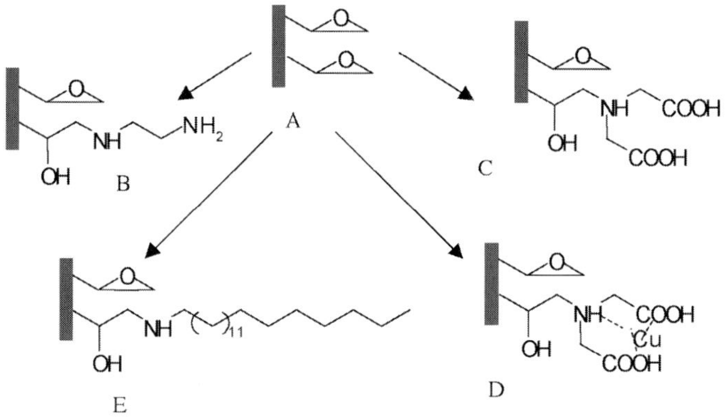

3.5 Factors Affecting Enzyme Performance

Scheme 3.51 Modification of epoxy carrier to give carriers with different microenvironments.

As shown in Scheme 3.52, immobilization via lysine residues of MML located far from the active centre might not disturb enzyme selectivity. However, immobilization via His residues that are close to the lid might affect the conformation and activity of the enzyme. Thus, the presence of ICM (immobilized chelating metal-Cu) in the proximity of epoxy ring will orientate these His residues to the carrier, resulting in a change in the conformation of the enzyme (the lid might be opened) and activation of the enzyme on the hydrophobic surface of Eupergit C. Unfortunately, only one type of carrier was used in this experiment. To discover whether the lid-opening event is influenced by the hydrophobic nature of the carrier it is worth using hydrophilic epoxy carriers such as epoxy Sepharose.

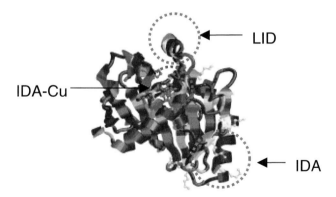

Scheme 3.52 Three-dimensional structure of MML with enzyme-binding functionality – lysine and histidine. Immobilization by means of lysine residues located far from the active centre might not affect enzyme selectivity. Immobilization by means of lysine residues close to the LID might affect enzyme conformation and activity.

Because the introduced CIG function as enzyme orientation groups on the carrier surface, it is also expected that variation or design of CAG of different nature and reactivity toward the active amino acid residues (AAR) might lead to the different orientation and thus different binding mode (position and number of bindings). Consequently enzyme selectivity can be definitively influenced by use of different CAG.

One such an example was recently clearly demonstrated by CRL immobilized on silica activated with 2,4,6-trichloro-1,3,5-triazine and on agarose activated by tosylation. The former had approximately seven times greater enantioselectivity than the soluble enzyme whereas CLR immobilized on agarose activated with tosylate had only four times the selectivity of the native enzyme [210].

3.5.3.2 CIG-controlled Selectivity

Selectivity is a powerful assets of enzymes in asymmetric synthesis [211]. The selectivity of enzymes usually refers to substrate selectivity, stereoselectivity, enantioselectivity, regioselectivity and functional group selectivity [212]. The native selectivity of an enzyme is dictated by its amino acid sequence. Thus, genetic engineering, e.g. site-directed mutagenesis or directed evolution, can improve the intrinsic selectivity by variation of the amino acid residues [213].

However, the apparent enantioselectivity of an immobilized enzyme, especially, a carrier-bound immobilized enzyme is influenced (or controlled) by many factors, e.g. the microenvironment of the carrier, the medium, the type of reaction [214], the temperature, the extent of protonation, diffusion constraints [215], and the method of immobilization. Therefore, at least two types of enantioselectivity can be distinguished – conformation-controlled and diffusion-controlled. Accordingly, linking the enzyme molecule to a suitable carrier or engineering the microenvironment in which an enzyme is accommodated can not only alleviate diffusion constraints and thus enhance the enantioselectivity and reaction rate [216], but also make a non-selective enzyme regioselective [218] or even reverse the enantioselectivity [214].

Although a change of enzyme selectivity (e.g. substrate selectivity, stereoselectivity, regioselectivity, functional group selectivity and reaction selectivity) as a result of different immobilization methods has been often observed, selectivity improvement by immobilization as a rational methodology is not well developed, because of a lack of systematic analysis of factors in immobilization which influence enzyme selectivity.

In general, CIG-controlled selectivity of carrier-bound immobilized enzymes is closely related to conformational change. Thus, selectivity changes by CIG can be ascribed to the following effects:

- improvement of the selectivity by the orientation effect of CIG,
- improvement of selectivity by inducing a new conformation,
- improvement of the selectivity as a result of a favourable microenvironment.

The effect of surface-pendant functionalities on the enantioselectivity of PFL (*Pseudomonas flurorescens*) lipase was studied with the use of three types of carrier differing only in pendant functional groups [219], as shown in Scheme 3.53.

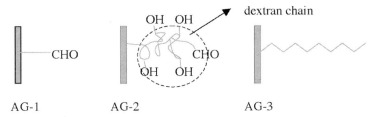

Scheme 3.53 Selective modification of carriers for introduction of CIG.

It was found that high enantioselectivity relative to the native enzyme was obtained with AG-1 and AG-3 whereas similar enantioselectivity was obtained with the AG-2, which binds the enzyme via a long hydrophilic spacer.

Obviously, the conformation of PFL was disturbed by the hydrophobic environment when AG-1 and AG-3 were used. However, compared with AG-1 (three times higher enantioselectivity relative to the native lipase), AG-3 was able to provide a compatible environment for lipase to achieve high enantioselectivity (10 times higher enantioselectivity relative to the native lipase). This result was actually consistent with the finding that PFL is lid-containing enzyme and that an enzyme molecule in an open conformation is essential for high enantioselectivity [220].

Like CAG, CIG also have a remarkable effect on enzyme enantioselectivity. For instance, controlled modification of an epoxy resin led to the preparation of carriers with different pendant CIG. When these were used for immobilization of lipase the enantioselectivity depended on the nature of the adjacent CIG [76], suggesting that the CIG can select (or recognize) the lysine residues to be bound (Scheme 3.54). This result also implies that selective modification of amino acid residues on the enzyme molecule surface might change enzyme selectivity (or enzyme conformation).

The important effect of CIG on the selectivity of the immobilized enzymes is also apparent for enzymes immobilized by non-covalent adsorption, because all the pendant groups can now be regarded as CIG. For instance, the enantioselectivity of PFL lipase adsorbed on hydrophobic carriers in the hydrolytic resolution of fully soluble (R,S)-2-hydroxy-4-phenylbutanoic acid ethyl ester was at least one order of magnitude better than that of the native enzymes.

Remarkably, the activity of the immobilized enzyme is also clearly related to the enantioselectivity and the hydrophobicity of the carrier. In general, the higher the hydrophobicity of the carrier, the higher the activity or selectivity, suggesting that

Scheme 3.54 Effect of carrier pendant groups on the enantioselectivity of lipase.

the hydrophobic carrier induced an active and selective enzyme conformation [216]. Interestingly, it was found that the enantioselectivity of the adsorbed CRL also strongly depended on the pH of the medium used, implying that engineering the medium is also a crucial step when evaluating the selectivity of the immobilized enzymes [217].

Similarly, it was found that the selectivity for two types of reaction concomitantly catalysed by one enzyme, for example in the kinetically controlled synthesis of β-lactam antibiotics, can be affected by the aquaphilicity of the carrier (i.e. the microenvironment of the carriers as discussed above) [221]. It is thus expected that the presence of more hydrophilic CIG might also affect the synthesis of β-lactam antibiotics, as confirmed by the use of new synthetic carriers of high hydrophilicity [222].

In general, improvement of enzyme selectivity by immobilization might be attractive with regard to simplicity and universal applicability; it also usually obviates the need for detailed structural information. However, combination of structural information about an enzyme with various immobilization techniques might expedite the design of more selective immobilized enzymes.

3.5.3.3 Spacer-controlled Enzyme Selectivity

The presence of spacer can, in the first instance, alter molecular flexibility and the microenvironment. A decrease of molecular mobility or rigidification of the enzyme molecule, relative to the native enzyme, is usually a direct reason for lower retention of activity, as revealed by the study performed by Clark et al. [163].

Study of the hydrolysis of (±)-mandelic acid methyl ester with immobilized PGA (penicillin G acylase) on AG-GL (agarose activated with gloxal) and AG-GA (agarose activated with glutaraldehyde) revealed that when the CAG content was controlled to the extent that it enabled one point attachment only the E values of the two immobilized enzymes were quite similar and significantly higher than for the native enzyme (Scheme 3.55). However, the E value decreased to that for the native enzyme when multipoint attachment occurred.

Although, the reason for the enhanced enantioselectivity of immobilized penicillin G acylase with controlled linkage is not clear, the fact that penicillin G acylase immobilized on AG-GA as carrier has similar selectivity to penicillin G acylase of controlled linkage on AG-GL carrier suggests that the enhanced selectivity is somehow related to the molecular mobility of the enzyme and the environmental effect of the carrier. It seems most likely that on the hydrophilic carrier the conformation

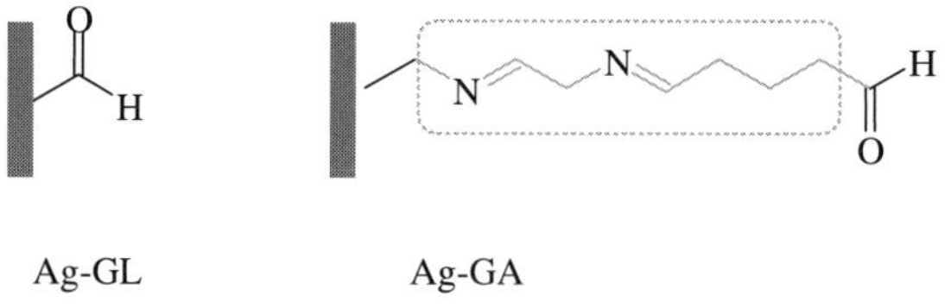

Ag-GL   Ag-GA

Scheme 3.55 Immobilized PGA on agarose-based carriers with different pendant groups: AG-GL, gloxal agarose; AG-GA, glutaraldehyde agarose.

of the enzyme is more compact and selective compared with on the relatively hydrophobic Eupergit C, which provided an immobilized enzyme with only half the selectivity of the free enzyme. In this instance the effect of diffusion constraints on enantioselectivity can be disregarded, because the loading of the enzymes was always much lower (0.001) [223].

It has also been shown that a spacer-mediated immobilized enzyme might have the same conformational flexibility as the native enzyme, with similar K_m and stability. Therefore, introduction of an appropriate spacer might be an efficient means of improving activity or selectivity.

3.5.3.4 Diffusion-controlled Selectivity

Despite many examples of diffusion constraints affecting the reaction selectivity of immobilized enzymes, for example the use of immobilized penicillin G acylase in the kinetically controlled synthesis of β-lactam antibiotics, in which diffusion constraints seriously reduce synthetic efficiency for the targeted products [222], there are few examples of diffusion constraints affecting the enantioselectivity of immobilized enzymes,.

Diffusion constraints can, however, change product selectivity. For instance, immobilization of cyclodextrin glucosyltransferase on Eupergit C led to a change of product spectrum – the enzyme selectivity was substantially shifted toward oligosaccharide production [148].

3.5.3.5 Aquaphilicity-controlled Selectivity

The effect of carrier aquaphilicity on the enantioselectivity of immobilized subtilisin in organic solvents might imply that the enzyme undergoes a change of conformation on attachment to the carrier [224]. The similar phenomenon was also observed by Guisan and his co-workers, who improved the enantioselectivity of immobilized lipase by hydrophilization of the enzyme microenvironment [214].

More recently, it has been found that synthesis of β-lactam antibiotics can be affected by the aquaphilicity of immobilized penicillin G acylase. High aquaphilicity usually favours a high S/H ratio. This result suggests that high carrier aquaphilicity favours an enzyme conformation related to a certain type of reaction (synthesis of semi-synthetic β-lactam antibiotics) or that the conformation change on use of hydrophilic carrier is less than that for hydrophobic carriers [15].

3.5.3.6 Conformation-controlled Enantioselectivity

It is known that the conformation and the conformational mobility of a carrier-bound immobilized enzyme can be affected by many factors, for example steric hindrance, the spacer, the number of attachments [209] the microenvironment, and the binding chemistry [91].

Accordingly, binding an enzyme molecule to a suitable carrier, engineering of the microenvironment, selection of suitable binding chemistry, or introducing spacer or arm [58,91] can not only alleviate the diffusion constraints but also make the enzyme conformation flexible, thus substantially enhancing enantioselectivity, make a non-selective enzyme regioselective [214], or even reversing enzyme enantioselectivity.

There have been few studies of the effect on enantioselectivity of covalent binding of enzyme to carriers. Sanchez and co-workers studied the enantioselectivity of CRL covalently immobilized on agarose and silica activated with tosylate and triazine, respectively. It was found that CRL immobilized on silica activated with 2,4,6-trichloro-1,3,5-triazine was approximately seven times more enantioselective than the soluble enzyme, whereas CLR immobilized on agarose activated with tosylate was only four times more selective than the native enzyme [210]. As noted above, although the difference in the enantioselectivity was explained as a result of the difference in the microenvironment, the difference in the enantioselectivity can be ultimately ascribed to the conformational difference in the obtained immobilized enzymes.

By combination of enzyme immobilization and medium engineering it has been demonstrated that the enantioselectivity of immobilized lipases can be modulated [225]. Although there is still no evidence that variation of the binding chemistry (for covalent binding) on the same carrier can lead to a change of the enzyme selectivity, an indirect example showed that different binding chemistry might lead to a change of enzyme selectivity. For instance, use of immobilized cellobiohydrolase I (CBH I) as a chiral stationary phase resulted in different separation behaviour depending on whether it was bound to aminopropyl silica via its carboxyl groups (peptide formation) or to aldehyde silica via its amino groups [226]; this suggested enzyme conformation depended on the binding chemistry.

It has been also widely demonstrated that immobilization can alter enzyme selectivity as a result of conformational change. For instance, immobilization of urokinase on aldehyde Sepharose 6B-CL led to a change in the specificity – the side-reaction of cryptic cleavage was significantly suppressed [227] compared with the native enzyme.

3.5.3.7 Selectivity and Particle Size

Diffusion limitation often results in reduction of apparent enzyme activity but also leads to the reversion of the product when the reaction is reversible. However, it has recently also been observed that reduction of enantioselectivity or change of reaction selectivity between two competing reactions that occurs in the same reaction systems (e.g. kinetically controlled peptide synthesis) can occur as a result of diffusion constraints. Consequently, it is might be necessary to control the particle size, to reduce diffusion constraints and thus improve the selectivity of the immobilized enzymes.

A diffusion-controlled product map was probably first observed in the 1970s, when it was found that the product maps of immobilized α-amylase and its soluble counterpart were different [228].

The reaction selectivity between two types of competing reaction (e.g. the kinetically controlled synthesis of peptides catalysed by protease or penicillin G acylase) was found to be largely dependent on the extent of diffusion constraints. Diffusion constraints generally reduce the synthetic efficiency of immobilized penicillin G acylase in the synthesis of semi-synthetic β-lactam antibiotics [14], necessitating control of the distribution of particle size to avoid heterogeneity of the performance of immobilized penicillin G acylase in the corresponding industrial processes [483].

It was recently found that the enantioselectivity of CAL-B is also largely dependent on the diffusion constraints. Intraparticle diffusion constraints usually reduce enantioselectivity severalfold [215].

3.6
Preparation of Active Carriers

Covalent enzyme immobilization usually refers to processes by which enzyme molecules are bound to carriers by formation of covalent bonds between the enzyme (native or modified) and ready-made carriers bearing active or inert functionality. Since the 1960s, covalent methods of enzyme immobilization have flourished into one of the most important methods of enzyme immobilization. The underlying principle is very simple – use of a pre-designed insoluble scaffold with active binding functionality that can react with amino acids under suitable conditions is obviously a simple means of providing the non-catalytic functions required by each specific enzyme and application.

Inspired by the discovery in 1960s that selected pre-designed carriers function not only as simple scaffolds for the immobilized enzymes but also substantially affect enzyme performance such as activity, stability and selectivity, many carriers of different physical and chemical nature, many coupling methods, and methods of activation of the carriers have been developed.

Usually, the active functionality is introduced to the carrier and subsequently the protein is bound to the carrier. Occasionally the proteins (or enzymes) are first activated and subsequently the active functionalities of the proteins are reacted with the reactive functionality of the carriers. One example is oxidized glycoenzymes, which either can be bound to a carrier bearing amino groups by Schiff base formation [229] or the trypsin and pepsin can be modified by treatment with 3-isothiocyanato-1-propylisocyanate which can then be bound to a carrier bearing NH_2 groups [230].

Covalent enzyme immobilization has gained increasing attention in recent decades, because of the simplicity of the methodology, because the binding is strongest among all the methods of immobilization, and because of the easy control of the geometric properties of the immobilized enzyme. The vast amount of coupling chemistry and the diversity of carrier structure are also a powerful asset for modulating the catalytic properties of the enzyme, e.g. activity, selectivity, and stability. Thus, selection or preparation of the carrier is of crucial importance with regard to the performance of the immobilized enzymes.

Currently, methods for preparation of such active carriers can be placed in the following groups:

- preparation by direct polymerization of monomers bearing active functionality;
- preparation by polymerization of monomers bearing inert functionality, followed by chemical modification;
- preparation of carriers by cross-linking of soluble polymers;
- preparation by modification of ready made carriers;
- preparation by coating of pre-polymers;
- preparation by grafting of the ready-made carrier;
- formation of composites of more than two different polymers.

Each method has advantages and disadvantages. For practical applications, especially for industrial applications, it is recommendable to pay attention to the following features of the selected carrier:

- the simplicity of the method,
- the potential loading of the enzyme,
- the type of application,
- the required specific activity,
- the structure of the carrier,
- the simplicity of the design,
- biodegradability,
- environmental impact,
- the economical viability,
- aquaphilicity,
- functionality ready for enzyme immobilization.

For detailed information, the reader can consult books and reviews [231, 232].

Since the 1950s, the use of the insoluble carriers with predetermined physical and chemical properties for enzyme attachment (by adsorption or covalent binding) has been regarded as one of the simplest methods of enzyme immobilization, because of its broad applicability and ease of operation. Thus, much effort has been devoted to the preparation of different solid pre-carriers of diverse defined chemical and physical nature for bioseparation and immobilization procedures. In particular, synthetic polymers are attracting increasing attention, because they can be reproducibly prepared from defined starting materials and thus the products will also have defined chemical and physical properties.

Many synthetic active polymeric beads or powders can be prepared directly by suspension or precipitation polymerization of the monomers bearing the appropriate activated groups; examples include epoxy resins e.g. Eupergit C [233], synthetic polymeric aldehyde and thiiranyl polymer [234], which can be used directly to bind proteins or enzymes via the activated functionality they contain.

In general, suspension polymerization in a two-phase system is one of the most popular means of controlling the physical nature of the carrier with regard to geometric properties such as size and shape and, especially, the internal structure, for example pore size, porosity and pore-size distribution. As a result, use of carriers with ready-made chemical and physical properties, especially tailor-made geometric properties, and active pendant groups ready for binding enzymes, is one of the simplest methods of enzyme immobilization.

One disadvantage of using this kind of fabricated polymer as enzyme carrier is probably that the "built-in" intrinsic functionality might be not the best for immobilization of the enzyme of interest, in terms of retention of activity and the performance of the resulting immobilized enzymes, thus necessitating modification of this type of carrier [235].

Carriers which bear relatively inert functionality such as hydroxyl, amide, carboxyl, amino, etc., can be prepared directly by suspension polymerization. For covalent enzyme immobilization, however, they must be activated before the covalent binding.

3.6.1
Synthetic Active Carriers

Active polymers are polymers which contain the active functionality and can be used directly to bind enzymes under appropriate conditions (pH, ionic strength, and temperature). Thus, at least one of the monomers used must contain the desired active functionality, which can react with one or more of the reactive amino acid (RAA) residues in the enzymes, as discussed in Chapter 2.

Many types of polymer bearing active functionality have been studied and tested for enzyme immobilization. They include carriers bearing the functionality acyl azide, acid anhydride, halogen, epoxy, isocyanate, thioisocyanate, cyclic carbonate, carbonyl, aldehyde, activated esters, and azalactone, etc. They can usually be prepared in one step by suspension polymerization in the presence of one or more co-monomer and a cross-linker. Alternatively, they can also be prepared as homo- or hetero-polymers, followed by cross-linking with a cross-linker which can react with the active functionalities. Here, discussion is focused on the first example.

3.6.1.1 Polymers Bearing Acyl Azide

Polymers bearing acyl azide were probably among the earliest active polymeric carriers used for covalent protein immobilization [236]. This type of polymer can be prepared by modification of several synthetic polymers [236] or derivatives of natural polymers [237–240].

For example, a composite of synthetic polymer and collagen membrane was activated to acyl azide for immobilization of urokinase and trypsin, leading to the formation of immobilized enzymes with catalytic properties comparable with those of the native enzymes [237]. PAAm-γ-alginic acid activated by esterification and subsequently converted to acyl azide was able to reach payloads of 0.236 and 0.269 for papain and trypsin, respectively [241].

Synthetic polymers such as polyacrylamide (PAAm) or other synthetic polyacrylic polymer bearing the acylhydrazine moiety [242], or composites [241, 243], can also be activated to bear active acyl azide as side-chains for binding enzymes. Direct

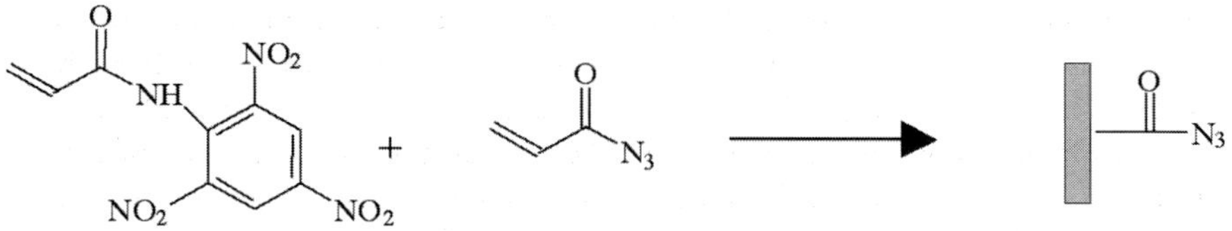

Scheme 3.56 Preparation of polymer bearing acyl azide groups.

preparation of polymers bearing acyl azide has been achieved by polymerization of unsaturated monomers bearing acyl azide functional groups, as illustrated in Scheme 3.56 [244].

This type of polymers is, however, currently rarely used for enzyme immobilization, owing to the lability of azide group and the heterogeneous nature of the binding, because acyl azide can react with several different amino acid residues, for example amino groups, hydroxyl, and mercapto groups, leading to the formation of immobilized enzymes with heterogeneous biding and, especially, low retention of activity.

3.6.1.2 Polymers Bearing Anhydrides

Polymers bearing anhydride functionality were probably the second type of synthetic polymer used for covalent enzyme immobilization. They can be prepared directly by polymerization of monomers bearing anhydride functionality, for example maleic anhydride [245], methyacrylic anhydride [246, 247], itaconic and citraconic acid anhydride as homopolymers, or heteropolymers in the presence of other comonomers, e.g. ethylene [248], acrylic acid, methylacrylic acid, butene, styrene, acrylonitrile [249, 252], acrylamide [253], and N-vinylpyrrolidone [252], or prepared as insoluble polymer in the presence of various bi-functional monomers such as bisacrylamide, or divinylbenzene, butanediol–divinyl ether (maleic anhydride) [254, 255] as cross-linkers.

Interestingly, variation of the comonomers or the cross-linkers can be used to tailor the structure and/or the hydrophobicity of the resulting carrier [117], as illustrated in Scheme 3.57.

In 1964, Levin and his co-workers reported for the first time the use of copolymers of maleic acid anhydride for immobilization of several proteolytic enzymes such as trypsin [248, 256]. With these negatively charged polymer–enzyme complexes they discovered that the microenvironment of the matrix can affect enzyme kinetics, because of unequal distribution of protons and the partition effect of the charged substrates.

At the beginning of 1970s polyanhydrides were intensively studied for enzyme immobilization [256]. Insoluble homo or heteropolymeric anhydrides can be fabricated in a variety of forms, for example powders, beads, and sheet. As illustrated in Scheme 3.57, water-insoluble copolymer powders with high enzyme loading were obtained by polymerization of the four components I-IX-VI-XVII in toluene [117].

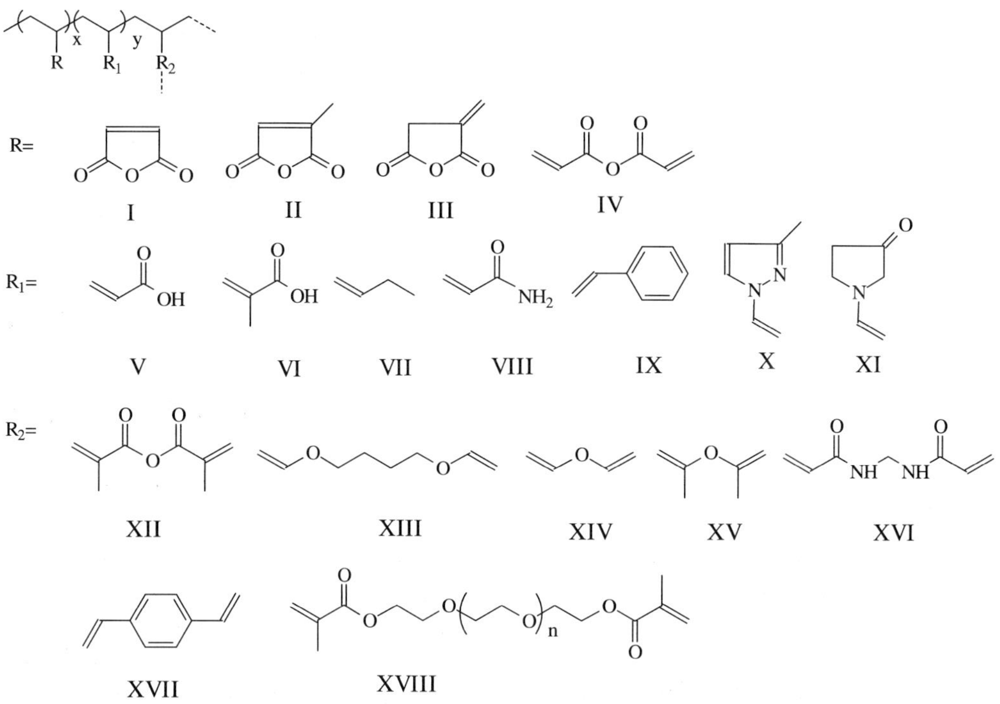

Scheme 3.57 Monomers used for preparation of polymeric anhydrides for enzyme immobilization.

Polymeric beads bearing anhydride groups have been prepared by dispersing vinyl ether-1,4-butanediol divinyl ether (XIII) and maleic anhydride (I) in heptane containing glycerol monooleate. The polymer obtained was of diameter 250 µm and average pore diameter 8000 Å. Unfortunately, the specific surface area was only 1.9 m² g⁻¹ [257].

The polarity of the polymers formed can be modulated by variation of the monomers used or the ratios of monomers such as maleic, itaconic, or citraconic anhydrides, of co-monomers such as aromatic monomers or vinylcarboxylic monomers, or cross-linkers such as hydrophobic monomers, e.g. aromatic divinyl monomers or vinylcarboxylic monomers as hydrophilic agents. For example, hydrophilic polymers such as copolymers of maleic anhydride, methacrylic acid, and tetraethylene glycol dimethacrylate can be formed by polymerizing the monomers in a suitable ratio in benzene. The resulting polymer has a surface area of only 8.6 m² g⁻¹ and a swelling factor of 12.4 mL g⁻¹. Approximately 10 U g⁻¹ penicillin G acylase can be immobilized on this type of polymer [258].

Enzyme immobilization on these polymers can be realized by many ways:

- Direct immobilization; for example, immobilization of the proteolytic enzymes, trypsin, chymotrypsin, and papain on cross-linked poly(methacrylic anhydride) resulted in only 3–20% activity toward casein and 40% toward low molecular weight substrates compared with the free enzyme [259].

- Enzymes can also be immobilized on ethylene–maleic anhydride copolymers (EMA) [262] with well defined and controlled anhydride content, achieved by controlled hydrolysis. The optimum recovery of immobilized enzyme activity and of insoluble material was achieved with EMA samples containing 50–70% anhydride [155].

- Anhydride-bearing polymer can be converted to other active functional groups for enzyme immobilization. For instance, the copolymer of ethylene–maleic anhydride can react with N_2H_4, 4,4′-methylenedianiline, or 1,6-hexanediamine, or with hydrazine followed by conversion to acyl azide [252]. The modified polymer obtained with hydrazine can be diazotized and used for enzyme immobilization [260].

- The polymeric anhydride modified with 1,6-diaminohexane (HMD) in the presence of dicyclohexylcarbodiimide can be used to bind enzyme through its carboxyl groups ($C-NH_2$ + HCOO-E) or activated with glutaraldehyde for immobilization of enzymes such as trypsin, by means of the enzyme's amino groups [261].

- Coating of heteropolymer on the other non-specific polymeric beads, for example the of methyl vinyl ether–maleic anhydride copolymer (MMAC) has been used to coat solid carriers for protein immobilization [247].

Interestingly, it has been demonstrated that polymeric anhydride partially substituted with pendant hydrophobic groups can be used to prepare amphipathic enzyme–polymer conjugates which could be dispersed on the surface of oil-in-water emulsions with reasonable enzyme activity [263].

Synthetic polymeric carriers bearing anhydride groups have several interesting features, for example ease of fabrication and mild conditions for binding enzyme. They also have several drawbacks, however, for example shifting of the pH profile of the immobilized enzyme by 1–2.5 pH units, toward alkaline pH, owing to the presence on the surface of excess carboxyl groups, produced by the reaction of anhydride with the protein or by hydrolysis by water [247], lower activity of the immobilized enzyme toward charged substrates, owing to the charged matrix [256], and lower surface area with beaded polymers [264].

As a result, interest in using these polymers for enzyme immobilization decreased at the end of the 1970s, because of the availability of other neutral microporous polymeric beads, for example acrylic polymers bearing epoxy rings, which are prepared by suspension polymerization, thus resulting in the formation of polymeric beads of controlled size, porosity and internal structure [265].

3.6.1.3 Polymers Bearing Halogen Atoms

The use of synthetic polymeric carriers bearing halogen atoms can be dated back to the beginning of the 1960s [266, 267]. These synthetic polymeric carriers bearing

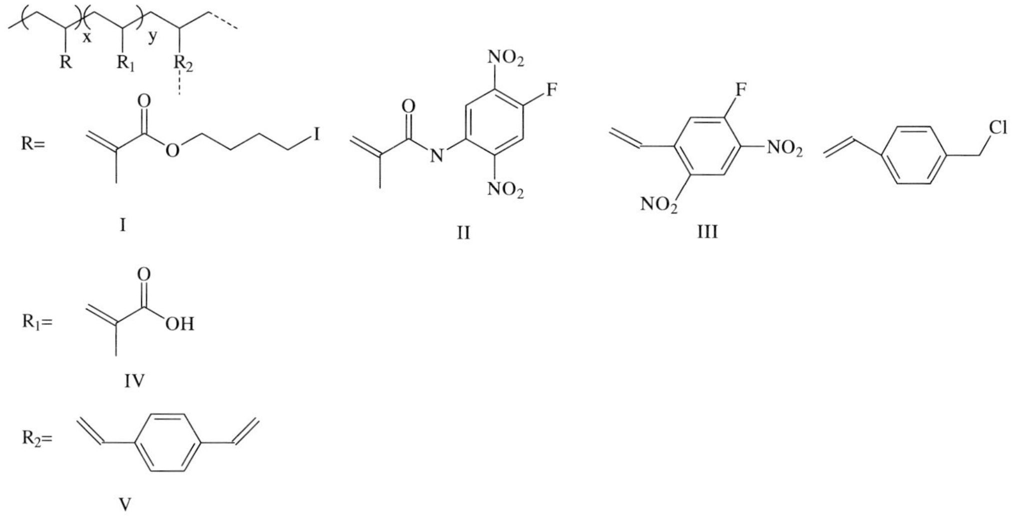

Scheme 3.58 Monomers used for preparation of active polymers bearing an active halogen atom for enzyme immobilization.

halogen atoms were synthesized from monomers bearing fluoro groups, for example I, II, III, and IV in Scheme 3.58, with co-monomers such as acrylic acid or divinylbenzene.

In 1970, Brown and co-workers used poly(4-iodobutyl) methacrylate, prepared by halogen exchange of poly(4-chlorobutyl methacrylate with sodium iodide, for immobilization of urease [268]. Remarkably, this study led to the important discovery that enzymes could be immobilized not only in buffer but also in organic solvents with low water content, for example dioxane [4, 268].

Following this work, other synthetic polymers, for example a copolymer of methacrylic acid *m*-fluoroanilide with methacrylic acid, divinylbenzene or fluorostyrene, have been synthesized since the 1960s [266-269]. In general, there has been almost no further interest in this type of polymer since the 1970s. The use of this kind of polymer has been reviewed by Manecke [270].

It was recently reported that conversion of non-porous poly(*p*-chloromethylstyrene) (PCMS) beads into a carrier by use of PEI, led to the preparation of carrier with a payload of 0.019 for invertase. Internal diffusion limitation is negligible and the thermostability of the immobilized invertase obtained was almost five times that of the native enzyme [271].

3.6.1.4 Oxirane Functional Polymers

The use of synthetic polymers bearing oxirane functional groups for enzyme immobilization can be dated back to the mid-1970s [272, 282]. Currently, this type of polymeric carrier is important for enzyme immobilization [178].

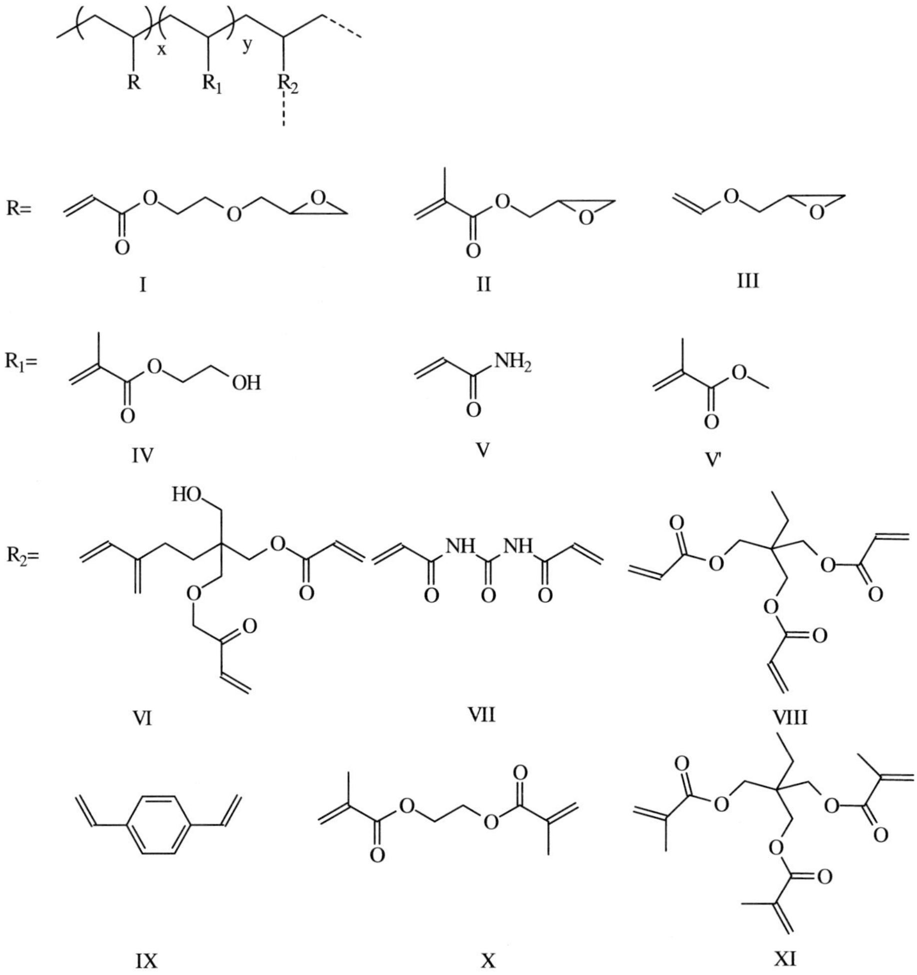

Scheme 3.59 Monomers used for preparation of epoxy polymers.

In general, as illustrated in Scheme 3.59, these polymeric beads can be prepared by reverse suspension polymerization of monomers containing oxirane groups (I, II, III), a co-monomer such as IV and V, and the cross-linkers such as VI, VII, VIII, IX, X and XI.

Depending on the nature of the active monomers or comonomers, the cross-linkers, and the reaction conditions, the polymers differ substantially from each other with regard to immobilization results such as activity retention, performance, and enzyme loading [15, 273]. They are usually regarded as neutral carriers and there is usually no pH shift with these polymers [274]. Occasionally the biocatalytic process might benefit from the pH shift or pH gradient [275].

Another characteristic of this type of carrier is that the nature of binding of the oxirane ring is heterogeneous, because the oxirane functionality is active toward several RAARs (Reactive amino acid residues) for example amino, carboxyl, hydroxyl, mercapto, phenol, imidazole, and indole groups in the protein. Moreover, although use of hydrophilic monomers, for example, hydrophilic acryl amide, as co-monomers or bis acrylamide as cross-linkers, can render the polymer more hydrophilic, this type of carrier usually has a remarkably hydrophobic nature overall.

For immobilization of enzymes, therefore, special measures should be taken, for example, increasing the ionic strength of the medium to facilitate enzyme binding or adding quenching agents after binding of the enzyme to eliminate excess epoxy groups, which might slowly react further with other relatively inert amino acids residues, leading to the deactivation of the immobilized enzyme obtained.

Active porous resins bearing oxirane groups can be synthesized by suspension polymerization of the monomers with cross-linkers such as II–XI [276], II–VIII, II–XI, II–VI [15], and II–IX [273]. The pore and pore-size distribution might be broad in the range of macroporous carriers [15], or close to the upper limit of the mesoporous carriers (10 nm) [273].

The enzyme loading is usually 10–100 mg protein g^{-1} carrier [276]. For penicillin G acylase, the activity (U g^{-1}) is usually in the range 150–205 U g^{-1} [276, 277] and activity retention is usually below 50%, depending on the nature of the polymer (monomer, comonomer or cross-linker used) [15], pore size [26] and the enzyme loading [277].

By comparing different macroporous beaded terpolymers bearing oxirane groups synthesized by use of II, IV, and V with IX or X as cross-linkers, Bahulekar found that the oxirane acrylic polymer cross-linked with IX (DVB: divinyl benzene) (Scheme 3.60) was not suitable for immobilization of penicillin G acylase [125], because of the incompatible interaction of the enzyme with the benzyl ring [279]. The same type of polymeric carrier might, however, be suitable for immobilization of other enzymes, for example lipase or amino acid oxidase [10].

These examples indicated that the desired carrier for enzyme immobilization might change from enzyme to enzyme and from application to application.

Fibrous polymeric carrier with oxirane groups can be prepared using photo- and radiation-induced grafting techniques. Glycidylmethacrylate (B) and 2-hydroxyethylmethacrylate (D) copolymer was grafted on to poly(acrylonitrile) fibres which could then be used to immobilize enzyme [125, 280, 281]. This type of carrier can be used directly for enzyme immobilization or after derivatization to introduce other functionality [281, 282].

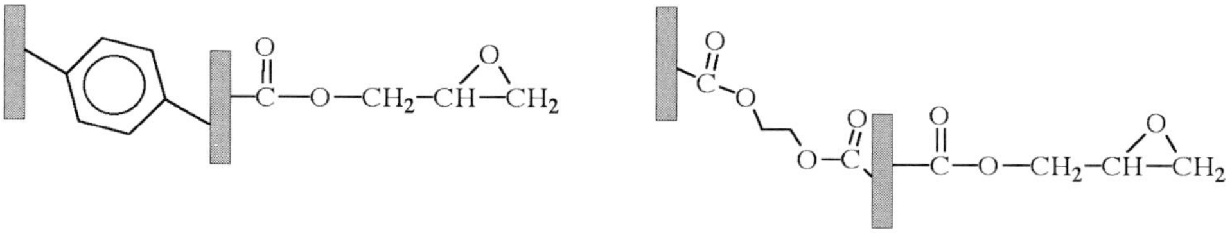

Scheme 3.60 Effect of chemical composition on the performance of immobilized enzymes.

Eupergit C was among the first commercialized synthetic oxirane polymers used for enzyme immobilization [125, 282, 283]. It is prepared from hydrophilic monomers such as methylene bis(methacrylamide), allyl glycidyl ether, acryl amide and glycidyl methacrylate and so is a relatively hydrophilic polymer with water retention capacity 3 g g^{-1} polymer.

The robustness of Eupergit C has been confirmed by many investigators (see [2] for a review). It is usually observed that retention of activity is low (usually below 40%) at high enzyme loading [277]. By expansion of the pore size to 100 nm (in Eupergit C 250 L) retention of activity can be increased when the loading is comparable [25], suggesting that a large pore size is essential for higher retention of activity, because of reduction of mass transfer limitation.

Penicillin G acylase immobilized on Eupergit C is an important industrial immobilized enzyme for production of 6-APA, a key intermediate in the synthesis of semi-synthetic β-lactam antibiotics such as ampicillin and amoxicillin [233]. It has, however, recently been demonstrated that penicillin G acylase immobilized on Eupergit C is not generally suitable for synthesis of semi-synthetic β-lactam antibiotics, because of precipitation of the product in the pores [284], or poor performance in the synthesis of semi-synthetic β-lactam antibiotics compared both with the native enzyme or with other forms of immobilized penicillin G acylase [14].

3.6.1.5 Isocyanate/Thioisocyante Functional Polymers

Direct polymerization of polymeric isocyanate was pioneered by Maneck at the end of the 1960s [285]. One of the earliest examples is the preparation of thioisocyanate from acrylic or of methacrylic acid with isothiocyanato styrene [267]. A copolymer was prepared by co-polymerization of acrylic acid (IV) with 3- or 4-isothiocyanatostyrene (III) and divinylbenzene (VII) as cross-linking agent [267]. Hydrophilic macroporous polymer bearing isocyanate groups was prepared by copolymerization of acrylic acid, divinylbenzene, and 4-isothiocyanatostyrene. However, increasing the amount of cross-linking led to a decrease of the binding enzyme. With papain as model enzyme it was found that immobilized papain had a lower K_m than the native enzyme [286].

Polymeric carriers containing isocyanate functional groups are usually prepared by polymerizing monomers bearing isocyanate groups with comonomers in an anhydrous organic solvent, as illustrated in Scheme 3.61.

For example, use of AIBN to copolymerize 20 g 2-isocyanatoethyl methacrylate (I) with 80 g trimethylolpropane trimethacrylate (VI) in heptane furnished polymeric powders with a surface area 134 m^2 g^{-1} [287]. Moreover, changing the cross-linkers to other monomers with two unsaturated moieties such as ethylene glycol dimethacrylate (EGDMA) could lead to the formation of a variety of isocyanate polymers with tailor-made properties for immobilization of different enzymes. It has also been found that the specific surface area of this polymer is usually reduced by increasing the concentration of cross-linker [287].

Because the practical use of direct polymerization to prepare polymers containing isocyanate functionality is not desired, this type of polymeric carrier is often ob-

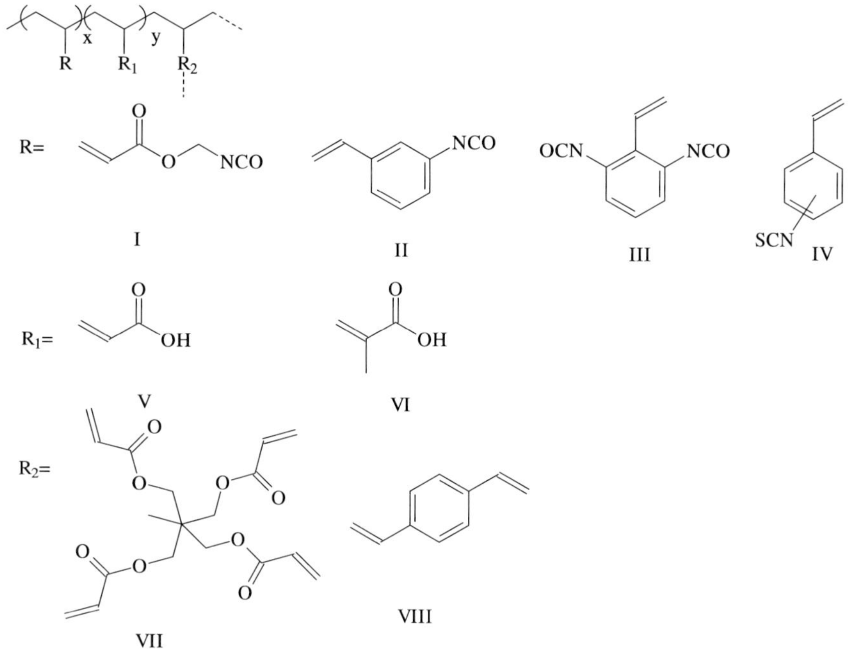

Scheme 3.61 Monomers used for preparation of synthetic polymers bearing isocyanate groups.

tained by modification of other polymers, for example polymers containing hydroxyl groups or amino groups, with di-isocyanate compounds, as discussed in the following sections.

3.6.1.6 Polycarbonate

The use of synthetic polycarbonate for enzyme immobilization can be dated back to the beginning of 1970s [288]. Kennedy and co-workers were probably the first to use this type of polymer for covalent enzyme immobilization. Several enzymes, for example α-amylase, d-glucosidase and trypsin were immobilized in these early days of enzyme immobilization.

Monomers bearing carbonate functionality, for example I, II, III, and IV, comonomers such as V, and cross-linkers such as VII, VIII, IX, and X can be used to prepare soluble or insoluble polycarbonates for enzyme immobilization.

It has been found that activity retention depends on pH, coupling time and enzyme concentration. It has also been found that coupling conditions differed from enzyme to enzyme and from carrier to carrier. With this type of polymer it was found for the first time that post-treatment of the immobilized enzyme by lyophilization of the insoluble immobilized enzyme in the presence of a polyol prevented instability of the immobilized enzyme on storage. In the succeeding 20 years, how-

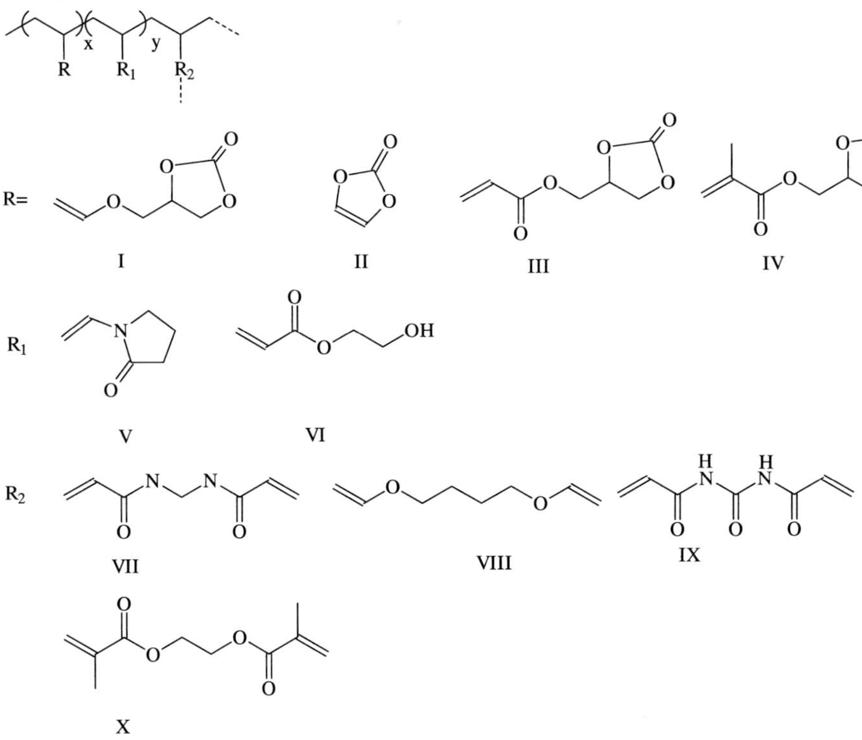

Scheme 3.62 Monomers used for preparation of polymeric carriers with carbonate groups.

ever, there have been few reports on the use of this type of polymer for enzyme immobilization.

Since the 1980s Mauz has introduced several new monomers, for example vinyl or allyl ethers (I) or (meth)acrylates of glycerol carbonate (III, IV) or vinylenecarbonate (II) and comonomers such as N-vinylpyrrolidone (V) and cross-linkers such as methylenebizacrylamide (VII) or butanediol divinyl ether and N,N'-divinylalkyleneurea (IX), for preparation of hydrophilic beaded polymers bearing carbonate groups for enzyme immobilization [289].

Using dispersion polymerization of vinyl or allyl ethers (I) or (meth)acrylates of glycerol carbonate (III, or IV) with N,N'-divinylalkyleneurea (IX), beaded polymer (10–600 µm) of pore size (5–1000 nm) with active carbonate groups was formed. Hydrophilic polymer prepared from glycerol carbonate methacrylate and N,N'-divinylalkyleneureas can be used to bind large amounts of penicillin G acylase (payload of dry beads, 1.28) with 51% retention of activity [290].

It was recently found that beaded hydrophilic polycarbonate prepared from vinylene carbonate (VCA), 2-hydroxyethylene acrylate, and N,N'-methylene bisacrylamide could also be used for immobilization of enzymes [291] such as trypsin [292] and glucoamylase [293].

Very recently soluble homo-polymer polycarbonate was cross-linked to form an insoluble beaded carrier which can be used to bind trypsin. Remarkably, the payload can reach 0.25 (dry carrier). Retention of activity is, however, less than 15% at high payload. This could be because of diffusion constraints or deactivation of enzyme by the high density of reactive groups. Taking the swelling factor (10) into account, the payload of wet beads is a maximum of 0.025 [294].

3.6.1.7 Activated Carbonyl Polymers

Activated carbonyl polymers can be prepared directly by polymerizing the corresponding monomers bearing activated leaving groups in the presence of a cross-linker, as shown in Scheme 3.63.

Although it is also possible to prepare these polymers by post-synthesis, direct preparation might provide an interesting means of studying the effect of individual leaving group on the performance of the polymers obtained (e.g. binding capacity and performance of the enzymes).

One such example was furnished by Lu and co-workers in 1981 [295]; they prepared four different polymers bearing activated carbonyl functionality – poly(N-p-

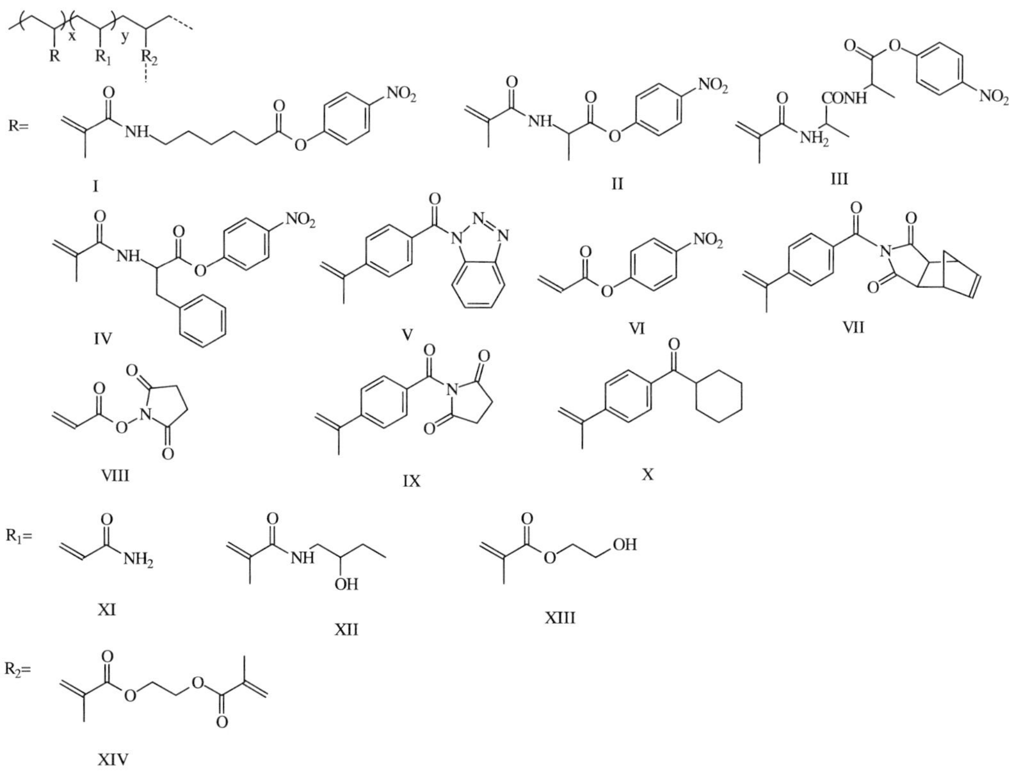

Scheme 3.63 Activated polycarbonyl polymers.

methacryloxybenzoylox-5-norbornene-2,3-dicarboximide), poly(N-p-methacryloxybenzoyloxysuccinimide), poly(N-p-methacryloxybenzoyloxy-4-oxo-3,4-dihydro-1,2,3-benzotriazine, and poly(N-p-methacryloxybenzoyloxybenzotriazole) from the corresponding monomers, as shown in Scheme 3.63.

Although the payloads of these polymers were quite high and very close for trypsin (147, 149, 138, and 142 mg g^{-1} carrier, respectively), retention of activity differed greatly (31, 25, 8, and 8%, respectively), suggesting that other factors which can affect enzyme immobilization might be the reactivity of the active functionality toward the amino groups of the enzyme, as mentioned above.

Other acylating polymers which have been developed for enzyme immobilization are prepolymers prepared from acryloxysuccinimide and acrylamide, for example PAN (poly(acrylamide-N-acryloxysuccinimide)) [296, 297]. This prepolymer was also used to entrap enzyme in the presence of a diamine such as cystamine as cross-linker. Cross-linked copolymers of N-acryloxysuccinimide can be also used as carriers for covalent enzyme attachment (covalent binding and entrapment occur concomitantly). Currently, this method is rarely used for preparation of active carriers for covalent enzyme immobilization.

3.6.1.8 Polyphenolic Polymers

These types of polymer are usually hydrophilic in nature, inexpensive, and easy to prepare [299]. They can be prepared by condensation of p- or o-benzoquinone with formaldehyde or glutaraldehyde (Scheme 3.64)

The highly porous materials obtained can be used to immobilize several enzymes, for example ovalbumin and acid and alkaline phosphatases, with high retention of activity [298].

Immobilized α-amylase, glucoamylase, and d-glucose isomerase can be used in packed bed reactors; they enable high flow rates and retain their activity in continuous use [299]. Immobilized penicillin G acylase on this type of carrier retains ca. 50% activity (2000 U g^{-1} dry carrier) [300].

Other commercially available phenolic polymeric resins, for example Duolite S-30 resin (phenol–formaldehyde resin), can be used for enzyme immobilization after modification by reaction with POCl$_3$ and DMF to obtain aldehyde groups [301], or introduction of diazonium salt groups [302].

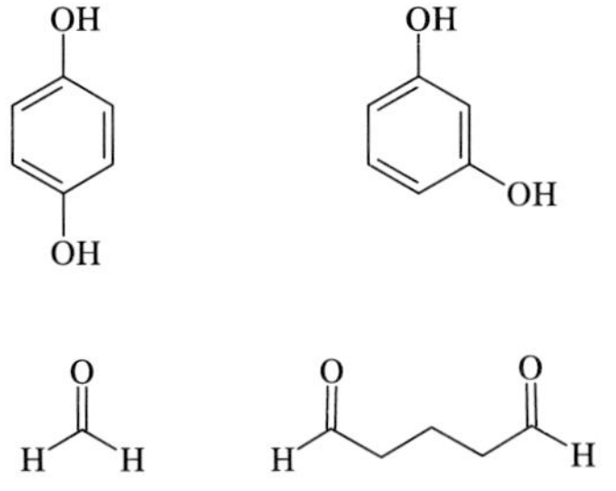

Scheme 3.64 Monomers used for preparation of polymeric p-quinone.

3.6.1.9 Polymeric Carriers Bearing Aldehyde Groups

The use of synthetic polymeric carriers bearing aldehyde groups for enzyme immobilization was pioneered by Epton and co-workers in 1971 [303], when monomers bearing protected aldehydes (as acetals) were used to prepare polymeric carriers bearing acetal functionality, which could then be activated into active aldehydes under acidic conditions [304].

This type of polymer can be prepared by polymerizing the protected aldehyde (III) or acrolein (I) in the presence of a comonomer such as styrene, allyl alcohol, acrylic acid, vinylpyrrolidone, or acryl amide for formation of soluble pre-polymers [305]; for formation of insoluble carriers, a cross-linker must be used [306].

Alternatively, polymers bearing aldehyde groups can be prepared by condensation of glutaraldehyde alone or glutaraldehyde in the presence of a diamine compounds such as xylenediamine, as shown in Scheme 3.65.

Synthetic polymeric aldehyde carriers, prepared by polymerization of monomers containing aldehyde groups, include polyacrolein prepared from acrolein [307–309], a water-soluble copolymer of 4-vinylpyridine with acrolein [305], which was used to prepare both immobilized cofactor-NADH and alcohol dehydrogenase, copolymer of acrolein with acrylic acid, protected acryl aldehyde derivatives with or without co-monomers such as styrene [310] or N-vinylpyrrolidone (H), a co-polymer of acryloyl vanillin (D) and ally alcohol (F), a copolymer of acrolein and sucrose, a copolymer of acrolein with acrylamide [306], a copolymer of vinyl ether and

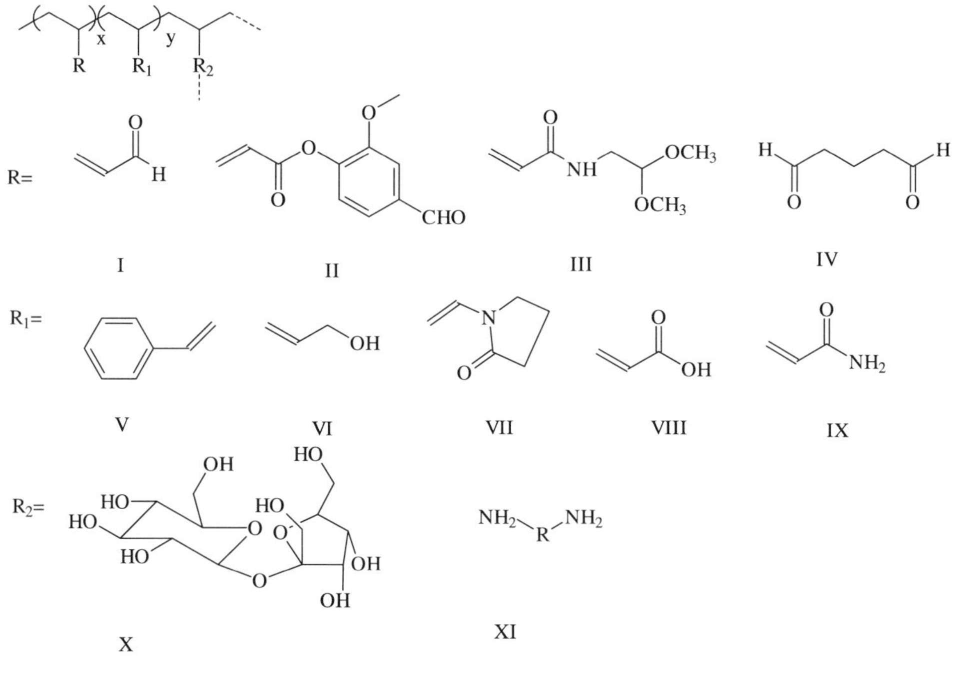

Scheme 3.65 Monomers used for preparation of polymers bearing aldehyde groups.

vinylbutyral [311], and a copolymer of acrolein and styrene with a sponge structure and particle size smaller than 10 μm which can be used to bind large amounts of enzyme, e.g. papain with payloads varying from 0.8 to 1.05, depending on the ratio of the two monomers [312].

A porous resin – acrolein–vinyl acetate–divinylbenzene polymer – has been synthesized by radical suspension polymerization of acrolein with vinyl acetate and divinylbenzene in the presence of a pore-creating agent, toluene [313]. The copolymers prepared from acrylic acid and acrolein or the condensation product of aliphatic dialdehyde or acrolein with diamines could be used to bind enzymes or proteins [314]. Comprehensive information on preparation of polyacrolein microspheres suitable for medical applications can be found in a review [309].

These polymers can be used to covalently bind enzymes [315], resulting in the formation of Schiff base which is usually unstable under acidic conditions. Reduction of the Schiff base with sodium borohydride often can enhance the covalent binding. Modification of this aldehyde polymeric carrier led to the introduction of new functionality for enzyme immobilization. For example, polyacrolein was used to immobilize enzyme (invertase) by ionic adsorption, by treating polyacrolein with diaminohexane. Interestingly, polyacrolein containing latexes and their assemblies on solid supports have been used for immobilization of human serum albumin, gamma globulins, horseradish peroxidase, and glucose oxidase [316].

Other synthetic polymers containing aldehyde groups can be prepared indirectly by reacting pyridine-containing polymers such as the polymer prepared from the polythiol with 4-vinylpyridine, then treatment with CNBr, followed by hydrolysis [316–318].

The enzymes immobilized on these synthetic polymers bearing aldehyde groups included lipoprotein lipase [319] and horseradish peroxidase on poly(styrene/acrolein) latexes [320]. Thermolysin from *Bacillus thermoproteolyticus rokko* was immobilized on acrolein-containing latex (AL, obtained by copolymerization of styrene and acrolein) [321].

Insoluble polymers can be obtained by cross-linking heteropolymers such as cross-linked acrolein (I)-2-hydroxyethyl methacrylate polymer [322]. It was found that the activity of horseradish peroxidase immobilized on cross-linked acrolein–2-hydroxyethyl methacrylate polymer particles depended on the acrolein content, reaching a maximum at 0.14 mol fraction and decreasing with further increases in the acrolein content of the particle.

Polystyrene (P(S)), poly(styrene/acrolein) (P(SA)), and polyacrolein (P(A)) latexes with varied proportions of polyacrolein in the surface layer (f_A = 0, 0.50, 0.63, 0.84, 1.00) have been used for attachment of horseradish peroxidase [320]. Polymerization of monomers bearing aldehyde functionality to prepare the polymeric carriers bearing aldehyde groups is, however, rarely explored. This is largely because a number of polymeric carriers bearing aldehyde groups can be prepared by derivatization of other ready-made polymers, for example oxidation of cross-linked beads of polysaccharides such as agarose and dextran, as will be discussed in the following sections.

Nowadays, the use of this type of polymers for enzyme immobilization gradually loses importance. This is not only because of instability of the formed the Schiff base linkage, but also because of the easy availability of carriers bearing aldehyde groups via the interconversion techniques as discussed in the following section.

3.6.1.10 Polymers Bearing Activated Ester

As illustrated in Scheme 3.66, polymeric activated esters can be prepared by direct polymerization of acryloyl derivatives containing activated ester groups (I, II, IV and V) [323] with co-monomers such as VI, VII and IX.

Immobilization of enzyme is usually performed under alkaline conditions to avoid the degradation of the active functionality (hydrolysis). Use of p-nitrophenyl esters of methacrylic acid and methacryloyl derivatives of glycine, β-alanine and ω-aminocaproic acid as the activated monomers enables preparation of active polymer with different spacers [324].

Disadvantages of these polymers is that nitrophenol can be released after the coupling reaction, thus some enzyme might be deactivated or inhibited.

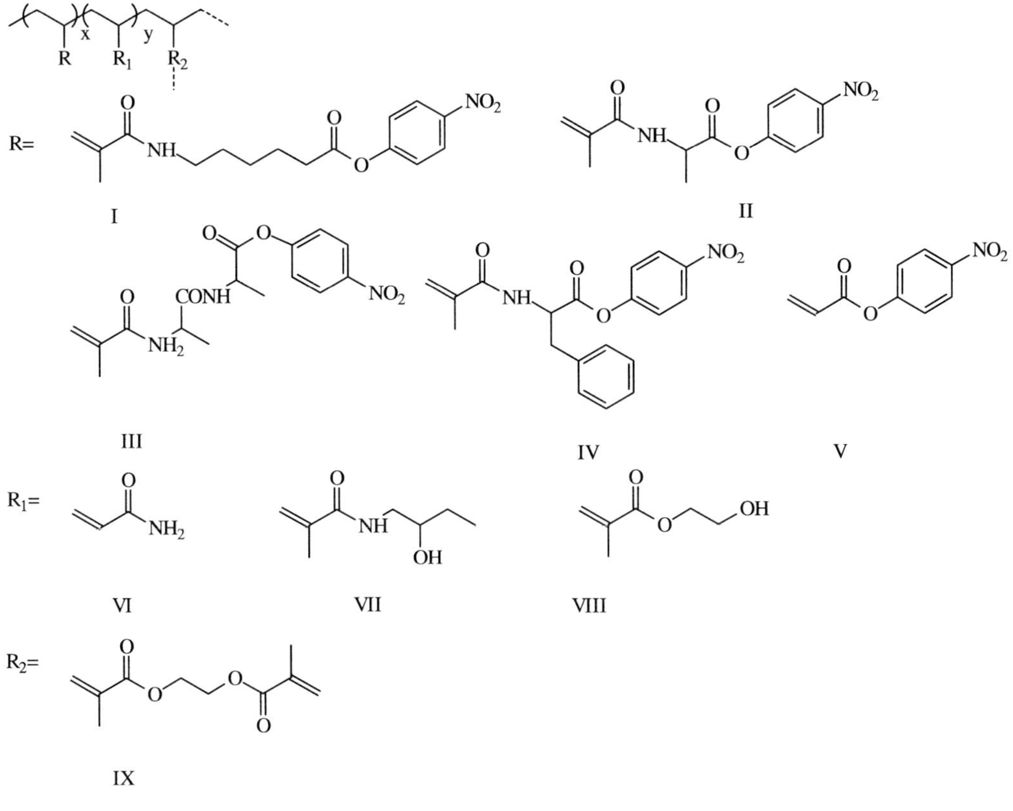

Scheme 3.66 Monomers used for preparation of activated esters.

3.6.1.11 Polymers Bearing Active Azalactone

Polymeric carriers bearing azalactone groups can be dated to the mid-1980s, when a group of synthetic chemists and biochemists at 3M began to collaborate on the use of azlactone-based polymers for covalent coupling of biomolecules to polymeric supports [325, 326, 328]. A comprehensive review was recently published by Heilmann and co-workers [327].

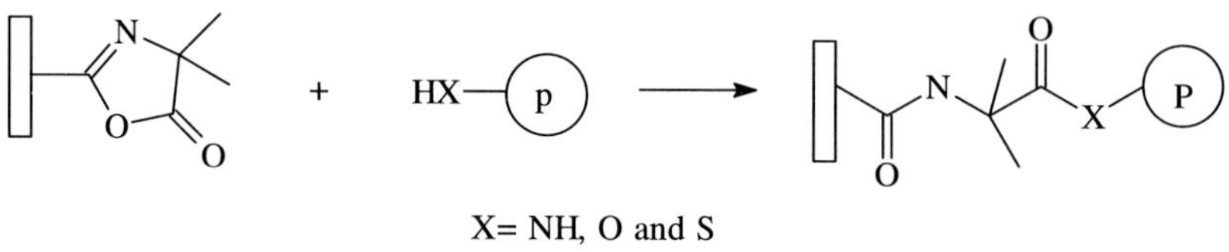

X= NH, O and S

Scheme 3.67 Polymeric carrier bearing azalactone.

Immobilization of biomolecules on this type active polymer can be performed under mild conditions. The functional groups can react with amino, sulphydroxyl, and hydroxyl groups. The monomers which can be used to prepare azlactone-bearing polymers are illustrated in Scheme 3.68

Porous polymeric carriers can be prepared by polymerization of vinylazalactone in the presence of cross-linkers such as bis-acrylamide or trimethylolpropane trimethacrylate (TMPTMA). Monolithic reactor can be also prepared with other co-monomers, for example ethylene dimethacrylate, and acrylamide or 2-hydroxye-

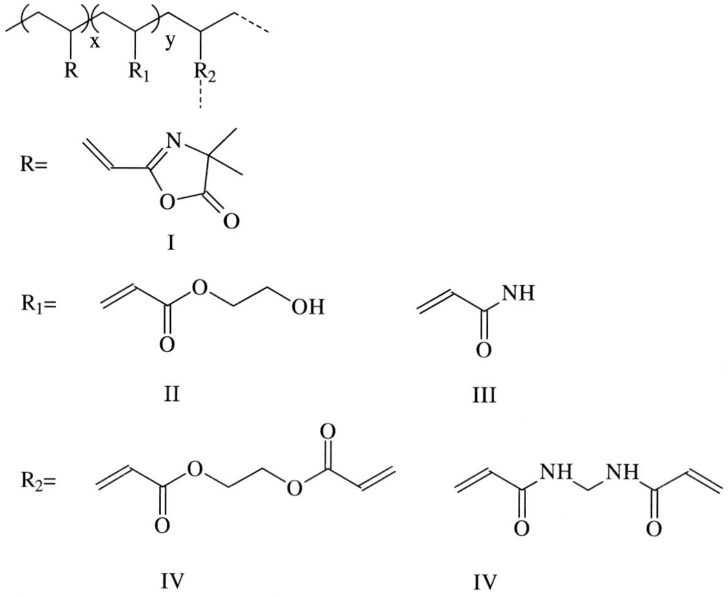

Scheme 3.68 Monomers used to prepare activated azalactone polymeric carriers.

thyl methacrylate. The activity of immobilized trypsin reached 260 U mL^{-1} (N-benzoyl-l-arginine ethyl ester as substrate) [329].

The hydrophilic/hydrophobic properties of this type of polymer can be modulated. The catalytic activity of the monolithic reactor is maintained even at a flow velocity of 180 cm min^{-1}, which substantially exceeds values reported in the literature for packed bed reactors [330]. However, these types of polymer resemble carriers bearing epoxy groups, with regard to the enzyme loading and immobilization conditions [331]. Thus, the characteristics of soybean lipoxygenases immobilized on EMD azlactone polymer were very similar to those on Eupergit C [332].

3.6.2
Inactive Pre-carriers

This class of polymer can be prepared from the inert monomers. In general, these polymers include polymers containing the functionality such as hydroxyl (-OH), carboxyl (-COOH), amino (-NH$_2$) and nitrile (-CN) groups. Often, they must be activated before covalent immobilization of the enzymes. Here, only the preparation of these synthetic polymers is covered, because the corresponding activation methods will be presented.

3.6.2.1 Hydroxyl Functionality

As illustrated in Scheme 3.69, synthetic polymers with hydroxyl groups can be prepared by polymerization of monomers bearing hydroxyl groups, for example 2-hy-

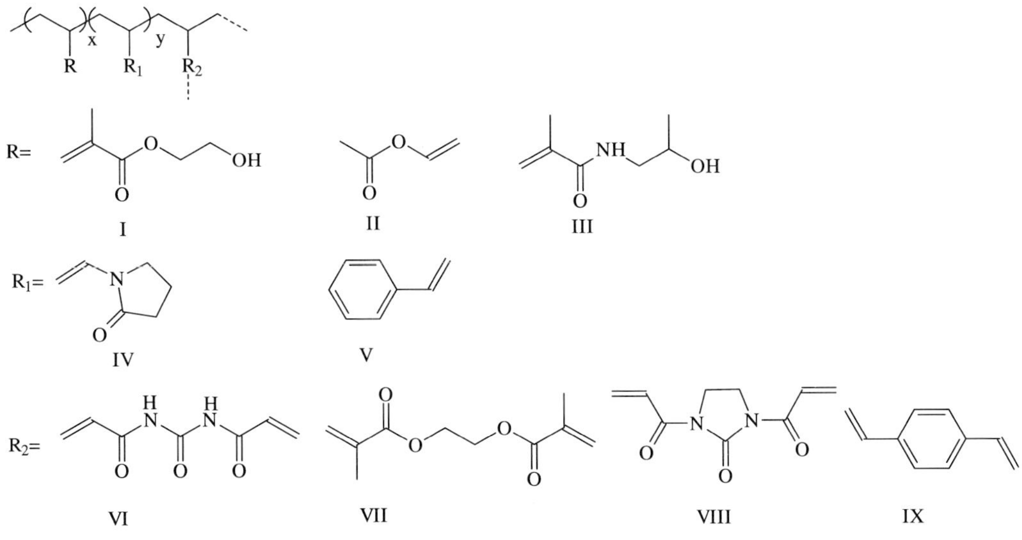

Scheme 3.69 Monomers used for preparation of synthetic carriers bearing hydroxyl groups.

droxyethyl methacrylate (I), with a cross-linker such as ethylene dimethacrylate (IV) [333], or by hydrolysis of the copolymer of vinyl acetate and acrylic acid.

Hydroxyl groups of synthetic polymers can also be activated by other general methods used for activation of hydroxyl groups. For example, functionalization of the copolymer of 2-hydroxyethyl methacrylate and ethylene dimethacrylate with epichlorohydrin furnishes a carrier with epoxy groups which can be used to immobilize several enzymes, for example aminoacylase, thermitase, pepsin, trypsin, chymotrypsin, elastase, subtilisin, penicillin amidase, carboxypeptidase A, and cystathionine synthase [333].

Bead-shaped macroporous carriers with hydroxyl groups can be obtained by suspension polymerization of vinyl acetate (II) and N,N'-divinylethyleneurea (VI) in water, followed by partial hydrolysis of the acetate groups to hydroxyl groups and subsequent reaction with epichlorohydrin [334]. Remarkably, carrier-bound penicillin acylase could be re-used 65 times without loss of activity. The carrier materials withstand steam sterilization without alteration of their properties [334]. Partially hydrolysed poly(VAc-co-DVB) can be activated with different activators. Interestingly the enzyme immobilized on this polymer had activator-dependent activity and stability [114].

Commercial synthetic polymers bearing hydroxyl groups, e.g. Spheron 300, have used to immobilize several enzymes for example chymotrypsin, trypsin, chymotrypsinogen, and tyrosinase, by using benzoquinone [335] or the CNBr method for activation [336]. Activation of the hydroxyl groups of a synthetic carrier can also be achieved with ECH (epichlorohydrin), leading to the formation of active oxirane polymer [337].

3.6.2.2 Polyacrylamide

Polyacrylamide (PAAm) gel, one of the earliest synthetic polymers used for entrapment of proteins or enzymes or resting whole cells [338-340], can also be used for

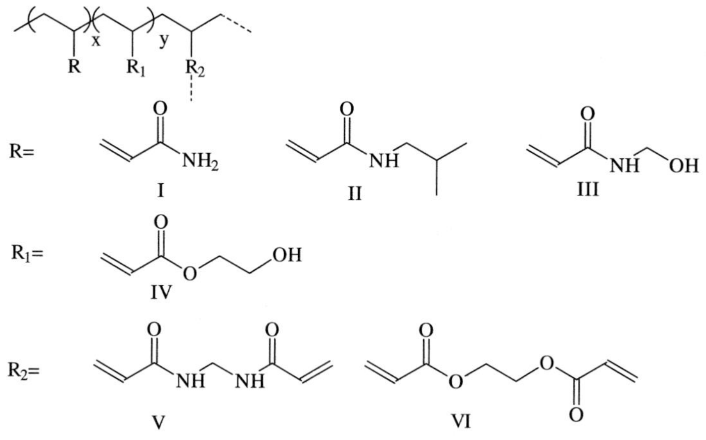

Scheme 3.70 Polyacrylamide.

covalent immobilization of enzymes [341, 342]. Beaded polyacrylamide gel with controlled size can be prepared in a two-phase system by use of chemical initiators (Scheme 3.70).

For covalent immobilization PAAm gel is usually activated by several methods, as discussed in the following section. For covalent enzyme immobilization, other monomers, for example glycidol dimethacrylate, bearing active functional groups can be co-polymerized to form an active carrier which can be directly used to bind enzymes; an example is the commercial carrier Eupergit C.

More interestingly, functional groups can also be formed by functionalization of dispersants such as PVA, used in suspension polymerization processes, and can be structurally incorporated into polymeric beads, for example copolymers of ethylene glycol dimethacrylate (EGDMA) with acrylic acid (AA), hydroxyethyl methacrylate (HEMA), and acrylamide (AAm), by suspension polymerization with or without toluene [343].

Polyacrylamide-based gels can also be prepared by precipitation polymerization with acrylamide, glycidyl methacrylate, N,N'-methylenebizacrylamide, methacrylic acid, and AIBN as monomers and cross-linkers at 60 °C [344].

3.6.2.3 Insoluble Polyacrylic Acid or Derivatives

Synthetic polymeric carriers bearing carboxylic or carboxylate groups are widely used as adsorbents. They can be prepared by use of monomers bearing unsaturated bonds, for example 2-hydroxylethylacrylate and methyacrylic acid, and copolymers of acrylamide and acrylic acid (Scheme 3.71)

Some polyacrylic polymers bearing special side groups, e.g. poly(4-methacryloxybenzoic acid), can, after activation with N-ethoxycarbonyl-2-ethoxy-1,2-dihydroquinoline, be used to bind enzymes. One advantage of this method is that the extent of polymer activation can be determined and monitored by analysis of quinoline, a by-product of the reaction [345]. The polymeric beads can be prepared by suspension polymerization. The physical and chemical nature of these polymers is largely dictated by the monomers, cross-linkers and the inert porogens used [346].

In an attempt to prepare acrylic polymers for immobilization of different enzymes, Kolarz and co-workers prepared several polymers from acrylonitrile or Bu acrylate as monomers and divinylbenzene, ethylene glycol dimethacrylate, trimethylolpropane trimethacrylate monomer (J), trimethylolpropane triacrylate monomer (I), pentaerythritol triacrylate (G) as cross-linkers. They found that immobilization of enzymes on these polymers could be achieved by aminolysis of the nitrile groups to N-alkylamino, for example for acrylonitrile (I)-vinyl acetate (II)-divinylbenzene (III) terpolymers, then activation of the amino groups with glutaraldehyde [344], or conversion of the ester with a diamino compound such as ethylenediamine for pentaerythritol triacrylate (PENTA)-Bu acrylate (BA) matrixes [111].

Compared with the copolymer prepared from acrylonitrile, polymers prepared from acrylate esters such as BA with trimethylolpropane trimethacrylate monomer (TMPMA), after aminolysis, were more suitable for enzyme immobilization [22].

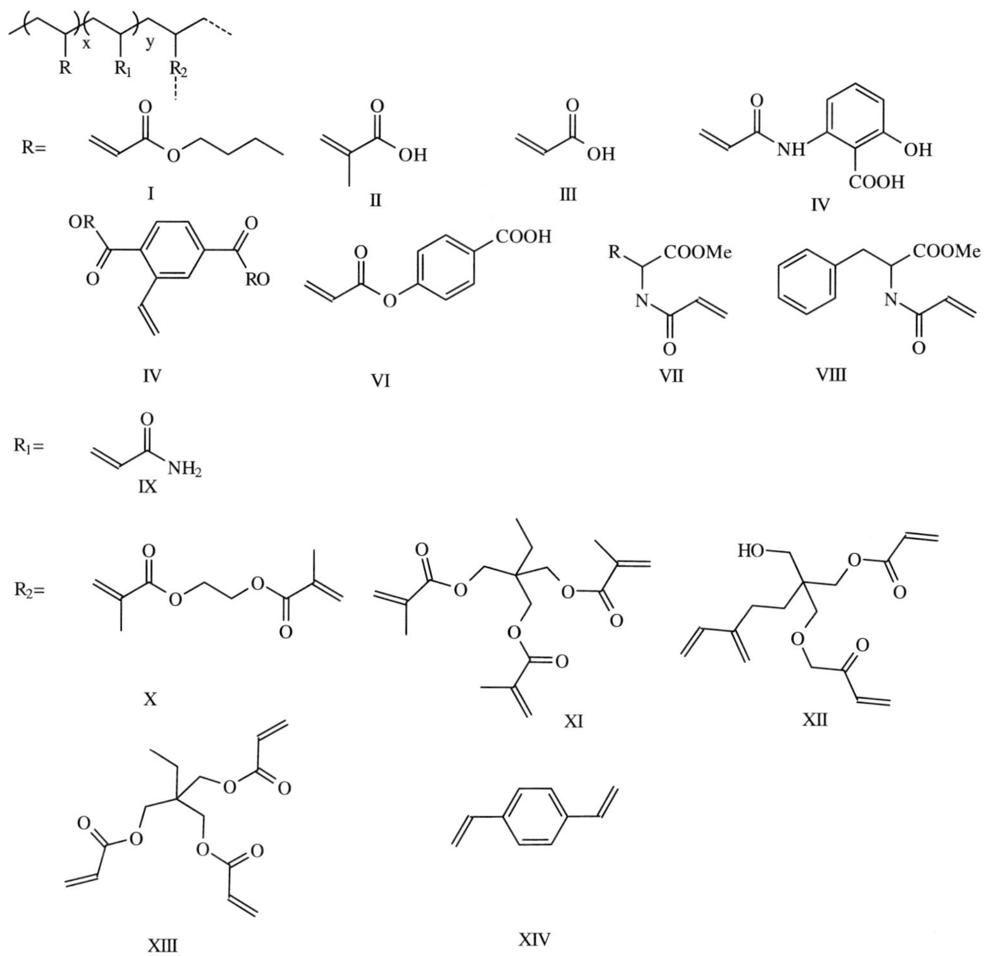

Scheme 3.71 Monomers bearing carboxyl groups.

Unfortunately, all these polymers have small pores (below 3 nm) [347] so enzyme loading is usually low.

Depending on the composition and nature of the monomer used, immobilization of enzymes on those polymers can be via adsorption (such as lipases), covalent binding (via the carboxyl groups), entrapment of enzymes or whole cells, or coordination binding if a coordination moiety is introduced into the polymers, for instance, polymer bearing 5-aminosalicylic acids can be used to prepare adsorbent which can bind an enzyme via the coordination binding [348]. On the other hand, polyacrylic polymers such as XAD-7 or methyl acrylate–divinyl benzene copolymer can, after aminolation, be used directly for enzyme immobilization by adsorption [349] or activated by glutaraldehyde for covalent immobilization of penicillin G acylase [350].

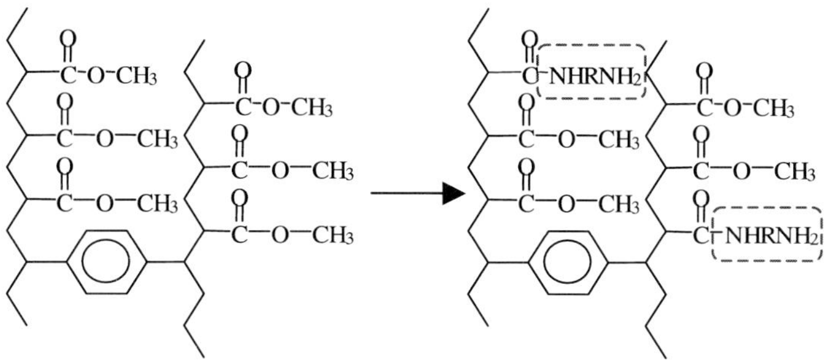

Scheme 3.72 Introduction of charged groups to synthetic polyacrylic polymers.

The surface ester functionality of poly(ethylene terephthalate) powder can be hydrolysed into a polymer containing carboxyl and hydroxyl groups. The former can be converted into isocyanide groups by reaction with 1,6-diisocyanohexane whereas the latter can be converted to aldehyde groups by reaction with dipyridyl chromium(VI) oxide [351]. Both of new functionality can be used to bind enzymes. The copolymer of vinyl acetate with acrylic acids can be converted into a hydrogel after hydrolysis of the acetate groups. Enzyme immobilization can be via carbodiimide activation of the carboxyl groups [352].

In principle N-acryolated amino acids can also be polymerized to form homopolymers or, in the presence of cross-linkers such as divinyl benzene, ethylene glycol dimethacrylate, or bis acrylamide, cross-linked polymers. For example, N-acrylated phenylalanine methyl ester can be used to prepare a beaded polymer which can, after hydrolysis of the ester group, be used to immobilize enzymes [353].

3.6.2.4 Polymers Bearing Nitrile Groups

Polymers bearing acrylonitrile can be prepared by polymerization of acrylonitrile or other unsaturated monomers bearing nitrile groups in the presence of comonomers such as acrylamide [354] and a cross-linker, e.g. divinylbenzene or bisacrylamide. Macroporous polymers bearing nitrile groups (pore size >120 nm) can be prepared by suspension polymerization in the presence of acetophenone or decalin as porogens and partially saponified poly(vinyl acetate) as dispersant. Enzymes such as urease, catalase, α-chymotrypsin, and pepsin could be adsorbed with 33, 73, 50, and 45% retention of activity, respectively [356].

Immobilization of enzyme on PAN can be performed by amidine, via the imido ester of polyacrylonitrile [355]. Uricase and peroxidase can be individually or simultaneously immobilized on to a copolymer of acrylonitrile and acrylamide [355]. Porous polyacrylonitrile can be reduced to polymers bearing free amino groups [357]. Immobilization of enzymes such as acylase from *Bacillus megaterium* on the polyacrylonitrile with free amino groups resulted in high capacity binding of the enzyme compared with the polymer without free amino groups [358].

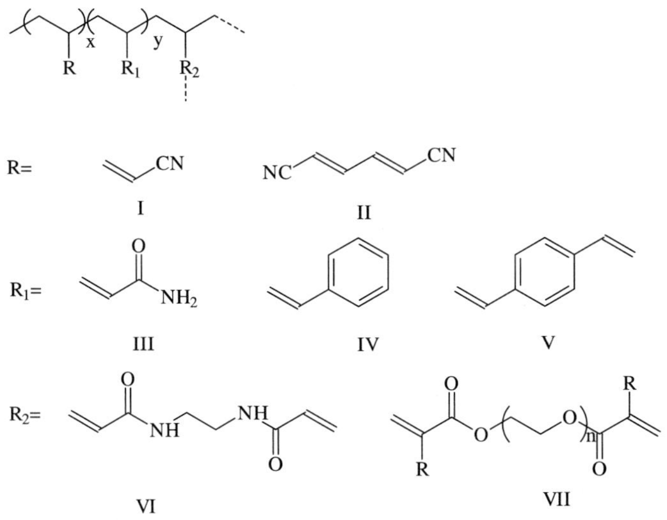

Scheme 3.73 Polymeric nitriles.

Similarly, poly(acrylonitrile) or acrylonitrile copolymers can be also converted into polymers bearing hydrazine and NH_2 groups, which can be subsequently diazotized by treatment with HCl and $NaNO_2$. Alternatively, the hydrazine-modified polymer can be activated by glutaraldehyde, p-benzoquinone, or cyanuric chloride [359].

Copolymers of acrylonitrile and 1,4-divinylbenzene or of acrylonitrile, acrylic acid, and 1,4-divinylbenzene can be activated by reaction with N-methylolmaleimide. The activated polymer can react with the thiolated enzymes such as α-chymotrypsin, RNase A and trypsin (reaction with N-acetyl-dl-homocysteine thiolactone).

It has been found that the nature of the copolymer can affect not only the physical properties of the polymers but also the performance of the immobilized enzymes. It has been found that a copolymer of acrylonitrile, acrylic acid, and 1,4-divinylbenzene had a higher swelling factor than acrylonitrile and 1,4-divinylbenzene, leading to high retention of activity of the immobilized enzyme. On the other hand, the presence of carboxyl groups also led to the pH shift toward alkaline pH [360].

3.6.2.5 Semi-synthetic Polysaccharides

It is well known that a variety of polysaccharides such as agarose, dextran and chitosan derivatives play important roles in bio-immobilization, including controlled drug-release. These linear polysaccharides can be cross-linked to form solid gel with defined structure and properties and widely used in bio-separation and immobilization. One characteristic of these polysaccharide derivatives is that the backbone of the polymer often consists of sugar units (Scheme 3.74).

It has long been known that synthesis of polymers bearing sugar moieties as side-chains is feasible by polymerization of monosaccharide bearing unsaturated

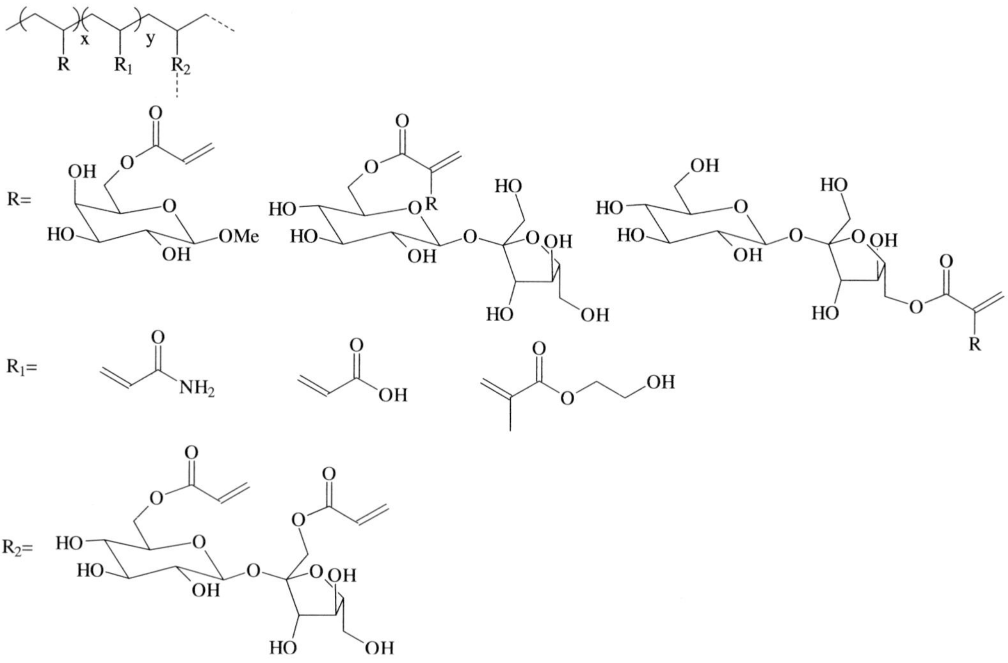

Scheme 3.74 Monomers and cross-linkers used for preparation of semi-synthetic polysaccharides.

bonds; lack of selectivity in the introduction of unsaturated bonds often necessitates the use of protection chemistry, however, thus undoubtedly limiting its industrial value. It has recently been shown that selective introduction of unsaturated bonds can be achieved by use of enzymes.

In line with this idea, it has been demonstrated that polymerization of these enzymatically synthesized semi-synthetic monomers in the presence of cross-linker can be used to produce hydrogels [361–365]. It has also been demonstrated that sustained release of the drugs can be achieved by use of macroporous polysucrose gel prepared with sucrose-1′,1-acrylate and bisacrylamide [366].

Interestingly, a series of degradable hydrogels based on different vinyl monomers, for example acrylamide, sucrose-1-acrylate, and acrylic acid, have been synthesized using sucrose-6,1-diacrylate (SDA) as cross-linking agent [367]. Highly swelling hydrogel can be prepared with monoacylated methylgalactoside [368].

3.6.2.6 Synthetic Polypeptide

Polymethylglutamate (PMG), a synthetic polypeptide, has been used as a carrier to immobilize glucose oxidase, uricase, peroxidase, trypsin, chymotrypsin, urease, and aminoacylase by the NH_3 method. The enzymes could be immobilized covalently on PMG coated on glass beads [369]. Remarkably, retention of activity by all immobilized enzymes was excellent (>90%). The amount of enzyme immobilized on the polymer varied markedly, depending on the nature of enzymes (trypsin 30 mg, chymotrypsin 27 mg, urease 5.8 mg, uricase 5.6 mg, aminoacylase 2.3 mg, glucose oxidase 1.8 mg, and peroxidase 2.3 mg/100 mg PMG).

It was also found that the enzyme loading is largely dependent on enzyme molecular weight and the amount of lysine residues. Also, enzymes immobilized on PMG were more stable because of both the increased hydrophilicity of the polymer and the multipoint binding mode, including covalent and ionic bonding.

3.6.2.7 Polymers Bearing Amino Groups

Direct preparation of polymers with amino functionality can be achieved by polymerization of monomers bearing amino groups functionality with a cross-linker. For instance, styrene bearing long spacer groups with active anchoring groups ($-NH_2$ or $-NH-$) has been used to prepare insoluble polymers for enzyme immobilization [370]. This method is not practically preferred, however, because amino groups can be easily introduced into ready-made polymers bearing other chemical functionality. This will be discussed in the next section.

Although covalent binding of an enzyme to a carrier is one of the simplest methods of enzyme immobilization, the lack of guidelines linking the nature of the carriers selected to the performance to be expected for a given application makes the rational design of immobilized enzymes very difficult.

Although many types of carrier are commercially available, or can be prepared easily on a laboratory scale, synthetic organic chemists often hesitate to spend their precious time optimizing the immobilization. Also, many methods, which work well on a laboratory scale might not be acceptable to industrial users, owing to economic and ecological considerations.

For this reason, it is hardly surprising that interest in the development of new carriers for enzyme immobilization are gradually switching to the creation of new types of carrier that can bind large amounts of enzyme with high retention of activity under mild conditions or development of carriers with ready-made chemical and physical properties which can easily be transformed into new carriers by chemical modification, as discussed in the following section.

3.6.3
Interconversion of Inert Carriers

Chemically inert carriers are carriers that cannot be directly used to bind enzyme. Thus, their functional groups must be activated to introduce functionality active toward the reactive amino acid residues on the enzyme surface.

Usually, many chemical routes are available to convert a chemically inert functionality into chemically active one. Thus, conversion of inert polymers by selective chemical modification not only increases the number of immobilization methods of choice but also provides an efficient means of modifying the carrier microenvironment by variation of the conversion method. Consequently, enzyme performance can be affected by introduction of different functionality such as CAG or CIG as mentioned above.

This chemically inert functionality for enzyme immobilization mainly include the hydroxyl groups of polysaccharides, for example Bio-Gel A (agarose), cellulose and its derivatives, dextran, Sephadex, Sepharose, synthetic polyacrylate and derivatives such as PVA cryogel, carboxyl or carboxylate groups of synthetic or semi-synthetic adsorbents such as XAD-7, amide groups of polyacrylamide gel or Nylon [144, 145, 438], amino groups of PEI or gelatine gel, nitrile groups of polyacrylonitrile or copolymers [359], and isonitrile groups.

Currently, many modifications or conversions of this ready-made functionality have been identified. Thus, an inert functionality can be easily converted to the new functionality. To date it is still very difficult to predict the most suitable functionality for a specific enzyme or application, because the performance of the enzyme obtained depends on many factors, for example the properties of the linkage, the physicochemical nature of the carrier, and the application conditions.

3.6.3.1 Polymers Containing Hydroxyl Groups

These carriers include polymers such as poly(vinyl alcohol) (I), poly(2-hydroxyethylmethylacrylamide), cross-linked polysaccharide beads such as agarose, chitosan, pectin, dextran, cellulose, and their derivatives, and inorganic carriers such as silica, etc.

As illustrated in Scheme 3.75, they contain two different types of hydroxyl group, separated hydroxyl groups and vicinal hydroxyl groups. Depending on the type, they can be converted into different new functionality ready for enzyme immobilization, as illustrated in Scheme 3.76 (more detailed information is given in [371]).

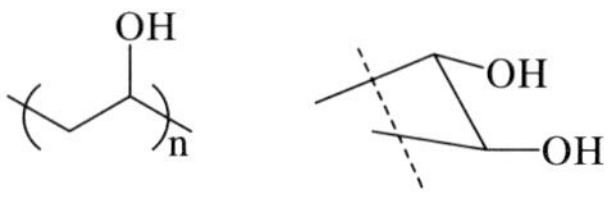

Scheme 3.75 Polymers containing hydroxyl groups.

Activation of Vicinal Hydroxyl Groups Carriers bearing vicinal hydroxyl groups include most cross-linked polysaccharide such as Sepharose and Sephadex. They can be activated by oxidation to aldehyde functionality, isourea bonds, amino groups, cyclic carbonate groups, as illustrated in Scheme 3.76.

Oxidation to Aldehyde Vicinal groups can be converted into aldehyde groups by oxidation with $NaIO_4$. This concept for enzyme immobilization was first proposed by Royer in 1975, when the sugar moieties of glycosylated enzymes were oxidized into aldehyde groups, which can subsequently be immobilized to carriers carrying amino groups [372, 373].

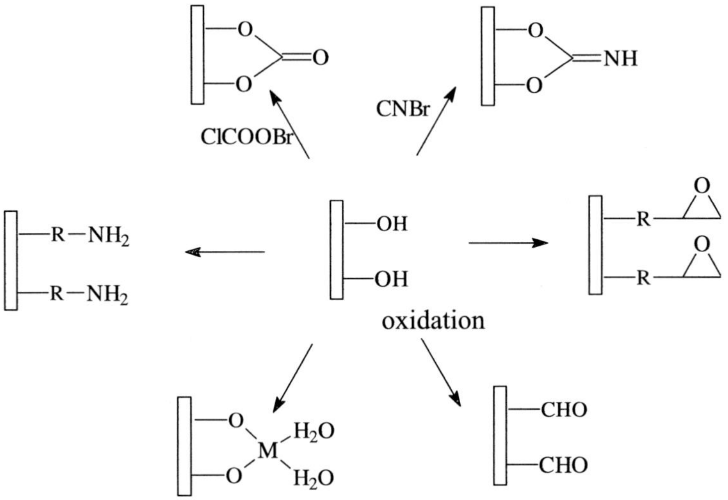

Scheme 3.76 Activation methods often used for vicinal hydroxyl groups.

Likewise, the vicinal hydroxyl functional groups in sugar moieties in polysaccharides such as agarose can also be oxidized to aldehydes or ketones that, on reaction with amines, lead to the formation of Schiff bases that can in turn be reduced to stable secondary amines [372].

Vicinal hydroxyl groups can also be introduced by reacting insoluble cross-linked polysaccharide beads, for example agarose beads, with glycidol, followed by hydrolysis and oxidation, as illustrated in Scheme 3.77.

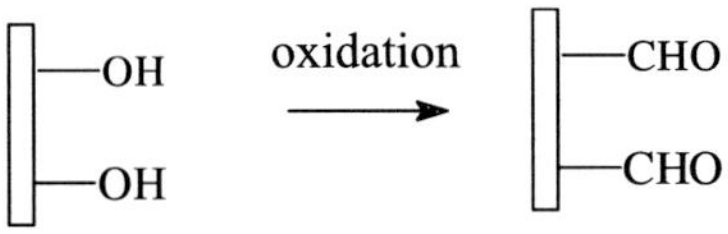

Scheme 3.77 Oxidation to aldehyde groups.

This last approach is widely used to introduce vicinal diols to polysaccharide carriers such as agarose; these are subsequently oxidized to aldehyde residues by use of periodate. Enzymes can be immobilized by formation of Schiff bases between the aldehyde groups of the carrier and the amino groups of the lysine residues in the enzyme molecules (Scheme 3.78) [374].

Several other carriers bearing hydroxyl groups, for example dextran-coated porous glass, Sepharose, and glass coated with a glyceryl silane, can be oxidized with $NaIO_4$ to introduce aldehyde groups for covalent binding of enzymes [375], as illustrated in Scheme 3.79.

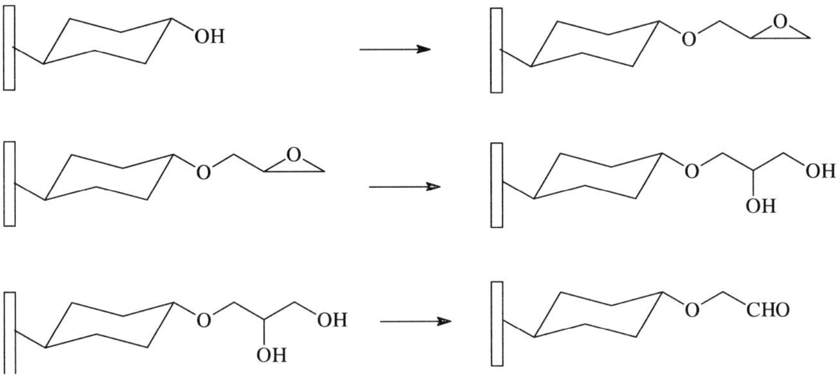

Scheme 3.78 Introduction of vicinal hydroxyl groups to polysaccharide carriers such as agarose.

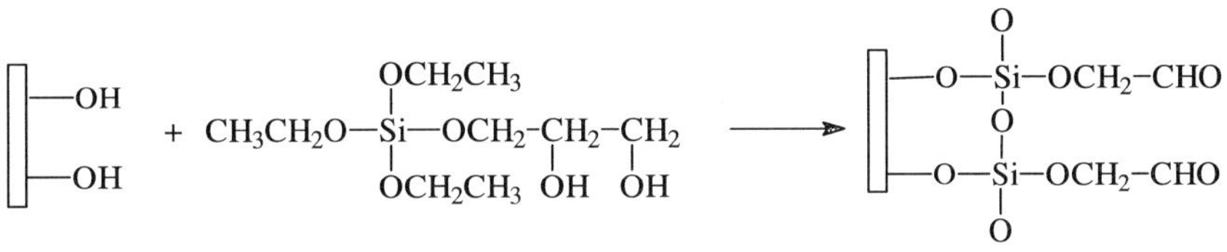

Scheme 3.79 Conversion of hydroxyl groups to aldehyde groups.

Activation with Cyanogen Bromide Since Axen [376] introduced this method in 1967, these methods have been increasingly used in enzyme immobilization using natural or synthetic polymers [377]. Cyanogen bromide is often used to activate the vicinal hydroxyl groups of polysaccharides, e.g. cellulose fibres [378] or synthetic polymers bearing hydroxyl groups [379], as shown in Scheme 3.80.

Activation of a polysaccharide carrier with CNBr should be performed at pH 11–12 and immobilization of enzymes on the activated carriers should be performed immediately after the activation, because of the instability of the activated bonds. Special care should also be taken to remove excess CNBr, which can react further with the protein, resulting in deactivation. The linkage formed (isourea bond) is only moderately stable, often leading to the slow leakage of the enzyme [380].

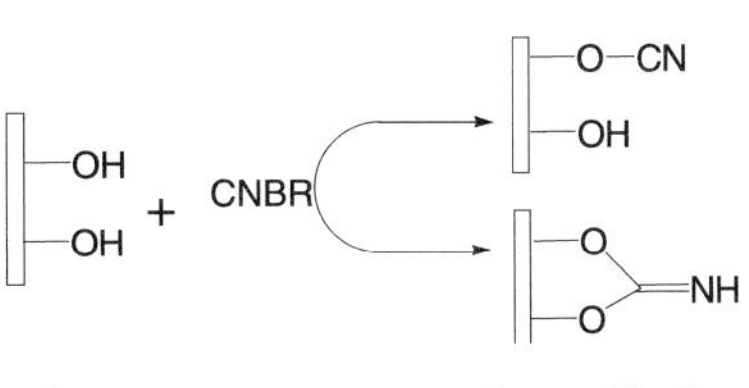

Scheme 3.80 Conversion of vicinal hydroxyl groups to active functionality.

Interestingly, it has been found that immobilization of subtilisins modified with dextran on aminosilochrome carriers by two methods, activation of the subtilisin–dextran complex with CNBr and oxidation of dextran tails, led to the discovery that the former method afforded an immobilized enzyme with higher stability than the latter method [381].

It has also been found that CNBr-activated polysaccharides might not be suitable for lipase immobilization because of steric hindrance and excessive loss of enzyme activity. In contrast, reaction of 4,4′-methylenedianiline with the CNBr-activated polysaccharides led to formation of a polysaccharide carrier with a spacer with terminal amino groups, which was subsequently used to bind enzyme by use of carbodiimide as coupling agent, resulting in the formation of immobilized enzyme with activity at least threefold that of the corresponding preparation without spacer [382].

Study of the immobilization of bacterial dextranase on cellulose-based carriers, for example CNBr-activated cellulose (ICC), acylazide CM cellulose (CMC), and cellulose carbonate (CC) (Scheme 3.81), found that coupling of dextranase to ICC cellulose not only afforded the highest enzyme loading (82 mg protein g^{-1} carrier) but also highest retention of activity (32.2%) and storage stability (only 5% activity loss during several months storage at 4 °C in buffer) [383].

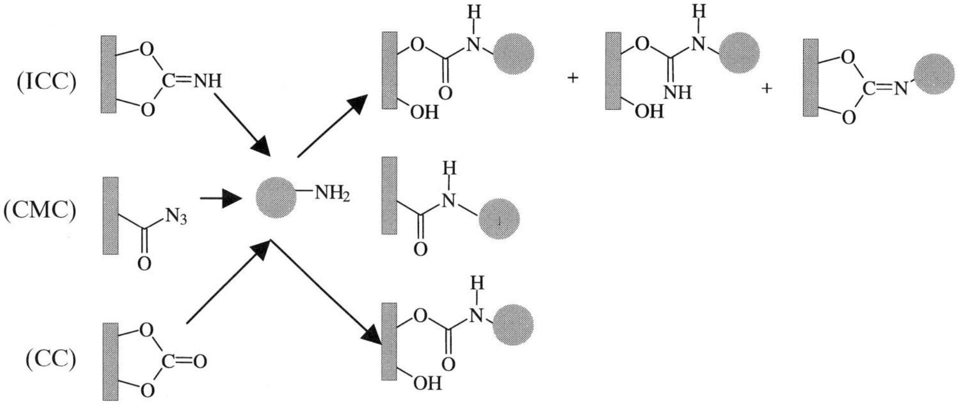

Scheme 3.81 Immobilization of dextranase on cellulose derivatives.

Conversion to Amino Groups   Vicinal hydroxyl groups on silica carriers or cross-linked polysaccharide beads such as Sepharose [384] can be converted directly to amino groups by use of 3-aminopropyltriethioxysilane (APTES) to modify the carrier and introduce the amino functionality [385], as illustrated in Scheme 3.82.

The resulting carriers can be used to immobilize the enzyme either directly by the glutaraldehyde method or indirectly by activation with a polymeric aldehyde such as aldehyde dextran, as mentioned above [381].

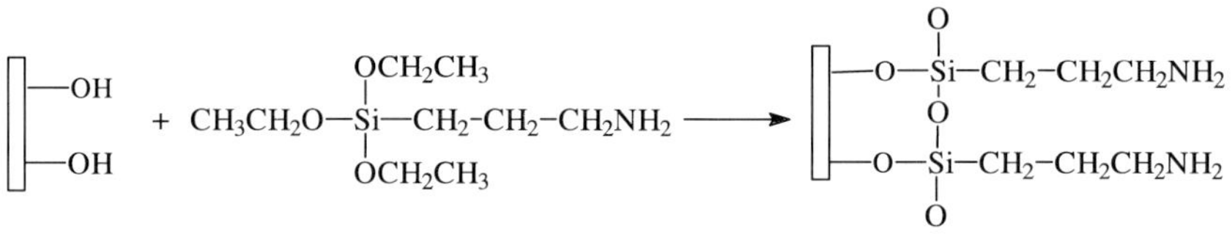

Scheme 3.82 Conversion of hydroxyl groups to amino groups.

Conversion to Oxirane Groups Hydroxyl groups can also be converted to oxirane functionality for enzyme immobilization, for instance, by reaction with epichlorohydrin [235]. 3-Glycidoxypropyltrimethoxysilane and diisopropyltrimethoxysilane can be used to convert hydroxyl groups on silica into epoxide groups [386].

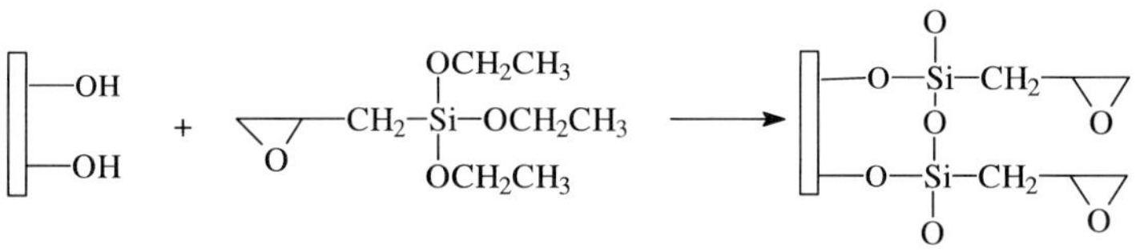

Scheme 3.83 Conversion of hydroxyl groups to epoxy groups.

Conversion to Isocyanate Silica or glass or organic polymers bearing vicinal hydroxyl groups can also be converted to carriers carrying isocyanate groups by use of 3-isothiocyanatopropyldiethoxysilane. With this method several enzymes, for example trypsin, glucoamylase, peroxidase, aminoacylase, and alkaline phosphatase, have been covalently bound to γ-isothiocyanatopropyldiethoxysilyl glass. Coupling reaction at pH 7.5 takes less than 1 h. Activity retention was 16–24% for trypsin, 68% for glucoamylase, 27% for horseradish peroxidase, 65% for alkyl phosphatase, and approximately 1% for kidney aminoacylase [387].

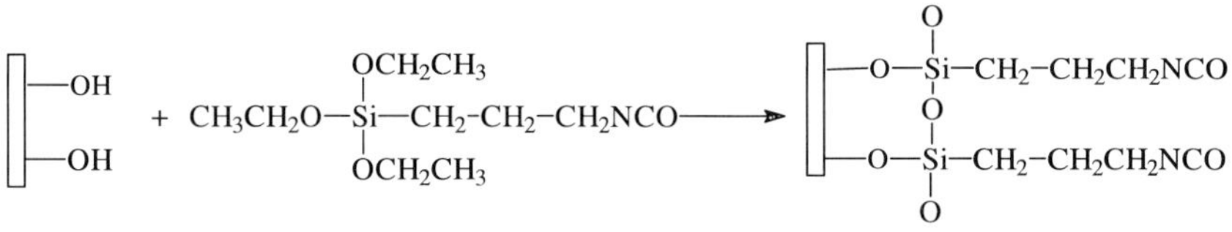

Scheme 3.84 Conversion of hydroxyl groups to isocyanate.

Conversion to Diazonium Salt Hydroxyl groups can be also converted to the diazonium salt by a series of reactions developed by Huelsmann and co-workers. Very high operational stability (1000 h constant activity) was observed for amino acylase from *Aspergillus oryzae* immobilized on silica activated and diazotized according to Scheme 3.85 [388].

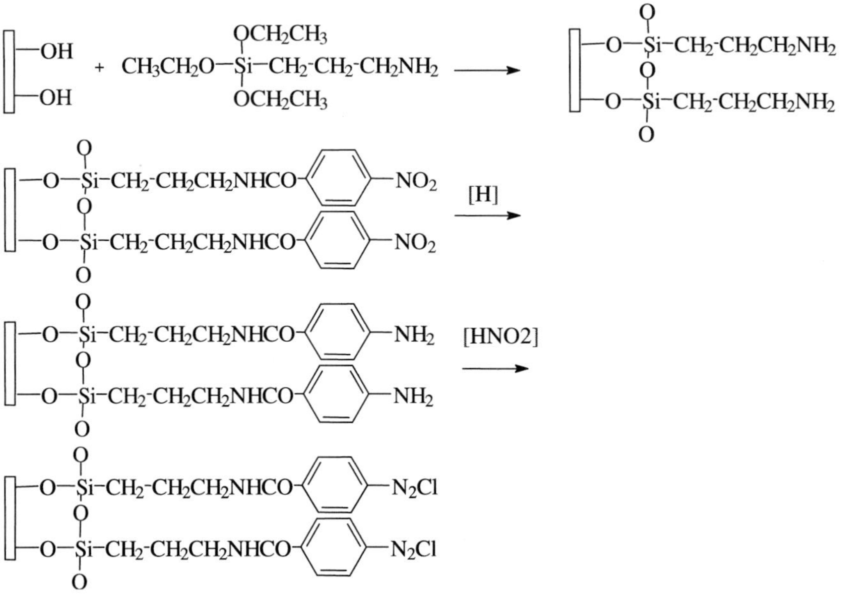

Scheme 3.85 Introduction of diazonium salt to silica or other hydroxyl group-bearing carrier.

In contrast, a simpler, two-step procedure, illustrated in Scheme 3.86, was developed by Messing et al. [388]. The half life of ABSP (alkaline *Bacillus subtilis* protease) immobilized on CPS (controlled pore silica) activated by use of this method was almost twice as high as the ABSP immobilized on the silanized CPS by adsorption at room temperature in H_2O [389].

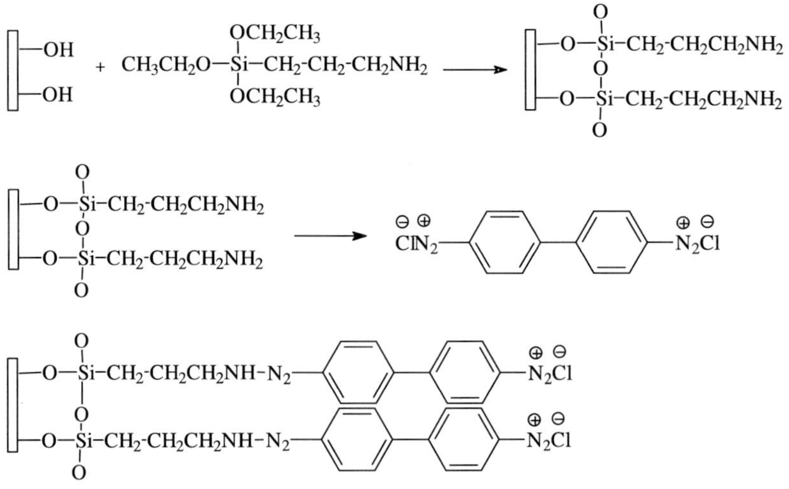

Scheme 3.86 Activation of silica to produce a carrier bearing a diazonium salt.

Retention of activity of glucose oxidase (GOD) immobilized on p-(β-ethylsulpho-nyl)aniline-activated agarose by the glutaraldehyde method is slightly better than when diazotization is used [390].

Activation with Metal Ions

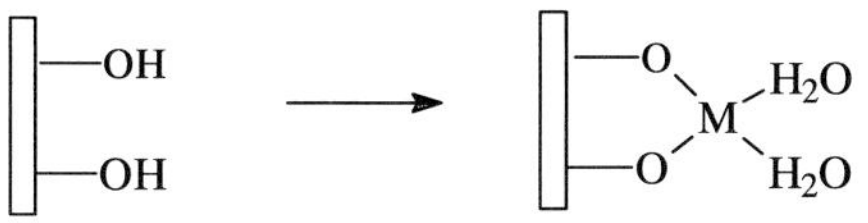

Scheme 3.87 Metal chelation for activation of a hydroxyl group.

Conversion into Carboxyl or Amino Groups Conversion of EVAL (ethylene vinyl alcohol) polymer bearing hydroxyl groups into an activated polymer can be achieved by several methods, for example conversion into COOH groups or into amino groups (via conversion into cyanate groups) [391].

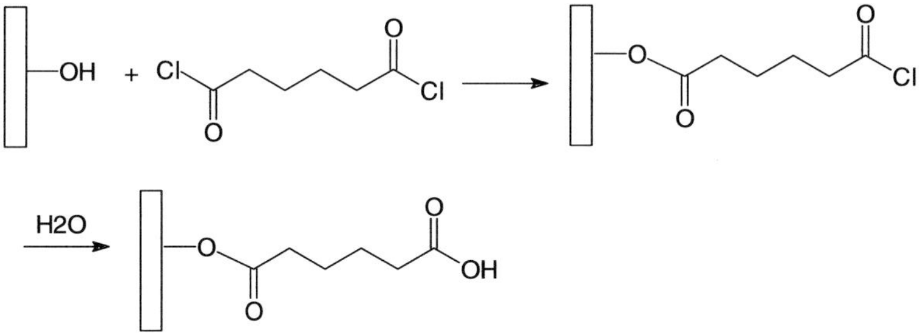

Scheme 3.88 Conversion into carboxyl groups.

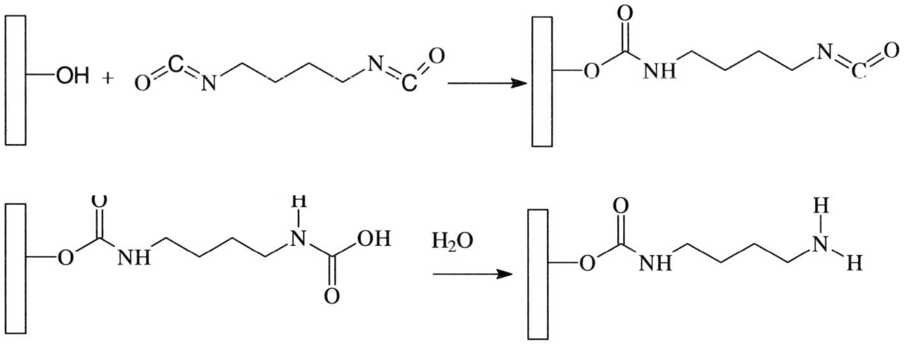

Scheme 3.89 Conversion into amino groups.

Conversion into Aldehyde Groups Silica based carriers can be also directly converted to aldehyde carriers with the use of triethoxysilylacetal. The aldehyde groups can be released by acid hydrolysis. With this type of carrier, ribonuclease A was immobilized, followed by reductive amination using NaBHCN [392].

In contrast to the widely used method of glutaraldehyde activated silanized carrier for covalent enzyme immobilization, this method necessitates the use of reducing agents, in order to stabilize the binding, because of the instability of the Schiff base linkage.

3.6.3.2 Activation of Separate Hydroxyl Groups

Polymeric carriers bearing separate hydroxyl groups can be converted into carriers bearing several activated species suitable for coupling to nucleophilic residues, as shown in Scheme 3.90. They can, for example, be activated as mesylates (A′–G) and tosylates (A′–H) but also as succinimido- or imidazolyl-carbonates, or as *p*-nitrophenylformates (A–C). All of these activated residues react readily with primary amines, leading to secondary amines from tosylates and mesylates or to stable carbamate bonds otherwise. Several other new types of active functionality, for example epoxy ring (A′–N), diazonium salt, triazine oxirane, *p*-benzoquinone, carboxyl groups, and acyl azide, etc., can be also introduced, as illustrated in Scheme 3.90.

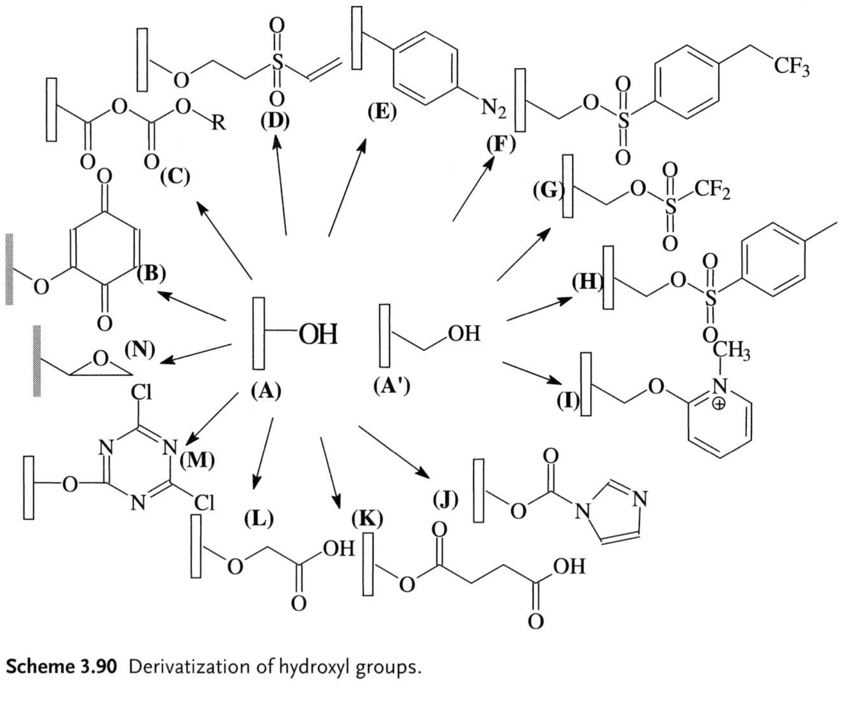

Scheme 3.90 Derivatization of hydroxyl groups.

Activated as Mesylates and Tosylates Carriers bearing primary hydroxyl can be converted into activated species such as mesylates and tosylates. Use of tosylated or mesylated polymers was pioneered by Nilsson and Mosbach, who immobilized peroxidase and alcohol dehydrogenase on Sepharose with activity retention 8% and 23% and payload 0.084 and 0.112, respectively [393, 394]. The reaction had been known for a long time before it was implemented in this coupling reaction for enzyme immobilization [395].

Tosylated cotton has been used to immobilize *Aspergillus oryzae* β-galactosidase, leading to immobilization of the enzyme with high loading, activity retention, and particularly high storage stability compared with the native enzyme [396].

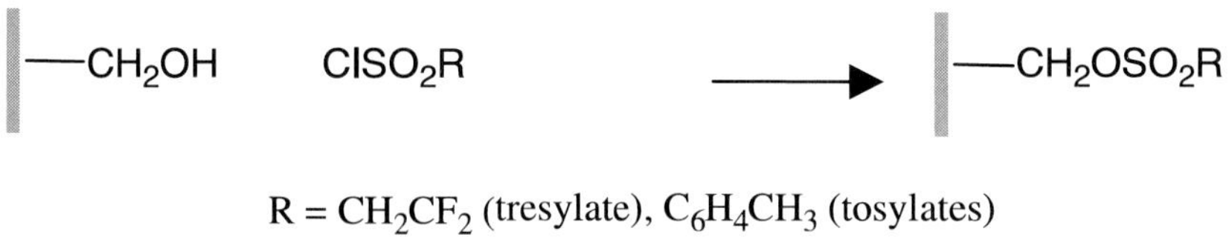

$$\text{—CH}_2\text{OH} \quad \text{ClSO}_2\text{R} \longrightarrow \text{—CH}_2\text{OSO}_2\text{R}$$

$$R = CH_2CF_2 \text{ (tresylate)}, C_6H_4CH_3 \text{ (tosylates)}$$

Scheme 3.91 Preparation of carriers containing tresylate and tosylate functionality.

Horse liver alcohol dehydrogenase has been immobilized on glycerylpropyl-silica (10 μm, 1000-Å pores) activated with 2,2,2-trifluoroethanesulphonyl chloride (tresyl chloride) with almost 100% coupling and activity yield [397].

Micrococcal endonuclease has been coupled to *p*-toluenesulphonyl chloride-activated agarose gel with loading 8 μg mL^{-1}, 100% coupling yield, and 50% retention of activity [398]. CRL has been immobilized on tosylated agarose with 24–82% retention of activity and 0.009 to 0.035 payload, depending on the pH used for coupling [399]. For lipase immobilized on tosylated silica in organic solvents synthetic activity in organic solvent was much higher than for lipase immobilized in aqueous medium [6].

Activation as Oxirane Polymers Activation with difunctional cross-linkers (homogeneous or heterogeneous bifunctional cross-linkers) will lead to the formation of new functional carriers. For example, use of bisoxirane compounds for activation of carrier with hydroxyl groups will lead to the formation of a carrier with oxirane groups (Scheme 3.92). Treatment of the synthetic copolymer of 2-hydroxyethylacrylate with epichlorohydrin also leads to the formation of a carrier with epoxy groups.

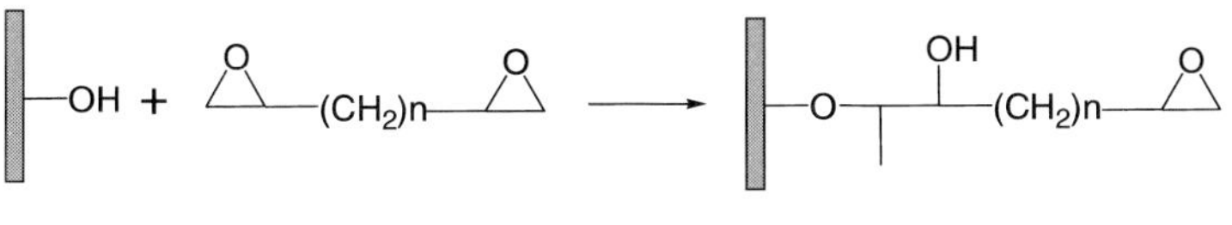

Scheme 3.92 Modification of carriers containing hydroxyl groups.

Cyanuric Chloride Method Another method of activation used for coupling of hydroxyl functions to primary amines is by use of trichlorotriazine (cyanuric chloride). This reagent, being tri-functional, can result in cross-linking and does not always lead to well-defined conjugates. It can, however, be used to activate synthetic polymers bearing hydroxyl groups, for example copolymers of 2-hydroxyethylacrylate [400, 401] or Duolite A7 [402] or natural polymers such as dextran [403].

Hydrolysed macroporous XAD-7 has been used to immobilize pectinlyase. Compared with the synthetic carrier Eupergit C the activity was approximately six times higher [404].

Immobilization of enzymes on synthetic carriers bearing hydroxyl groups can be achieved by activation of a carrier bearing hydroxyl groups then use of interconversion techniques to produce other active functionality. For example, nonporous poly(HEMA-co-EDMA) microspheres activated with 2,4,6-trichloro-1,3,5-triazine can be converted into carriers bearing adipohydrazide, which can bind oxidized enzymes bearing aldehyde functionality [405].

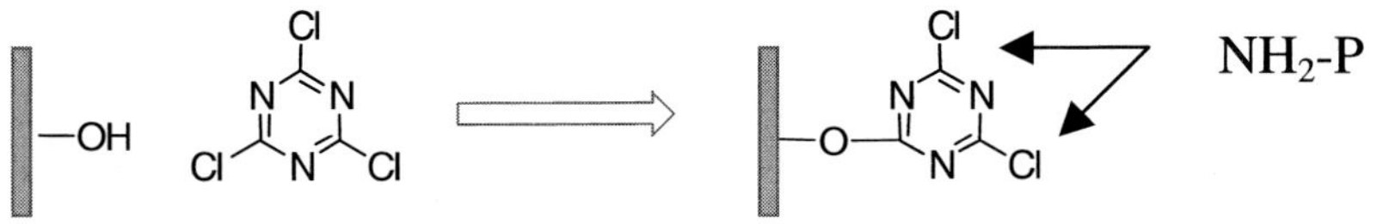

Scheme 3.93 Activation of hydroxyl groups with triazine.

It has been reported that this reagent lacks selectivity and can react with nucleophiles other than the lysines or primary amines such as tyrosine residues. This behaviour, and the potential toxicity of the reagent, explain why it is now used only for conjugation of enzymes designed for bioconversion in organic solvents and is almost ignored for drug modification.

Divinylsulphone Method Divinylsulphone was introduced by Porath and Sunberg in 1972 for the activation of Sepharose for preparation of affinity adsorbent [406].

In an extended work, α-chymotrypsin was immobilized by covalent attachment to activated Sepharose with high enzyme loading, on the basis of the dry weight and 41% retention of activity [406].

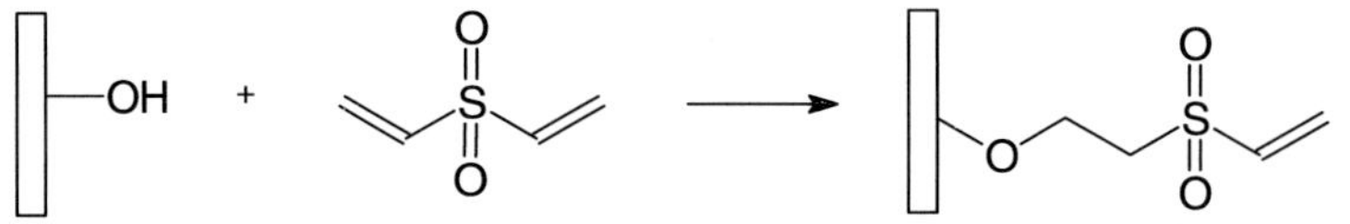

Scheme 3.94 Activation of carrier containing hydroxyl groups with divinylsulphone.

Benzylquinone Method Activation of polymeric carriers bearing hydroxyl groups by use of benzoquinone can be dated back to the mid 1970s. The use of this technique enables the preparation of active carriers which can bind enzymes in a broad pH range with higher retention of activity [407].

Natural polymers or synthetic polymers bearing hydroxyl groups, for example poly(hydroxyalkyl methacrylate) gel, PVA beads, or partially hydrolysed copolymer, i.e. poly(vinyl-co-divinylbenzene), can be activated by this method [114].

Activation of the hydroxyl groups of Spheron 300 with benzoquinone at pH 8 led to formation of an activated carrier with maximum binding capacity at pH 7–8. Enzyme activity after activation of the Spheron 300 carrier with benzoquinone was twice that after activation with BrCN, although the latter could bind five times more enzyme [408]. This suggests that the nature of the linkage plays a very important role in determining enzyme activity retention and other characteristics such as stability. Because of the insolubility of benzoquinone in aqueous medium, activation of the carrier often necessitates the presence of an organic solvent such as ethanol or dioxane.

This method was also recently used to activate hydrolysed synthetic polymer poly(vinyl-co-divinylbenzene) for immobilization of penicillin G acylase. Remarkably, not only high specific activity (almost twice as high as for the diazonium salt and 30% higher than for the glutaraldehyde method) but also high retention of activity was obtained. The operational stability, however, seemed lower than for use of the other methods [114].

Activation of poly(hydroxyethyl methacrylate)-grafted pectins with *p*-benzoquinone then coupling of diaminohexane to the copolymer has been used to immobilize trypsin with 0.369 payload when glutaraldehyde was used for coupling [409]. Among the different carriers investigated, for example CNBr-activated Sepharose 4B, PBQm-activated Sepharose 4 B, CPG-COOH or CPG-NH$_2$ (with carbodimide as activating agent for coupling), and TiCl$_4$-activated silica, it was found that wheat shoot phosphotransferase (WSP) immobilized on benzoquinone-activated Sepharose gave the best result [410].

Poly(maleic anhydride-vinyl acetate-ethylene), hydrolysed to release the hydroxyl group from the poly(vinyl acetate) component of the grafted chains, has been used to immobilize enzymes via tosylation (by use of *p*-toluenesulphonyl chloride) and the *p*-benzoquinone method. It was found that the *p*-benzoquinone method was suitable for coupling BSA and acid phosphatase whereas the tosylation method was very effective for immobilizing trypsin [247].

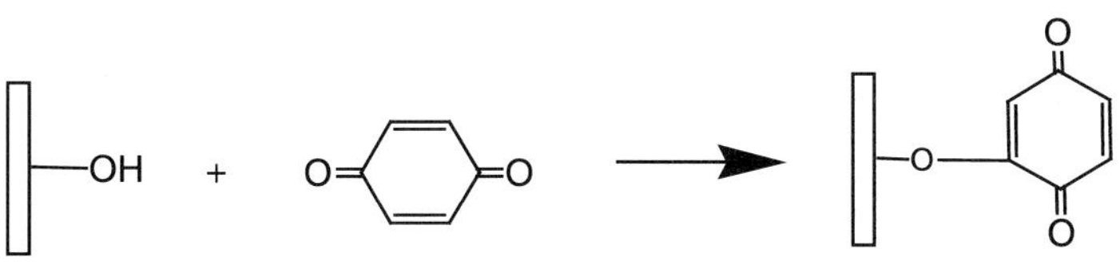

Scheme 3.95 Activation of hydroxyl groups for immobilization of penicillin G acylase.

Activation with Chloroformate Hydroxyl groups on carriers can be activated with chloroformate, resulting in the formation of activated carrier, as illustrated in Scheme 3.96.

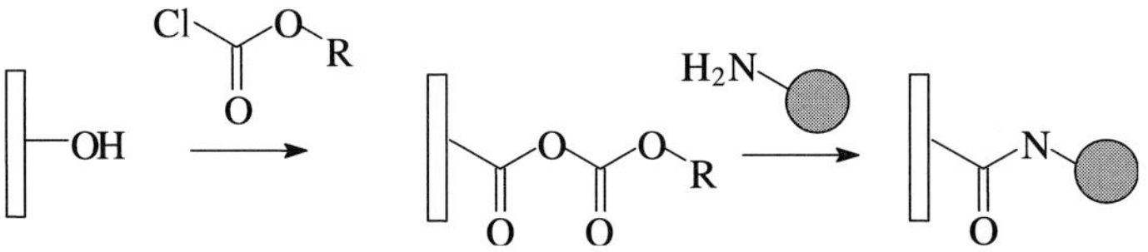

Scheme 3.96 Activation of hydroxyl groups with chloroformate.

For example, trisacryl GF 2000, a new synthetic gel support can be activated with N-hydroxysuccinimide, chloroformate and p-nitrophenyl chloroformate in organic solvents, providing the best activation yield and subsequent coupling [411].

Activation with Carbonyldiimidazole Hydroxyl groups can be easily activated with carbonyldiimidazole, leading to the formation of activated carrier which can react with the amino groups of the enzymes, as shown in Scheme 3.97.

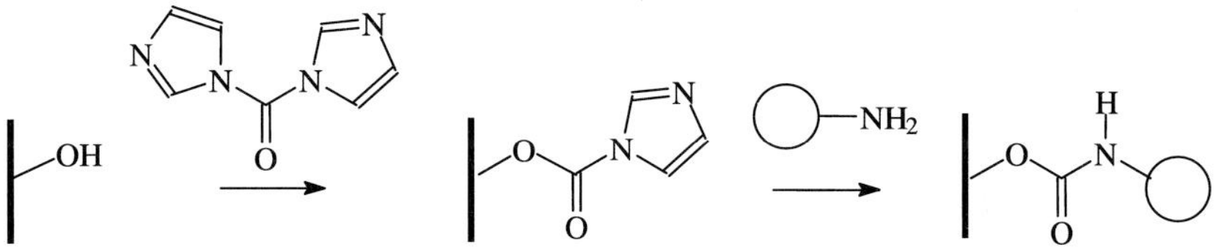

Scheme 3.97 Activation of hydroxyl groups with carbonyldiimidazole.

It has been shown that GAH-cross-linked PVA microspheres activated with carbonyldiimidazole can bind invertase with 74% retention of the native activity. The thermostability and storage stability of the immobilized enzyme were, moreover, better than for the native enzyme. The extent of thermostabilization depends on the temperature, varying from 1.5–2.5-fold in the range 50–70 °C [412].

Watermelon β-galactosidase immobilized on chitin activated with 1,1′-carbonyldiimidazole displayed better storage stability than that immobilized on Amberlite IRA-938 [413].

3.6.3.3 Polymers Containing Carboxylic or Ester Groups

Although direct use of carboxylic groups for enzyme immobilization is possible in the presence of a carbodiimide, use of this technique is rare, because careful control of the coupling conditions such as pH, ionic strength are often requested, in order to ensure higher coupling yield [414]. Thus, they are often converted into other functionality that is easier to handle.

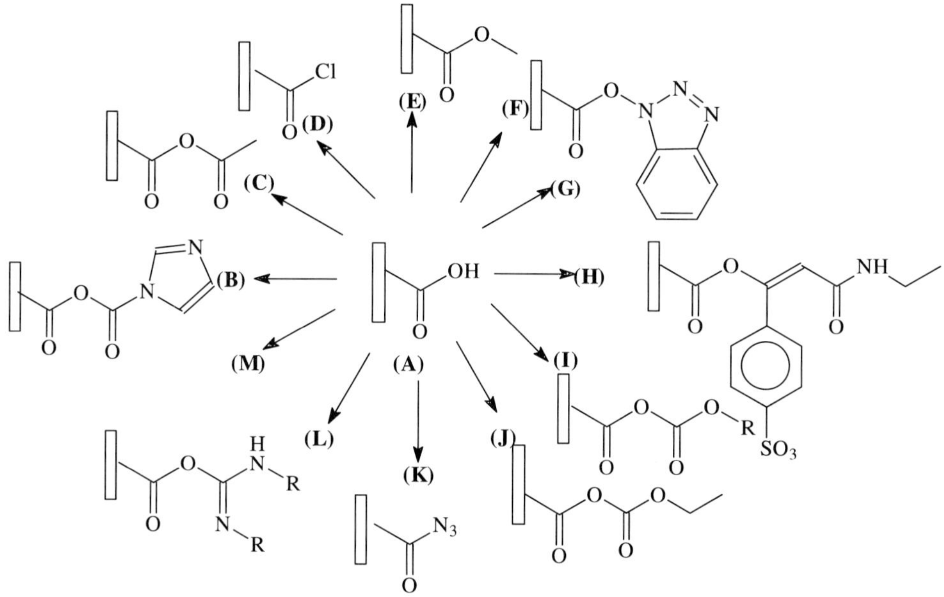

Scheme 3.98 Activation of carriers containing carboxyl groups: (AB) activation with carbonyldiimidazole, (AC) mixed anhydride, (AD) thionyl chloride activation, (AE) esterification, (AF) activation with N-hydroxybenzotriazole, (AH) activation with Woodward reagent K, (AI) activation with ethyl 2-ethoxy-1,2-dihydro-1-quinolinecarboxylate, (AJ) activation with chlorocarbonate, (AK) acyl azide via three steps, i.e. esterification and further reaction with hydrazine and nitric acid, (AL) activation with various carbodiimides.

As illustrated in Scheme 3.98, carboxyl groups can be converted into numerous new groups such as activated (N-hydroxysuccinimidyl) esters, amino groups, mixed anhydride, and chloride, etc.

Activated Esters The most common method of activation for carboxyl functions (Scheme 3.99) is via their N-hydroxysuccinimidyl esters, by use of N-hydroxysuccinimide (NHS) and a carbodiimide. These active esters are suitable for coupling primary amines and, less frequently, hydroxyl functionality, or sulfhydryl, leading to stable amides in the former instance and to hydrolytically unstable esters in the latter [415, 416].

Carboxyl groups can also be coupled directly to amines without NHS, with use of the condensation agents such as carbodiimides. Either water-soluble or organic-

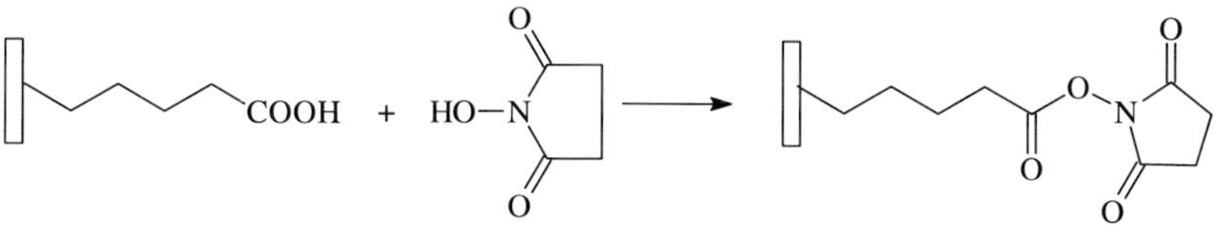

Scheme 3.99 Activation of carboxyl groups.

soluble carbodiimides can be used, depending on the reaction environment or the solubility of the reagents. Low pH (about 4.5) is usually required for good coupling.

Activation with Woodward Reagent and Water-soluble Carbodiimide Polymeric carriers with carboxyl groups can be activated by the use of Woodward's reagent K. This method was pioneered by Patel and co-workers in 1967 [417]. It has been found that the enzyme payload depends on the surface area of the carriers whereas activity retention depends on the texture of the carrier [418].

Like other types of covalent enzyme immobilization, the presence of substrate in the immobilization medium has proved to be essential, regarding the higher activity retention. For instance, coupling of pullulanase in the presence of substrate-pullulan to crosslinked poly (acrylamide-co-acrylic acid) (Biogel CM-100) led to 5-fold more activity than that without substrate involved in the coupling [419].

This type of polymer is often prepared by polymerization of acrylamide with monomers bearing carboxyl. Before binding of the enzyme, the carriers must be activated by use of the activators mentioned above [421–423]. Many enzymes have been immobilized by use of these methods; for example, aldolase, aminoacylase, arginase, carboxypeptidase B, cholinesterase, cyclodextrin glycosyltransferase and glucoamylase have been immobilized by covalent coupling on polyacrylamide bead polymers with carboxyl functional groups activated by water-soluble carbodiimides [420]. D-galacturonan digalacturonohydrolase has been immobilized on Bio-Gel CM 100 activated by water-soluble carbodiimide [424].

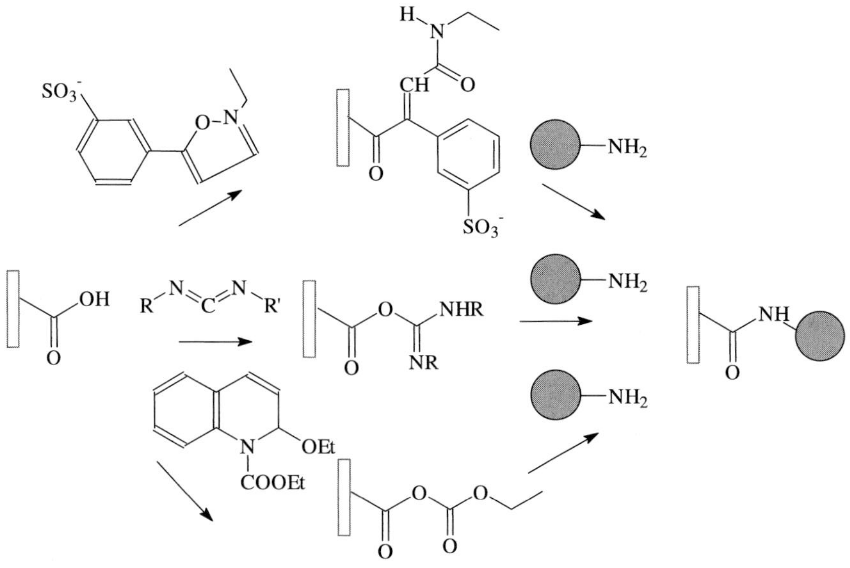

Scheme 3.100 Activation of carboxyl groups with carbodiimide and Woodward's reagent K.

Remarkably, it has been found that PPL immobilized on polyacrylamide beads with carboxyl functional groups activated by a water-soluble carbodiimide had higher volume activity (2,187 U g^{-1} solid) than the native enzyme, and that immobilization also stabilized the enzyme against heat and urea treatment. Cross-linking of the immobilized enzyme with glutaraldehyde or 3,5-difluoronitrobenzene further improved the thermal stability [425].

Similarly, polyethylene-g-AAc has been used to immobilize bovine serum albumin with 0.065 payload in the presence of ethyl 2-ethoxy-1,2-dihydro-1-quinolinecarboxylate as activating agent of the carboxylic groups [426].

Although a peptide bond is always formed between the enzyme and the carrier, enzyme performance – activity, activity retention and stability – were found to be largely dependent on the nature of the activation agents, suggesting that the leaving groups might also affect the binding mode of the enzyme, for example binding position or number of bonds , as validated by the observation of Szajani et al. who found that the loading, activity retention and the performance of the immobilized enzymes are largely influenced by the nature of the CDI used for the activation of the carboxylic groups of the carriers [419, 420].

More interestingly, it was found that formation of electrostatic complexes between the carrier and the enzyme before immobilizing chymotrypsin on synthetic copolymers of acrylamide and acrylic acid, by use of water-soluble carbodiimide techniques, was essential for high retention of activity and high thermostability, suggesting that the mode of interaction of the enzyme with the carrier (for example orientation) depends on ionic strength and pH [427, 428].

Although these two methods have not been widely used for enzyme immobilization, CDI (carbodiimide) is the most popular agent for activation of the carboxyl groups of the carrier or enzyme in enzyme immobilization.

In the activation of the carboxyl groups of proteins before conjugation with amino-carrying polymers, the risk of reaction with the amino groups of the same protein, leading to intra molecular cross-linking, cannot be overlooked. An original procedure for avoiding this problem is use of a water-soluble carbodiimide to activate the carboxyl functions in the protein at pH 4–5, in the presence of a hydrazide-containing polymer. Under these conditions only the polymeric hydrazide residues, which have a very low pK_a, will react whereas the primary amines will remain unaffected.

Immobilization of enzymes such as glucose isomerase, urease, glucamylase, trypsin and glucose oxidase to chitosan bearing amino groups with the aid of a water-soluble carbodiimide led to different activity yields: I, 32%; II, 44%; III, 8%; IV, 10%; and V, 37% [429].

In contrast, it was found that immobilization of carbodiimide activated enzyme on alkyl or arylamino groups at the end of longer substituent-bearing carriers preserved most of the activity [125]. It is very likely steric hindrance plays an important role in activity retention.

Conversion to Aldehyde Functionality Along with the activated pendant functional groups that can be used for direct coupling of the enzyme (Scheme 3.13), carboxyl groups can be converted into other active pendant groups which are more suitable for immobilization of some enzymes. For instance, carboxyl groups can be easily converted into terminal primary amino groups ready for enzyme immobilization via the glutaraldehyde or carbodiimide methods. To do this the carboxylic acid can be first converted to an ester; aminolysis of the ester with a diaminoalkane then leads to the formation of the desired aminolated carriers, as illustrated in Scheme 3.101. By use of this methodology penicillin G acylase was immobilized on commercial XAD-7 with ester groups which were further derivatized by use of diamino-compounds such as 1,4-diaminobutane. The pendant amino groups were further activated by use of glutaraldehyde for immobilization of the enzyme [350], in accordance with Scheme 3.101.

After conversion of the carboxylic group into an amino group, there are two possibilities for coupling the enzyme to carriers bearing amino groups, namely coupling the enzyme to the aminolated carrier in the presence of a carbodiimide compound or activating the carrier by glutaraldehyde [260].

Remarkably, although the non-converted parent carrier bearing carboxylic groups and the converted carrier bearing amino groups can both be directly used for enzyme immobilization in the presence of a suitable carbodiimide compound as activating agent, an interesting study revealed that α-amylase immobilized on carboxymethylated cellulose displayed higher activity than that immobilized on AE (aminoethyl) cellulose, although they were both prepared in the presence of a carbodiimide [250].

This might suggest that the carboxylic residues of amylase are not situated such that attachment through them can stabilize the enzyme. In α-amylase the number of lysine residues is significantly higher than the number of Glu residues. Besides, the 3D structure showed that the distribution of highly exposed Lys residues is more uniform than of the highly exposed Glu residues. Glu residues are present on the surface of α-amylase mainly as cluster. Thus, higher stability of the enzyme immobilized via lysine residues can be expected [251].

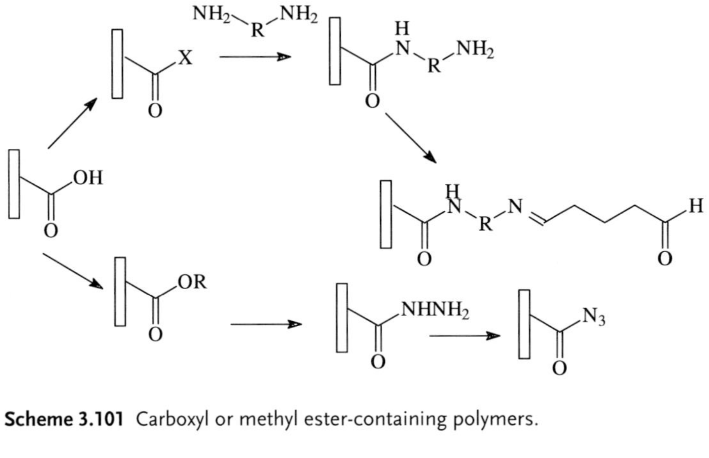

Scheme 3.101 Carboxyl or methyl ester-containing polymers.

Conversion to Acyl Azide Polymers such as cross-linked acrylic acid-1,4-divinyl-benzene-1-vinyl-2-pyrrolidinone copolymer can be converted to polymers bearing acyl azide functionality (see Scheme 3.101) [252].

Urokinase and trypsin has been immobilized on a collagen-synthetic polymer composite via activated acyl azide, formed from carboxyl groups; the catalytic properties were comparable with those of the native enzyme [240].

PAAm-γ-alginic acid activated by esterification and subsequent conversion to the acyl azide achieved payloads of 0.236 and 0.269 for papain and trypsin, respectively [241].

Synthetic polymers such as polyacrylamide (PAAm), or other synthetic polyacrylic polymers or composites bearing acylhydrazine moieties, can also be activated to give acyl azide side-chains for binding enzymes [242, 243].

Natural polymers such as CM-cellulose can be derivatized to acyl azide which can bind enzymes under certain conditions. Glucoamylase has been covalently linked to dialdehyde cellulose resulting in an immobilized enzyme with 0.0098 payload and 46% retention of activity whereas binding to acyl azide-CM-cellulose led to an immobilized enzyme with much lower retention of activity (10%). Obviously, the active centre was modified by the pendant acylazide groups [430], because protection of the active centre led to higher retention of activity [431].

Conversion to Amino Groups The conversion of carrier-bearing carboxyl groups into amino groups opened many possibilities for the coupling chemistry between the enzyme and the carrier. This is because amino groups can be converted into active groups such as aldehyde groups obtained by glutaraldehyde activation, s-triazine, benzoquinone, imidoester, etc [432].

3.6.3.4 Polymers Containing Amino Groups

Amino groups of the polymeric carriers can be used directly to bind enzymes activated by different carbodiimides. Often, however, amino groups must be activated by use of a bifunctional cross-linker for binding enzyme via the terminal functional group of the cross-linker. A frequently used method is the glutaraldehyde method, because of its stability in aqueous medium. Other methods used for activation of amino groups include the s-triazine method, and use of difunctional isocyanate, succinic anhydride, benzoquinone [408, 407], bifunctional diazonium salts, and cyanogen bromide, etc. [433]. The frequently activation methods are depicted in Scheme 3.102.

Zero Cross-linking Amine-containing polymers can be coupled directly with the carboxyl functionality of the enzyme in the presence of carbodiimide reagents (Scheme 3.103). In this reaction the enzyme is activated before immobilization. Coupling of activated enzyme (with CDI) to the amino bearing carriers proved to be advantageous because higher retention of activity was usually obtained, as compared with other methods [125].

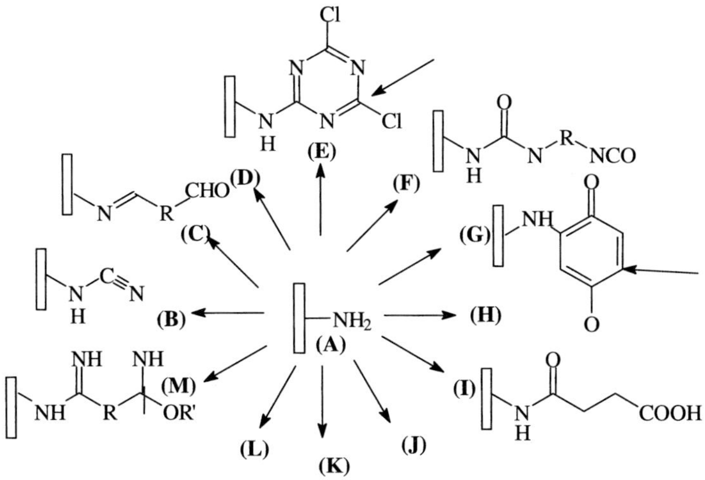

Scheme 3.102 Activation of carriers containing amino groups.

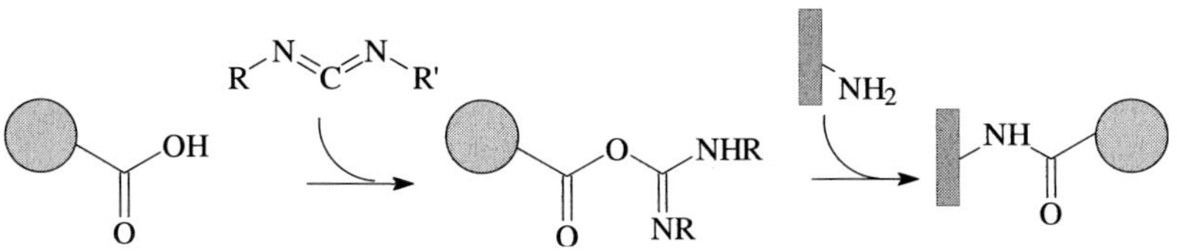

Scheme 3.103 Zero coupling of protein with carriers.

It has also been discovered that the β-glucosidase immobilized by peptide bond formation between carboxyl groups of the enzyme and amino groups of the carrier is about twice as stable as when bound via Schiff base formation and four times more stable than the free enzyme. Obviously, the underlying mechanism is the same as the previously mentioned example – because these enzymes are much richer in carboxyl groups than lysine residues, enhanced multipoint attachment via the carboxyl groups is likely to be the major reason for the enhanced stability [434].

Aldehyde Conversion of carriers carrying amino groups into aldehyde functionality has become one of the most popular methods of enzyme immobilization (Scheme 3.104), because the method is simple and the reagent glutaraldehyde is relatively cheap [45].

Currently, many types of carrier, for example natural polymer such as proteins, gelatine, and chitosan, and some synthetic polymers, for example Amberlite CG 50 [434], polyaminostyrene [435], polystyrene anion-exchange resin (Indion 48-R) [436], PAP IV (-NH$_2$), DAP III/IV (-NH-(CH$_2$)$_2$-NH$_2$), Lewait 258 K (-NH$_2$) [437], hydrolysed Nylon [438], and silanized silica bearing amino groups [10], or PEI-coat-

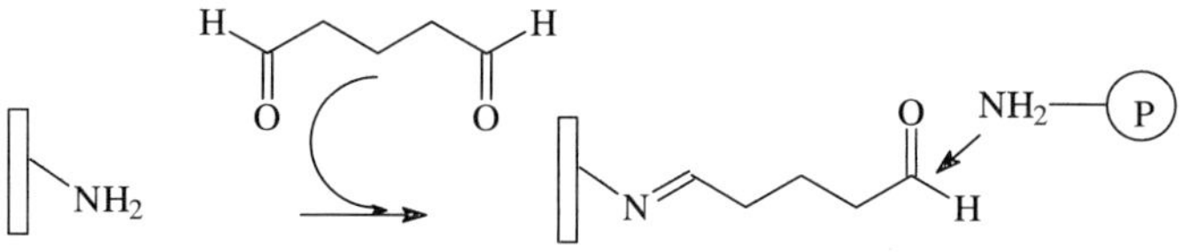

Scheme 3.104 Activation of amino groups with difunctional aldehyde.

ed macroporous beaded polymers [278] can be activated by dialdehyde, e.g. glutaraldehyde, to give aldehyde pendant groups.

Inorganic silica, after silanization with 3-aminopropyltriethoxylsilane then glutaraldehyde activation, has also been used for immobilization of enzymes, for example amino acid oxidase [439, 440].

Isocyanates/Thioisocyanates Some macroporous basic anion exchangers with amino functional groups, for example Diaion CR-20, have been used for enzyme immobilization after activation by diisocyanate (Scheme 3.105). The primary amino groups can be activated by use of hexamethylene diisocyanate [441] or 2,4-toluylenediisocynate [250]. Aromatic amines can be also activated as isocyanates or isothiocyanates by using phosgene or thiophosgene, respectively.

To avoid the hydrolysis of hexamethylene diisocyanate, the carrier is usually washed with DMF and the reaction performed in DMF. Remarkably, penicillin G acylase immobilized on Diaion CR 20 activated by hexamethylene diisocyanate can be used 300 times with 70% of the original activity [441].

The advantages of using diisocyanate-activated aminolated carrier have been demonstrated in an early work in the 1970s, when α-amylase was immobilized onto aerosol derivatives activated by different methods, either by direct coupling of this carrier with the enzyme or the enzyme with the carboxymethylated carrier in the presence of a carbodiimide, or activated by a diisocyanate compound. It was found that α-amylase immobilized onto amino aerosol activated by N,N′-dicyclohexylcarbodiimide gave the higher enzyme loading, activity retention and stability [182].

These activated carriers bearing active isocyanate groups can subsequently react with primary amines, leading to isoureas or isothioureas, respectively, and with hydroxyl-carrying molecules, leading to carbamate or thiocarbamate bonds. Isoureas can also be obtained by activation of the primary amine with disuccinimidylcarbonate and triethylamine, to give a succinimidyl carbamate, and further reaction of this activated reagent with the second amine-carrying molecule.

Scheme 3.105 Activation of amino group with diisocyanate.

Cross-linked polyethylenimine (PEI) can be activated with thiophosgene to introduce thioisocyanate for binding several enzymes, for example glucose oxidase, glucoamylase, acetylcholinesterase, and butyrylcholinesterase with high retention of activity [442].

s-Triazine Chloro-s-triazine was introduced by Kay and Crrok in 1967 for activation of hydroxyl-carrying carriers, for example polysaccharide-based carriers, for enzyme immobilization [443]. Soon after, use of chloro-s-triazines was extended to activation of amino-bearing carriers for enzyme immobilization [444]. The mechanism of enzyme immobilization on amino carriers activated with s-triazine is similar to that of hydroxyl carriers, as illustrated in Scheme 3.106.

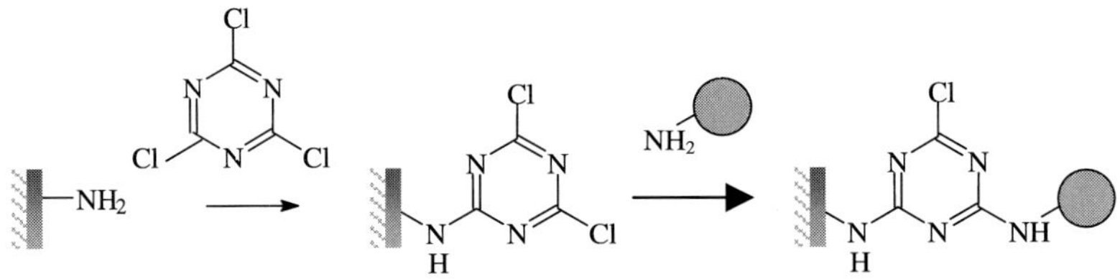

Scheme 3.106 Immobilization of enzymes on amino carriers activated with s-triazine.

Compared with enzyme immobilization on s-triazine-activated hydroxyl carriers, however, there are few reports dealing with comparison of the performance of enzymes immobilized on these s-triazine-activated amino carriers. Interestingly, it has been reported that invertase immobilized on aminopolystyrene by the triazine method was less thermally stable than the soluble enzyme [444]. Remarkably, it has been found that enzymes immobilized on chlorotriazine-activated carriers are more resistant to hydrolysis than those activated with CNBr.

Acid oxidation of polyethylene beads generated surface -CO_2H groups which react with excess ethylenediamine via carbodiimide-promoted reactions. Glucose oxidase has been covalently immobilized on amine-substituted beads by using glutaraldehyde, triazine trichloride, or dimethylsuberimidate as cross-linkers. The kinetic properties, pH-activity profile, and stability of the immobilized enzymes were reported [445].

Cyanogen Bromide Activation Similar to the activation of hydroxyl groups, amino carriers can be also activated with cyanogen bromide [446], leading to the formation of a guanidine linkage, as shown in Scheme 3.107. Although there are few examples on the use of this method for enzyme immobilization, several examples suggest that enzymes immobilized via guanidine linkage are often much more stable than those immobilized by use of other types of linkage. For example, among a variety of carriers (underivatized polystyrene beads, cyanogen bromide-activated Sepharose, aminopropyl controlled-pore glass, oxirane acrylic beads, aminoethyl biogel, and polymertrager VA epoxy Biosynth), it was found that alkyl

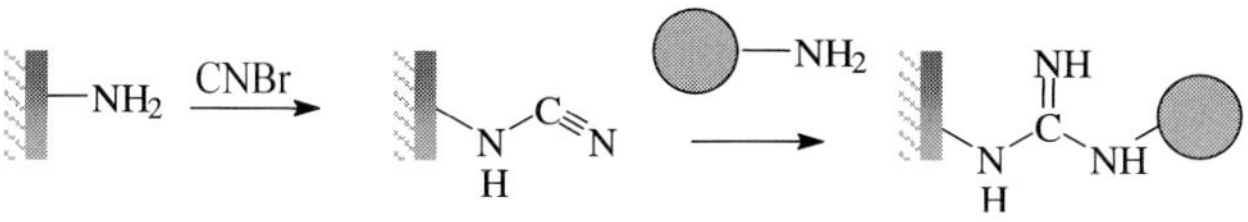

Scheme 3.107 Immobilization of enzymes via guanidine linkage.

phosphatase immobilized on cyanogen bromide activated aminopropyl controlled-pore glass was by far the most stable [447].

It was also wonderfully demonstrated recently that glucoamylase immobilized by guanidine linkage via a spacer-arm attached to magnetic poly(methyl methacrylate) microspheres was not only 40% more stable than that attached by amide linkage but activity retention was almost 50% higher, even though the carrier and spacer used were the same for both (Scheme 3.108). The pH profile of the immobilized glucoamylase was not significantly different from that of the native enzyme [181].

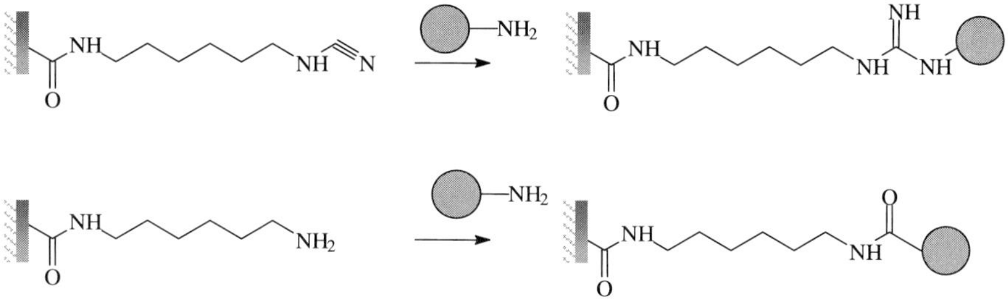

Scheme 3.108 Immobilization of enzymes on amino carriers by use of the CDI or CNBr methods [181].

Conversion to Carboxyl Groups As discussed above, immobilization of enzymes on a carrier bearing amino groups by peptide formation can be realized by attaching the activated enzyme with carbodiimide methods [372]. Although it has been observed that immobilization of enzymes on to the carriers bearing terminal amino groups at the end of a long spacer results in better retention of activity [125], other studies have revealed that activation of carriers bearing carboxylic functionality, then attachment of the enzyme through the amino group to the activated carrier bearing carboxylic groups led to formation of more stable immobilized enzymes. Thus, it is necessary to convert the amino functionality into carboxyl groups. The method developed by Cuatrecasas and Parikh can be used, as shown in Scheme 3.109 [448].

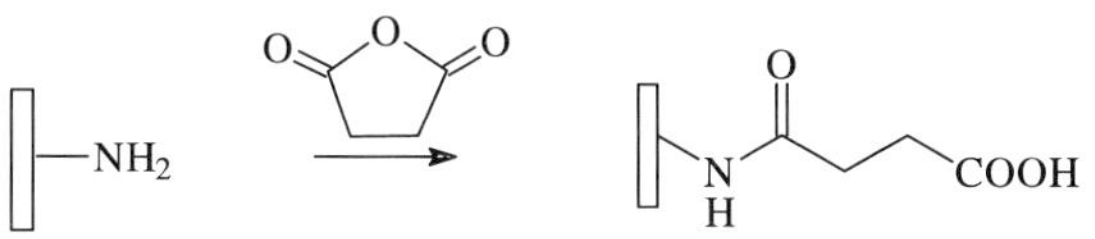

Scheme 3.109 Conversion of an amine support to a carboxylic acid support.

Activation with Benzoquinone Amino-carrying carriers such as AH-Sepharose 4B activated with 1,4-benzoquinone and 1,2-naphthoquinone has been used to immobilize trypsin; the native carrier has no capacity for physical adsorption [433].

There are many examples showing that use of benzoquinone-activated aminolated carriers for enzyme immobilization is advantageous compared with other carriers or methods of activation [449, 450, 452], with regard to the performance, for example activity and stability, of the immobilized enzymes obtained.

In contrast, there are also examples of benzoquinone-coupled carrier-bound immobilized enzymes such as α-amylase with very poor binding stability [453] or immobilized enzymes with lower thermal stability than the soluble enzyme, depending on the properties of the carriers used [451]; this suggests the properties of the immobilized enzymes are dictated by many factors e.g. the binding properties and the nature of the carriers, as discussed in the introduction.

The presence of a spacer between the polymeric matrix and p-benzoquinone can enhance enzyme loading and retention of activity [454].

Comparison of different compounds such as glutaric dialdehyde, benzoquinone, and s-trichlorotriazine for activation of porous glass beads for immobilization of α-amylase revealed that payloads were always quite similar but that the activity of the glutaric dialdehyde-linked enzyme was lower than that of the benzoquinone and s-trichlorotriazine-linked preparations, implying that the binding chemistry or the microenvironment of the carrier largely dictated the performance of the immobilized enzymes, as shown in Scheme 3.110 [455].

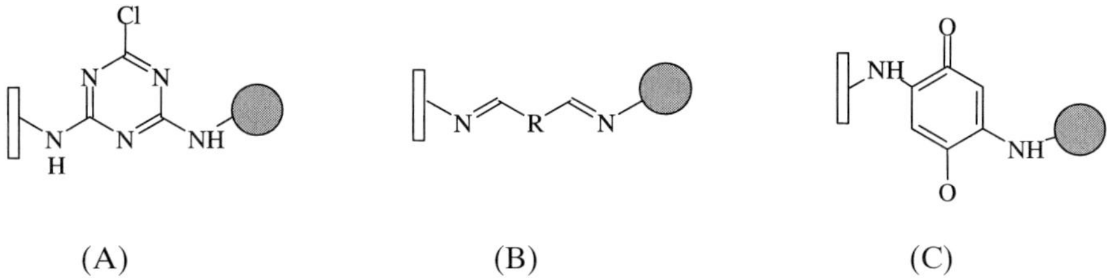

Scheme 3.110 Immobilization of a-amylase on amino silica activated with (A) cyanuric chloride, (B) glutaraldehyde, and (C) p-benzoquinone.

Conversion to Imidoester Although difunctional imidoesters are often used to cross-link the enzymes, with the aim of increasing the stability, the use of difunctional imidoesters for activation of amine carriers is not well recognized. A few examples have, however, shown that enzymes coupled to carriers via diimidate esters usually had higher activity and stability [456]. For instance, protein kinase CK2 and calf intestine alkaline phosphatase CIP immobilized on a resin bearing secondary amino and thiol groups via the imidoester di-Me pimelimidate hydrochloride had constant storage stability and activity for at least 3 months [456].

3.6.3.5 Polymers Containing Amide Groups

Polyacrylamide is one of the most popular synthetic carriers with amide functionality. It is usually prepared by polymerization of acrylamide in the presence of N,N'-methylenedipropyleneamide as cross-linker. Variation of the ratio of the cross-linker leads to the formation of matrix with controlled pore size.

Advantages of polyacrylamides are ease of preparation, excellent stability over a broad pH range (1–10), and non-biodegradability (thus it is resistant against microbial attack). Polyacrylamides are, however, usually of low mechanical stability. For enzyme immobilization they can be used as entrapment matrix or activated by glutaraldehyde or hydrazine or benzoquinone. A similar polyacrylamide gel is trisacryl, prepared from acroyl-tris.

Compared with PAAm (polyacrylamide), trisacryl has high water-retention capacity, chemical stability, and high mechanical strength. It can also be easily functionalized for different applications. For enzyme immobilization, however, less information is available, although the amide functionalities can be activated into other functionality such as isocyanate, acyl azide, aldehyde, amino, and triaznyl as illustrated in Scheme 3.111.

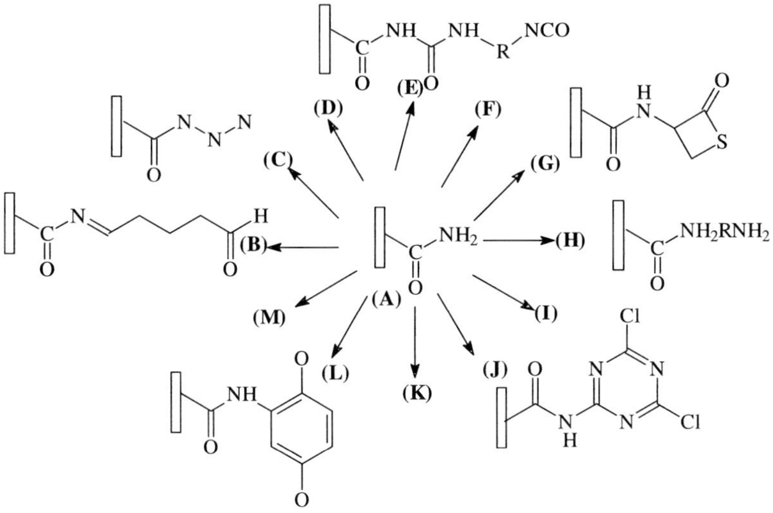

Scheme 3.111 Polymers with amide functionality.

Activation to Acyl Azide Conversion of amide to acylazide functionality has been a popular method for activation of polyacrylamide since the end of the 1960s. The reaction requires two steps, initial reaction with hydrazine then reaction with HNO_2, resulting in the formation of active azide which can be directly used to bind enzymes.

Ribonuclease T1 has been immobilized on a water-insoluble cross-linked polyacrylamide (Enzacryl AH) bearing acid azide with 45% and 77% retention of the original activity toward yeast RNA and 2′,3-cyclic GMP, respectively, as substrates.

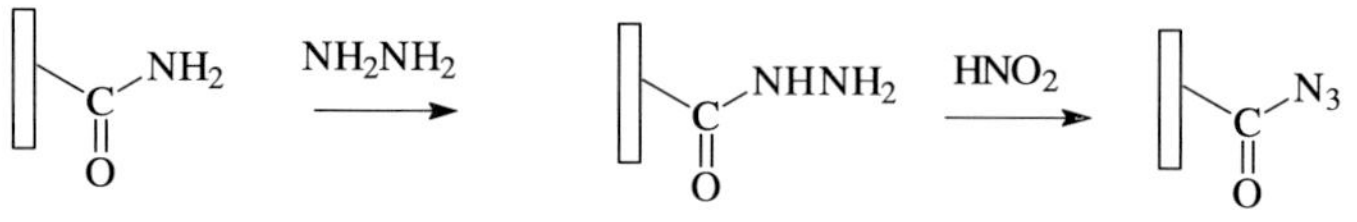

Scheme 3.112 Activation of polyacrylamide to polyacylazide.

The immobilized enzyme was found to be far more stable to heat and extremes of pH than the native enzyme [432].

β-Galactosidase and glucose oxidase have been immobilized on cellulose-polyacrylamide (C-PAM) graft copolymers, using the azide method or by use of glutaraldehyde. Binding of β-galactoside was more successful than binding of glucose oxidase. A relationship between the level of immobilized enzyme activity and the extent of grafting in the copolymer has been established for the β-galactosidase system [457].

In general, enzymes coupled to polymers bearing active acyl azide groups often display lower activity than other covalent coupling. For instance, thermolysin immobilized to polyamide nonwovens via azide coupling displayed lower activity than other coupling methods such as glutaraldehyde and adipimidate [458].

Activation to Amine with Diamino Compounds Conversion of amide into more active amine with the aim of exploring other types of binding chemistry was pioneered by Axen and co-workers in 1971, when aldehyde agarose (RCOR) and other polymers such as aminoethylated polyacrylamide, (p-$CH_2CH_2NH_2$), poly(methyl methacrylate, -CH_2-CH_2NH_2), and a cross-linked carboxymethylated dextran (CH_2COOH) were used to immobilize enzyme by means of the Ugi reaction [72].

The conversion can be realized by reacting polymeric amide with a diaminoalkane, resulting in the formation of a carrier with terminal primary amino groups as illustrated in Scheme 3.113. The resulting amino groups can be used to immobilize enzymes by means of the chemistry generally applicable to carriers bearing amino functionality [459].

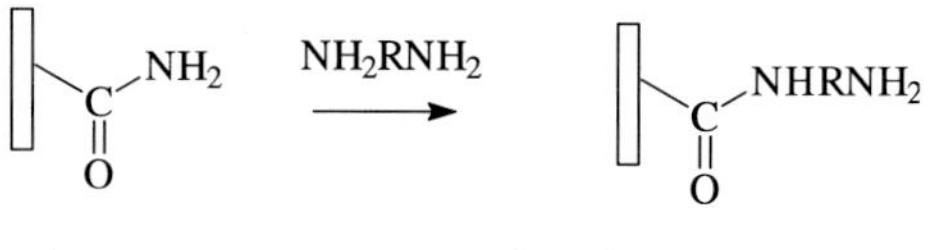

Scheme 3.113 Conversion of amide to primary amine.

Polyamide carriers such as Nylon 66 can be hydrolysed. Thus, the original amino groups and the carboxyl groups that form the amide bond can be partially released. The released amino or carboxylic groups can be used to bind enzyme by means of the chemistry generally applicable to carriers bearing carboxylic groups or amino groups as mentioned above. Besides, the partially hydrolysed nylon bearing free amino groups can be used to bind the oxidized glycoenzymes bearing aldehyde groups [481].

Activation with Bifunctional Aldehyde Activation of amide functionality with glutaraldehyde can be dated back to 1970s [460]. Enzymes can be directly immobilized on carriers bearing the amide function, after activation of the amide functionality with glutaraldehyde [423]. Alternatively, amide functionality can be first converted into acylhydrazine which can react with a protected glutaraldehyde with one terminal aldehyde functionality, resulting in the formation of a polymeric carrier with protected aldehyde functionality.

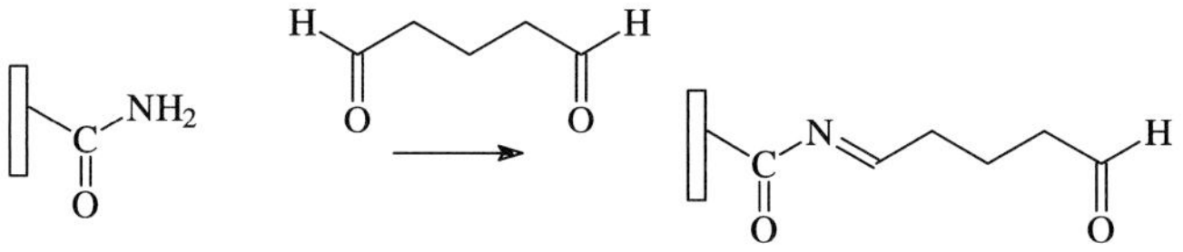

Scheme 3.114 Activation of amide to aldehyde functionality.

Subsequent hydrolysis will convert the protected aldehyde to the free aldehyde groups for coupling enzymes. It has been found that protecting the aldehyde group during the linking reaction increased the activity of the immobilized enzyme [461].

Activation with Triazine

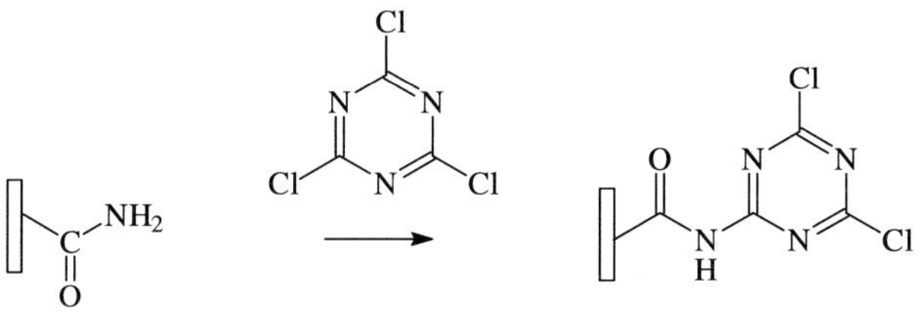

Scheme 3.115 Activation of amide to triazyl.

Activation with p-Benzoquinone Polyacrylamide gel such as Akrilex P-100 xerogel (an acrylamide-N,N'-methylene-bis(acrylamide) copolymer) can be activated with p-benzoquinone in buffer containing dioxane. The activated polymer can be used to bind enzymes or proteins such as bovine serum albumin, glucose oxidase, and catalase [462].

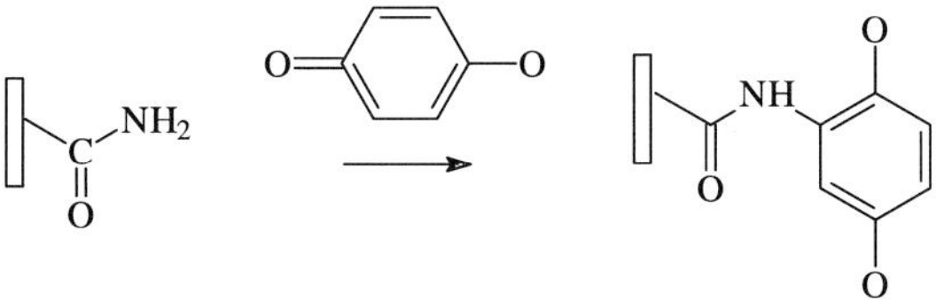

Scheme 3.116 Activation of polyacrylamide by p-benzoquinone.

3.6.3.6 Polymers Containing Nitrile Groups

Nitrile groups on polymers can be converted into several other new functional groups, for example amide, carboxyl, imidoester, amino groups etc., as shown in Scheme 3.117.

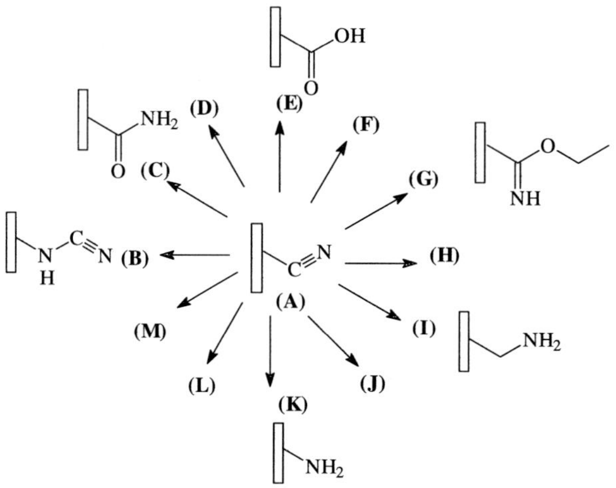

Scheme 3.117 Activation of nitrile by different methods.

Activation to Imidoester Polyacrylonitrile can be converted into imidoester-containing polymers for enzyme immobilization, according to the method shown in Scheme 3.117 [359]. The coupling reaction is usually performed at high pH to prevent deactivation of the active functionality. Advantages of using this coupling method are that the net charge of the enzyme will not be changed. Using this method Zaborsky immobilized several enzymes including trypsin and α-chymotrypsin. It was found that the immobilized enzyme obtained has high retention of activity, enhanced stability against lyophilization, and high storage stability (no loss of activity after storage for 1 year at 5 °C in 0.1 M phosphate buffer, pH 6.7) [463].

In a recent example, it was shown that α-amylase (AA) from *Bacillus subtilis* immobilized on alternating acrylonitrile–butadiene copolymer (ABC) by amidination after imidoesterification of the resin had higher operational stability (it was re-used more than ten times without significant loss of activity) [464].

Conversion to Acyl Azide Poly(acrylonitrile) or acrylonitrile copolymers can be converted into polymers bearing hydrazine and NH_2 groups which can subsequently be activated by other methods to different functionality. By use of this method polyacrylonitrile was diazotized for coupling of trypsin, leading to 52% and 4% retention of activity for the small amido substrates and casein. The immobilized enzyme also retained 80% of its activity after storage for 3 months at 4 °C [360].

By means of azido transfer with di-Ph phosphoryl azide, PAN was converted into acylazide for coupling amyloglucosidase to the PAN membrane, leading to an immobilized enzyme with activity similar to that of the native enzyme [64].

Activation to Acid   Polymers bearing nitrile groups can be also selectively hydrolysed and used to immobilize enzyme by the carbodiimide method [465].

Activation to Amino Groups   Carriers with CN groups e.g. Nylon-g-CAN can be reduced to form carriers carrying primary amino groups; these can then be used to immobilize enzymes by the glutaraldehyde activation or carbodiimide method.

Aminolysis of the nitrile groups of acrylic copolymers cross-linked with divinylbenzene or ethylene dimethacrylate leads to the formation of carriers with amino groups, which can be activated by glutaraldehyde for enzyme immobilization. Remarkably, conversion of the nitrile groups into amino groups leads to formation of polymeric gel with high swelling factor [357].

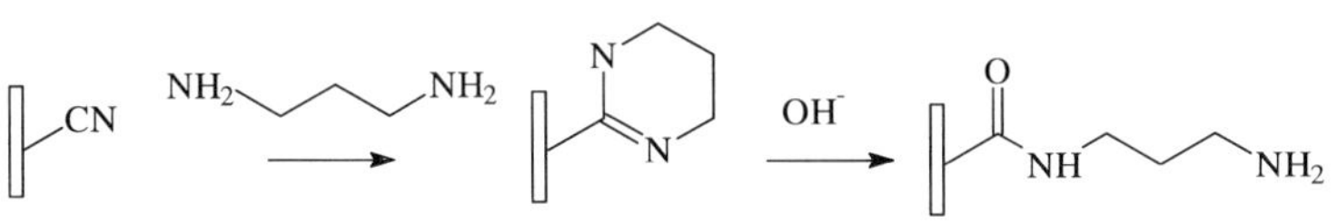

Scheme 3.118 Conversion to amine functionality.

3.6.3.7 Polymers Bearing Isonitrile Functional Groups

The use of polymers bearing isonitrile functional groups for binding enzymes to the carriers is based on the Ugi reaction [70], a method which can be dated back to the 1970s [71–73]. The first use of Ugi reaction for enzyme immobilization was pioneered by Axen, who used oxidized agarose (aldehyde group) and 3-(dimethylamino)propylisonitrile to bind chymotrypsin with a payload of 0.04 (40 mg g^{-1} agarose) and 40% retention of activity. Other polymers used for this type of enzyme immobilization include aminoethylated polyacrylamide, poly(methyl methacrylate) and a cross-linked carboxymethylated dextran [71].

Use of Ugi method for binding enzyme to carrier involves four components – amine, carboxyl, isonitrile, and aldehyde moieties [74] – in the binding chemistry, as illustrated in Scheme 3.119. Thus, in addition to polymers bearing isocyanate groups, other polymers bearing amino, carboxyl, and aldehyde groups can also be used and functional groups of the enzymes participating in the binding can be carboxylic group or amino groups.

One of the advantages of the method is that the protein can be linked to a polymer containing any of the functionality amine, carboxyl, isonitrile, and aldehyde. However, one disadvantage might be that the optimization of the conditions is relatively difficult compared with other covalent binding methods.

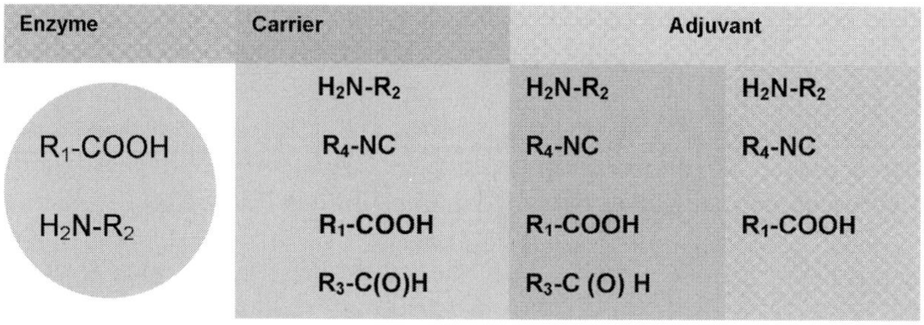

Enzyme	Carrier		Adjuvant
	H_2N-R_2	H_2N-R_2	H_2N-R_2
R_1-COOH	R_4-NC	R_4-NC	R_4-NC
H_2N-R_2	R_1-COOH	R_1-COOH	R_1-COOH
	R_3-C(O)H	R_3-C(O)H	

Scheme 3.119 Enzyme immobilization by the Ugi method.

3.6.4
Interconversion of Active Functionality

Although some active carrier functionality, for example the epoxy group in Eupergit C [57, 148, 468], the aldehyde [466] group in oxidized dextran, or isourea functionality of cyanogen bromide activated polysaccharide [377] can be used for direct binding of enzyme molecules, direct binding of enzymes to these active carriers often leads to undesired results, for example poor retention of activity, lower stability, or poor selectivity, for reasons which include unmatched surface chemistry, steric hindrance, unstable linkage, inappropriate microenvironment, and, particularly, modification of the essential amino acid residues involved in the active centre – for example modification of the active centre of pectin esterase on Enzacryl AA (diazotization) led to complete loss of activity [467].

Fortunately, the different techniques for interconversion of different active functionality provide powerful means of designing an appropriate covalent linkage between a selected enzyme and the chosen carrier. In principle, any of the active carriers discussed in the previous section can be converted to other types of active functionality by use of the methods presented in the Section 3.6.3. Because only a few types of active polymer are commercially available, however, it might be useful for readers who are interested in the use of commercial carriers or who do not have the resources to prepare their own carriers to use the most frequently used interconversion techniques as discussed below.

Currently, the most frequently used carriers for covalent enzyme immobilization are those with aldehyde and oxirane ring functionality. Thus, interconversion of active functionality is mainly based on these two moieties. More detailed Information can be found in the following sections.

Recently, it has been increasingly found that modification of a carrier's microenvironment or introduction of spacers or some orientation groups can bring about unexpected results, for example enhancement of retention of activity, selectivity, and stability, simply by converting the original active functionality into another active functionality.

3.6.4.1 Converting Epoxy Groups

As already noted in this book, polymeric beads bearing oxirane groups are among the most important carriers for enzyme immobilization [178]. Occasionally, however, unsatisfactory results have been obtained with this type of carrier, because of modification of the essential amino residues or the heterogeneity of the binding, thus justifying the use of interconversion approaches.

Converting Epoxy Groups to Amino Groups It has often been observed that immobilization of enzymes on an active oxirane-bearing carrier leads to very low retention of activity, because of the heterogeneous nature of the binding [468]. To overcome this drawback, epoxy groups can be converted to other functionality, for example amino groups [469], which can subsequently be activated by use of the appropriate chemistry and used for enzyme immobilization, as illustrated in Scheme 3.120.

It has been found that modification of the epoxy groups not only enabled introduction of a spacer [468–470], but also several other chemistries for binding enzymes, with improved results. For example, tyrosinase was immobilized on poly(MMA–GMA–DVB) (poly(methyl methacrylate–glycidyl methacrylate–divinyl benzene) by converting the epoxy groups into amino groups by reaction either with ammonia or 1,6-diaminohexane, with glutaric dialdehyde as coupling agent. The results showed that the presence of the spacer increased the activity of the enzyme 35 % compared with carriers aminated with ammonia (the absolute activity was 51 % that of the native enzyme) [469].

Among these different coupling chemistries with aminoalkylated carrier, it was often found that the best result was obtained by coupling the carbodiimide-activated enzyme to a carrier bearing alkyl or arylamino groups at the end of a longer substituent [125].

Retention of activity and binding yield were higher for glutaryl-7-aminocephalosporanic acid (ACA) acylase immobilized on aminoalkylated epoxy silica by glutaraldehyde coupling than for the enzyme coupled directly to the epoxy carrier [468]. Polymers bearing epoxy groups can also be converted to amino groups, and can

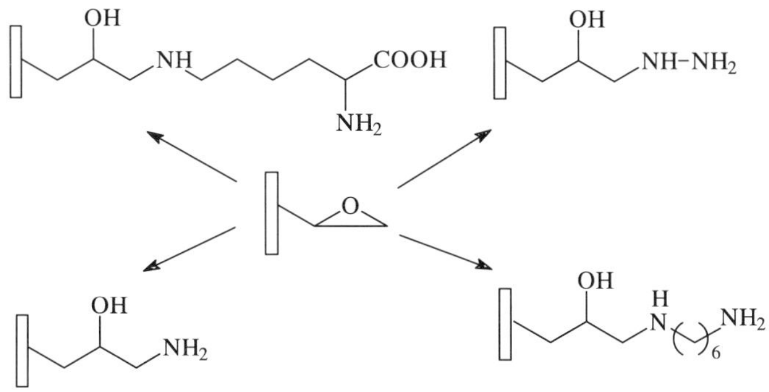

Scheme 3.120 Converting an epoxy carrier to other functionality.

thus be used as ionic adsorbents for non-covalent enzyme immobilization. By use of this method a hollow-fibre grafted with glycidyl methacrylate was converted into an ionic carrier by reaction with diethylamine. The adsorbed enzyme was then cross-linked with glutaraldehyde. High enzyme loading was achieved [471].

Polymer beads grafted with 2,3-epoxypropyl methacrylate can be converted into adsorbents containing amino groups by reacting the oxirane groups with a diamino compound. Compared with native carriers bearing oxirane groups, the new derivatives afforded higher enzyme loading and activity retention when used for immobilization of peroxidase, glucoamylase and urease [58].

Invertase, a glycoenzyme, can, after oxidation with periodate, be conveniently immobilized on carriers bearing amino groups introduced by reaction of oxirane polymers with diamino compounds. High activity was obtained by using the modified Ugi reaction for coupling [472].

Converting Epoxy Groups to Hydrazide Although Eupergit C carrier can be used to bind enzymes directly, this carrier is usually suitable only for the enzymes that are rich in lysine. When the enzymes are poor in these residues, for example the glycosylated enzymes, it is desirable to exploit other possibilities. Interconversion techniques provide the only possibility of making use of carriers with defined chemical and physical nature. Thus, glucosylated enzymes, after introduction of aldehyde groups by oxidation of the carbohydrate moieties, can be coupled to converted Eupergit C carrier bearing hydrazine functionality [473].

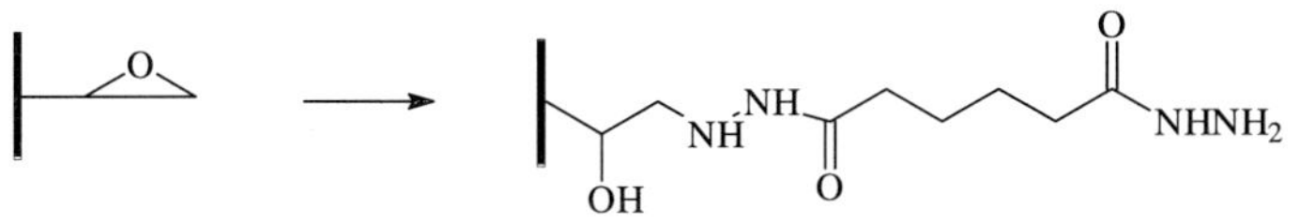

Scheme 3.121 Conversion of an epoxy group to acylhydrazine.

Conversion of Oxirane to Aldehyde Groups Oxirane groups can be converted into aldehyde groups by oxidation of the vicinal hydroxyl groups obtained by hydrolysis of the oxirane functionality. This method was first pioneered in 1986 by Lenfeld, who used periodate to oxidize poly(2,3-epoxypropyl methacrylate) bound to spherical silica gel, leading to the formation of a polymeric carrier bearing aldehyde groups [474, 475]. It is essential to control the reaction time, reactant concentration and temperature to avoid overoxidation.

This technology was later extended by other research groups for preparation of aldehyde agarose, obtained by reacting epichlorohydrin with agarose then by following the procedure as introduced by Lenfeld et al. [177, 374].

3.6.4.2 Converting Anhydride to New Functionality

Anhydride also can be converted into many other new types of functionality, e.g. amino groups, acyl hydrazine, and diazonium salt.

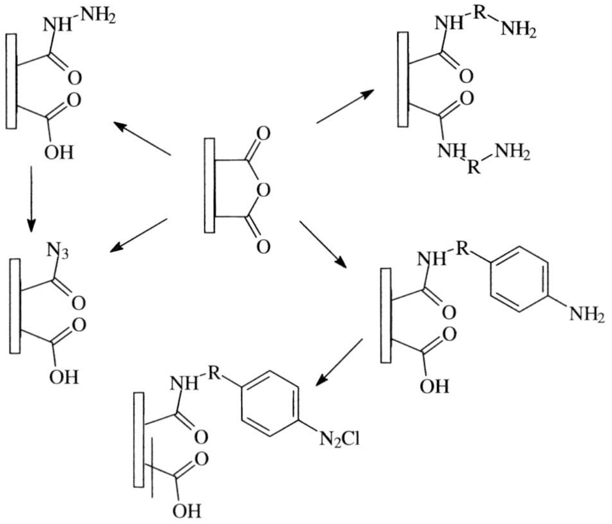

Scheme 3.122 Conversion of a polymeric anhydride to other functionality.

Conversion to an Amine Ethylene–maleic anhydride copolymer can react with N_2H_4, 4,4′-methylenedianiline, 1,6-hexanediamine, or hydrazine followed by conversion to the acyl azide [252]. The modified polymer obtained with hydrazine can be diazotized and used for immobilization of trypsin with 40% retention of activity [260].

Modification of a polymeric anhydride to an aminolated carrier with 1,6-diaminohexane (HMD) in the presence of dicyclohexylcarbodiimide led to the formation of carriers bearing amino groups which could then be used to immobilize enzymes by the glutaraldehyde activation method or the carbodiimide method by means of the enzyme's amino groups [261]. The latter method is often advantageous in that higher retention of activity can be achieved.

Conversion to a Diazonium Salt Coupling of a copolymer of ethylene and maleic anhydride (EMA) with *p,p*′-diaminodiphenylmethane (MDA) or with hydrazine led to high-capacity anionic arylamine (EMA-MDA) or acylhydrazide (EMA-hydrazide) resins [476] which were further converted into polymeric diazonium salts or acylazides, respectively. Enzymes such as trypsin, chymotrypsin, subtilisin Novo, subtilisin Carlsberg, and papain could be coupled to these active polymers with high retention of activity and enhanced temperature and lyophilization stability. The anionic derivatives were more stable in the alkaline pH range than the native enzymes whereas the cationic derivatives were more stable in the acidic pH range.

Conversion to an Acylazide Reaction of polymeric anhydrides with hydrazine leads to the formation polymeric acylhydrazine, which can be further converted to acylazide.

3.6.4.3 Aldehyde

Aldehyde functional groups can be easily converted into many other groups for example amino, hydroxyl, carboxyl, and aromatic amino (for diazotization), which can be used to explore other types of binding chemistry to immobilize the enzymes, as shown in Scheme 3.123.

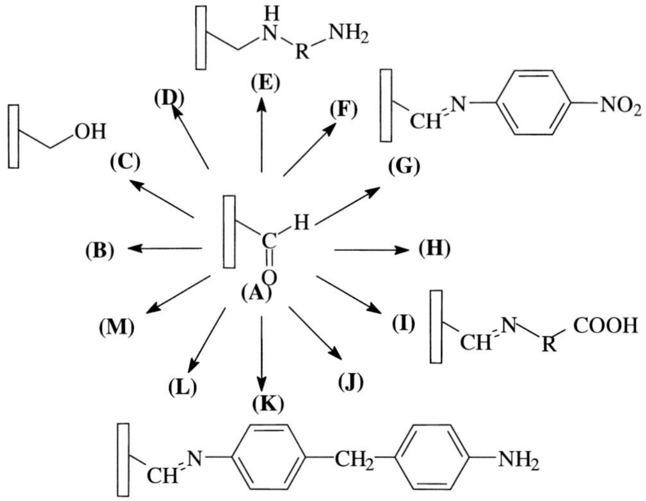

Scheme 3.123 Converting active functionality into new active functionality.

Conversion to Amino Groups Conversion of aldehyde groups into amino groups is wanted for two reasons. First, the aldehyde functionality of the parent carrier may be not ideal functional groups, because of low retention of activity. Second, the oxidized glycosylated enzyme, which bears aldehyde functionality, can be easily coupled to amine supports. Remarkably, the oxidized enzymes can be coupled to the aminolated carriers in an oriented manner [478, 479].

In line with this idea, collagenase was immobilized on oxidized cellulose acetate membranes, via a spacer, with high retention of activity, after derivatization with a diaminoalkane [466]. In contrast, the oxidized enzyme carrying aldehyde groups can also be immobilized on amino carriers. For example, β-d-glucopyranosidase has been immobilized on amino agarose prepared from the aldehyde agarose, with high retention of activity and stabilization [434]. Similarly, poly(styrene/acrolein) (P (SA)), and polyacrolein (P(A)), after derivatization with ethylenediamine, has been used to immobilize oxidized HRP (horseradish peroxidase) with higher retention of activity than the native enzyme immobilized on to the carriers bearing aldehyde groups [477].

Conversion into Carboxyl Groups Aldehyde can be converted to carboxyl groups by reaction with ethylenediamine and, subsequently, with succinic acid anhydride to introduce a spacer [466].

Conversion into Diazonium Salts Polymeric carriers with aldehyde groups can be reduced to hydroxyl functionality, or converted into polymers bearing free amino groups, by use of diaminoalkane or aromatic diamines [372, 374]. Thus, immobilization of enzyme can be achieved by diazonium chemistry or isocyanate chemistry [480].

Reaction of aldehyde polymer, e.g. aldehyde starch, with an aromatic diamine led to the formation pendant aromatic amines which can be subsequently diazotized. Enzymes can be coupled to the active polymers via the tyrosine moiety. By use of this technique Katchalski et al. were first to develop an interesting method for coupling chemically modified trypsin to the active carrier [480].

Similarly, a porous resin, acrolein–vinyl acetate–divinylbenzene polymer has been synthesized by radical suspension polymerization of acrolein with vinyl acetate and divinylbenzene in the presence of a pore-creating agent, toluene, and further reacted with *p*-nitroaniline to form a porous polymer carrier bearing anilino and hydroxy groups. The amino groups introduced by reduction of the nitro group can be used to immobilize chymotrypsin by the glutaraldehyde-cross-linking method or by the diazo-coupling method [313].

Conversion into Polyacrylhydrazide Carriers bearing aldehyde functionality, for example oxidized agarose can be converted into carriers bearing acylazide by reaction with polyacrylhydrazide [482]. The resulting carrier could be used to immobilize enzymes with higher thermostability and stability against 6 M urea than either the native enzymes or enzyme immobilized on cyanogen bromide-activated Sepharose.

3.7
References

1 Zarbosky O: *Immobilised enzymes*, CRC Press, Cleveland, 1973
2 Gemeiner P: Materials for enzyme engineering. In: Gemeiner P (Ed) *Enzyme Engineering*. Ellis Horwood, New York, 1992, pp 13–119
3 Glazer AN, Bar-Eli A, Katchaski E: Preparation and characterization of polystyrosyl trypsin. *J Biol Chem* 1962, 237:1832–1838
4 Monsan et al., Nouvelle Methode de Preparation d'Enzymes Fixes sur des Supports Mineraux. *CR Acad Sci Paris* 1971, 273:33–36
5 Bartling et al., Synthesis of a matrix-supported enzyme in non-aqueous conditions. *Nature* 1973, 243:342–344
6 Stark MB, Kolmberg K: Covalent immobilisation of lipase in organic solvents. *Biotechnol Bioeng* 1988, 34:942–950
7 Xu K, Griebenow K, Klibanov AM: Correlation between catalytic activity and secondary structure of subtilisin dissolved in organic solvents. *Biotechnol Bioeng* 1997, 56:485–491
8 Aparicio J, Sinisterra JV: Influence of the chemical and textural properties of the support in the immobilisation of penicillin G-acylase from *Kluyvena citrophila* on inorganic supports. *J Mol Catal* 1993, 80:269–276
9 Kotha A, Raman RC, Ponrathnam S, Kumar KK, Shewale JG: Beaded reactive polymers, 2. Immobilisation of penicillin G acylase on glycidyl methacrylate–divinyl benzene copolymers of differing pore size and its distribution, *React Funct Polym* 1996, 28:235–242
10 Mujawar SK, Kotha A, Rajan CR, Ponrathnam S, Shewale JG:

Development of tailor-made glycidyl methacrylate–divinyl benzene copolymer for immobilisation of d-amino acid oxidase from Aspergillus species strain 020 and its application in the bioconversion of cephalosporin C. *J Biotechnol* 1999, 75:11–22

11 Porath J: General methods and coupling procedures. *Methods Enzymol* 1974, 34:13–30

12 Woodley JM: Immobilised biocatalysts. In: K. Smith (Ed.) *Solid supports and catalysts in organic synthesis*, Ellis Horwood,1992

13 Messing RA: Immobilised enzymes for industrial reactors. In: Messing RA (Ed.) *Immobilised enzymes for industrial reactors*. Academic Press, New York, 1975

14 Bruggink A, Roos EC, de Vroom E: Penicillin acylase in the industrial production of β-lactam antibiotics. *Org Process Res Dev* 1998, 2:128–133

15 Kotha A, Raman RC, Ponrathnam S, Kumar KK, Shewale JG: Beaded reactive polymers. 3. Effect of triacrylates as crosslinkers on the physical properties of glycidyl methacrylate copolymers and immobilisation of penicillin G acylase. *Appl Biochem Biotechnol* 1998, 74:191–203

16 Hossain MM, Do DD: Immobilisation of multi-enzyme in porous solid supports – a theoretical study. *Chem Eng Sci* 1987, 42:255–264

17 Messing RA: Adsorption and inorganic bridge formation. *Methods Enzymol* 1976, 44:148–169

18 Messing RA: Potential applications of molecular inclusion to beer processing. *Brewers Digest* 1971, 46:60–63

19 Monan P: Optimisation of glutaraldehyde activation of a support for enzyme immobilisation, *J Mol Catal* 1977, 3:371–384

20 Haynes R, Walsh KA: Enzyme envelopes on colloidal particles. *Biochem Biophys Res Commun* 1969, 36:235–242

21 Spagna G, Pifferi PG, Tramontini M: Immobilisation and stabilisation of pectinlyase on synthetic polymers for application in the beverage industry. *J Mol Catal A: Chemical* 1993, 101:99–105

22 Al-Duri B, Yong YP: Lipase immobilisation: an equilibrium study of lipases immobilised on hydrophobic and hydrophilic/hydrophobic supports. *Biochem Eng J* 2000, 4:207–215

23 Blanco RM, Terreros P, Fernandez-Perez M, Otero C, Diaz-Gonzalez G: Fictionalisation of mesoporous silica for lipase immobilisation. Characterisation of the support and the catalysts. *J Mol Catal B Enzymatic* 2004, 30:83–93

24 Hannibal-Friedrich O., Chun M. Sernertz M: Immobilization of galactosidase, albumin and globulin on epoxy-activated acrylic beads. *Biotechnol Bioeng* 1980, 22:157–175

25 Hernaiz MJ, Crout DHG: Immobilisation/stabilization on Eupergit C of the β-galactosidase from *B. circulans* and an α-galactosidase from *Aspergillus oryzae*. *Enzym Microb Technol* 2000, 27:26–32

26 Tischer W, Kasche V: Immobilised enzymes: crystals or carriers? *Trends Biotechnol* 1999, 17:326–335

27 Gabel D, Porath J: Molecular properties of immobilised proteins. *Biochem J* 1972, 127:13–14

28 Chang M, Colvin M, Rembaum A: Acrolein and 2-hydroxyethyl methacrylate polymer microspheres. *J Polym Sci C: Polym Lett* 1986, 24:603–606

29 Koilpillai L, Gadre RA, Bhatnagar S, Raman RC, Ponrathnam S, Kumar KK: Immobilisation of penicillin-G-acylase on methacrylate polymer. *J Chem Technol Biotechnol* 1990, 49:173–182

30 Mandel M, Lepp E, Siimer E: Effect of monomers on penicillin amidase under incorporation in polyacrylamide gel. *USSR Tr Tallin Politekh Inst* 1979, 465:55–63

31 Kotha A, Selvaraj L, Rajan CR, Ponrathnam S, Kumar KK, Ambekar GR, Shewale JG: Adsorption and expression of penicillin G acylase immobilised on to methacrylate polymers generated with varying pore generating solvent volume. *Appl Biochem Biotechnol* 1991, 30:297–302

32 Carenza M, Lora S, Palma G, Boccu E, Largajolli R, Veronese FM: Influence of matrix porosity on the immobilisation of penicillin acylase by radiation-induced

polymerisation. *Radiat Phys Chem* 1988, 31:657–662
33. Kil'deeva NR, Krasovskaya SB, Tsarevskaya IY, Virnik AD, Filippova OV: Effect of the structure of polymeric materials on properties of immobilised enzymes. *Prikl Biokhim Mikrobiol* 1988, 24:375–379
34. Ferreira L, Ramos MA, Dordick JS, Gil MH: Influence of different silica derivatives in the immobilisation and stabilization of a *Bacillus licheniformis* protease (subtilisin Carlsberg). *J Mol Catal B: Enzymatic* 2002, 835:1–11
35. Ison AP, Macrae AR, Smith CG, Bosley J: Mass transfer effects in solvent-free fat interesterification reactions: influences on catalyst design. *Biotechnol Bioeng* 1994, 43:122–130
36. Fonseca LP, JP Cardoso JP, Cabral JMS: Immobilization Studies of an Industrial Penicillin Acylase on a Silica Carrier. *J Chem Tech Biotechnol* 1993, 58:27–37
37. Bosely JA, Clayton JC: Blueprint for a lipase support: use of hydrophobic controlled-pore glass as model systems. *Biotechnol Bioeng* 1994, 43:934–938
38. Artemova AA, Voroshilova OI, Nikitin YS, Khokhlova TD: Macroporous silica in chromatography and immobilisation of biopolymers. *Adv Colloid Interface Sci* 1986, 25:235–248
39. Al-Duri B, Robinson E, MacNerlan S, Bailie P: Hydrolysis of edible oils by lipases immobilised on hydrophobic supports: effect of internal support structure. *J Am Oil Chem Soc* 1995, 72:1351–1359
40. Romero CSS, Manjon S, Iborra JL: Optimisation of the pectinesterase/endo-d-polygalacturonase coimmobilisation process. *Enzym Microb Technol* 1989, 11:837–843
41. Hayashi S, Hayashi T, Kinoshita J, Takasaki Y, Imada K: Immobilisation of β-fructofuranosidase from Aureobasidium sp. ATCC 20524 on porous silica. *J Ind Microbiol* 1992, 9:247–250
42. Kasche V, Galunsky B: Enzyme catalysed biotransformations in aqueous two-phase systems with precipitated substrate and/or product. *Biotechnol Bioeng* 1995, 45:261–267
43. Ettalibi M, Baratti JC: Sucrose hydrolysis by thermostable immobilised inulinases from *Aspergillus ficuum*, *Enzym Microb Technol* 2001, 28:596–601
44. Muller J, Pfleiderer G: Factors affecting the activity of immobilized enzymes. I. Diffusional limitation. *Hoppe-Seyler's Z Phys Chem* 1980, 361: 675–680
45. Nehete PN, Kothari RM, Shankar V: Immobilisation of amyloglucosidase on polystyrene anion exchange resin. I. Preparation and properties. *Food Biotechnol* 1987, 1:107–116
46. Lalonde J: Practical catalysis with enzyme crystals. *Chemtech* 1997, 27:38–45
47. Margolin AL: Novel crystalline catalysts. *Trends Biotechnol* 1996, 14:223–230
48. Quiocho FA, Richards FM: The enzyme behaviors of carboxypeptidase-A in the solid state. *Biochemistry* 1966, 5:4062–4076
49. Varlan AR, Sansen W, Van Loey A, Hendrickx M: Covalent enzyme immobilization on paramagnetic polyacrolein beads. *Biosens Bioelectr* 1996, 11:443–448
50. Fan CH, Lee CK: Purification of d-hydantoinase from adzuki bean and its immobilisation for N-carbamoyl-d-phenylglycine production. *Biochem Eng J* 2001, 8:157–164
51. Kumakura M, Adachi S, Kaetsu I: Flexible porous disks by radiation polymerisation method for enzyme immunoassay of α-fetoprotein. *Arch Immunol Ther Exp* 1984, 32:135–141
52. Cao L, Fischer A, Bornscheuer UT, Schmid RD: Lipase-catalysed solid phase preparation of sugar fatty acid esters. *Biocatal Biotransform* 1997, 14:269–283
53. Kobayashi Y, Matsuo R, Oya T, Yokoi N: Enzyme-entrapping behaviors in alginate fibers and their papers. *Biotechnol Bioeng* 1987, 30:451–457
54. Kondo A, Murakami F, Kawagoe M, Hagashitani K: Kinetic and circular dichroism studies of enzymes adsorbed on ultrafine silica particles. *Appl Microbiol Biotechnol* 1993, 39:726–731
55. Zoungrana T, Findenegg GH, Norde W: Structure, stability and activity of adsorbed enzymes. *J Colloid Interface Sci* 1997, 190:437–448

56 Bahulekar RV, Ponrathnam S, Uphade BS, Ayyangar NR, Kumar KK, Shewale JG: Immobilisation of penicillin G acylase on to alumina: effect of hydrophilicity. *Biotechnol Tech* 1991, 5:401–404

57 Park SW, Choi SY, Chung KH, Hong SI, Kim SW: New method for the immobilisation of glutaryl-7-aminocephalosporanic acid acylase on a modified epoxy/silica gel hybrid. *J Biosci Bioeng* 2002, 94:218–224

58 Wojcik A, Lobarzewski J, Blaszczynska T: Immobilisation of enzymes to porous-bead polymers and silica gels activated by graft polymerisation of 2,3-epoxypropyl methacrylate. *J Chem Technol Biotechnol* 1990, 48:287–301

59 Kawakita H, Sugita K, Saito K, Tamada M, Sugo T, Kawamoto H: Optimisation of reaction conditions in production of cycloisomaltooligosaccharides using enzyme immobilised in multilayers on to pore surface of porous hollow-fiber membranes. *J Membr Sci* 2002, 205:175–182

60 Solomon BA, Chen CC, Colton CK: Immobilisation of acetate kinase on functionalized solid-core polymeric beads. *Enzym Eng* 1978, 4:105–108

61 Bahulekar RV, Prabhune AA, SivaRaman H, Ponrathnam S: Immobilisation of penicillin G acylase on functionalized macroporous polymer beads. *Polymer* 1993, 34:163–166

62 Watanabe K, Royer GP: Polyethlenimine/silica gel as enzyme support. *J Mol Catal* 1983, 22:145–152

63 Torres-Bacete J, Arroyo M, Torres-Guzman R, De la Mata I, Castillon MP, Acebal C: Stabilization of penicillin V acylase from *Streptomyces lavendulae* by covalent immobilisation. *J Chem Technol Biotechnol* 2001, 76:525–528

64 Hicke HG, Boehme P, Becker M, Schulze H, Ulbricht M: Immobilisation of enzymes on to modified polyacrylonitrile membranes: application of the acyl azide method. *J Appl Polym Sci* 1996, 60:1147–1161

65 Guisan JM: Aldehyde-agarose gels as activated supports for immobilisation-stabilization of enzymes. *Enzym Microb Technol* 1988, 10:375–382

66 Taylor JB, Swaisgood HE: Kinetic study on the effect of coupling distance between insoluble trypsin and its carrier matrix. *Biochim Biophys Acta* 1972, 284:268–277

67 Bowers LD, Carr PW: XXX, Biotechnol Bioeng 1976, 18:1331

68 Spagna G, Barbagallo RN, Pifferi PG, Blanco RM, Guisan JM: Stabilization of a β-glucosidase from *Aspergillus niger* by binding to an amine agarose gel. *J Mol Catal B: Enzymatic* 2000, 11:63–69

69 Bleha M, Plichta Z, Votavova E, Kalal J: Redox, polymers containing p-phenylenediamine groups. *Angew Makromol Chem* 1978, 70:173–178

70 Ugi I, Steinbrückner C: Isonitrile 2. Reaktionen von Isonitrilen mit Carbonylverbindungen, Aminen und Stickstoffwasserstoffsäure. *Chem Ber* 1961, 94:734–742

71 Vretblad P, Axen R: Use of isocyanides for the immobilisation of biological molecules. *Acta Chem Scand* 1973, 27:2769–2780

72 Axen R, Vretblad P, Porath JO: Fixed polymeric products, especially adsorption materials and polymer-bound enzymes. *Ger Offen, DE 2,061,009*, 1971

73 Goldstein L, Freeman A, Granot R, Sokolovsky M: Polymers containing isonitrile functional groups as supports for the covalent fixation of biologically active molecules. In: Marekov N, Ognyanov I, Orahovats A (Eds) *Symp Pap – IUPAC Int Symp Chem Nat Prod*, 11th 4(Part 1), 1978:42–53

74 Goldstein L: Polymers bearing isonitrile functional groups as supports for enzyme immobilisation. *Methods Enzymol* 1987, 135:90–102

75 Voivodov K, Chan WH, Scouten W: Chemical approaches to oriented protein immobilisation. *Makromol Chem Macromol Symp* 1992, 275–283

76 Palomo JM, Munoz G, Fernandez-Lorente G, Mateo C, Fuentes M, Guisan JM, Fernandez-Lafuente R: Modulation of *Mucor miehei* lipase properties via directed immobilisation on different hetero-functional epoxy resins. Hydrolytic resolution of (R,S)-2-butyroyl-2-phenylacetic acid. *J Mol Catal B: Enzymatic* 2003, 21:201–210

77 Fernadez-Lafuente R, Rosell CM, Rodriguez V, Santana C, Soler G, Bastida A, Guisan JM: Preparation of activated supports containing low pK amino groups. A new tool for the protein immobilisation via the carboxyl coupling methods. *Enzym Microb Technol* 1993, 15:546–550

78 Liu JG, Wei C, Ouyang F, Xu GZ, Han H: Oxirane copolymer of 2-hydroxyethyl methacrylate and its application in the immobilisation of penicillin acylase. *Huagong Yejin* 2000, 21:278–282

79 Cardias HCT, Grininger CC, Trevisan HC, Guisan JM, Giordano RLC: Influence of activation on the multipoint immobilisation of penicillin G acylase on macroporous silica. *Braz J Chem Eng* 1999, 16:141–148

80 Kobayashi K, Kageyama B, Yagi S, Sonoyama TA: Novel method of immobilising penicillin acylase on basic anion exchange resin with hexamethylene diisocyanate and *p*-hydroxybenzaldehyde. *J Ferment Bioeng* 1992, 74:10–12

81 Hipwell MC, Harvey MJ, Dean PDG: Affinity chromatography on a homologous series of immobilised N6-ω-aminoalkyl AMP. Effect of ligand–matrix spacer length on ligand–enzyme interaction. *FEBS Letters* 1974, 42:355–359

82 Steers E Jr, Cuatrecasas P, Pollard HB: Purification of β-galactosidase from *Escherichia coli* by affinity chromatography. *J Biol Chem* 1971, 246:196–200

83 Cuatrecasas P, Wilchek M, Anfinsen CB: Selective enzyme purification by affinity chromatography. *Proc Natl Acad Sci USA* 1968, 61:636–643

84 Lowe CR: *An introduction to affinity chromatography*. Elsevier Biomedical, Amsterdam, 1979

85 Bulmus V, Ayhan H, Piskin E: Modified PMMA monosize Microbeads for glucose oxidase immobilisation. *Chem Eng J* 1997, 65:71–76

86 Hayashi T, Ikada Y: Spacer effects on enzymic activity immobilised on to polymeric substrates. In Gebelein CG (Ed) *Biotechnol Polym Proc Am Chem Soc Symp Polym Biotechnol*, 1991: 321–332

87 Marcus SL, Balbinder E: Use of affinity matrices in determining steric requirements for substrate binding: Binding of anthranilate 5-phosphoribosylpyrophosphate phosphoribosyltransferase from Salmonella typhimurium to sepharose-anthranilate derivatives. *Anal Biochem* 1972, 48:448–459

88 Hayashi T, Ikada Y: Spacer effects on enzymic activity immobilised on to polymeric substrates. *Polym Mater Sci Eng* 1990, 62:512–516

89 Nouaimi M, Moschel K, Bisswanger H: Immobilisation of trypsin on polyester fleece via different spacers. *Enzym Microb Technol* 2001, 29:567–574

90 Mosbach K, Larsson PO: Preparation of a NAD(H)-polymer matrix showing coenzymic function of the bound pyridine nucleotide. *Biotechnol Bioeng* 1971, 13:393–398

91 Taylor KE, Boss SC: Immobilisation of proteins and cells using hydrophilic high molecular weight polyether spacer arms. *Eur Pat Appl, EP345789*, 1989

92 Garcia CM, Singhal RP: Immobilisation of urease on poly(*N*-vinylcarbazole)/ stearic acid Langmuir–Blodgett films for application to urea biosensor. *Biochem Biophys Res Commun* 1979, 86:697–703

93 Clark DS, Bailey JE: Structure–function relationships in immobilised chymotrypsin catalysis. *Biotechnol Bioeng* 1983, 25:1027–1047

94 Maneck G, Polakowski D: Some carriers for the immobilisation of enzymes based on derivatized poly(vinyl alcohol) and on copolymers of methacrylates with different spacer lengths. *J Chromatogr* 1981, 215:13–24

95 Dhal P, Babu GN, Sudhakaran S, Borkar SP: Immobilisation of penicillin acylase by covalent linkage on vinyl copolymers containing epoxy groups. *Makromol Chem Rapid Commun* 1985, 6:91–95

96 Gemeiner P, O'Carra P, Barry S, Griffin T: In: Jakoby WB, Wilched M (Eds) *Methods Enzymology*, Academic Press, New York, 1974, pp 108–126

97 Toshifumi S, Tamura N, Yasui M, Fujimoto K, Kawaguchi H: Enzyme

98 Penzol G, Armisen P, Fernandez-Lafuente R, Rodes L, Guisan JM: Use of dextrans as long and hydrophilic spacer arms to improve the performance of immobilised proteins acting on macromolecules. *Biotechnol Bioeng* 1998, 60:518–523

99 Anzai J, Lee S, Osa T: Enzyme sensors based on an ion-sensitive field effect transistor coated with Langmuir–Blodgett membranes. Use of polyethyleneimine as a spacer for immobilising α-chymotrypsin. *Chem Pharm Bull* 1989, 37:3320–3322

immobilisation on thermosensitive hydrogel microsphere. *Colloid Surf B: Biointerfaces* 1995, 4:267–274

100 Ciaran OF: Understanding and increasing protein stability. *Biochim Biophys Acta* 1995, 1252:1–14

101 Hsiao HY, Royer GP: Immobilisation of glycoenzymes through carbohydrate side-chains. *Arch Biochem Biophys* 1979, 198:379–385

102 Marek M, Valentova O, Kas J: Invertase immobilisation via its carbohydrate moieties. *Biotechnol Bioeng* 1984, 26:1223–1226

103 Duggleby KG, Kaplan HA: A competitive labeling method for the determination of the chemical properties of solitary functional groups in proteins. *Biochemistry* 1975, 14: 5168–5175

104 Shewale JG, Brew K: Effect of Fe^{3+} binding on the microenvironments of individual amino groups in human serum transferring as determined by different kinetic labeling. *J Biol Chem* 1982, 257:9406–9415

105 Hasegawa M, Kitano H, Nishida R, Kobashi T: Macroporous and hydrophilic polymer resins modified with isothiocyanate groups for immobilisation of enzymes. *Biotechnol Bioeng* 1990, 36:219–223

106 Mateo C, Fernandez-Lorente G, Abian O, Fernandez-Lafuente R, Guisan JM: Multifunctional epoxy supports: a new tool to improve the covalent immobilisation of proteins. The promotion of physical adsorptions of proteins on the supports before their covalent linkage. *Biomacromolecules* 2000, 1:739–745

107 Skaria S, Rao ES, Ponrathnam S, Kumar KK, Shewale JG: Porous thiiranyl polymers: newer supports for immobilisation of penicillin G acylase. *Eur Polym J* 1997, 33:1481–1485

108 Anspach FB, Altmann-Haase G: Immobilised-metal-chelate regenerable carriers: (I) adsorption and stability of penicillin G amidohydrolase from *Escherichia coli*. *Biotechnol Appl Biochem* 1994, 20:313–322

109 Warmuth W, Wenzig E, Mersmann A: Selection of a support for immobilisation of a microbial lipase for the hydrolysis of triglycerides. *Bioprocess Eng* 1995, 12:87–93

110 Scharer R, Hossain MM, Do DD: Determination of total and active immobilised enzyme distribution in porous solid supports. *Biotechnol Bioeng* 1992, 39:679–687

111 Kolarz BN, Wojaczynska M, Bryjak J, Lobarzewski J, Pawlow B: Influence of pentaerythritol triacrylate acrylic carrier swelling on immobilisation of enzymes. *J Appl Polym Sci* 1995, 58:1317–1323

112 Barker SA, Somers PJ: Cross-linked polyacrylamide derivatives (enzyacryls) as water-insoluble carriers of amylolytic enzymes. *Carbohydr Res* 1970, 4:287–296

113 Taylor RF: A comparison of various commercially available liquid chromatographic supports for immobilisation of enzymes and immunoglobulins. *Anal Chim Acta* 1985, 172:241–248

114 Liu JG, Cong W, Wang SA, Ouyang F: Studies of poly(vinyl acetate-co-divinyl benzene) beads as a carrier for the immobilisation of penicillin acylase and the kinetics of immobilised penicillin acylase. *React Funct Polym* 2001, 48:75–84

115 Seip JE, Faber JE, Gavagen DL, Anton, R de Cosimo: Glyoxylic acid production using immobilised glycolate oxidase and catalase, *Bioorg Med Chem* 1994, 2:371–378

116 Wu CW, Lee JG, Lee WC: Protein and enzyme immobilisation on non-porous microspheres of polystyrene. *Biotechnol Appl Biochem* 1998, 27:225–230

117 Batkai L, Horvath I, Horvath-Feher E, Boross L, Li VP: Insoluble enzymes. *Fr. Demande, FR 74-6448* 1974

118 Kolarz BN, Bryjak J, Wojaczynska M, Pawlow B: Carriers from triacrylate for penicillin acylase immobilisation. *Polymer* 1996, 37:2445–2449

119 Mauz O, Noetzel S, Sauber K: New synthetic carriers for enzymes. *Ann NY Acad Sci* 1984, 434:251–253

120 Kondo K, Kimura N, Miyamori S: Divinyl benzene polymer, its manufacture and use in enzyme immobilisation for water-insoluble substrate reaction. *Jpn Kokai Tokkyo Koho, JP 1986-260,733* 1988

121 Lu ZZ, Yang JJ, Leng LC, Feng XD, Li DC: Studies on biologically active *p*-(methacrylamido) benzoic acid esters. *Kexue Tongbao* (Chinese Edition) 1981, 23:1433–1435

122 Diaz JF, Balkus KJ Jr: Enzyme immobilisation in MCM-41 molecular sieve. *J Mol Catal B: Enzym* 1996, 2:115–126

123 Goldstein L: Immobilised enzymes: coupling of biologically active proteins to ethylene-maleic anhydride copolymers of different hydride content. *Anal Biochem* 1972, 50: 40–46

124 Xu H, Li M, He B: Immobilisation of *Candida cylindracea* lipase on methyl acrylate-divinyl benzene copolymer and its derivatives. *Enzym Microb Technol* 1995, 17:194–199

125 Drobnik J, Saudek V, Svec F, Kalal J, Vojtisek V, Barta M: Enzyme immobilisation techniques on poly(glycidyl methacrylate-co-ethylene dimethacrylate) carrier with penicillin amidase as model. *Biotechnol Bioeng* 1979, 21:317–332

126 Dhal PK, Babu GN, Sudhakaran S, Borkar PS: Immobilisation of penicillin acylase by covalent linkage on vinyl copolymers containing epoxy groups. *Makromol Chem Rapid Commun* 1985 6:91–95

127 Karyekar SK, Hegde MV: The hydrophobic domain of penicillin acylase is negatively charged. *Enzym Microb Technol* 1991, 13:139–141

128 Prieto I, Martin J, Arche R, Fernandez P, Perez-Aranda A, Barbero JL: Penicillin acylase mutants with altered site-directed activity from *Kluyvera citrophila*. *Appl Microb Biotechnol* 1990, 33:553–559

129 Shtilman MI, Torchilin VP, Kozlov AA: *High-Molec Compd* 1984, 26:916–920

130 Yamskov IA, Budanov MV, Davankov VA, Nys PS, Savitskaya EM: Enantioselective hydrolysis of *N*-phenacetyl-d,l-C-phenylglycine by native and immobilised penicillin amidase from *E. coli*. *Bioorg Khim* 1979, 5:604–610

131 Kasche V, Galusky B: Ionic strength and pH effects in the kinetically controlled synthesis of benzylpenicillin by nucleophilic deacylation of free and immobilised phenyl-acetyl-penicillin amidase with 6-aminopenicillanic acid. *Biochem Biophys Res Commun* 1982, 104:1215–1222

132 Al-Adhami AJH, Bryjak J, Greb-Markiewicz B, Peczuska-Czoch W: Immobilisation of wood-rotting fungi laccases on modified cellulose and acrylic carriers. *Process Biochem* 2002, 37:1387–1394

133 Yodoya S, Takagi T, Kurotani M, Hayashi T, Furuta M, Oka M, Hayashi T: Immobilisation of bromelain on to porous copoly(γ-methyl-l-glutamate/l-leucine) beads. *Eur Polym J* 2002, 39:173–180

134 Hayashi T, Ikada Y: Protease immobilisation on to polyacrolein microspheres. *Biotechnol Bioeng* 1990, 35:518–524

135 Rexova-Benkova L, Mrackova-Dobrotova M: Effect of immobilisation of *Aspergillulus Niger* extracellulor endo-D-galacturonanase on kinetics and action pattern, *Carbohydrate Research* 1981, 98:115–122

136 Itoyama K, Tanibe H, Hayashi T, Ikada Y: Spacer effects on enzymatic activity of papain immobilised on to porous chitosan beads. *Biomaterials* 1994, 15:107–112

137 Itoyama K, Tokura S, Hayashi T: Lipoprotein lipase immobilisation on to porous chitosan beads. *Biotechnol Prog* 1994, 10:225–229

138 Kitano H, Ise N: Pressure effects on catalysis by thermolysin immobilised on polymer lattices. *Biotechnol Bioeng* 1988, 31:507–510

139 Arica MY, Yavuz H, Denizli Al: Immobilisation of glucoamylase on the plain and on the spacer arm-attached poly(HEMA-EGDMA) microspheres. *J Appl Polym Sci* 2001, 81:2702–2710

140 Bulmus V, Ayhan H, Piskin E: Modified PMMA monosize microbeads for glucose oxidase immobilisation. *Chem Eng J* 1997, 65:71–76

141 Bulmus V, Kesenci K, Piskin E: Poly(EGDMA/AAm) copolymer beads: a novel carrier for enzyme mmobilisation. *React Funct Polym* 1998, 38:1–9

142 Szajani B, Boross L: Factors influencing covalent coupling of enzymes on derivatized polyacrylamide gel beads: A survey. *Hung J Ind Chem Veszprem* 1999, 27: 125–130

143 Nogues MV, Guasch A, Alonso J, Cuchillo CM: Affinity chromatography study of the interaction of ribonucleotides with bovine pancreatic ribonuclease covalently bound to Sepharose 4B. *J Chromatogr* 1983, 268:255–64

144 Isgrove FH, Williams RJH, Niven GW, Andrews AT: Enzyme immobilisation on nylon– optimisation and the steps used to prevent enzyme leakage from the support. *Enzym Microb Technol* 2001, 28:225–232

145 Andrews AT, Mbafor W: Immobilisation of enzymes to Nylon film. *Biochem Soc Trans* 1991, 19:271

146 Bayramoğlu G, Kaya B, Arıca MY: Immobilization of Candida rugosa lipase onto spacer-arm attached poly(GMA-HEMA-EGDMA) microspheres. *Food Chem*, in press

147 Melius P, Wang BC: Immobilization of lipase to cyanogen bromide activated polysaccharide carriers. *Adv Experim Med Biol* 1974, 42:339–343

148 Martin MT, Plou FJ, Alcade M, Ballesteros AM: Immobilisation on Eupergit C of cyclodextrin glucosyltransferase (CGTase) and properties of the immobilised biocatalyst. *J Mol Catal B: Enzymatic* 2003, 21:299–308

149 Martin MT, Plou FJ, Alcade M, Ballesteros A: Covalent immobilisation of cyclodextrin glucosyltransferase (CGTase) in activated silica and Sepharose. *Indian J. Biochem Biophys* 2002, 39:229–234

150 Katalin I, Szajani B, Seres G: Immobilisation of starch-degrading enzymes. I. A comparative study on soluble and immobilised cyclodextrin glycosyltransferase. *J Appl Biochem* 1983, 5:158–164

151 Steighardt J, Kleine R: Production and immobilisation of a proteinase-reduced cyclodextrin glycosyltransferase preparation. *Appl Microbiol Biotechnol* 1993, 39:63–68

152 Venkataraman S, Horbett TA, Hoffman AS: The reactivity of α-chymotrypsin immobilised on radiation grafted hydrogel surfaces. *Polym Preprints (Am Chem Soc, Div Polym Chem)* 1975, 16:197–202

153 Koch-Schmidt AC, Mosbach K: Studies on conformation of soluble and immobilized enzymes using differential scanning calorimetry. 2. Specific activity and thermal stability of enzymes bound weakly and strongly to Sepharose CL 4B. *Biochemistry* 1977, 16:2105–2109

154 Pashova V, Dakov V, Georgiev G: Immobilisation of penicillin-amidase on poly(glycidylmethacrylate) or on a copolymer of glycidylmethacrylate and 2-hydroxyethylmethacrylate both grafted on to poly(acrylonitrile) fibers. *Biotekhnol Biotekh* 1992, 4:21–24

155 Goldstein L: Immobilised enzymes. Coupling of biologically active proteins to ethylene-maleic anhydride copolymers of different hydride content. *Anal Biochem* 1972, 50:40–46

156 Venkataraman S, Horbett TA, Hoffman AS: The reactivity of α-chymotrypsin immobilised on radiation-grafted hydrogel surfaces. *J Biomed Mater Res* 1977, 11:111–123

157 Ferreira l, Ramos MA, Dordick JS, Gil MH: Influence of different silica derivatives in the immobilization and stabilization of a Bacillus licheniformis protease (subtilisin Carlsberg). *J Mol Catal B: Enzymatic* 2002, 835:1–11

158 Bahar T, Tuncel A: Immobilization of α-chymotrypsin onto newly produced poly(hydroxypropyl methacrylate–co-methacrylic acid) hydrogel beads. *React Funct Polym* 2000,44:71–78

159 Lobarzewski J, Wolski T: The function of free carboxyl groups in the action of peroxidase and indole-3-acetic acid oxidase. *Phytochemistry* 1985, 24: 2231–2213

160 Kjellen KG, Neujahr HY: Immobilization of phenol hydroxylase. *Biotechnol Bioeng* 1979, 21:715–719

161 Gouda MK, Abdel-Naby MA: Catalytic properties of the immobilised *Aspergillus tarmarii* xylanase. *Microbiol Res* 2002, 157:275–281

162 Chae HJ, Kim EY: Optimisation of protease immobilisation by covalent binding using glutaraldehyde, *Appl Biotechnol* 1998, 73:195–204

163 Clark DS, Bailey JE: Structure–function relationships in immobilised chymotrypsin catalysis. *Biotechnol Bioeng* 1983, 25:1027–1047

164 Dravis BC, Swanson PE, Russell AJ: Haloalkane hydrolysis with an immobilised haloalkane dehalogenase. *Biotechnol Bioeng* 2001, 75:416–423

165 Klibanov AM: Enzyme stabilization by immobilisation. *Anal Biochem* 1979, 93:1–25

166 Mozhaev VV, Berezin IV, Martinek K: Structure-stability relationship in proteins: fundamental tasks and strategy for the development of stabilised nzyme catalysts for biotechnology. *CRC Crit Rev Biochem Mol Biol* 1988, 23:235–281

167 Germain P, Slagmolen T, Crichton RR: Relation between stabilization and rigidification of the three-dimensional structure of an enzyme *Biotechnol Bioeng* 1989, 33:563–569

168 Colacino F, Vrichton RR: Enzyme Thermstabilisation. *Biotechnol Genet Eng Rev* 1996, 14:243–277

169 Mozhaev VV, Sergeeva V, Belova AB, Khmelnitski YL: Multipoint attachment to a support protects enzyme from inactivation by organic solvents: α-chymotrypsin in aqueous solutions of alcohols and diols. *Biotechnol Bioeng* 1990, 35:653–659

170 Blanco RM, Calvete JJ, Guisan JM: Immobilisation-stabilisation of enzymes: Variables that control the intensity of the trypsin (amine) agarose (aldehyde) multi-point covalent attachment. *Enzym Microb Technol* 1988, 11:353–359

171 Gabel D, Steineberg IZ, Katchalski E: Changes in conformation of insolubilised trypsin and chymotrpsin, followed by fluorescence. *Biochemistry* 1971, 10:4661–4665

172 Jaworek D: New immobilisation techniques and supports. In: Pye EK, Wingard LB Jr (Eds) *Enzym Eng (Pap Res Rep Eng Found Conf, 2nd*, 1974, pp 105–114

173 Martinek K, Klibanov AM, Goldmacher VS, Berezin IV: The principles of enzyme stabilisation, I. Increase in thermostability of enzymes covalently bound to a complementary surface of a polymer support in a multipoint fashion. *Biochem Biophys Acta* 1977, 485:1–12

174 Lenders JP, Crichton RR: Thermal stabilization of amylolytic enzymes by covalent coupling to soluble polysaccharides. *Biotechnol Bioeng* 1984, 26:1343–1351

175 Bryjak J, Noworyta A, Trochimczuk A: Immobilisation of penicillin acylase on acrylic carriers. *Bioprocess Eng* 1989, 4:159–162

176 Muronets VI, Cherednikova TV, Nagradova NK: Use of immobilisation for investigation of glyceraldehyde 3-phosphate dehydrogenase. Immobilised Tetramers. *Biokhimiia* 1981, 46:1731–1739

177 Pedroche J, Yust MM, Girón-Calle J, Vioque J, Alaiz M, Mateo C, Guisán JM, Millán F: Stabilization–immobilisation of carboxypeptidase A to aldehyde–agarose gels, A practical example in the hydrolysis of casein. *Enzym Microb Technol* 2002, 31:711–718

178 Boller T, Meier C, Menzler S: Eupergit oxirane acrylic beads: how to make the enzyme fit for biocatalysts. *Org Process Res Dev* 2002, 6:509–519

179 Monsan P, Combes D: Enzyme stabilization by immobilisation. *Methods Enzymol* 1988, 137:584–598

180 Cabral JMS, Kennedy JF: Immobilisation techniques for altering thermal stability of enzymes. In: Gupta MN (Ed) *Thermostability of enzymes.* Springer, Berlin, 1993, pp 163–179

181 Arica MY, Yavuz H, Patir S, Denizli A: Immobilisation of glucoamylase on to spacer-arm attached magnetic poly-(methylmethacrylate) microspheres: characterization and application to a continuous flow reactor. *J Mol Catal B: Enzymatic* 2000, 11:127–138

182 Kolesnik LA, Galich IP, Kovachuk TA: Properties of amylase immobilised on

aerosol derivatives. *Ukr Biokhim Zh* 1979, 51:369–373
183 Giacomini C, Irazoqui G, Batista-Viera F, Brena BM: Influence of the immobilisation chemistry on the properties of immobilised β-galactosidases. *J Mol Catal B Enzymatic* 2001, 11:597–606
184 Ovsejevi K, Brena B, Batista-Viera F, Carlsson J: Immobilisation of β-galactosidase on thiolsulphonate-agarose. *Enzym Microb Technol* 1995, 17:151–156
185 Yang YG, Chase HA: Immobilisation of α-amylase on poly(vinyl alcohol)-coated perfluoropolymer supports for use in enzyme reactors. *Biotechno. Appl Biochem* 1998, 2:145–154
186 Barkers SA, Somers PJ: Preparation and stability of exo-amylolytic enzymes chemically coupled to microcrystalline cellulose. *Carbohydr Res* 1969, 9:257–263
187 Mansfeld J, Ulbrich-Hofmann R: Site-specific and random immobilisation of thermolysin-like proteases reflected in the thermal inactivation kinetics. *Biotechnol Appl Biochem* 2000, 32:189–195
188 Nouaimi M, Moschel K, Bisswanger H: Immobilisation of trypsin on polyester fleece via different spacers. *Enzym Microb Technol* 2001, 29:567–574
189 Watanabe T, Mori T, Tosa T, Chibata I: Immobilisation of aminoacylase by adsorption to tannin immobilised on aminohexyl cellulose. *Biotechnol Bioeng* 1979, 21:477–486
190 Timasheff SN: In disperse solution; "osmotic stress" is a restricted case of preferential interactions. *Proc Natl Acad Sci USA* 1998, 95:7363–7367
191 Tyagi R, Gupta MN: Chemical modification and chemical crosslinking. In: Gupta MN (Ed) *Thermostability of Enzymes.* Springer, Berlin, 1993
192 Schmid RD: Stabilised soluble enzymes, *Adv Biochem Eng* 1979, 12:41–118
193 Aksoy S, Tumturk H, Hasirci N: Stability of α-amylase immobilized on poly(methyl methacrylate-acrylic acid) microspheres. J Biotechnol 1998, 60:37–46
194 Demers AG, Wong SS: Increased stability of galactosyltransferase on immobilisation. *J Appl Biochem* 1985, 7:122–125
195 Jayakumari VG, Pillai VNR: Immobilisation of papain on cross-linked polymer supports: role of the macromolecular matrix on enzymic activity. *J Appl Polym Sci* 1991, 42:583–590
196 Hasegawa M, Kitano H, Nishida R, Kobashi T: Macroporous and hydrophilic polymer resins modified with isothiocyanate groups for immobilisation of enzymes. *Biotechnol Bioeng* 1990, 36:219–223
197 Nguyen DT, Smit M, Dunn B, Zink JI: Stabilization of creatine kinase encapsulated in silicate sol-gel materials and unusual temperature effects on its activity. *Chem Mater* 2002,14:4300–4306
198 Bismuto E, Martelli PL, De Maio A, Mita DG, Irace G, Casadio R: Effect of molecular confinement on internal enzyme dynamics: Frequency domain fluorimetry and molecular dynamics simulation studies. *Biopolymers* 2002, 67:85–95
199 Horigome T, Kasai H, Okuyama T: Stability of Taka-amylase A immobilised on various sizes of matrix. *J Biochem* 1974, 75:299–307
200 Mody HM, Mody KH, Jasra RV, Shin HJ, Ryong R: Catalytic activity of an immobilised α-amylase on mesoporous silicas. *Indian J Chem Sect A: Inorganic, Bio-inorg, Phys Theor Anal Chem* 2002, 41:1795–1803
201 Takahashi H, Li B, Sasaki T, Miyazaki C, Kajino T, Inagaki S: Catalytic activity in organic solvents and stability of immobilised enzymes depend on the pore size and surface characteristics of mesoporous silica. *Chem Mater* 2000, 12:3301–3305
202 Messing RA: Immobilised catalase-glucose oxidase preparations *Ger Offen, DE 2,405,352 19,740,822,* 1974
203 Messing RA: Simultaneously immobilised glucose oxidase and catalase in controlled-pore Titania. *Biotechnol Bioeng* 1974, 16:897–908
204 Abian O, Wilson l, Mateo C, Fernandez-Lorente G, Palomo JM, Fernandez-Lafuente R, Guisan JM, Re D, Tam A, Daminatti M: Preparation of artificial

hyper-hydrophilic microenvironments (polymeric salts) surrounding enzyme molecules. new enzyme derivatives to be used in any reaction medium. *J Mol Catal B: Enzymatic* 2002, 19:295–303

205 Guisan JM, Sabuquillo PP, Fernandez-Lafuente R, Fernandez-Lorente G, Mateo C, Halling PJ, Kennedy D, Miyata E, Re D: Preparation of new lipase derivatives with high activity stability in anhydrous media: adsorption on hydrophobic supports plus hydrophilization with polyethylenimine. *J Mol Catal B: Enzymatic* 2001, 11:817–824

206 Torres-Bacete J, Arroyo M, Torres-Guzman R, De la Mata I, Castillon MP, Acebal C: Stabilization of penicillin V acylase from Streptomyces lavendulae by covalent immobilization. *J Chem Technol Biotechnol* 2001, 76:525–528

207 Park SW, Lee JW, Hong SI, Kim SW: Enhancement of stability of GL-7-ACA acylase immobilised on silica gel modified by epoxide silanization. *Process Biochem* 2003, 39:359–366

208 Hwang S, Lee KT, Park JW, Min BR, Haam SJ, Ahn IS, Jung JK: Stabilisation of Bacillus stearothermophilus lipase immobilised on surface-modified silica gels. *Biochem Eng J* 2004, 17:85–90

209 Tosa T, Mori T, Chibata I: Studies on continuous enzyme reactions Part VI. Enzymatic properties of DEAE-Sephadex-aminoacylase complex. *Agric Biol Chem* 1969, 33:1053–1059

210 Sanchez EM, Bello JF, Roig MG, Burguillo FJ, Moreno JM, Sinisterra JV: Kinetic and enantioselective behaviour of the lipase from *Candida cylindracea*. A comparative study between the soluble enzyme and the enzyme immobilised on agarose and silica gels. *Enzym Microb Technol* 1996, 18:468–476

211 Klibanov AM: Asymmetric transformations catalysed by enzymes in organic solvents. *Acc Chem Res* 1990, 23:114–120

212 Rozzell JD: Commercial scale biocatalysis: myths and realities. *Bioorg Med Chem* 1999, 7:2253–2261

213 Reetz MT, Zonta A, Schimossek K, Liebeton K, Jaeger K-E: Creation of enantioselective biocatalysts for organic chemistry by *in vitro* evolution. *Angew Chem Int Ed* 1997, 36:2830–2832

214 Palomo JM, Fernandez-Lorente G, Mateo C, Ortiz C, Fernandez-Lafuente R, Guisan JM: Modulation of the enantioselectivity of lipases via controlled immobilisation and medium engineering: hydrolytic resolution of mandelic acid esters. *Enzym Microb Technol* 2002, 31:775–783

215 Rotticci D, Norin T, Hult K: Mass transport limitations reduce the effective stereospecificity in enzyme-catalysed kinetic resolution. *Org Lett* 2000, 2:1373–1376

216 Fernandez-Lorente G, Terreni M, Mateo C, Bastida A, Fernandez-Lafuente R, Dalmases P, Huguet J, Guisan JM: Modulation of lipase properties in macro-aqueous systems by controlled enzyme immobilisation: enantio-selective hydrolysis of a chiral ester by immobilised Pseudomonas lipase. *Enzym Microb Technol* 2001, 28:389–396

217 Fernandez-Lorente G, Fernández-Lafuente R, Palomo JM, Mateo C, Bastida A, Coca J, Haramboure T, Hernández-Justiz O, Terreni M, Guisán JM: Biocatalyst engineering exerts a dramatic effect on selectivity of hydrolysis catalyzed by immobilized lipases in aqueous medium. *J Mol Catal B: Enzymatic* 2001, 11: 649–656

218 Aoun S, Baboulene M: Regioselectivity bromohydroxylation of alkenes catalysed by chloroperoxidase: advantages of the immobilisation of enzyme on talc. *J Mol Catal B Enzymatic* 1998, 4:101–109

219 Schrag JD, Li Y, Cygler M et al.: The open conformation of a Pseudomonas lipase. *Structure* 1997, 5:187–202

220 Margolin AL: Novel crystalline catalysts. *Trends Biotechnol* 1996, 14:223–230

221 Terreni M, Pagani G, Ubiali D, Fernandez-Lafuente R, Mateo C, Guisan JM: Modulation of penicillin acylase properties via immobilisation techniques: one-pot chemoenzymatic synthesis of cephamandole from cephalosporin C. *Bioorg Med Chem Lett* 2001, 11:2429–2432

222 Sheldon RA, van Langen LM, Cao LQ, Janssen MHA: Biocatalysts and biocatalysis in the synthesis of β-lactam antibiotics. In: Bruggink A (Ed) *Synthesis of β-lactam antibiotics*, Kluwer Academic, 2001

223 Silvia R, Vicente UAS, Pregnolato M, Tagliani A, Guisán JM, Fernández-Lafuente R, Terreni M: Influence of the enzyme derivative preparation and substrate structure on the enantioselectivity of penicillin G acylase. *Enzym Microb Technol* 2002, 60/62:1–6

224 Orsat B, Drtina GJ, Williams MG, Klibanov AM: Effect of support material and enzyme pretreatment on enantioselectivity of immobilised subtilisin in organic solvents. *Biotechnol Bioeng* 1994, 44:1265–1269

225 Goldstein L, Katchalski E: The use of water-insoluble enzyme derivatives in biochemical analysis and separation. *Z Anal Chem* 1965, 243:375–396

226 Marle I, Joensson S, Isaksson R, Pettersson C, Pettersson G: Chiral stationary phases based on intact and fragmented cellobiohydrolase I immobilised on silica. *J Chromatogr* 1993, 648:333–347

227 Suh CW, Choi GS, Lee EK: Enzymic cleavage of fusion protein using immobilised urokinase covalently conjugated to glyoxyl-agarose. *Biotechnol Appl Biochem* 2003, 37:149–155

228 Linko Y, Saarinen RL, Linko M: Starch conversion by soluble and immobilised α-amylase. *Biotechnol Bioeng* 1975, 17:153–165

229 Latyshko NV, Gudkova LV, Degtiar RG, Gulyi MF: Immobilisation of *Penicillium vitale* catalase on aminoethyl cellulose and properties of the obtained preparations. *Ukr Bochim Zhur* 1981, 53:48–52

230 Gemeiner P, Halak P, Polakova K: Two-step covalent immobilisation of enzymes as a way for study of effects influencing catalytic activity. *J Sol-Phase Biochem* 1980, 5:197–209

231 White CA, Kennedy JF: Popular matrices for enzyme and other immobilisations. *Enzym Microb Technol* 1980, 2:82–90

232 Taylor RF: Commercially available supports for protein immobilisation. In: Taylor RF (Ed) *Protein Immobilisation*. Marcel Dekker, New York, 1991, pp 139–160

233 Katchalski-Katzir E, Kraemer DM: Eupergit C, a carrier for immobilisation of enzymes of industrial potential. *J Mol Catal B Enzymatic* 2000, 10:157–176

234 Skaria S, Rao ES, Ponrathnam S, Kumar KK, Shewale JG: Porous thiiranyl polymers: newer supports for immobilisation of penicillin G acylase. *Eur Polym J* 1997, 33:1481–1485

235 Mauz O, Wernicke R: Derivatization experiment in the polymer carrier VA-Hydroxy-BIOSYNTH. *BioTec* 1992, 4:30, 32–35

236 Micheel F, Evers J: Synthesis of cellulose-bound proteins. *Macromol Chem* 1949, 3:200–209

237 Goldstein L: Immobilised enzymes: Synthesis of a new type of polyanionic and polycationic resins and their utilization for the preparation of water-insoluble enzyme derivatives. *Biochim Biophys Acta* 1973, 315:1–17

238 Coulet, PR, Godinot C, Gautheron DC: Surface-bound aspartate aminotransferase on collagen films. Properties compared with native enzyme. *Biochim Biophys Acta* 1975, 391:272–281

239 Loeffler LJ, Pierce JV: Acyl azide derivatives in affinity chromatography immobilisation of enzymically active trypsin on beaded agarose and porous glass. *Biochim Biophys Acta* 1973, 317:20–27

240 Watanabe S, Shimizu Y, Teramatsu T, Murachi T, Hino T: The in vitro and in vivo behaviour of urokinase immobilised on to collagen-synthetic polymer composite material. *J Biomed Mater Res* 1981, 15:553–563

241 Kumaraswamy MDK, Rao K, Panduranga JK, Thomas SM: Immobilisation of enzymes on alginic acid-polyacrylamide copolymers. *Biotechnol Bioeng* 1981, 23:1889–1892

242 Epton R, Marr G, Ridley RG: Triethoxysilane-substituted acrylate copolymers as reagents for the derivatization of porous silica beads. *Polymer* 1979, 20:1447–1448

243 El Sherif H, Di Martino S, Travascio P, De Maio A, Portaccio M, Durante D, Rossi S, Canciglia P, Mita DG: Advantages of using non-isothermal bioreactors in agricultural waste water treatment by means of immobilised

urease. Study on the influence of spacer length and immobilisation method. *J Agric Food Chem* 2002, 50:2802–2811

244 Epton R, Marr G and Morgan GJ: Soluble polymer-protein conjugates: 1. Reactive N-(sym-trinitroaryl) polyacrylamide/acrylhydrazide copolymers and derived carbonic anhydrase conjugates. *Polym* 1977, 18:319–323

245 Lai TS, Cheng PS: A new type of enzyme immobilization support derived from styrene-maleic anhydride copolymer and the (-chmotrypsin with which it immobilizes. *Biotechnol Bioeng* 1978, 20:773–779

246 Horvath C: Peculiar immobilised enzymes. *Biochim Biophys Acta* 1974, 358:164–177

247 Beddows CG, Gil HG, Guthrie JT: The immobilisation of enzymes and cells of Bacillus stearothermophilus on to poly(maleic anhydride-styrene-ethylene) and poly(maleic anhydride-vinyl acetate-ethylene). *Biotechnol Bioeng* 1985, 27:579–584

248 Levin Y, Pecht M, Goldstein L, Katchalski E: A water-insoluble polyanionic derivative of trypsin. *Biochemistry* 1964, 3:1905–1913

249 Abel C, Malsch G, Lehmann I, Ziegler HJ, Scharnagl N, Becker M, Hicke HG: Acrylic reactive copolymer membranes. *Angew Makromol Chem* 1995, 226:71–87

250 Kolesnik LA, Galich IP: Stability of α-amylase with immobilization through its different functional groups. *Ukr Biokhim Zh (Ukr. Biochem J)* 1979, 51:154–159 (in Russian)

251 Goldstein L: New polyamine carrier for the immobilization of proteins. Water-insoluble derivatives of pepsin and trypsin. *Biochim Biophys Acta* 1973, 327:132–137

252 Manecke G, Korenzecher R: Reaktive Copolymere des N-Vinyl-2-pyrrolidons zur Immobilisierung von Enzymen. *Makromol Chem* 1977, 178:1729–1738

253 Jaworek D: New methods for covalent binding of proteins to synthetic polymers. In: Salmona M, Saronio C, Garattini S (Eds) *Insolubilized Enzymes*, Raven, New York, 1974, p 65

254 Bruemmer W, Hennrich N, Klockow M, Lang H, Orth HD: Preparation and properties of carrier-bound enzymes. *Eur J Biochem* 1972, 25:129–135

255 Alsen C, Bertram U, Gersteuer T, Ohnesorge FK, Delin S: Studies on acetylcholinesterase and cholinesterase covalently bound to polymaleinic anhydride. *Biochim Biophys Acta* 1975, 377:297–302

256 Goldstein L, Levin Y, Pecht M, Katchalski E: A water-insoluble polyanionic derivatives of trypsin, effect of the polyelectrolyte carrier on the kinetic behaviour of the bound trypsin. *Biochemistry* 1964, 3:1914–1919

257 Cross-linked copolymers for the covalent bonding of substances containing reactive groups. (Merck GmbH Pat). *Br Pat, GB 71-22503*, 1973

258 Hueper F, Rauenbusch E, Schmidt-Kastner G, Boemer B, Bartl H: Water-insoluble protein. *Ger Offen, DE 72–2215687*, 1973

259 Conte A, Lehmann K: Fixation of proteolytic enzymes on poly(methacrylic anhydride). *Hoppe-Seyler's Z Physiol Chem* 1971, 352:533–541

260 Goldstein L, Katchalski E, Levin Y, Blumberg S: Polymer carrier for binding biologically active proteins. *Ger Offen, DE 74–2,420,747*, 1974

261 Goldstein L: New polyamine carrier for the immobilisation of proteins. Water-insoluble derivatives of pepsin and trypsin. *Biochim Biophys Acta* 1973, 327:132–137

262 Isosaki K, Seno N, Matsumoto I, Koyama T, Moriguchi H: Immobilisation of protein ligands with methyl vinyl ether-maleic anhydride copolymer. *J Chromatogr* 1992, 597:123–128

263 Smith RAG: Preparation and properties of amphipathic enzyme-polymer conjugates. *Biochem J* 1979, 181:111–118

264 Merck GmbH: Crosslinked copolymers for the covalent bonding of substances containing reactive groups. *Br Pat, GB 71–22,503*, 1973

265 Kraemer K, Lehmann HP, Plainer H, Reisner W, Sproessler BG: *J Polym Sci Symp* 1974, 47:77–89

266 Manecke G, Forster HJ: Reaktionsfähige Hochpolymere auf Polystyrol-Basis als Träger von Proteinen und Enzymen. *Makromol Chem* 1966, 91:136–154

267 Manecke G, Gunzel G: Verwendung eines nitrierten Copolymerisates aus Methacrylsäure und Methacrylsäure-m-fluoranilid zur Darstellung von Enzymharzen sowie zu Racematspaltungs- und Gerbungsversuchen. *Makromol Chem* 1962, 51:199–216

268 Brown E, Racois A, Gueniffey H: Preparation and properties of urease derivatives insoluble in water. *Tetrahedron Lett* 1970, 25:2139–2142

269 Manecke G, Singer S: Umsetzungen an Copolymerisaten des Methacrylsäure-fluoranilids. *Makromol Chem* 1960, 39:13

270 Manecke G: Immobilisation of enzymes by various synthetic polymers. *Biotechnol Bioeng Symp* 1972, 3:185–187

271 Bahar T, Tuncel A: Immobilisation of invertase on to cross-linked poly(p-chloromethylstyrene) beads. *J Appl Polym Sci* 2002, 83:1268–1279

272 Kramer DM, Lehmann K, Pennewiss M, Plainer H: Photo-beads and oxirane beads as solid supports for catalysis and bio-specific adsorption. In: Peeters H (Ed) *Protides of Biological Fluids 23rd Colloquium.* Pergamon Press, Oxford, 1975, pp 505–511

273 Kotha A, Raman RC, Ponrathnam S, Kumar KK, Shewale JG: Beaded reactive polymers, 2. Immobilisation of penicillin G acylase on glycidyl methacrylate-divinyl benzene copolymers of differing pore size and its distribution. *React Funct Polym* 1996, 28:235–242

274 Erarslan A, Kocer H: Thermal inactivation kinetics of penicillin G acylase obtained from a mutant derivative of *Escherichia coli* ATCC 11105. *J Chem Technol Biotechnol* 1992, 55:79–84

275 Spiess AC, Kasche V: Direct measurement of pH profiles in immobilised enzyme carriers during kinetically controlled synthesis using CLSM. *Biotechnol Prog* 2001, 17:294–303

276 Bigwood MP, Naples JO: Oxirane-bearing polymer carriers for immobilising enzymes and other materials bearing active hydrogen, and oxirane-bearing polymer carriers. *Eur Pat Appl, EP 84–308,588,* 1985

277 Janssen MHA, van Langen LM, Pereira SRM, van Rantwijk F, Sheldon RA: Evaluation of the performance of immobilised penicillin G acylase using active-site titration. *Biotechnol. Bioeng* 2002, 78:425–432

278 Bahulekar RV, Prabhune AA, Sivaraman H, Ponrathnam S: Immobilisation of penicillin G acylase on functionalized macroporous polymer beads. *Polymer* 1993, 34:163–166

279 Bahulekar RV, Ponrathnam S, Ayyangar NR, Kumar KK, Shewale JG: Effect of microenvironment of oxirane groups on the immobilisation of penicillin G acylase. *J Appl Polym Sci* 1992, 45:279–284

280 Pashova V, Dakov V, Georgiev G: Immobilisation of penicillin-amidase on poly(glycidylmethacrylate) or on a copolymer of glycidylmethacrylate and 2-hydroxyethylmethacrylate both grafted on to poly(acrylonitrile) fibers. *Biotekhnol Biotekh* 1992, 4:21–24

281 Pashova VS, Georgiev GS, Dakov VA: Photoinitiated graft copolymerisation of glycidyl methacrylate and 2-hydroxyethyl methacrylate on to polyacrylonitrile and application of the synthesized graft copolymers in penicillin-amidase immobilisation. *J Appl Polym Sci* 1994, 51:807–813

282 Kraemer DM, Lehmann K, Pennewiss H, Plainer H: Oxirane-acrylic beads, preparation 2878-C. *Enzyme Eng* 1978, 4:153–154

283 Kraemer D, Pennewiss H, Plainer H, Schnee R: A process for preparing oxirane-containing polymer carriers for the immobilisation of enzymes and other materials Hydrophilic carrier polymer beads for immobilisation of enzymes and other proteins. *Eur Pat Appl, EP 81-109,694 19,811,114,* 1982

284 Kasche V, Galunsky B: Enzyme catalysed biotransformations in aqueous two-phase systems with precipitated substrate and/or product. *Biotechnol Bioeng* 1995, 45:261–267

285 Manecke G, Günzel G: Polymere Isothiocyanate zur Darstellung hochwirksamer Enzymharze, *Naturwissenschaften* 1967, 54:531–533

286 Manecke G, Pohl R, Schluensen J, Vogt HG: Some reactive carriers and immobilised enzymes. *Enzym Eng* 1978, 4:409–412

287 Heilmann SM, Drtina GJ, Haddad LC, Moren DM, Hyde FW, Pranis RA: Crosslinked isocyanate-functional polymer supports for catalysts, their preparation and use. *Eur Pat Appl*, EP 607963 A1, 1994

288 Kennedy JF, Barker SA, Rosevear A: Use of a poly(allyl carbonate) for the preparation of active, water insoluble derivatives of enzymes. *J Chem Soc Perkin Trans II* 1972, 1:2568–2573

289 Mauz O, Noetzel S, Sauber K: New synthetic carriers for enzymes. *Ann NY Acad Sci* 1984, 434:251–253

290 Mauz O, Noetzel S, Sauber K: Preparation of cross-linked polymers by dispersion polymerisation. *Eur Pat Appl*, EP 87-112,388, 1988

291 Ding LH, Li Y, Jiang Y, Cao Z, Huang J: New supports for enzyme immobilisation based on copolymers of vinylene carbonate and β-hydroxyethylene acrylate. *J Appl Polym Sci* 2002, 83:94–102

292 Ding LH, Jiang Y, Huang L, Li YG, Huang JX: New supports for enzyme immobilisation based on copolymers of vinylene carbonate and acrylamide. *Appl Biochem Biotechnol* 2001, 95:11–21

293 Huang JX, Li Y, Huo YL, Liu T, Yang Y, Yuan Z: Synthesis of new supports containing cyclic carbonate and immobilisation of glucoamylase. *Xuexiao Huaxue Xuebao* 2002, 23:1605–1609

294 Ding LH, Qu BJ: New supports for enzyme immobilisation based on the copolymers of poly(vinylene carbonate) and α-(2-aminoethylene amino)-ω-(2-aminoethylene amino)–poly(ethylene oxide). *React Funct Polym* 2001, 49:67–76

295 Lu ZZ, Yang JJ, Leng LC, Feng XD, Li DC: Studies on biologically active p-(methacrylamido)benzoic acid esters. *Kexue Tongbao (Chinese Edition)* 1981, 23:1433–1435

296 Pollak A, Blumenfeld H, Wax M, Baughn RL, Whitesides GM: Enzyme immobilisation by condensation copolymerisation into crosslinked polyacrylamide gels. *J Am Chem Soc* 1980, 102:6324–6336

297 Seip JE, Fager SK, Grosz R, Gavagan JE, DiCosimo R, Anton DL: Enzymic synthesis of cytidine 5′-diphosphate using pyrimidine nucleoside monophosphate kinase. *Enzym Microb Technol* 1990, 12:361–366

298 Ponnuchamy NP, Gupta MN: The polymeric p-quinone as a matrix for enzyme immobilisation. *Biotechnol Bioeng* 1989, 33:927–931

299 Chaplin MF, Kennedy JF: Use of some new poly-phenolic resins for fractionation of carbohydrates and immobilisation of carbohydrate hydrolases and isomerases. *J Chem Soc Perkin Trans 1: Org Bio-Org Chem* 1979:2144–2153

300 Ye QL, Li YX: Studies on a phenolic resin as a carrier for the immobilisation of penicillin G acylase. *Polym Adv Technol* 1996, 8:727–730

301 Masri MS, Randall VG, Stanley WL: Insolubilization of enzymes on modified phenolic polymers. *US Pat Appl*, US 77–827,659, 1978

302 Masri MS, Randall VG, Stanley WL: Insolubilization of enzymes on modified phenolic polymers. *US Pat Appl*, US 76-712,298, 1976

303 Epton R, Mclaren JV, Thomas TH: Enzyme insolubilisation with crosslinked polyacryloylaminoacetalaldehyde dimethylacetate. *Biochem J* 1971, 123:21–22

304 Epton R, Mclaren JV, Thomas TH: Water-insolubilisation of glucoside hydrolase with crosslinked poly(acryolaminoacetalaldehyde) dimethyl acetal. *Carbohydr Res* 1972, 22:301–306

305 Egorov AM, Osipov AP, Ovchinnikov AN, Dikov MM: Synthesis and properties of the catalytically active NAD analog-formate dehydrogenase conjugate containing the coenzyme immobilised in the vicinity of the active site of the enzyme. *Chem Biotechnol Biol Act Nat Prod Proc, 1st*, 1981, 3:357–361

306 Flemming C, Gabert A, Gomoll M, Roth P: Synthesis and properties of immobilised enzymes. VIII. Kinetic studies on the binding rate of enzymes to macroporous carriers. *Acta Biol Med Ger* 1977, 36:1007–1018

307 Manecke G, Pohl R: Reaktive Trager zur Immobilisierung von Enzymen ausgehend von Polyacrolein. *Makromol Chem* 1978, 179:2361–2377

308 Slomkowski S, Basinska T, Miksa B: New types of microspheres and microsphere-related materials for medical diagnostics. *Polym Adv Technol* 2002, 13:906–918

309 Slomkowski S: Polyacrolein-containing microspheres: synthesis, properties and possible medical applications. *Prog Polym Sci* 1998, 23:815–874

310 Basinska T, Slomkowski S: Attachment of horseradish peroxidase (HRP) on to the poly(styrene/acrolein) latexes and on to their derivatives with amino groups on the surface; activity of immobilised enzyme. *Colloid Polym Sci* 1995, 273:431–438

311 Zhuang P, Butterfield DA: Structural and enzymatic characterizations of papain immobilised on to vinyl alcohol/vinyl butyral copolymer membrane. *J Membr Sci* 1992, 66:247–257

312 Tarhan L, Pekin B: Immobilisation of papain with acrolein-styrene copolymers containing carbonyl groups. *Biotechnol Bioeng* 1983, 25:2777–2783

313 Tao GL, Furusaki S: Synthesis of porous polymer carrier and immobilisation of α-chymotrypsin. *Polym J* 1995, 27:111–121

314 Krasnobaev V, Boeniger R: Aromatic polyamine-dialdehyde resins for fixation of proteins. *Ger Offen* 2529604, 1976

315 Varian AR, Sansen W: Covalent enzyme immobilisation on paramagnetic polyacrolein beads. *Biosens Bioelectron* 1996, 11:443–448

316 Slomkowski S, Miksa B, Kowalczyk D, Basinska T, Wang FW: Proteins at surfaces of poly(styrene/acrolein) latexes and latex assemblies. *Book of Abstracts, 214th ACS National Meeting, Las Vegas, NV*, 1997, pp 7–11

317 Pittner F, Miron T, Pittner G, Wilchek M: Enzyme immobilisation on pyridine containing polymers. *Enzym Eng* 1980, 5:447–449

318 Pittner F, Miron T, Pittner G, Wilchek M: Pyridine-containing polymers: new matrices for protein immobilisation. *J Am Chem Soc* 1980, 102:2451–2452

319 Hayashi T, Ikoda Y: Lipoprotein lipase immobilised on to polyacrolein microsphere. *Biotechnol Bioeng* 1990, 36:593–600

320 Basinska T, Slomkowski S: Attachment of horseradish peroxidase (HRP) on to the poly(styrene/acrolein) latexes and on to their derivatives with amino groups on the surface; activity of immobilised enzyme. *Colloid Polym Sci* 1995, 273:431–438

321 Kitano H, Ise N: Pressure effects on catalysis by thermolysin immobilised on polymer lattices. *Biotechnol Bioeng* 1988, 31:507–510

322 Chang M, Colvin M, Rembaum A: Acrolein and 2-hydroxyethyl methacrylate polymer microspheres. *J Polym Sci C: Polym Lett* 1986, 24:603–606

323 Dattagupta N, Buenemann H: New type of polymeric carrier for immobilisation of biologically specific molecules. *J Polym Sci* 1973, 11:189–192

324 Coupek J, Labsky J, Kalal J, Turkova J, Valentova O: Reactive carriers of immobilised compounds. *Biochim Biophys Acta* 1977, 481:289–296

325 Rasmussen JK, Heilmann SM, Krepski LR, Jensen KM, Mickelson J, Johnson K: Crosslinked, hydrophilic, azlactone-functional polymeric beads: a two-step approach. *React Polym* 1992, 16:199–212

326 Johnson PR, Stern NJ, Eitzmann PD, Rasmussen JK, Milbrath DS, Gleason RM, Hogancamp RE: Reproducibility of physical characteristics, protein immobilisation and chromatographic performance of 3M Emphaze biosupport medium AB. *J Chromatogr A* 1994, 667:1–9

327 Heilmann SM, Rasmussen JK: Chemistry and technology of 2-alkenyl azlactones. *J Polym Sci A: Polym Chem* 2001, 39:3655–3677

328 Drtina GJ, Heilmann SM, Moren DM, Rasmussen JK, Krepski LR, Smith HK, Iranis RA, Turek TC: Highly crosslinked azlactone functional supports of tailorable polarity. *Macromolecules* 1996, 29:4486–4489

329 Xie SF, Svec F, Frechet JMJ: Monolithic poly(2-vinyl-4,4-dimethylazlactone-co-acrylamide-co-ethylene dimethacrylate) support for design of high throughput bioreactors. *Polym Preprints (Am Chem Soc, Div Polym Chem)* 1997, 38:211–212

330 Xie SF, Svec F, Frechet JMJ: Design of reactive porous polymer supports for

high throughput bioreactors: poly(2-vinyl-4,4-dimethylazlactone-co-acrylamide-co-ethylene dimethacrylate) monoliths. *Biotechnol Bioeng* 1999, 62:30–35

331 Coleman PL, Milbrath DS, Walker MM: Biologically active material covalently immobilised on to azlactone-functional polymeric supports and method for preparing it. *PCT Int Appl, WO 9,207,879*, 1992

332 Chikere AC, Galunsky B, Schuhlnemann V, Kasche V: Stability of immobilised soybean lipoxygenases: influence of coupling conditions on the ionization state of the active site Fe. *Enzym Microb Technol* 2001, 28:168–175

333 Smalla K, Turkova J, Coupek J, Hermann P: Influence of salts on the covalent immobilisation of proteins to modified copolymers of 2-hydroxyethyl methacrylate with ethylene dimethacrylate. *Biotechnol Appl Biochem* 1988, 10:21–31

334 Burg K, Mauz O, Noetzel S, Sauber K: New synthetic supports for immobilisation of enzymes. *Angew Makromol Chem* 1988, 157:105–121

335 Stambolieva N, Turkova J: Covalent attachment of proteins to Spheron by means of benzoquinone. *Collect Czech Chem Commun* 1980, 45:1137–1143

336 Brynda E, Bleha M: Reversible thermal denaturation of immobilised chymotrypsinogen. *Collect Czech Chem Commun* 1979, 44:3090–3101

337 Arica MY, Hasirci V, Alaeddinoglu NG: Covalent immobilisation of α-amylase on to pHEMA microspheres: preparation and application to fixed bed reactor. *Biomaterials* 1995, 16:761–768

338 Wieland T, Determann H, Buenning K: Insoluble enzymes fixed in polyacrylamide gel. *Z Naturforsch B* 1966, 21:1003

339 Hicks GP, Updick SJ: The preparation and characterization of lyophilised polyacrylamide enzyme gels for chemical analysis. *Anal Chem* 1966, 38:726–730

340 Mosbach K, Mosbach R: Entrapment of enzymes and microorganisms in synthetic crosslinked polymers and their application in column techniques. *Acta Chem Scand* 1966, 20:2807–2811

341 Inman JK, Dintzis HM: The derivatization of crosslinked polyacrylamide beads. Controlled introduction of functional groups for the preparation of special purpose. *Biochem Adsorb Biochem* 1969, 8:4074–4082

342 Weston PD, Avrameas S: Protein coupled to polyacrylamide beads using glutaraldehyde, *Biochem Biophys Res Commun* 1971, 45:1574–1580

343 Bulmus V, Kesenci K, Piskin E: Poly(EGDMA/AAm) copolymer beads: a novel carrier for enzyme immobilisation. *React Funct Polym* 1998, 38:1–9

344 Hasegawa J, Oikawa H, Kobayashi O, Kataoka Y, Sekiya M: Manufacture of hydrophilic fine gel particles with narrow particle distributions. *Eur Pat Appl, EP 335,703 A2*, 1989

345 Bartling GJ, Chattopadhyay SK, Barker CW, Brown HD: Preparation and properties of matrix-supported horseradish peroxidase. *Can J Biochem* 1975, 53:868–874

346 Kolarz BN, Trochimczuk A, Bryjak J, Wojaczynska M, Dziegielewski K, Noworyta A: A search for optimum acrylic carriers for immobilisation of penicillin acylase. *Angew Makromol Chem* 1990, 179:173–183

347 Kolarz BN, Wojaczynska M, Bryjak J, Lobarzewski J: Synthesis and properties of porous carriers from acrylonitrile and trimethylolpropane triacrylate. *Macromol Rep A* 1993, 30:201–209

348 Kennedy JF, Epton J: Poly(*N*-acryloyl-4 and 5-aminosalicylic acids) Part III. Uses as their titanium complexes for the insolubilisation of enzymes. *Carbohydr Res* 1973, 27:11–20

349 Xu H, Li M, He B: Immobilisation of *Candida cylindracea* lipase on methyl acrylate-divinyl benzene copolymer and its derivatives. *Enzym Microb Technol* 1995, 17:194–199

350 Bianchi D, Golini P, Bortolo R, Cesti P: Immobilisation of penicillin G acylase on aminoalkylated polyacrylic supports. *Enzyme Microb Technol* 1996, 18:592–596

351 Blassberger D, Freeman A, Goldstein L, George S: Chemically modified polyesters as supports for enzyme immobilisation: isocyanide, acylhydrazide, and

aminoaryl derivatives of poly(ethylene terephthalate). *Biotechnol Bioeng* 1978, 20:309–315

352 Popa M, Bajan N, Sunel V, Daranga M: Catalase immobilised on poly(acrylic acid-co-vinyl alcohol). *Eurasian Chem Technol J* 2002, 4:199–206

353 Melamed O, Margel S: Poly(N-vinylphenylalanine) microspheres: synthesis, characterization, and use for immobilisation and microencapsulation. *J Colloid Interface Sci* 2001, 241:357–365

354 Ivanov IP, Yotova LK: Simultaneous immobilisation of uricase and peroxidase to copolymer of acrylonitrile with acrylamide. *Biotechnol Biotechnol Equip* 2002, 16:104–110

355 Oosawa T: Immobilisation of glucoamylase. *Jpn. Kokai Tokkyo Koho, JP 52,034,979*, 1977

356 Miyake T, Takeda K, Ikeda A, Yokohama K, Mizuno M: Adsorption agent for proteins. *Ger Offen, DE 2746275*, 1978

357 Kolarz BN, Lobarzewski J, Trochimczuk A, Wojaczynska M: Acrylic carriers for the immobilisation of enzymes. *Angew Makromol Chem* 1989, 171:201–211

358 Matsumoto K, Izumi R, Seijo H, Mizuguchi H: Immobilisation of biologically active substances. *Jpn Kokai Tokkyo Koho, JP 55,048,392*, 1980

359 Gomoll M, Langhammer G: Immobilisation of biologically active substances on acrylonitrile polymers. *Ger. (East), DD 157342 Z 19821103*, 1982

360 Manecke G, Middeke HJ: Reactive carriers with maleimide groups for the immobilisation of enzymes. *Angew Makromol Chem* 1984, 121:27–39

361 Chen X, Dordick JS, Rethwisch DG: Chemoenzymatic synthesis and characterization of poly(α-methyl galactoside 6-acrylate) hydrogels. *Macromolecules* 1995, 28:6014–6019

362 Chen X, Martin BD, Neubauer TK, Linhardt RJ, Dordick JS, Rethwisch DG: Enzymatic and chemoenzymatic approaches to synthesis of sugar-based polymers and hydrogels. *Carbohydr Polym* 1995, 28:15–21

363 Martin BD, Ampofo SA, Linhardt RJ, Dordick JS: Biocatalytic synthesis of sugar-containing poly(acrylate) based hydro-gels. *Macromolecules* 1992, 25:7081–7085

364 Patil DR, Dordick JS, Rethwisch DG: Chemoenzymatic synthesis of a sucrose-containing polymers. *Macromolecules* 1991, 24:3462–3463

365 Patil DR, Rethwisch DG, Dordick JS: Enzymatic synthesis of a sucrose-containing linear polyester in nearly anhydrous organic media. *Biotechnol Bioeng* 1991, 37:639–646

366 Patil NS, Dordick JS, Rethwisch DG: Macroporous poly(sucrose acrylate) hydrogel for controlled release of macromolecules. *Biomaterials* 1996, 17:2343–2350

367 Patil NS, Li YZ, Rethwish DG, Dordick JS: Sucrose diacrylate: a unique chemically and biologically degradable crosslinker for polymeric hydrogels. *J Polym Sci A Polym Chem* 1997, 35:2221–2229

368 Martin BD, Lindhardt RJ, Dordick JS: Highly swelling hydrogels from ordered galactose-based polyacrylates. *Biomaterials* 1998, 19:69–76

369 Minamoto Y, Yugai Y: Preparation and properties of various enzymes covalently immobilised on polymethylglutamate. *Chem Pharm Bull* 1980, 28:2052–2058

370 Narita T, Hirano T: New polymer containing aminoalkyl group and its application to enzyme- and coenzyme-support. *Polym Preprints (Am Chem Soc, Div Polym Chem)* 1979, 20:878–879

371 Cabral JMS, Kennedy JF: Covalent and coordination immobilisation of proteins. In: Taylor RF (Ed) *Protein immobilisation: fundamentals and applications.* Marcel Dekker, New York, 1991

372 Liberatore FA, McIsaac JE, Royer GP: CL-Sepharose 4B with hexamethylene diamine added after oxidation to aldehyde groups. *FEBS Lett* 1976, 66:45–48

373 Royer GP, Liberatore FA: Immobilisation of enzymes by reductive alkylation. *Enzym Eng* 1978, 3:43–49

374 Blanco RM, Guisan JM: Protecting effect of competitive inhibitors during very intense insolubilized enzyme-activated support multipoint attachments: trypsin (amine)-agarose (aldehyde) system. *Enzym Microb Technol* 1988, 10:227–232

375 Royer GP, Liberatore FA, Green GM: Immobilisation of enzymes on aldehydic matrixes by reductive alkylation. *Biochem Biophys Res Commun* 1975, 64:478–484

376 Axen R, Porath J, Ernback S: Chemical coupling of peptides and proteins to poly-saccharides by means of cyanogen halides. *Nature* 1967, 214:1302–1304

377 Axen R, Erback S: Chemical fixation of enzymes to cyanogen halide activated polysaccharide carriers. *Eur J Biochem* 1971, 18:351–356

378 Tosa T, Sano R, Chibata I: Immobilised d-amino acid oxidase. Preparation, some enzymic properties, and potential uses. *Agric Biol Chem* 1974, 38: 1529–1534

379 Brynda, E, Bleha M: Reversible thermal denaturation of immobilised chymo-trypsinogen. *Collect Czech Chem Commun* 1979, 44:3090–3101

380 Comfort AR, Mullon CJP, Langer R: The influence of bond chemistry on immo-bilised enzyme systems for ex vivo use. *Biotechnol Bioeng* 1988, 32:554–563

381 Nakhapetyan LA, Akparov VK: Thermo-stability of soluble and immobilised subtilisins after their modification by dextrans and dextrins. *Enzym Eng* 1980, 5:423–426

382 Melius P, Wang B-C: Immobilisation of lipase to cyanogen bromide activated polysaccharide carriers. *Adv Exp Med Biol* 1974, 42:339–343

383 Cheetham NWH, Richards GN: Studies on dextranases Part III. Insolubilisation of a bacterial dextranase. *Carbohydr Res* 1973, 30:99–107

384 Groff L, Cherniak R: The incorporation of amino groups into cross-linked Sepharose by use of (3-aminopropyl)-triethoxysilane. *Carbohydr Res* 1980, 87:302–305

385 Lomako OV, Menyailova II, Nakhapetian LA, Kozlovskaya LI, Rodionova NA: Immobilisation of β-glucosidase on an inorganic carrier. *Acta Biotechnol* 1982, 2:179–185

386 Subramanian A, Kennel SJ, Oden PI, Jacobson KB, Woodward J, Doktycz MJ: Comparison of techniques for enzyme immobilisation on silicon supports. *Enzym Microb Technol* 1999, 24:26–34

387 Flemming C, Gabert A, Wand H: Synthesis and properties of insolubilized enzymes. V. Covalent coupling of trypsin, glucoamylase, peroxidase, aminoacylase, and alkaline phosphatase to isothiocyanatopropyldiethoxysilyl glass. *Acta Biol Med Ger* 1974, 32:135–141

388 Huelsmann HL, Janssen P, Renckhoff G, Vahlensieck HJ: Fixing soluble acylases on to inorganic carrier material *Ger Offen, DE 2619571,* 1977

389 Messing RA, Stinson HR: Covalent coupling of alkaline *Bacillus subtilis* protease to controlled-pore silica with new simplified coupling technique. *Mol Cell Biochem* 1974, 4:217–220

390 Li LX, Chen CZ, Yu YT: Reactivity and stability improvement of immobilised glucose oxidase. *Biomat Artif Cell Artif Organ* 1989, 17:183–188

391 Marconi W, Faiola F, Piozzi A: Catalytic activity of immobilised fumarase. *J Mol Catal B Enzymatic* 2001, 15:93–99

392 Marty JL, Coste CM: Application of reponse surface methodlgy to optimisa-tion of immobilisation of ribonuclease A on aldehyde silica. *J Mol Catalysis* 1985, 32:275–283

393 Nilsson N, Mosbach K: *p*-Toluene-sulphonyl chloride as an activating agent of agarose for the preparation of immo-bilised affinity ligands and proteins. *Eur J Biochem* 1980, 112:397–402

394 Nilsson K, Mosbach K: Immobilisation of enzymes and affinity ligands to various hydroxyl group carrying supports using highly reactive sulphonyl chlorides. *Biochem Biophys Res Commun* 1981, 102:449–457

395 Crossland RK, Wells WE, Shiner Jr VJ: Sulphonate leaving groups, structure and reactivity. 2,2,2-trifluoroethane-sulphonate. *J Am Chem Soc* 1971, 93:4217–4219

396 Albayrak N, Yang ST: Immobilisation of *Aspergillus oryzae* β-galactosidase on tosylated cotton cloth. *Enzym Microb Technol* 2002, 31:371–383

397 Nilsson K, Larsson PO: High-perform-ance liquid affinity chromatography on silica-bound alcohol dehydrogenase. *Anal Biochem* 1983, 134:60–72

398 Ballesteros A, Sanchez-Montero JM, Sinisterra JV: *p*-Toluenesulphonyl chloride activation of agarose as exemplified by the coupling of lysine and micrococcal endonuclease. *J Mol Catal* 1986, 38:227–236

399 Arroyl M, Moreno JM, Sinisterra JV: Immobilisation/stabilisation on different hydroxylic support of lipase from *Candida rugosa*. *J Mol Catal* 1993, 83:261–271

400 Laane A, Chytry V, Haga M, Sikk P, Aaviksaar A, Kopecek J: Covalent attachment of chymotrypsin to poly[*N*-(2-hydroxypropyl)methacrylamide. *Collect Czech Chem Commun* 1981, 46:1466–1473

401 Horak D, Rittich B, Sÿafar J, Lenfeld AS Jr, Benes MJ: Properties of RNase A immobilised on magnetic poly(2-hydroxyethyl methacrylate) microspheres. *Biotechnol Prog* 2001, 17:447–452

402 Morikawa Y, Tezuka T, Teranishi M, Kimura K, Fujimoto Y, Samejima H: Dichloro-*s*-triazinyl resin as a carrier of immobilised enzymes. *Agric Biol Chem* 1976, 40:1137–1142

403 Wykes JR, Dunnill P, Lilly M: Immobilisation of α-amylase by attachment to soluble support materials. *Biochem Biophys Acta* 1971, 250:522–529

404 Spagna G, Pifferi PG, Tramontini M: Immobilisation and stabilisation of pectinlyase on synthetic polymers for application in the beverage industry. *J Mol Catal A Chemical* 1993, 101:99–105

405 Horak D, Karpisek M, Turkova J, Benes M: Hydrazide-functionalized poly(2-hydroxyethyl methacrylate) microspheres for immobilisation of horseradish peroxidase. *Biotechnol Prog* 1999, 15:208–215

406 Coombe RG, George AM: An alternative coupling procedure for preparing activated Sepharose for affinity chromatography of penicillinase. *Aust J Biol Sci* 1976, 29:305–316

407 Brandt J, Andersson LO, Porath J: Covalent attachment of proteins to polysaccharide carriers by means of benzoquinone. *Biochim Biophys Acta* 1975, 386:196–202

408 Stambolieva N, Turkova J: Covalent attachment of proteins to Spheron by means of benzoquinone. *Collect Czech Chem Commun* 1980, 45:1137–1143

409 Beddows CG, Gil MH, Guthrie JT: Investigation of the immobilisation of bovine serum albumin, trypsin, acid phosphatase and alkaline phosphatase to poly(hydroxyethyl acrylate)-co-cellulose and poly(hydroxyethyl acrylate)-co-pectin. *Polym Bull (Berlin, Germany)* 1984, 11:1–6

410 Ademola JI, Hutchinson DW: Preparation and properties of an insolubilised phosphotransferase. *Biotechnol Bioeng* 1980, 22:2419–2424

411 Miron T, Wilchek M: Activation of trisacryl gels with chloroformates and their use for affinity chromatography and protein immobilisation. *Appl Biochem Biotechnol* 1985, 11:445–456

412 Akgol S, Kacar Y, Denizli A, Arcab MY: Hydrolysis of sucrose by invertase immobilised on to novel magnetic polyvinylalcohol microspheres. *Food Chem* 2001, 74:281–288

413 Onal S, Telefoncu A: Comparison of chitin and amberlite IRA-938 for β-galactosidase immobilization. *Artifi Cel Blood Substit Biotechnol* 2003, 31:19–33

414 Papisov MI, Maksimenko AV, Torchilin VP: Optimization of reaction conditions during enzyme immobilization on soluble carboxyl-containing carriers. *Enzym Microbial Technol* 1985, 7:11–16

415 Miyazaki JH, Croteau R: Immobilization of cyclase enzymes for the production of monoterpenes and sesquiterpenes. *Enzym Microb Technol* 1990, 12: 841–845

416 Martensson K, Mosbach K: Covalent coupling of pullulanase to an acrylic copolymer using a water-soluble carbodimide. *Biotechnol Bioeng* 1972, 14:715–724

417 Patel RP, Lopiekes DV, Brown SR, Price S: Derivatives of proteins II. Coupling of α-chymotrypsin to carboxyl containing polymers by use of *N*-ethyl-5-phenyl-isoxazolium-3′-sulphonate. *Biopolymer* 1967, 5:577–582

418 Ohtsuka Y, Kawaguchi H, Yamamoto T: Immobilisation of α-amylase on polymeric carriers having different structures. *J Appl Polym Sci* 1984, 29:3295–3306

419 Szajani B, Sudi P, Klamar G, Jaszay ZsM, Petnehazy I, Toke L: Effect of carbodiimide structure on the immobilisationof enzyme. *Appl Biochem Biotechnol* 1991, 30:225–231

420 Szajani B, Boross L: Factors influencing covalent coupling of enzymes on derivatized polyacrylamide gel beads. A survey. *Hung J Ind Chem* 1999, 27:125–130

421 Pizarro C, Gonzalez-Saiz JM, Sanchez-Jimenez JJ, Martinez CB: Application and optimisation of polyacrylamide gels as the support for enzyme immobilisation. *Recent Res Develop Biotechnol Bioeng* 1999, 2:19–36

422 Capet-Antonini FC, Tamenasse J: Kinetic studies on soluble and insoluble urokinases. *Can J Biochem* 1975, 53: 890–894

423 Ito H, Hagiwara M, Takahashi K, Ichikizaki I: The structure and function of ribonuclease T1 XXIV. Preparation and properties of a stable water-insoluble polyacrylamide derivative of ribonuclease T1. *J Biochem* 1977, 82:877–883

424 Heinrichova K, Dzurova M, Ziolecki A, Wojciechowicz M: d-Galacturonan digalacturonohydrolase covalently bound to a polyacrylamide-type support. *Lett Appl Microbiol* 1989, 8:105–107

425 Bagi K, Simon LM, Szajani B: Immobilisation and characterization of porcine pancreas lipase. *Enzym Microb Technol* 1997, 20:531–535

426 Beddows CG, Gil H, Guthrie JT: The use of graft copolymers as enzyme supports. The preparation and use of polyethylene-co-acrylic acid supports. *Polym Bull (Berlin)* 1980, 3:645–653

427 Torchilin VP, Tishchenko EG, Smirnov VN: Covalent immobilisation of enzymes on ionogenic carriers. Effect of electrostatic complex formation prior to immobilisation. *J Solid-Phase Biochem* 1977, 2:19–29

428 Papisov MI, Maksimenko AV, Torchilin VP: Optimisation of reaction conditions during enzyme immobilisation on soluble carboxyl-containing carriers. *Enzym Microb Technol* 1985, 7:11–16

429 Kasumi T, Tsuji M, Hayashi K, Tsumura N: Preparation and some properties of chitosan bound enzymes. *Agric Biol Chem* 1977, 41:1865–1872

430 Roth P, Feist U, Flemming C, Gabert A, Taufel A: Synthesis and properties of carrier-fixed enzymes. VII. Linking of glucoamylase to dialdehyde cellulose, carboxymethylcellulose hydrazide and carboxymethylcellulose azide. *Acta Biol Med Ger* 1977, 36:179–183

431 Miwa N, Ohtomo K: Enzyme immobilisation in the presence of substrates and inhibitors. *Jpn Kokai Tokkyo Koho, JP 56,045,591*, 1975

432 Ngo TT, Ivy J, Lenhoff HM: Polyethylene beads as supports for enzyme immobilization. *Biotechnol Lett* 1980, 2:429–434

433 Kojima H, Suzuki O, Iwamoto K, Okamoto Y: Immobilisation of biologically active substances. *Jpn. Kokai Tokkyo Koho, JP 78–71,770*, 1980

434 Spagna G, Barbagallo RN, Pifferi PG, Blanco RM, Guisan JM: Stabilization of a β-glucosidase from *Aspergillus niger* by binding to an amine agarose gel. *J Mol Catal B Enzymatic* 2000, 11:63–69

435 Bihari V: Diffusional behavior of immobilised glucose oxidase. *J Chem Technol Biotechnol* 1985, 35B:83–93

436 Beitz J, Schellenberger A, Lasch J, Fischer J: Catalytic properties and electrostatic potential of charged immobilised enzyme derivatives. Pyruvate decarboxylase attached to cationic polystyrene beads of different charge densities. *Biochim Biophys Acta* 1980, 612:451–454

437 Chae HJ, Kim EY, In MJ: Improved immobilisation yields by addition of protecting agents in glutaraldehyde-induced immobilisation of protease. *J Biosci Bioeng* 2000, 89:377–379

438 Alkorta I, Garbisu C, Lima ML, Serra JL: Immobilisation of pectin lyase from *Penicillium italicum* by covalent binding

438 to nylon. *Enzym Microb Technol* 1996, 18:141–146
439 He F, Zhuo RX, Liu LJ, Xu MY: Immobilisation of acylase on porous silica beads: preparation and thermal activation studies. *React Funct Polym* 2000, 45:29–33
440 Anton O, Crichton R, Lenders JP: Immobilised enzymes and their use. *Eur Pat Appl, EP 158909 A2*, 1985
441 Kobayashi K, Kageyama B, Yagi S, Sonoyama T: A novel method of immobilising penicillin acylase on basic anion exchange resin with hexamethylene diisocyanate and p-hydroxybenzaldehyde. *J Ferment Bioeng* 1992, 74:410–412
442 Zemek J, Kuniak L, Gemeiner P, Zamocky J, Kucar S: Crosslinked polyethylenimine: an enzyme carrier with spacers of various lengths introduced in crosslinking reaction. *Enzym Microb Technol* 1982, 4:233–238
443 Kay G, Crrok EM: Coupling of enzymes to cellulose using chloro-s-triazines. *Nature* 1967, 216:514–515
444 Mansfeld J, Schellenberger A: Thermostability of immobilized invertase. *Acta Biotechnol* 1986, 6:89–99
445 Ngo TT, Ivy J, Lenhoff HM: Polyethylene beads as supports for enzyme immobilisation. *Biotechnol Lett* 1980, 2:429–434
446 Schnapp J, Shalitin Y: Immobilisation of enzymes by covalent binding to amine supports via cyanogen bromide activation. *Biochem Biophys Res Commun* 1976, 70:8–14
447 Tham SY: A comparison of the different supports for immobilising alkaline phosphatase for enhancing its activity. *J Biosci (Penang, Malaysia)* 1998, 9:51–60
448 Cuatrecasas P, Parikh I: Adsorbents for affinity chromatography. Use of N-hydroxysuccinimide ester of agarose. *Biochemistry* 1972, 11:2291–2299
449 Di Gregorio F, Morisi F: Chemical modification of protein materials by reaction with quinones. *Ger Offen, DE 2,615,349*, 1976
450 Laszlo K, Szava A, Simon LM: Stabilization of various α-chymotrypsin forms in aqueous-organic media by additives. *J Mol Catal B Enzymatic* 2001, 16:141–146
451 Abraham M, Alexin A, Szajani B: Immobilised triosephosphate isomerases. A comparative study. *Biochem Biotechnol* 1992, 36:1–12
452 Fischer J, Ulbrich R, Schellenberger A: The influence of charged matrix surfaces on the thermostabilizing effect of calcium ions on immobilised fungal α-amylase. *Acta Biol Med Ger* 1978, 37:1413–1424
453 Ademola JI, Hutchinson DW: Preparation and properties of an insolubilized phosphotransferase. *Biotechnol Bioeng* 1980, 22:2419–2424
454 Ulbrich R, Golbik R, Schellenberger A: Protein adsorption and leakage in carrier-enzyme systems. *Biotechnol Bioeng* 1991 37:280–287
455 Manecke G, Beier W: Polymers containing quinone groups as carriers for immobilisation of enzymes. *Angew Makromol Chem* 1981, 97:23–33
456 Chernukhin IV, Klenova EM: A method of immobilisation on the solid support of complex and simple enzymes retaining their activity. *Anal Biochem* 2000, 280:178–181
457 Abdel-Hay FI, Guthrie JT, Morrish CEJ, Beddows CG: The use of graft copolymers as enzyme supports. II. The immobilisation of β-galactosidase and glucose oxidase on cellulose-polyacrylamide graft copolymers. *Polym Bull (Berlin, Germany)* 1979, 1:755–761
458 Moeschel K, Nouaimi M, Steinbrenner C, Bisswanger H: Immobilization of thermolysin to polyamide nonwoven materials. *Biotechnol Bioeng* 2003, 82:190–199
459 Bulmus V, Kesenci K, Piskin E: Poly(EGDMA/AAm) copolymer beads: a novel carrier for enzyme immobilisation. *React Funct Polym* 1998, 38:1–9
460 Fiddler MB, Gray GR: Immobilisation of proteins on aldehyde-activated polyacrylamide supports. *Anal Biochem* 1978, 86:716–724
461 Chen C, Yu YT: Immobilised enzyme activity improvement in the glutaraldehyde method. *Biomater Artif Cell Artif Organ* 1989, 17:329–334
462 Antal Z, Beller E, Boross L, Daroczi I, Kalman M, Suto I, Szajani B: Immobilisation of compounds with one or

more nucleophilic groups and preparation of an activated polymer carrier for this immobilisation. *Ger Offen, DE 3,330,573 A1*, 1984

463 Zaborsky OR: Immobilisation of enzymes with imido ester-containing polymers. In: Olson AC, Cooney CL (Eds) *Immobilised Enzymes Food Microb. Processes*, 1974, pp 187–203

464 Mineki S, Yajima H, Goto S, Nakazato K, Ishii T: Properties of bacterial α-amylase immobilised on alternating acrylonitrile-butadiene copolymer by amidination reaction (2000) *Mater Technol* 2000, 18:393–399

465 Abdel-Hay FI, Beddows CG, Guthrie JT: The use of graft copolymers as enzyme supports. IV. The immobilisation of invertase, pepsin, acid and alkaline phosphatases and bovine serum albumin to hydrolysed and reduced nylon-co-acrylonitrile graft copolymers. *Polym Bull* 1980, 2:607–612

466 Chen Y, Mason NS, Sparks RE, Scharp DW, Ballinger WF: Collagenase immobilised on cellulose acetate membranes. *Adv Chem Ser* 1982, 199:483–491

467 Markovic O, Machova E: Immobilisation of pectin esterase from tomatoes and *Aspergillus foetidus* on various supports. *Collect Czech Chem Commun* 1985, 50: 2021–2027

468 Park SW, Choi SY, Chung KH, Hong SI, Kim SW: New method for the immobilisation of glutaryl-7-aminocephalosporanic acid acylase on a modified epoxy/silica gel hybrid. *J Biosci Bioeng* 2002, 94:218–224

469 Arica MY, Bayramoglu G, Biçak N: Characterisation of tyrosinase immobilised onto spacer-arm attached glycidyl methacrylate-based reactive microbeads. *Process Biochem* 2004, 39:2007–2017

470 Dhal PK, Babu GN, Sudhakaran S, Borkar PS: Immobilisation of penicillin acylase by covalent linkage on vinyl copolymers containing epoxy groups. *Makromol Chem Rapid Commun* 1985, 6:91–95

471 Kawai T, Saito K, Sugita K, Sugo T, Misaki H: Immobilisation of ascorbic acid oxidase in multilayers on to porous hollow-fiber membrane. *J Membr Sci* 2001, 191:207–213

472 Marek M, Valentova O, Kas J: Invertase immobilisation via its carbohydrate moiety. *Biotechnol Bioeng* 1984, 26:1223–1226

473 Solomon B, Koppel R, Schwartz F, Fleminger G: Enzymic oxidation of monoclonal antibodies by soluble and immobilised bifunctional enzyme complexes. *J Chromatogr* 1990, 510:321–329

474 Lenfeld J, Svec F, Kalal J: Reactive polymers. LII. Periodate oxidation of poly(2,3-epoxypropyl methacrylate) bound on to porous glass. *Acta Polym* 1986, 37:31–36

475 Lenfeld J, Svec F, Kalal J: Reactive polymers. LIII. Periodate oxidation of poly(2,3-epoxypropyl methacrylate) bound to silica gel. *Acta Polym* 1986, 37:377–381

476 Goldstein L: Immobilised enzymes. Synthesis of a new type of polyanionic and polycationic resins and their utilization for the preparation of water-insoluble enzyme derivatives. *Biochim Biophys Acta* 1973, 315:1–17

477 Basinska T, Slomkowski S: Attachment of horseradish peroxidase (HRP) on to the poly(styrene/acrolein) latexes and on to their derivatives with amino groups on the surface; activity of immobilised enzyme. *Colloid Polym Sci* 1995, 273: 431–438

478 Zuzana B, Marcela S, Antonin L, Daniel H, Jiri L, Jaroslava T, Jaroslav C: Oriented immobilisation of galactose oxidase to bead and magnetic bead cellulose and poly(HEMA-co-EDMA) and magnetic poly(HEMA-co-EDMA) microspheres. *J. Chromatogr B Anal Technol Biomed Life Sci* 2002, 770:25–34

479 Royer GP: Immobilisation of glycoenzymes through carbohydrate chains. *Methods Enzymol* 1987, 135:141–146

480 Katchalski E, Goldstein L, Levin Y, Blumberg S: Water-insoluble enzyme derivatives. *US Pat 3,706,633*, 1972

481 Braun B, Klein E: Immobilisation of *Candida rugosa* lipase to Nylon fibers using its carbohydrate groups as the chemical link. *Biotechnol Bioeng* 1996, 51:327–341

482 Miron T, Wilchek M: Polyacryl-hydrazido-agarose: preparation via periodate oxidation and use for enzyme immobilisation and affinity chromatography. *J Chromatogr* 1981, 215:55–63

483 Bozhinova D, Galunsky B, Yueping G, Franzreb M, Koster R, Kasche V: Evaluation of magnetic polymer microbeads as carriers of immobilized biocatalysts for selective and stereoselective transformations. *Biotechnol Lett* 2004, 26: 343–350

4
Enzyme Entrapment

4.1
Introduction

As the word implies, entrapment of enzymes means that the enzyme molecules or enzyme preparations are confined in a matrix formed by dispersing the catalytic component (a soluble/insoluble enzyme preparation) in a fluid medium (polymer solution), followed by formation of a insoluble matrix with confined enzymes by chemical or physical methods, as illustrated in Scheme 4.1.

The entrapment technique is one of the simplest methods for immobilization of enzymes and whole cell-based immobilized enzymes. Historically, entrapment of enzyme was probably the third developed enzyme immobilization method after covalent enzyme immobilization and adsorption. This method is also characterized in that more than one enzyme can be immobilized simultaneously [2].

Early in the mid 1950s it was reported for the first time that enzymes can be physically entrapped in an inorganic gel matrix (glass) with retention of biological activity [1]. However, the potential of entrapment techniques for enzyme immobilization was given significant attention only in the early 1960s, when Bernfeld and Wan demonstrated that the water-soluble monomer acrylamide could be cross-linked in the presence of the cross-linker N,N-bis-acrylamide and the dissolved enzyme, leading to the formation of entrapped immobilized enzymes with retention of activity [3]. In this case, the enzyme molecules are physically entrapped in the gel matrix formed by polymerization.

Soon after, this method was extended by Mosbach for the entrapment of whole cells and the preparation of beaded immobilized enzymes [5]. Obviously, the ma-

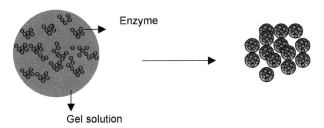

Scheme 4.1 Entrapment of biocatalysts.

Carrier-bound Immobilized Enzymes: Principles, Application and Design. Linqiu Cao
Copyright © 2005 WILEY-VCH Verlag GmbH & Co. KGaA, Weinheim
ISBN: 3-527-31232-3

Table 4.1 Entrapment techniques

Methods	Approach	Remarks	Ref.
Entrapment-template leaching	Soluble additives such as polymers or small compounds such as sugars or surfactants were used	Intended to increase the porosity of the matrix	7
Double entrapment	Two types of entrapment can be consecutively combined	The first matrix for entrapment can be selectively leached, in order to increase the porosity of the matrix	7, 93
Physical entrapment	Non-covalent binding is formed between the enzyme and the matrix	Enzyme molecules are not usually damaged	212
Post-loading-cross-linking entrapment	In contrast with the conventional entrapment technique, enzyme entrapment and preparation of the matrix are separated	Because the enzyme is loaded in a consecutive step, it is possible to control the payload. Diffusion constraints might also be lower than with conventional entrapment	9, 231
Entrapment-in-situ cross-linking	After normal physical entrapment, a cross-linking procedure is applied	Avoiding enzyme leakage and reinforcement of the matrix can be achieved	9
Covalent entrapment	Analogous to physical entrapment, a monomer bearing active functional groups that can form covalent bonds with the enzyme is used	One-step combination of covalent enzyme immobilization and entrapment	62, 184, 185
Cross-linking entrapment	Use of an active pre-polymer than can form an insoluble matrix with enzyme cross-linked in the matrix	Use of an active pre-polymer can avoid the toxicity of the monomer	10, 11
Entrapment encapsulation	After normal entrapment, a permeable layer is formed around the original matrix	Reducing enzyme leakage and increasing the mechanical stability of the matrix	12, 245

trix formed should be permeable to the small substrate molecules. The matrix can be prepared in the shape of beads, membranes, films, disks and fibres.

In those original entrapment techniques the enzymes (or whole cells) had first to be dispersed in a solution and followed by a solidification process, with the aim of forming a physical barrier. Today, formation of the matrix can be realized by several other methods, for example as chemical means (e.g. cross-linking, polymerization) or physical means (physical gelation) or a combination of these two methods. The enzyme molecules can also be physically or covalently confined in the matrix. The latter approach often mitigates leakage of enzyme, as compared with physical entrapment.

Noticeably, many variations of entrapment techniques have appeared in the last few decades, with the aim of solving the intrinsic drawbacks associated with the original entrapment techniques such as covalent entrapment, double entrapment, post-loading entrapment, entrapment and crosslinking, etc., as shown in Table 4.1, or to solve problems that cannot be solved by any of the available immobilization techniques.

Depending on the interaction of the materials used for entrapment with the enzymes to be entrapped, new variations of entrapment method can be classified into: double entrapment; template-leaching; covalent entrapment; cross-linking entrapment; adsorption entrapment; attachment-entrapment; and entrapment-coating.

Currently, there is an increasing trend toward the use of entrapment techniques to immobilize not only whole cells but also isolated enzymes, because entrapment is a simple and easy technique. In addition, the entrapped enzyme in the gel matrix is often stabilized by confinement effects. Also, selection of the appropriate system or variation of the gel components can provide a simple and efficient means of modulating enzyme activity and stability [6].

4.2
Definition of Entrapment

By definition, entrapment of enzymes refers to the processes by which the enzymes are embedded in a matrix formed by chemical or physical means such as cross-linking or gelation. In general, the entrapment matrix is generally formed during the immobilization process. Thus, it is not surprising that the precursors of the gel matrix and the conditions used for the formation should be compatible with the enzyme molecules. The enzyme molecules can be physically embedded or covalently linked to the matrix. Thus, entrapment can be also classified as covalent entrapment and chemical entrapment.

Depending on the method chosen, the precursor used for preparation of the matrix is also different. In most cases of polymerization entrapment, unsaturated monomers and co-monomers are used as cross-linker and the polymerization can be irradiation-initiated [67] or photo [60] or chemically initiated. In physical gelation a solution containing the enzyme (or whole cells) and dissolved polymer is usually gelified under at low temperature by use of poly(vinyl alcohol) (PVA)–cryogel beads [13], salts (e.g. alginate-Ca^{2+} system) [144], or phase inversion by solvent removal, etc.

Many enzymes, hydrophilic and hydrophobic have been entrapped in various gel matrices, for example lipase from *Candida rugosa* in alginate or chitosan [14, 182], whole cells with PGA (penicillin G acylase) activity in gelatin [128] or polyacrylamide gel (PAAm), glucose oxidase (GOx) and glutamate oxidase (GlutOx) in pH-responsive hydrogel [15], glucose oxidase (GOx) entrapped in p(HEMA)-hydrogel microspheres [16], yeast alcohol dehydrogenase (YADH) [59] in poly(acrylamide-co-hydroxyethyl methacrylate) copolymer, cellulose bead-entrapped whole-cell glucose isomerase [17], microbial whole cell in cross-linked ENTP-2000 (synthesized from poly(propylene glycol)-2000, hydroxyethylacrylate, and isophorone diisocyanate) [18].

Remarkably, the geometric properties of the entrapped enzymes can be easily adapted in various forms such as beads, film, fibre, etc., depending on the application and the method of entrapment.

Although it was found that entrapment can lead to serious diffusion limitation, thus reducing the apparent activity of the enzyme [144], there are also many examples of enzymes entrapped in matrix which show that such entrapment can result in retention of activity as good as other enzyme immobilization methods such as covalent [59, 144] or even better [213], because the entrapment method is usually very mild compared with covalent enzyme immobilization. Thus, this method might be extremely useful for the enzymes, which can be easily deactivated by covalent enzyme immobilization [59].

Conventionally, the enzyme is usually entrapped in the polymeric matrix by dispersing the enzyme–polymer solution in an immiscible medium, followed by chemical or physical gelation. Thus, the formation of the non-catalytic function of the entrapped enzyme (in this case, the matrix) and the immobilization occur concomitantly. However, the enzyme can also be entrapped in the ready-made matrix such as Sepharose or Sephadex beads, which are often used as chromatography stationary phases, via a two-step method in which the enzyme is first sucked into the beads and then chemically cross-linked with glutaraldehyde [9].

The original idea of this methodology is to improve the utility of cross-linked enzymes, because of the difficulty of controlling the geometric properties. Compared with the classic method for entrapment of enzyme [20], retention of activity might be high, because the enzyme can freely enter the matrix and the accessibility of the enzyme might be higher than for the enzyme that is homogeneously dispersed in the gel matrix.

In a similar approach, the enzyme can also be selectively partitioned in a suitable hydrogel matrix for the purpose of controlled release [232]. Remarkably, high payload (up to 0.4) can be reached with this method. While this method is specifically designed for the drug delivery, for enzyme immobilization it is not emphasized, because these gels usually have a high swelling factor (~10), and thus the specific activity based on wet weight is normally 10 times reduced. They are, also, unsuitable for use in organic solvents, because the gel will collapse in hydrophilic solvents.

4.3
Requirement of the Carriers

As already discussed, any immobilized enzyme has to contain two functional components, its catalytic and the non-catalytic functions. Thus, the entrapped enzyme, without exception, must contain these two functional components to be suitable for specific applications.

In general, the carrier of entrapped enzymes is prepared concomitantly during the immobilization. Because the process of preparation of the carriers cannot be separated from the immobilization process, the whole immobilization process must render into the immobilized enzyme its two desired functions – the catalytic and the non-catalytic components. Thus, it is often difficult to find a compromise when preparing an immobilized enzyme that not only has the desired catalytic activity but also the desired non-catalytic functions required for control of processes such as re-use and separation of the immobilized enzymes [213]. This is largely because the conditions to achieve good catalytic and non-catalytic properties are often not compatible. For example, the condition for preparation of beaded entrapped enzyme by suspension-polymerization in organic solvents often does not favour high retention of activity.

4.3.1
Physical Requirements

The physical requirements of the carrier include the pore size, particle size, nature of the pore (open or closed), morphology, shape and mechanical stability.

4.3.1.1 Pore Size

Because entrapping is often conducted on the basis of gel forming precursors such as monomers used for the synthesis of synthetic gels, e.g. acrylamide for the preparation of polyacrylamide (PAAm), or by cross-linking prepolymers such as PAAm hydrazine [17], or by gelation of gel-forming polymers such as the synthetic polymer PVA or natural polymers such as gelatin [127], these gel matrix cannot in the strict sense be regarded as porous carriers, because they lack the permanent pores of macroporous synthetic carriers.

In the swollen state, however, they can be regarded as a porous carrier, because these pores are filled with solvent and substrate, as shown in Scheme 4.2

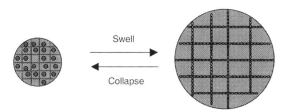

Scheme 4.2 Enzyme molecules immobilized in a gel matrix.

When entrapping enzyme in synthetic polymers such as polyacrylamide, the size of the matrix depends on the concentration of the monomers and the cross-linker. Moreover, the swelling factor is also crucial.

4.3.1.2 Porosity

Like the carriers used for adsorption and covalent binding, an increase in carrier porosity generally leads to an increase in enzyme activity. Consequently, inert templates are often used to reduce the diffusion limitation by increasing carrier porosity.

Often, the diffusion constraints are closely related to the porosity of the gel matrix. For instance, the apparent K_m of acid phosphatase (AP) immobilized in bovine serum albumin-poly(ethylene glycol) (PEG) was found to be increased two to five-fold after immobilization, depending on the chain length of PEG and the bead size. Remarkably, it was found that the half life of the immobilized enzyme was 85 to 200 h, depending on the molecular mass of the PEG used, compared to a half-life of 72 min for the soluble enzyme [10]

4.3.1.3 Geometry

Without exception, selection of the shape/size of an immobilized enzyme depends not only on the method and the carrier but also on the peculiarities of the application. In general, the geometries of the entrapped enzymes can be classified as beaded carrier, fibrous carrier, film, foams, disks, etc. (Scheme 4.3)

It has been observed that each type of geometry is better suited for one application than another. Some types of geometries, for example films are mainly used to prepare specific immobilized enzyme devices such as biosensors (electrodes). Membrane-bound immobilized enzyme is suitable for in-situ biotransformation and separation.

In practice, beaded carriers are widely accepted as conventional carriers, because they usually have good flow characteristics. However, other shapes, for example fibrous carriers, which generally has the advantages of high volume activity, has been also preferred, presumably because of the high surface available for the enzyme immobilization.

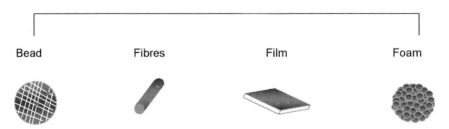

Scheme 4.3 Classification of carriers according to geometry.

4.3.1.4 Particle Size

Although the sizes of entrapped enzymes might vary from method to method, the size is usually severalfold larger than for covalent enzyme immobilization, because entrapped enzyme is generally prepared by dropping the enzyme-gel precursor into an immiscible solution.

It is, on the one hand, conceivable that retention of activity might be generally lower, because of the serve diffusion limitation. On the other hand, this phenomenon can be positively used to improve the non-catalytic properties of some immobilized enzymes. For instance, immobilized enzymes on carriers of small size can be further enlarged by entrapping these pre-immobilized enzymes in another suitable matrix, resulting in improvement of the non-catalytic performance. Thus, the doubled immobilized enzyme can be used in the sieve-plate reactor, in which immobilized enzymes of large particle size are needed to facilitate the separation of the immobilized enzyme from the reactors [19].

4.3.2 Chemical Requirements

The chemical nature of carriers is largely dictated by the precursors used. Like other type of carrier-bound immobilized enzymes, the chemical nature of the carriers of entrapped enzymes can be described by the properties:

- hydrophilicity/hydrophobicity,
- aquaphilicity,
- nature of the active functionality,
- nature of the inactive functionality.

In general, the gel precursors selected are hydrophilic, because they are often compatible with the enzymes [92]. Occasionally, however, hydrophobic gels may afford good performance with regard to enzyme activity and stability [77].

4.3.2.1 Nature of the Active Functionality

The nature of the active functionality varies from less active to very active, depending on the precursors used. In general, the nature of the active functionality refers to the functional groups, which can be cross-linked with each other but also react with the enzymes during the fabrication of the entrapped enzymes. They can be classified into several groups.

Inert Functionality   The functionality in the precursor is inert, which means there is no chemical reaction between the enzyme and the gel matrix, the enzyme is physically entrapped in a physically set inactive gel [128].

Active Functionality   Active functionality not only reacts with itself to form a gel matrix (cross-linked) but can also react with enzymes. The entrapped enzyme can now be regarded as an analogue of a covalently bound immobilized enzyme. Thus, enzyme leakage can be reduced to minimum [186]. However, the enzyme may be deactivated by the chemical modification. Discrepancies in the nature of the active functionality, for example size, charge, and polarity might influence the nature of the binding, for example binding position and number of bonds formed between the enzyme and the carriers.

Studies have revealed that the nature of inert functionality also has a large effect on enzyme activity and stability. For instance, the activity of enzyme entrapped in a gel matrix formed by a sol–gel process depends largely on the nature of the precursor, namely the side chains, suggesting the interaction of the enzyme molecules with this inert functionality might have a large effect on enzyme conformation or orientation [204].

4.3.2.2 Aquaphilicity of the Carriers

Like other types of immobilized enzyme, the hydrophilicity of the matrix of the entrapped enzyme has also large effect on enzyme activity, stability, and selectivity. A systematic study on effect of aquaphilicity is, however, still lacking, because it is very difficult to obtain a set of matrices with variable aquaphilicity.

Interestingly, this was beautifully demonstrated by the observation that the ratios of the two polymers used for immobilizing enzymes such as invertase and glucose isomerase, and yeast cells, in PEG-HEA/PPG-HEA dictated the activity of the entrapped biocatalysts, suggesting that the aquaphilicity does have great influence on the enzyme activity [60].

In the same way as for lipase covalently immobilized on hydrophobic carriers, lipase entrapped in a hydrophobic matrix also displayed preference for the hydrophobic matrix with regard to retention of activity [77].

4.4
Effect of Entrapment

Depending on the method selected, the matrix formed also has a substantial effect on the performance of the immobilized enzymes, for example activity, selectivity, and stability. Thus, it is essential to select the appropriate method, according to the peculiarities of each enzyme and application.

4.4.1
Activity of the Entrapped Enzyme

Retention of activity of the immobilized enzymes is a dimensionless quantity and denotes usually the ratio of the specific activity of the immobilized enzyme (U mg^{-1} protein immobilized) to that of the free enzyme (U mg^{-1} protein).

This term is only valid when the activity of the free enzyme and the immobilized enzyme are measured in the same reaction and medium. When the free enzyme is in lyophilized form or dissolved in aqueous solution, it is difficult to measure the retention of activity in organic solvents. Thus, it is hardly surprising that retention of activity is often based on measurement of activity in aqueous medium. As a result, it often fails to predict the catalytic behaviour of the immobilized enzymes in non-conventional medium (organic solvents), because the activity measured in aqueous medium (based on certain type of reaction such as hydrolytic reaction) does not correlate well with the activity in organic solvents [73]. For this reason, retention of activity is restricted solely to the activity measured on the basis of aqueous reactions.

For entrapment of enzymes in the matrix, retention of activity is found to depend on many factors such as nature of the matrix, particle size (diffusion controlled), enzyme loading, additives, immobilization method, and immobilization conditions.

4.4.1.1 Loading-dependent Activity

Like many other matrix-based immobilized enzyme, the activity of the matrix-entrapped enzyme also depends on the diffusion constraints. For instance, loading-dependent activity was observed for the lipase CRL entrapped in matrix formed by ENTP/ENT prepolymer [77] or in the sol–gel matrix [226].

Remarkably, it was found that CRL entrapped in ENTP 4000 matrix had an activity–loading profile completely different from that of enzyme immobilized in the porous carriers by adsorption or covalent attachment, because the enzyme activity is linearly related to the amount of enzyme added to the matrix [77].

Similarly, the activity of the same enzyme entrapped in cross-linked PVA was strongly dependent on enzyme loading [78].

4.4.1.2 Matrix-dependent Activity

It is often found that one matrix is better-suited than others. For instance, among the various matrices such as cross-linkable albumin and the photo-cross-linked ENT matrix, it was found that the stability of alcohol oxidase entrapped in albumin was greater than that entrapped in ENT, whereas mitochondria were active in the photo-cross-linked ENT matrix only [60].

Compared with ENT-3400 (prepolymer based on polyethylene glycol 3400), lipase *Candida rugosa* immobilized in ENTP-4000 (prepolymer based on poly(propylene glycol) by entrapment gave better overall performance with regard to activity and stability [77].

Obviously, the microenvironment effect has to be taken into consideration. In enzyme entrapment the nature of the matrix can be also divided into different groups, for example hydrophobic, hydrophilic, charged, or neutral.

However, the dependence of enzyme activity on hydrophilicity differs from enzyme to enzyme. For instance, it was found that the activity of invertase immobilized in PUF increased with increasing amount of ethylene oxide hydrophilic segments in the polyols and with increased molecular weight of the polyols [95].

The effect of the hydrophilicity of the matrix on enzyme activity was excellently demonstrated by immobilization of *Mucor miehei* lipase (MML) in alginate and photo-crosslinkable resins of the hydrophilic type ENT 1000. Although the lipase entrapped in alginate had much lower activity, MML immobilized in crosslinked ENT retained 25% esterification activity of the native enzyme [27]. Similarly, the activity of *Candida rugosa* lipase entrapped in the matrix formed by poly(N-vinyl-2-pyrrolidone-co-2-hydroxyethyl methacrylate) (poly[VP-co-HEMA]) hydrogel was also very dependent on the hydrophilicity of the matrix. Lipase immobilized on VP(%):HEMA(%) 90:10 had the highest activity [49]

Retention of activity by cholesterol oxidase immobilized in polyacrylamide gels was four times higher than that in alginate gels [60]. Because polyacrylamide gel is neutral and alginate gel is negatively charged it is suggested that the hydrophobicity of the gel or electrostatic interaction might be crucial for retention of activity. Similarly, it has been found that the ratios of the two polymers (PEG-HMA/PPG-HMA) dictated the activity of entrapped biocatalysts such as invertase, GI, and yeast cells [92].

β-Glucosidase immobilized in calcium alginate (3% alginate, 0.2 M $CaCl_2$) had 66% of native activity whereas 55% of native activity was achieved by entrapment in polyacrylamide (20% AAm and 1.2% bis-AAm) [28]. Moreover, K_m for β-glucosidase immobilized in PAAm is very similar to that of the free enzyme, suggesting that this carrier is neutral and the diffusion limitation in the matrix is less [284].

The effect of the matrix on the activity of immobilized enzyme was excellently demonstrated by entrapment of hydrophobic lipases in PVA matrix. It was found that PVA-entrapped pegylated *Candida antarctica* lipase B was only 10% as active as the native enzyme [284] whereas for lipase entrapped in hydrophobic sol–gel matrix retention of activity was high (50–100%) [205]. In special cases more than 100% retention of activity relative to the native enzyme was achieved, because of the interfacial activation effect for enzymes such as lipase [29].

4.4.1.3 Diffusion-controlled Enzyme Activity

For enzyme entrapped in the gel matrix the guidelines of diffusion constraints apply. Thus, retention of activity is also closely related to the size of the particles [30, 68], porosity [163], and pore size [31]. Often, activity is low because of the presence of diffusion limitation [32].

Immobilization of CRL (*Candida rugosa* lipase) by sol–gel-entrapment led to formation of immobilized lipase with 10–20% of the activity of the native lipase [223], However, the immobilized enzyme was more enantioselective for the synthesis of (S)-ibuprofen ester from racemic ibuprofen in isooctane than the native lipase. The stability to repeated use was also improved by immobilization.

Immobilization of microbial β-D-glucosidase in alginate beads also led to the discovery that the remaining activity of the matrix-entrapped enzymes were below 30% of the original activity [32].

Studies of sol–gel-entrapped bovine carbonic anhydrase has led to the discovery that the specific activity of the encapsulated enzyme is only 1–2% of the value for the enzyme in solution, suggesting that the activity of the immobilized enzyme is mainly determined by diffusion constraints [33].

Several interesting strategies have been developed, to mitigate diffusion constraints; these include the template-leaching technique developed by Prabhune et al. to reduce diffusion constraints for penicillin G acylase immobilized in PAAm matrix with Na alginate as template, which can be leached out with potassium phosphate. Less measurable diffusion limitation was observed for immobilized penicillin G acylase in stirred reactors and the enzyme was highly stable in the batch mode over 90 cycles without any apparent loss in hydrolytic activity [64].

Another widely used technique is the use of inert additives such as sugar, surfactant or PEG; these are applicable to any type of entrapment technique, for example the sol–gel process or physical gel entrapment [34].

4.4.1.4 Conformation-controlled Enzyme Activity

The conformation of the immobilized enzyme is often dictated by the conformation induced during immobilization.

This is particularly true, when the enzyme is entrapped in rigid sol–gel matrix, because the mobility of the immobilized enzyme is extremely low. Thus, it is understandable that selection of a matrix of suitable nature or the presence of the additives (conformation selectors) might improve enzyme activity.

The effect of the nature of the matrix on enzyme activity was recently demonstrated by studying the activity of *Pseudomonas* sp. lipase in PEO matrix [35]. It was found that the reaction rate catalysed by the lipase PS entrapped in PEO is dependent on the nature of the solvents used. The reaction rate was, however, usually enhanced compared with that of free enzyme powders (Scheme 4.4). More interestingly, it was found that the enantioselectivity of the immobilized lipase was also enhanced (see Scheme 4.5, see below), suggesting that the enzyme entrapped in the

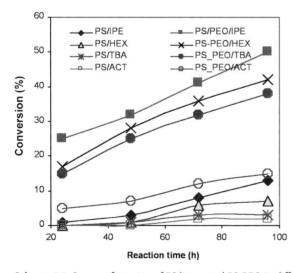

Scheme 4.4 Course of reaction of PS lipase and PS-PEO in different organic solvents.

PEO matrix might adopt a new conformation, which is responsible for the high enantioselectivity.

Remarkably, by use of additives the specific activity of enzymes such as BcL or PfL lipases entrapped in sol–gel matrix can be increased to five times that of the enzyme powder [204].

4.4.1.5 Additives-controlled Enzyme Activity

Like other types of immobilized enzyme, the activity of many immobilized enzymes obtained by entrapment also depends on the additives [103, 202, 209, 218]. This is especially true for enzymes entrapped in sol–gel matrix [203, 294]. Additives added during the enzyme-entrapment process usually have the following functions:

- mitigation of diffusion constraints [294],
- increase of the porosity (the additives function as templates for the formation of pores) [7],
- effect on enzyme activity [218],
- modulation of enzyme conformation, for instance by imprinting [55, 338].

Some additives might enhance the effective diffusion coefficient by promoting a highly porous matrix structure, as demonstrated by the immobilization of urease in a matrix prepared with HEMA (2-hdroxyethylmethyacrylate) and/or N-vinyl-2-pyrrolidone (VP) [57].

Many compounds can be used as additives for the sol–gel process, for example crown ethers, polysaccharides, cell-free extract, PEG, surfactants, metal salts [209, 210, 212, 215, 218], in order to influence activity of the corresponding entrapped enzymes.

The additives selected are, however, largely dictated by the method of immobilization. For entrapment of enzyme in soft polymeric matrix, the function of additives might be mainly limited to mitigation of the diffusion constraints [7, 294]. Selective leaching off of the additives apparently increases the porosity of the matrix, thus leading to high retention of activity [7, 128].

4.4.2
Stability

Although it has very often been found that covalently immobilized enzymes or carrier-adsorbed enzymes could be less stable than the native enzyme [36], few entrapped enzymes are less stable than the native enzymes [37, 179].

In contrast, many examples have shown that entrapped enzymes in the gel matrix might be more stable than the adsorbed enzymes [38] or the covalently carrier-bound enzymes. For example, the glucoamylase adsorbed on PEI-coated non-porous glass followed by cross-linking was less thermostable than the free enzyme [39] whereas the enzymes entrapped in the polymeric matrix formed by polarization phenomena were found to be more stable than the native enzymes.

In contrast, *Candida rugosa* lipase entrapped in alginate gel was found to be more stable than the covalently bound enzyme on Eupergit C or the encapsulated enzyme in a sol–gel matrix [40]. Immobilized glucoamylase entrapped in polyacrylamide gels was more stable than that bound to SP-Sephadex C-50 [38].

Another striking example is that pronase and CT covalently attached to PDMS film are less stable than the entrapped enzymes [41].

It has recently been demonstrated that simple entrapment of α-amylase in PAAm matrix enhanced the thermal stability of the enzyme approximately fivefold at 60 °C compared with the native enzyme [42].

Remarkably, lipases entrapped in sol–gel matrix are usually more stable than the native enzyme. For instance, the half-life of lipase in sol–gel matrix was 51 times longer than that of the native enzyme [335].

These examples clearly demonstrate that might be common mechanism of stabilization of the enzyme in these entrapment techniques.

4.4.2.1 Confinement-determined Stability

Analysis of sol–gel-entrapped proteins has revealed that the crowding effect, hydration effect [43], and confinement effect [45, 214] play a crucial role in stabilization of the entrapped enzyme. Indeed, these observations are also valid for other types of immobilized enzyme.

It has been found that sol–gel-entrapped enzymes could be several orders of magnitude more thermally stable than the free enzyme or the same enzyme immobilized on the surface of a sol–gel sphere [212, 218, 219, 225, 226]. Although the degree of crowding and confinement might differ from method to method, these effects are obviously important for all entrapped enzymes.

That the stability of glucoamylase immobilized in crosslinked polyacrylamide gel was highly dependent on the degree of crosslinking suggested that the pore might become smaller as the degree of crosslinking increases [47].

A similar effect of stabilization by confinement has also been found for other types of entrapped enzyme. For instance, mesophilic β-galactosidase from *Aspergillus oryzae* entrapped in agarose gel also had enhanced thermal stability, because of the reduced freedom of the peptide chains [46]. Compared with other types of immobilized enzyme, the entrapped enzyme might have the advantage of high stability [47].

4.4.2.2 Matrix-nature-dependent Stability

However, stability is also found to be dependent on the properties of the monomers or the properties of the matrix [60, 49] or even on the temperature [39]. For instance, matrix-dependent thermostability was observed for β-D-glucosidase immobilized in polyacrylic polymeric matrix [61]. Pegylated OPH entrapped in PGA-BSA based matrix was more stable than the soluble enzyme preparation [79].

Interestingly, it was found that at high temperatures the thermostability behaviour of the most stabilized entrapped enzyme lipase CRL in alginate gel was similar to that of the free enzyme [39], suggesting that the stabilization mechanism and scope might be temperature-dependent.

The matrix used is often hydrophilic in nature; that this also is an important stabilizing factor was revealed by study of the effect of different monomers on the stability of the entrapped glucosidase in synthetic gel matrices [61]. Also, oxidase entrapped in an albumin-based matrix was more stable than that entrapped in a synthetic gel matrix prepared from PEG-HMA/PPG-HMA [92].

The techniques used for preparation of stable entrapped enzyme preparations can be classified into the following groups:

- enhancement of the confinement effect using the sol–gel techniques (the confinement is maximized),
- stabilization-entrapment,
- double-immobilization technique,
- entrapment-cross-linking.

In sol–gel-entrapment maximum confinement can be achieved because of shrinkage of the gel (the pore size decreases from 20 nm to 2–5 nm) during ripening of the gel. Thus, the unfolding of the enzyme can be greatly reduced. Stabilization-entrapment is based on the fact that an enzyme stabilized, e.g., by chemical modification can be further immobilized by suitable entrapment techniques. In double immobilization a ready-made immobilized enzyme with improved stability can also be entrapped. Finally the entrapped enzymes can be subjected to further chemical cross-linking, with the aim of improving enzyme stability [50].

Sol–gel-entrapment of bovine carbonic anhydrase, a very labile enzyme, revealed that encapsulation in a sol–gel matrix significantly changed the deactivation pattern. It was found that encapsulated enzyme is more thermostable than the free enzyme [33].

The stability and properties of mushroom tyrosinase entrapped in alginate, polyacrylamide and gelatin gels was found to be strongly dependent on the nature of the matrix. The enzyme in gelatin had 1–2-fold higher storage stability and thermal stability at 40 °C compared with the other preparations [134]

4.4.2.3 Enzyme-dependent Stability

It has been found that the stability of three flavoprotein oxidases, glucose oxidase, lactate oxidase, and glycolate oxidase is enzyme-dependent. The half-life of glucose entrapped in a sol–gel matrix oxidase at 63 °C increased 200-fold on immobilization; the half-lives of lactate oxidase and of glycolate oxidase were not extended beyond those of the water-dissolved enzymes [51].

If lactate oxidase (IP pH 4.6) was electrostatically complexed with the weak base poly(N-vinylimidazole) before its immobilization, most of its activity was retained and its half-life at 63 °C increased 150-fold. Lactate oxidase was also stabilized when electrostatically complexed with the stronger base poly(ethyleneimine) before immobilization. Glycolate oxidase (IP pH 9.6) was not stabilized by poly(N-vinylimidazole) but was stabilized by poly(ethyleneimine) complexation before immobilization. The complexed enzyme retained its initial activity on immobilization, and its half-life at 60 °C also increased 100-fold. The results show that encaging an oxidase in a silica gel can lead either to a gain in stability or to loss of activity and that electrostatic complexing is required for stabilization by encapsulation of some, but not all, flavoprotein oxidases.

Activation or stabilization of the enzymes by chemical modification before entrapment has attracted some attention during recent decades. di-Me adipimidate (DMA)-cross-linked *Escherichia coli* β-galactosidase entrapped in PAAM was 50% more active than the native enzyme entrapped in PAAm [290].

4.4.2.4 Enzyme Structure-dependent Stability

The stabilizing effect of the method of immobilization is ultimately dictated by any change in the structure of the enzyme when the application changes. If the enzyme structure under application conditions is stabilized by the immobilization, enhanced stability can be expected.

In contrast, the immobilized enzyme might be not significantly stabilized. For instance, amyloglucosidase entrapped in PVA matrix prepared by photo-cross-linking of PVA bearing styrylpyridinium groups had only slightly better pH and temperature stability than the native enzyme, suggesting the immobilized enzyme is only slightly stabilized by this method of immobilization [78].

4.4.3 Selectivity

Like other enzyme immobilization techniques such as the covalent enzyme immobilization and adsorption, physical entrapment of enzyme is also able to affect enzyme selectivity even without involving chemical modification [161], suggesting that the enantioselectivity of an enzyme is not only determined by the sequence of the amino acid residues but also by the microenvironment (which can be changed by the use of additives or the carriers).

Consequently, improvement of enzyme enantioselectivity by entrapment immobilization might be attractive, because of its simplicity and universal applicability. In general, the improvement of enzyme selectivity by entrapment can be ascribed to the following effects:

- microenvironment,
- diffusion constraints,
- conformational change induced by the conformer selectors.

4.4.3.1 Microenvironment-dependent Selectivity

The microenvironment of the entrapped enzymes is closely related to enzyme conformation. Thus, 2-mannosidase entrapped in sodium alginate [52] had a different product spectrum. Similarly, the product selectivity of glucoamylase entrapped in gelatin matrix was also changed compared with the enzyme immobilized on the surface, probably because of diffusion limitation [53].

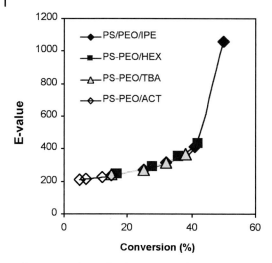

Scheme 4.5 Relationship between E and conversion of PS lipase entrapped in PEC (data from [35]).

Entrapment of RML (*Rhizopus miehei* lipase) improved enantioselectivity compared with the native enzyme [338] and increased threefold the enantioselectivity of pegylated PCL (*Pseudomonas cepacia* lipase) in Ca-alginate gel beads [54].

Interestingly, it was recently found that Pseudomonas sp. lipase (PSL) immobilized in poly(ethylene oxide) (PEO) was more active and selective than the native enzymes for esterification of R,S-methyl mandelate whereas the enzyme entrapped in agar had no activity [35].

Remarkably, the selectivity is increased with increasing conversion as shown in Scheme 4.5. Also, all the data obtained from different solvents seem to fit the same trend and the highly active enzyme is also more selective. This suggests that the structure of the enzyme might be changed during the reaction or the diffusion constraints of the immobilized enzyme were changed. Otherwise the E value should be constant.

Remarkably, sol–gel-entrapment is also applicable to the imprinting of enzymes, as in sol–gel-entrapped lipase deposited on celite [338], because the sol–gel matrix can trap the conformation induced under certain conditions, for instance in the presence of inhibitor or substrate or substrate analogues or conformer selectors [55]. Thus, addition of additives such as surfactants or crown ethers is found to affect enantioselectivity and enzyme activity [204].

4.4.3.2 Conformation-dependent Selectivity

As with the conformation change induced by conformer selectors, it has been found that PFL (*Pseudomonas fluorescens* lipase) entrapped in sol–gel matrix in the presence of the 18-crown ether-6 was almost twice as enantioselective as without the additive, suggesting the conformation of the enzyme might be altered by the presence of additive [204].

This effect of additives on enantioselectivity is, however, largely dependent on the enzymes used. In the same publication it was demonstrated that the additive had almost no effect on CaLB or BcL lipases, although the activity of the enzyme (U mg^{-1} protein) was enhanced severalfold, suggesting that the role of the additive is complex [204].

Most likely, the presence of the crown ether or surfactant can alter not only the structure of the sol–gel matrix but also the conformational flexibility of the enzyme. The additives might improve the porosity of the carrier, thus mitigating diffusion constraints and improving diffusion-controlled enantioselectivity [56].

4.4.3.3 Carrier Nature-dependent Selectivity

It has been found that lipase immobilized in a sol–gel matrix formed by DMDMOS/TMOS on Celite 545 was 3–10 times more active than that disposed on Celite, depending on the reaction temperature between menthol and butyric acid. More interestingly, it was found that the enantioselectivity increased with increasing chain length of the alkyl groups, suggesting that the nature of the carrier, for example surface pendant groups, might affect enzyme conformation [332].

4.5
Preparation of Various Entrapped Enzymes

As noted above, many methods are known for preparation of entrapped enzymes. Compared with other types of immobilized enzyme, for example immobilized enzymes prepared by covalent attachment to the ready-made carrier or non-covalent adsorption on functionalized adsorbents, preparation of entrapped enzyme is usually not straightforward. In particular, it is very difficult to realize the desired catalytic functions (activity, stability, and selectivity) and non-catalytic functions (geometric properties such as shape and size, etc.) concurrently.

It is, therefore, hardly surprising that one characteristic of a robust method of enzyme entrapment is often that many related methods are combined, with the objective of solving the problems that are unattainable with a single entrapment method. For instance, many conventional entrapment methods might have the drawback of enzyme leakage. However, modification of the enzyme with a polymer might enhance the interaction of the entrapped enzyme molecules with the matrix, thus reducing the possibility of enzyme leakage [274]. Similarly, sol–gel-entrapped enzymes are usually very stable. The particle size of these resulting inorganic sol–gel-entrapped enzymes is, however, usually below the low limit for a carrier-bound immobilized enzyme (100 µm). This limits their application as robust immobilized enzymes in many processes in which the particle size might be crucial for the separation of the enzyme from the heterogeneous reaction mixture.

To his end, impregnation was combined with sol–gel-entrapment techniques, i.e. the pre-gel solution bearing the sol–gel precursors and enzymes is sucked into the pores of ready made carriers such as resin plate or beaded porous carriers, then was gelified by conventional sol–gel processes [355, 336].

Examples of these "combi-methods" of enzyme entrapment are:
- double entrapment,
- modification/entrapment,
- entrapment/crosslinking,
- concomitant entrapment/covalent attachment,
- impregnation/entrapment,
- supported entrapment, and
- entrapment and coating.

The reader should be aware there is no universally applicable method of enzyme entrapment and judicious choice of method requires knowledge about the peculiarity of a given application.

4.5.1
Conventional Entrapment Process

Conventional entrapment of enzymes is based on the inclusion of enzyme molecules in a matrix formed by physical or/and chemical gelation of the pregel components, which can be monomers such as acrylamide, used in the first example of enzyme entrapment in a crosslinked polyacrylamide matrix [5], or a pregel polymer, e.g. gelatin, BSA (bovine serum albumin), PVA, alginate, chitosan, etc. [178, 283].

In general, the enzyme molecules are only physically or mechanically included in the formed matrix. Thus, no covalent bonds between the enzyme and matrix are formed. However, some entrapment conditions such as irradiation might be hazardous to the enzyme, leading to enzyme deactivation by the entrapment processes [24].

4.5.1.1 Formation of Entrapment Matrix by Chemical Cross-linking
Classic entrapment technology is the technique by which the enzyme is dispersed in solution containing the gel-forming precursors, followed by a dispersion of the enzyme-gel precursor solution in a second immiscible phase and consecutive solidification of the dispersed solution by a chemical or a physical means, as illustrated in Scheme 4.6.

Depending on the means used for the formation of solid particles, traditional entrapment technology can be classified as either chemical cross-linking and or physical cross-linking, for example gelation.

Polymerization of Monomers Classical entrapment is based on polymerization of the enzyme solution containing monomers and co-monomers bearing double bonds. The monomers most often used are acrylamide, acrylic acid with glycidy-

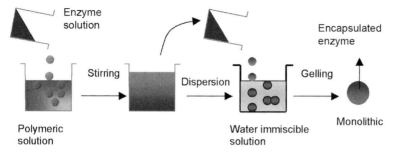

Scheme 4.6 Conventional encapsulation of an enzyme. The immobilized enzyme preparation is actually a monolithic composite (non-porous).

lacrylate, 2-hydroxyethylmethylacrylate, *N*-vinylpyrrolidone, 2-hydroxylpropylamide, 2-hydroxypropylacrylate, poly(ethylene glycol) methylacrylate, butanediolacrylate, and ethyleneglycoldiacrylate, as illustrated in Scheme 4.7.

The combinations of monomers and cross-linkers most often encountered are:
- AAAm-BisAAm
- AAAm-2-HEMA-BisAAm
- HEMA-DMAEMA-BisAAm
- NIPAAm-BisAAm
- NIPAAm-AAc-BisAAm
- HEMA-BisAAm

Occasionally other specially designed monomers, for example 1-acryloylpiperidine-4-spiro-2′-(1′,3′-dioxacyclopentane) can be used, with the objective of modulating the enzyme properties. Examples of these polymeric gels, which are used for entrapped enzyme are poly(hydroxyethyl) methacrylate-co-dimethylaminoethyl methacrylate, (p(HEMA-DMEMA)) hydrogel for entrapment of glucose oxidase

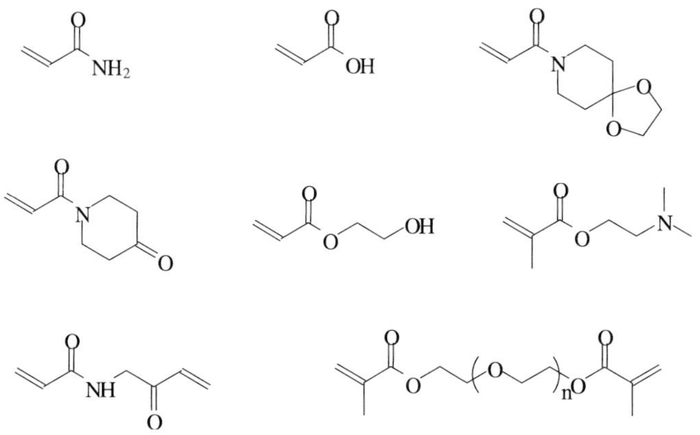

Scheme 4.7 Monomers used for polymerization entrapment.

[16], poly(acrylamide-co-hydroxyethyl methacrylate) copolymer for entrapment of yeast alcohol dehydrogenase (YADH) [21], copolymer of acrylamide and 2-hydroxyethyl methacrylate for α-chymotrypsin [59] and arginase [22], polyacrylamide for α-amylase [42].

Enzyme performance such as activity, stability, or selectivity are highly dependent on the nature of the monomers, the concentration of the crosslinker, the amounts of monomer and comonomer, and initiator/radical concentration [23].

Occasionally additives such as surfactants or inert polymers such as PEG and PVA [25] are needed to improve enzyme activity by protection of the enzyme molecules from deactivation or by enhancing the porosity of the matrix and thus reducing the diffusion constraints [57] or enlarging the pores. For example, it has been found that poly(ethylene glycol) (PEG) 1500, PEG 6000, and poly(N-vinylpyrrolidone) (PVP 10,000) are effective additives for entrapment of acid phosphatase in poly(2-hydroxyethyl methacrylate) [26].

Cross-linking of Prepolymers In a new variation of classic entrapment, a variety of soluble prepolymer (e.g. polyacrylamide, PVA, PEI, PAA, and poly(vinylpyrrolidone) (PVP)) or prepolymers bearing active functionality (e.g. PVA-sbQ (styrylpyridinium), PEG derivatives bearing photocrosslinkable groups [76], or other crosslinkable groups (e.g. amide groups, cross-linked with transglutaminase) or polyurethane prepolymers) can be also used to entrap enzymes; the prepolymers are used instead of the monomers, because the latter might poison the entrapped enzyme, owing to the higher reactivity of monomers such as acrylic amide.

To form an insoluble matrix, these prepolymers must be cross-linked to form the desired matrix by addition of a chemical cross-linker or by light [80] or by use of a radical and irradiation, depending on the properties of the prepolymers used, as illustrated in Scheme 4.8.

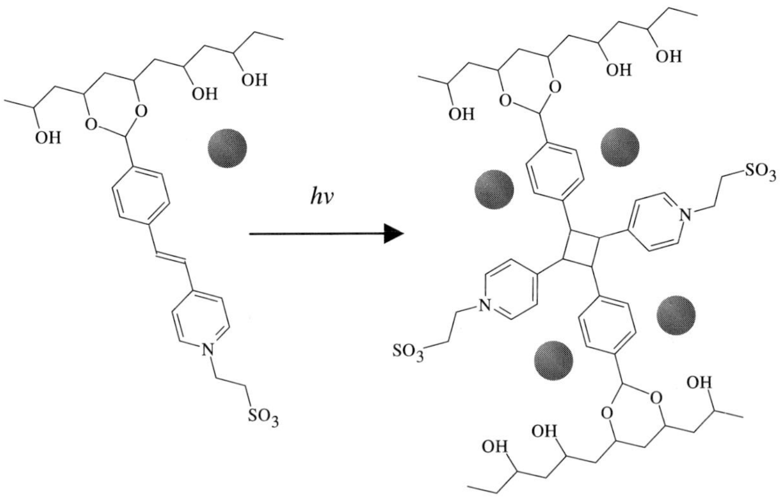

Scheme 4.8 Entrapment of enzyme in a matrix formed by photo-initiated polymerization.

4.5 Preparation of Various Entrapped Enzymes

Table 4.2 Enzyme entrapment with synthetic polymeric matrix

Synthetic carrier	Method	Comment	Enzyme	Ref.
Poly(HEMA) or poly (VP) gel with PEG or PPG	Entrapment	The presence of PEG or PPG enhanced the enzyme activity due to the formation of highly porous structure	Urease	57
PAAm	Entrapment	Polymerizable NAD derivatives	Formate dehydrogenase	58
Poly(AAm–co-HEMA)	Entrapment	90% retention of activity; improved thermal and storage stability	YADH	59
Polyacrylamide gels	Entrapment	Fourfold higher retention of activity than alginate gels	Cholesterol oxidase	60
Poly(1-APSDCP-BisAAm)	Entrapment	Enhanced thermostability; stability is dependent on the properties of the monomers	β-d-Glucosidase	61
PAAm	Entrapment	Photo-polymerization	α-Amylases and glucoamylase	62
PAAm	Entrapment	Among alginate, agarose, agar, carrageenan, polyurethane, and polyacrylamide, polyacrylamide was the best	B. stearothermophilus G-82 cells	63
PAAm	Entrapment	Use of alginate as template facilitates diffusion of the substrate	Penicillin G acylase	64
Poly(NIPAAm-co-AAc)	Entrapment	Glucose-responsive gel, because of the production of gluconic acid	GOD	65
Poly(HEMA) gel	Entrapment	47% retention of activity	Urate oxidase	66
Poly(2-HEMA-co-DMA)	Entrapment	Enzyme activity and leakage depend on irradiation strength.	GAL, INT, and GAD	67
Poly(2-HMA-co-BisAAm)	Entrapment	35% retention of activity	β-Galactosidase	68
NIPA-acrylamide copolymer	Entrapment	The enzyme was entrapped within a thermo-sensitive gel membrane	Asparaginase	69
Poly(NIPA)-based microspheres	Entrapment	The activity of the enzyme–gel system was completely "shut-off" at 50°C, the temperature at which the activity of the free enzyme was maximum	β-d-Galactosidase	70
Poly(2-hydroxyethyl methacrylate)	Entrapment	Similar K_m to the native enzyme; slight leakage	l-Asparaginase	75

4 Enzyme Entrapment

Using prepolymers instead of monomers for entrapment of enzymes is advantageous in that the poisoning effect of the monomers on the enzymes and, in particular, the toxic effect on whole cells, can be mitigated [78, 178]. For instance, immobilization of amyloglucosidase in gel cross-linked with PVA-styrylpyridinium groups does not damage the enzyme molecules [80, 297].

It is essential to point out that the nature of the gel, for example its hydrophilicity, is also of crucial importance with regard to enzyme performance such as activity and stability. Usually, for lipases, interface-active enzymes, hydrophobic gels such as ENTP-based gels often give better results than the hydrophilic gels [77].

Mixing the enzyme solution with a solution of macromolecular monomer that is able to react with enzyme to form insoluble matrix with the enzyme to be entrapped is another variation of this technology. This method is, however, rarely used to prepare bulky immobilized enzymes. It is often used to fabricate renewable surface electro-biosensors. The polymeric systems used for gel formation include hydrophobic two-component epoxy-amine resins derived from bisphenol-A–epichlorohydrin; glycidylacrylate/methacrylate-containing polymers, glycidyl-terminated poly(ethylene glycol) (PEG) or glycerol propoxylate.

In-situ formation of epoxy-amine resins can be used to entrap lipase in the presence of surfactant. For instance, lipase or other enzymes have been entrapped in the gel matrix by mixing sorbitol-ethylene oxide 80, epichlorohydrin, PEI and lipase (or alkaline protease + tris buffer) [71].

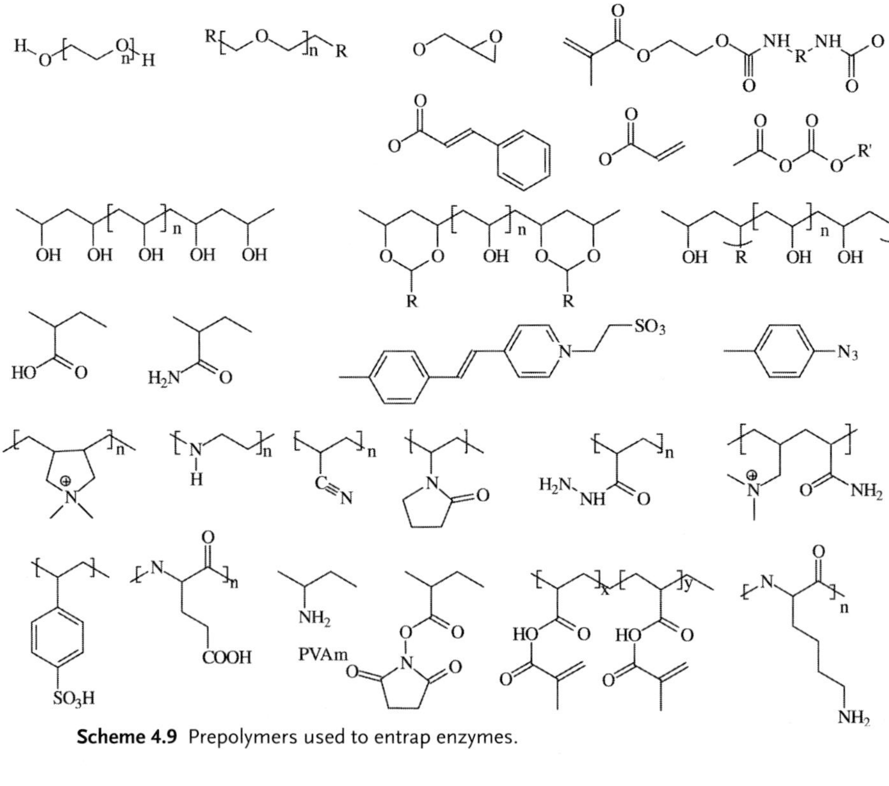

Scheme 4.9 Prepolymers used to entrap enzymes.

4.5 Preparation of Various Entrapped Enzymes

Table 4.3 Entrapment of enzymes by cross-linking the prepolymers via photocrosslinking

Prepolymer	Method	Remarks	Enzyme	Ref.
ENTP-4000/ENT-3400	Photo-cross-linking	Immobilization led to enhancement of enzyme activity up to 2.8-fold, improvement of ester yield, and alleviation of alcohol inhibition of enzyme activity. Moreover, the enzyme activity is also loading-dependent	Lipase CRL	77
PVA-styrylpyridinium	Photo-cross-linking	Enzyme activity and immobilization efficiency depends on the degree of cross-linking and the molecular size of the enzyme	Enzymes	78
PEG-CA gel	Photo-curing	IME was more stable than soluble enzyme preparations	Pegylated OPH	79
Poly(2-CEVE-co-NVP)-stilbazole photosensitive prepolymer	Entrapment by photo-cross-linking	Enzyme molecules remain intact. The activity of IVT, GA and CAT was 30, 60, and 35%, respectively	Invertase, GA, and catalase	80
ENT-4000	Photo-cross-linking	Enzyme films were stored in buffer at 4 °C without significant loss of enzyme activity for up to 2 years	β-Glucosidase	81
ENT-2000	Photo-cross-linking	The immobilization techniques using DEAE-Cellulofine and ENT-2000 were suitable for the bromoperoxidase reaction	Bromoperoxidase	82
PVP	γ-Ray irradiation	30% retention of activity, no leakage at low enzyme concentrations	β-Galactosidase	83
PVA	γ-Ray irradiation	Payload is 0.046 and 0.03 for amylase and invertase respectively (enzyme leakage?)	Amylase/invertase	84
Photocross-linkable polyurethane (FPCPU) with double bonds on ends	Entrapment	Covalent coating on the fibre-micelle surface	BSLC producing α-amylase	85

Table 4.4 Formation of entrapment matrix by reactive prepolymers

Prepolymer	Methods	Remarks	Enzyme	Ref.
Urethane prepolymer	Entrapment	Double entrapment; enhanced activity and stability	BYLC entrapped in alginate gel	93
Hydrophilic urethane prepolymer: Hypol HFP2002/Hypol FHP8190H	Entrapment	The immobilized enzyme was equally active for the large substrates up to 200 kD or for the small molecules, suggesting that the diffusion limitation is less. The immobilized enzyme was stabilized	Amyloglucosidase	94
Urethane prepolymer (PEG)	Entrapment	Activity decreased with increasing NCO content; Immobilized invertase was stable over >30 batch reactions without loss of activity	Invertase	95
Prepolymer HYPOL 3000	Entrapment	Highest activity; covalent attachment	β-d-Galactosidase	96
A polyurethane hydrogel prepolymer, or polymethylene isocyanate	Entrapment	Enzyme molecules are either covalently bound to the carrier formed by polymerization of the prepolymer or entrapped in the pores	Enzymes	97
Urethane prepolymer	Entrapment	Entrapped	Cytosine deaminase	98
Polyurethane (PU-6) prepolymers	Entrapment	Water-induced polymerization	Glucose oxidase	99
Urethane resin	Entrapment	Hydrophobic gels are suitable for use in organic solvents	Enzymes	100
Acrylated-polyurethane photocurable polymer	Entrapment	–	Glucose oxidase?	101
Epikote DX 255 (Epoxy prepolymers)	Casamide (curing agent)	Epoxy beads can be prepared with the use of Na alginate/CaCl$_2$	PGA activity of *Escherichia coli* cells	102
PUF	Reactive entrapment	Internal mass transfer dictated the enzyme activity. Addition of surfactant affected the structure of the carrier	Diisopropyl-fluorophosphatase	103

4.5 Preparation of Various Entrapped Enzymes

Table 4.5 Formation of entrapment matrix by cross-linking of the prepolymers with use of a bi-functional cross-linker

Prepolymer/cross-linker	Method	Remarks	Enzymes	Ref.
PAAm-hydrazide	Entrapment-cross-linking	Prepolymer instead of monomer can avoid the toxic effects of the monomer	Cells, organelles, and enzymes	104
PAAm-hydrazide	Entrapment-cross-linking	Prepolymer instead of monomer can avoid the toxic effects of the monomer	Cells, organelles, and enzymes	105
Poly(AAm-co-MAAm) hydrazides	Entrapment-cross-linking	pH sensor	ACE	106
PVP	Entrapment-cross-linking	A matrix cross-linked with γ-ray irradiation	ASA, UA, LDH, GOD, GDH, POX	107
PCS	Entrapment-cross-linking	Entrapment-coating on the surface of electrode	OPH	108
Polyazetidine	Entrapment	Polymeric membrane coated with prepolymer–enzyme mixture	BCE	109
Tri-functional polyaziridines	Entrapment	Prepolymers enable both covalent immobilization and entrapment. Immobilization can be performed either in water or organic solvent	Enzymes	110
Polyfunctional aziridines	Entrapment	Cells may be formed as a coating on a solid inert carrier	Microbial cells	111
Polyazetidine (PA)		Grafting carboxylic Nylon membrane?	l-Glutamate oxidase (GAO)	112
Prepolymer	Entrapment-cross-linking	Self-mounted enzyme membranes, directly coating the surface of glass pH electrodes	ACE, urease, penicillinase	113
Epoxy-diacrylate prepolymer	Encagement	Grafting of prepolymer on membrane (cellulose)	Catalase	114
Condensation product of CD-GAH	Entrapment		Glucose oxidase (GOD)	115
MEEP hydrogel	Entrapment	80% retention of activity	Urea amidohydrolase (urease)	116
PAAm hydrazide	Chemical cross-linker	Cross-linked with glyoxal	Enzymes, whole cells	117
Polyacrylamide-hydrazide (PAAH)	Chemical cross-linker	—	Enzymes	118
PAAm hydrazide	Chemical cross-linker	Cross-linked with glyoxal	Monooxygenase activity of RLM	119
Cellulose acetate-DMF/Pt wire	Entrapment	The indicator electrode was stable for a period of at least 30 days	LDH	120

4.5.1.2 Physical Entrapment

Another widely used approach for entrapping enzyme in gel matrix was developed in the 1970s. It is based on entrapment of enzyme or whole cells in the physical gels formed by the natural polymers e.g. alginate, gelatin, synthetic polymers, and the semi-synthetic hybrid gels.

The often used naturally occurring polymers include polypeptide (albumin, gelatin, collagen, casein), polysaccharide (chitosan, agar or agarose, starch, kappa-carrageenan, cellulose, pectins, galactomanans, xanthan) and their derivatives such as ethylcellulose, propyl aginate, whereas synthetic polymers often used for this purpose include PVA, PEI, PAA and other synthetic polyelectrolytes.

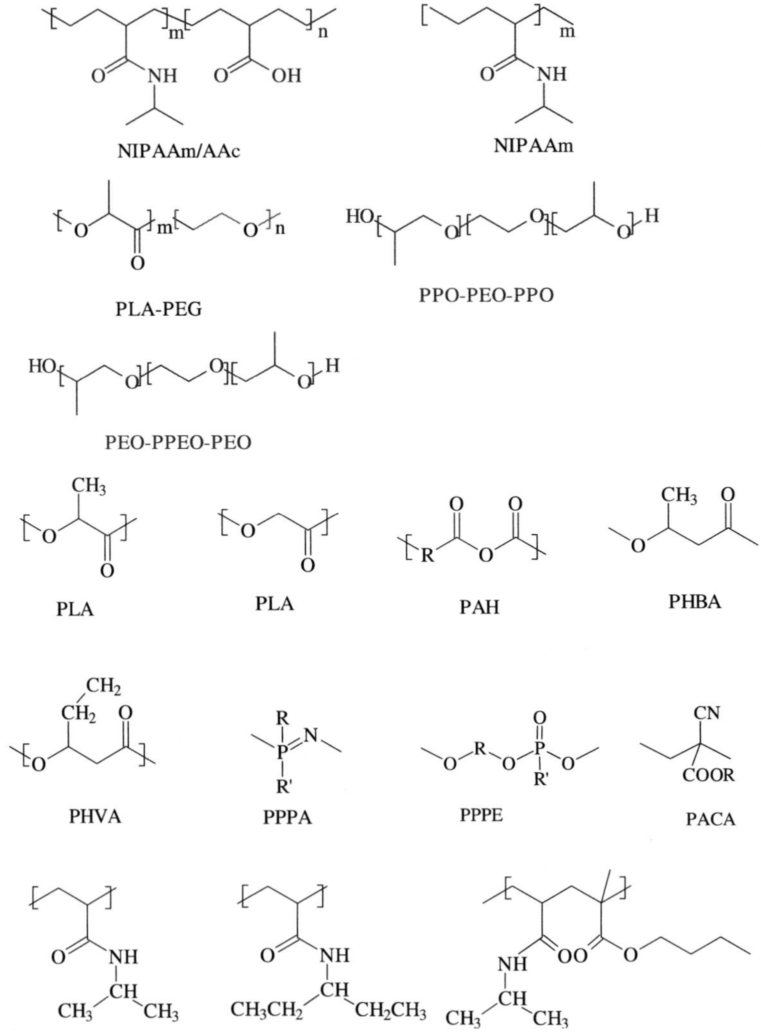

Scheme 4.10 Polymeric smart polymers for entrapment of enzymes.

Other types of synthetic gel, for example biodegradable block copolymers such as block copolymer PLA-PEG [70, 121], poly(N-vinylcaprolactam) (PNVCl), N-alkylated polyacrylamides, e.g. poly(N-isopropylacrylamide [122], block co-polymers, e.g. poly(methyl vinyl ether) [123], polymers of poly(N-isopropylacrylamide-co-acrylic acid) (NiPAAm/AAc), triblock PEO–PPO–PEO copolymers (Pluronics), or polyoxamers [126], are very often used in the field of controlled release and delivery of drugs.

In contrast to conventional entrapment in which the enzyme is entrapped in the matrix and used in the insoluble form during the catalytic action, the soluble–insoluble enzyme–polymer complexes are only entrapped in the gel formed by the polymers on changing the temperature, pH or ionic strength. Here, the enzymes are usually covalently bound to the smart polymers as illustrated in Scheme 4.10.

In general, under the influence of pH, temperature, salts, or solvents, these soluble polymers can be jellified to form insoluble matrix. If the enzyme is present in the solution or bound to the polymer backbone during the gelation it can be physically and covalently entrapped, as illustrated in Scheme 4.11.

For physical gelation, hybrid gels such as macroporous composite poly(N-vinylcaprolactam)–calcium alginate (PVCL–CaAlg) hydrogels can be used to entrap animal cells and enzymes. Remarkably, immobilization of enzymes such as proteases, for example trypsin, α-chymotrypsin, carboxypeptidase B, and thrombin, led to enhancement of the thermal stability of the immobilized enzyme and enabled their use at 65–80 °C, while the native enzymes were completely inactivated at 50–55 °C [125].

Although the enzymes are usually physically embedded in the matrix, enhancement of the stability of the enzyme can often achieved, because of the confinement effect of the gel matrix on the molecular mobility (conformation rigidification) [127]. Interaction of the enzyme with the matrix, for example formation of hydrogen bonds, salt bridges, and hydrophobic interaction, must, moreover, also be

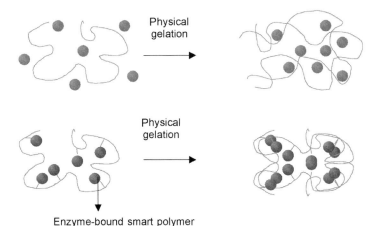

Scheme 4.11 Physical gelation of polymer with enzyme entrapped in the matrix.

taken into consideration when interpreting the thermostability of the entrapped enzymes obtained.

Diffusion limitation often occurs in the entrapped gel–enzyme systems. However, like the encapsulation of enzyme in the sol–gel system, dopants can be used to enhance the porosity of the gel matrix [128], with the aim of mitigating diffusion constraints. On the other hand, leakage of enzyme might occur as a result of selecting the inappropriate gel precursors [182]. In this case, extra cross-linking or coating is often required to avoid the enzyme leakage, as discussed below [262].

Gelation by the Action of Temperature The solubility of many natural or synthetic polymers can be changed by changing the temperature. This technique was first used for entrapment of enzymes or whole cells by use of natural cold-set gel precursors such as gelatin, agarose and agar [129–134]. Later, by use of cryogel techniques, the PVA cryogel was introduced for the immobilization of enzymes [283] or whole cells [128].

The activity of the immobilized enzymes is probably dictated strongly by diffusion limitation and the matrix effect (or the microenvironmental effect), as demonstrated by use of hydrophobic monomers for entrapment of lipase [205]. The stability of the immobilized enzymes is also matrix-dependent, because often it is found that one matrix gives better stability than others [134]. Often, the confinement effect enhances the stability of the immobilized enzyme, especially for sol–gel-entrapped enzymes [127]. This is mainly ascribed to restriction of the enzyme molecules in a confined space, which reduces the possibility of unfolding [124, 127, 214].

Another drawback associated with physical entrapment is that the enzyme might leak slowly from the gel matrix [288]. An elegant technique was developed in the 1970s to mitigate enzyme leakage. It is based on entrapment of the modified enzyme by means of a compound that can be used both as a gelling precursor and a modifier. Because of the introduction of a gel precursor on the protein surface, interaction of the enzyme molecules with the gel matrix is enhanced, leading to the formation of more robust entrapped enzymes [131, 288].

Some thermally reversible gels, for example Pluronics and Tetronics-based polymers and thermally reversible poly(isopropylacrylamide), also undergo phase transition on increasing or reducing the temperature. Thus, they can also be used to entrap enzymes or proteins for delivery [143].

For enzyme immobilization, however, they are less suitable, because they are usually not chemically and physically stable. The proteins and enzymes entrapped in these matrices are usually more stable, however [143], which might have important implications for the design of stable carrier-based immobilized enzymes. The thermo-sensitive properties of gels such as PVME (poly(vinyl methyl ether)) also provide an interesting means of controlling reactions by temperature adjustment [138].

Table 4.6 Enzyme entrapment by physical gelation effect of temperature

Gel/precursors	Method of gelation	Comments	Enzyme	Ref.
Agarose	Cold-set gelation	Confinement increases enzyme thermostability by increasing its molecular rigidity	Mesophilic β-galactosidase from *Aspergillus oryzae*	127
Gelatin	Cold-set gelation	The cells were immobilized in an open-pore gelatin matrix which was prepared by selectively leaching alginate/dopant	*K. marxianus* cells with inulinase	128
Agarose	Cold-set gelation	Among the methods tested (e.g. adsorption, cross-linking, and entrapment), entrapment proved to be best method of immobilization	P 450 monooxygenase containing chloroplasts and yeast microsomes	129
Agarose	Cold-set gelation	10% retention of activity	Cells of pseudomonas species	130, 131
Agar	Cold-set gelation	Compared with adsorption on Dowex 1, the activity with entrapment is eight times higher, with improved stability	Rhodococcus AJ270 containing amidase activity	132
Agar	Cold-set gelation		Whole cell lactase of *E. coli*	133
Gelatin	Cold-set gelation	Retention of activity is approximately 88%, which is higher than in Cu alginate and PAAm gel (67 and 57%). Enzyme stability is also higher	Tyrosinase	134
	Iterative freezing-thawing	The presence of Con A facilitates the diffusion of the substrates through the pores	Glucoamylase	135
PVA	Freezing/thawing	The PVA concentration did not affect the kinetic behaviour of the immobilized cells. The pH of optimum enzyme activity was not changed by immobilization, although the activity profile was broader than that of the free cells	*E. coli* whole cells with β-galactosidase	136
Agarose	Cold-set gelation	The stability of the sensors and their linear response range was strikingly improved by casting a cellulose acetate membrane on top of the agarose–enzyme gel	Glucose oxidase and lactate oxidase	137
Poly(vinyl Me ether) gel	Thermally induced reversible swelling-collapsing	The hydrolysis of maltose proceeds at temperatures below 37 °C and stops above this temperature; the initiation and termination of this enzymatic process can be repeated by a rapid temperature switching, e.g. from 32 to 42 °C and back to 32 °C	exo-1,4-β-d-Glucosidase	138
PEO-PPO-PEO	Heat-set gels	Temperature-sensitive polyurethanes are obtained by reacting trifunctional isocyanate with poly(ethylene oxide)-poly(propylene oxide)-poly(ethylene oxide) (Pluronic L 122)	Proteins	139

Gelation by pH Adjustment pH-responsive gels are prepared from polymers bearing pendant acidic (e.g. carboxylic or sulphonic acids) or basic (e.g. ammonium salts) groups. Their solubility depends on their state of ionization. Basically, two types of polyelectrolyte, cationic and anionic, can be used to prepare pH sensitive gels.

The polyacids or polybases often used for enzyme immobilization are synthetic polymers, e.g. PAAc, PAAc-co-HEMA, poly(MAAc-co-PEG), PEI, PMA, poly(MAAc-MA-MMA), and Eudragit S. Semi-synthetic polymers such as hydroxypropylmethylcellulose acetate succinate (AS) can be also used for this purpose [145].

It is, however, relatively unusual to use these polymers to entrap enzymes. In contrast, to make use of pH-controlled solubilization behaviour covalent or non-covalent enzyme complexes are often prepared, with the aim of using these conjugates in the soluble state, at which the activity of the insoluble substrates is maximum; the enzymes can be recovered as insoluble particles by changing the pH.

Enzymes can occasionally be immobilized in pH-sensitive gel for the purpose of controlled drug release. For instance, glucose oxidase has been immobilized in a pH sensitive gel which swells and releases the entrapped drug as a result of the pH decrease in the environment of the gel caused by production of glucuronic acid by glucose oxidase-catalysed oxidization of glucose [140].

Gelation by Wetting Spin Techniques Mixing of a polymer solution with a solvent which is miscible with the solvent that dissolves the polymer but does not dissolve the polymer itself leads to the gelation of some polymers. This is ascribed to the different solubility of the polymer in the two solvents and is used as the basis for the wetting spin technique for enzyme entrapment.

In general, the spinning medium is crucial to the rheological properties of the fibres and their geometric properties, for example the diameter and internal structure, e.g. porosity, which critically affect the performance of immobilized enzyme [6].

Entrapping cells or enzyme in fibres such as cellulose derivative fibres was intensively studied in the 1970s. Initially, the wet spinning technique was used for the purpose of entrapping enzyme in the fibrous shape. The most popular polymers used for this purpose are cellulose derivatives such as triethylcellulose or diethylcellulose and the Na alginate/$CaCl_2$ system [178].

Gelation by Formation of Complexes Insoluble complexes can be formed by polymer–polymer, polymer–salt, or polymer–inorganic metal oxide interactions. In general, under effect of pH, temperature, salts, or solvents, these soluble polymers can be jellified to form an insoluble matrix. If the enzyme is present in the solution or bound to the polymer backbone during the gelation, enzyme can be physically and covalently entrapped.

The most popular method is use of the alginate–calcium chloride system. The enzyme or whole cells can be easily embedded in the matrix formed by dropping the solution of enzyme with the alginate into $CaCl_2$ solution for hardening [176].

4.5 Preparation of Various Entrapped Enzymes

Table 4.7 Enzyme entrapment by physical gelation (pH-sensitive gel)

Gel/precursors	Method of gelation	Comments	Enzyme	Ref.
Hydroxypropyl methyl-cellulose acetate succinate (AS)	Physical gelation	A reversibly soluble–insoluble transition occurred at pH 4.0; a sharp response of solubility to slight changes of pH without decrease in enzyme activity	Amylase	145
pH Responsive poly(MAAc-MA-MMA)	Physical gelation	High retention of activity; enhanced thermostability and stability in water-miscible organic solvent	Papain, chymotrypsin	146
Poly(ACr-co-AAc)	Physical gelation	Easily precipitated at pH 4–4.5 and re-dissolved at pH 7.0 with complete retention of activity	Trypsin	147
Polyethylenimine (PEI)	Physical gelation	Broad pH optimum in alkaline region	Invertase	148
Polyacrylic acid (PAAc)	Physical gelation	Broad pH optimum in acidic region	Invertase	148
PEI/Eudragit	Physical gelation	96% retention of activity on Eudragit and more than 100% activity was found with PEI; the payload is approximately 0.02–0.05	α-Amylase	149
AA poly(ethyleneimine) (CEPEI)	Physical gelation	Re-dissolved in organic solvent with full restoration of catalytic activity and remarkably high storage stability in the dry state	Chymotrypsin laccase	150
PVNEPB-PAAc (1:3)	Physical gelation	Slight change in pH and ionic strength can reverse the solubility	Penicillin G acylase	151
Eudragit L	Physical gelation	Reversibly soluble–insoluble, depending on pH	TV cellulase	152
Poly(N-vinyl-caprolactam)	Physical gelation	Increasing the NaCl concentration from 0.01 to 1.0 m shifted the half-precipitation temperature maximum from 34.5 to 24.5 °C	Penicillin amidase chymotrypsin	153
Eudragit L100-55	Physical gelation	Reversibly soluble–insoluble, depending on pH; similar performance to NE; and enhanced stability at 25–45 °C	endo-Pectinlyase	154
Eudragit S-100	Physical gelation	Retention of activity depends on the pH of the medium	AP, GL, GA, Trypsin xylanase	155
Eudragit S-100	Physical gelation	64% retention of activity; 30% reduced K_m	Trypsin	156

Table 4.8 Enzyme entrapment by physical gelation via wet spinning

Gel/precursors	Method of gelation	Comments	Enzyme	Ref.
Hydrophilic cellulose fibres	Entrapment	Dry enzyme powders; entrapment in organic solvent (N-EPC and DMF) with 40–60% retention of activity	Glucose isomerase β-galactosidase	157
Diacetyl cellulose	Solvent-set gels	Mixing with the polymer in dimethylacetamide (I) and precipitate the enzyme–polymer solution	Enzymes	158
Diacetyl cellulose	Lyophilization	Adding the enzyme solution to the polymer solution dissolved in dioxane, followed by lyophilization	Enzymes	159
Cellulose triacetate fibres	Wet spinning	–	Penicillin acylase	160
Cellulose triacetate fibres	Wet spinning	Pyridoxal 5′-phosphate was cofactor	Tryptophan synthetase	161
Cellulose triacetate fibres	Wet spinning	The entrapped enzyme has better storage stability under refrigeration (for at least 6 months)	Tryptophan synthetase	162
Fibre	Wet spinning	Diffusion-controlled activity	Invertase	163
Cellulose triacetate fibres	Wet spinning	The small diameter of the fibres guaranteed low mass resistances	Invertase	164
Cellulose triacetate fibres	Entrapment	Diffusion limitations were severe	Esterase	165
Cellulose triacetate fibres	Entrapment	No need for extra coenzyme in the external reaction mixture	LDH, ADH; high-mol.-wt. NAD derivatives	166
Cellulose triacetate fibres	Entrapment	In the presence of platelet anti-aggregating agent	Phenylalanine ammonia-lyase	167

Table 4.9 Enzyme entrapment by physical gelation (PEC)

Gel/precursors	Method of gelation	Comments	Enzyme	Ref.
Cellulose acetate-TiO_2 gel fibre	Complexation	Improved enantioselectivity compared with native enzymes	RML	168
P(TM-co-AAm)/PAA	Symplex	Inter-polymer complex	Growing cells of Serratia marcescens	169
CS/PDMDAAC	Symplex	High storage stability and operational stability were obtained, at low substrate concentrations. pH and temperature optima slightly changed. The K_m was increased by encapsulation as a result of diffusion limitation	Invertase/PS-bound invertase	170
Poly(N-ethyl-4-pyridinium bromide) and poly(methylic acid)	Symplex	pH-sensitive phase transition; thermostability of the enzyme depends on pH, varying from 7 to 300 times that of the native enzyme	Penicillin amidase, α-chymotrypsin, urease	171
Polycation polybrene	Non-covalent complex	Enhanced stability in organic solvents	α-Chymotrypsin	172
KPVS-TGCl	Symplex	After 18 runs, no serious leakage or loss of enzyme activity was found	α-Amylase	173
CS/PDMDAAC	Symplex	35–38% retention of activity; enhanced stability	Urease	174
CS/PDMDAAC	Symplex	Complete immobilization can be achieved by inclusion flocculation with a symplex formed by an anionic and a cationic linear polyelectrolyte or by immobilization in symplex microcapsules	Invertase	175

Table 4.9 Continued

Gel/precursors	Method of gelation	Comments	Enzyme	Ref.
Chitosan-PVP/hexametapolyphosphate	Symplex	Immobilization of whole cell-bound lipase in Chitosan-PVP cross-linked with hexametapolyphosphate gave the most mechanically durable immobilized enzyme	Intracellular *Mucor circinelloides* lipase	176
Na+ Alginate/Ca^{2+}Cl$^-$	Salt-set gels	Increased thermostability (~40% increase in thermal stability), broad pH optimum	Naringinase	177
Alginate fibres	Salt-set gels	Divalent metal ions greatly affected the entrapment of glucoamylase in alginate fibres, the order of which approximately followed the ionotropic series of Thiele	GA, CDGT, EPG protease	178
Alginate	Salt-set gel	IME were less thermally stable than the native enzyme	β-Glucosidases	179
Cellulose acetate-TiO$_2$ gel fibre	Complexation	The activity of IME was not particularly affected by changes in reaction conditions relative to free enzyme and Novozyme 435	CAL-B	180
Alginate/CaCl$_2$	Salt-set gels	The gelation conditions (concentration of CaCl$_2$, gelation time, concentration of Alginate) has large effect on the thickness of the capsules, on enzyme leakage, and on encapsulation efficiency	Glucose oxidase	181
Chitosan/tripolyphosphate	Salt-set gels	Undesirable swelling of agarose beads was observed during leaching. Chitosan proved much better than alginate	Lipase	182

Another technique, developed in the mid-1980s, was use of symplex (polyelectrolyte–polyelectrolyte complex) formation for preparation of immobilized enzymes. This is based on the formation of an insoluble symplex with entrapped enzyme [174]. Occasionally the symplex might undergo a phase transition on change of pH. Remarkably, it has been found that the enzyme immobilized by the symplex is usually more thermostable. One such example is penicillin G acylase immobilized by the water-soluble complex formed by poly(N-ethyl-4-vinylpyridinium bromide) and poly(methylacrylic acid) [171].

Gelation by Solvent Removal Water-soluble or organic-soluble prepolymers can be used to entrap enzyme by solvent evaporation. For example, urea amidohydrolase (urease) has been immobilized within poly(di(methoxyethoxyethoxy)phosphazene) (MEEP) hydrogels by solvent-evaporation methods then gamma-irradiation cross-linking, leading to 80% retention of its native activity [171]. In contrast with water-soluble polymers, other organic soluble prepolymers can be used to entrap the organic soluble enzymes by this method.

4.5.1.3 Covalent Entrapment

Occasionally, smart polymers, for example poly(NIZAAm), or copolymers can be used to prepare reversible soluble–insoluble polymer–enzyme complexes; these are advantageous when sparingly soluble substrates or macromolecular substrates are used.

Unlike physical entrapment, in the covalent entrapment process enzymes can be covalently bound to the matrix formed simultaneously. Thus drawbacks associated with classic entrapment as developed by Bernfeld and Wan [3], for instance, enzyme leakage, can be efficiently avoided by use of this technique.

Covalent enzyme entrapment techniques can currently be classified into two groups—cross-linking of modified enzymes bearing unsaturated bonds and cross-linking of enzymes with an active pre-polymer bearing reactive functionality.

Polymerization of Unsaturated Enzymes This technique can be dated back to the early 1970s, when Jaworek copolymerized an enzyme bearing an unsaturated bond to entrap the enzyme in gel matrix. The advantages of the process were high specific activity, retention of activity, no swelling or shrinkage of the gel, no adsorption of charged substrates or reactants, unchanged kinetic properties, and high activity yield of proteins consisting of subunits [186].

Following this work, several publications on the preparation of such entrapped enzymes appeared in the mid 1970s. By this end of the 1970s this technology was becoming the basis of complementary multipoint attachment [186]. Detailed discussion of this can be found in Chapter 3 of this book.

The introduction of polymerizable functional groups into protein will enable the preparation of spherical granule enzymes [187], as shown in Scheme 4.12.

By use of this method enzyme plastic was formed by polymerizing AMP deaminase and cross-linking with glycidyl methacrylate in the presence of cross-linker

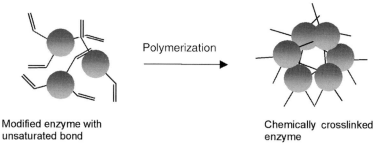

Scheme 4.12 Copolymerization of enzyme modified with unsaturated compounds.

such as methylene bisacrylamide; retention of activity of the plastic immobilized enzyme was ca. 15%. It has been found that the immobilized enzymes formed by this method are usually stable compared with the native enzymes or with other immobilized forms, because of rigidification of the conformation and multipoint attachment [190].

Penicillin G acylase immobilized in a similar spherical granule form by, first, modification with maleic anhydride then co-polymerization with acrylamide via a cross-linking agent, is not only more resistant to heating but also has a lower affinity for benzylpenicillin. It is also less inhibited than the native enzyme by phenylacetate. Its substrate specificity and optimum pH, however, remain unchanged [191].

Although the unsaturated monomer is small, long tails can be integrated into the monomers by use of poly(ethylene glycol) (PEG, MW 3400) with an acrylate group at one terminus and an active ester at the other; this enabled the preparation of cross-linked-entrapped subtilisins with longer half-lives (>100 days) than other immobilized forms. Besides, the enzyme's tolerance of both heat and a miscible organic solvent was also enhanced [192].

As well as long PEG arms, other arms such as polypeptide arms can also be introduced into the enzymes. For instance, a polypeptide chain with unsaturated terminal functionality was introduced into alcohol dehydrogenase and this was then polymerized in the presence of acrylamide [200]. Variation of the nature of the polypeptide chains might be a good method of studying the effect of the microenvironment on enzyme activity [200].

Recently this methodology has also been successfully implemented for imprinting of several enzymes, for example protease [192, 194] and epoxide hydrolase [195].

In situ Entrapment/Covalent Enzyme Immobilization Another technology developed in the 1970s [183] was based on the use of an active monomer in the presence of an inert monomer. By use of this technique enzymes can be covalently bound to the matrix during or after formation of the matrix by polymerization.

Instead of the inert monomer, a monomer bearing active groups which can react with enzyme molecules can also be used during polymerization. Covalent immo-

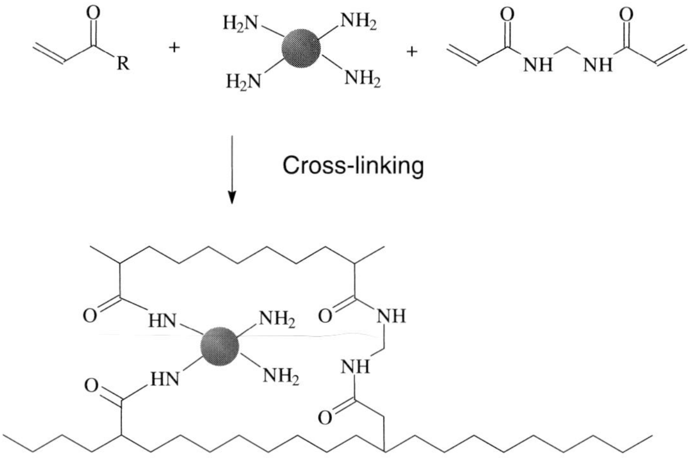

Scheme 4.13 Entrapment of an enzyme in a matrix by covalent bonding.

bilization and entrapment can therefore occur concomitantly during entrapment [196], as shown in Scheme 4.13.

This technique was pioneered by Polak et al., who immobilized adenylate kinase, acetate kinase, and horseradish peroxidase in a copolymer of acrylamide, N,N'-methylenebisacrylamide, and N-acryloxysuccinimide, with retention of activity in the range 20–80% [183, 184].

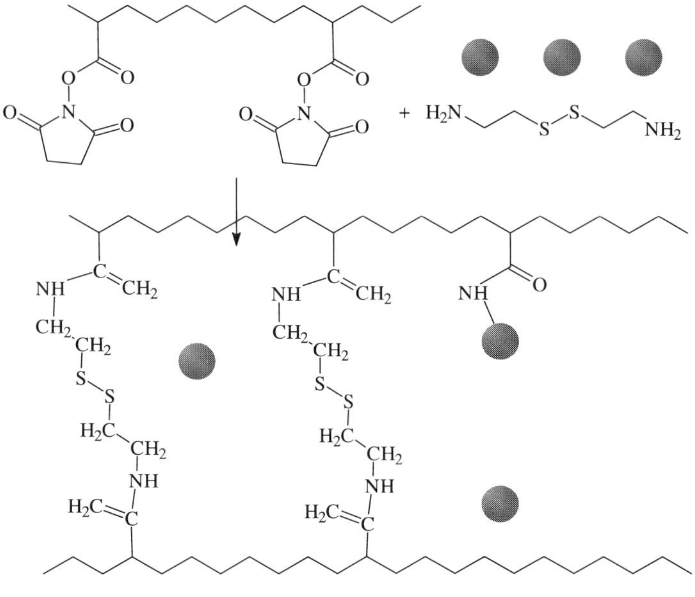

Scheme 4.14 Covalent entrapment.

354 4 Enzyme Entrapment

Table 4.10 Covalent entrapment

Matrix/monomers	Method	Comments	Enzyme	Ref.
Poly(AAm-co-NAS-co-BAAm)	Entrapment	Covalently in good yields (20–90%)	ADK, ACK, HRP	183
A mixture of poly(acrylamide-co-N-acryloxysuccinimide) and triethylenetetraamine	Entrapment	A variety of enzymes were covalently immobilized in high (20–80%) yield. The resulting insoluble enzyme-containing gels have adequate physical properties for use in organic synthetic procedures	–	184
PAN, poly(acrylamide-N-acryloxysuccinimide/cystamine	Entrapment	The gels are too soft to be used alone in columns, but by mixing them with filter aides or by forming them on glass beads, columns with excellent flow characteristics can be assembled	Enzymes	185
AAm/BisAAm/active monomer	Entrapment	The heat stability of immobilized glucoamylase gels is improved?	α-Amylases and glucoamylase	186
AAm/BisAAm/active monomer	Entrapment	Complimentary multipoint attachment	Enzymes	187
Glycidyl methacrylate/methylene bisacrylamide	Entrapment	Ca 15% activity retained; the immobilized enzyme obtained is more stable than the native enzyme	AMP deaminase	188
Polyacrylamide/AAm	Entrapment	The stabilizing effect increased with increasing the number of bonds formed	Chymotrypsin	189
AAm/BisAAm	Entrapment	The immobilized enzymes can be used repeatedly in the form of suspensions or columns without appreciable loss of activity	Modified penicillin G acylase and other enzymes	190
AAm/BisAAm	Entrapment	–	Maleic anhydride-modified penicillin G acylase	191

Table 4.10 Continued

Matrix/monomers	Method	Comments	Enzyme	Ref.
Incorporated into polyacrylates	Entrapment	The biopolymer had lower K_m and k_{cat} than the subtilisin, but the specificity constant (k_{cat}/K_m) was only reduced to one-ninth. Enzyme stability toward both heat and a miscible organic solvent was enhanced	Subtilisin and thermolysin modified with a poly(ethylene glycol) (PEG, MW 3400)	192
Modified enzyme bearing unsaturated bonds/acrylate monomer	Entrapment	The presence of the conformer inducer is able to imprint the enzyme during cross-linking	Modified proteases bearing double bonds	193
EGDMA/cyclohexane	Entrapment	The cross-linked imprinted enzyme can accept d-configured substrates	Lyophilized imprinted protease	194
EGDMA/water-free solvent	Entrapment	Sevenfold higher stability was obtained; enhanced selectivity can be obtained, depending on the substrates and imprinters	Lyophilized imprinted epoxide hydrolase powder	195
Poly(AAm-NAS)	Entrapment	Entrapment and covalent immobilization of enzyme were combined	Adenylate kinase, acetate kinase	196
Poly(acrylamide)	Entrapment	The enzyme first modified with acrylamide and chemically polymerized to form insoluble immobilized enzyme	Penicillin amidase from *Proteus rettgeri*	197
Poly(NIPAAm-co-NAS)-EDA (cross-linker)	Entrapment	Raising or lowering temperature through the LCST (lower critical solution temperature) is able to influence activity	Chymotrypsin	198
P(DEAEMA)/PEG-Diamine	Entrapment	The hydrogels were prepared in the form of disks and microparticles. Gel swelling was responsive to glucose concentration	Functionalized enzyme glucose oxidase and catalase	199
PAAm/enzyme derivatives	Entrapment	The enzyme was covalently bound to the matrix via a polypeptide spacer. Variation of the spacer influenced the microenvironment	Alcohol dehydrogenase grafted with polypeptide chains	200

A new variation of this technique is based on the use of functionalized pre-polymer (PAN: poly(acrylamide-N-acryloxysuccinimide) instead of the monomers [185]. Interestingly, it was found that use of cystamine led to the formation of a matrix that can be re-dissolved by addition of a reducing thiol under mild conditions.

Similarly, functionalized prepolymer poly(acrylamide-N-acryloxysuccinimide) was also used to entrap enzyme in the presence of a diamine such as triethylentetramine in pH neutral aqueous solution [185], as illustrated in Scheme 4.14.

Advantages of this method are that it is simple and the entrapped enzyme can be coated on various solid carriers thus various reactor configurations are possible.

Similarly, adenylate kinase, acetate kinase, and horseradish peroxidase have also been immobilized by free-radical copolymerization of acrylamide, N,N'-methylenebisacrylamide, and N-acryloxysuccinimide, in a solution containing enzyme and enzyme-protecting reagents (substrates, dithiothreitol) in a suspension system [183].

4.5.1.4 Sol–Gel Process

Introduction The use of inorganic sol–gel physical entrapment techniques for preparation of immobilized enzymes has, since the beginning of the 1990s, become a very important area of enzyme immobilization [206]. Although the method has been known since early in the mid-1950s, when Dickey showed for the first time that some enzymes could be entrapped in silicic acid-derived glasses with partial retention of biological activity [1], in the succeeding three decades there have been few reports of the use of this methodology for preparation of entrapped protein/enzymes [201, 207, 334].

However, several early observations revealed that activity retention in sol–gel processes might be comparable with that in other types of enzyme-immobilization processes; the enzyme might be stabilized; the matrix might be not permeable to high-molecular-weight substrates [208]; enzymes can be imprinted because of molecular confinement [334].

Because of its simplicity and the potential to increase enzyme stability by stabilization of protein tertiary structure by the rigid gel network, entrapment of enzyme in hybrid sol–gels is very attractive [202]. Moreover, variation with monomers of different properties, for example different hydrophilicity, also enables modulation of enzyme performance such as activity and stability or selectivity, by altering the proximity of the enzyme molecules entrapped or addition of conformer selectors such as surfactants or crown ethers [204], or other additives such as acetylated dextran or polysaccharides [209] or cell-free extract [210], polymers, and metal salts [212, 215] as illustrated in Scheme 4.15.

The stability of the sol–gel-entrapped enzyme was, however, also found to be highly dependent on enzyme concentration, pH, ionic strength, and other immobilization conditions [210]. Increasing the concentration of enzyme entrapped in the matrix led to increased enzyme thermostability [210]. Remarkably, reduced

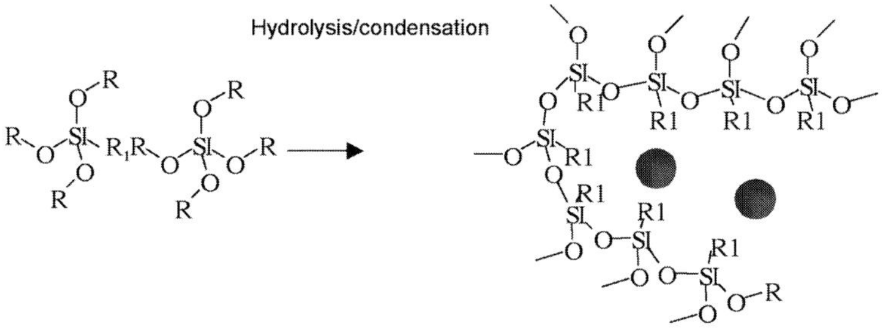

Scheme 4.15 Enzyme encapsulation in a sol–gel matrix

thermostability relative to the native enzyme was also obtained with sol–gel matrix entrapped enzymes [210], suggesting that stabilization of enzymes by immobilization might be governed by many factors. The pre-immobilization conformation of the enzyme is also probably very important. Thus, the conditions selected should be sufficient to confer a more stable conformation on the enzyme molecules.

Remarkably, enzymes immobilized in inorganic gels are often stabilized [211]. This stabilization effect was recently interpreted as the result of molecular confinement [335]. Often, surprisingly high stabilization effects can be achieved compared with the non-immobilized counterpart or other immobilized enzymes [218], as shown in Table 4.11

Conventional Sol–Gel Process Conventional sol–gel process refers to the process by which the enzyme is mixed with the sol–gel solution, followed by a gelation process under the influence of the pH and an aging process.

Because, after aging, the pore sizes of the gel formed are usually in the range 2–20 nm, the materials are, strictly, categorized as mesoporous materials. Their pore sizes are smaller than that needed to overcome diffusion constraints. As a result, serious diffusion can occur when the enzymes are confined in the matrix formed by the sol–gel process, thus leading to significant loss of the original activity [219].

Because the pore size of the sol–gel falls into the same range as the protein crystals, much insight into the performance of sol–gel-entrapped enzymes can be obtained from cross-linked enzyme crystals (CLEC) [216]. For instance, CLEC often need to adopt a suitable size and shape to be maximally active by overcoming diffusion constraints. Similarly, sol–gel-entrapped enzymes (SGEE) are usually prepared in small sizes (around 5–10 μm).

To increase retention of activity of enzymes it is usually necessary to use the templates such as PEG and sugars or surfactants to modulate the pore size and increase the permeability of the substrate through the pores and the accessibility of the enzyme [220, 221]. Nonetheless, enzyme activity expression is dictated by many factors, for example the nature of the precursor for the formation of a gel matrix [204], the type of template [220], the drying process [222], and the properties of the substrates [225].

Table 4.11 Entrapment of enzyme by the sol–gel process

Carrier	Method	Remark	Enzyme	Ref.
Sol–gel matrix formed from silicic acid-derived glass	Entrapment	Dickey demonstrated for the first time that some enzymes were entrapped with reasonable retention of biological activity	AMP deaminase	1
Sol–gels	Entrapment	Addition of surfactants or crown ether can be used to improve the enantioselectivity and activity of the immobilized lipases, depending on the enzymes	Lipases	204
Hydrophobic sol–gels	Entrapment	Modulation of the nature of the sol–gel by variation of the monomer enabled improvement of enzyme activity	Lipase	205
TMOS	Entrapment	The immobilized purified ALP from bovine intestinal mucosa had 30% activity yield and improved stability to thermal deactivation compared with a solution	Alkaline phosphatase (ALP)	206
Silica gel derived from TKDEAES	Entrapment	By hydrolysing an aqueous solution of tetrakis(diethylamino-ethoxy)silane HCl–enzyme the immobilized enzyme was obtained	Cholinesterase	207
TMOS	Entrapment	The presence of peracetylated β-cyclodextrin improved enzyme selectivity and activity	*Pseudomonas cepacia* lipase	209
TMSO	Entrapment	The stability of the immobilized enzyme is much lower than that of the native enzyme	d-Galactosidase	210
TMOS and alkyltrimethoxysilane	Entrapment	Urease encapsulated in gels made from a mixture of TMOS and alkyltrimethoxysilane retained 60% of its native activity	Urease	211
TMOS/PEG 6000	Entrapment	Almost no diffusion was observed for the small substrate. The sol–gel-entrapped enzyme was stable for several months when incubated at ambient temperature (pH 7.5) whereas the enzyme immobilized by adsorption on surface of sol–gel glass was completely autodigested overnight	Trypsin and acid phosphatase	212
TEOS	Entrapment	Lipase entrapped in the sol gel matrix has more or less sevenfold higher specific activity (mol h^{-1} kg^{-1} lipase) than lipase covalently immobilized on silica beads	Lipase	213

4.5 Preparation of Various Entrapped Enzymes

Table 4.12 Stabilization with a sol–gel matrix

Methods	Methods	Stabilization factors	Enzyme	Ref.
Tetrakis(2-hydroxyethyl) orthosilicate (THEOS)	Entrapment	Entrapment of enzymes can be performed at any pH compatible with their structural integrity and functionality; the enzymes are stabilized 100-fold; for some enzymes, activity depends on the additives	d-Glucanase, d-galactosidase	218
Sol–gel glass matrix	Entrapment	The entrapped enzyme was almost two orders of magnitude less active than FE	Parathion hydrolase	219
Pure silica or organically modified silica matrices	Entrapment	Glucose was used as template. Enzyme activity can be increased threefold by improvement of pore size	HRP, GOD	220
Nano-porous silica (sol–gel reactions of TMOS)	Entrapment	No detectable leakage of ACP; improved thermostability; the presence of DG as non-surfactant template is important with regard to the enhanced activity	ACP, Trypsin	221
Silica and aluminosilicate gels	Entrapment	Supercritical drying avoids compression of the gel network, hence improving activity	PCL	222
tetramethoxy-silane-based glass	Entrapment	The stability of ACP entrapped in the sol–gel matrix was about 120 times stabilized than the free enzyme in solution	ACP	223
Cellulose acetate-titanium isopropoxide gel	Entrapment	The activity of the IME was approximately 10–20% that of the native lipase. The reaction was more enantioselective compared with the native lipase. The operational stability was improved	Lipase (*Candida rugosa*)	224
Ca silicate microcapsules	Entrapment	Hydrophilicity of the substrate is a key factor; the half-life was enhanced 23-fold	Chymotrypsin	225
Celite-supported sol–gel (TMOS or TrMMOs/TMOS)	Entrapment	The entrapped enzyme was 100 and 200 times stable, respectively, than that deposited on Celite. Retention of activity was 28 and 52%, respectively	Lipase	226
TMOS	Entrapment	The thermostability of the entrapped enzyme is temperature-dependant but an order of magnitude higher than that of the native enzyme can usually be obtained	Creatine kinase	227

The intrinsic drawbacks of the classic entrapment technique are probably that not all the entrapped enzyme molecules are catalytically active [217]. Some of the enzyme molecules within the matrix might be deactivated during the entrapment process. Some of the enzyme molecules are, moreover, physically inaccessible to the substrate molecules, despite their native structures, as illustrate in Scheme 4.16.

The precursors used for preparation of silica via sol–gel processors are generally denoted $(RO)_4Si$ (e.g. tetramethoxysilane (TMOS) or tetraethoxysilane TEOS) and organically modified siloxane precursors $R''(R'O)_3Si$. They can be classified into the following groups:

- Group I: simple tetraalkoxysilanes.
- Group II: functional trialkoxysilanes, namely simple alkyl (methyl, ethyl, propyl, butyl, pentyl, cyclohexyl).
- Group III: aryl (phenyl, benzyl, phenethyl) and long alkyl chains $(R''=(C_nH_{2n+1})Si(OMe)_3;\ n = 6, 8, 10, 12, 14, 18)$.
- Group IV: inert functional groups such as those bearing γ-amino groups (3-aminopropyl, 3-carboxylpropyl, 3-mercaptopropyl, etc.), which can be easily activated for covalent bonding of the enzyme or can be used as ionic adsorbent.
- Group V: functional trialkoxysilanes bearing active functional groups (isocyanyo, epoxy, photoactive double bonds).
- Group VI: inert functional groups, which are usually used to modulate the microenvironment of the gel matrix.
- Groups VII: poly(alkoxylsilanes) and their derivatives bearing different functional groups, as mentioned above.

The structures of the monomers most frequently used in sol–gel processes are depicted in Scheme 4.17.

To obtain an immobilized enzyme with suitable properties, for example retention of activity, stability and selectivity, it is often necessary to select the appropriate precursors and optimize their ratio. As with other carrier-bound immobilized

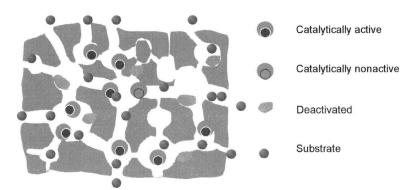

Scheme 4.16 Molecular accessibility of an enzyme entrapped in a gel matrix (adapted from [217]).

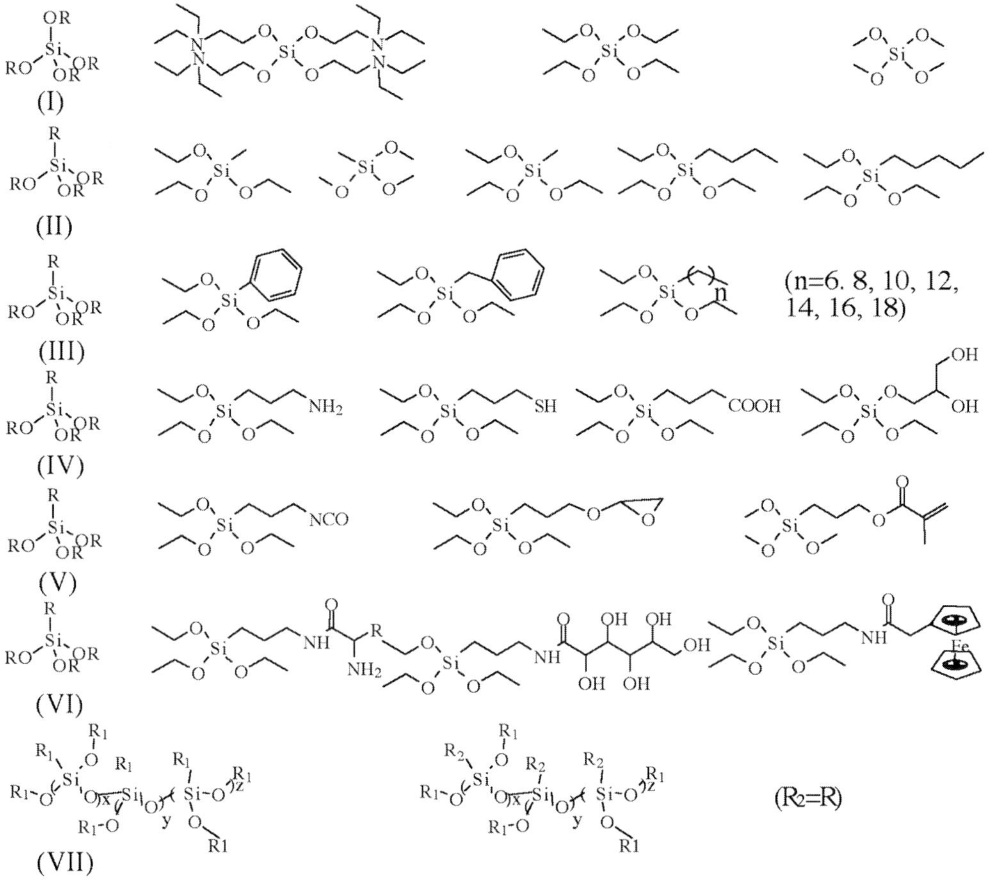

Scheme 4.17 Precursors often used for entrapment of enzymes by the sol–gel process.

enzymes, the optimum conditions for entrapment change from enzyme to enzyme and from application to application. The availability of diverse precursors with different chemical properties (hydrophilic/hydrophobic positively/negatively charged, active/or inert functionalities), however, is a unique means of modulating enzyme properties by varying the precursor used.

Researchers have found that not only do the precursors play an important role in the performance of the immobilized enzymes obtained; the additives used also have strong effect on enzyme properties. For example, immobilization of 1,3-d-glucanase in tetrakis(2-hydroxyethyl)orthosilicate (THEOS) revealed that 1,3-d-glucanase was usually immobilized without loss of its activity whereas the activity of α-d-galactosidase in the immobilized state depended on the type of polysaccharide material [218].

4.5.2
Non-conventional Entrapment

4.5.2.1 Post-loading Entrapment (PLE)

In addition to the methods described above, which usually involve entrapment of the enzyme during formation of the matrix, enzymes can be also separately entrapped in the ready-made matrices, which have larger pore sizes to enable the enzyme molecules to pass thorough and be accommodated inside the pores. In general, this technique is termed "post-loading entrapment" (PLE).

This technique already featured at the end of 1960s, when glycerol dehydratase was adsorbed in PAAm and Sephadex, followed by drying, leading to the formation of entrapped enzyme with high storage stability [228]. In the last 30 years, however, there have been few examples based on this technology. For instance aminoacyl-tRNA synthetase was immobilized in PAAm gel or Sephadex gel by the PLE technique, leading to the discovery that the immobilized enzymes obtained were much more thermostable then the native enzymes [228].

Currently, post-loading entrapment techniques can be placed in two groups, diffusion-controlled entrapment and partition-controlled entrapment. The former has been mainly used to improve the performance of cross-linked enzymes, whereas the latter is mainly used for controlled drug release.

Diffusion Entrapment Diffusion entrapment is based on the concept that the enzyme molecules can diffuse into the preformed gel matrix or a pre-designed carrier [229]. Consequently, the enzyme can be irreversibly insolubilized in the gel matrix by cross-linking or used only in a medium, for example organic solvents, in which the enzyme cannot diffuse from gel matrix.

An advantage of these techniques is that deactivation of the enzymes during the conventional entrapment process can be avoided, because preparation of the matrix and loading of enzyme are separated into two steps. This technology is also advantageous in that the enzyme molecules might be more accessible to the substrate than enzyme molecules entrapped in a single step, because the voids in the matrix accessible to the enzyme molecules might be also accessible to the small substrates. Finally, the resulting enzyme composite had the strength and rigidity of the prefabricated support. Thus, an immobilized enzyme with desired non-catalytic performance can be designed at will.

On the basis of this concept, glucose isomerase was immobilized within macroporous Celite or titania beads (0.4–0.8 mm) with high mechanical stability. In a gel type carrier, laccase was immobilized for the first time simply in Sepharose 4B beads, which enabled the enzyme molecules to diffuse freely [231]. It was found that the immobilized laccase obtained in this way was very stable in organic solvents. This work was extended by Khare who, instead, entrapped enzymes e.g. β-galactosidase, acid phosphatase, and trypsin, in Sepharose beads than performed cross-linking with a bifunctional cross-linker such as glutaraldehyde [9], as illustrated in Scheme 4.18.

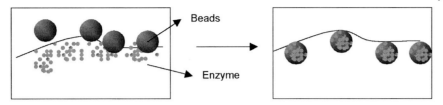

Scheme 4.18 Entrapment and cross-linking of an enzyme in ready-made gel beads.

Similarly, water-soluble enzymes can be entrapped in gel-film by diffusion of the enzyme into the gel and subsequently cross-linking by physical or chemical means (e.g. drying, freeze-drying, or treatment with electrolytes), with the aim of reducing the porosity of the gel [9] with retention of activity inside the matrix. The principle is illustrated in Scheme 4.19.

The same method was also used to entrap enzyme by incubation of enzyme solution with Con A-coated Sephadex beads, followed by cross-linking [232].

Selective Protein Partition Entrapment   Although PLE has not been recognized as a powerful technique for preparation of robust immobilized enzymes, it has recently been implemented for drug-delivery. Examples are amylase and lysozyme entrapped in poly(2-hydroxyethyl methacrylate) by soaking in solutions of these enzymes [232], or bovine serum albumin in a polyacrylamide gel [232], or lysozyme, bovine serum albumin and immunoglobulin G entrapped in a glycidyl methacrylate-derivatized dextran gel [233].

These diffusion-controlled entrapment techniques generally have the drawback that loading of the enzyme is lower than 0.1% of the dry weight. It has recently been found that the payload can be enhanced by selective partitioning of the enzymes in the gel matrix (Scheme 4.20). Remarkably, it has been found that high payload can be obtained (~0.1) by use of Sepharose-based beads [236].

This technique can have several advantages compared with classic entrapment. First, entrapment of the protein is separated from the process of preparation of the gel matrix, thus avoiding protein deactivation. Second, it is simple and generally applicable, like covalent binding of enzymes on ready-made carriers. For enzyme immobilization, however, this technique is still poorly developed. Another drawback is that the gel used often swells substantially in aqueous medium, resulting in very low mechanical stability.

Scheme 4.19 Entrapment of an enzyme by diffusion of the enzyme into the gel then reduction of the pore size.

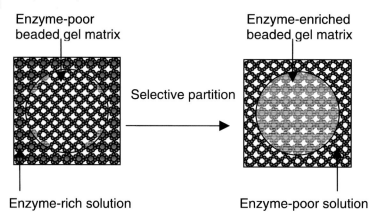

Scheme 4.20 Entrapment of an enzyme by selective partition.

Impregnation Entrapment Lai and Shin immobilized acid phosphatase on montmorillonite and chitosan to improve the phosphorus content of soil. In this work, chitosan was first integrated with activated clay, to form a composite bead, and then cross-linked with glutaraldehyde [237].

The robustness of this technology has been demonstrated by tyrosinase, which was immobilized by adsorption on Fuller's-earth then entrapment in gelatin. The immobilized enzyme had greater thermal and operational stability than the conventionally entrapped enzyme [238]

In a new approach, impregnation with an enzyme can be combined with the sol–gel process. For instance, sol–gel precursor solution containing lipase *Candida antarctica* B has been used to impregnate the pores of the prefabricated carriers. Subsequent gelation of the mixture in the pores of the resin plate led to the formation of immobilized lipase not only with high stability (51-fold increase in operational stability after entrapment) and high activity in organic solvents (from the sol–gel process) but also controlled geometry (from the poly(vinyl formaldehyde) resin plate) and enhanced mechanical stability [335]. In an extension of this work, it was further demonstrated that sol–gel-entrapped enzymes can also be formed in other porous bead carriers, for example porous chitosan beads or synthetic polymeric beads such as Amberlite [336, 337].

4.5.2.2 Entrapment-based Double-immobilization Technique

Entrapment–Coating/Cross-linking To improve mechanical stability, the operational stability of the bead-entrapped enzyme, and to avoid enzyme leakage [241], the beads can be further coated with a second layer formed by cross-linking a coating of a soluble polymer, for example chitosan, polyethyleneimine, or polyethyleneimine-glutaraldehyde. Formation of a dense layer around the beads often increases the diffusion constraint, however, leading to reduced enzyme activity [242].

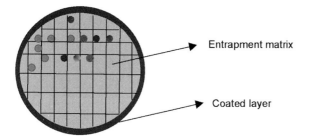

Scheme 4.21 Entrapment and coating.

As noted above, the enzymes entrapped in the gel matrix might slowly leach from the gel matrix, because they are not covalently bound to the matrix [241]. To avoid enzyme leakage, the enzymes entrapped in the gel matrix of natural or synthetic polymers can subsequently be encapsulated in a semi-permeable capsule which allows only the substrate to cross, as shown in Scheme 4.21

In a modified approach, the enzyme can be entrapped in a ready-made gel matrix such as DEAE Sepharose (by so-called post-loading entrapment), followed by encapsulation by coating techniques [250]. In addition to mitigation of enzyme leakage, this method has several other advantages, for example enhanced mechanical stability of the gel matrix [244] and improved operational stability of the immobilized enzymes. As with the one-step encapsulation technique mentioned above, it is essential to control properties of the capsules such as thickness, pore size, aquaphilicity, and the permeability to the substrates [243, 247].

Entrapment and Cross-linking Just as when other gels are used for entrapment, biocatalysts (mainly free enzymes) entrapped in the form of beads or film can be cross-linked by addition of chemical cross-linkers with the aim of reducing enzyme leakage [263] and improving mechanical stability [257] and thermostability [128].

α-Amylase entrapped in a BSA matrix and cross-linked with glutaraldehyde was more thermostable than the native form, with a remarkable fourteen fold increase in catalytic half-life [260].

In both cases, cross-linking between the enzymes, between the enzyme and matrix, or between the molecules of the matrix can occur. High payloads (up to 0.2) can usually be achieved by entrapment. An increase in the payload might, however, lead to leakage of the immobilized enzymes [261].

Retention of activity differs greatly, depending on the matrix used, the type of cross-linker, the concentration, and the enzyme used [261]. Interestingly, it was found that when PCL (*Pseudomonas cepacia* lipase) or polyethylene glycol (PEG)-modified PCL was entrapped in Ca alginate gel the enantioselectivity was at least threefold selective than that of the native enzyme, suggesting that the conformation of the enzyme had been changed on entrapment in the alginate gel [54].

Table 4.13 Entrapment-coating techniques

Matrix (core)/shell	Method	Comments	Enzyme	Ref.
Alginate microcapsules/ PEC multilayers	Entrapment-coating	Mitigation of activity loss by use of double layers rather than a single layer	Cytochrome C	243
Alginate beads/chitosan or PMG membrane	Entrapment-coating	Exclude α-chymotrypsin and other proteases	Urease	244
Alginate microcapsules/ silica	Entrapment-coating	Enhanced mechanical stability, protein diffusion and enabling enzyme immobilization	Urease	245
PAAM beads/a polypeptide membrane	Entrapment-coating	The coating significantly (fivefold) increased the half-life	Amino acid oxidase	246
Alginate beads/poly-(methylene co-guanidine) membranes	Entrapment-coating	70% mass yield and 31% activity yield resulted from encapsulation of urease	Urease	247
Alginate beads/ polyamide	Entrapment-coating	Enzyme leakage was reduced	Amino acylase	248
Gelatin gels hardened by in-situ polymerization of TEOS	Entrapment-encapsulation	Stable in both aqueous and organic solvents; good activity in cyclohexane	Lipase CV (*Chromobacterium viscosum*)	249
DEAE-Sepharose beads/ PVA-SbQ	Adsorption/ Entrapment	Enzyme is immobilized first in DEAE-Sepharose, followed by entrapment in PVA gels	Choline oxidase	250
Poly(methylsiloxane)–SiO_2/Alginate–SiO_2	Enclosed	–	Lipoxygenase type I	251
Alginate-silicate sol–gel matrix	Entrapment	The yeast was immobilized in small glasslike beads of alginate-silicate sol–gel matrix	*Saccharomyces cerevisiae*	252
Photocrosslinkable polymer/metal-chelate substituted beads or anion-exchanger beads	Entrapment-coating	–	Histidine-modified peroxidase/choline oxidase	253

4.5 Preparation of Various Entrapped Enzymes

Table 4.14 Entrapment of enzyme in natural gel by entrapment and cross-linking

System	Method	Remark	Enzyme	Ref.
PVA gel	Entrapment-cross-linking	Enzyme leakage was avoided by coating the gel-entrapped enzyme with 0.2% glutaraldehyde solution. Enzyme activity was increased by glutaraldehyde treatment	Penicillin G acylase, aldolase and lactase	254
Gelatin pellet	Entrapment-cross-linking	40% Retention if activity; negligible internal diffusion limitation	Microbial cell with glucose isomerase	255
Cobalt alginate	Entrapment-cross-linking	Highest retention of activity (83%) was obtained. Subsequent cross-linking eliminated enzyme leakage	β-d-Galactosidase	256
Chitosan-formaldehyde gel	Entrapment-cross-linking	67% Retention of activity; higher operational stability	β-d-Galactosidase	257
Xanthan-alginate spheres coated with gelatin cross-linked with GAH	Entrapment-cross-linking	Slight decrease in activity but increased stability; 75% retention of activity	Urease	258
Gelatin thin layers on activated glass, polyester, and Al foil	Entrapping-cross-linking	Immobilization did not result in changes of pH and temperature optima	Glucoamylase	259
Gelatin(or BSA)-matrix and cross-linked with glutaraldehyde	Entrapment-cross-linking	IME were more thermostable than the native form, with a remarkable fourteenfold increase in catalytic half-life	β-Amylase	260
Gelatin matrix insolubilized by HCHO	Entrapment-cross-linking	Retention of activity, 9.6–48.0%, mainly because of diffusion resistance	Enzymes	261
Collagen fibrils	Entrapment-cross-linking	Payload should be kept below 0.2	Glucoamylase	262
Gelatin beads	Entrapment-cross-linking	Both entrapment and cross-linking played their role	AP, INT and GLD	263
Erythrocytes	Entrapment-cross-linking	–	Pegylated urease	264
Cellulose acetate was in contact with a metal alkoxide solution	Entrapment-cross-linking	The immobilized enzyme is stable for a long period and can be used in a packed column	Enzymes	265

4.5.2.3 Modification and Entrapment

It has been widely recognized that entrapment of the enzyme molecules is a relatively simple method of enzyme immobilization. However, the entrapped enzyme molecules often tend to leach, even if they are large [241]. One efficient solution to this problem is to modify the enzyme, thus enhancing interaction of the enzyme with the matrix or making the enzyme insoluble. Often used approaches are:

1. preparing enzyme–polymer conjugates;
2. cross-linking the enzymes by means of a bi-functional cross-linker; and
3. cross-linking the enzyme in the presence of exotic protein.

Entrapment of Enzyme–polymer Covalent Conjugates   It has long been known that enzyme–polymer conjugates can be formed by reacting enzyme with activated water-soluble polymer, for example dextran, gelatin, and PEG [284], and use of this technique for enzyme immobilization can be dated back to the beginning of 1970s [273].

In this way, enzymes could first be covalently bound to gelatin or water-soluble dextran and starch. The resulting polymer–enzyme conjugates were subsequently entrapped in a gel matrix, as illustrated in Scheme 4.22.

The advantages of this method might be stabilization, because of the formation of gelatin–enzyme conjugate, and less diffusion limitation, because conjugation of gelatin with the enzyme molecules might increase the void of the matrix, resulting in less diffusion limitation. Moreover, less or no leakage of the enzyme from the gel beads might be expected from this method, because of enhanced interaction of the protein molecules with the gel matrix [274].

In contrast with post-treatment of the immobilized enzymes, the enzyme can be modified before immobilization, with the aim of enhancing the binding of the enzyme to the carrier, thus stabilizing the enzyme. Immobilization of soluble modified enzyme derivatives was pioneered by Glazer [275].

Originally, the soluble modified enzyme was covalently immobilized to another insoluble activated polymeric carrier. Nowadays, however, several other techniques have been developed to make use of enzyme derivatives, for example entrapment [276] or adsorption of these modified enzymes. Moreover, soluble enzyme derivatives attached to active soluble polymers can also be used directly in the

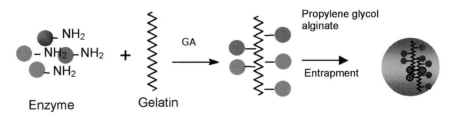

Scheme 4.22 Enzyme conjugation with a soluble polymer before formation of gel beads.

membrane reactor. Because of the increase in size the enzyme cannot pass through the membrane [276].

These modified enzymes are often immobilized by entrapment methods. For instance, modified penicillin G acylase has been immobilized in a variety of fibres and films as support carriers [278]. Native bovine serum amine oxidase (BSAO) and poly(ethylene glycol) (PEG)-treated ("PEGylated") BSAO have been immobilized in a hydrogel during synthesis of the latter, with improved operational stability in the presence of the substrate [279]. Lipid-coated enzyme dissolved in an organic solvent [280] can be entrapped in hydrophobic gel formed by cross-linking of the photo-cross-linkable prepolymer such as ENTP-polyethylene glycol derivatives with photo-cross-linkable terminal functionality [92].

Novick and Dordick have developed a similar procedure for entrapment of hydrophilic enzymes in hydrophobic polymer film. To render the enzyme soluble and homogeneously dispersed in the polymer solution, surfactant–enzyme ionic pairs soluble in organic solvents were used [299]. The principle is illustrated in Scheme 4.23.

When lipase B from *Candida antarctica* (MW 35000) was pegylated with monomethoxypoly(ethylene glycol)-norleucine (PEG-Nle) 5 kDa hydroxysuccinimide ester and entrapped in PVA cryogel it had only 10% of the native activity of PEG-enzymes [284]. When bovine pancreas α-chymotrypsin covalently modified with homo-bifunctional poly(ethylene glycol) derivatives was entrapped in Ca-alginate gel particles its thermostability and operational stability in the hydrolysis of *N*-acetyl-l-phenylalanine Me ester (APME) were improved both in batch mode and in a continuous fixed bed reactor [292], suggesting that the first step stabilized the enzyme.

To prevent enzyme leakage β-galactosidases were co-cross-linked with bovine serum albumin then entrapped in agarose beads [288]. Leakage of entrapped enzyme from a gel matrix is often ascribed to the fact that the enzyme can re-dissolved in an aqueous medium. In a recent report, enzyme was first cross-linked and then entrapped in polyacrylamide [281]. Remarkably, for immobilized penicillin G acylase preparation retention of activity was generally in the range 62–76%, indicating that diffusion limitation was not significant compared with many other enzymes immobilized on the same carrier. Unfortunately, the specific activity of the immobilized penicillin G acylase obtained in this way is only 45.5 U g^{-1} (dry), i.e. much lower than the most commercialized penicillin G acylase

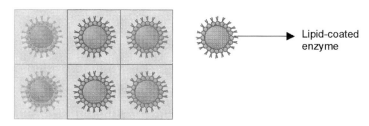

Scheme 4.23 Entrapment of modified enzymes.

Table 4.15 Modification-entrapment

Gel for entrapment	Methods	Comments	Enzyme derivatives	Ref.
PVA matrix	Entrapment by Freezing/thawing	Activity is one tenth that of the enzyme-PEG in the free form	Pegylated enzymes and proteins	283
PVA-cryogel	Entrapment	Leakage of lipase was minimized by using high mol. wt. PVA and by previous conjugation of the enzyme to PEG, diffusion limitation exists (10% retention of activity relative to PGE lipases) and can be reduced by cutting into small pieces	Pegylated lipase	284
Cross-linked acrylamide beads	Grafting	The polymeric beads showed little dependence on diffusion; less change in K_m; high storage stability	Glycidyl methacrylate-treated horseradish peroxidase	285
Poly(diethylaminoethyl methacrylate-g-ethylene glycol) gels	Entrapment	By copolymerization of the constituent monomers and the functionalized enzyme solutions	Functionalized enzyme, glucose oxidase and catalase	286
Poly(3,4-ethylenedioxythiophene) (PEDT) films	Entrapped	This modified enzyme was entrapped afterwards within poly(3,4-ethylenedioxythiophene) (PEDT) films electro-generated on glassy carbon (GC) electrodes. The composite (PEG-GOD/PEDT) film is more porous than the film without enzyme (PEDT + PEG)	PEG-GOD	287
Agarose beads	Entrapment	The use of co-cross-linked enzyme aggregates avoided enzyme leakage in the agarose beads or gellan gum beads	Co-cross-linking of β-galactosidases with bovine serum albumin	288
PLY(N-vinylcaprolactam) gel	Entrapment	After immobilization, approximately 90 and 75% of original trypsin and COB activity, respectively, were retained	Poly(vinylpyrrolidone-co-acrolein proteases	289
Entrapped in PAAm gel lattice	Entrapment	The DMA-cross-linked preparation, entrapped in PAAm in the presence of BSA, lactose, and Cysteine, was a significantly better catalyst and hydrolysed 47% milk lactose, compared with 31% hydrolysis by entrapped native enzyme, in 6 h	Cross-linked derivatives of *Escherichia coli* β-galactosidase	290
PVA-cryogel	Entrapment	–	PEG-modified glucose oxidase	291
Ca-alginate	Entrapment	85% modification reduced the enzyme activity to 50% of the native activity. However, the long term stability was improved because of reduced enzyme leakage and enzyme stability	α-Chymotrypsin modified with PEG	292

4.5 Preparation of Various Entrapped Enzymes

(ca. 100–200 U g^{-1} wet), when a free soluble penicillin G acylase with a specific activity of 4.55 U mg^{-1} was used.

However, it should be pointed out that this, the main drawback of the method, could be overcome by using a purer enzyme preparation. It might, therefore, constitute a potentially useful method of immobilization for penicillin G acylase; if semi-purified penicillin G acylase (20–50 IU mL^{-1}) can be used for the immobilization, an immobilized penicillin G acylase preparation with high specific activity (5–10 times) can possibly be obtained.

Entrapment of Cross-linked Enzymes Enzymes can be cross-linked before entrapment with the aim of improving the enzyme activity. For instance, *Aspergillus niger* glucosidase cross-linked with glutaraldehyde and entrapped in calcium alginate was more thermally stable than the free enzyme (half-life 23 days, at 50 °C). The free enzyme had a 9-day half life at the same temperature [282].

Entrapment of Enzyme–polymer Non-covalent Complexes Non-covalent enzyme–polymer complexes can also be entrapped in a gel matrix [274].

4.5.2.4 Supported Entrapment

Enzymes can also embedded in film-like polymer matrices by normal entrapment techniques, for example:

- entrapment in the gel matrix formed by the sol–gel process [43, 44],
- entrapment in the matrix formed by electro-polymerization [295, 296],
- entrapment in the polymer matrix formed from photosensitive resin [293],
- entrapment in the polymer matrix formed by cross-linking a solution containing enzyme and cross-linkable polymer with a macro cross-linker.

Usually, the enzyme films are supported, as illustrated in Scheme 4.24. Therefore, enzyme-containing coating (ECC) can be formed by casting enzyme–polymer solution followed by drying, electrochemical, or photo-initiated polymerization or deposition of the polymeric solution containing enzymes. By analogy with normal entrapment techniques, retention of activity is usually dictated by the properties of the enzyme, the matrix, and the method used and is below 50% [80, 297, 298].

Several types of matrix film have been developed, for example sol–gel matrix prepared by coating of the solution containing a precursor of variable hydrophilicity

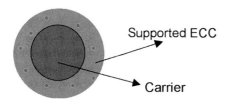

Scheme 4.24 Supported enzyme-containing coating.

followed by gelation, hydrophobic matrix formed by casting a solution of hydrophobic polymer in organic solvent, followed by evaporation [299], and hydrophilic cross-linked film matrix formed by cross-linkers of different length [302]. Thus, enzyme activity in the films can also be modulated, principally by selecting a suitable matrix and the condition for entrapment.

Supported Entrapment by Chemical Gelation Supported enzyme entrapment can be achieved by polymerizing the solution (containing enzyme and monomer) and casting on a support (flat sheet, electrode) [299, 300]. In this way, enzyme membranes which not only have a separation function but also catalytic function, can be prepared in one step. In general, supported entrapment is often used for fabrication of electrodes.

This technique has recently been used for preparation of supported sol–gel-entrapped enzymes, with the objective of overcoming the drawbacks of sol–gel-entrapped enzymes, for example lower activity and smaller particles. Use of extra supports for preparation might reduce diffusion constraints, thus enhancing enzyme activity.

It was recently found that sol–gel polymer lipase can be also fabricated within the pores of different porous supports, for example chitosan beads, Amberlite, and non-woven polyester, with the aim of improving its mechanical stability. Of all the supports tested, non-woven fibres of medium pore size were found best for this purpose. The thermal stability of lipase was increased 55-fold on entrapment in sol–gel materials [301].

The enzymes can be immobilized by entrapment in foams formed by proteins [308, 311] or urethane prepolymers [308], as shown in Scheme 4.25. Usually, entrapment of enzyme in PUF (Polyurethane foam) leads to improvement of the longevity of the immobilized enzyme compared with the native enzyme [315]. High volume activity (U g^{-1}) can also be obtained by delicate control of the amount of pre-polymer used [313].

Scheme 4.25 Formation of foam as entrapment carrier for enzyme immobilization.

Supported Immobilized Enzyme by Physical Gelation Physical gelation of the enzyme-polymer solution in or on the porous carrier can be dated to the beginning of 1980s, when Jancsik and co-workers immobilized several enzymes such as lactase, aldolase, and penicillin acylase in a poly(vinyl alcohol) matrix by several methods, for example drying of enzyme–PVA solution on the glass plate, drying the PVA-enzyme polymer solution impregnated in the pores of filter paper by air, or contact of the above supported enzyme–polymer solution with concentrated salt solution [254].

Table 4.16 Entrapment of enzyme in supported films

Polymer/support	Method	Remarks	Enzyme	Ref.
Photosensitive methyacrylated poly(HEMA)	Entrapment	–	LOD	293
Sol–gel glass prepared from APTES, 2-(3,4-ECH)-ETMS	Entrapment	Additives, e.g. PEG, GP, or PEG-GP, mitigate diffusion constraints	GOD	294
PMBQ film prepared by electro-polymerization of MHQe	Entrapment	Deposition in the presence of the enzyme	GOD	295
Polypyrrole	Deposition-entrapment	Electro-polymerization of pyrrole in the presence of the enzyme	GOD	296
Film of PVA bearing aromatic azido groups	Entrapment	~40% of its native enzyme activity	β-Glucosidase	297
Water-borne polyurethane coatings	Covalent-entrapment	Less hydrophilic PIC enhanced the activity; 39% retention of activity	DFPase	298
Poly(MMA), polystyrene, and PVAc	Entrapment in solvent	High operational stability of the immobilized enzyme	Chymotrypsin and trypsin	299
Poly(1-vinylimidazoyl)-Os(bipyridine)$_2$Cl)+/2+-PEG-diglycidyl ether (DGE)	Entrapment	–	GOD	302
PET FILM coated with PAAc, enzymes and substrates	Entrapment	Ca^{2+} ion-responsive drug release	Amylase, GOD	303
Poly(1-vinylimidazoyl)-dmeOs)-PEG-DGE	Entrapment	–	GOD or LOD	304
Polyester film-layered-photo-sensitive resin, PEG-DMA	Entrapment	Activity increased with increasing chain length of the PEG moiety	Invertase	305
Photocross-linkable PVA bearing stilbazolium	Entrapment	The presence of more than 1 mol % cross-linking units is essential to avoid enzyme leakage	Enzymes	306
Electropolymerized poly(1,3-diaminobenzene)	Adsorption-cross-linking	–		307

Table 4.17 Supported Entrapment by physical gelation

Pregel component/carrier	Method	Remark	Enzyme	Ref.
Water-borne polyurethane coatings	Supported entrapment	Less hydrophilic PIC enhanced the activity; 39% retention of activity	DFPase	317
PVA/filter paper	Supported entrapment	The thermostability of the immobilized enzyme was twice that of the free enzyme	Adolase	318
Poly(vinyl alcohol) cryogel membrane	Supported entrapment	A bienzymic sensor for detection of acetylcholine was prepared by physical co-immobilization of acetylcholinesterase in a PVA cryogel membrane obtained by a cyclic freezing–thawing process	Acetylcholinesterase/ PEG-modified choline oxidase	319
Platinum electrode poly(vinyl stearate)	Supported entrapment	The presence of hydrophilic domains or pores can make the membrane permeable	PEG-modified GOD	320
Vinnapas M54/25C/an electrode	Supported entrapment	Covering the enzyme membrane with a second membrane, which can prevent non-specific interferences	GOD	32
Biodegradable polymer	Supported entrapment	Resulting in the release of the targeting compounds	Protein	322
Redox gel membrane by poly(VF-co-BIS-AAm)	Supported entrapment	Enzyme redox gel electrode is suitable for use with dehydrogenases that require NADP$^+$ as a cofactor	Glutathione reductase	323
PAAM or poly(AAm-coMAAm)	Supported entrapment	Cross-linked *in situ* with glyoxal, resulting in a thin film which adheres firmly to the surface of the electrode	ACE	324
Poly(MMA), polystyrene, and poly(vinyl acetate)	Supported entrapment	The enzymes were prepared in the hydrophobic ionic pairing form and soluble in hydrophobic solvents	α-Chymotrypsin and trypsin	325

Table 4.17 Continued

Pregel component/carrier	Method	Remark	Enzyme	Ref.
Photo-crosslinking polymers are cyclic formals prepared by treating poly(vinyl alcohol)/silica gel	Supported entrapment	The loading and activity retention were very low	Invertase	326
Gelatin-glutaraldehyde/solid substrate	Supported entrapment	Similar coatings can be prepared on thermocouples, thermistors, or pH electrodes. The coating was stable at RT in the air for a month	Glucose oxidase	326
Polyhydroxyethyl methyacrylate/electrode	Supported entrapment	–	Glucose oxidase	327
Polyvinyl alcohol and carboxymethyl hydroxyethyl cellulose/electrode		The obtained electrode can be used in non-aqueous medium due to the use of hydrated highly hydrophilic gels	Horseradish peroxidase	328
Poly(ester-sulphonic acid)-glucose oxidase solutions	Supported entrapment	The cation exchanger simultaneously performs the entrapment, charge permselectivity, and antifouling functions	Glucose oxidase	329
Hydroxyethylcellulose, and 1.5% polyethyleneimine/glutaraldehyde/glass filaments	Supported entrapment	The obtained immobilized enzymes are characterized in that they are easy and inexpensive to prepare, durable, and highly active	Diazyme	330

Several types of supported matrix film have been developed, for example:

- sol–gel matrix prepared by coating of the solution containing the precursor of variable hydrophobicity, then gelation;
- hydrophobic matrix formed by casting solution of hydrophobic polymer in organic solvent, followed by evaporation [299];
- heating [299, 300].

It has been found that a hydrophilic cross-linked film matrix can be formed by use of cross-linkers of different length [302]. Thus, enzyme activity in the films can also be principally modulated by selecting a suitable matrix and conditions for entrapment.

Supported Immobilized Enzyme by Entrapment/cross-linking In another approach, enzyme entrapment in the film formed from synthetic or natural polymers, followed by cross-linking, can be achieved by chemically cross-linking the pre-polymer (or polymer)–enzyme solution [72, 83], or crosslinking the polymer matrix by irradiation [75, 82, 83].

Obviously, diffusion constraints depend on the thickness of the film. Thus, reduction of the thickness can lead to high retention of activity. The cross-linking reaction can make the enzyme insoluble and increase the mechanical stability of the film. Enzyme stability can also be increased [74, 75, 82].

Supported Sol–gel-entrapped Enzymes As noted above, sol–gel-entrapped enzymes usually suffer from a serious drawback – the intrinsic size of the particles are often too small to be used in processes in which separation of the enzyme from the reaction medium might be an important issue in term of process control and recycling of the costly enzymes, for example the synthesis of semi-synthetic β-lactam antibiotics.

Also, conditions for in-situ entrapment are often not compatible with the enzyme selected. To avoid this intrinsic drawback, a new concept was introduced in the mid of 1980s, when Glad et al. prepared sol–gel-entrapped enzyme on a preformed carrier-silica [334].

It has recently been reported that these intrinsic drawbacks can be overcome by use of a support with a large pore volume in which the sol–gel-entrapment process can be performed or the enzyme can be entrapped in a mesoporous zeolite-MCM-41 with the aid of alginate–$CaCl_2$ gels (see Scheme 4.26) [341].

As shown in Scheme 4.26, the enzyme is first entrapped in the sol–gel and this is subsequently confined in the pores of the selected support [338]. In the other process the mesoporous materials used function only as a host for the enzymes entrapped in another gel system. The sol–gel material functions as an extra support, with the aim of improving the geometric properties of the IME.

Alternatively, the sol–gel particles can be also formed on the surface of another solid, for example Celite [226, 338]. Remarkably, the stability of enzymes entrapped in this way is at least two orders of magnitude greater than for enzymes deposited on the Celite (Scheme 4.26) [226]. Similarly, glucose oxidase was covalently immo-

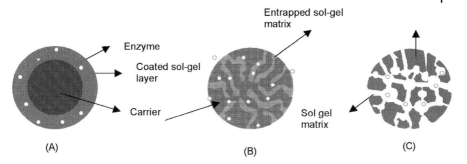

Scheme 4.26 Types of supported entrapped enzyme: (A) sol–gel-entrapped enzyme coated on another carrier; (B) sol–gel-entrapped enzyme confined to the pores of another host carrier; (C) gel-entrapped enzyme hosted in the pores of a sol–gel matrix.

bilized in Sepabeads and the preimmobilized enzyme was subsequently entrapped in sol–gel matrix with high retention of enzyme activity [340].

This technology was applied by Glad and co-workers as early as the 1980s, mainly for the enzyme imprinting. They applied the sol–gel process for the preparation of substrate-selective polymers and the entrapment of enzymes [334]. Originally, they polymerized organic silane monomers in the presence of enzyme or substrate on the silica surface, leading to the formation of entrapped enzyme with high mechanical and operational stability.

Although, in general, this technology is not well developed, the demand of more robust immobilized enzymes both with preserved high catalytic performance and broadly applicable non-catalytic functions, e.g. suitable geometry (approx. 100 µm), might provide inspiration for this technology [336].

As discussed above, homopolymers such as PAAH and PEI have been used to encage enzymes in sandwich or tentacle-brush structures. A similar method of enzyme encagement, which we call adsorption–encapsulation, had been developed a long time previously. This is exemplified by adsorption of urease on activated charcoal granules which were then coated with an ultrathin layer of cellulose nitrate [266]. Enzyme activity was 28.1 % of the activity of urease in solution. The enzyme did not leak out after immobilization and worked efficiently.

Table 4.18 Entrapment of enzyme by supported sol–gel process

Carrier	Method	Remark	Enzyme	Ref.
DMDMOS/TMOS/Celite	Entrapment	Activity was 3–10 times higher than that of the enzyme deposited on the Celite. Enantioselectivity for the reaction of menthol and butyric acid depended on the length of the alkyl chain	*Candida rugosa* lipase	332
TMSO/alginate beads	Entrapment	Enzyme-containing calcium alginate beads were soaked in a solution of TMSO in hexane, leading to the formation of rigid and robust silica-alginate beads that do not deform	β-Glucosidase	333
Bis(2-Hydroxyethyl)amino-propyltriethoxysilane and tetraethoxysilane/silica	Entrapment	To make the entrapped enzyme preparation easier to handle and apply, especially in column experiments, the entrapment was, alternatively, left to proceed on the surface of porous silica	Glucose oxidase	334
Polyvinylformal (PVF) resin plate-supported silicate	Entrapment	Fifty-onefold enhancement of operational stability was obtained; highest activity was obtained at 0.06 payload; enzyme activity was influenced by the water/silane molar ratio and the ratio of TMSO/PTMS	Lipase	335
Propyltrimethoxysilane (PTMS)/TMOS/Non-woven polyester sheet and other porous beads	Entrapment	The mechanical stability of the entrapped enzyme was improved. Entrapment on non-woven polyester sheet and other porous beads proved to be more efficient with regard to enzyme activity	*Candida rugosa* lipase	336
ATMOS)/TMOS/Non-woven polyester sheet	Entrapment	The activity of the immobilized enzyme was increased 2.6-fold and the stability of the immobilized enzyme was also increased more than 50-fold	*Candida rugosa* lipase	337
Celite-supported sol–gel (TMOS/organic silane)d	Entrapment	The presence of the substrates changed the selectivity of the immobilize lipase	Lipase	338
Celite/PTMOS/TMOS	Entrapment	The hybrid gel-entrapped protease on Celite 545 had up to eight times the activity of the protease deposited on Celite 545 in the temperature range of 35 to 85 °C	Protease	339
TMSO/Sepabaeds	Binding and Entrapment	High retention of activity was obtained, due to the avoidance of the direct contact of enzyme with the sol–gel matrix. Besides, no enzyme leakage was observed	Glucose oxidase	340
Silica (MCM 41)-Ca alginate	Entrapment	Enzyme-alginate was deposited in the pores of the silica (~3 nm). Enhanceds stability was obtained, compared with the enzyme entrapped in Alginate bead	β-Glucosidase	341

4.6
References

1. Dickey FH: Specific adsorption. *J Phys Chem* 1955, 59:695–707
2. Wei Y, Dong H, Xu JG, Feng QW: Simultaneous immobilization of horseradish peroxidase and glucose oxidase in mesoporous sol–gel host materials. *Chem Phys Chem* 2002, 3:802–808
3. Bernfeld P, Wan J: Antigens and enzymes made insoluble by entrapping them into lattice of synthetic polymers. *Science* 1963, 142:678–679
4. Sato T, Tosa T, Chibata I: Continuous production of 6-aminopenicillanic acid from penicillin by immobilized microbial cells. *Eur J Appl Microbiol* 1976, 2:153–160
5. Mosbach K, Mosbach R: Entrapment of enzymes and microorganisms in synthetic cross-linked polymers and their applications in column techniques. *Acta Chem Scand* 1966, 20:2807–2810
6. Kil'deeva NR, Krasovskaia SB, Tsarevskaia II, Virnik AD, Filippova OV: Effect of the structure of polymeric materials on the properties of immobilized enzymes. *J Prikl Biokhim Mikrobiol* 1968, 24:375–379
7. Singh D, Goel R, Johri BN: Production of 6-aminopenicillanic acid through double entrapped *E. coli* NCIM 2563. *Curr Sci* 1988, 57:1229–1231
8. Kawakami K, Abe T, Yoshida T: Silicon-immobilised biocatalyst effective for bioconversions in non-aqueous media. *Enzyme Microb Technol* 1992, 14:371–375
9. Khare SK, Vaidya S, Gupta MN: Entrapment of proteins by aggregation within Sephadex beads. *Appl Microb Biotechnol* 1991, 27:205–216
10. D'Urso EM, Fortier G: Albumin-poly(ethylene glycol) hydrogel as matrix for enzyme immobilisation: biochemical characterization of crosslinked acid phosphatase. *Enzym Microb Technol* 1996, 18:482–488
11. Seip JE, Fager SK, Grosz R, Gavagan JE, DiCosimo R, Anton DL: Enzymic synthesis of cytidine 5′-diphosphate using pyrimidine nucleoside monophosphate kinase. *Enzym Microb Technol* 1990, 125:361–366
12. Lee KH, Lee PM, Siaw YS: Studies of l-phenylalanine production by immobilized in stabilized calcium alginate beads. *J Chem Technol Biotechnol* 1992, 54:375–382
13. Szczesna-Antczak M, Antczak T, Rzyska M, Bielecki S: Catalytic properties of membrane-bound Mucor lipase immobilized in a hydrophilic carrier. *J Mol Catal B Enzym* 2002, 19/20:261–268
14. Alsarra Ibrahim A, Betigeri SS, Zhang H, Evans BA, Neau SH: Molecular weight and degree of deacetylation effects on lipase-loaded chitosan bead characteristics *Biomaterials* 2002, 23:3637–3644
15. Guiseppi-Elie A, Sheppard NF Jr, Brahim S, Narinesingh D: Enzyme microgels in packed-bed bioreactors with downstream amperometric detection using microfabricated interdigitated microsensor electrode arrays. *Biotechnol Bioeng* 2001, 75:475–484
16. Brahim S, Narinesingh D, Guiseppi-Elie A: Kinetics of glucose oxidase immobilized in p(HEMA)-hydrogel microspheres in a packed-bed bioreactor. *J Mol Catal B Enzym* 2002, 18:69–80
17. Linko YY, Viskari R, Pohjola L, Linko P: Cellulose bead entrapped whole cell glucose isomerase in fructose syrup production. *Enzym Eng* 1978, 4:345–347
18. Kawashima J, Izumida H, Hirama T: Adjustment of specific gravity of granules for enzyme or microorganism immobilization. *Jpn Kokai Tokkyo Koho*, JP 10,139,803, 1998
19. Fan CH, Lee CK: Purification of d-hydantoinase from adzuki bean and its immobilization for *N*-carbamoyl-d-phenylglycine production. *Biochem Eng J* 2001, 8:157–164
20. Salleh AB, Mohamed S: Protein-alginate gel for enzyme immobilisation. *Biotechnol Lett* 1982, 4:587–592

21 Soni S, Desai JD, Devi S: In situ entrapment of α-chymotrypsin in the network of acrylamide and 2-hydroxyethyl methacrylate copolymers. *J Appl Polym Sci* 2000, 77:2996–3002

22 Veronese FM, Boccu E, Caliceti P, Lora S, Carenza M, Palma G: Arginase immobilization on poly(hydroxyethyl acrylate) matrix beads. *J Bioact Compat Polym* 1989, 4:42–50

23 Schulz B, Riedel A, Abela PU: Influence of polymerization parameters and entrapment in poly(hydroxyethyl methacrylate) on activity and stability of GOD. *J Mol Catal B: Enzymatic* 1999, 7:85–91

24 Kumakura M, Kojima T, Kaetsu I: Takasaki Radiat. Chem. Res. Establ. Fibrous polymeric composites immobilizing glucose oxidase. *Acta Chim Hung* 1986, 122:65–71

25 Higa OZ, Del Mastro NL.; Castagnet AC:Immobilization of cellulase and cellobiase by radiation-induced polymerization. *Radiat Phys Chem* 1986, 27:311–16

26 Cantarella M, Cantarella L, Alfani F:Entrapping of acid phosphatase in poly(2-hydroxyethyl methacrylate) matrixes: preparation and kinetic properties *Brit Polym J* 1988, 20:477–85

27 Frings K, Koch M, Hartmeier W: Kinetic resolution of 1-phenylethanol with high enantioselectivity with native and immobilized lipase in organic solvents. *Enzyme Microb Technol* 1999, 25:303–309

28 Ortega N, Busto MD, Perez-Mateos M: Optimisation of β-glucosidase entrapment in alginate and polyacrylamide gels. *Bioresour Technol* 1998, 64:105–111

29 Hsu AF, Foglia TA, Shen SY: Immobilization of *Pseudomonas cepacia* lipase in a phyllosilicate sol–gel matrix. Effectiveness as a biocatalyst. *Biotechnol Appl Biochem* 2000, 31:179–183

30 Ikeda Y, Kurokawa Y, Nakane K, Ogata N: Entrap-immobilization of biocatalysts on cellulose acetate-inorganic composite gel fiber using a gel formation of cellulose acetate-metal (Ti, Zr) alkoxide. *Cellulose* 2002, 9:369–379

31 Xu JG, Dong H, Feng QW, Wei Y: Direct immobilization of horseradish peroxidase in hybrid mesoporous sol–gel materials. *Polym Prepr (Am Chem Soc, Div Polym Chem)* 2000, 41:1044–1045

32 Busto MD, Ortega N, Perez-Mateos M: Studies on microbial β-d-glucosidase immobilized in alginate gel beads. *Process Biochem (Oxford)* 1995, 30:421–426

33 Badjic JD, Kostic NM: Effects of encapsulation in sol–gel silica glass on esterase activity, conformational stability, and unfolding of bovine carbonic anhydrase II. *Chem Mater* 1999, 11:3671–3679

34 Marani A, Bartoli F, Bendoricchio G, Morisi F: Leakage of enzyme from cellulose triacetate fibers. *Chim Ind Milan* 1977, 59:243–247

35 Queiroz N, Nascimento MG, Pseudomonas sp. lipase immobilized in polymers versus the use of free enzyme in the resolution of (R,S)-methyl mandelate. *Tetrahedron Lett* 2002, 43:5225–5227

36 Kimura K, Yokote Y, Fujita M, Samejima H: Production of high-fructose syrup using glucoamylase and glucose isomerase immobilized on phenol–formaldehyde resin. *Enzym Eng* 1978, 3:531–536

37 Bunting PS, Laidler KJ: Properties of α-chymotrypsin and β-galactosidase supported in polyacrylamide gels. *Can J Biochem* 1973, 51:1598–1603

38 Moriyama S, Noda A, Nakanishi K, Matsuno R, Kamikubo T: Thermal stability of immobilized glucoamylase entrapped in polyacrylamide gels and bound to SP-Sephadex C-50. *Agric Biol Chem* 1980, 44:2047–2054

39 Wasserman BP, Burke D, Jacobson BS: Immobilization of glucoamylase from *Aspergillus niger* on polyethylenimine-coated nonporous glass beads. *Enzym Microb Technol* 1982, 4:107–109

40 Matsumoto M, Ohashi K: Effect of immobilization on thermostability of lipase from *Candida rugosa*. *Biochem Eng J* 2003, 14:75–77

41 Kim Y, Dordick J, Clark D: Siloxane-based biocatalytic films and paints for use as reactive coatings. *Biotechnol Bioeng* 2001, 72:475–482

42 Raviyan P, Tang JM, Rasco BA: Thermal stability of α-amylase from Aspergillus oryzae Entrapped in polyacrylamide gel. *J Agri Food Chem* 2003, 51:5462–5466

43 Eggers DK, Valentine JS: Crowding and hydration effects on protein conformation: a study with sol–gel encapsulated proteins. *J Mol Biol 2001*, 314:911–922

44 Mizutani F, Sato Y, Yabuki S, Iijima S: Amperometric urea-sensing electrode based on a tri-enzyme/polydimethyl-siloxane-bilayer membrane. *Chem Sensor* 2001, 17:151–153

45 Liu DM, Chen IW: Encapsulation of protein molecules in transparent porous silica matrices via an aqueous colloidal sol–gel process. *Acta Mater* 1999, 47:535–4544

46 Bismuto E, Martelli PL, De Maio A, Mita DG, Irace G, Casadio R: Effect of molecular confinement on internal enzyme dynamics: frequency domain fluorometry and molecular dynamics simulation studies. *Biopolymers* 2002, 67:85–95

47 Haraguchi T, Hatanaka C, Ide S, Nishimiya K, Kajiyama T: Immobilization of glucoamylase by plasma-initiated polymerization and evaluation of enzyme activity. *J Appl PolymSci: AppliPolym Symp* 1990, 46:385–97

48 Goncalves AP, Cabral JM, Aires-Barros MR: Immobilization of a recombinant cutinase by entrapment and by covalent binding. Kinetic and stability studies. Appl Biochem Biotechnol 1996, 60:217–28

49 Basri M, Harun A, Ahmad MB, Razak CAN, Salleh AB: Immobilization of lipase on poly(N-vinyl-2-pyrrolidone-co-styrene) hydrogel. *J Appl Polym Sci* 2001, 82:1404–1409

50 Emi S: Microporous polyalkylene-terephthalate fiber and its preparation and use for immobilization of sustained-release enzymes. *Jpn Kokai Tokkyo Koho, JP 03,175,986 A2*, 1991

51 Chen Q, Kenausis GL, Heller A: Stability of oxidases immobilized in silica gels. *J Am Chem Soc* 1998, 120:4582–4585

52 Suwasono S, Rastall RA: Synthesis of oligosaccharides using immobilised 1,2-mannosides from *Aspergillus phoenicis*, Immobilisation-dependent modulation of product spectrum. *Biotechnol Lett* 1998, 20:15–17

53 Kennedy JF, Cabral JMSM, Kalogerakis B: Comparison of action patterns of gelatin-entrapped and surface-bound glucoamylase on an α-amylase degraded starch substrate: a critical examination of reversion products. *Enzym Microb Technol* 1985, 7:22–28

54 Mohapatra SC, Hsu JT: Optimizing lipase activity, enantioselectivity, and stability with medium engineering and immobilization for β-blocker synthesis. *Biotechnol Bioeng* 1999, 64:213–220

55 McIninch JK, Kantrowitz ER: Use of silicate sol–gels to trap the R and T quaternary conformational states of pig kidney fructose-1,6-bisphosphatase. *Biochim Biophys Acta* 2001 1547:320–328

56 Rotticci D, Norin T, Hult K: Mass transport limitations reduce the effective stereospecificity in enzyme-catalyzed kinetic resolution. *Org Lett* 2000, 2:1373–1376

57 Demircioglu H, Beyenal H, Tanyolac A, Hasirci N: Entrapment of urease in glycol-containing polymeric martrices and estimation of effective diffusion cofficient of urea. *Polymer* 1995, 36:4091–4096

58 Yamazaki Y, Maeda H: The co-immobilization of NAD and dehydrogenases and its application to bioreactors for synthesis and analysis. *Agric Biol Chem* 1982, 46:1571–1581

59 Soni S, Desai JD, Devi S: Immobilization of yeast alcohol dehydrogenase by entrapment and covalent binding to polymeric supports. *J Appl Polym Sci* 2001, 82:1299–1305

60 Duarte JMC, Lilly MD: The use of free and immobilized cells in the presence of organic solvents: the oxidation of cholesterol by *Nocardia rhodochrous*. *Enzym Eng* 1980, 5:363–367

61 Epton R, Marr G, Shackley AT: New hydrophilic acrylate monomers suitable for the gel-bead entrapment of enzymes. *Polymer* 1981, 22:553–557

62 Walton HM, Eastman JE: Insolubilized amylases. *Biotechnol Bioeng* 1973, 15:951–962

63 Manolov RJ, Kambourova MS, Emanuilova EI: Immobilization of *Bacillus stearothermophilus* cells by entrapment in various matrices. *Process Biochem* (Oxford) 1995, 30:141–144

64 Prabhune A, SivaRaman H: Immobilization of penicillin acylase in porous beads of polyacrylamide gel. *Appl Biochem Biotechnol* 1991, 30:265–272

65 Matsuda K, Orii H, Hirata M, Kokufuta E: Construction of a biochemo-mechanical system using enzyme-loaded polyelectrolyte gels. *Polym Gel Network* 1994, 2:299–305

66 Hinberg I, O'Driscoll KF: Preparation and kinetic properties of gel entrapped urate oxidase *Biotechnol Bioeng* 1975, 17:1435–1441

67 Maeda H, Suzuki H, Yamauchi A, Sakimae A: Preparation of immobilized enzymes from acrylic monomers under γ-ray irradiation. *Biotechnol Bioeng* 1975, 17:119–128

68 Hinberg I, Korus R, O'Driscoll KF: Gel entrapped enzymes. Kinetic studies of immobilized β-galactosidase. *Biotechnol Bioeng* 1974, 16:943–963

69 Dong LC, Hoffman AS: Thermally reversible hydrogels III. Immobilization of enzymes for feedback reaction control. *J Controlled Release* 1986, 4:223–227

70 Park TG, Hoffman AS: Immobilization and characterization of α-galactosidase in thermally reversible hydrogel beads. *J Biomed Mater Res* 1990, 24:21–38

71 Kokufuta E: Novel applications for stimulus-sensitive polymer gels in the preparation of functional immobilized biocatalysts. In: Dusek K (Ed.) *Advances in Polymer Science*, vol 110, Springer, Berlin, 1993, pp 157–177

72 Elcin, Y. Murat; Sungur, Sibel; Akbulut, Ural. Glucose oxidase immobilized on gelatin by various crosslinkers. *Macromol Rep* 1993, A30 (Suppl. 1–2): 137–147

73 Stark MB, Kolmberg K: Covalent immobilisation of lipase in organic solvents, *Biotechnol Bioeng* 1989, 34:942–950

74 Ge, Shi-Jun; Zhang, Long-Xiang: The immobilized porcine pancreatic exopeptidases and its application in casein hydrolyzates debittering. Applied Biochem Biotechnol 1996, 59:159–165

75 O'Driscoll KF, Korus RA, Ohnuma T, Walczack IM: Gel entrapped L-asparaginase. Kinetic behavior and antitumor activity. *J Pharmacol Experim Therapeut* 1975, 195:382–388

76 Sperinde JJ, Griffith LG: Synthesis and characterization of enzymatically-cross-linked poly(ethylene glycol) hydrogels. *Macromolecules* 1997, 30:5255–5264

77 Chen JP: Production of ethyl butyrate using gel-entrapped *Candida cylindracea* lipase. *J Ferment Bioeng* 1996, 82:404–407

78 Uhlich T, Ulbricht M, Tomaschewski G: Immobilization of enzymes in poly(vinyl alcohol). *Enzym Microb Technol* 1996, 19:124–131

79 Andreopoulos FM, Roberts MJ, Bentley MD, Harris JM, Beckman EJ, Russell AJ: Photoimmobilization of organophosphorus hydrolase within a PEG-based hydrogel. *Biotechnol Bioeng* 1999, 65:579–588

80 Ichimura K, Watanabe S: Immobilization of enzymes with use of photosensitive polymers having the stilbazolium group. *J Polym Sci Polym Chem Ed* 1980, 18:891–902

81 Yeoh HH: Cassava β-glucosidase electrode prepared with photocrosslinkable prepolymer resins. *Biotechnol Appl Biochem* 1992, 15:221–225

82 Itoh N, Cheng LY, Izumi Y, Yamada H: Immobilized bromoperoxidase of *Corallina pilulifera* as a multifunctional halogenating biocatalyst. *J Biotechnol* 1987, 5:29–38

83 Maeda H: Preparation of immobilized β-galactosidase by poly(vinylpyrrolidone) and the continuous hydrolysis of lactose in acid whey. *Biotechnol Bioeng* 1975, 17:1571–1589

84 Maeda H, Suzuki H, Yamauchi A: Preparation of immobilized enzymes

using poly(vinyl alcohol). *Biotechnol Bioeng* 1973, 15:607–610
85 Sun QH, Jia DM: Studies on the fiber-containing photocrosslinkable polyurethane as the carrier of immobilized cells. *Gaofenzi Cailiao Kexue Yu Gongcheng* 1993, 9:38–42
86 Leca B, Morelis RM, Coulet PR: Bienzyme electrode compatible behaviour for water-soluble and -insoluble substrates. *Anal Lett* 1996, 29:661–672
87 Tanaka A, Hagi N, Yasuhara S, Fukui S: Immobilization of catalase with photo-crosslinkable resin prepolymers. *J Ferment Technol* 1978, 56:511–515
88 Tanaka A, Yasuhara S, Fukui S, Iida T, Hasegawa E: Immobilization of invertase using photocrosslinkable resin oligomers and properties of the immobilized enzyme. *J Ferment Technol* 1977, 55:71–75
89 Fukui S, Tanaka A, Iida T, Hasegawa E: Application of photo-crosslinkable resin to immobilization of an enzyme. *FEBS Lett* 1976, 66:179–182
90 Kajiwara S, Maeda H: The improvement of a droplet gel-entrapping method: the coimmobilization of leucine dehydrogenase and formate dehydrogenase. *Agric Biol Chem* 1987, 51:2873–2879
91 Fukunaga K, Minamijima N, Sugimura Y, Zhang ZZ, Nakao K: Immobilization of organic solvent-soluble lipase in nonaqueous conditions and properties of the immobilized enzymes. *J Biotechnol* 1996, 52:81–88
92 Fukui S, Tanaka A, Gellf G: Immobilization of enzymes, microbial cells, and organelles by inclusion with photo-crosslinkable resins. *Enzym Eng* 1978, 4:299–306
93 Kanda T, Miyata N, Fukui T, Kawamoto T, Tanaka A: Doubly entrapped baker's yeast survives during the long-term stereoselective reduction of ethyl 3-oxobutanoate in an organic solvent. *Appl Microbiol Biotechnol* 1998, 49:377–381
94 Storey KB, Ducan JA, Chakrabarti AC: Immobilisation of amyloglucosidase using to forms of polyureathane polymer. *Appl Biochem Biotechnol* 1990,23:221–236
95 Fukushima S, Nagai T, Fujita K, Tanaka A, Fukui S: Hydrophilic urethane prepolymers: convenient materials for enzyme entrapment. *Biotechnol Bioeng* 1978, 20:1465–1469
96 Hu ZC, Korus RA, Stormo KE: Characterization of immobilized enzymes in polyurethane foams in a dynamic bed reactor. *Appl Microb Biotechnol* 1993, 39:289–295
97 Wood LL, Calton GJ: Immobilized microbial cells and their use. *Eur Pat Appl, EP 89,165 A2*, 1983
98 Katsuragi T, Sakai T, Tonomura K: Implantable enzyme capsules for cancer chemotherapy from bakers' yeast cytosine deaminase immobilized on epoxy-acrylic resin and urethane prepolymer. *Appl Biochem Biotechnol* 1987, 16:61–69
99 Alva S, Phadke RS, Govil G: Glucose oxidase immobilized in polyurethane matrix: a biosensor for amperometric estimation of glucose. *J Indian Chem Soc* 1993, 70:403–408
100 Tanaka A, Iida T: Entrapment of biocatalysts by prepolymer methods. *Methods Biotechnol* 2001, 15:19–30
101 Puig-Lleixa C, Jimenez C, Bartroli J: Acrylated polyurethane – photopolymeric membrane for amperometric glucose biosensor construction. *Sens Actuators B Chem* 2001, 72:56–62
102 Klein J, Eng H: Immobilization of microbial cells in epoxy carrier systems. *Biotechnol. Lett* 1979, 1:171–176
103 Drevon GF, Russell AJ: Irreversible immobilization of diisopropylfluorophosphatase in polyurethane polymers. *Biomacromolecules* 2002, 1:571–576
104 Freeman A, Blank T, Haimovich B: Gel entrapment of enzymes in cross-linked prepolymerised PAAm-hydrazide. *Ann NY Acad Sci* 1983, 413:557–559
105 Freeman A, George S: Gel entrapment of whole cells and enzymes in cross-linked, prepolymerised PAAm hydrazide. *Ann NY Acad Sci* 1984, 434:418–426

106 Tran-Minh C, Pandey PC, Kumaran S: Studies on acetylcholine sensor and its analytical application based on the inhibition of cholinesterase. *Biosens Bioelectron* 1990, 5:461–471

107 Denti E: Enzyme trapping in polyvinyl-pyrrolidone. *Biomater Med Dev Artifi Organ* 1974, 2:293–294

108 Gaberlein S, Knoll M, Spener F, Zaborosch C: Disposable potentiometric enzyme sensor for direct determination of organophosphorus insecticides. *Analyst* 2000, 125:2274–2279

109 Campanella L, Colapicchioni C, Favero G, Sammartino MP, Tomassetti M: Organophosphorus pesticide (paraoxon) analysis using solid state sensors. *Sens Actuators B Chem* 1996, 33:25–33

110 Wood LL, Cobbs CS, Lantz L 2nd, Peng L, Calton GJ: Immobilization of enzymes with polyaziridines. *J Biotechnol* 1990, 13:305–314

111 Wood LL, Calton GJ: Immobilizing microbial cells with polyfunctional aziridines. *US Ser No 465,551*, 1987

112 Botre F, Botre C, Lorenti G, Mazzei F, Porcelli F, Scibona G: Determination of l-glutamate and l-glutamine in pharmaceutical formulations by amperometric l-glutamate oxidase based enzyme sensors. *J Pharm Biomed Anal* 1993, 11:679–686

113 Tor R, Freeman A: New enzyme membrane for enzyme electrodes. *Anal Chem* 1986, 58:1042–1046

114 Selli E, D'Ambrosio A, Bellobono IR: Enzymatic activity under tangential flow conditions of photochemically grafted membranes containing immobilized catalase. *Biotechnol Bioeng* 1993, 41:474–478

115 Kutner W, Wu HH, Kadish KM: Condensation β-cyclodextrin polymer membrane with covalently immobilized glucose oxidase and molecularly included mediator for amperometric glucose biosensor. *Electroanalysis* 1994, 6:934–944

116 Allcock HR, Pucher SR, Visscher KB: Activity of urea amidohydrolase immobilized within poly[di(methoxyethoxyethoxy)phosphazene] hydrogels. *Biomaterials* 1994, 15:502–506

117 Freeman A: Gel entrapment of whole cells and enzymes in crosslinked, prepolymerised PAAm hydrazide. *Ann NY Acad Sci* 1984, 434:418–426

118 Freeman A: Crosslinked polyacrylamide-hydrazide (PAAH) as matrix for the gel entrapment of cells and enzymes – an overview. *Makromol Chem Macromol Sym* 1988, 19:125–131

119 Yawetz A, Perry AS, Freeman A, Katchalski-Katzir E: Monooxygenase activity of rat liver microsomes immobilized by entrapment in a crosslinked prepolymerised PAAm hydrazide. *Biochim Biophys Acta* 1984, 798:204–209

120 Dubinin AG, Li FC, Li YR, Yu JT: A solid-state immobilized enzyme polymer membrane microelectrode for measuring lactate-ion concentration. *Bioelectrochem Bioenergy* 1991, 25:31–35

121 Molina I, Li SM, Martinez MB, Vert M: Protein release from physically crosslinked hydrogels of the PLA/PEO/PLA triblock copolymer-type, *Biomaterials* 2001, 22:363–369

122 Galaev IYu, Mattiasson B: Thermoreactive water-soluble polymers, nonionic surfactants, and hydrogels as reagents in biotechnology. *Enzym Microb Technol* 1993, 15:354–366

123 Park TG, Hoffman AS: Immobilization of arthrobacter simplex in a thermally reversible hydrogel: effect of temperature cycling on steroid conversion. *Biotechnol Bioeng* 1990, 35:152–159

124 Sotiropoulou S, Vamvakaki V, Chaniotakis NA: Stabilization of enzymes in nanoporous materials for biosensor applications. Biosens Bioelectron 2005, 20:1674–9

125 Markvicheva E, Kuptsova SV, Mareeva TY, Vikhrov AA et al.: Immobilized enzymes and cells in poly(N-vinyl caprolactam)-based hydrogels: preparation, properties, and applications in biotechnology and medicine. *Appl Biochem Biotechnol* 2000, 88:145–157

126 Kabanov AV, Batrakova EV, Melik-Nubarov NS, Fedoseev NA, Dorodnich TYu, Alakhov VYu, Chekhonin VP, Nazarova IR, Kabanov VA: A new class of drug carriers: micelles of

127 Bismuto E, Martelli PL, De Maio A, Mita DG, Irace G, Casadio R: Effect of molecular confinement on internal enzyme dynamics: frequency domain fluorimetry and molecular dynamics simulation studies. *Biopolymers* 2002, 67:85–95

128 Bajpai P, Margaritis A: Immobilization of *Kluyveromyces marxianus* cells containing inulinase activity in open pore gelatin matrix. 1. Preparation and enzymic properties. *Enzym Microb Technol* 1985, 7:373–376

129 Hara M, Iazvovskaia S, Ohkawa H, Asada Y, Miyake J: Immobilization of P450 monooxygenase and chloroplast for use in light-driven bioreactors. *J Biosci Bioeng* 1999, 87:793–797

130 Agarwal M, Roy U, Roy PK, Shukla OP: DOPA decarboxylase activity and dopamine synthesis by Pseudomonas sp. immobilized in different matrices. *Enzym Microb Technol* 1991, 6:28–34

131 Dominguez E, Nilsson M, Hahn-Hagerdal B: Carbodiimide coupling of β-galactosidase from *Aspergillus oryzae* to alginate. *Enzym Microb Technol* 1988, 10:606–610

132 Colby J, Snell D, Black GW: Immobilization of Rhodococcus AJ270 and use of entrapped biocatalyst for the production of acrylic acid. *Monatsh Chem* 2000, 131:655–666

133 Toda K: Interparticle mass transfer study with a packed column of immobilized microbes. *Biotechnol Bioeng* 1975, 17:1729–1747

134 Munjal N, Sawhney SK: Stability and properties of mushroom tyrosinase entrapped in alginate, polyacrylamide and gelatin gels. *Enzym Microb Technol* 2002, 30:613–619

135 Kokufuta E, Jinbo E: A hydrogel capable of facilitating polymer diffusion through the gel porosity and its application in enzyme immobilization. *Macromolecules* 1992, 25:3549–3552

136 Ariga O, Kato M, Sano T, Nakazawa Y, Sano Y: Mechanical and kinetic properties of PVA hydrogel immobilizing β-galactosidase. *J Ferment Bioeng* 1993, 76:203–206

137 Schneider BH, Daroux ML, Prohaska OJ: Microminiature enzyme sensors for glucose and lactate based on chamber oxygen electrodes. *Sens Actuators B Chem* 1990, 1/6:565–570

138 Kokufuta E, Ogane O, Ichijo H, Watanabe S, Hirasa O: Poly(vinyl methyl ether) gel for the construction of a thermosensitive immobilized enzyme system exhibiting controllable reaction initiation and termination. *J Chem Soc Chem Commun* 1992, 5:416–418

139 Lev B: Temperature-sensitive star-branched poly(ethylene oxide)-b-poly(propylene oxide)-b-poly(ethylene oxide) network, *Polymer* 1998, 39: 5663–5669

140 Hoffman AS: Intelligent Polymers. In: Park K (Ed.) *Controlled drug delivery: challenge and strategies*. American Chemical Society, Washington, DC, 1997, pp 485–497

141 Allcock HR, Ambrosio A: Synthesis and characterization of pH-responsive poly(organophosphazene) hydrogels. *ACS Symposium Series*, 2003, 833: 82–101

142 Futs KA, Johnston TP: Sustained-release of urease from a poloxamer gel matrix. *J Parent Sci Technol* 1990, 44:58–65

143 Park TG, Cohen S, Langer R: Controlled protein release from polyethyleneimine-coated poly(l-lactic acid)/pluronic blend matrices. *Pharm Res* 1992, 9:37–39

144 Omata T, Tanaka A, Yamane T, Fukui S: Immobilization of microbial cells and enzymes with hydrophobic photo-crosslinkable resin prepolymers. *Eur J Appl Microb Biotechnol* 1979, 6:207–215

145 Hoshino K, Taniguchi M, Marumoto H, Fujii M: Repeated batch conversion of raw starch to ethanol using amylase immobilized on a reversible soluble-autoprecipitating carrier and flocculating yeast cells. *Agric Biol Chem* 1989, 53: 1961–1967

146 Fujimura M, Mori T, Tosa T: Preparation and properties of soluble-insoluble immobilized proteases. *Biotechnol Bioeng* 1987, 29:747–752

147 van Leemputten E, Horisberger M: Soluble-insoluble complex of trypsin: immobilised on acrolein-acrylic copolymer. *Biotechnol Bioeng* 1976, 18:587–590

148 Gianfreda L, Pirozzi D, Greco G Jr: Microenvironmental effect of stabilizing polyelectrolytes in ultrafiltration membrane enzymic reactors. *Biotechnol Bioeng* 1989, 33:1067–1071

149 Cong L, Kaul R, Dissing U, Mattiasson B: A model study on Eudragit and polyethyleneimine as soluble carriers of α-amylase for repeated hydrolysis of starch. *J Biotechnol* 1995, 42:75–84

150 Vakurov AV, Gladilin AK, Levashov AV, Khmelnitsky YL: Dry enzyme– polymer complexes: stable organosoluble biocatalysts for nonaqueous enzymology. *Biotechnol Lett* 1994, 16:175–178

151 Margolin AL, Izumrudov VA, Svedas V, Zezin AB, Kabanov VA, Berezin IV: Preparation and properties of penicillin amidase immobilized in polyelectrolyte complexes. *Biochim Biophys Acta* 1981, 660:359–365

152 Taniguchi M, Kobayashi M, Fujii M: Properties of a reversible soluble-insoluble cellulase and its application to repeated hydrolysis of crystalline cellulose. *Biotechnol Bioeng* 1989, 34:1092–1097

153 Kirsh YE, Galaev IY, Karaputadze TM, Margolin AL, Svedas V: Thermo-precipitating polyvinylcaprolactam–enzyme conjugates. *Biotekhnologiya* 1987, 3:184–189

154 Dinnella C, Lanzarini G, Ercolessi P: Preparation and properties of an immobilized soluble-insoluble pectinlyase. *Process Biochem (Oxford)* 1995, 30:151–157

155 Tyagi R, Roy I, Agarwal R, Gupta MN: Carbodiimide coupling of enzymes to the reversibly soluble insoluble polymer Eudragit S-100. *Biotechnol Appl Biochem* 1998, 28:201–206

156 Fujimura M, Mori T, Tosa T: Preparation and properties of soluble-insoluble immobilized proteases. *Biotechnol Bioeng* 1987, 29:747–752

157 Linko YY, Pohjola LA: Simple entrapment method for immobilizing enzymes within cellulose fibers. *FEBS Lett* 1976, 62:77–80

158 Kaputskii FN, Bil'dyukevich AV, Khlyustov SV, Rytik PG et al.: Enzyme immobilization on cellulose. *SU, 1,567,625 A1*, 1987

159 Nakamura T, Ono Y: Immobilized enzyme. *Japan. Kokai, JP 76–19,374*, 1977

160 Marconi W, Bartoli F, Cecere F, Galli G, Morisi, F: Synthesis of penicillins and cephalosporins by penicillin acylase entrapped in fibers, *Agric Biol Chem* 1975, 39:277–279

161 Marconi W, Bartoli F, Cecere F, Morisi F: Synthesis of l-tryptophan from indole and dl-serine by tryptophan synthetase entrapped in fibers. II. Reactor studies. *Agric Biol Chem* 1974, 38:1343–1349

162 Zaffaroni P, Vitobello V, Cecere F, Giacomozzi E, Morisi F: Synthesis of l-tryptophan from indole and dl-serine by tryptophan synthetase entrapped in fibers. I. Preparation and properties of free and entrapped enzyme. *Agric Biol Chem* 1974, 38:1335–1342

163 Marconi W, Gulinelli S, Morisi F: Properties and use of invertase entrapped in fibers. *Biotechnol Bioeng* 1974, 16:501–511

164 Marani A, Scaltriti G, Bartoli F, Morisi, F: Fixed beds of fiber-entrapped enzymes. *J Ferment Technol* 1979, 57:357–363

165 Konecny J: The immobilization of a stable esterase by entrapment, covalent binding and adsorption. *Enzym Eng* 1978, 3:11–18

166 Marconi W, Prosperi G, Giovenco S, Morisi F: Entrapment of co-enzymically active NAD+ polymers in fibers. *J Mol Catal* 1976, 1:111–120

167 Marconi W, Bartoli F, Gianna R, Morisi F, Spotorno G: Phenylalanine ammonia-lyase entrapped in fibers. *Biochimie* 1980, 62:575–580

168 Ikeda Y, Kurokawa Y: Hydrolysis of 1,2-diacetoxypropane by immobilized lipase on cellulose acetate-TiO_2 gel fiber derived from the sol–gel method. *J Sol–Gel Sci Technol* 2001, 21:221–226

169 Wang HZ, Liu SY, Wang Y: Alkaline protease production by immobilized growing cells of *Serratia marcescens* with interpolymer complexes of P(TM-co-AAm)/PAA. *J Appl Polym Sci* 2002, 84:178–183

170 Mansfeld J, Forster M, Schellenberger A, Dautzenberg H: Immobilization of invertase by encapsulation in polyelectrolyte complexes. *Enzym Microb Technol* 1991, 13:240–244

171 Margolin AL, Sherstyuk SF, Izumrudov VA, Zezin AB, Kabanov VA: Enzymes in polyelectrolyte complexes. The effect of phase transition on thermal stability. *Eur J Biochem* 1985, 146:625–632

172 Levitsky V, Lozano P, Gladilin A, Iborra JL: Stability of immobilized enzyme-polyelectrolyte complex against irreversible inactivation by organic solvents. *Prog Biotechnol* 1998, 15:417–422

173 Kokufuta E, Shimizu N, Tanaka H, Nakamura I: Use of polyelectrolyte complex-stabilized calcium alginate gel for entrapment of α-amylase. *Biotechnol Bioeng* 1988, 32:756–759

174 Ristau O, Pommerening K, Jung C, Rein H, Scheler W: Activity of urease in microcapsules. *Biomed Biochim Acta* 1985, 44:1105–1111

175 Dautzenberg H, Koetz J, Philipp B, Rother G, Schellenberger A, Mansfeld J: Interaction of invertase with polyelectrolytes. *Biotechnol Bioeng* 1991, 38:1012–1019

176 Szczesna-Antczak M, Antczak T, Rzyska M, Moderzejewska Z, Patura J, Kalinowska H, Bielecki S: Stabilisation of an intracellular *Mucor circinelloides* lipase for application in non-aqueous media, *J Mol Catal B Enzym* 2004, 29:163–171

177 Zhao L: Studies on the immobilized naringninase and its debittering the orange juices. *Shipin Yu Fajiao Gongye* 1987:25–34

178 Kobayashi Y, Matsuo R, Oya T, Yokoi N: Enzyme-entrapping behaviors in Alginate fibers and their papers. *Biotechnol Bioeng* 1987, 30:451–457

179 Busto MD, Ortega N, Perez-Mateos M: Effect of immobilisation on the stability of bacterial and fungal β-d-glucosidase 1997, *Process Biochem* 32:441–449

180 Ikeda Y, Kurokawa Y: Synthesis of geranyl acetate by lipase entrap-immobilized in cellulose acetate-TiO_2 gel fiber. *J Am Chem Soc* 2001, 78:1099–1103

181 Blandino A, Macias M, Cantero D: Immobilization of glucose oxidase within calcium alginate gel capsules. *Process Biochem (Oxford)* 2001, 36:601–606

182 Betigeri SS, Neau SH: Immobilization of lipase using hydrophilic polymers in the form of hydrogel beads. *Biomaterials* 2002, 23:3627–3636

183 Adalsteinsson O, Lamotte A, Baddour RF, Colton CK, Pollak A, Whitesides GM: Preparation and magnetic filtration of polyacrylamide gels containing covalently immobilized proteins and a ferrofluid. *J Mol Catal* 1979, 6:199–225.

184 Pollak A, Baughn RL, Adalsteinsson O, Whitesides GM: Immobilization of synthetically useful enzymes by condensation polymerisation. *J Am Chem Soc* 1978, 100:302–304

185 Pollak A, Blumenfeld H, Wax M, Baughn RL, Whitesides GM: Enzyme immobilization by condensation copolymerisation into crosslinked polyacrylamide gels. *J Am Chem Soc* 1980, 102:6324–6336

186 Jaworek D: New immobilization techniques and supports. In: Pye EKl, Wingard LB Jr (Eds) *Enzyme Eng Pap. Res. Rep. Eng. Found. Conf. 2nd*, 1973:105–114

187 Martinek K, Klibanov AM, Coldmacher VS, Berezin LV: The principles of enzyme stabilisation. 1. Increase in thermostability of enzymes covalently bound to a complimentary surface of a polymer support in a multipoint fashion. *Biochim Biophys Acta* 1977, 485:1–12

188 Suzuki S, Hirano K, Takagi Y: Plastic immobilized enzyme. *Japan Kokai, JP 51,070,872*, 1976

189 Mozhaev VV, Sergeeva, MV, Belova, AB, Khmel'nitskii YL: Multipoint attachment to a support protects enzyme from inactivation by organic solvents:

α-chymotrypsin in aqueous solutions of alcohols and diols. *Biotechnol Bioeng* 1990, 35: 653–9
190 Hamsher JJ: Immobilized enzymes and intermediates for their preparation. *Fr. Demande, FR 2212340*, 1974
191 Szewczuk A, Ziomek E, Mordarski M, Siewinski M, Wieczorek J: Properties of penicillin amidase immobilised by copolymerisation with acrylamide. *Biotechnol Bioeng* 1979, 21:1543–1552
192 Yang Z, Mesiano AJ, Venkatasubramanian S, Gross SH, Harris JM, Russell AJ: Activity and stability of enzymes incorporated into acrylic polymers. *J Am Chem Soc* 1995, 117:4843–4850
193 Fischer L: Immobilized bioimprinting. An innovative preparation for the commercial utilization of "tailor-made" biocatalysts. *Initiativen Umweltschutz* 1999, 14:91–102
194 Peiûker F, Fischer L: Crosslinking of imprinted proteases to maintain a tailor-made substrate selectivity in aqueous solutions. *Bioorg Med Chem* 1999, 7:2231–2237
195 Kronenburg NAE, de Bont JAM, Fischer L: Improvement of enantioselectivity by immobilized imprinting of epoxide hydrolase from *Rhodotorula glutinis*. *J Mol Catal B Enzym* 2001, 16:121–129
196 Whitesides GM, Lamotte A, Adalsteinsson O, Colton CK: Covalent immobilization of adenylate kinase and acetate kinase in a polyacrylamide gel: enzymes for ATP regeneration. *Methods Enzymol* 1976, 44:887–897
197 Robak M, Szewczuk A: Penicillin amidase from *Proteus rettgeri*. *J Acta Biochim Pol* 1981, 28:275–278
198 Liu F, Zhuo RX: A convenient method for the preparation of temperature-sensitive hydrogels and their use for enzyme immobilization. *Biotechnol Appl Biochem* 1993, 18:57–65
199 Podual K, Doyle FJ 3rd, Peppas NA: Glucose-sensitivity of glucose oxidase-containing cationic copolymer hydrogels having poly(ethylene glycol) grafts. *J Controlled Release* 2000, 67:9–17

200 Bille V, Plainchamp D, Lavielle S, Chassaing G, Remacle J: Effect of the microenvironment on the kinetic properties of immobilized enzymes. *Eur J Biochem* 1989, 180:41–47
201 Venton DL, Cheeseman KL, Chatterton RT, Anderson TL: Entrapment of highly specific antiprogesterone antiserum using polysiloxane copolymers. *Biochim Biophys Acta* 1984, 797
202 Avnir D, Braun S, Lev O, Ottolenghi M: Enzymes and other proteins entrapped in sol–gel materials. *Chem Mater* 1994, 6:1605–1614
203 Soares CM, dos Santos OA, de Castro HF, de Moraes FF, Zanin GM: Studies on immobilised lipase in hydrophobic sol–gel. *Appl Biochem Biotechnol* 2004, 113–116:307–319
204 Reetz MT, Tielman P, Wiesenhoefer W, Koenen W, Zonta A: Second generation sol–gel encapsulated lipases: robust heterogeneous biocatalyst, *Adv Synthesis* 2003, 345:717–728
205 Reetz MT, Zonta A, Simpelkamp J: Efficient heterogeneous biocatalyst by entrapment of lipases in hydrophobic sol–gel materials. *Angew Chem Int Ed Engl* 1995, 34:301–303
206 Braun S, Rappoport S, Zusman R, Avnir D, Ottolenghi M: Biochemically active sol–gel glasses; the trapping of enzymes. *Mater Lett* 1990, 10:1–5
207 Janczarski I, Mazur A, Witkowski K, Lubaszka E: Activity of cholinesterase included on silica gel. *Acta Physiol Pol* 1976, 27:301–306
208 Johnson P, Whateley TL: On the use of polymerizing silica gel systems for the immobilization of trypsin. *J Colloid Interf Sci* 1971, 37: 557–563
209 Ghanema A, Schurig V: Entrapment of *Pseudomonas cepacia* lipase with peracetylated β-cyclodextrin in sol–gel: application to the kinetic resolution of secondary alcohols. *Tetrahedron Asymmetry* 2003, 14:2547–2555
210 Arica O, Suzuki T, Sano Y, Murakami Y: Immobilization of a thermostable enzyme using a sol–gel preparation method. *J Ferment Bioeng* 1996, 82:341–345

211 Finnie KS, Bartlett JR, Woolfrey JL: Encapsulation of biological species in sol–gel matrices. *J Aust Ceram Soc* 2000, 36:109–113

212 Shtelzer S, Rappoport S, Avnir D, Ottolenghi M, Braun S: Properties of trypsin and acid phosphatase immobilized in sol–gel matrices. *Biotechnol Appl Biochem* 1992, 15:227–235

213 Sato S, Murakata T, Ochifuji M, Fukushima M, Suzuki T: Development of immobilized enzyme entrapped within inorganic matrix and its catalytic activity in organic medium. *J Chem Eng Jpn* 1994, 27:732–736

214 Daryl K, Eggers A, Joan S: Valentine molecular confinement influences protein structure and enhances thermal protein stability. *Protein Sci* 2001, 10:250–261

215 Uo M, Yamashita K, Suzuki M, Tamiya E, Karube I, Makishima A: Immobilization of yeast cells in porous silica carrier with sol–gel process. *J Ceram Soc Jpn* 1992, 100:426–429

216 Vilenchik LZ, Griffith JP, St Clair NL, Navia MA, Margolin AL: Protein crystals as novel microporous materials. *J Am Chem Soc* 1998, 120:4290–4294

217 Jin W, Brennan JD: Properties and applications of proteins encapsulated within sol–gel derived materials. *Analytica Chimica Acta* 2002, 461:1–36

218 Shchipunova YA, Karpenkoa TY, Bakuninab IY, Burtsevab YV, Zvyagintsevab TN: A new precursor for the immobilization of enzymes inside sol–gel-derived hybrid silica nanocomposites containing polysaccharides. *J Biochem Biophys Methods* 2004, 58:25–38

219 Dosoretz C, Armon R, Starosvetzky J, Rothschild N: Entrapment of parathion hydrolase from Pseudomonas sp. in sol–gel glass. *J Sol–Gel Sci Technol* 1996, 7:7–11

220 Dong H, Xu JG, Feng QW, Wei Y: Simultaneous immobilization of oxidase/peroxidase in the mesoporous sol–gel silicate matrix. *Abstracts of Papers Am Chem Soc, 220th PMSE-243,* 2000

221 Wei Y, Xu JG, Feng QW, Lin MD, Dong H, Zhang WJ, Wang C: A novel method for enzyme immobilization: direct encapsulation of acid phosphatase in nanoporous silica host materials. *J Nanosci Nanotechnol* 2001, 1:83–93

222 Buisson P, Hernandez C, Pierre M, Pierre AC: Encapsulation of lipases in aerogels. *J Non-Crystal Solid* 2001, 285:295–302

223 Braun S, Rappoport S, Shtelzer S, Zusman R, Druckmann S, Avnir D, Ottolenghi M: Design and properties of enzymes immobilized in sol–gel glass matrices. In: Ed(s): Kamely D, Chakrabarty AM, Kornguth SE. *Proc US-Isr Res Conf Adv Appl Biotechnol* 1991:205–18

224 Ikeda Y, Kurokawa Y: Enantioselective esterification of racemic ibuprofen in isooctane by immobilized lipase on cellulose acetate-titanium iso-propoxide gel fiber. *J Biosci Bioeng* 2002, 93:98–100

225 Matsumoto M, Kondo K: Enhanced thermostability of α-chymotrypsin enclosed in inorganic microcapsules. *J Biosci Bioeng* 2001, 92:197–199

226 Kawakami K, Yoshida S: Thermal stabilisation of lipase by sol–gel-entrapment in organically modified silicates formed on kieselguhr. *J Ferment Bioeng* 1996, 82:239–245

227 Nguyen DT, Smit M, Dunn B, Zink JI: Stabilization of creatine kinase encapsulated in silicate sol–gel materials and unusual temperature effects on its activity. *Chem Mater* 2002, 14:4300–4306

228 Norris RD: Stabilisation of aminoacyl–tRNA synthetase by Sephadex and polyacrylamide gels. *Phytochem* 1975, 14:1701–1706

229 Rosevear A, Kent CA, Thomson AR, Bucke C: Studies of insoluble organic/inorganic composites of glucose isomerase. *Enzym Eng* 1978, 4:415–416

230 Jin W, Brennan JD: Properties and applications of proteins encapsulated within sol–gel derived materials. *Anal Chim Acta* 2002, 461:1–36

231 Milstein O, Nicklas B, Huettermann A: Oxidation of aromatic compounds in organic solvents with laccase from

Trametes versicolor. Appl Microbiol Biotechnol 1989, 31:70–74

232 Husain Q, Saleemuddin M: An inexpensive procedure for the immobilization of glycoenzymes on Sephadex G-50 using crude concanavalin A. *Biotechnol Appl Biochem* 1989, 11:508–512

233 Hennick WE, Talsma H, Borchert JCH, De Smedt SC, Demeester J: Controlled release of proteins from dextran hydrogels. *J Controlled Release* 1996, 39:47–55

234 Shalaby A, Abdallah AA, Park H, Park K: Loading of bovine serum albumin into hydrogels by an electrophoretic process and its potential applications to protein drugs, *Pharm Res* 1993, 10:457–460

235 Antonsen K, Bohnert J, Naveshima Y, Sheu MS, Wu XS, Hoffman A: Controlled release of proteins from 2-hydroxyethyl methacrylate copolymer gels, *Biomater Artif Cells Immob Biotechnol* 1993, 21:1–22

236 Stevin H, Gehrke L, Uhden H, McBride JF: Enhanced loading and retention of activity of bioactive proteins in hydrogel delivery systems. *J Controlled Release* 1998, 55:21–33

237 Lai CM, Shin CY: Stability of immobilized acid phosphatase added to soils. *J Biomass Energy Soc Chin* 1993, 12:31–32

238 Sharma NM, Kumar S, Sawhney SK: A novel method for the immobilization of tyrosinase to enhance stability. *Biotechnol Appl Biochem* 2003, 38:137–141

239 Gotoh T, Shidara M, Iwanaga T, Kikuchi KI, Hozawa M: immobilisation of γ-glutamyl transpeptidase, a membrane enzyme, in gel beads via liposome entrapment. *J Ferment Bioeng* 1994, 77:268–273

240 Banerjee S, Premchandran R, Tata M, John VT, McPherson GL, Akkara J, Kaplan D: Polymer precipitation using a micellar nonsolvent: the role of surfactant-polymer interactions and the development of a microencapsulation technique. *Ind Eng Chem Res* 1996, 35(9):3100–3107

241 Tanaka H, Kurosawa H, Kokufuta E, Veliky IA: Preparation of immobilized glucoamylase using Ca-alginate gel coated with partially quaternized poly(ethyleneimine). *Biotechnol Bioeng* 1984, 26:1393–1394

242 Lee KH, Lee PM, Siaw YS: Studies of l-phenylalanine production by immobilized in stabilized calcium alginate beads. *J Chem Technol Biotechnol* 1992, 54:375–382

243 Blandino A, Macias M, Cantero D: Immobilization of glucose oxidase within calcium alginate gel capsules. *Process Biochem (Oxford)* 2001, 36(7):601–606

244 Rilling P, Walter T, Pommersheim R, Vogt W: Encapsulation of cytochrome C by multilayer microcapsules. A model for improved enzyme immobilization. *J Membr Sci* 1997, 129:283–287

245 DeGroot AR, Neufeld RJ: Encapsulation of urease in alginate beads and protection from α-chymotrypsin with chitosan membranes. *Enzym Microb Technol* 2001, 29:321–327

246 Nippon Kayaku Co: Preparation of immobilized enzymes and microorganisms *Jpn Kokai Tokkyo Koho, JP 58,116,683 A2*, 1983

247 Hearn E, Neufeld RJ: Poly(methylene co-guanidine) coated alginate as an encapsulation matrix for urease. *Process Biochem (Oxford)* 2000, 35:1253–1260

248 Konishiroku Photo Industry Co: Preparation of reactive carriers. *Jpn Kokai Tokkyo Koho, JP 59,063,186 A2*, 1984

249 Schuleit M, Luisi PL: Enzyme immobilization in silica-hardened organogels. *Biotechnol Bioeng* 2001, 72:249–253

250 Leca B, Blum LJ: Luminol electrochemiluminescence with screen-printed electrodes for low-cost disposable oxidase-based optical sensors. *Analyst* 2000, 125:789–791

251 Hsu AF et al.: Immobilized lipoxygenase in a packed-bed column bioreactor: continuous oxygenation of linolenic acid. *Biotechnol Appl Biochem* 1999, 30:245–250

252 Bressler E, Pines O, Goldberg I, Braun S: Conversion of fumaric acid to l-malic by sol–gel immobilized *Saccharomyces cerevisiae* in a supported liquid

membrane bioreactor. *Biotechnol Prog* 2002, 18:445–450

253 Tsafack VC, Marquette CA, Pizzolato F, Blum LJ:Chemiluminescent choline biosensor using histidine-modified peroxidase immobilised on metal-chelate substituted beads and choline oxidase immobilised on anion-exchanger beads co-entrapped in a photocrosslinkable polymer. *Biosens & Bioelectr* 2000 15:125–133

254 Rill-Jancsik V, Nagy M, Keleti T, Wolfram E: Immobilization of water-soluble enzymes by embedding into poly(vinyl alcohol) gel. *Hung Teljes* 1977, HU 75–2662

255 Park YH, Chung TW, Han MH: Studies on microbial glucose isomerase. 4. Characteristics of immobilized whole-cell glucose isomerase from Streptomyces spp. *Enzym Microb Technol* 1980, 2:227–233

256 Ates S, Mehmetoglu U: A new method for immobilisation of β-galactosidase and its utilization in a plug flow reactor. *Process Biochem* 1997, 32:433–436

257 Kusaoke H, Suzuki K, Nihei T, Kimura K: Utilization of gels prepared from chitosan as supports for enzyme and microorganism immobilization. Ed(s): Kennedy JF, Phillips GO, Williams PA, *Cellulose* 1990:501–506

258 Elcin YM: Encapsulation of urease enzyme in xanthan-alginate spheres. *Biomaterials* 1995, 16:1157–1161

259 Krauze J, Wawrzyniak B: Properties of gel-entrapped glucoamylase. *Starch/Staerke* 1988, 40:314–319

260 Ray RR, Jana SC, Nanda G: Biochemical approaches of increasing thermostability of α-amylase from *Bacillus megaterium* B, *FEBS Lett* 1994, 356:3–32

261 De Alteriis E, Parascandola P, Salvadore S, Scardi V: Enzyme immobilization within insolubilized gelatin. *J Chem Technol Biotechnol* 1985, 35:60–64

262 Gondo S, Koya H: Solubilized collagen fibril as a supporting material for enzyme immobilization. *Biotechnol Bioeng* 1978, 20:2007–2010

263 De Alteriis E, Parascandola P, Pecorella MA, Scardi V: Effect of gelatin-immobilization on the catalytic activity of enzymes and microbial cells. *Biotechnol Tech* 1988, 2:205–210

264 Baysal SH, Uslan AH: Encapsulation of Urease and PEG-Urease in erythrocyte. *Artif Cell Blood Substitutes Immobilization Biotechnol* 2000, 28:263–271

265 Kurokawa Y, Ohta H, Okubo M, Takahashi M: Formation and use in enzyme immobilization of cellulose acetate-metal alkoxide gels. *Carbohydr Polym* 1994, 23:1–4

266 Piskin K, Chang TMS: A new combined enzyme-charcoal system formed by enzyme adsorption on charcoal followed by polymer coating. *Int J Artif Organs* 1980, 3:344–346

267 Nitto Electric Industrial Co: Preparation of immobilized enzymes *Jpn Kokai Tokkyo Koho, JP 82-154,856*, 1984

268 Mitsubishi Petrochemical Co: Immobilization of an enzyme: *Jpn Kokai Tokkyo Koho, JP 59,102,393 A2*, 1984

269 Mizutani F, Yabuki S, Hirata Y: Amperometric biosensors using poly-l-lysine/poly(styrenesulfonate) membranes with immobilized enzymes. *Denki Kagaku Oyobi Kogyo Butsuri Kagaku* 1995, 63:1100–1105

270 Sakurada Y, Yamane T: Fibrous substance entrapping an enzyme. *Jpn Kokai, JP 50,029,728*, 1975

271 Miyawaki O, Nakamura K, Yano T: Mass transfer and reaction with microcapsules containing enzyme and adsorbent. *Enzym Eng* 1978, 3:79–84

272 Secundo F, Chilin A, Guiotto A: Pegylated enzyme entrapped in poly(vinyl alcohol) hydrogel for biocatalytic application. *Il. Farmaco* 2001, 56:541–547

273 Hixson HF: Water-soluble enzyme-polymer grafts. Thermal stabilization of glucose oxidase. *Biotechnol Bioeng* 1973, 15:1011–1016

274 Husain Q, Iqbal J, Saleemuddin M: Entrapment of concanavalin A-glyco-enzyme complexes in calcium alginate gels. *Biotechnol Bioeng* 1985, 27:1102–1107

275 Glazer AN, Bar-Eli A, Katchalski E: Preparation and characterization of polytyrosyl trypsin. *J Biol Chem* 1962, 237:1832–1838

276 Broun GB: Crosslinked enzymes. In: *Methods Enzymol* 1976, 44:263–280

277 Marshall JJ, Rabinowitz ML: *J Biol Chem* 1976, 251:1081–1087

278 Virnik AD, Krasovskaya SB, Kil'deeva NR, Biber BL, Solomon ZG: Enzyme immobilization in fibers and film structures, *Antibiot Med Biotekhnol* 1986, 31:117–122

279 Demers N, Agostinell E, Averill-Bates DA, Fortier G: Immobilization of native and poly(ethylene glycol)-treated ('PEGylated') bovine serum amine oxidase into a biocompatible hydrogel, *Biotechnol Appl Biochem* 2001, 33:201–207

280 Okahata Y, Ijiro K: A lipid–coated lipase as a new catalyst for triglyceride synthesis in organic solvents. *J Chem Soc Chem Commun* 1988, 1392–1394

281 Welwardova A, Gemeiner P, Michalkova E, Welward L, Jakubova A: Gel-entrapped penicillin G acylase optimized by an enzyme thermistor. *Biotechnol Tech* 1993, 7:809–814

282 Bello Magalhaes D, da Rocha-Leao M, Helena M: Immobilization of β-glucosidase aggregates in calcium alginate. *Biomass Bioenergy* 1991, 1:213–216

283 Veronese FM, Mammucari C, Caliceti P, Schiavon O: Influence of PEGylation on the release of low and high molecular-weight proteins from PVA matrices. *J Bioact Compat Polym* 1999, 14:315–330

284 Veronese FM, Mammucari C, Schiavon F, Schiavon O, Lora S, Secundo F, Chilin A, Guiotto A: Pegylated enzyme entrapped in poly(vinyl alcohol) hydrogel for biocatalytic application. *Farmaco* 2001, 56:541–547

285 Cremonesi P, Mazzola G, Focher B, Vecchio G: Peroxidase immobilized on acrylic copolymer beads. *Angew Makromol Chem* 1975, 48:17–27

286 Podual K, Doyle FJ 3rd, Peppas NA: Glucose-sensitivity of glucose oxidase-containing cationic copolymer hydrogels having poly(ethylene glycol) grafts. *J Controlled Release* 2000, 67:9–17

287 Piro B, Dang LA, Pham MC, Fabiano S, Tran-Minh C: A Glucose biosensor based on modified-enzyme incorporated within electro polymerised poly(3,4-ethylenedioxythiophene) (PEDT) films. *J Electroanal Chem* 2001, 512:101–109

288 Berger JL, Lee BH, Lacroix C: Immobilization of β-galactosidases from *Thermus aquaticus* YT-1 for oligo-saccharides synthesis. *Biotechnol Tech* 1995, 9:601–606

289 Markvicheva EA, Bronin AS, Kudryavtseva NE, Rumsh LD, Kirsh YE, Zubov VP: Immobilization of proteases in composite hydrogel based on poly(N-vinylcaprolactam). *Biotechnol Tech* 1994, 8:143–148

290 Khare SK, Gupta MN: Immobilization of *E. coli* β-galactosidase and its derivatives by PAAm gel. *Biotechnol Bioeng* 1988, 31:829–833

291 Doretti L, Ferrara D, Gattolin P, Lora S, Schiavon F, Veronese FM: PEG-modified glucose oxidase immobilized on a PVA cryogel membrane for amperometric biosensor applications. *Talanta* 1998, 45:891–89

292 Mohapatra, SC, Hsu, JT: Immobilization of α-chymotrypsin for use in batch and continuous reactors. *J Chem Technol Biotechnol* 2000, 75:519–525

293 Yu LS, Urban G, Moser I, Jobst G, Gruber H: Photolithographically patternable modified poly(HEMA) hydrogel membrane. *Polym Bull* 1995, 35:759–765

294 Pandey PC, Upadhyay S, Pathak HC: A new glucose sensor based on encapsulated glucose oxidase within organically modified sol–gel glass. *Sens Actuator B Chem* 1999, 60:83–89

295 Arai G, Masuda M, Yasumori I: Glucose sensor of poly(mercapto-p-benzoquinone) films containing immobilized glucose oxidase. *Chem Lett* 1992, 9:1791–1794

296 Umana M, Waller J: Protein-modified electrodes. The glucose oxidase/polypyrrole system. *Anal Chem Sci Anal Chem* 1986, 58:2979–2983

297 Miyairi S: An enzyme-polymer film prepared with the use of poly(vinyl alcohol) bearing photosensitive aromatic azido groups. *Biochim Biophys Acta* 1979, 571:374–377

298 Drevon GF, Danielmeier K, Federspiel W, Stolz DB, Wicks DA, Yu PC, Russell AJ: High-activity enzyme-polyurethane coatings. *Biotechnol Bioeng* 2002, 79:785–794

299 Novick SJ, Dordick JS: Protein-containing hydrophobic coatings and films. *Biomaterials* 2001, 23:441–448

300 Fortier G, Belanger D, Basavant L, Wilson C, Lawrence MF: Glucose biosensor based on immobilization of glucose oxidase in an anionic ion exchange polymer blend. *Anal Lett* 1992, 25:1835–1842

301 Fortier G, Chen JW, Belanger D: Biosensors based on entrapment of enzymes in a water-dispersed anionic polymer. *ACS Symposium Series* 1992, 487:22–30

302 Csoeregi E, Quinn CP, Schmidtke DW, Lindquist SE, Pishko MV, Ye L, Katakis I, Hubbell JA, Heller A: Design, characterization, and one-point in vivo calibration of a subcutaneously implanted glucose electrode. *Anal Chem* 1994, 66:3131–3138

303 Kaetsu I, Sindo H, Uchida K, Sutani K: Ca^{2+} ion responsive controlled release systems using porous PET film. *Proc Int Symp Controlled Release Bioact Mater* 1996, 23:469–470

304 Ohara TJ, Rajagopalan R, Heller A: Wired enzyme electrodes for amperometric determination of glucose or lactate in the presence of interfering substances. *Anal Chem* 1994, 66:2451–2457

305 Fukui S, Tanaka A, Iida T, Hasegawa E: Application of photo-crosslinkable resin to immobilization of an enzyme. *FEBS Lett* 1976, 66:179–182

306 Ichimura K: A convenient photochemical method to immobilize enzymes. *J Polym Sci Polym Chem Ed* 1984, 22:2817–2828

307 Reynolds ER, Geise RJ, Yacynych AM: Electropolymerised films for the construction of ultramicrobiosensors and electron-mediated amperometric biosensors. *ACS Symp Ser* 1992, 487:186–200

308 Freedman HH: Process for crosslinking polyamines and gelatin foams with polyisocyanates. *Can, US 84-585,744*, 1990

309 Klug JH: Poly(ureaurethane) foams containing immobilized active. *US Pat 3,905,923*, 1975

310 Legoy MD, Kim HS, Thomas D: Use of alcohol dehydrogenase for flavor aldehyde production. *Eur Congr Biotechnol 3rd* 1984, 2:497–501

311 Marolia KZ, D'Souza SF: A simple technique for the immobilization of lysozyme by cross-linking of hen egg white foam. *J Biochem Biophys Methods* 1993, 26:143–147

312 Goodson LH, Jacobs WB: Monitoring of air and water for enzyme inhibitors. *Methods Enzymol* 1976, 44:647–658

313 Hartdegen FJ, Swann WE: Process for immobilizing proteins. *US Pat 4,195,127*, 1980

314 Allinson BT, LeJeune KE: Organophosphorus pesticide decontamination and detoxification using environmentally friendly enzyme polymers. *Am Chem Soc Div Environ Chem* 2002, 42:380–386

315 LeJeune KE, Hetro AD, Russell AJ: Stabilizing nerve agent hydrolyzing enzymes. *Book of Abstracts, 213th ACS National Meeting 1997*, ACS, Washington, DC

316 Kutney JP, Berset JD, Hewitt GM, Singh M: Biotransformation of dehydroabietic, abietic, and isopimaric acids by *Mortierella isabellina* immobilized in polyurethane foam. *Appl Environ Microbiol* 1988, 54:1015–1022

317 Drevon GF, Danielmeier K, Federspiel W, Stolz DB, Wicks DA, Yu PC, Russell AJ: High-activity enzyme-polyurethane coatings. *Biotechnol Bioeng* 2002, 79:785–794

318 Jancsik V, Beleznai Z, Keleti T: Enzyme immobilization by poly(vinyl alcohol) gel entrapment. *J Mol Catal* 1982, 14:297–306

319 Doretti L, Ferrara D, Lora S, Schiavon F, Veronese FM: Acetylcholine biosensor involving entrapment of acetylcholinesterase and poly(ethylene glycol)-modified choline oxidase in a poly(vinyl alcohol) cryogel membrane. *Enzym Microb Technol* 2000, 27:279–285

320 Lee S, Takahashi T, Anzal J, Suzuki Y, Osa T: Preparation of enzyme membranes by use of organic solvent. A composite membrane of poly(vinyl stearate) and poly(ethyleneglycol)-modified glucose oxidase for enzyme sensors. *Pharmazie* 1994, 49:620–621

321 Hampp N, Scholze J, Braeuchle C: Immobilization of organic macromolecules or biopolymers in a polymer membrane for biosensors. *Ger Offen, DE 4,027,728*, 1992

322 Ashley SL, McGinity JWM: Enzyme-mediated drug release from poly(dl-lactide) matrices. *5th Congr Int Technol Pharm* 1989, 5:195–204

323 Bu HZ, Mikkelsen SR, English AM: NAD(P)H sensors based on enzyme entrapment in ferrocene-containing polyacrylamide-based redox gels. *Anal Chem* 1998, 70:4320–4325

324 Freeman A, Tor R: Membranes and films. *UK Pat Appl GB 84-21,588*, 1985

325 Novick SJ, Dordick JS: Protein-containing hydrophobic coatings and films. *Biomaterials* 2002, 23:441–448

326 Agency of Industrial Sciences and Technology, Japan: Preparation of immobilized enzymes. *Jpn. Kokai Tokkyo Koho 1981*, JP 56137889 A2

327 Quennesson JC, Thomas D: Films and biochemically-active coatings. *Fr Demande* 1978, FR 2391254

328 Shimada K, Yano M, Shibatani K, Makimoto T:Enzyme electrode. *Jpn. Kokai Tokkyo Koho 1978*, JP 53149397

329 Dong SJ, Guo YZ: A novel enzyme electrode for the water-free organic phase. *J Electroanal Chem* 1994, 375:405–7

330 Wang J, Leech D, Ozsoz M, Martinez S, Smyth MR: One-step fabrication of glucose sensors based on entrapment of glucose oxidase within poly(estersulfonic acid) coatings. *Anal Chim Acta* 1991, 245:139–43.

331 Lee DM, Nishioka GM, Swann WE, Nolf CA: Enzyme immobilization on non-porous glass fibers. U S 1988, US 4749653 A

332 Furukawa SY, Kawakami K: Characterization of *Candida rugosa* lipase entrapped into organically modified silicates in esterification of menthol with butyric acid. *Ferment Bioeng* 1998, 85:240–242

333 Heichal-Segal RS, Braun S: Immobilization in alginate-silicate sol–gel matrix protects β-glucosidase against thermal and chemical denaturation: enzyme stabilization for use in e.g. wine aroma improvement. *Biotechnology* 1995, 13:798–800

334 Glad M, Norrloew O, Sellergren B, Siegbahn N, Mosbach K: Use of silane monomers for molecular imprinting and enzyme entrapment in polysiloxane-coated porous silica. *J Chromatogr* 1985, 347:11–23

335 Chen JP, Hwang YN: Polyvinylformal resin plates impregnated with lipase entrapped sol–gel polymer for flavour ester synthesis, *Enzym Microb Technol* 2003, 33:513–519

336 Chen JP, Lin WS: Sol–gel powders and supported sol–gel polymers for immobilization of lipase in ester synthesis. *Enzym Microb Technol* 2003, 32:801–811

337 Chen JP, Lin WS, Chang MF: Synthesis of geranyl acetate by esterification with lipase entrapped in hybrid sol–gel formed within nonwoven fabric. *J Am Chem Soc* 2002, 79:309–331

338 Furukawa S, Ono T, Ijima H, Kawakami K: Effect of imprinting sol–gel immobilized lipase with chiral template substrates in esterification of (R)-(+)- and (S)-(–)-glycidol. *J Mol Catal B Enzym* 2002, 17:23–28

339 Furukawa SY, Ono T, Ijima H, Kawakami K: Activation of protease by sol–gel-entrapment into organically modified hybrid silicates. *Biotechnol Lett* 2002, 24:13–16

340 Betancor L, Lopez-Gallego F, Hidalgo A, Fuentes M, Podrasky O, Kuncova G, Guisan JM, Fernandez-Lafuente R: Advantages of the pre-immobilization of enzymes on porous supports for their entrapment in sol–gels. *Biomacromolecules* 2005 6:1027–30

341 Coradin T, Livage J: Mesoporous alginate/silica biocomposites for enzyme immobilization. *CR Chimie* 2003, 6:147–152

5
Enzyme Encapsulation

5.1
Introduction

Encapsulation of an enzyme is the formation of a membrane-like physical barrier around an enzyme preparation. To distinguish entrapment techniques from encapsulation techniques, we restrict our discussion to the preparation of immobilized enzymes enclosed in a spherical membrane and to the immobilization techniques derived from these encapsulation techniques.

Thus, other encapsulation techniques such as the sol–gel process are put into the category entrapment, because here the enzyme is enclosed in a matrix not in the membrane, although the microscopic structure of the enzyme in the sol–gel matrix may resemble that of the encapsulated enzyme.

As discussed previously, although there are many methods of preparation of encapsulated enzymes, for example in-situ encapsulation, encapsulation-cross-linking, immobilization and encapsulation, and post-loading encapsulation, methods for preparation of the membrane can be generally categorized into the classes:

- interfacial processes for the formation of solid shell around the liquid droplet of the enzymes,
- phase inversion methods for the preparation,
- template leaching,
- post-loading encapsulation.

Compared with other enzyme-immobilization techniques, characteristics of encapsulation techniques are that many enzymes can be immobilized simultaneously [1] and the conditions used are often mild [2].

Scheme 5.1 Encapsulated enzymes – defined as enzymes physically enclosed in a semi-permeable membrane.

Carrier-bound Immobilized Enzymes: Principles, Application and Design. Linqiu Cao
Copyright © 2005 WILEY-VCH Verlag GmbH & Co. KGaA, Weinheim
ISBN: 3-527-31232-3

5.1.1
General Considerations

Micro-encapsulation of enzymes refers to the process, by which an enzyme preparation such as dissolved enzymes, or lyophilized enzymes, or whole-cell catalysts are enclosed physically or chemically within spherical semi-permeable polymer membranes with diameters in 1–100 μm range [4].

Although, usually, enzymes within the capsules are not chemically modified, it might be desirable to insolubilize the enzyme molecules to lessen enzyme leakage. Although the enzyme molecules are physically confined in the interior by the membrane formed around the drops of enzyme solution, the substrates or products are able to diffuse freely across the membrane, depending on the pore size of the membranes, as illustrated in Scheme 5.2. Thus, the membrane functions as a physical barrier to the enzyme molecules only.

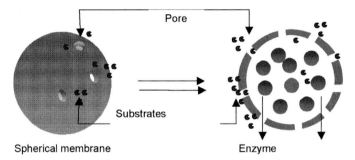

Scheme 5.2 Micro-encapsulation of enzymes in microcapsules.

5.1.2
An Historical Overview

Use of this concept for enzyme immobilization was pioneered by Chang in 1964. Initially, he dispersed aqueous droplets containing monomer, 1,6-hexanediamine, and erythrocyte haemolysate in a mixture of chloroform and cyclohexane and added sebacoyl chloride to the organic phase, leading to the formation of the first artificial cells [1].

One characteristic of this method is that the enzymes within the capsules are free enzymes, and are thus not stabilized. It is laborious to control the thickness and permeability of the membrane to avoid enzyme leakage and mitigate the diffusion limitation. For this reason, this technology is less used for preparation of industrially useful immobilized biocatalysts.

Currently, many variations of encapsulation techniques have been developed, with the aim of overcoming the drawback of this original encapsulation technique and meeting the needs of different applications. These new variations include:

- Encapsulation-cross-linking in which the enzyme is first encapsulated in the capsules then cross-linked with a bifunctional agent to insolubilize the enzymes.
- Immobilization and encapsulation in which the enzyme is first immobilized by other techniques such as covalent, adsorption, entrapment and cross-linking techniques, then encapsulated by coating. The aim is to further stabilize the enzyme.
- The post-loading encapsulation technique in which the enzyme solution is charged into the ready-made hollow microsphere and other techniques, for example cross-linking or aggregation/cross-linking, are then used to insolubilize the enzymes.

Among these variations post-loading encapsulation is attractive with regard to simplicity and broad applicability.

5.1.3
Pros and Cons of Micro-encapsulation

Encapsulation of an enzyme is the process by which enzyme molecules are immobilized in a capsule of size ranging from a few micrometers to hundreds of micrometers.

Usually, enzyme molecules cannot pass through the membrane whereas substrates can pass freely [2]. Thus, the pores of the membrane must be smaller than the size of the enzyme molecules. Consequently, diffusion limitation might be serious compared with other enzymes immobilized on porous carriers.

Advantages of this technique are:
- It enables the preparation of multi-enzyme systems and the sequential enzyme reaction [3, 70].
- There is no chemical modification during the entrapment process, thus selecting the appropriate entrapment process might avoid the enzyme deactivation that occurs in some encapsulation processes.

5.2
Classification of Encapsulation

5.2.1
Conventional Encapsulation

Conventional encapsulation is the process by which the enzyme dissolved or dispersed in a solution is spontaneously encapsulated in a membrane formed around the drops of enzyme solution. Thus, the enzyme molecules are not immobilized in the liquid core of the microcapsules.

The method for formation of the membrane usually utilizes the concept of interfacial processes e.g. polymerizing the monomers located on the interfaces between the aqueous enzyme solution and the immiscible organic phase [42] or deposition

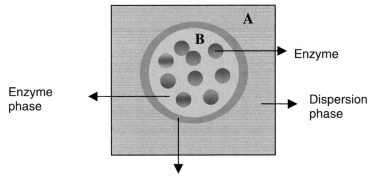

Scheme 5.3 Conventional encapsulation by an interfacial process.

of the polymer on the interface or formation of insoluble complex (A–B) at the interface of two aqueous solutions as in the formation of a symplex (a polyelectrolyte–polyelectrolyte complex) [43]. The A and B components are dissolved separately in two aqueous phases and interaction A–B leads to formation of the membrane around the droplet, as illustrated in Scheme 5.3.

5.2.1.1 Encapsulation by an Interfacial Process

The interfacial process for enzyme encapsulation is the process by which a membrane or film is formed around the enzyme solution by physical or chemical transformation of the liquid phase around the enzyme solution into the solid phase. The methods used can be classified into two groups, the chemical method, i.e. crosslinking or polymerization, and physical means, for example physical gelation of the liquid phase around the enzyme drops or complexation of the components in the liquid phase. In both cases, the enzyme solution should be not miscible with the liquid phase that is to be solidified.

According to this definition, Interfacial encapsulation can be classified into several groups:

- interfacial gelation, for example encapsulation of enzyme in alginate capsules (A is alginate; B is $CaCl_2$) [57];
- reverse interfacial gelation (A is $CaCl_2$, B is Alginate solution) [7, 8, 57];
- interfacial cross-linking [44];
- interfacial polymerization (A is water-soluble monomer; B is a water-insoluble monomer) [41];
- interfacial deposition (A is liposome which functions as a template; B is a polymer) [93];
- interfacial complexation i.e. formation of PEC (polyelectrolyte complex) (A and B are oppositely charged polyelectrolytes) [9].

Among these methods, interfacial gelation techniques were widely used for encapsulation of enzymes or whole-cell-associated enzymes [10]. Activity retention and leakage of the immobilized enzyme obtained depend on the thickness of the membrane [11, 55] and its pore size [12]. For cellulase entrapped in the capsules formed by cross-linking 2-hydroxyethyl methacrylate, and tetraethylene glycol diacrylate it was found that the optimum thickness in terms of the activity is approximately 1% of the capsule size (500–2000 µm) [41].

As with the interfacial polymerization process, the enzyme can be deactivated by the process and the enzyme might be incorporated into the wall of the membrane.

5.2.1.2 Encapsulation by Phase Inversion

Phase inversion refers to the method by which the liquid phase around the enzyme solution is solidified because of the change of solubility stimulated by diffusion or evaporation of the solvents of the components to be integrated into the membrane. They often involve methods such as:

- coacervation,
- double emulsion (W1/O/W2) process, where A is water and B is water-insoluble polymer dissolved in an organic phase [13],
- liquid drying.

5.2.2
Non-conventional Encapsulation

Non-conventional processes combine conventional microencapsulation techniques with other enzyme immobilization techniques. For instance, encapsulation–cross-linking, covalent encapsulation, encapsulation–coating and immobilization–encapsulation are the most useful variants which can be used to solve problems that cannot be solved by conventional encapsulation techniques.

5.2.2.1 Encapsulation in Liquid Membrane

Enzyme immobilization in semi-permeable membranes can be dated back to 1970s, when several enzymes were encapsulated by use of liquid-surfactant membranes for preparation of encapsulated enzyme [14, 15]. In principle, the mechanisms for encapsulation are similar to those of conventional encapsulation. The aqueous solution is encapsulated in a hydrocarbon membrane which must be permeable to the substrate and the product but not to the enzyme [16].

5.2.2.2 Encapsulation–Reticulation

Microcapsules of immobilized invertase have been prepared by cross-linking of the enzyme protein itself [45].

5.2.3
Double Immobilization Based on Encapsulation

5.2.3.1 Encapsulation–Cross-linking

One of the intrinsic drawbacks of the conventional encapsulation technology mentioned above is that it is laborious to control the pore size and thickness of the membrane and to find a compromise between activity retention and the leakage of the enzymes. To overcome this dilemma the encapsulation–cross-linking technique was designed. With this technique enzyme leakage was mitigated and enzyme stability was improved [17]. The technique is simple and large amounts of enzymes can be entrapped or many enzymes can be entrapped simultaneously [3]. In particular, enzyme molecules can be further stabilized by subsequent cross-linking techniques.

Moreover, unlike the conventional enzyme encapsulation technique, it is not necessary to use a permeable membrane with a pore size far below the size of the enzyme (5 nm), because the size of the cross-linked enzyme in the interior of the capsules is much bigger than the native enzyme, because of the cross-linking, as shown in Scheme 5.6.

Thus, a membrane with a pore size being much larger than the enzyme size can be used. Consequently, diffusion limitation can easily be overcome by use of microporous or macroporous capsules.

Other advantages of using this technology for preparation of cross-linked enzymes, instead of cross-linking the dissolved enzyme in bulky solution are obvious. It is possible to overcome the intrinsic drawback associated with conventional cross-linked enzymes such as gelatinous properties and difficulties in handling [18]. It is, moreover, possible to control the size of the cross-linked enzymes by selecting appropriate capsules with defined geometries. Unfortunately, this technology has not yet been exploited further for preparation of cross-linked enzymes.

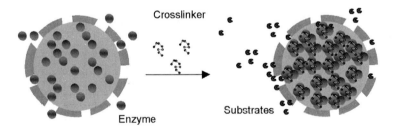

Scheme 5.4 Encapsulation cross-linking.

5.2.3.2 Encapsulation and Coating

The microcapsules with enzymes in the interior can be further coated with polyelectrolytes with the aim of altering the substrate permeability by the use of pH-conditioned configuration changes in the adsorbed polyion layer [19, 20, 57, 178] or enhancing the stability of the enzymes [57], avoiding enzyme leakage.

For instance, enzyme capsules prepared by interfacial gelation of poly(bis-carboxylatophenoxy)phosphazene Ca cationic solution, was coated with poly(L-lysine), resulting in the formation of encapsulated enzyme with permeability properties highly dependent on the molecular weight of the polycation [21].

Although this technique is less important for preparation of immobilized enzymes for industrial applications, it might be a very interesting approach for the controlled drug delivery.

5.2.3.3 Entrapment and Coating

The enzyme entrapped in a gel matrix can be further encapsulated in a capsule which can be formed by coating. For instance, the enzyme bovine carbonic anhydrase (BCA) was first immobilized in alginate beads followed by chitosan coating. SEM studies have clearly shown the difference between chitosan-coated beads and conventional alginate beads (Scheme 5.5).

Control of the weight cut-off of the beads was accomplished by adjusting the cross-linking conditions, coating components, and molecular weight of the system [22].

Scheme 5.5 Entrapment and encapsulation.

5.2.3.4 Immobilization and Encapsulation

Although micro-encapsulation was initially developed for encapsulation of enzyme solutions, many variations of encapsulation techniques have been developed in recent decades; these include carrier-bound immobilization–encapsulation, cross-linking–encapsulation, and entrapment–encapsulation, as illustrated in Scheme 5.6.

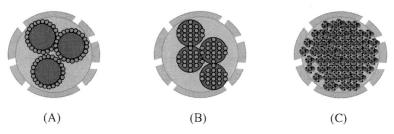

(A)   (B)   (C)

Scheme 5.6 Immobilization–encapsulation: (A) entrapped carrier-bound enzyme; (B) encapsulated entrapped enzyme; (C) encapsulated cross-linked enzyme.

Often, consecutive encapsulation is primarily designed to improve the encapsulated enzymes, for example:

- mitigation of enzyme leakage for enzymes adsorbed on fine carriers;
- enhancement of activity retention, because of leakage [22], or enhanced activity retention because of improvement of dispersion [22];
- improvement of stability such as mechanical stability and stability in organic solvents [177, 196] or operational stability;
- enhancement of selectivity [195];
- mitigating substrate or substrate inhibition [197];
- enlargement of size [197].

In entrapment–encapsulation, enzymes or whole cells are first entrapped in the gel matrix, followed by encapsulation with polymeric capsules obtained by formation of a symplex [25] such as γ-carrageenan and chitosan [26].

Other carrier-bound immobilized enzymes can be also further entrapped in capsules. For example, the rate of interesterification palm oil with stearic acid in hexane was higher for lipase from *Rhizopus arrhizus* adsorbed on Celite and then encapsulated in lecithin reverse micelles [27]

5.2.4
Post-loading Encapsulation

Analogous to the post-loading entrapment of dissolved enzymes in the ready-made gel matrix such as Sepharose 4 B [28], dissolved enzyme can also be charged into the interior of suitable preformed capsules. In this way, the ready-made polymeric microcapsules with predetermined pore sizes and particles size can be used as a micro-container, as illustrated in Scheme 5.7.

Entrapment of enzyme in a pre-designed matrix can be dated backed to the 1980s. Laccase was adsorbed on to Sepharose CL-6B and followed by freeze-drying the gel beads to give small spherical particles containing 104 µg protein mg^{-1} gel, which could be used in organic solvents [31]. Surprisingly, it was shown that the immobilized laccase in organic solvents was very stable and had high tolerance of elevated temperatures [32].

Encapsulation of enzyme in pre-designed capsules can probably be traced back to the beginning of 1970s, when Rony reported that alternative microencapsulation could be achieved by use of pre-designed hollow fibres containing enzyme solution within the hollow core, thus preventing the enzyme from diffusing from the microcapsules. In this case, the hollow microsphere functions only as a microdevice [33].

5.2.4.1 Encapsulation in Non-swellable Microcapsules
As shown in Scheme 5.7, the encapsulation of enzyme in ready-made capsules can be performed either in swellable or non-swellable capsules.

Because only dissolved enzyme molecules can be loaded into the capsules, it is conceivable that the enzyme loading in the ready-made capsules depends on

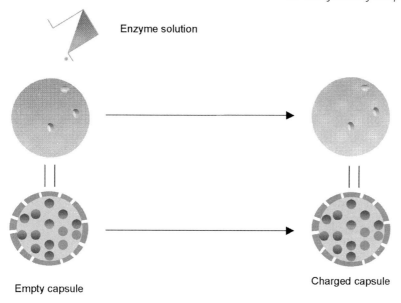

Scheme 5.7 Loading the interior of microcapsules with an enzyme.

whether the capsules are swellable or non-swellable. For non-swellable capsules the enzyme loading per capsule can be calculated as the volume of the void of the capsule multiplied by the concentration of enzyme. In contrast, for swellable capsules the loading of the enzyme is SF (swelling factor) times larger than for non-swellable capsules. It is, therefore, obviously advantageous to use swellable rather than non-swellable capsules, because of the high enzyme loading in swellable capsules.

The use of non-swellable capsules for enzyme encapsulation was reported for the first time by Parthasarathy and Martin in 1996 [34]. Several enzymes such as glucose oxidase, catalase, trypsin, subtilisin, and alcohol dehydrogenase were loaded into polypyrrole microcapsules prepared by a template-dependent method, followed by a sealing procedure.

Encapsulation of enzyme or whole cells can be also performed in glass capsules with pore sizes in the millimetre range, although serious diffusion limitation may hamper practical application as a robust immobilized enzyme.

5.2.4.2 Encapsulation in Soft Microcapsules

The pore size and internal volume of swellable microspheres can be controlled by changing the pH, temperature, or degree of hydration. For example, α-chymotrypsin was loaded into ready-made capsules prepared by the layer-by-layer (LBL) assembly of the non-biodegradable polyelectrolytes poly(allylamine) and poly(styrenesulphonate) on melamine formaldehyde microcores followed by the core decomposition at low pH [35].

To encapsulate the enzyme, the so prepared capsules should be able to shrink or swell upon pH changes. Otherwise, the enzyme might not enter the interior or leak from the capsules. Thus, it is expected that diffusion constraints might occur, when the enzyme is included in the interior with a membrane having pore size small than the size of the enzyme entrapped.

Admittedly, encapsulation of enzyme in the ready made microdevices for the purpose of biocatalysis has not attracted much interest. This technique might, however, be of interest for DCR (drug-controlled release), because high enzyme loading can be achieved. Preparation and encapsulation of the enzyme can also be separated, probably leading to high retention of activity.

5.3
Effect of Encapsulation

Encapsulation of enzyme in the capsules can also have a large influence on the enzyme performance, depending on the individual encapsulation method.

5.3.1
Activity of the Encapsulated Enzymes

The activity of the encapsulation enzyme has been found to be dictated by:

- The nature of the monomer A (water-soluble) used in the formation of the membrane: for instance, the activity of urease encapsulated in microcapsules formed by a symplex membrane from cellulose sulphate and polydimethyldiallyl ammonium chloride was reduced to 35 to 38%, because of the negative interaction of urease with cellulose sulphate [29].

- Diffusion constraints: like other types of carrier-bound immobilized enzymes, encapsulated enzymes also suffer from diffusion constraints. Thus, broken enzyme-containing microcapsules were more active than the native enzyme-containing capsules. The pH profile can be also altered [42] and K_m also depends on the thickness of the membrane [30].

- The nature of the process: studies have found that one third of these proteins were lost during the overall preparation procedure and a further fraction was attached to the membranes of the microcapsules, when encapsulating the enzyme in semi-permeable polyamide capsules.

The retention of activity of the encapsulated enzymes depends on many factors such as the thickness of the membrane [11, 41, 55], the pore size [12], the processes used, and the properties of the enzymes. Typical activity retention is usually in the range 20–90%.

It has been found that loss of the activity is not only attributed to diffusion constraints [8] but also to deactivation of the enzymes, depending on the methods used. For instance, in coacervation processes using organic solvents more than

80% of the native activity can often be lost, because of inactivation of the enzyme by use of the organic solvents [37, 38] or deactivation by the toxic monomers used in the interfacial cross-linking process. In contrast, use of interfacial gelation techniques such as the formation of a polyelectrolyte complex often leads to high retention of activity [56, 179].

5.3.2
Stability of the Encapsulated Enzymes

Stability of the encapsulated enzyme covers thermal stability, operational stability, etc. Usually, the enzyme is not only physically encapsulated but also entrapped in the membrane or adsorbed on the membrane. Thus, the entrapped enzyme might be heterogeneous in nature with regard to enzyme performance. Consequently, the stability of the encapsulated enzymes varies to a great extent, from less stabilized to greatly stabilized, relative to the free enzymes, depending on the methods used.

Subsequent cross-linking of the formed encapsulated enzymes often leads to enhancement of enzyme stability [56, 92]. Subsequent coating has been found to efficiently enhance enzyme stability [92].

The entrapment–encapsulation technique (or coating) has significantly enhanced the operational stability and eliminated the enzyme leakage of an encapsulated enzyme [57].

5.3.3
Enantioselectivity

There are a few examples of enhancement of the enantioselectivity of the encapsulated enzymes.

Lipid-coated lipase can be regarded as a dehydrated encapsulated enzyme. In this sense, it has been found that lipase OF coated with lipid in the presence of a substrate enhanced the enantioselectivity by at least one order of magnitude [40].

5.4
Processes for Preparation of Encapsulated Enzymes

5.4.1
Interfacial Processes

Interfacial processes are based on the physical and chemical reactions that occur at the interface formed by two immiscible phases. If reactant A is soluble in phase I and reactant B in phase II, as illustrated in Scheme 5.8, reaction of A and B can lead to the formation of a membrane between the phases. Depending on the properties and the methods, the enzyme dispersed in phase I (or II) can be encapsulated.

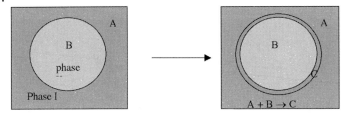

Scheme 5.8 Basic interfacial process for preparation of a spherical membrane.

5.4.1.1 Interfacial Cross-linking/Polymerization

Encapsulation of biologically active components in capsules formed by interfacial cross-linking and polymerization was the first reported method of encapsulation of enzymes.

Usually, B (precursor of the membrane) is dissolved in aqueous phase with the enzyme and is thus a water-soluble monomer, for example 1,6-hexanediamine, piperidine, polyphenol, 2,2-bis(4-hydroxyphenyl)propane or a water-soluble polymer such as PAA, PEI, chitosan, etc., while A should be an organic solvent-soluble monomer, for example sebacoyl chloride, terephthaloyl chloride, bischloroformate, 2,2-dichloroether, 1.6-hexane diisocyanate, 2,4-TDI, etc. (Scheme 5.9).

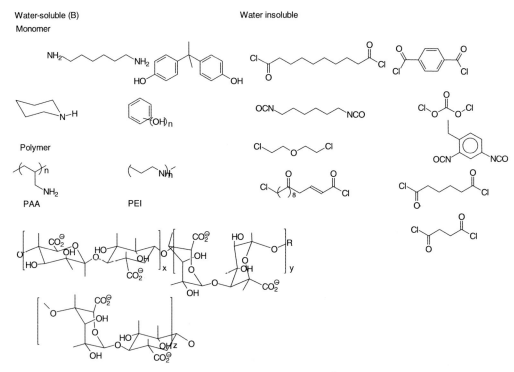

Scheme 5.9 Monomers and polymers for hollow sphere membrane (HSM) by interfacial cross-linking.

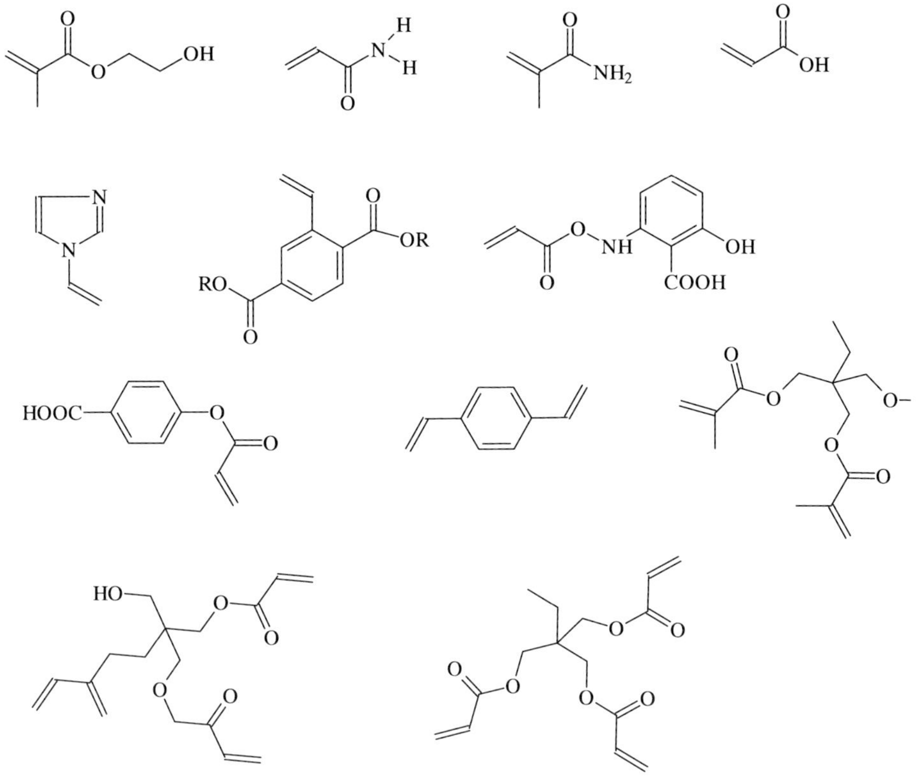

Scheme 5.10 Monomers for interfacial polymerization.

Reaction of A and B leads to the formation of insoluble polymers that deposit at the interface. Consequently, the catalysts (enzymes or whole cell-associated enzymes) present in the phase II during the formation of the membrane will be trapped in the interior with a liquid core.

Apart from interfacial condensation, spherical membranes can be formed by interfacial polymerization. Here, the monomer used for phase II should be water-soluble, for example 2-hydroxyethylmethylacrylate, 3-hydroxypropylmethylacrylate, and acrylamide. For phase I the monomers B can be organic solvent-soluble cross-linkers such as tetraethylene glycol diacrylate [41], as shown in Scheme 5.10.

However, the organic solvents used (phase I), for example chloroform and toluene often toxic to the enzymes or whole cells. Alternatively, use of a non-toxic polymer such as chitosan or PEI might reduce the risk of deactivating enzymes and whole cells [53].

5 Enzyme Encapsulation

Table 5.1 HSM prepared by interfacial cross-linking and polymerization

Process	Material	Remark	Enzyme	Ref.
Interfacial polymerization	2-Hydroxyethyl methacrylate, and tetraethylene glycol diacrylates	The enzyme activity of the immobilized enzyme capsule varied with the thickness of the membrane and with the size of the capsule	Enzymes	41
Interfacial polymerization	Polyamide membrane	Less than 40% activity retention/pH optimum shifted up 1 unit; 33% protein is lost during the encapsulation procedure/67%protein was bound to the membrane	α-Chymotrypsin	42
Interfacial polymerization	Semi-permeable polyamide membranes	Less than 40% activity retention because of the diffusion constraints, because broken capsule gave higher activity	Histidase	43
Interfacial cross-linking	Toluene-2,4-diisocyanate/chitosan		Lactococci cells	44
Interfacial polymerization	Polyamide microcapsules	pH, duration of polymerization, surfactant concentration, stirring rate, improvements in the isolation procedure, effect of lyophilization can effect encapsulation efficiency	Invertase	45
Interfacial polymerization	Semi-permeable polyamide membranes	Enhanced storage stability	l-Asparaginase	46
Interfacial polymerization	Spherical microcapsules of nylon	Lower pH optimum/increase K_m, depending on the size	Enclosing aqueous solution of urease	47
Interfacial polymerization	Capsules of poly(2-hydroxyethyl methacrylate	The activity of the immobilized enzyme decreased with the increasing degree of cross-linking	Urokinase	48
Interfacial polymerization	Polyamide membranes	62.3 mg mL^{-1} with 92.5% retention of activity	Urease	49
Interfacial cross-linking	PEI/acid dichloride	PEI capsules were formed in non-polar solvent; The mean diameter and size distribution of the PEI microcapsules were similar to those observed with nylon membranes	–	50
Interfacial polymerization	Polyamide	Increase in stirring rate was associated with a proportional increase in the activity of urease; suggestive evidence for a diffusion barrier	Urease	51

5.4.1.2 Interfacial Physical Gelation

The application of polymers that can be gelled at the interface is an important method for preparation of encapsulated drugs, for controlled release, and encapsulated enzymes.

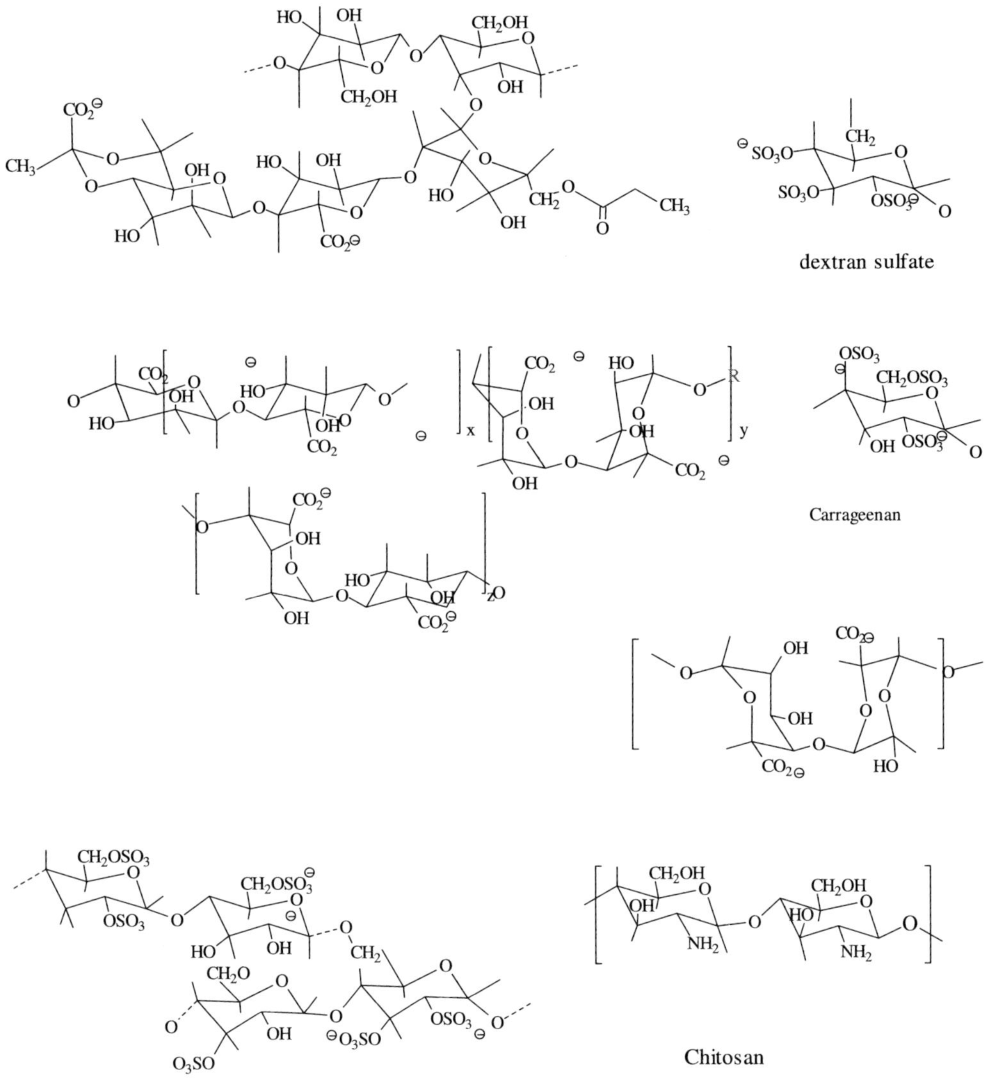

Scheme 5.11 Natural polyelectrolyte for formation of a PEC (polyelectrolyte complex).

On contact of A and B, the following events might occur, depending on the properties of A and B.

- Formation of insoluble polyelectrolyte complex; in this case, A and B are polyelectrolytes of opposite charge, for example chitosan and alginate [64].
- Formation of an insoluble complex can occur in the presence of salts, for example gelation of alginate with use of $CaCl_2$ [66–72]. In this case, the location of A and B can be changed – for instance, a capsule with liquid core can be formed directly when $CaCl_2$ is mixed with the enzyme or whole cell. In contrast, a solid core might be formed when the alginate is located with the catalysts.

The pre-gel materials used for preparation of membrane can be classified into two groups:

- natural polymers, e.g. chitosan, alginate, cellulose sulphate, dextran sulphate, pectin, carrageenan and Xantan;
- synthetic polyelectrolytes, e.g. poly(L-lysine) (PLL), polydimethyldiallyl ammonium chloride (PDMDAAC), poly(methylene co-guanidine) (PMG), PLLAA (poly(L-aspartic acid), PLGA (poly(L-glutamic acid), gamma-PGA, PEI, PPI (poly(propylenimine), K poly(vinyl alcohol) sulphate (KPVS), trimethylammonium glycol chitosan I (TGCl), etc.

The membrane can be also prepared by the use of single polyelectrolyte, for example alginate, in combination with salts such as $CaCl_2$ [55], or prepared as a PEC (polyelectrolyte complex) complex between PA^+ and PB^- (two oppositely charged polyelectrolytes).

The most important factors with regard to enzyme performance, enzyme leakage, and efficiency are often the nature of the gelation conditions and the thickness of the membrane [54, 56].

5.4.2
Surfactant-related Hollow Microsphere

5.4.2.1 Microemulsion-based Encapsulated Enzymes

Emulsion based liquid hollow microspheres have been long pursued as micro-containers for enzymes which can be used to immobilize enzymes [73] or as controlled-release device [74], because enzymes encapsulated in capsules formed by lipids are usually protected from denaturation and also from proteolysis in the microemulsion [75, 76, 105]. Thus, increased activity can also be obtained occasionally [77].

Methods for preparation and application of these entrapped enzymes have been review intensively by many people since the 1970s [78, 79].

In the same way as encapsulation in a membrane formed by the polymers, this type of entrapped enzyme can be regarded as entrapped enzyme in a hollow spherical membrane with a liquid cores, as shown in Scheme 5.12.

5.4 Processes for Preparation of Encapsulated Enzymes

Table 5.2 HSM prepared by PEC

Capsule	Method	Remarks	Enzyme	Ref.
Alginate/$CaCl_2$	Interfacial gelation	Capsule characteristics such as thickness, percentage of enzyme leakage, and encapsulation efficiency were effected by gelation conditions; the optimum conditions selected for effective encapsulation of glucose oxidase were 1 % (w/v) sodium alginate, 5.5 % (w/v) $CaCl_2$ and 1 h gelation time, diffusion limitations exist	Glucose oxidase	55
Alginate/$CaCl_2$	Interfacial gelation	Immobilization resulted in 87 % relative activity and 36 days without decrease in activity. Invertase is also stabilized	Invertase from *Saccharomyces cerevisiae*	56
Ca alginate beads stabilized with poly-l-lysine	Gelation	Enhanced stability, operational stability	Aminoacylase	57
Cellulose sulphate-poly-DADMAC	Symplex formation	The exclusion limit of the membrane was low enough to secure irreversible entrapping of the enzyme	Bovine muscle lactate dehydrogenase.	58
Cellulose sulphate-poly-DADMAC	Symplex formation	The pH and temperature optima were slightly changed, but K_m was changed because of the diffusion limitation. Encapsulation enhanced the storage and operational stability	Free or PS-bound invertase	59
Interfacial gelation	Ca alginate beads	The immobilized enzyme was more thermally stable than the extracellular β-glucosidase produced by this organism	Mycelial-associated β-glucosidase	60
Chitosan (QTS)-poly(vinyl alcohol) (PVA)	Interfacial gelation	Shift of pH optimum to lower pH	Urease	61

414 | *5 Enzyme Encapsulation*

Table 5.2 Continued

Capsule	Method	Remarks	Enzyme	Ref.
Alginate/PEI or PLL/alginate/leaching	Interfacial gelation	Liquid core is formed by the use of buffer sodium citrate solution	Islets	62
Alginate?	Interfacial gelation	Alginate concentration, air flow rate, and liquid flow rate are critical for obtaining microcapsules with the desired characteristics and permselectivity	Engineered cells	63
PVA/Boric acid	Interfacial gelation	150–800 µm capsules with 80–85% activity retention at 38 °C for 10 weeks	Protease (API 21)	64
Alginate polycation microcapsules	Interfacial gelation	Alginate/poly-l-lysine and alginate/chitosan were used. The molecular weight cut off could be increased by increasing M_w of PLL or prolonging the reaction time	Cells	65
Alginate/CaCl$_2$	Interfacial gelation	Encapsulation of lactic acid bacteria with alginate/starch capsules	Lactic acid bacteria	66
Alginate/CaCl$_2$	Interfacial gelation	Recombinant *Saccharomyces cerevisiae* cells with invertase activity were encapsulated in liquid core alginate capsules	Recombinant *Saccharomyces cerevisiae* cells with invertase	67
Alginate/CaCl$_2$	Interfacial gelation	–	Glucose oxidase	68
		Substrate diffusion into the capsules was not rate-limiting; PEG (35,000)-NADH was retained within the microcapsules. Activity retention was approximately 15–25%	Formate dehydrogenase (FDH)/PEG-NADH	69
Alginate/CaCl$_2$		The hydrogen peroxide formed in the oxidation reaction deactivates catalase first	GOD-CAT enzymic system	70
Alginate/CaCl$_2$	Interfacial gelation	–	Glucose oxidase	71
Alginate/CaCl$_2$	Interfacial gelation	The density of cells immobilized in the capsule increased with batches whose cycle time is 24 h and the cells leak out of the capsule during the fourth cycle	*E. coli* cells	72

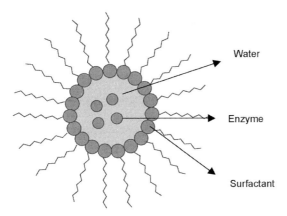

Scheme 5.12 Emulsion-based encapsulated enzymes.

Many enzymes, for example polyphenol-oxidase [81], various lipases [81], chintase [82], trypsin, and α-chymotrypsin [83, 84, 99], protease [85], hydrogenase [86], laccase [87], etc., can be entrapped in the microemulsion formed by surfactants such as ionic surfactants, anionic surfactants, or non-ionic surfactants, with retention of catalytic activity, as illustrated in Scheme 5.13.

Surfactants often used include AOT, TOMAC, SDS, CTAB, DCP, DDAB, and TTAB, as illustrated in Scheme 5.13.

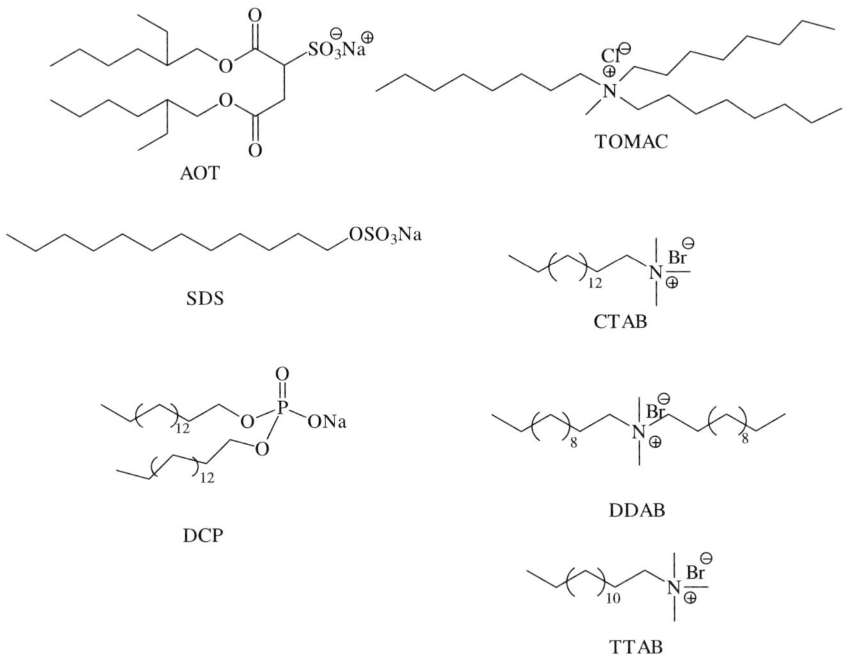

Scheme 5.13 Surfactants used for preparation of micelles.

5.4.2.2 Polymeric Micelles/Liposomes

The emulsion membrane formed by monomer surfactants can be polymerized to form a stable polymeric capsule. Some polymeric surfactants can, moreover, spontaneously form capsules and can be used to encapsulate enzymes [93].

Remarkably, the capsules formed can be coated with a polymer with the aim of stabilizing the liposome or changing the properties of the capsule for drug targeting. Subsequent removal of the micelle can lead to the formation of polymeric nanocapsules [95, 96], which might open a new route for preparation of nanocapsules.

5.4.2.3 Liposome capsules

Lipid vesicles, which are formed by dispersion of bi-layer-forming amphiphilic molecules in aqueous solution (see Scheme 5.14), has long been pursued as a method of encapsulation of enzymes or proteins for different applications, e.g. biomedical, analytical, and food applications, since the beginning of the 1970s [101]. More often, liposomes have been used as controlled-release devices for drug delivery [102]. The use of liposomes for enzyme encapsulation has been intensively reviewed in Ref. [78].

The performance of the encapsulated enzymes depends, however, on the properties of the liposome, for example size and stability, and the method used for preparation of the liposome [102]. Usually, the size of the liposome varies from 20 to 1000 nm, depending on the properties of the lipid used or the method of preparation of the liposome.

The entrapment efficiency and the number of entrapped enzymes are also highly dependent on the properties of the liposome and the size of the enzymes [121].

In specific cases enzymes can be immobilized on the surface of the liposome. For example, α-chymotrypsin has been immobilized on photopolymer liposomes prepared from polymerizable phospholipids [103]. Immobilization of α-chymotrypsin has also been achieved in a sucrose stearate–palmitate containing liposome

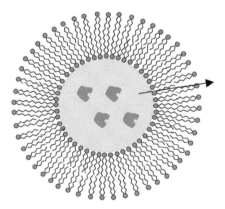

Scheme 5.14 Liposome capsule for enzyme encapsulation.

5.4 Processes for Preparation of Encapsulated Enzymes | 417

Table 5.3 Micelle-aided encapsulation of enzymes

Capsule	Method	Remarks	Enzyme	Ref.
Detergentless microemulsion	Micelle	Alterations of the catalytic properties of polyphenol-oxidase in detergent-less microemulsion and ternary water–organic solvent mixtures	Polyphenol oxidase	80
Micelles of AOT/isooctane	Micelle	The enzyme-containing micelle was used in a dialysis stirred cell to integrate reaction and product recovery. The resistance of the dialysis membrane to reversed micelles was controlled by the water content	Candida rugosa lipase	81
Phosphatidylcholine/ isooctane	Micelle	Esterification of butyric acid was performed in organic solvent	Chintase	82
Bis(2-ethylhexyl) sodium sulphosuccinate in isooctane	Micelle	The intermicellar diffusion and the intramicellar diffusion were used to describe the distribution of the micelle system	α-Chymotrypsin	83
–	Micelle	100 to 1000-fold increase in K_m and simultaneous decrease in k_{cat} by a factor of 2–5 was observed	α-Chymotrypsin	84
Lecithin reverse micelles	Micelle	The water activity in the micelle was controlled	Trypsin and α-protease	85
	Micelle		Hydrogenase	86
Detergentless microemulsion	Micelle	The emulsions are obtained with the use of hydrocarbon/alcohol/water, in defined zones of the phase diagram.	Laccase	87
Liquid membrane emulsions	Micelle		Cellulase	88
Lecithin reverse micelles	Adsorption–encapsulation	A 2.8-fold increase in the productivity of the interesterification when compared with shake-flask experiments	Celite–Rhizopus arrhizus lipase	89
Detergentless microemulsion	Micelle	The emulsions are obtained with the use of hydrocarbon/alcohol/water, in defined zones of the phase diagram	α-Chymotrypsin	90
Reversed micelles of sodium lauryl sulphate and sodium tauroglycholate	Micelle	–	Invertase	91

5 Enzyme Encapsulation

Table 5.4 Polymer-stabilized liposome-entrapped enzymes

Capsule	Method	Remarks	Enzyme	Ref.
Surfactant nanocapsules in cyclohexane	The reversed hydrated micelles from N,N-diallyl-N,N-didodecyl ammonium bromide (DDAB)	Cross-linking enhanced the thermostability of the enzyme molecules	α-Chymotrypsin	92
Polymer-coated liposome	Poly(1,4-pyridinium diethylene salt) was coated on the liposome	Liposome of dicetyl phosphate (DCP), or of a 7:2:1 (mole ratio) mixture of phosphatidylcholine, DCP and cholesterol	Glucose oxidase	93
Polymeric micelles	Polymeric reversed micelles were based on modified polyethyleneimine	Successive alkylation of polyethyleneimine with cetyl bromide and ethyl bromide, thus being able to solubilize considerable amounts of water in the benzene/BuOH mixture	α-Chymotrypsin and laccase	94
Surface-modified polymeric nanogranules (SMPN)	Polymerization of an AAm/ N,N'-MBAm mixture in a mixed reversed of AOT and surfactant Pluronic F-108	Chromatographic removal of the auxiliary surfactant and AOT was necessary	α-Chymotrypsin	95
Surface-modified polymeric nanogranules (SMPN)	One of the surfactants bearing double bonds such as Pluronic F-108 was polymerized	Activity and stability of α-chymotrypsin entrapped in SMPN strongly depended on conditions of preparation of SMPN	α-Chymotrypsin	96
Polymeric surfactant	PEI modified with cetyl bromide and ethyl bromide	Polymeric reverse micelles of 20–50 nm were capable of solubilizing enzymes in non-polar solvents with retention of catalytic activity	α-Chymotrypsin and laccas	97
VET200	Polymeric liposome	Enzyme is entrapped inside polymerized PMOXA-PDMS-PMOXA-triblock copolymer vesicles mean diameter 250 nm; the membrane was 10 nm thick	Lactamase	98
AOT reverse micelles/gel	Gel-entrapped enzymes in reversed micelles	The enzymes are first gel-entrapped then "equilibrated" with a reversed micellar solution	α-Chymotrypsin	99
Poly(MA-co-MVE), poly(MA-co-St)-coated emulsion	Depositing the polymer around emulsified aqueous droplets	The permeability of the former responded to pH changes in the range pH 5–60	Invertase	10
FAT-VET200	POPC/polymer-stabilized POPC-based vesicles	Enzyme is entrapped inside POPC or polymer-stabilized POPC-based vesicles (FAT-VET200)	Lactamase	101

Table 5.5 Liposome-based capsules

Method of preparation of liposome	Type of liposome	Remark	Enzyme	Ref.
DRV	–	AChE can be stabilized in porin-functionalized liposome against denaturation by proteolysis	AChE	106
FAT-VET100	POPC vesicles	No activity against the larger substrate Suc-Ala-Ala-Pro-Phe-pNA or casein. Inhibition of externally present enzyme by an inhibitor protein	α-Chymotrypsin	107
FAT-VET100	POPC vesicles	87 enzyme molecules per vesicle with a diameter of 125 nm was obtained in FAT-VET100	α-Chymotrypsin	108
DDV	DPPC–cholesterol–stearylamine 6:3:1 or 5:3:2	Liposome is around 61 ± 18 nm and 29 ± 11 nm, respectively. The enzyme was partially localized on the external vesicle surface, particularly with ratio of DPPC–cholesterol–stearylamine 6:3:1	Carbonic anhydrase	109–111
REV	Egg PG–egg PC–cholesterol, 1:4:5	Vesicles are characterized by a relatively large internal aqueous space (0.2–1 μm)	Alkaline	112
DDV	Egg PC–lyso PC: SM–PE, 72.5:4.6:3.8:19.1	The liposome is ~30–125 nm and enabled the permeation of (N-benzoyl-dl-arginine-4-nitroanilide) the interior of the vesicles	Trypsin	113
DDV	PC–cholesterol (2:1)	Entrapment in vesicles yielded about 150 enzyme molecules per vesicle	Alkaline phosphatase	114
VET/REV	Dipalmitoylphosphatidyl-choline (DPPC)/phosphatidylinositol	~20% encapsulation efficiency with the use of 1 mg mL^{-1} enzyme content	Chloroperoxidase/lactoperoxidase/glucose oxidase	115
VET/REV	Dipalmitoylphosphatidyl-choline (DPPC) and phosphatidylinositol (PI)	The antibacterial activity of these "reactive" liposomes arising from hydrogen peroxide and oxyacids in the presence of the substrates glucose and iodide ions	Glucose oxidase (GO) and GO in combination with horseradish peroxidase (HRP)	116
Dry-rehydration with enzyme–buffer solution/extrusion	(SL) Palmitoyloleoyl-phosphatidylcholine (POPC) and dipalmitoyl phosphatidylethanol-amine-N-PEG 2000/cholesterol	There is no significant difference in K_m values between free and encapsulated phosphotriesterase; therefore paraoxon readily penetrates the membrane of the carrier cells	A recombinant phosphotriesterase	117

Table 5.5 Continued

Method of preparation of liposome	Type of liposome	Remark	Enzyme	Ref.
VET	DMPC and CS-incorporating dimethyldioctadecylammonium bromide and DMPC incorporating PTI	On addition of glucose, H_2O_2 is produced, which is toxic to the bacterial	Glucose oxidase/ horseradish peroxidase (HRP)/ lactoperoxidase (LPO)	118
Rehydration	PC liposome	An increase in the number of stearoyl residues attached to the enzyme results in a dramatic decrease of ATEE binding to the active centre (K_m increase)	Hydrophobized α-chymotrypsin derivatives	119
SUV, "Millipore-filtered MLV", and "ether-injection method"	Egg PC-egg PA, 15 : 2.12	Entrapment experiments indicated that at least part of the enzyme is localized inside the vesicles	Aminolevulinate dehydratase	120
MLV	PC-cholesterol-dicetylphosphate, 5 : 5 : 1	The enzyme-containing vesicles obtained was 2–3.5, depending on the lipid	Trypsin	12
FAT-VET400	POPC vesicles	The diameter is ~150 nm and addition of cholate enabled the passage of substrate across the membrane	Phosphorylase	122
REV or DRV	Soybean phosphatidylcholine–cholesterol vesicles	Enzyme entrapment in soybean phosphatidylcholine–cholesterol vesicles (REV or DRV) decreased with increasing cholesterol content. The enzyme was localized both inside and on the surface of the vesicles	β-Galactosidase	123
MLS	Egg-phosphatidylcholine lipics (eggPC),1-palmitoyl-2-oleoyl-sn-glycero-3-[phosphoserine] (sodium salt) (POPS) etc.	Liposomes insert a barrier between the enzyme and the external environment and protects the enzyme	Acetylcholinesterase	124

Abbreviations: DDV, vesicles prepared by the detergent dialysis method; SUV, small (sonicated) unilamellar vesicles; FAT-VET200, vesicles prepared by means of freeze–thaw cycles then repeated extrusions; MLV, multilamellar vesicles; VET100, vesicles prepared by the extrusion method (without freezing–thawing cycles); REV, vesicles prepared by the reverse-phase evaporation; DRV, vesicles prepared by the dehydration–rehydration method; MLS, multilamellar spherulites prepared by shearing a lamellar phase by oxidation of the hydroxyl groups in the sugar moieties [104]. Alternatively, derivatized enzymes with artificial hydrophobic tails can be immobilized on the surface of liposomes [105].

5.4.2.4 Colloidal Liquid Aphrons (CLA)

Polyaphrons are spherical droplets (~5–10 mm) of solvent stabilized by a mixture of ionic and non-ionic surfactants (also called oil-in-water microemulsions) which contain a higher proportion of internal solvent phase (f) than that associated with the hexagonal close packing of spheres [123].

Thus, the enzymes can be encapsulated in the aqueous shell, resulting in the formation of encapsulated enzymes as illustrated in Scheme 5.15. Although this phenomenon was observed early in 1971, exploitation of these techniques has since remained mainly in the field of protein separation.

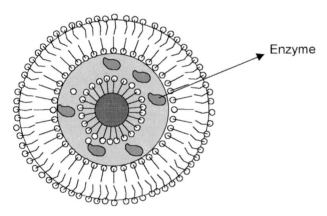

Scheme 5.15 CLA-encapsulated enzymes.

Recently, the technique was extended for immobilization of enzymes, especially hydrophobic enzymes [126, 127]. Interestingly, it was found that variation of the solvent core and the surfactant can lead to the formation of immobilized enzymes with 100% immobilization efficiency [127]. By studying the immobilization of several classes of enzyme such as α-amylase, β-galactosidase, lipase, lysozyme, trypsin and ribonuclease A it was found that the results of immobilization, for example efficiency, were related to pH, molecular weight, adiabatic compressibility, and hydrophobicity [126].

Compared with microemulsion or liposome methods, entrapment of enzymes in CLA has many advantages, for example high entrapment efficiency and broad applicability.

Table 5.6 Encapsulation of enzyme in CLA

Surfactants	Method	Remark	Enzyme	Ref.
Softanol 30 (organic phase)/ SDS (aqueous phase)	CLA-entrapment	The activity of α-amylase was increased more than 15 times. The increasedenzyme activity may be attributed to the loosening of the enzyme structure	α-Amylase	125
Softanol 30 (organic phase)/ SDS (aqueous phase)	CLA-entrapment	Enzyme activity was increased sixfold when immobilized in CLA	β-Galactosidase	126
Softanol (organic phase)/ SDS (aqueous phase)	CLA-entrapment	The enzyme retained in CLA was independent of the pH and ionic strength of the bulk aqueous phase indicating that immobilization was primarily due to hydrophobic interactions	β-Galactosidase	127
Softanol 30 (organic phase)/ SDS (aqueous phase)	CLA-entrapment	80% of the lipase was retained under certain conditions, and activity was comparable with that of the free enzyme for the hydrolysis of p-nitrophenylacetate	Candida cylindracea lipas	126, 128
Softanol 30 (organic phase)/ SDS (aqueous phase)	CLA-entrapment	–	Lysozyme	126
Softanol 30 (organic phase)/ SDS (aqueous phase)	CLA-entrapment	25% activity retention was obtained	Ribonuclease A	128
Softanol 30 (organic phase)/ SDS (aqueous phase)	CLA-entrapment	–	Trypsin	126
Softanol 30 (organic phase)/ SDS (aqueous phase)	CLA-entrapment	Although over 90% of the enzyme could be successfully immobilized on CLA, activity retention was less than 1%	α-Chymotrypsin	129
Softanol 30 (organic phase)/ Synperonic A20 in RO water (aqueous phase)	CLA-entrapment	Efficiency of P450 immobilization was greater than 85%, and in this state enzymatic activity could be measured for more than 24 h at 15 °C	Cytochrome P450	130

5.4.3
Phase Inversion

The phase-inversion method is usually based on a change in the solubility of the polymeric phase around droplets containing drugs.

5.4.3.1 Coacervation

Use of these techniques was pioneered by Chang in 1966. This technique is characterized in that it is very simple and broadly applicable. The solvents used are, however, usually toxic to the enzyme entrapped, often leading to lower retention of activity [133]. In addition, not all the enzyme can be entrapped [135].

The drugs to be encapsulated are usually dissolved in an aqueous solution and then dispersed in a water-immiscible solvent containing the dissolved polymer. Subsequent evaporation of the solvents leads to the deposition of the polymer on the surface of the aqueous droplets as illustrated in Scheme 5.16.

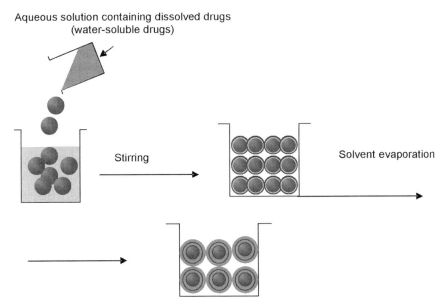

Scheme 5.16 Preparation of monolithic microspheres.
(This method is not suitable for entrapment of highly water-soluble drugs.)

5.4.3.2 The Double-emulsion Method

Phase-inversion techniques (PIT) for encapsulation of biologically active compounds such as proteins, enzymes, or peptides are mainly used for the purpose of controlled drug release.

Table 5.7 Coacervation process for preparation of HSM

Material	Method	Remarks	Enzymes	Ref.
Cellulose nitrate?	Interfacial coacervation	Haemoglobin preserves the physical stability of the enzyme during microcapsule manufacture and also enhances the structural integrity of the microcapsules	–	131
Cellulose nitrate	Interfacial coacervation	Enzyme was immobilized in semipermeable microcapsules which retain the enzyme and allow asparaginase to diffuse in	Asparaginase	132
Cellulose nitrate	Interfacial coacervation		Asparaginase	133
Cellulose nitrate (collodion artificial cells)	Interfacial coacervation	Microencapsulated PAL reportedly loses 80% of its activity. V_{max} is about 20% of the free enzyme	Phenylalanine ammonia-lyase	134
Cellulose nitrate	Interfacial coacervation	Lower retention of activity was obtained because of deactivation of the enzyme during encapsulation. 70% encapsulation efficiency of which 21% is active	Phenylalanine ammonia-lyase	135
–	Interfacial coacervation	–	Phenylalanine ammonia-lyase	136
Collodion artificial cells	Interfacial coacervation	20% activity retention was obtained	Phenylalanine ammonia-lyase	137
Collodion artificial cells	Interfacial coacervation	70% entrapment efficiency and 20% activity retention were obtained, because of deactivation of the enzyme by the organic solvents	Phenylalanine ammonia-lyase	138
Cellulose nitrate	Interfacial coacervation	Lost 88% of its activity because of denaturation of the enzyme after contact with organic solvent	Arginase	138
Cellulose nitrate	Interfacial coacervation	K_m was not changed but V_{max} decreased	Catalase	140
Cellulose nitrate	Interfacial coacervation	K_m was not changed but V_{max} decreased	Histidase	141
Cellulose nitrate	Interfacial coacervation	–	Lactase	142
NH$_4$ polyacrylate-enzyme/liquid paraffin	Interfacial coacervation	–	Alkaline protease	143
Cellulose nitrate	Interfacial coacervation	The artificial cells can be used to remove intestinal amino acids in diseases where they accumulate	Asparaginase, glutaminase, and tyrosinase	144

Use of these concepts for preparation of immobilized enzymes has not been intensively studied, because:

- the materials used are often expensive,
- enzymes are often deactivated,
- the encapsulation efficiency is often very low,
- leakage of the enzyme from the micro-sphere occurs (this is the intrinsic nature of this technology).

Direct use of this type of entrapped enzyme for biotransformation is not recommended. After cross-linking of the entrapped enzymes, however, it is obviously possible to irreversibly insolubilize the enzyme so they can be used as normal immobilized enzymes.

Currently, several double-emulsion techniques are available:
- W/O/W (water/oil/water)double-emulsion technique [145, 147],
- S/O/W (solid/oil/water) [148],
- W/O/O (water/oil/oil) [149],
- O/W1/W2 (oil/water/water) [150],
- O/W1/O (for encapsulation of oil) [147].

In general, an aqueous phase containing water-soluble drugs can first be emulsified within a polymer-solvent phase (W/O; polymer 1) and the emulsion is re-emulsified in a large volume of aqueous phase (containing probably another water-soluble polymer such as PVA) to produce the (W/O/W) double emulsion (Scheme 5.17).

Here, organic solvent (O) is usually volatile and evaporates quickly at the water/air interface and precipitation will also occur at the interface. A skin layer is therefore formed. The drugs dispersed in the first aqueous phase will be entrapped. The entrapment efficiency and the drug-release kinetics are determined by the structure and porosity of the microsphere.

The polymers used for preparation can be water-soluble or organic solvent-soluble, depending on the method used. For the solvent-evaporation method it is desirable to use organic solvent-soluble polymers such as PMM (for encapsulation of ovalbumin [146]), PPP (for encapsulation of insulin) [151], Eudragit S [152], PLGA [150], or block copolymers such as copolymers consisting of poly(L-lactide-co-glycolide) A blocks attached to central poly(ethylene oxide) (PEO) B blocks [147].

The methods used also affect the structure of the microsphere. For example, encapsulation of peptides in PLA-PLGA microspheres by means of solvent evaporation, spray-drying, and solvent extraction produced different types of microsphere with different structures [153].

The polymers used are usually organic solvent-soluble and the enzyme is normally encapsulated in the liquid core. Depending on the properties of the polymer used the enzyme can be slowly released by diffusion or erosion.

Nevertheless, for enzyme immobilization the capsules used should be chemically and physically stable. It is also obvious that diffusion limitation might be serious for the encapsulated enzyme, because the pore size of the membrane should be

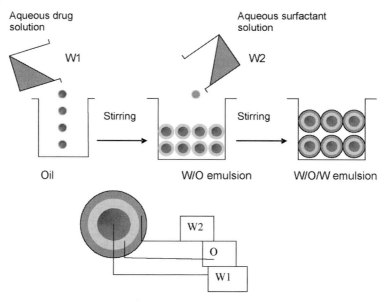

Scheme 5.17 W/O/W emulsion for preparation of microspheres with in-situ encapsulation of drugs.

close to the size of the enzyme or even smaller. In addition to this intrinsic drawback, the process used often deactivates the enzymes owing to interfacial deactivation phenomena. Addition of additives such as sugars might improve activity retention [154].

5.4.3.3 Modified Double-emulsion Methods
Conventionally, entrapment of drugs by use of W/O/W double emulsion technology involves the use of a stabilizer, for example a polymer or a surfactant, to stabilize the emulsion in the second step. Consequently, several variants have appeared, with the objective of improving the encapsulation, especially encapsulation efficiency and size homogeneity.

Method without stabilizer It has been found that encapsulation of bovine serum albumin (BSA) in poly(L-lactide) (PLLA) nanocapsules or polycaprolactone–poly(ethylene oxide) block copolymer (PCE) can be achieved by the modified double-emulsion method. In the second step a glycerine–water mixture was used instead of a polymer solution. Remarkably, more than 50% encapsulation efficiency can be obtained with PLLA [174]

Use of the induced phase-separation method An aqueous surfactant solution was slowly added to a W1 (protein)/O (organic solvent +polymer) emulsion to obtain a W/O/W emulsion. The solvent was removed under reduced pressure [158, 175].

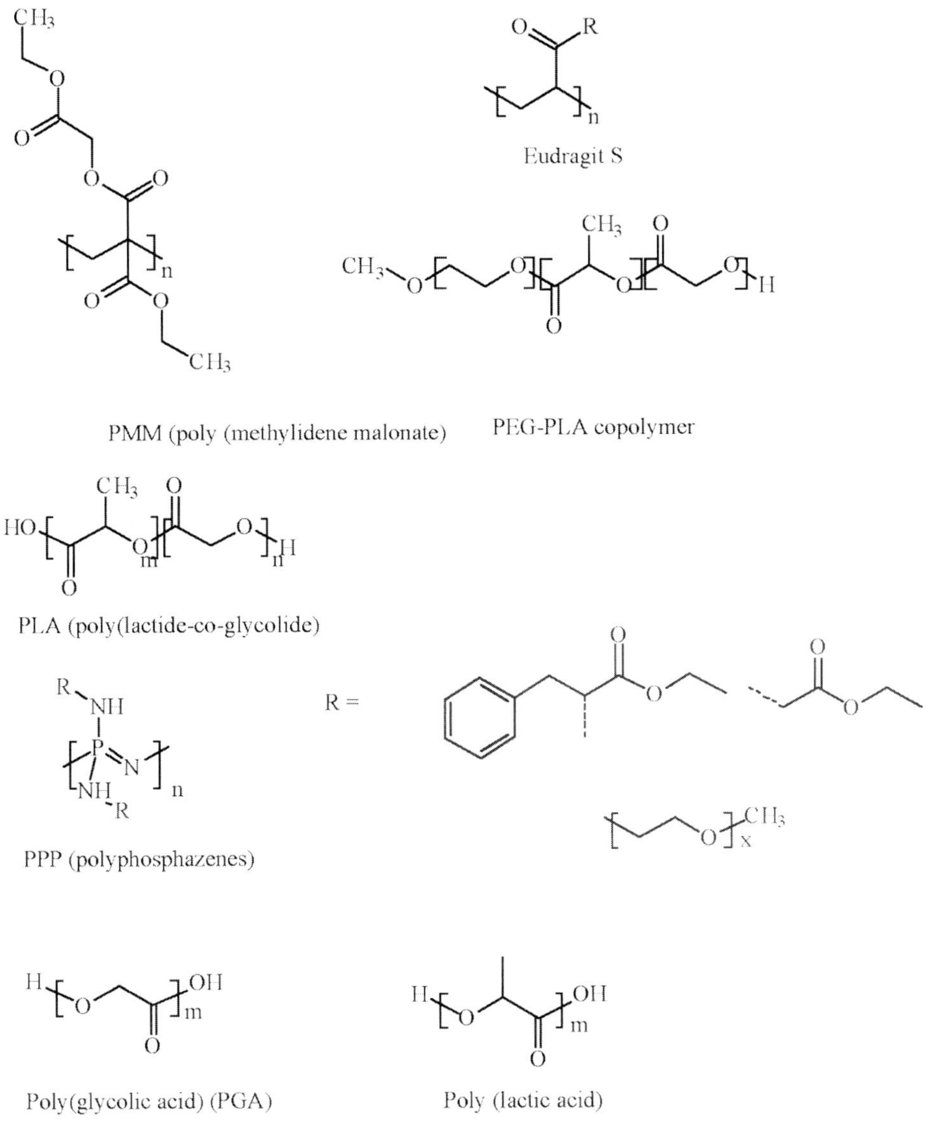
Eudragit S
PMM (poly (methylidene malonate)
PEG-PLA copolymer
PLA (poly(lactide-co-glycolide)
R =
PPP (polyphosphazenes)
Poly(glycolic acid) (PGA)
Poly (lactic acid)

Scheme 5.18 Polymers used for preparation of polymeric capsules for entrapment of drugs.

Table 5.8 HSM and encapsulated enzymes prepared by double-emulsion methods

Double emulsion	Materials	Remarks	Enzyme	Ref.
O/W1/O	Linear and star-branched block copolymers	The size, and the entrapment efficiency were influenced both by the composition of the polymer and temperature	Human erythropoietin	147
S/O/W	Eudragit S100 microspheres	Activity loss was mainly due to aggregation; addition of sugar or PEG prevented structural change	γ-Chymotrypsin	148
W/O1/O2	Eudragit S100 microspheres	The profile of drug release was pH dependent.	Water soluble drugs	149
O/W1/W2	PLGA (poly(d,l-lactide-co-glycolide)	Variation of O/W1 ratio and solvent removal rate enables the fabrication of different type microspheres with different size distributions	–	150
W/O/W Solvent diffusion	Polyphosphazene	Double emulsion and solvent evaporation	Insulin	151
	Eudragit RS 100 polymer	Eudragit RS 100 polymer were dissolved in ethanol	Ibuprofen	152
W/O/W (solvent evaporation)	Poly(lactic acid) and poly (lactic acid-co-glycolic acid)	The mean diameter and encapsulation efficiency are influenced by molecular weight and the glycolic acid content in the polymer	Peptide	153
W/O/W	Poly(lactide-co-glycolide) microspheres	Addition of sugars enhanced the entrapment efficiency; the loading is below 5%, activity retention is variable from 30–90%, depending on the additives used	Urease	154
W1/O/W2	Polystyrene (PS) and/or styrene-butadiene rubber (SBR)	A composition of 2:1 PS-SBR yielded a homogeneous and tough wall structure, resilient to the impact and tight confinement of enzyme macromolecules	*Pseudomonas fluorescens* lipase	155
W/O/W	PLGA-Pluronic L-121/DCM/PVA-buffer/DCM	–	Human growth hormone aggregates	156
W/O/W	Poly-dl-lactide–poly(ethylene glycol)	Activity retention is approximately 40 %	Glucose oxidase (GOD)	157

5.4.3.4 Other Methods

Along with the double-emulsion methods mentioned above, other methods such as spray-drying, solvent extraction, freeze-drying, in-situ solidification, etc., have been also frequently used to prepare capsulated protein drugs. Among these, spray-drying was probably the first method developed for encapsulation of drugs [172]. The enzyme is usually dispersed in a polymer solution, followed by spray-drying.

In in-situ solidification of the polymeric materials the polymer shell will be quickly formed, because of dissipation of solvents to the external aqueous medium, whereas in the solvent-extraction technique partially water-miscible solvents are used [159–161].

5.4.4
Pre-designed Capsules for Post-loading Encapsulation

5.4.4.1 Introduction

Conventionally, enzymes are usually trapped in the capsule formed during encapsulation. Although it has recently been demonstrated that post-loading techniques also apply to encapsulation of enzymes for certain applications, for example controlled release, implementation of these techniques for preparation of robust immobilized enzyme remains to be further explored.

As discussed in Chapter 4, enzymes can be loaded into ready-made insoluble matrix. Similarly, it is conceivable that the ready-made capsules can also serve as a micro-container for enzyme encapsulation.

In line with this concept, the enzyme has been enclosed in the interior of the capsule. On changing the pH, it was found that the pores of some capsules could be opened and closed [162, 163], thus enabling the release of some drugs in response to the pH change. On the other hand, the enzyme can also be permanently immobilized inside the capsules by chemical cross-linking.

Although a similar concept was used as early as 1979, encapsulation of an enzyme in a ready-made hollow spherical membrane was first reported by Parthasarathy and Martin in 1996 [163], when several enzymes such as glucose oxidase, catalase, trypsin, subtilisin, and alcohol dehydrogenase were loaded into the polypyrrole microcapsules prepared by the template method, followed by a sealing procedure.

In a similar approach, capsules prepared by the so-called coating and template-leaching technique have also been used for encapsulating enzymes [164]. Briefly, an insoluble core such as melamine formaldehyde micro/cores or polystyrene with defined particles was used as template for preparation of HSM by layer-by-layer assembly of non-biodegradable polyelectrolytes, e.g. poly(allylamine) (PAH) and poly(styrenesulphonate) (PSS).

Subsequent etching of the solid core will result in the formation of hollow microspheres, which can be used to immobilize enzymes by two mechanisms – adsorption of protein on to the capsule shells and pH-dependent opening and closing of capsule wall pores [162–164].

It is obvious that this technique reduces the risk of deactivating the enzyme by the in-situ method mentioned above, because preparation of the hollow microspheres and loading of the enzyme are separated [164].

5.4.4.2 Theoretical Considerations

Intuitively, this method is relatively simple. A central question, however, is: what kind of capsule is best-suited for preparation of encapsulated cross-linked enzymes? Before we try to address this question, let us first consider the essential requirement for the capsules required for this process:

- Appropriate pore size: the pore size of the capsules should enable the enzyme to freely enter the interior of the capsules during the loading process but also mitigate diffusion constraints after cross-linking of the enzyme inside the capsules.
- Suitable physical and chemical nature: the capsules should be mechanically and chemically stable.
- The capsules should be stable under the conditions used to aggregate the enzyme in the interior of the capsules by use of non-denaturing protein-aggregation methods.
- Highly swelling factor: the capsules prepared should be able to swell, in order to enable high enzyme loading. Because the maximum amount of the enzyme that can be encapsulated in the interior is equal to the volume of the capsule multiplied by the concentration of enzyme of the enzyme solution, it is obviously advantageous to prepare capsules which swell immediately in the aqueous enzyme solution so the enzyme molecules can be easily loaded into the interior of the capsules.
- The capsules should be readily dehydrated to remove the water.
- The capsules should not swell significantly after enzyme immobilization when immersed in aqueous medium. Otherwise the volume of the immobilized enzyme will be increased.

5.4.5
Non-conventional Encapsulation Processes

Along with the conventional processes mentioned above, in recent decades several other encapsulation processes, for example encapsulation/coating, encapsulation/cross-linking, and immobilization and encapsulation, have also appeared, usually with the objective of improving the properties of the pre-immobilized enzymes.

5.4.5.1 Encapsulation/Coating

Encapsulated enzymes often suffered from enzyme leakage, as a result of diffusion of the enzyme molecules from the interior of the capsules into the application mix-

5.4 Processes for Preparation of Encapsulated Enzymes

Table 5.9 Encapsulation/coating techniques

Hollow microsphere	Coating	Remark	Enzymes	Ref.
Alginates/BaCl$_2$	PNVA/PAAc/CS	Cellulose derivatives are better than polyacrylic acid and single-layer capsule is not able to retain the enzyme	Cytochrome C	177
Alginate	Poly(methylene co-guanidine)	The coating led to 35% loss of activity relative to the uncoated enzyme. Addition of glucose prevents deactivation of the enzyme during coating	Urease	178
Alginate-CaCl$_2$	K poly(vinyl alcohol) sulphate, trimethylammonium glycol	The immobilized system remained stable without leading to serious loss of activity or to a large leakage of the enzyme from the support, because of formation of the PEC-coating	β-Amylase	179
Alginate	PVAm	–	Mouse erythro-leukaemia cell	180
Alginate/cellulose sulphate microcapsules	Poly(methylene co-guanidine)	–	Living cells	181
Alginate/cellulose sulphate microcapsules/coating	Poly(methylene co-guanidine)	–	Living cells	182
Alginate/cellulose sulphate microcapsules	PMG	–	Living cells	183
Alginate	PEI/PAAc	The activity of the enzyme encapsulated in the capsules depended on membrane composition (no. and sequence of layers) and storage time	Acidic phosphatase	184
Alginate capsules	Chitosan	Microcapsules were prepared by interfacial polymerization of chitosan, and alginate microspheres formed by emulsification/internal gelation. Diameters ranged from 20 to 500 μm, depending on the formulation conditions	DNA	185
Alginate	Chitosan	–	Proteins	186
Alginate	Polylysine	–	Pancreatic islets	187
Alginate	Polylysine	–	–	188
Alginate-polylysine microcapsules	Polylysine	–	Mammalian cell	189

ture. This could lead to serious problems in the practical application of the immobilized enzymes as a reusable catalyst; for example, lower operational stability and contamination of the product with the enzyme used.

Consequently, coating of the capsules with a layer or with multiple layers of polycation/polyanion complexes was found to mitigate he loss of the enzyme activity [177]. The microcapsules with the enzymes in the interiors can be coated with polyelectrolytes with the objective of altering permeability by use of pH-conditioned changes in the configuration of the adsorbed polyion layer [20]. Although this technique is of less importance for preparation of immobilized enzymes for industrial application, it is might be a very interesting approach for controlled drug delivery.

Alginate gel beads coated with a membrane consisting of a single layer of synthetic polymers are in extensive use for immobilization of living cells and enzyme systems [182–184] and coating of alginate-based microcapsules with silica gel has been found to enhance mechanical resistance and reduce protein leakage by diffusion [186].

5.4.5.2 Encapsulation/Cross-linking

As remarked in Chapter 4, enzyme can leach from the interior of the entrapping matrix. Thus, if the enzymes that are only physically encapsulated in the interior of the capsules are to be useful as biocatalysts they must be fixed in the interior of the capsules.

To this end, in the same way as for other types of immobilized enzyme, encapsulated enzyme preparations also can be cross-linked with a diffusion cross-linker, for example glutaraldehyde, thus overcoming this intrinsic problem. Although this technique was pioneered by Chang et al. in 1970s, it was not further developed for the purpose of preparing immobilized enzymes, as compared with enzyme entrapment and cross-linking [190].

Alternatively, addition of cross-linkable component such as gelatin or albumin to the enzyme solution might lead to an encapsulated enzyme with enhanced mechanical stability and operational stability. For example, urease-containing xanthan–alginate spheres were prepared by gelling enzyme–xanthan–alginate–gelatin solution, followed by subsequent cross-linking with GAH, leading to the formation of encapsulated enzyme with very high stability [190].

It is obvious that use of this technique obviates the use of capsules with a pore size much below the size of the enzyme (5 nm), because the size of cross-linked enzyme in the interior of the capsules is much bigger than that of the native enzyme, as shown in Scheme 4.3. Thus, a membrane with a pore size being much larger than that of the native enzyme can be used. Consequently, the diffusion limitation is readily overcome by use of microporous or macroporous capsules.

Another advantages of using this technology for preparation of cross-linked enzymes, instead of cross-linking the dissolved enzyme in the bulk solution are obvious. It is possible to overcome the intrinsic drawbacks associated with conventional cross-linked enzymes, for example gelatinous properties and difficulties in han-

5.4 Processes for Preparation of Encapsulated Enzymes | 433

Table 5.10 Encapsulation and subsequent cross-linking methods

Hollow microsphere	Cross-linking	Remark	Enzymes	Ref.
Xanthan-alginate spheres	Gentle glutaraldehyde cross-linking with amine groups of gelatin present in the initial mixture	Cross-linking reduced the enzymatic activity but increased the stability (75 % activity retention after 20 cycles). Enzyme loading influenced the enzyme activity retention; because of diffusion constraints, the enzyme inside the capsules is much more stable than the native enzyme	Urease	190
Gelatin		Enzyme loading affected enzyme activity retention because of diffusion constraints	Urease	191
	Cross-linking		Urease	192
Ca-alginate gel beads	Entrapment-cross-linking	Threefold increase in enantioselectivity after immobilization compared with the native enzyme	Pegylated PCL	193
Ca-alginate beads coated with chitosan, PEI, and PEI-GAH	Entrapment-coating	Enhanced operational stability	Aminoacylase	194

dling. It is, moreover, possible to control the size of the cross-linked enzymes by selecting appropriate capsules with defined geometric properties. Unfortunately, this technology has not yet been exploited for preparation of cross-linked enzymes.

5.4.5.3 Immobilization and Encapsulation

It is often found that immobilized enzymes have one or more of the following drawbacks of enzyme leakage [20], low activity retention, poor stability [198, 200], poor selectivity or serious substrate inhibition [200], or small size [197]. Use of encapsulation techniques with consecutive immobilization might, however, solve these problems, as illustrated in Scheme 5.19.

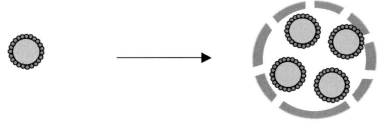

Scheme 5.19 Immobilization–encapsulation.

Often, consecutive encapsulation is primarily designed to improve the properties of encapsulated enzymes, for example:

- mitigation of enzyme leakage (for enzymes adsorbed on fine carriers) [198]
- enhancement of activity retention, by reducing leakage [197], or enhancing activity retention by improvement of dispersion [22]
- improvement of stability, for example mechanical stability, stability in organic solvents [177, 196] or operational stability
- enhancement of selectivity [195]
- mitigating substrate or substrate inhibition [197]
- enlargement of size [197]

5.4.5.4 Immobilization and Encagement

Absorbed enzymes on different carriers can be further encapsulated in a gel matrix formed on the top of immobilized enzyme particles. The characteristics of this technology, which is similar to immobilization and encapsulation, are that the enzymes can be further stabilized and enzyme leakage is avoided. The principle is depicted in Scheme 5.20. Remarkably, encagement of immobilized enzymes substantially enhanced their stability [205].

Table 5.11 Immobilization and encapsulation

Systems	Method	Comment	Enzyme	Ref.
Alginate beads coated with chitosan, PLL or PMG membrane	Entrapment-coating	Protect from attack by α-chymotrypsin	Urease	196
–	–	Mitigating substrate or substrate inhibition	Glucose oxidase	197
Calcium alginate/PEI bead	Immobilization-encapsulation	Enlarging the size	Hydantoinase on fine PGL	198
Sandwich structure	Entrapment-coating	Encagement can be realized by reacting the enzyme connected with polyaldehyde	Urease	199
Within cross-linked propylene glycol alginate/bone gelatin gel spheres	Adsorption-encapsulation	Mitigation of substrate inhibition at high concentration	β-Glucosida CAS complex	200
Alginate beads, coated with polyamide	Entrapment-encapsulation	Reduced enzyme leakage	Amino acylase	201
Coating of the particles with a semi-permeable poly(vinyl acetate) film	Adsorption-coating	Silica gel mixture of enzyme > pellet > coating	Invertase	202
–	Adsorption-coating	–	–	203
BCP electrode by coating the enzyme-loaded surface with an NC film	Coating	–	Glucose oxidase	204
Dowex 50W-X8 within a capsule	Entrapment-coating	Artificial kidney	Urease	205
Adsorbed on ACG, followed by coating with ultrathin cellulose nitrate	Entrapment-coating	28.1% of the activity of urease in solution	Urease	206

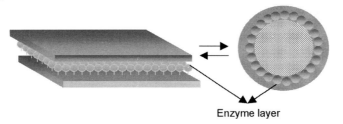
Enzyme layer

Scheme 5.20 Encagement of adsorbed enzyme.

For example, α-amylase adsorbed on the particles of copolymers of methyl methacrylate, trimethylolpropane trimethacrylate, and acrylonitrile was encapsulated in the polymer layer formed by cross-linking of amines with, e.g., glutaraldehyde, resulting in the formation of an immobilized enzyme with higher stability [216].

Urease has been adsorbed on activated charcoal granules which were then coated with ultrathin cellulose nitrate [215]. Its activity was 28.1% of the activity of urease in solution. The enzyme did not leak after immobilization and worked efficiently.

Encapsulation of enzymes can be achieved by encagement in a sandwich structure; this can be achieved by reacting the enzyme with polyaldehyde polymer, then reacting with polyamine, leading to the formation of a bilayer-encaged immobilized enzyme [208].

Encagement of enzymes in the layer form by use of a symplex was proposed by Decher and Hong at the beginning of the 1990s [208] and later extended by Ivov et al. [209] and Ge et al. [210]. It has been reviewed by Decher et al. [207].

Encagement of the enzyme on the carrier surface is relatively simple – by casting an enzyme solution containing poly-*l*-lysine on an electrode, followed by addition of a solution of the counter-polyelectrolyte, for example poly(4-styrenesulphonate). Several enzymes, for example lactate oxidase, choline oxidase, and glucose oxidase, can be immobilized on electrodes with high operational stability [212]. Similarly, enzymes can also be encaged in a membrane or film by formation of a polyelectrolyte complex.

Fibres can be soaked in enzyme solution then coated with a semi-permeable membrane. For example, by use of this method asparaginase was immobilized on rayon filament; the filament containing the enzyme was passed through a 15% ethanolic solution of a copolymer of 50% 2-hydroxyethylmethacrylate, 40% ethoxyethylacrylate, and 10% hydroxypropylacrylate then heated at 50 °C to evaporate the ethanol [214].

The enzyme activity is governed by two steps, the adsorption and entrapment steps. Thus, the principles that govern the activity of adsorbed immobilized enzymes and entrapped enzymes are also applicable here. In general, the following factors should be taken into consideration:

- the nature of the carriers,
- binding functionality,
- the properties of the entrapment matrix,
- the extent of diffusion limitation.

Table 5.12 Immobilization of enzyme by adsorption and engagement

Polymer	Method	Comments	Enzyme	Ref.
Polyaldehyde polymer	Polyacrylamide hydrazine	Engagement enhanced enzyme stability	Enzymes	207
PGAH/PAAm	Sandwich PAAm-PGAH-E-PGAH-PAAm	18 to 96% residual activity with higher values in the presence of cofactors, depending on the engagement conditions	l-LDH, penicillinase	208
Poly-l-lysine	The counter polyelectrolyte such as poly(4-styrene-sulphonate) was used	The poly-l-lysine was dissolved in the enzyme solution and coated on the solid surface, followed by immersing the coated enzyme in the solution containing the counter polyelectrolyte such as poly(4-styrenesulphonate)	Lactate oxidase, choline oxidase, and glucose oxidase	209
PVC film	Blocked-isocyanate prepolymer	–	Glucose oxidase	213
Fibres	Poly(2-HEMA-co-EEA-co-HPA)	Soaked in fibres, followed by coating	Asparaginase	214
Activated charcoal	Ultrathin cellulose nitrate	Activity retention is 28.1% of the activity of urease in solution	Urease	215
Copolymers of MMA-TMP and acrylonitrile	1% Xylylene diamine aqueous solution in the presence of 1% glutaraldehyde	Total activity retention was about 60%. The immobilized enzyme was much more stable than the enzyme without the top layer	α-Amylase	216
Celite	ENTP 2000	The immobilized enzyme has higher operational stability than the adsorbed lipase on Celite	Lipase	217

Immobilization of α-amylase on polymeric granules resulted in 35% loss of activity whereas formation of a surface layer by cross-linking xylene diamine with glutaraldehyde resulted in 5% loss of activity [216], suggesting that the loss of activity is mainly governed by the adsorption step. Immobilizing urease by adsorption on to activated charcoal and subsequent coating with cellulose nitrate resulted in 72% loss of activity, suggesting that the loss of activity might be ascribed to diffusion constraints within the cellulose nitrate layer [215].

The occurrence of diffusion constraints as a result of the formation of the extra layer might, however, constitute an extra benefit of activity retention. For instance, β-glucosidase was immobilized on con A Sepharose then encapsulated in the gel, leading to the formation of immobilized enzyme with size-dependent activity. Thus, the diffusion limitation might be the dominant factor affecting enzyme activity. The immobilized enzyme did not, however, suffer from any substrate inhibition effect, compared with the free or adsorbed enzyme for substrate concentrations of 10–180 mM.

A striking effect is worth mentioning – formation of the outer layer usually improves the stability of the enzyme, for example thermostability and operational stability. The stabilization effect might depend on the method used to form the outer layer. Formation of an extra layer by cross-linking of the monomers in the proximity of the enzymes usually substantially stabilizes the enzyme [216, 217]. The presence of the extra layer might also prevent enzyme leakage, because of the weak interaction between the enzyme and the carrier, thus increasing operational stability [217].

5.5
References

1 Chang TMS, McIntosh FC, Mason FG: Semipermeable microcapsules: preparation and properties. *Can J Physiol Pharmacol* 1996, 44:115–119
2 Jen AC, Wake MC, Milos AG: Review: hydrogels for cell immobilization. Biotechnol Bioeng 1995, 50:357–364
3 Cousineau J, Chang TMS: Formation of amino acid from urea and ammonia by sequential enzyme reaction using a microencapsulated multi-enzyme system. *Biochem Biophys Res Commun* 1977, 79:24–31
4 Kennedy JF, Roig MG: Principles of immobilisation of enzymes. In: Wiseman A (Ed) *Handbook of enzyme Biotechnology*, 3rd edn. Ellis Horwood, 1991
5 Chang TMS: Semipermeable microcapsules. *Science* 1964, 146:524–525
6 Stefuca VGP, Kurillova L, Dautzenberg H, Polakovic M, Bales V: Polyelectrolyte complex capsules as a material for enzyme immobilization: catalytic properties of encapsulated lactate dehydrogenase. *Appl Biochem Biotechnol* 1991, 30:313–324
7 Klein J, Stock J, Vorlop KD: Pore size and properties of spherical Ca-alginate biocatalysts. *Eur J Appl Microbiol Biotechnol* 1983, 18:86–91
8 Chang HN, Seong GH, Yoo IK, Park JK, Seo JH: Microencapsulation of recombinant *Saccharomyces cerevisiae* cells with invertase activity in liquid-core alginate capsules. *Biotechnol Bioeng* 1996, 51:157–162
9 Stefuca V, Gemeiner P, Kurillova L, Dautzenberg H, Polakovic M, Bales V: Polyelectrolyte complex capsules as a

material for enzyme immobilization: catalytic properties of encapsulated lactate dehydrogenase. *Appl Biochem Biotechnol* 1991, 30:313–324

10 Siso MIG, Lang E, Carreno-Gomez B, Becerra M, Espinar FO, Mendez JB: Enzyme encapsulation on chitosan microbeads. *Process Biochem (Oxford)* 1997, 32:211–216

11 Lee KS, Choi MR, Lim HS: Development of coencapsulating technology for the production of chitosanoligosaccharides. *Biotechnol Bioprocess Eng* 2000, 5:345–349

12 Yuan ZY, Li X, Li S, Wu Y: Microencapsulation of biocatalysts with ALG-PLL technique. *Shengwu Gongcheng Xuebao* 1990, 6:81–84

13 Iso M, Shirahase T, Hanamura S, Urushiyama S, Omi S: Immobilization of enzyme by microencapsulation and application of the encapsulated enzyme in the catalysis. *J Microencap* 1989, 6:165–176

14 Shrier AL: Liquid membranes in enzyme research. *Biotechnol Bioeng Symp* 1972, 3:323–326

15 May SW, Li NN: Encapsulation of enzymes in liquid membrane emulsions. *Enzym Eng* 1974:77–82

16 Scheper T, Halwachs W, Schuegerl K: Production of *l*-amino acids by continuous enzyme-catalyzed *d,l*-amino ester hydrolysis using liquid membrane emulsions. *Chemie Ingenieur Technik* 1982, 54:696–697

17 Chang TMS: Stabilisation of enzymes by microencapsulation with a concentrated protein solution or by microencapsulation followed by cross-linking with glutaraldehyde. *Biochem Biophys Res Commun* 1971, 44:1531–1536

18 Zarborsky O: *Immobilised Enzymes*. CRC Press, Cleveland, 1973

19 Cohen S, Allcock HR, Langer R: Cell and enzyme immobilization in ionotropic synthetic hydrogels. In: Hincal AA, Kas HS (Eds) *Recent Adv Pharm Ind Biotechnol, Minutes 6th Int Pharm Technol Symp 1992*, 1993, pp 36–48

20 Kokufuta E: Polyelectrolyte-coated microcapsules and their potential applications to biotechnology. *Bioseparation* 1998, 7:241–252

21 Lee KH, Lee, Pat M: Effect of pH on the preparation of poly-*l*-lysine-stabilised calcium alginate beads for immobilization of aminoacylase. *Biotechnol Tech* 1993, 7:131–136

22 Simsek-Ege FA, Bond GM, Stringer J: Matrix molecular weight cut-off for encapsulation of carbonic anhydrase in polyelectrolyte beads. *J Biomater Sci Polym Ed* 2002, 13:1175–1187

23 Furukawa S, Ono T, Ijima H, Kawakami K: Effect of imprinting sol–gel immobilized lipase with chiral template substrates in esterification of (*R*)-(+)- and (*S*)-(–)-glycidol. *J Mol Catal B: Enzym* 2002, 17:23–28

24 Suwasono S, Rastall RA: Synthesis of oligosaccharides using immobilised 1,2-mannosides from *Aspergillus phoenicis*. Immobilisation-dependent modulation of product spectrum. *Biotechnol Lett* 1998, 20:15–17

25 Dautzenberg H, Foerster M, Hoffmann T, Schellenberger A, Doepfer KP, Mansfeld J, et al. Immobilization of microorganisms and co-immobilization of enzymes by two-step microencapsulation. *Ger Offen (East) DD 285,370 A5*, 1990

26 Bader H, Rueppel D, Walch A. Immobilization of biologically active material in polysaccharide capsules. *Ger Offen DE 3,615,043*, 1987

27 Mojovic L, Siler-Marinkovic S, Kukic G, Bugarski B, Vunjak-Novakovic G. Enzymic interesterification of the palm oil midfraction in an air-lift reactor by immobilized lipase. *Hemijska Industrija* 1993, 47:158–162

28 Khare SK, Vaidya S, Gupta MN: Entrapment of proteins by aggregation within Sephadex beads. *Appl Biochem Biotechnol* 1991, 27:205–216

29 Ristau O, Pommerening K, Jung C, Rein H, Scheler W: Activity of urease in microcapsules. *Biomed Biochim Acta* 1985, 44:1105–1111

30 Sundaram PV. Kinetic properties of microencapsulated urease. *Biochim Biophys Acta* 1973, 321(1),319–328

31 Huettermann A, Milstein O, Nicklas B. Immobilization of lignin-modifying

enzymes for use in organic solvents. *Ger Offen DE 3,827,001*, 1989

32 Milstein O, Nicklas B, Huettermann A: Oxidation of aromatic compounds in organic solvents with laccase from *Trametes versicolor*. *Appl Microb Biotechnol* 1989, 31:70–74

33 Rony PR: Multiphase catalysis. II. Hollow fiber catalysts. *Biotechnol Bioeng* 1971, 13:431–447

34 Parthasarathy RV, Martin CR. Enzyme and chemical encapsulation in polymeric microcapsules. *J Appl Polym Sci* 1996, 62:875–886

35 Tiourina OP, Antipov AA, Sukhorukov GB, Larionova NI, Lvov Y, Mohwald H: Entrapment of α-chymotrypsin into hollow polyelectrolyte microcapsules. *Macromol Biosci* 2001, 1:209–214

36 Shellenberg NN, Nys PS, Savitskaya EM: Penicillin amidase from *Escherichia coli*. Some physicochemical properties of the enzyme incorporated in polyacrylamide gel. *Antibiotiki (Moscow)* 1977, 22:125–130

37 Yuan ZY, Li X, Li SY, Wu Y: Microencapsulation of biocatalysts with ALG-PLL technique. *Shengwu Gongcheng Xuebao* 1990, 6:81–84

38 Habibi-Moini S, D'mello AP: Evaluation of possible reasons for the low phenylalanine ammonia lyase activity in cellulose nitrate membrane microcapsules. *Int J Pharm* 2001, 215:185–196

39 Kondo T, Muramatsu N: Enzyme inactivation in microencapsulation. In: Nixon JR (Ed) *Microencapsulation*. Marcel Dekker, New York, 2001, pp 67–75

40 Okahata Y, Hatano A, Ijiro K: Enhacing enantioselectivity of a lipid-coated lipase via imprinting method for esterification in organic solvents. *Tetrahedron Asymmetry* 1995, 6:1311–1322

41 Kumakura M, Kaetsu, TI: Encapsulation of enzymes by a radiation technique. *Biotechnol Lett* 1984, 6:409–412

42 Wood DA, Whateley TL: A study of enzyme and protein microencapsulation – some factors affecting the low apparent enzymic activity yields. *J Pharm Pharmacol* 1982, 34:552–557

43 Ristau O, Pommerening K, Jung C, Rein H, Scheler W: Activity of urease in microcapsules. *Biomed Biochim Acta* 1985, 44:1105–1111

44 Groboillot AF, Champagne CP, Darling GD, Poncelet D: Membrane formation by interfacial cross-linking of chitosan for encapsulation of *Lactococcus lactis*. *Biotechnol Bioeng* 1993, 42:1157–1163

45 Rambourg P, Levy J, Levy MC: Microencapsulation. III: Preparation of invertase microcapsules. *J Pharm Sci* 1982, 71:753–758

46 Chang TMS: l-Asparaginase immobilized within semipermeable microcapsules. In vitro and in vivo stability. *Enzyme* 1973, 14:95–104

47 Sundaram PV: Kinetic properties of microencapsulated urease. *Biochim Biophys Acta* 1973, 321:319–328

48 Liu LS, Ito Y, Imanishi Y: Biological activity of urokinase immobilized to crosslinked poly(2-hydroxyethyl methacrylate). *Biomaterials* 1991, 12:545–549

49 Monshipouri M, Neufeld RJ: Activity and distribution of urease following microencapsulation within polyamide membranes. *Enzym Microb Technol* 1991, 13:309–313

50 Poncelet D, Alexakis T, de Smet BP, Neufeld RJ. Microencapsulation within crosslinked polyethyleneimine membranes. *J Microencap* 1994:11:31–40

51 Monshipouri M, Neufeld RJ: Kinetics and activity distribution of urease coencapsulated with hemoglobin within polyamide membranes. *Appl Biochem Biotechnol* 1992, 32:111–126

52 Takahashi T, Takayama K, Machida Y, Nagai T: Characteristics of polyion complexes of chitosan with sodium alginate and sodium polyacrylate. *Int J Pharm* 1990, 61:35–41

53 Shao Wen, Leong KW: Enzymically degradable synthetic polymers. *Mater Res Soc Symp Proc* 1995, 394:199–204

54 Dashevsky A: Protein loss by the microencapsulation of an enzyme (lactase) in alginate beads. *Int J Pharm* 1998, 161:1–5

55 Blandino A, Macias M, Cantero D: Immobilization of glucose oxidase within calcium alginate gel capsules. *Process Biochem (Oxford)* 2001, 36:601–606

56 Tauriseren A, Dogan S: Immobilisation of invertase with Ca alginate gel capsules. *Process Biochem* 2001, 36:1081–1083

57 Lee KH, Lee PM, Siaw YS: Immobilization of aminoacylase by encapsulation in poly-*l*-lysine-stabilised calcium alginate beads. *J Chem Technol Biotechnol* 1993, 57:27–32

58 Stefuca V, Gemeiner P, Kurillova L, Dautzenberg H, Polakovic M, Bales V: Polyelectrolyte complex capsules as a material for enzyme immobilization: catalytic properties of encapsulated lactate dehydrogenase. *Appl Biochem Biotechnol* 1991, 30:313–324

59 Mansfeld J, Foerster M, Schellenberger A, Dautzenberg H: Immobilization of invertase by encapsulation in polyelectrolyte complexes. *Enzym Microb Technol* 1991, 13:240–244

60 Matteau PP, Saddler JN: Glucose production using immobilized mycelial-associated β-glucosidase of Trichoderma E58, *Biotechnol Lett* 1982, 4:513–518

61 Miguez MJB, Rodrigues BC, Sanchez, NM, Laranjeira MCM: Preparation and scanning electronic microscopy study of chitosan/poly(vinyl alcohol)-encapsulated crude urease extract. *J Microencap* 1997, 14:639–646

62 Lim F, Sun AM: Microencapsulated islets as bioartificial endocrine pancrease. *Science* 1980, 210:908–910

63 Chang TMS, Prakash S: Therapeutic uses of microencapsulated genetically engineered cells. *Mol Med Today* 1998, 4:221–227

64 Onouchi T, Sugai H, Sekiguchi K, Hosoda Y, Yoshida E: Water-soluble microcapsules containing enzymes. *Eur Pat Appl EP 266796*, 1988

65 Goosen MEA, King GA, McKnight CA, Marcotte N: Animal cell culture engineering using alginate polycation microcapsules of controlled membrane molecular weight cutoff. *J Membr Sci* 1989, 41:323–343

66 Jankowski T, Zielinska M, Wysakowska A: Encapsulation of lactic acid bacteria with alginate/starch capsules. *Biotechnol Tech* 1997, 11:31–34

67 Chang HN, Seong GH, Yoo IK, Park JK, Seo JH: Microencapsulation of recombinant *Saccharomyces cerevisiae* cells with invertase activity in liquid core alginate capsules. *Biotechnol Bioeng* 1996, 51:157–162

68 Blandino MM, Cantero D: Formation of calcium alginate, gel capsules: influence of sodium alginate and $CaCl_2$ concentration on gelation kinetics. *J Biosci Bioeng* 1999, 88:686–689

69 Stengelin M, Patel RN: Phenylalanine dehydrogenase catalyzed reductive amination of 6-(1′,3′-dioxolan-2′-yl)-2-keto-hexanoic acid to 6-(1′,3′-dioxolan-2′-yl)-2*S*-aminohexanoic acid with NADH regeneration and enzyme and cofactor retention. *Biocatal Biotransform* 2000, 18:373–400

70 Blandino A, Macias M, Cantero D: Kinetic behaviour of glucose oxidase-catalase enzymatic system co-immobilized within calcium alginate gel capsules. *Mededelingen – Faculteit Landbouwkundige en Toegepaste Biologische Wetenschappen (Universiteit Gent)* 2000, 65:251–256

71 Blandino MM, Cantero D: Glucose oxidase release from calcium alginate gel capsules. *Enzyme Microb Tech* 2000, 27:319–324

72 Prakash S, Chang TMS: Preparation and in vitro analysis of microencapsulated genetically engineered *E. coli* DH5 cells for urea and ammonia removal. *Biotech Bioeng* 1995, 46:621–626

73 Gregoriadis G: Medical applications of liposome-entrapped enzymes. *Methods Enzymol* 1976, 44:698–709

74 Steger LD, Desnick RJ: Enzyme therapy. VI: Comparative in vivo fates and effects on lysosomal integrity of enzyme entrapped in negatively and positively charged liposomes. *Biochim Biophys Acta* 1977, 464:530–546

75 Sada E, Katoh S, Terashima M, Tsukiyama, K: Entrapment of an ion-dependent enzyme into reverse-phase evaporation vesicles. *Biotechnol Bioeng* 1988, 32:826–830

76 Madeira VMC: Incorporation of urease into liposomes. *Biochim Biophys Acta* 1977, 499:202–211

77 Yamada Y, Kuboi R, Komasawa I: Increased activity of *Chromobacterium viscosum* lipase in aerosol OT reverse

micelles in the presence of nonionic surfactants. *Biotechnol Prog* 1993, 9:468–472

78 Walde P, Ichikawa S: Enzymes inside lipid vesicles: preparation, reactivity and applications. *Biomol Eng* 2001, 18:143–177

79 Carvalho CML, Cabral JMS: Reverse micelles as reaction media for lipases. *Biochimie* 2000, 82:1063–1085

80 Vulfson EN, Ahmed G, Gill I, Kozlov IA, Goodenough PW, Law BA: Alterations of the catalytic properties of polyphenoloxidase in detergentless microemulsions and ternary water–organic solvent mixtures. *Biotechnol Lett* 1991, 13:91–96

81 Tsai SW, Chiang CL: Kinetics, mechanism, and time course analysis of lipase-catalyzed hydrolysis of high concentration olive oil in AOT-isooctane reversed micelles. *Biotechnol Bioeng* 1991, 38:206–211

82 Pinto-Sousa AM, Cabral JMS, Aires-Barros MR: Ester synthesis by a recombinant cutinase in reversed micelles of a natural phospholipids. *Biocatalysis* 1994, 9:169–179

83 Maestro M, Walde P: Application of a simple diffusion model for the enzymatic activity of α-chymotrypsin in reverse micelles. *J Colloid Interface Sci* 1992, 154:298–302

84 Fletcher PDI, Freedman RB, Mead J, Oldfield C, Robinson BH: Reactivity of α-chymotrypsin in water-in-oil microemulsions. *Colloids Surf* 1984, 10:193–203

85 Peng Q, Luisi PL: The behavior of proteases in lecithin reverse micelles. *Eur J Biochem* 1990, 188:471–480

86 Hoppert M, Braks I, Mayer F: Stability and activity of hydrogenases of *Methanobacterium thermoautotrophicum* and *Alcaligenes eutrophus* in reversed micellar systems. *FEMS Microb Lett* 1994, 118:249–254

87 Khmelnitsky YL, Gladilin AK, Neverova IN, Levashov AV, Martinek K: Detergentless microemulsions as media for enzymatic reactions: catalytic properties of laccase in the ternary system hexane-2-propanol-water. *Collect Czech Chem Commun* 1987, 55:555–563

88 Meyer ER, Scheper T, Hitzmann B, Schuegerl K: Immobilization of enzymes in liquid membranes for enantioselective hydrolysis. *Biotechnol Techn* 1988, 2:127–132

89 Mojovic L, Silermarinkovic S, Kukic G, Bugarski B, Vunjak-Novakovic G: *Rhizopus arrhizus* lipase-catalyzed interesterification of palm oil midfraction in a gas-lift reactor. *Enzym Microb Technol* 1994, 16:159–162

90 Khmelnitsky YL, Zharinova IN, Berezin IV, Levashov AV, Martinek K: Detergentless microemulsions: a new microheterogeneous medium for enzymatic activity. *Ann NY Acad Sci* 1987, 501:161–164

91 Madamwar DB, Bhatt JP, Ray RM. Activation and stabilisation of invertase entrapped into reversed micelles of sodium lauryl sulphate and sodium tauroglycholate in organic solvents. *Enzyme Microb Technol* 1988, 10:302–305

92 Shapiro YE, Pykhteeva EG: Immobilization of α-chymotrypsin into poly(N,N-diallyl-N,N-didodecyl ammonium bromide)/surfactant nanocapsules. *Appl Biochem Biotechnol* 1998, 74:67–84

93 Ozden MY, Hasirci VN: Enzyme immobilization in polymer-coated liposomes. *Br Polym J* 1990, 23:229–234

94 Gladilin AK, Khmelnitsky YL, Roubailo VL, Martinek K, Levashov AV: Polymeric reversed micelles based on modified polyethyleneimine: formation and usage as nonaqueous media for enzymic reactions. *Bioorganicheskaya Khimiya* 1992, 18:1170–1185

95 Khmelnitsky YL, Neverova IN, Gedrovich AV, Polyakov VA, Levashov AV, Martinek K: Catalysis by α-chymotrypsin entrapped into surface-modified polymeric nanogranules in organic solvent. *Eur J Biochem* 1992, 210:751–757

96 Khmelnitsky YL, Neverova IN, Momtcheva R, Yaropolov AI, Belova AB Levashov AV, Martinek K Surface-modified polymeric nanogranules containing entrapped enzymes, a novel biocatalyst for use in organic media, *Biotechnol Lett* 1989, 3:275–280

97 Khmelnitsky YL, Gladilin AK, Roubailo VL, Martinek K, Levashov AV: Reversed micelles of polymeric surfactants in nonpolar organic solvents: a new microheterogeneous medium for enzymatic reactions. *Eur J Biochem* 1992, 206:737–745

98 Nardin C, Thoeni S, Widmer J, Winterhalter M, Meier W: Nano reactors based on polymerized ABA-triblock copolymer vesicles. *Chem Commun* 2000:1433–1444

99 Fadnavis NW, Luisi PL: Immobilized enzymes in reversed micelles: studies with gel-entrapped trypsin and α-chymotrypsin in AOT reverse micelles. *Biotechnol Bioeng* 1989, 33:277–1282

100 Kokufuta E, Shimizu N, Nakamura I: Preparation of polyelectrolyte-coated pH-sensitive poly(styrene) microcapsules and their application to initiation-cessation control of an enzyme reaction. *Biotechnol Bioeng* 1988, 32:289–294

101 Sessa G, Weissmann G: Incorporation of lysozyme into liposomes. A model for structure-linked latency. *J Biol Chem* 1970, 245:3295–3301

102 Gregoriadis G, Leathwood PD, Ryman BE: Enzyme entrapment in liposomes. *FEBS Lett* 1971, 14:95–99

103 Walde P, Goto A, Monnard P-A, Wessicken M, Luisi PL: Oparin's reactions revisited: enzymatic synthesis of poly(adenylic acid) in micelles and self-reproducing vesicles. *J Am Chem Soc* 1994, 116:7541–7547

104 Regen SL, Singh M, Sameul NK: Functionalized polymeric liposomes, efficient immobilisation of α-chymotrypsin. *Biochem Biophys Res Commun* 1984, 119:646–651

105 Bogadanov AA, Klibanov AL, Torchillin VP: Immobilisation of α-chymotrypsin on sucrose stearate-palmitate containing liposomes. *FEBS Lett* 1984, 175:178–182

106 Nasseau M, Boublik Y, Meier W, Winterhalter M, Fournier D: Substrate-permeable encapsulation of enzymes maintains effective activity, stabilises against denaturation, and protects against proteolytic degradation. *Biotechnol Bioeng* 2001, 75: 615–618

107 Niedermann G, Weissig V, Sternberg B, Lash J: Carboxyacyl derivatives of cardiolipin as four-tailed hydrophobic anchors for the covalent coupling of hydrophilic proteins to liposomes. *Biochim Biophys Acta* 1991, 1070:401–408

108 Walde P, Marzetta B: Bilayer permeability-based substrate selectivity of an enzyme in liposomes. *Biotechnol Bioeng* 1998, 57:216–219

109 Blocher M, Walde P, Dunn IJ: Modeling of enzymatic reactions in vesicles: the case of α-chymotrypsin. *Biotechnol Bioeng* 1999, 62:36–43

110 Ramundo-Orlando A, Morbiduzzi U, Mossa G, D'Inzeo G: Effect of low frequency, low amplitude magnetic fields on the permeability of cationic liposomes entrapping carbonic anhydrase I. Evidence for charged lipid involvement. *Bioelectromagnetics* 2000, 21:491–498

111 Annesini MC, Di Marzio L, Finazzi-Agro A, Serafino AL, Mossa G: Interaction of cationic phospholipid vesicles with carbonic anhydrase. *Biochem Mol Biol Int* 1994, 32:87–94

112 Szoka F Jr, Papahadjopoulos D: Procedure for preparation of liposomes with large internal aqueous space and high capture by reverse-phase evaporation. *Proc Natl Acad Sci USA* 1978, 75:4194–4198

113 Formelova J, Breier A, Gemeiner P, Kurillova L: *Collect Czech Chem Commun* 1991, 56:712–717

114 Freytag JW: Large unilamellar lipid vesicles for use in therapeutic and diagnostic medicine. *J Microencapsul* 1985, 2:31–38

115 Jones MN, Hill KJ, Kaszuba M, Creeth JE: Antibacterial reactive liposomes encapsulating coupled enzyme systems, *Int J Pharm* 1998, 162:107–117

116 Hill KJ, Kaszuba M, Creeth JE, Jones MN: Reactive liposomes encapsulating a glucose oxidase-peroxidase system with antibacterial activity. *Biochim Biophys Acta: Biomembr* 1997, 1326:37–46

117 Petrikovics I, Hong K, Omburo G, Hu QZ, Pei L, McGuinn WD, Sylvester D,

Tamulinas C, Papahadjopoulos D, Jaszberenyi JC, Way JL: Antagonism of paraoxon intoxication by recombinant phosphotriesterase encapsulated within sterically stabilised liposomes. *Toxicol Appl Pharmacol* 1999, 156:56–63

118 Kaszuba M, Jones MN: Hydrogen peroxide production from reactive liposomes encapsulating enzymes. Antagonism of paraoxon intoxication by recombinant phosphotriesterase encapsulated within sterically stabilised liposomes. *Biochim Biophys Acta* 1999, 141:221–228

119 Kozlova NO, Bruskovskaya IB, Melik-Nubarov NS, Yaroslavov AA Kabanov VA: Catalytic properties and conformation of hydrophobized α-chymotrypsin incorporated into a bilayer lipid membrane. *FEBS Lett* 1999, 461:141–144

120 Espinola LG, Wider EA, Stella AM, Batlle AM, Del C: Enzyme replacement therapy in porphyrias – II: Entrapment of δ-aminolaevulinate dehydratase in liposomes. *Int J Biochem* 1983, 15:439–445

121 Lariviere B, El Soda M, Soucy Y, Trepanier G, Paquin P, Vuillemard JC: Microfluidized liposomes for acceleration of cheese making. *Int Dairy J* 1991, 1:111–124

122 Oberholzer T, Meyer E, Amato I, Lustig A, Monnard P-A: Enzymatic reactions in liposomes using the detergent-induced liposome loading method. *Biochim Biophys Acta* 1999, 1416:57–68

123 Matsuzaki M, McCafferty F, Karel M: *Int J Food Sci Technol* 1989, 24:451–460

124 Chaize B, Fournier D: Sorting out molecules reacting with acetylcholinesterase by enzyme encapsulation in liposome. *Biosens Bioelectron* 2004, 20:628–632

125 Lissant KJ, Peace BW, Wu SH, Mayhan KG: Structure of high-internal phase-ratio emulsions. *Colloid Interface Sci* 1974, 47:416–423

126 Lamb SB, Stuckey DC: Enzyme immobilisation on colloidal liquid aphrons CLAs: the influence of protein properties. *Enzyme Microb Technol* 1999, 24:541–548

127 Lamb SB, Stuckey DC: Enzyme immobilization on colloidal liquid aphrons (CLAs): the influence of system parameters on activity. *Enzyme Microb Technol* 2000, 26:574–581

128 Lye GJ, Pavlou OP, Rosjidi M, Stuckey DC: Immobilisation of *Candida cylindracea* lipase on colloidal liquid aphrons CLAs and development of a continuous CLA-membrane reactor. *Biotechnol Bioeng* 1996, 51:69–78

129 Lye GJ: Stereoselective hydrolysis of dl-phenylalanine methyl ester and separation of l-phenylalanine using aphron-immobilised α-chymotrypsin. *Biotechnol Lett* 1997, 11:611–616

130 Lamb SB, Lamb DC, Kelly SL, Stuckey DC: Cytochrome P450 immobilisation as a route to bioremediation/biocatalysis, *FEBS Lett* 1998, 431:343–346

131 Chang TMS, Bourget L, Lister C: A new theory of enterorecirculation of amino acids and its use for depleting unwanted amino acids using oral enzyme artificial cells, as in removing phenylalanine in phenylketonuria. *Artif Cells Blood Subs Immob Biotech* 1995, 23:1–21

132 Chang TMS: l-Asparaginase immobilized within semipermeable microcapsules. In vitro and in vivo stability. *Enzyme* 1973, 14:95–104

133 Mori T, Sato T, Matuo Y, Tosa T, Chibata I: Preparation and characteristics of microcapsules containing asparaginase. *Biotechnol Bioeng* 1972, 14:663–673

134 Bourget L, Chang TMS: Phenylalanine ammonia lyase immobilized in semipermeable microcapsules for enzyme replacement in phenylketonuria. *FEBS Lett* 1985, 180:5–8

135 Habibi-Moini S, D'mello AP: Evaluation of possible reasons for the low phenylalanine ammonia lyase activity in cellulose nitrate membrane microcapsules. *Int J Pharm* 2001, 215:185–196

136 Chang TMS, Macintosh FC, Mason SG: Semipermeable aqueous microcapsules. I. Preparation and properties. *Can J Physiol Pharmacol* 1966, 44:115–128

137 Bourget L, Chang TM: Artificial cell-microencapsulated phenylalanine

ammonia-lyase. *Appl Biochem Biotechnol* 1984, 10:57–59

138 Habibi-Moini S, D'mello AP: Evaluation of possible reasons for the low phenylalanine ammonia lyase activity in cellulose nitrate membrane microcapsules, *Int J Pharm* 2001, 215:185–196

139 Kondo T, Muramatsu N: Enzyme inactivation in microencapsulation. In: Nixon JR (Ed) *Microencapsulation*. Marcel Dekker, New York, 1976, pp 67–75

140 Poznansky MJ, Chang TMS: Comparison of the enzyme kinetics and immunological properties of catalase immobilized by microencapsulation and catalase in free solution for enzyme replacement. *Biochim Biophys Acta* 1974, 334:103–115

141 Khanna R, Chang TMS: Characterization of l-histidine ammonia lyase immobilized by microencapsulation in artificial cells: preparation, kinetics, stability and in vitro depletion of histidine. *Int J Artif Organs* 1990, 13:189–195

142 Wang XL, Shao JY: New preparation for oral administration of digestive enzymes. Lactase complex microcapsules. *Biomater Artif Cells Immob Biotech* 1993, 21:637–646

143 Langley J, Symes KC: Enzyme-containing particles for liquid detergent compositions. *Eur Pat Appl* 1990, 6

144 Chang TMS, Lister C: Plasma/ intestinal concentration patterns suggestive of enteroportal recirculation of amino acids: effects of oral administration of asparaginase, glutaminase, and tyrosinase immobilized by microencapsulation in artificial cells. *Biomater Artif Cells Artif Organs* 1989, 16:915–926

145 Crotts G, Park TC: Preparation of porous biodegradable polymeric hollow microspheres. *J Controlled Release* 1995, 35:91–105

146 Visage CL, Quaglia F, Dreux M, Ounnar S, Breton P, Bru N, Couvreur P, Fattal E: Novel microparticulate system made of poly(methylidene malonate). *Biomaterials* 2001, 22:2229–2238

147 Pistel KF, Bittner B, Koll H, Winter G, Kissel T: Biodegradable recombinant human erythropoietin loaded microspheres prepared from linear and star-branched block copolymers: Influence of encapsulation technique and polymer composition on particle characteristics. *J Controlled Release* 1999, 59:309–325

148 Castellanos IJ, Cruz G, Crespo R, Griebenow K: Encapsulation-induced aggregation and loss in activity of γ-chymotrypsin and their prevention. *J Controlled Release* 2002, 81:307–319

149 Lee J-H, Park TG, Choi H-K: Effect of formulation and processing variables on the characteristics of microspheres for water-soluble drugs prepared by w:o:o double emulsion solvent diffusion method. *Int J Pharm* 2000, 196:75–83

150 Sah H: Microencapsulation techniques using acetate as dispersed solvent effect of its extraction rate on the characteristics of PLGA microspheres. *J Controlled Release* 1997, 47:233–245

151 Caliceti P, Veronese FM, Lora S: Polyphosphazene microspheres for insulin delivery. *Int J Pharm* 2000, 211:57–65

152 Perumal D: Microencapsulation of ibuprofen and Eudragit® RS 100 by the emulsion solvent diffusion technique. *Int J Pharm* 2001, 218:1–11

153 Witschi C, Doelker E: Influence of the microencapsulation method and peptide loading on poly(lactic acid) and poly(lactic-co-glycolic acid) degradation during in vitro testing. *J Controlled Release* 1998, 51:327–341

154 Sturesson C, Carlfors J: Incorporation of protein in PLG-microspheres with retention of bioactivity. *J Controlled Release* 2000, 67:171–178

155 Iso M, Shirahase T, Hanamura S, Urushiyama S, Omi S: Immobilization of enzyme by microencapsulation and application of the encapsulated enzyme in the catalysis. *J Microencap* 1989, 6:165–176

156 Kim HK, Park TG: Microencapsulation of dissociable human growth hormone aggregates within poly(d,l-lactic-co-glycolic acid) microparticles for sustained release. *Int J Pharm* 2001, 229:107–116

157 Li XH, Zhang YH, Yan RH, Jia WX, Yuan ML, Deng XM, Huang ZT: Influence of process parameters on the protein stability encapsulated in poly-*dl*-

lactide–poly(ethylene glycol) microspheres. *J Controlled Release* 2000, 68:41–52

158 Lamprecht NU, Hombreiro Pérez M, Lehr CM, Hoffman M, Maincent P: Influences of process parameters on nanoparticle preparation performed by a double emulsion pressure homogenization technique. *Int J Pharm* 196:177–182

159 Dunn RL, Tripton AJ, Southard GL: Biodegradable polymer composition. *Euro Pat Appl, EP 0,539,751 A1*, 1992

160 Lambert WJ, Peck KD: Development of an in-situ forming biodegradable polylactide-co-glycolide system for the controlled release of proteins. *J Controlled Release* 1995, 33:189–195

161 Sah H, Smith MS, Chern RT: A novel method of preparing PLGA microcapsules utilizing methyl ethyl ketone. *Pharm Res* 1996, 13:360–367

162 Caruso F, Caruso RA, Donath Ed, Mohwald H, Sukhorukov G: Nanocomposite multilayers on decomposable colloidal templates. *PCT Int Appl, WO 9,947,253*, 1999

163 Donath E: From polyelectrolyte capsules to drug carriers and artificial cells. *Abstracts of Papers, 223rd ACS National Meeting, Orlando, FL, USA*, April 7–11, 2002

164 Tiourina OP, Sukhorukov GB, Antipov AA, Donath E, Mohwald H, Larionova NI: Encapsulation of α-chymotrypsin on to the hollow polyelectrolyte microcapsules. *Proc 28th Int Symp Controlled Release of Bioactive Materials and 4th Consumer & Diversified Products Conference, San Diego, CA, USA*, vol 2, 2001, pp 1400–1401

165 Cochran JK: Ceramic hollow spheres and their applications. *Curr Opin Solid State Mater Sci* 1998, 3:474–479

166 Hurysz KM et al: Steel and titanium hollow sphere foams. In: Schwarz D, Shih DS, Evans A, Wadley HNG (Eds) *MRS Symp Proc, vol 521: Porous and cellular materials for structural applications*, 1998, pp 191–203

167 Sun L, Crooks RM, Chechik V: Preparation of polycyclodextrin hollow spheres by templating gold nanoparticles. *Chem Commun* 2001, 359–360

168 Hotz JMW: Vesicle-templated polymer hollow spheres. *Langmuir* 1998, 14:1031–1036

169 Yang XM, Chaki TK: Hollow lead zirconate tiranate microspheres prepared by sol–gel/emulsion technique. *Mater Sci Eng B* 1996, 39:123–128

170 Buchholz CM: R: Verfahrenstechnische Auslegung einer Apparatur zur Herstellung microverkapselter Biokatalysatoren. *Chem-Tech* 1993, 65:1221–1223

171 Donath E, Sukhorukov GB, Caruso F, Davis SA, Möhwald H: Novel hollow polymer shells by colloid-templated assembly of polyelectrolytes *Angew Chem Int Ed* 1998, 37:2201–2205

172 Narayan P, Wheatley MAC: Preparation and characterization of hollow microcapsules for use as ultrasound contrast agents. *Polym Eng Sci* 1999, 39: 2242–2255

173 Rengel RG, Barisic K, Pavelic Z, Grubisic TZ, Cepelak I, Filipovic-Grcic J: High efficiency entrapment of superoxide dismutase into mucoadhesive chitosan-coated liposomes. *Eur J Pharm Sci* 2002, 15:441–448

174 Lu Z, Bei JZ, Wang SG: A method for the preparation of polymeric nanocapsules without stabiliser, *J Controlled Release* 1999, 61:107–112

175 Hildebrand GE, Tack JW: Microencapsulation of peptides and proteins. *Int J Pharm* 2001, 196:173–176

176 Lasic DD: *Liposomes: from physics to applications*. Elsevier, Amsterdam, 1993

177 Rilling P, Walter T, Pommersheim R, Vogt W: Encapsulation of cytochrome C by multilayer microcapsules. A model for improved enzyme immobilization. *J Membr Sci* 1997, 129:283–287

178 Hearn E, Neufeld RJ: Poly(methylene co-guanidine) coated alginate as an encapsulation matrix for urease, *Process Biochem* 2000, 35:1253–1260

179 Kokufuta E, Shimizu N, Tanaka H, Nakamura I: Use of polyelectrolyte complex-stabilised calcium alginate gel for entrapment of α-amylase. *Biotechnol Bioeng* 1988, 32:756–759

180 Wang FF, Wu CR, Wang YJ: Preparation and application of poly(vinylamine)/ alginate microcapsules to culturing of a

mouse erythroleukemia cell line. *Biotechnol Bioeng* 1992, 40:1115–1118

181 Lacik L, Brissova M, Anilkumar AV, Powers AC, Wang T: New capsule with tailored properties for the encapsulation of living cells. *J Biomed Mater Res* 1998, 39:52–60

182 Brissova M, Lacik, Powers AC, Anilkumar AV, Wang T: Control and measurement of permeability for design of microcapsule cell delivery system. *J Biomed Mater Res* 34 1988, 61–70

183 Tsou CL: Inactivation precedes overall molecular conformation changes during enzyme denaturation. *Biochim Biophys Acta* 1995, 1253:151–162

184 Pommersheim R, Schrezenmeir J, Vogt W: Immobilization of enzymes by multilayer microcapsules. *Macromol Chem Phys* 1994, 195:1557–1567

185 Alexakis T, Boadi DK, Quong D, Groboillot A, O'Neill E, Poncelet D, Neufeld RJ: Microencapsulation of DNA within alginate microspheres and crosslinked chitosan membranes for in vivo application. *Appl Biochem Biotechnol* 1995, 50:93–106

186 Polk AE, Amsden B, Scarratt DJ, Gonzal A, Okhamafe AO, Goosen MFA: Oral delivery in aquaculture: controlled release of proteins from chitosan-alginate microcapsules. *Aquacultural Eng* 1994, 13:311–323

187 Lim F, Sun AM: Microencapsulated islets as bioartificial endocrine pancreas. *Science* 1980, 210:908–910

188 Ma K, Vacek L, Sun A: Generation of alginate poly-lysine-alginate (APA) biomicrocapsules: the relationship between membrane strength and the reaction conditions. *Artif Cells Blood Subs Immob Biotech* 1994, 22:43–49

189 King GA, Daugulis AJ, Faulkner P, Goosen MEA: Alginate-polylysine microcapsules of controlled membrane molecular weight cut-off for mammalian cell culture engineering. *Biotech Prog* 1987, 3:231–240

190 Elcin YM: Encapsulation of urease enzyme in xanthan-alginate spheres. *Biomaterials* 1995, 16:1157–1161

191 Elcin YM, Sungur S, Akbulut U. Urease immobilization into poly(acrylamide)-gelatin gels. *Bioorg Med Chem Lett* 1992, 2:433–438

192 Sungur S, Elcin YM, Akbulut U. Studies on immobilization of urease in gelatin by crosslinking. *Biomaterials* 1992, 13: 795–800

193 Mohapatra SC, Hsu JT: Optimizing lipase activity, enantioselectivity, and stability with medium engineering and immobilization for β-blocker synthesis. *Biotechnol Bioeng* 1999, 64:213–220

194 Lee KH, Lee KH, Lee PM, Siaw YS: Studies of *l*-phenylalanine production by immobilized in stabilised calcium alginate beads. *J Chem Technol Biotechnol* 1992, 54:375–382

195 Suwasono S, Rastall RA: Synthesis of oligosaccharides using immobilised 1,2-mannosidase from *Aspergillus phoenicis*. Immobilisation-dependent modulation of product spectrum. *Biotechnol Lett* 1998, 20:15–17

196 DeGroot AR, Neufeld RJ: Encapsulation of urease in alginate beads and protection from α-chymotrypsin with chitosan membranes. *Enzym Microb Technol* 2001, 29:321–327

197 Bao J, Furumoto K, Fukunaga K, Nakao K: A kinetic study on air oxidation of glucose catalyzed by immobilized glucose oxidase for production of calcium gluconate. *Biochem Eng J* 2001, 8:91–102

198 Fan CH, Lee CK: Purification of *d*-hydantoinase from adzuki bean and its immobilization for *N*-carbamoyl-*d*-phenylglycine production. *Biochem Eng J* 2001, 8:157–164

199 Tor R, Dror Y, Freeman A: Enzyme stabilisation by bilayer "encagement". *Enzym Microb Technol* 1989, 11: 306–312

200 Woodward J, Clarke KM: Hydrolysis of cellobiose by immobilized β-glucosidase entrapped in maintenance-free gel spheres. *Appl Biochem Biotechnol* 1991, 28/29:277–283

201 Konishiroku Photo Industry: Preparation of reactive carriers. *Jpn Kokai Tokkyo Koho*, JP 59,063,186 A2, 1984

202 Pichko VB, Elchits SV, Dobrolinskaya GM: Characteristics of mechanically immobilized *d*-fructofuranosidase.

Ukrainskii Biokhimicheskii Zhurnal 1980, 52:732–736

203 El'chits SV, Pichko VB, Chugui VA, Tikhomirova AS, Serova YZ: Method for the immobilization of enzyme preparations which catalyze the splitting of low-molecular substrates. *Ukrainskii Biokhimicheskii Zhurnal* 1979, 51:378–381

204 Ikeda T, Hamada H, Miki K, Senda M: Glucose oxidase-immobilized benzoquinone-carbon paste electrode as a glucose sensor. *Agric Biol Chem* 1985, 49:541–543

205 Miyawaki O, Nakamura K, Yano T: Mass transfer and reaction with microcapsules containing enzyme and adsorbent. *Enzym Eng* 1978, 3:79–84

206 Piskin K, Chang TMS: A new combined enzyme-charcoal system formed by enzyme adsorption on charcoal followed by polymer coating. *Int J Artif Organs* 1980, 3:344–346

207 Decher G, Eckle M, Schmitt J, Struth B: Layer-by-layer assembled multicomposite films. *Curr Opin Colloid Interface Sci* 1998, 3:32–39

208 Lehn C, Freeman A, Schuhmann W, Schmidt HL: Stabilisation of NAD(+)-dependent dehydrogenases and diaphorase by bilayer encagement. *J Chem Technol Biotechnol* 1992, 54:215–221

209 Decher G and Hong JD:Buildup of ultrathin multilayer film by a self-assembly process; Consecutive adsorption of anionic and cationic bipolar amphiphiles. *Macromol Chem Macromol Symp* 1991, 46:321–327

210 Lvov Y, Decher G, Möhwald H: Assembly structural characterization and thermal behavior of layer-by-layer deposited ultrathin films of poly-(vinylsulfate) and poly(allylamine). *Langmuir* 1993, 9:481–486

211 Ge YB, Burmaa B, Zhang S, Wang SY, Zhou BL, Li W: Co-immobilisation of cellulase and glucose isomerase by molecular deposition technique. *Biotechnol Techniques* 1997, 11:359–361.

212 Mizutani F, Yabuki S, Hirata Y: Amperometric biosensors using poly-*l*-lysine/poly(styrenesulfonate) membranes with immobilized enzymes. *Denki Kagaku Oyobi Kogyo Butsuri Kagaku* 1995, 63:1100–1105

213 Mitsubishi Petrochemical, Japan: Immobilization of an enzyme. *Jpn Kokai Tokkyo Koho, JP 59,102,393 A2*, 1984

214 Sakurada Y, Yamane T: Fibrous substance entrapping an enzyme. *Jpn Kokai, P 500,29,728*, 1975

215 Piskin K, Chang, TMS: A new combined enzyme-charcoal system formed by enzyme adsorption on charcoal followed by polymer coating. *Int J Artif Organs* 1980, 3:344–346

216 Nitto Electric Industrial: Preparation of immobilized enzymes *Jpn Kokai Tokkyo Koho*, JP 59045884 A2

217 Ajinomoto Japan, Kansai Paint: Preparation of immobilized enzymes for use in a nonaqueous reaction system. *Jpn Kokai Tokkyo Koho, JP 57118792*, 1982

218 Coradin T, Mercey E, Lisnard L, Livage J: Design of silica-coated microcapsules for bioencapsulation. *Chem Commun* 2001, 7:2496–2497

6
Unconventional Enzyme Immobilization

6.1
Introduction

Since the first industrial application of immobilized amino acylase in 1967 for resolution of amino acids [1], enzyme immobilization technology has attracted increasing attention and considerable progress has been made in recent decades, because of progress in organic chemistry, protein chemistry, polymer chemistry, and materials science.

In general, methods of enzyme immobilization can be simply classified into attachment, embedding, and encapsulation. However, the author feels that this oversimplified classification cannot be used to exactly describe this rapidly growing field, because many new immobilization methods that are often characterized in that more than two methods of enzyme immobilization are involved. Moreover, combinatorial approaches are also increasingly used to solve the specific problems that cannot be solved by one of these basic immobilization techniques [2]. In this context we felt it was necessary to deal with these novel combined immobilization methods separately in a new chapter. The aim of this chapter is to provide special hints about how to design a desired immobilized enzyme, when conventional methods fail.

In general, these unconventional methods can be classified into the following eight subgroups:

- adsorption-based methods of enzyme immobilization,
- coating-based methods of enzyme immobilization,
- entrapment-based methods of enzyme immobilization,
- site-specific methods of enzyme immobilization,
- immobilization in non-aqueous media,
- imprinting immobilization methods,
- stabilization immobilization,
- modification-based methods of enzyme immobilization.

Most of the methods, listed in Table 6.1, are characterized in that these immobilized enzymes are prepared by multiple steps or by combination of at least two different methods of immobilization. Some of these methods such as the new variations based on adsorption and entrapment have already been discussed in previous

Carrier-bound Immobilized Enzymes: Principles, Application and Design. Linqiu Cao
Copyright © 2005 WILEY-VCH Verlag GmbH & Co. KGaA, Weinheim
ISBN: 3-527-31232-3

chapters. Accordingly, emphasis will be given to the rest of the immobilization methods listed in Table 6.1, although it is might be important to point out that it is difficult to draw a distinguished boundary between the different methods.

Table 6.1 Unconventional methods of enzyme immobilization.

Unconventional enzyme immobilization	Technical peculiarities
Entrapment-based	Native enzyme, enzyme derivatives or immobilized enzyme are entrapped in a gel matrix formed by various techniques
Adsorption-based	Native enzyme or enzyme derivatives are adsorbed on a carrier, followed by other immobilization techniques, for example covalent binding, cross-linking, encagement, or encapsulation
Modification-based	Native enzyme is modified, with the aim of improving the properties of the enzyme, followed by appropriate enzyme immobilization techniques
Coating-based enzyme immobilization	Enzyme or enzyme derivatives are coated on a carrier, followed by other-enzyme immobilization techniques
Immobilization on smart polymers	Enzyme–polymer conjugates formed which are reversibly soluble under the influence of pH or temperature
Site-specific enzyme immobilization	Enzyme or enzyme derivatives are immobilized via various site-specific immobilization techniques with the aim of improving the activity of the enzyme
Immobilization in non-aqueous medium	Enzyme is immobilized by covalent bonding or entrapment in organic solvents
Stabilization-immobilization	Enzyme is stabilized by various stabilization techniques, followed by other appropriate enzyme immobilization techniques
Imprinting-based enzyme immobilization	Enzyme is imprinted, with the aim of improving activity retention, stability and selectivity, followed by other enzyme immobilization techniques

6.2
Coating-based Enzyme Immobilization

These types of method of enzyme immobilization refer to the immobilization processes by which the enzyme molecules are coated on the surface of a solid carrier in different geometries without covalent bonding between the enzyme and the carrier. In general, the methods used for coating enzymes on to solid carrier surfaces can be categorized into five different groups–monolayer coating of the enzyme on

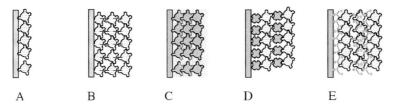

Scheme 6.1 Coating of enzymes on the solid carriers: monolayer coating can be reached by simple adsorption (A); multiple layer coating can be obtained by crosslinking (B), gelation (C), affinity coating (D), and simplex formation between polyelectrolyte and enzyme (E).

the solid surface (A), multilayer coating (B), gelation coating (C), affinity coating (D), and simplex coating (E), as shown in Scheme 6.1.

Except for monolayer coating, a characteristic of all these methods of coating enzymes on carriers is that multiple layers of enzyme molecules are coated to the carriers by chemical or physical methods. However, the nature of the interaction between the enzyme and the carrier are all exclusively physical.

6.2.1
Monolayer Enzymes

The first immobilized enzyme in monolayer form was reported by Haynes and Walsh in 1969. It was prepared by adsorption of enzyme on solid particles followed by the intermolecular cross-linking with a bi-functional cross-linker, i.e. glutaraldehyde [3]. The loading of the enzyme by adsorption is generally dictated by the available surface. Although the loading of the enzyme was quite high (~0.25 g enzyme/g carrier), the nano-sized silica particles makes it very difficult to use them as useful industrial immobilized biocatalysts. This type of immobilized enzyme was intensively studied in the 1960s [4]. Since the 1970s the method has gradually been abandoned, because many active carriers with tailor-made physical and chemical properties have been developed and are commercially available [5].

6.2.2
Phase-inversion Coating

A solution or mixture containing enzyme and other coating components, e.g. film-forming polymers such as hydroxyethylcellulose and polyethyleneimine, can be coated on a ready-made carrier surface. The gel can be cross-linked. Thus, the enzyme can be entrapped in the gel layer, as illustrated in Scheme 6.1. Urease has been successfully immobilized in radiation-cross-linked poly(vinyl alcohol) gel, by use of a post-irradiation immobilization procedure [6]. With this method, glucoamylase was coated on to glass fibres with a half life of 40 days at 60 °C [7].

In a similar approach, the soluble enzyme or insoluble enzyme or intact whole cell [8] can be coated on to a number of insoluble carriers. The coated layer can be

5–50% of the core, in term of the radius of the insoluble carrier. Enzyme can be also coated on to a polymer matrix by casting the enzyme–polymer solution that is used as the cross-linker and matrix-forming component. For instance, water-borne polyurethane coatings containing diisopropylfluorophosphatase have been coated on to a polymer matrix by use of a less hydrophilic polyisocyanate, leading to the formation of a stable matrix–coated enzyme layer [9].

6.2.3
Multiple Enzyme Coating by Physical Adsorption

Enzymes can be coated to the carriers in multiple layers simply by physical adsorption. This can be achieved by use of so-called tentacle carriers, as illustrated in Scheme 6.2. In this case, the enzyme molecules are immobilized on the tentacle chains attached to the solid carrier, which is equivalent to a multilayer on the solid surface (Scheme 6.2).

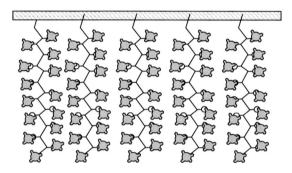

Scheme 6.2 Tentacle carriers for enzyme immobilization.

With this technique, several enzymes such as trypsin, urease, β-galactosidase, ClTase, α-amylase, glucose oxidase, β-galactosidase and urokinase have been immobilized with very high enzyme loadings. Retention of activity is usually very high, owing to the low diffusion limitation [10–17]. Some these examples are listed in Table 6.2.

6.2.4
Mediated Formation of Multiple Enzyme Layers

Enzymes can be also coated on to a film by use of the symplex concept for encagement of enzyme in multilayers. For instance, horseradish peroxidase was recently coated on a thin film by alternate formation of enzyme–polyelectrolyte composite mediated by bipolar quaternary ammonium, as illustrated in Scheme 6.3 [18].
Similarly, enzyme layers of β-glucosidase (β-GL) separated by oppositely charged polyelectrolyte (polystyrenesulphonate, PSS) have been deposited on to polystyrene (PS) latex particles using the layer-by-layer adsorption technique [19].

Table 6.2 Adsorption of enzyme on tentacle carriers.

Tentacle carrier	Method	Remark	Enzyme	Ref.
Ozonized PET fibres-g-PAA	Adsorption	The adsorbed trypsin was inhibited more easily than the covalently immobilized enzyme	Trypsin	10
Hollow-fibre membrane-g-GMA-diethylamino (DEA) group	Ionic adsorption	A novel "tentacle-type" porous membrane with 50 times more enzyme loading than the native membrane	Urease	11
Cotton cloth-enzyme-PEI aggregates	Adsorption	250 mg g^{-1} support with approximately 90–95% efficiency	β-Galactosidase	12
Anion-exchange PHFM-g-GMA-diethylamine	Adsorption?	The enzyme loading is 38 and 110 mg-CITase g^{-1} carrier, which is equivalent to ca 7–9 layers of enzymes	CITase	13
A porous hollow-fibre membrane-g-GMA opened with EA and Ph	Adsorption	Negligible mass-transfer resistance of the starch to the α-amylase because of convective flow	α-Amylase	14
Poly(AN-co-MMA-co-SVS)membranes-g-AMPSA (or DMAEM)	Adsorption	Glucose oxidase was immobilized on modified acrylonitrile copolymer membranes with DMAEM and AMPSA and had high relative activity	Glucose oxidase	15
PP-g-GMA-AAs	Adsorption	Payload increased with the increasing content of the amino acid group with the order: l-Phe > d,l-Phe > d,l-Try > l-Cys but is pH-dependent: Loading at pH 7.4 was higher than that at pH 9.0	Urokinase	16
Sepabeads-coated with PEI	Adsorption	The adsorption strength was much higher for PEI-Sepa beads than on DEAE-supports at both pH 5 and 7. No activity loss after incubation at 50 °C for weeks	β-Galactosidase	17

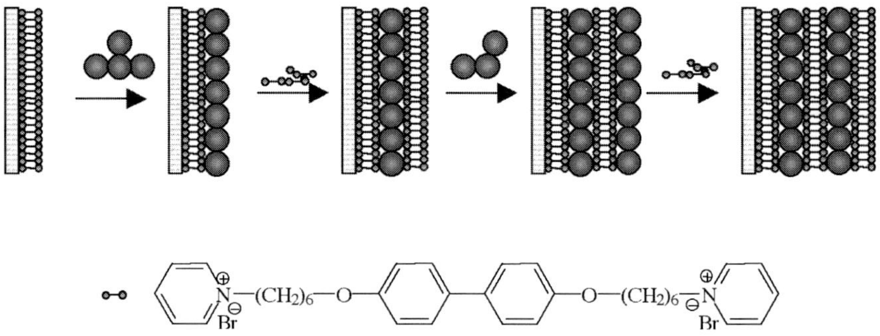

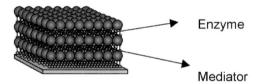

Scheme 6.3 Illustration of enzyme immobilisation by layer-by-layer methods. (Adapted from [18]).

Instead of using oppositely charged polyelectrolyte, horseradish peroxidase HRP-containing multilayer films have been fabricated by alternate deposition with the bipolar pyridine salt PyCBPCPy on the surface of a gold electrode derivatized with a 3-mercapto-1-propanesulphonate monolayer. The first layer was formed by the salt PyCBPCPy, which can adsorb negatively charged HRP. In turn, a second layer of the salt PyCBPCPy can be formed on the enzyme layer. By repeating the process, the desired amount of enzyme can be adsorbed (Scheme 6.4) [20].

Scheme 6.4 Enzyme coated on the carrier surface as a multilayer.

By symplex formation, organophosphorus hydrolase (OPH) was entrapped in polyelectrolyte multilayers with branched poly(ethylene imine) (BPEI) and polystyrenesulphonate (PSS) as two priming layers and glass beads as the hard core. Interestingly, it was found that OPH multilayers coated on glass beads had catalytic activity very similar to that of free enzyme [21]. The multiple enzyme layer (MEL) prepared in this way can be encapsulated by forming a polymer layer, by polymerizing the anchored monomers, 1,2-dihydroxypropyl methacrylate (DHPM), 1,2-dihydroxypropyl-4-vinylbenzyl ether (DHPVB), and N-[3-(trimethoxysilyl)propyl]ethylenediamine (TMSED) to the deposited poly(acrylic acid) (PAA) as an outer priming layer. Enhanced stability was obtained with TMSED.

Interestingly, the layer-by-layer enzyme-immobilization technique has been also used to prepare covalently bound layers of enzymes. With this method, multilayers of lactozyme were immobilized on activated silica gel bearing the free amino groups (A) with the aid of dialdehyde as the cross-linker and activator of the first enzyme layer [22], as illustrated in Scheme 6.5.

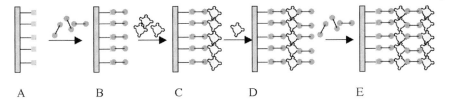

Scheme 6.5 Covalent multiple enzyme layers by alternating cross-linking.

6.2.5
Affinity-ligand-mediated Formation of Enzyme Coatings

Similarly, the enzyme can be attached to a carrier bearing multivalent affinity ligands. When the enzyme is bound to the carrier, addition of the ligand can lead to the formation of a new layer of ligand non-covalently bound to the enzyme, which can further bind enzyme. By repeating the above process, multilayers of enzymes can be formed on the carrier surface. In line with this idea, it was recently demonstrated that horseradish peroxidase could be immobilized layer-by-layer on polystyrene beads via biotinylated enzyme [23], as illustrated in Scheme 6.6.

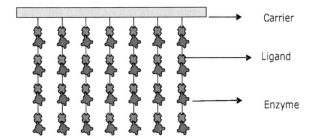

Scheme 6.6 Illustration of the affinity layer-by-layer immobilization technique.

It has been reported that loading and activity can be enhanced several times. Unfortunately, the surface area of the polymeric beads is only 775 $\mu m^2\, g^{-1}$, thus resulting in a lower enzyme loading. However, high enzyme loading could be expected by using the principle and application of a carrier with a high surface area.

By the use of the affinity of Con A for glycol enzyme, a layer-by-layer deposition technique was used to construct an enzyme film for an enzyme electrode, as illustrated in Scheme 6.7 [24]. Similarly, glycoenzyme invertase was immobilized on cellulose beads and on poly(glycidyl methacrylate) beads by use of the affinity of ConA for glycol enzyme; a tenfold increase in catalytic activity of the immobilized invertase was achieved [25].

It was specifically claimed by the author that the layer-by-layer deposition technique was useful for optimizing the loading and geometry of the enzyme layers in the multilayer film, by changing the number of layers and the sequence of deposition.

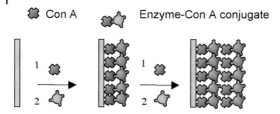

Scheme 6.7 Illustration of affinity coating of immobilized enzymes (adapted from [25]).

6.2.6
Coating of Soluble Enzyme–Polymer Complexes

Soluble enzyme–polymer conjugates can be coated on another solid support by casting followed by UV irradiation. For example, linking of urease to oxidized poly(2-hydroxyethyl methacrylate) by use of $(NH_4)_2Cr_2O_7$ then coating on a polyethylene support by radiation afforded an immobilized enzyme with 85% of the activity of urease after storage for 1 month at 23 °C [26].

Direct ionotropic gelation of the polycationic biopolymer chitosan (CHIT) with the polyanionic enzyme lactate oxidase (LOx) has been used to form thin biopolymer–enzyme films on the surface of platinum electrodes [27]. Remarkably, the use of enzyme–polymer complexes as precursors of immobilized enzymes often enhances enzyme stability

Enzymes can be also coated on carrier surfaces by in-situ reticulation in the presence of an active copolymer which can covalently bind the enzymes. With this method several enzymes can be bound to carriers such as glass beads, alumina, or carbon beads. The immobilized enzymes obtained have higher mechanical stability and can be used in packed beds [28].

6.2.7
Enzymatically Gelified Multienzyme Layer

Transglutaminase can catalyse the formation of intermolecular peptide bonds, leading to enzyme cross-linking. Taking advantage of this mechanism, a layer of enzyme can be created on a solid surface by adsorption and an enzyme-cross-linking technique. This technique resembles the chemical cross-linking for preparation of multiple enzyme layers developed by Ho et al. [29], as illustrated in Scheme 6.8.

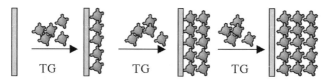

Scheme 6.8 Multilayer enzyme coating with transglutaminase.

Multilayers of cycloisomaltooligosaccharide glucanotransferase (CITase) can be formed on anion-exchange porous hollow-fibre membranes (graft polymerization of an epoxy group-containing monomer led to formation of chemically active functionality which reacted with diethylamine, with the aim of introducing the ionogenic groups). More specifically, the enzyme is first adsorbed on the membrane and gelified by the action of transglutaminase. In this way, a multilayer of enzyme is formed with a payload of 0.038 to 0.11 [30]. In the same way, urease was immobilized on a porous hollow-fibre membrane to which an anion-exchange polymer chain had been grafted by adsorption; cross-linking of the urease was achieved with transglutaminase [31]. Interestingly, the payload was 1.2, which means that 1.2 g urease was bound to 1 g of a matrix. Obviously, multi-layers of enzyme were formed.

Trypsin and β-amylase have been immobilized on ion exchangers (QAE-Sephadex A-25 and SP-Sephadex C-25) by using bacterial transglutaminase as cross-linker without any damage to enzyme activity [32], suggesting that transglutaminase treatment was milder than glutaraldehyde treatment.

6.2.8
Sol–Gel Coating and Covalent Attachment

A sol–gel layer containing entrapped enzyme can be coated on an electrode by casting enzyme solution with soluble sol–gel precursors. The enzyme–gel precursor solution was applied to the wall of a Pt disc electrode, followed by cross-linking using glyoxal [33].

6.2.9
Electrochemical Deposition

Native enzyme or modified enzyme in polyelectrolyte solutions containing monomers such as ferrocene aldehyde, pyrrole, N-methylpyrrole, and o-phenylenediamine [34] and electrolytes such as Na alginate can be electronically polymerized and deposited on the electrodes under the influence of a potential. For example, choline oxidase can be electronically deposited on a Pt electrode in the presence of aniline, leading to the formation of enzyme polymer membrane with enzyme activity and permeability to O_2 and H_2O_2 [35].

Similarly, glucose oxidase [36] or modified glucose oxidase (conjugated with a polyanion, poly(2-acrylamido-2-methylpropane sulphonic acid) via a poly(ethylene oxide) spacer [37] can be immobilized on to the electrode.

Glucose oxidase, lactate oxidase, l-amino acid oxidase and alcohol oxidase have been immobilized on new films on Pt electrodes based on 2,6-dihydroxynaphthalene (2,6-DHN) co-polymerized with 2-(4-aminophenyl)ethylamine (AP-EA). The electropolymerization was performed by cyclic voltammetry [38].

6.2.10
Enzyme Coating by Use of Small Pore-size Carriers

A simple method of enzyme immobilization based on adsorption on the surface of colloidal particles with small pores has been achieved by using ion exchangers as the support materials. Due to the exclusion of enzyme molecules, enzyme molecules are only coated on the surface of the carriers bearing pore size that is smaller than the enzyme to be coated. With trypsin, the best results were obtained with Sephadex C-25-type cation-exchangers. The optimum pH of immobilized trypsin on the cation exchangers was shifted toward alkaline pH (10–11) from that of the free enzyme (pH 8.5) [39].

6.3
Site-specific Immobilization

As discussed previously in this book, an enzyme molecule contains many similar and different amino acid residues which are usually located in different domains of the enzyme molecule (Scheme 6.9). Thus, one characteristic of an immobilized enzyme might be that the binding nature is heterogeneous, because the same amino acids located in different domains or surfaces might have the same possibility linkage to the carriers, which bear the same active functionality spread over the whole structure. Moreover, some of these amino acid residues potentially involved in binding to the carriers might be essential for the catalytic function. Thus, modification of these essential residues might lead to the loss of the catalytic activity.

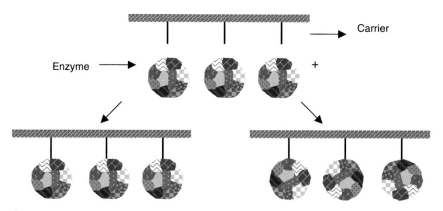

Scheme 6.9 Randomly immobilized enzyme (RIE) and site-directed immobilized enzyme (SDIE).

Table 6.3 Coating based enzyme immobilization.

Carrier	Coating method	Comment	Enzyme	Ref.
Nano sized silica	Monolayer	Enzyme envelope was firmed by adsorption-cross-linking technique. Monolayer was formed	Trypsin	3
Glass fibres	Gelation coating	Enzyme is immobilized on the fibre-coated with cross-linked gel	Urease	6
Glass fibres	–	–	Glucoamylase	7
Thin film	Symplex	Bipolar quaternary ammonium	Horseradish peroxidase	18
Polystyrene (PS) latex	Symplex	Charged polyelectrolyte (poly(styrenesulphonate), PSS), were deposited on to particles using the layer-by-layer adsorption technique	β-Glucosidase (β-GL)	19
Surface of gold electrode derivatized by 3-mercapto-1-propanesulphonate monolayer		Alternate deposition with bipolar pyridine salt PyC BPC Py on the surface of gold electrode derivatized by 3-mercapto-1-propanesulphonate monolayer	Horseradish peroxidase HRP	20
Glass beads	Branched poly(ethylene imine) (BPEI) and polystyrenesulphonate (PSS) as two priming layers	Interestingly, it was found that OPH multilayers coated on glass beads had catalytic activity very similar to that of free enzyme	Organophosphorus hydrolase (OPH)	21
Activated silica gel	Covalent	Enzymes are cross-linked by glutaraldehyde	Lactozyme	22
Sephadex C-25	Glutaraldehyde cross-linking	Enzyme layers are formed on small size ion exchanger resin. The fact that the enzyme molecules cannot enter the pore size suggested that diffusion limitation might be overcome	Trypsin	39

Table 6.3 Continued

Carrier	Coating method	Comment	Enzyme	Ref.
Polystyrene beads	–	The enzyme loading was increased severalfold	Horseradish peroxidase	23
Enzyme electrode, cellulose beads, and other polymer beads	Affinity of Con A	Tenfold increase in catalytic activity of the immobilized invertase	Glycosylated enzyme	24, 25
Polyethylene support	UV-radiation cross-linking	Enzymes are entrapped and cross-linked	Urease-oxidized poly(2-hydroxy-ethyl methacrylate)	26
Platinum electrodes	Ionotropic gelation	Symplex formation	Lactate oxidase (LOx)	27
Glass beads	Cross-linking-coating	A copolymer of maleic anhydride and vinyl Me ester was used as cross-linking agent	α-Chymotrypsin, papain, RNase, l-asparaginase and alkyl phosphatase	28
–	Chemical cross-linking	–		29
Anion-exchange porous hollow-fibre membranes	Enzymatically cross-linking	Transglutaminase-mediated cross-linking of the enzyme on anion-exchange porous hollow-fibre membranes led to the formation of immobilized enzymes with high enzyme loading from 0.038–0.01	Cycloisomaltooligosaccharide glucanotransferase (CITase)	30
On a porous hollow-fibre membrane	Enzymatically cross-linking	Surprisingly, it was found that the payload can be 1.2, which suggested that the enzyme is in the multilayer form	Urease	31
QAE-Sephadex A-25 and SP-Sephadex C-25	Enzymatically cross-linking	No damage to enzyme activity was observed, suggesting that transglutaminase treatment was milder than with glutaraldehyde	Trypsin and β-amylase	32

Table 6.3 Continued

Carrier	Coating method	Comment	Enzyme	Ref.
Pt disc electrode	Sol–gel coating-covalent attachment	A layer of sol–gel containing entrapped enzyme can be coated on the electrode casting enzyme solution with soluble sol–gel precursors. The enzyme–gel precursor solution was applied to the well of the Pt disc electrode, followed by cross-linking using glyoxal	Glucose oxidase	33
Electrode	Electrochemical deposition	Electronically deposited on to a Pt electrode in the presence of aniline	Choline oxidase	35
Electrode	–	Similarly, glucose oxidase or modified glucose oxidase (conjugated with a polyanion, poly(2-acrylamido-2-methylpropane sulphonic acid via a poly(ethylene oxide) spacer can be immobilized on to the electrode	Glucose oxidase	36
Pt electrodes	Electrochemical deposition	2,6-Dihydroxynaphthalene (2,6-DHN) was co-polymerized with 2-(4-aminophenyl)ethylamine (AP-EA) and deposited onto the electrode	Glucose oxidase, lactate oxidase, l-amino acid oxidase	37

Consequently, the immobilized enzyme might be less active than the free enzyme. The reduced performance can possibly be ascribed to one or more of the following factors:

- active site modification,
- structure rigidification by multiple-point binding,
- denaturation of the enzyme because of the wrong orientation [40, 41], and
- blockage of active site from substrate accessibility.

This is often ascribed to non-specific chemical cross-linking of protein to the surface or carrier [42].

In general, immobilization of an enzyme without controlling the binding mode (binding site, number of bonds) is regarded as random enzyme immobilization, which often results in heterogeneity of the immobilized derivatives, as mentioned above [43].

In contrast, it was observed early in 1972 that site-specific enzyme orientation could improve the performance, for example activity, of the immobilized enzyme [44]. It is, therefore, hardly surprising that site-specific enzyme immobilization attracted much research interest in the mid 1970s [45, 46], although the concept "site-specific immobilization" was not mentioned. In general, site-specific enzyme immobilization refers to the technique by which the enzyme molecule is bound covalently or by affinity to the carrier via a specific binding site of the enzyme molecules (see Scheme 6.10).

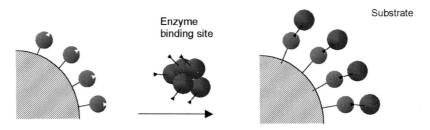

Scheme 6.10 Biospecific oriented enzyme immobilization

Site-specific enzyme-immobilization techniques have several advantages, including controlled orientation of enzyme molecules and thus good accessibility of the enzyme active centre; and avoidance of chemical modification of essential amino acids, thus increasing retention of activity compared with randomly immobilized enzymes [47]. Other benefits include improved thermal stability [48], stability against denaturation [49], reuse of support matrix [49], and enhancement of enzyme loading [50], compared with randomly immobilized enzyme. The improved performance is attributed mainly to exclusion of multipoint attachment, multiple orientations, and steric hindrance.

Several site-specific enzyme-immobilization techniques have been developed and can be placed in the following groups:

- carrier-bound antibody–antigen mode [51],
- protein–ligand interaction (avidin–biotin) [52, 59],
- use of molecular biology to introduce bio-tags to proteins for site-specific binding [42],
- introduction of chemical tags [60], and
- carrier-bound substrate (or inhibitor or substrate analogue)–enzyme mode [61–63].

These technologies have been also used for other purposes such as identification of essential amino acids involved in substrate binding and catalysis, to facilitate protein purification, and understanding protein interactions [64].

6.3.1
Site-specific Immobilization via Biospecific Ligand–Enzyme Interaction

Biospecific ligands are ligands which can bind a specific group of proteins or enzymes or specific carriers. The ligands, which are often used for site-specific immobilization, include:

- Con A (for binding glycosylated enzymes) [45, 46],
- avidin (for binding biotinylated enzymes) [65],
- biotin (for binding proteins with fused avidin) [52], and
- antibodies (for binding the corresponding proteins) [54].

Wallace and Lovenberg were probably among the first to prepare this type of immobilized enzyme in the mid of 1970s. They used Sepharose-bound Con A (as selective ligand) for affinity binding of dopamine β-hydroxylase [45]. Another type of oriented immobilized laccase from *Neurospora crassa* was soon obtained, based on the same principle, i.e. via the carbohydrate moiety of laccase on the same type of carrier i.e. Sepharose-bound Con A [46].

Other biospecific ligands used for oriented enzyme immobilization is the biotin–avidin interaction. In detail, the enzyme can be biotinylated and subsequently bound to a carrier bearing immobilized avidin. For example, bintinylated papain was immobilized on a poly(ether)sulphone (PES) membrane via avidin–biotin complex formation. Compared with papain directly immobilized on PES, papain immobilized on PES by avidin–biotin interaction had higher retention of activity and better stability [65]. Avidin can also be fused with the targeted enzyme molecules by genetic approaches [52].

Some special tags, for example poly(his) tag, poly(arg) or protein A domains, or cellulose-binding domains can be also genetically fused into the targeting enzymes. Thus, the enzyme can be oriented to the carriers which can selectively interact with these special tags, resulting in site-specific orientation of the enzyme molecules.

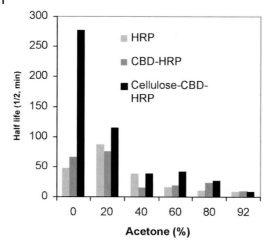

Scheme 6.11 Stability of different HRP preparations. HRP denotes free horseradish peroxidase; CBD-HRP denotes HRP with a cellulose-binding domain; cellulose-CBD-HRP denotes CBD-HRP immobilized on cellulose.

Immobilization of HRP bearing a cellulose domain led to the formation of immobilized enzymes with more stability in organic–water mixtures, as shown in Scheme 6.11 [66].

Some examples of this type of immobilized enzyme are listed in Table 6.4. Remarkably, the stability and activity of the immobilized enzyme are often better than those of the native enzyme (Table 6.4), suggesting that the bio-specific interaction between the enzyme and the carrier might stabilize the enzymes.

6.3.2
Introduction of Chemical Tags

In contrast with bio-affinity immobilization, site-directed immobilization can also be achieved by chemical methods. A native functional group specifically located on the enzyme surface can be activated by use of chemical methods. Subsequently,

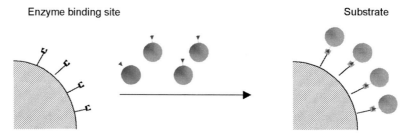

Scheme 6.12 Introduction of chemical tags into the enzymes.

Table 6.4 Site-specific enzyme immobilization via biospecific anchoring ligands.

Carrier/anchor	Binding site of enzyme	Comment	Enzyme	Ref.
Sepharose/Concanavalin A	Sugar moiety	Con A binds to the carbohydrate moiety of the glycolated enzymes	Dopamine β-hydroxylase	45
Sepharose/Concanavalin A	Sugar moiety	Enzyme binding to the lectin had no effect on enzyme activity. Thus, the carbohydrate moiety of enzyme is not close to the active centre	Laccase	46
Sepharose/Concanavalin A	Sugar moiety	Increased thermal stability	Chicken brain arylsulphatase A	47
Nonporous copolymer of HEEMA-EDMA/anti-chymotrypsin antibodies	Antibody–antigen	166.7 μg chymotrypsin g^{-1} dry carrier with 100% of its native proteolytic activity	Chymotrypsin	51
Immobilized on controlled pore glass	Avidin	Although the immobilized enzyme had similar properties to the native enzyme in term of K_m, the apparent optimum pH for activity was shifted 0.5 units to the acid region and the optimum temperature was 5 °C lower	Streptavidin-β-galactosidase	52
Immobilized ConA	Sugar	96% Activity retention; very good operational stability	Carboxypeptidase Y	53
Hydrazide cellulose anti-chymotrypsin antibodies antibody (oxidized)	Antibody–antigen	—	Chymotrypsin	54
Cobalt-charged imino-diacetate-Sepharose-IgGs	Antibody–antigen	High enzyme loading was obtained. Thermal stability and stability toward water-miscible organic solvents was enhanced	*Aspergillus niger* glucose oxidase	55
Eupergit C/A mouse monoclonal antibody (mAb 100) to CPA	Antibody–antigen	EC-mAb-CPA with full catalytic activity whereas the conventional covalently immobilized enzyme did not have full activity. Enhanced stability was obtained	Carboxypeptidase A (CPA)	56
Sepharose-protein A (SPA)	Antibody–antigen	SPAmAb-CPA, with full catalytic activity. Enzyme stability was also improved	Carboxypeptidase A (CPA)	56
AH Sepharose-GAH-anti-CHT IgG	Antibody–antigen	High activity than CHT covalently bound to AH-Sepharose	Chymotrypsin	57
Protein A bound MPS(aldehyde-modified polyethersulphone)	Antibody–antigen	The immobilized BAP had more than 80% of the free enzyme activity. Interestingly, tolerance of product inhibition was three times higher than for the free enzyme	Bacterial alkaline phosphatase (BAP)	58

the activated enzyme molecules can be covalently or non-covalently immobilized on carriers bearing a functional group that can react with the active enzyme molecules. Thus, enzyme immobilization and orientation can be combined in one step, as shown in Scheme 6.12.

6.3.2.1 Oxidation of the Sugar Moiety of Enzymes

Oxidation of a carbohydrate moiety in an enzyme can create a specific active bonding site in the enzyme. This method was first introduced by Royer in 1979 [67], when the sugar moiety of glycosylated enzymes such as glucoamylase, peroxidase, glucose oxidase, and carboxypeptidase Y were oxidized to aldehyde groups and subsequently converted into amino functionality and covalently bound to activated Sepharose, for example the aminocaproate adduct of CL-Sepharose via an *N*-hydroxysuccinimide ester or to CNBr-activated Sepharose.

Soon after, oxidized *Penicillium vitale* catalase bearing aldehyde groups was directly bound to aminoethyl cellulose (AE-cellulose), leading to the formation of immobilized enzyme with orderly orientation on the carrier surface. High retention of activity was obtained, compared with the same carrier activated with glutaraldehyde methods [60]. Immobilization of *Candida rugosa* lipase via its carbohydrate moiety also resulted in relatively high activity [68].

Hydrazide derivatives of non-magnetic and magnetic cellulose beads and of magnetic and non-magnetic poly(HEMA-co-EDMA) microspheres have been used to attach sodium periodate oxidized galactose oxidase [69].

6.3.2.2 Introduction of a Cofactor into the Enzyme

Several proteases have been immobilized on alumina by a two-step procedure. The first step converted them into semi-synthetic phosphoproteins which, in the second step, spontaneously bonded to alumina through their phosphate function. The immobilized enzymes thus obtained had physical properties typical of the inorganic carrier and a high activity on low-molecular-weight substrates [70].

6.3.2.3 Orientation and Covalent Binding

Voivodov et al. proposed an interesting chemical approach for orientated enzyme immobilization based on the use of chloroformates of nitrophenyl derivatives bearing different moieties, as illustrated in Scheme 6.13.

In this way, the orientation of the enzyme molecule is governed by the properties of the side chain of the carrier. Because the binding mode between the enzyme and the carrier is same, different enzyme activity can be ascribed to different enzyme orientation only. In other words, the properties of the immobilized enzymes are expected to be influenced by variation of the side chain. Although different enzyme activity was achieved with this method, the homogeneity of orientation of the enzyme was unknown [71].

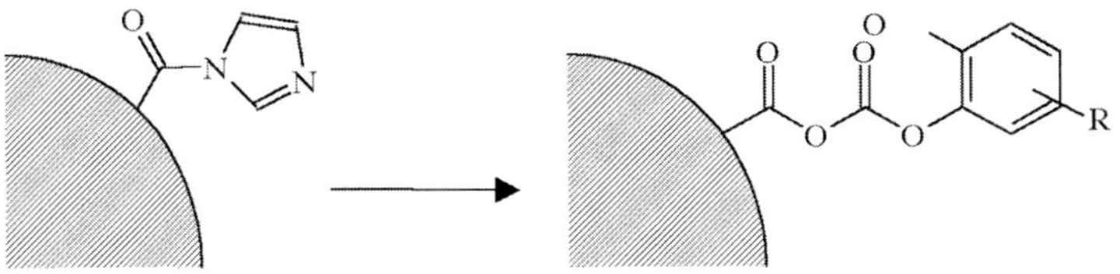

Scheme 6.13 Orientation of enzyme by the leaving groups of the activated carriers. R denotes ortho nitrophenyl derivatives containing ionic and/or hydrophobic groups [72].

In fact, the effect of the side chain of the activated carrier bearing the same backbone was previously also observed by Lu and co-workers in 1981 [72]. Four different polymers bearing different activated carbonyl functionalities, poly(N-p-methacryloxybenzoylox-5-norbornene-2,3-dicarboximide), poly(N-p-methacryloxybenzoyloxysuccinimide), poly(N-p-methacryloxybenzoyloxy-4-oxo-3,4-dihydro-1,2,3-benzotriazine and poly(N-p-methacryloxybenzoyloxybenzotriazole) were prepared from the corresponding monomers as shown in Scheme 6.14.

It was found the payloads of these polymers were quite similar and high for trypsin (147, 149, 138, and 142 mg g^{-1} carrier, respectively) whereas recovery of activity was 31, 25, 8, and 8% respectively, suggesting that other factors which can influence enzyme immobilization might govern the reactivity of the active functionality toward the amino groups in different domains of the enzyme, thus leading to the different activity retention.

It is most probable that the microenvironment of the binding functionality governs enzyme orientation and activity retention. The different activity retention

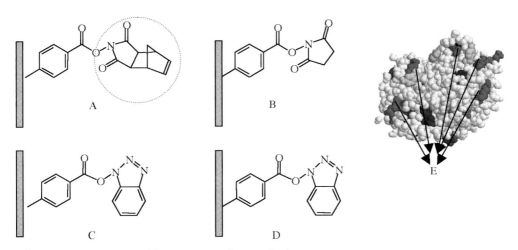

Scheme 6.14 Monomers used for preparation of activated polymeric carbonyl carriers, poly(N-p-methacryloxybenzoylox-5-norbornene-2,3-dicarboximide) (A), poly(N-p-methacryloxybenzoyloxysuccinimide) (B), poly(N-p-methacryloxybenzoyloxy-4-oxo-3,4-dihydro-1,2,3-benzotriazine) (C), and poly(N-p-methacryloxybenzoyloxybenzotriazole) (D).

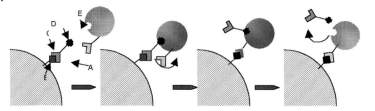

Scheme 6.15 Site-specific covalent enzyme immobilization via an orientation group that leaves after covalent enzyme immobilization. B is pedant functionality of the carrier, C is the leaving group bearing an orientation group D, E is the binding site of the enzyme, which can recognize D, and A is the active amino acid residue close to the recognition site of the enzyme.

might reflect the fact that the enzyme was bound to the carrier via different domains. Lower activity retention might be because essential amino acid residues close to the active centre were linked to the carrier. This hypothesis is confirmed by the fact that the active centre of trypsin is negatively charged. Thus, the exposed lysine residues near the active centre could react with carriers C and D (Scheme 6.14), resulting in very low retention of activity. In contrast, carriers A and B bearing relatively small neutral leaving groups were probably active toward the amino groups of the lysine residues located far from the active centre, leading to high retention of activity.

On the basis of these observations it is surmised that a novel oriented enzyme immobilization method might be designed on the basis of linking a chemical orientation group to the leaving groups of the activated functionality of the carrier. The advantages of this method might be that covalent enzyme immobilization and orientation can be combined into one step. In particular, this method might be extremely useful for orientation immobilization of enzyme with immobilized ligand (substrate or substrate analogue), because occasionally the active centre might be occupied by the ligand, thus the activity might be inhibited by the presence of immobilized substrate in the active centre. Thus, this problem can be circumvented by using the leaving ligand as orientation groups, as illustrated in Scheme 6.15. Immobilization and purification might also be possible.

In addition to advantages such as ease of immobilization, lack of chemical modification, and enhancement of stability, affinity-based site-oriented enzyme immobilization may also lead to favourable orientation of enzyme molecules on the carriers, thus leading to the enhancement of enzyme activity.

6.3.3
Immobilized Ligand (Substrate Analogue) Enzyme-binding

The idea of using an immobilized ligand for binding enzyme stems from affinity chromatography [61, 62]. As shown in Scheme 6.16, the substrate (or inhibitor or cofactor) is immobilized on a suitable carrier.

Table 6.5 Site-specific enzyme immobilization via chemical tags.

Carrier/anchor	Tags	Comment	Enzyme	Ref.
Biotinylated aminopropyl glass	Biotin	Both biotinylation of enzyme amino groups and association of the biotinylated enzyme with soluble avidin caused some loss of enzyme activity; reduction in the kcat value was observed	Biotinylated transglutaminase	59
Aminoethyl cellulose	Sugar moiety (oxidized)	–	*Penicillium vitale* catalase	60
Poly(ether)sulphone membranes	Biotin	Immobilization via avidin-biotin complex as a non-covalent spacer increased V_{max}(app), and reduced K_m (app), relative to directly immobilized papain. However, IME stability and reusability was significant increased	Biotinylated papain	65
Nylon fibres	Sugar moiety (oxidized)	–	*Candida rugosa* lipase	68
Hydrazide derivatives of non-magnetic and magnetic bead cellulose	Sugar moiety (oxidized)	Oriented immobilized galactose oxidase has high storage stability and lower susceptibility to inappropriate microenvironmental conditions	Galactose oxidase	69
Magnetic and non-magnetic poly(HEMA-co-EDMA) microspheres	Sugar moiety (oxidized)	Oriented immobilized galactose oxidase has high storage stability and lower susceptibility to inappropriate microenvironmental conditions	Galactose oxidase	69
Alumina		The immobilized enzymes thus obtained had physical properties typical of the inorganic carrier and high activity toward low-molecular-weight substrates	Proteases	70
Membranes (nitrocellulose, nylon, and positively charged nylon): streptavidin	Biotinylated HRP	Enhanced activity and operating stability, because of more favourable enzyme orientation	Horseradish peroxidase (HRP)	74

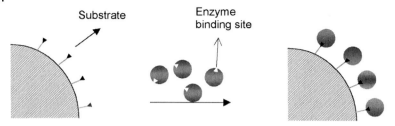

Scheme 6.16 Immobilized ligands as orientation groups.

The enzyme can be spontaneously oriented and bound to the immobilized substrate. In principle, this technology is universally applicable. The only requirement is that a substrate or a substrate analogue has to be identified and bound to the carrier.

6.3.3.1 Immobilized Substrate or Substrate Analogues

An early example of oriented enzyme immobilization by the use of an immobilized ligand, described in a patent [63], was based on affinity adsorption of enzymes requiring pyridoxal 5′-phosphate as coenzyme, for example immobilization of tryptophanase on carrier-bound pyridoxal 5′-phosphate on p-aminobenzamidohexyl Sepharose 4B.

After this work several other site-specific enzyme immobilization techniques featured in the literature. For example, Mosbach and co-workers immobilized lactate dehydrogenase and alcohol dehydrogenase (NAD^+ oxidoreductase) on agarose beads to which a bis-NAD analogue had been attached. Thus the enzyme can be site-specifically immobilized. Subsequent cross-linking of the adsorbed enzyme with glutaraldehyde might enhance the binding of the enzyme on the carrier [75].

Several triazine dyes such as Cibacron blue F3G-A have been immobilized on functionalized alkanethiol self-assembled mono-layers as anchored ligand to bind lactate dehydrogenase to gold electrodes. Interestingly, the so immobilized enzyme could catalyse the electro-oxidation of lactate only when the biological cofactor (NAD^+) was present in the reaction mixture, suggesting that the NAD^+-binding pocket used to anchor the enzyme to the monolayer is not involved in the enzymatic reaction [76]. Similarly, substrate analogues such as artificial pyridine-nucleotide analogue can be introduced instead of Cibacron blue F3G-A [77].

In principle, this method is generally applicable, because it is always possible to find a suitable substrate or substrate analogue for every enzyme.

6.3.3.2 Immobilized Non-substrate Ligand

In addition to substrate analogues, other non-substrate ligands such as catechol, 2-(diethylamino)carbonyl-4-bromomethylphenylboronate [92], aminophenylboric acid and [98], triazine scaffold bis-substituted with 5-aminoindan [102], which can selectively bind to the sugar moiety of the glycoenzymes, have been used to orient

6.3 Site-specific Immobilization

Table 6.6 Immobilized substrate or substrate analogues or other ligands.

Carrier/ligand	Ligand	Comment	Enzyme	Ref.
Agarose beads	NAD analogue	Subsequent cross-linking with glutaraldehyde enhances the binding of the enzyme	Lactate dehydrogenase	75
Supported functionalized alkanethiol self-assembled monolayers	Triazine dyes such as Cibacron blue F3G-A	NAD$^+$-binding pocket is not involved in the binding of the dyes	Lactate dehydrogenase	76
Self-assembled monolayers with biospecific affinity	An artificial pyridine-nucleotide analogue as pendant ligand group	The enzymes were site-specifically bound to metallic electrodes with a particular orientation which enables complexation of the ligand into the enzyme's NAD$^+$-binding pocket	NAD(H)-dependent dehydrogenases	77
Cellulose beads	Catechol (2-(diethylamino)carbonyl-4-bromomethyl)phenylboronate	90.12% activity retention; reversibility of the binding as the pH is reduced	Horseradish peroxidase (HRP)	78
CNBr-activated Sepharose 4B, Enzacryl AA, CM-Sephadex G-50	p-Aminobenzoic acid	The enzyme was absorbed on the carrier at pH 4.7 with 0.1 m acetate buffer	Dopa oxidase	79
Sepharose 4B	p-Aminobenzamido-hexyl/pyridoxal 5′-phosphate	–	Tryptophanase	80
Sepharose	Substrate analogues	For purification	Penicillin G acylase	81
Sepharose	Substrate analogues	–	Penicillin G acylase	82

Table 6.6 Continued

Carrier/ligand	Ligand	Comment	Enzyme	Ref.
Sepharose	Substrate analogues	–	Penicillin G acylase	83
CNBr-activated-Sepharose 4B	Aminolevulinate	Ionic strength	Bovine liver I dehydratase	84
Sepharose 4B	Derivative of pyridoxal 5′-phosphate	–	Tryptophanase and tyrosine phenollyase	85
Porous glass	p-Aminophenyl-β-d-thiogalactopyranoside	–	*Aspergillus niger* β-galactosidase	86
Sepharose CL-4B	Analogue of a Phe-Arg dipeptide	The enzyme can be selectively adsorbed to the carriers bearing this peptide ligand, resulting in 110-fold purification	Pancreatic kallikrein	87
Sepharose	Soybean trypsin inhibitor (Kunitz)	–	Peptidylarginine deiminase	88
Non-porous monodisperse silicas	Triazine dyes such as procion red HE3B, procion red MX5B, and Cibacron blue F3G-A	The selectivity for NADH-dependent enzymes was higher than with the two procion dyes	Lactate dehydrogenase and malate dehydrogenase aldehyde reductase	89
–	Ligand C4/6 (2-alanyl-alanyl-4-tryptamino-6-(lysyl)-s-triazine)	Biomimetic affinity ligand	Elastase	90
Agarose	–	–	Adenosine deaminase	91

Table 6.7 Immobilized non-substrate ligands.

Carrier	Ligand	Comment	Enzyme	Ref.
Agarose	Cibacron blue F3G-A	This method has the advantages of simplicity, reproducibility, and rapidity	Sulphurtransferase	93
Sepharose 4B	p-(N-acetyl-l-tyrosine azo) benzamidoethyl	The carrier is regenerable	Proteins and enzymes (e.g. lactase)	94
CL-Sepharose 4B	p-(N-acetyl-l-tyrosine azo) benzamidoethyl	Upon immobilization there was also a dramatic increase in the apparent thermal stability of the lactase, and the mean half-life at 50 °C was increased from 7.2 to 13 days at pH 4.5 and from 3.8 to 16 days at pH 6.5	*Aspergillus oryzae* lactase	95
–	Procion red H 8BN, procion yellow H-A, and Cibacron blue F3G-A	For example, Zn^{2+} promotes binding	CP G2, AK and yeast hexokinase	96
–	Cibacron blue F3G-A and Procion orange MX-G	The binding of ovalbumin to the carrier is selectively enhanced in the presence of Al^{3+}	Ovalbumin	96
γ-Glycidoxypropyl-trimethoxysilane-activated silica	Procion red H-8BN, Procion yellow H-A and Cibacron blue F3G-A and 6-aminohexyl derivatives of dyes	The effect of divalent metal ions such as Mg^{2+} and Zn^{2+} was able to promote the adsorption of metalloenzymes on triazine dye adsorbents	LDH, hexokinase, AK, CP G2, and tryptophanyl-tRNA synthetase	97
Agarose matrixes	Aminophenylboronic acid (APBA-NADP conjugate)	–	6-Phosphogluconate and alcohol dehydrogenases	98
Sepharose 4B	Calmodulin tryptic fragments	–	cAMP-PDE, cAMP-PK, Calcineurin	99
Gel	Pyridinium ligands	Mimics of protein A or G	Immunoglobulin G	100
Biotinylated aminopropyl glass	–	The presence of soluble avidin is necessary for the immobilization. Activity retention was substrate-dependent	Biotinylated transglutaminase	101
Amine-derivatized agarose	Triazine scaffold bis-substituted with 5-aminoindan	The adsorbed enzyme was quantitatively eluted with 0.5 m α-d-methyl-mannoside and to a lesser extent with the equivalent glucoside	Glucose oxidase	102

Abbreviations: LDH, lactate dehydrogenase; AK alkaline phosphatase; CP, carboxypeptidase G2; PDE, phosphodiesterase; PK, protein kinase

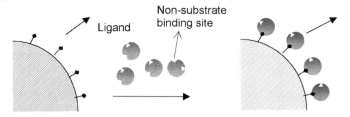

Scheme 6.17 Immobilization of non-substrate ligands.

enzymes on to carriers. Synthetic mimics for protein A [102] or Con A [100] can be also used for the site-specific immobilization (Scheme 6.17).

6.3.4
Genetically Engineered Tags

If the enzyme lacks the innate ability to bind to affinity supports, genetically engineered tags such as polypeptides/domains, for example the cellulose binding domain, protein A, histidine-rich peptides, arginine-rich peptides, and single-chain antibodies can be attached to the targeting enzymes. The simplest genetically engineered tag is the single cysteine residue.

The first oriented enzyme immobilization technique via this approach was proposed by Persson and his co-workers, and was based on genetically introducing an "affinity tag", for example a single cysteine residue, to an enzyme. Thus, this technique not only simplifies purification of the enzyme but also enables site-specific immobilization of glucose dehydrogenase on Thiopropyl-Sepharose [102].

In general, these oriented immobilized enzymes have high retention of activity compared with the randomly immobilized enzyme, because of avoidance of covalent binding, modification of the amino acid residues close to the active centre, and correct orientation of enzyme. They are, however, only applicable for genetically engineered enzymes.

6.3.4.1 Non-covalent Oriented Enzyme Immobilization
When His-tag, a polycationic hexa-arginine, protein A, AG-tag, or Streptag are introduced to the corresponding enzyme, the enzyme can be non-covalently adsorbed by the carriers, which can recognize these orientation groups. For example, enzyme-poly His-tag can be bound to a carrier bearing chelated metals or metal-oxide composites (e.g. Ni–TiO$_2$); enzymes bearing a polycationic hexa-arginine tag can be bound to a polyanionic carrier; enzyme-Streptag can be bound to a carrier bearing Strepavidin, etc. (Table 6.8).

All immobilizations via bio-affinity supports are characterized in that they can be reversibly removed from the carrier. Thus regeneration of the carrier is possible.

Table 6.8 Genetically engineered tags.

Carrier	Tags	Comment	Enzyme	Ref.
Carrier-bound/streptavidin	Streptag-linker peptide (Gly-Ser)$_5$	8.4-fold higher activity was obtained in the presence of linker	Streptag-linker peptide (Gly-Ser)$_5$-Enzyme	42
Thiopropyl-Sepharose/Carrier-CH_2-CH_2-CH_2-SH	Cysteine	Purification-immobilization	Glucose dehydrogenase	103
Electrode with monolayer of thiols/NTA complexed with metal transition ions	Genetically histidine pair (His-X3-His)	Direct electrical communication between the electrode and the enzyme, because of the proper orientation	NADP$^+$ reductase bearing-genetically introduced histidine pair	104
IgG-coated Sepharose matrixes/protein A	Cellulose binding domain, protein A	Higher specific activity and lower K_m values relative to covalently immobilized β-lactamase	β-Lactamase	105
Thermo-sensitive polymeric latex Human γ-globulin (HGb)	Affinity tag AG	75% of its activity in solution and the binding is stable enough to allow reasonable operational stability	AGβgal	106
Conjugated agarose beads/avidin	Fused segment of biotin	Only little loss of activity was observed	Oxidoreductase and luciferase	107
Magnetic beads coated with protein A/MAb	FLAG	Highest activity retention was obtained relative to the randomly carrier-bound cross-linked enzyme or to the randomly immobilized MAb	OPH-FLAG	108
Ni-TiO$_2$ film	His-tag	Same loading but activity twice as high relative to randomly immobilized enzyme	His-tag alkaline phosphatase (ALP)	109

Table 6.8 Continued

Carrier	Tags	Comment	Enzyme	Ref.
Polyanionic matrixes/electrostatic interaction	A polycationic hexa-arginine fusion peptide	The stability of the fusion protein is not affected by pH, urea, or thermal denaturation	A fusion protein of yeast β-glucosidase	110
Thiol-activated Sephadex G-10	Free sulphydryl	—	EcoRI endonuclease	111
	Single cysteine residues	Higher catalytic efficiency compared with subtilisin that was randomly immobilized	Subtilisin	112
Chelating supports	Poly-His-tags	A strong affinity interaction between the poly-His tail and a single chelate moiety is mainly responsible for the adsorption of poly-His GA	Poly(his)$_x$ tagged Glutaryl acylase	113
A biotinylated porous glass	—	Selective adsorption resulted in a one-step purification and immobilization of TRYPSA from crude cell lysate	Trypsin–streptavidin	114
Cu(II)-loaded chelating Sepharose/Agarose	His	Trapping of Cu^{2+} by the enzyme deactivated it. However, this drawback can be overcome by means of a polishing step by chromatography on Mono-Q in the presence of the chelator, EDTA	*Actinoplanes missouriensis* d-xylose isomerase	115
Maleic anhydride-Me vinyl ether copolymers	Six contiguous lysines/a hexa-histidine tag	The double tags enabled a high degree of purification and coupling yield. Also, covalent site-specific binding can be realized in one-step	Proteins	116

6.3.4.2 Covalent Orientation in Enzyme Immobilization

Enzymes, which bear a genetically introduced single cysteine residue can be covalently bound to thioactivated carriers. Thus, orientation, purification, and binding can be realized in one-step.

Alternatively, orientation of the enzyme on a specifically designed carrier can be realized by adsorption of the enzyme via orientating groups such as His-tag on a carrier bearing immobilized IDA-Co^{2+} (or other metal ions). The subsequent covalent binding is completed by means of a neighbouring active functionality that is able to covalently bind enzyme via the corresponding binding chemistry, for example alkylation of a protein with an oxirane functionality bound previously to the carrier, as illustrated in Scheme 6.18.

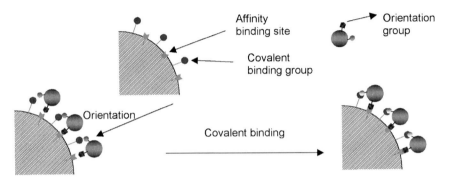

Scheme 6.18 Illustration of site-directed enzyme immobilization followed by covalent enzyme immobilization.

This method was recently exemplified by Cesar and co-workers [117]. It can also be modified to be generally applicable, if a substrate or substrate analogue is immobilized on a suitable carrier. In this way, any enzyme can be immobilized in oriented manner, without any genetic manipulation.

6.4
Immobilization in Organic Solvents

Even today, enzyme immobilization is predominantly performed in aqueous media in which the enzyme is completely soluble. The preference for aqueous media often necessitates the design of carriers and the corresponding binding functionality that are effective in aqueous medium, with regard to stability, activity of the functionality, and the resulting linkage.

Undoubtedly, the chemistry with regard to coupling (binding) conditions or carrier design is severely restricted by the aqueous chemistry available. As a result, in recent decades there has been interest in developing non-aqueous enzyme immobilization techniques.

6.4.1
Covalent Attachment in Organic Solvents

Early in 1970, Brown and co-workers used poly(4-iodobutyl methacrylate) [118], prepared by halogen exchange of poly(4-chlorobutyl methacrylate) with sodium iodide, to immobilize urease. During this work it was discovered that enzymes could be immobilized not only in buffer but also in organic solvents such as dioxane.

In the years after this pioneering work, the feasibility of immobilizing enzymes in organic solvents such as DMF (hydrophilic or hydrophobic) was studied by the same group [118, 135–137]. Consequently, it was further established that fixing of enzymes to a mineral, organic, or organomineral support could be performed in anhydrous organic liquids at temperatures from 60 to 120 °C. Astonishingly, this method enables preparation of support–enzyme complexes with rapid fixation reaction speed at high temperatures without specific pH conditions [136]. Unfortunately, this interesting work did not attract much attention in the following 30 years.

Nevertheless, although the potential of immobilization of enzymes in organic solvents was not fully recognized until the end of the 1980s, several valuable insights were obtained. For instance, it was appreciated that the scope of available chemistry for coupling enzyme to carriers can be extended in organic solvents. Moreover, modulation of the enzyme conformation and activity can be also achieved.

At the end of the 1980s, Stark and Holmberg found that the medium used for the immobilization of lipase could have a large effect on the activity of the lipase in organic solvent but not in aqueous reaction medium. In their study, *Rhizopus sp* lipase was immobilized on two types of Celite-based carrier activated with tresylate, i.e. with/or without PEG as spacer, and in three types of media – aqueous, hexane, or micro emulsion of hexane-buffer.

It was found that although there was no significant difference between the hydrolytic activity in aqueous or organic solvents of the enzyme immobilized in the aqueous medium, lipase immobilized in buffer had no transesterification activity in the organic solvents. In contrast, lipase immobilized in organic solvent was not only fully active in aqueous medium but also in organic solvent, suggesting that the immobilization chemistry was different in two cases [138].

The authors suggested that the enzyme binding functionality that is involved in the coupling with the active tresylate groups of the carriers are mainly phenolic hydroxyl groups of tyrosine and the thiol groups of cysteine, because of their activity in either the protonated or de-protonated form, and the activity of lipase immobilized in organic solvents is probably ascribed to enlargement of the active site by binding of the amino acids residues near the active site.

It was recently revealed that *Rhizomucor miehei* lipase can be crystallized in the open conformation from PEG solution after formation of enzyme–inhibitor complexes or enzyme–cofactor complexes in the presence of organic solvents or detergents [139, 140]. Obviously, the high activity of lipases in organic solvent obtained

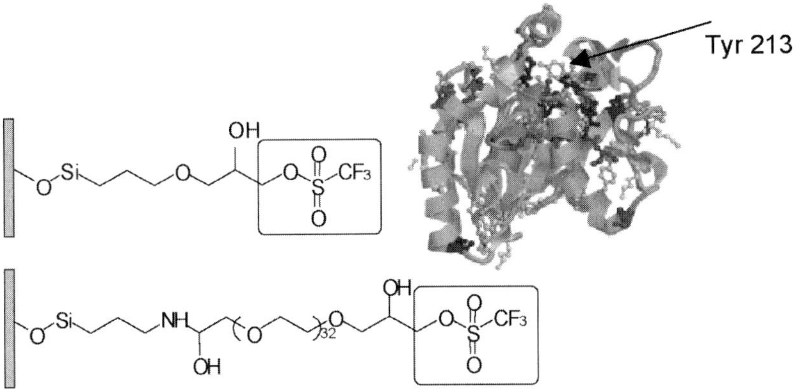

Scheme 6.19 Immobilization of lipase in organic solvent on the activated carrier.

by immobilization of lipases in organic solvents can be explained as a result of opening of the lid of the lipases in organic solvents during the immobilization. In subsequent coupling of the enzyme to the carrier, the conformation with the opened lid could be restricted to this open state by binding of the buried amino acids near the active site. In contrast, the lipase immobilized in aqueous buffer via the lysine groups exposed on the protein surface might have difficulty opening the lid in organic solvents. Moreover, it was demonstrated that introduction of a spacer significantly improved enzyme activity and stability.

As shown in Scheme 6.19, *Rhizomucor miehei* lipase contains a tyrosine residue located on the lid which is only exposed to the medium when the lipase is in the open conformation. Thus, it was most likely that lipase from *Rhizopus sp* was immobilized in an irreversible lid-open conformation and was thus active both in aqueous medium and organic solvents.

6.4.2
Entrapment of Enzyme in Organic Solvent

Immobilization of enzyme in organic solvents other than by covalent methods was also reported in the middle 1976 by Linko and co-workers [141]. Dried Glucose isomerase and β-galactosidase were entrapped in hydrophilic cellulose fibres by dispersing dried enzyme preparations in a mixture of *N*-ethylpyridinium chloride and dimethylformamide with dissolved cellulose.

Similarly, entrapment of enzyme in ethyl cellulose gel can be achieved by mixing with the polymer in dimethylacetamide and precipitating the enzyme–polymer solution, or adding the enzyme solution to the polymer solution dissolved in dioxane [142], followed by lyophilization. It was found that the enzyme immobilized in this way had enhanced thermostability [143].

6.4.3
Immobilization of Organic-soluble Enzyme Derivatives

In recent decades, non-aqueous enzymology has spurred the development of the organo-soluble enzyme, which can be dissolved in organic solvents, with the aim of improving the dispersion of the enzyme and reducing diffusion limitation. These organic-soluble enzymes include lipid coated [144, 145] and acrylated enzymes [146] or PEG or poly(styrene) [147] modified enzymes [148], or non-covalent enzyme–amphiphilic polymer complexes [149].

It was usually found that the modified enzymes were not only more active but also more stable. Consequently, these organic-soluble enzymes were further immobilized by entrapment in organic solvents by the solvent-evaporation method or by cross-linking of acrylated PEG-enzyme conjugates and other monomers or cross-linkers, for example methyl methacrylate (MMA)–trimethacrylate (TMA) in a homogeneous organic solutions [149, 150].

Advantages of these methods are that improvement of the enzyme in terms of activity, stability, and selectivity can be combined. Immobilization of enzyme in organic solvents also has several other obvious advantages, for example:

- many types of reaction that are not favourable for the binding enzyme in aqueous medium can be used;
- modulation of enzyme activity and structure is possible; and
- polymers that are not water soluble can be dissolved in organic solvent for binding to the enzyme, and enzymes immobilized in this way can also be used in aqueous media.

It has recently been discovered that native enzymes can be dissolved in some organic solvents with retention of enzyme structure and activity [152, 153]. This method can be used to immobilize enzymes in organic media in which enzyme activity and conformation can be modulated [152].

6.4.4
Adsorption of Enzyme on to the Carrier in Organic Solvents

It has recently been reported that *Candida rugosa* lipase (CRL) has been immobilized on poly(styrene–divinylbenzene) in heptane with higher enzyme loading than in aqueous medium. CRL immobilized in organic solvent not only had higher activity in both aqueous media (for hydrolysis) and organic solvent (for synthesis) but also had higher operational stability in organic solvents [154].

Similarly, CRL has been immobilized by physical adsorption on several inorganic supports using hexane as coupling medium [155].

6.5
Imprinting–Immobilization

Imprinting enzyme immobilization (IEI) is an enzyme immobilization process in which the geometry of the enzyme activity centre is perturbed by additives such as ligands, substrate, substrate analogues or other components (salts and polymer) and the induced conformation is subsequently frozen by the corresponding immobilization processes, for example covalent attachment, encapsulation, cross-linking or lyophilization). In this way, the activity or selectivity can be improved or changed.

The root of molecular imprinting technique can be dated back to early 1970s. The fact that enzyme conformation is usually affected by changes of the environmental factors such as pH, temperature, ionic strength, and binding ligands lays the foundation for molecular imprinting enzyme immobilization.

In fact, ligands, substrate, and inhibitors [155] are widely used to prevent 3D conformation change during processing such as purification and immobilization, thus protecting the enzyme from deactivation, owing to the tightening effect of these substances. Similarly, imprinting enzyme immobilization, on the other hand, takes advantage of the tightening effect of these substances to create immobilized enzyme with tailor-made properties by freezing the induced conformation. Subsequently, the induced new conformation is immobilized or imprinted by the selected immobilization process.

Along with the imprinting effect exerted by the substrate or substrate analogues, other non-substrates or analogues, for example as salts or polymers, can be also used to alter the activity or selectivity of the enzyme.

6.5.1
Imprinting-Multipoint Attachment

In general, the so-called ligands including substrate, substrate analogues, and inhibitors can be added to the aqueous solution before the enzyme is subjected to covalent binding to the carrier. Consequently, the pre-imprinted enzyme molecules can react with carriers bearing activated function groups. Because multipoint attachment occurs during the immobilization, the induced conformation will be "imprinted", because of the conformation-tightening effect of multipoint attachment. The first immobilized enzyme imprinted in this way was probably reported in the mid-1970s, when Miwa and Ohtomo immobilized glucoamylase on polyacrylic acid azide in the presence of substrates, substrate analogues, and inhibitors and followed this by cross-linking the polymer-enzyme conjugate with hydroxylamine [157]. The presence of the substrate in the active centre often prevents the enzyme from conformation change during covalent binding, thus leading to high retention of activity. For instance, the activity loss (57%) of α-amylase coupled to an oxidized dextran by reductive alkylation in the presence of Na cyanoborohydride was interpreted as the result of steric hindrance near the catalytic site [316]. In contrast, higher activity retention was obtained in the presence of a substrate which protects the catalytic site.

Similarly, a polymeric inhibitor – a water-soluble acrylic polymer with spacer-arms bearing benzamidine groups – was used to protect trypsin during immobilization. The trypsin–VINAC-S (VINAC-S is a polyvinyl matrix with aldehyde groups) conjugates prepared in the presence of this polymeric inhibitor were twice as active toward N^2-benzoyl-*l*-arginine ethyl ester and four times as active toward Hb compared with the corresponding conjugates prepared without the inhibitor [158].

Surprisingly, the activity of invertase immobilized on aminoalkylate magnetic particles by adsorption and cross-linking was 79% of the initial activity in the presence of 1% sucrose, whereas the enzyme immobilized in the absence of the substrate has almost no activity, suggesting that the active site is indeed protected by the substrate [159, 160].

Apart from the protection of the active centre, it was recently found that the kinetic properties of the immobilized penicillin acylase from *K. citrophila*, which was prepared by multipoint covalent immobilization on agarose gel activated by oxidation of the hydroxyl groups to aldehyde groups, was largely dependent on the inhibitors used during covalent binding [161]. This example clearly demonstrated that the presence of the inhibitor induced an inhibitor-dependent conformation, which was frozen by the covalent binding. In other word, the conformation of individual immobilized enzymes was imprinted. The imprinted penicillin G acylase was, moreover, 10,000 times more stable than the common soluble enzyme from *E. coli*. Several inhibitors of penicillin G acylase (penicillin sulphoxide, phenylacetic acid, mandelic acid, and phenylglycine) were used to induce conformational changes. The author suggested that this technique provides a broad spectrum of enzyme derivatives with a range of activity/stability depending on the inhibitors used in their stabilization. The resulting choice offers considerably increased potential for the use of the enzyme, because one can select a derivative which will specifically catalyse the reaction of interest [161].

6.5.2
Imprinting–Cross-linking

Although polymerization of vinylated enzymes bearing unsaturated bonds has been well-known since the mid 1970s [162] for preparation of spherical granule plastic enzymes [163] or for the enzyme stabilization via the complimentary surface attachment proposed by Martinek [164], use of this technology for imprinting the enzyme in the presence of a conformation inducer was not reported until almost two decades later [166]. The principle of this technology was depicted in Scheme 6.20.

It had previously been reported by Staal et al. that α-chymotrypsin can be converted to accept the d amino acid ester as substrate by precipitation, in PrOH, of the enzyme-inhibitor complex between chymotrypsin and *N*-acetyl-d-tryptophan [165]. Following this work, vinylated α-chymotrypsin or subtilisin Carlsberg (with use of itaconic anhydride) were precipitated by 1-propanol in the presence of

6.5 Imprinting–Immobilization

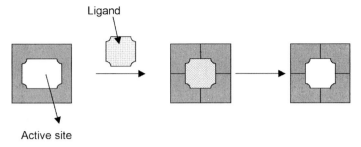

Scheme 6.20 Illustration of imprinting enzyme immobilization

N-acetyl-d-tryptophan and followed by cross-linking with ethylene glycol dimethacrylate in cyclohexane, leading to the formation of immobilized enzymes with imprinted properties (acceptance of d-configured substrates) which do not lose their induced "new" property in the presence of water [166].

Recently, several other enzymes such as epoxide hydrolase [167] were also imprinted with better selectivity toward some substrates. Moreover, the stability of these enzymes was also improved like other similarly cross-linked enzymes without "imprinters". Obviously, this method is able to combine imprinting, immobilization, and stabilization and might thus be useful for preparation of imprinted immobilized enzymes with altered specificity and with high stability.

The size of this cross-linked enzyme preparation is quite small (around 10 µm), as illustrated in Scheme 6.21. Thus, their application might be limited for some applications, for which the size of the immobilized enzyme must exceed a lower limit for carrier-bound immobilized enzyme (100 µm) in order to be used in a sieve-plate reactor [168].

Scheme 6.21 Cross-linked imprinted enzymes (from [166]).

6.5.3
Entrapment–Imprinting

The use of non-covalent sol–gel entrapment techniques for imprinting enzymes was developed in 1985 by Glad et al. [170], who polymerized organic silane monomers in the presence of enzyme or substrate on a silica surface, leading to the formation of an entrapped enzyme with high mechanical stability. Study on the effect of various additives on the performance of the immobilized enzyme using sol–gel entrapped enzyme can, however, be dated back to the end of the 1970s [169].

It was recently also demonstrated by Furukawa and co-workers that lipase adsorbed on celite in the presence of R-2-octanol and further entrapped in a gel formed by n-butyltrimethoxysilane (n-BuTrMOS) and tetramethoxysilane (TMOS) had enhanced activity and selectivity in the esterification of R-(+)-glycidol-n-butyric acid in organic solvents [171]. More recently, it was also discovered that the R and T quaternary conformational states of pig kidney fructose-1,6-bisphosphatase can be trapped by encapsulation in silicate sol–gels [172].

Obviously, confinement of enzyme molecules in a rigid gel matrix can freeze the conformation induced by the individual conformation inducer. This technique is not yet fully developed, however. Further development might be directed toward enzymes with poor selectivity, or combination with other imprinting techniques.

6.5.4
Crystallization and Cross-linking

Because it is known that the changes of crystallization conditions, for example ionic strength and pH, often lead to the formation of different crystals form with different enzyme conformations [173]. Substrate or substrate analogues are, moreover, often used as conformer selectors during crystallization [174], suggesting that the presence of substrate or substrate analogues induces a corresponding conformation in the whole crystal lattice. Thus cross-linking of each specifically-made enzyme crystals might potentially constitute an important enzyme imprinting technique. It has, for example, been demonstrated that many lipases which bear a lid on the top of the active centre can be crystallized in an open or closed conformation, depending on the properties of the crystallization broth or presence of surfactants [175] and it has also been found that the selectivity and activity of the cross-linked crystals of some lipases depend on whether the conformation is open or closed [176, 177]. These examples strongly suggest that enzyme conformation in the crystal lattice is readily imprintable. Combination of imprinting techniques with crystallization-cross linking, however, must still be further exploited to prepare best-suited CLEC for a given application. They also imply that the properties of each enzyme crystal depend on the broth used for crystallization.

This technology is obviously laborious, requiring skill and the time to find appropriate crystallization conditions and to obtain the desired activity and selectivity.

6.5.5
Aggregation and Cross-linking

Analogous to protein crystallization, non-denaturing protein aggregation techniques have recently been used to prepare cross-linked enzyme aggregate with tailored properties [178]. It has been found that enzyme selectivity depends not only the precipitants but also on the additives added during the cross-linking [179]. For example, CLEA of penicillin G acylase, precipitated by ammonium sulphate, had synthetic behaviour similar to that of the native enzyme in the synthesis of ampicillin, whereas CLEA prepared using *tert*-butanol as precipitant had relatively stable performance during the course of the reaction [178]. It seems likely that different precipitants could induce different enzyme conformations and these are subsequently frozen by the cross-linking.

In line with this thought, CLEA of seven lipases e.g. *Candida antarctica* lipase A (CAL-A), *Candida antarctica* lipase B (CAL-B), *Candida rugosa* lipase (CRL), etc., have been prepared, usually with high retention of activity, by varying the precipitant or addition of additives such as SDS, triton, and crown ether, etc. [179].

6.5.6
Intra-molecular Cross-linking – Imprinting

The imprinting enzyme immobilizations (IEI) discussed above are all based on the formation of insoluble chemically aggregated enzymes by inter-molecular cross-linking or confinement of enzyme in a matrix, leading to the formation of solid enzyme preparations. Consequently, a question arises: it is possible to imprint the enzyme molecules by the intramolecular cross-linking, as shown in Scheme 6.22.

Although there is little information available on the possibility of imprinting the enzyme by single molecular imprinting, a clue to the solution to this question can be found in work performed by Royer more than 25 years ago, when he immobilized trypsin and papain on CL-Sepharose-NH(CH$_2$)$_6$-NHCO(CH$_2$)$_2$S-S(CH$_2$)$_2$CO-(*N*-hydroxysuccinimide) and subsequently cross-linked the immobilized enzyme

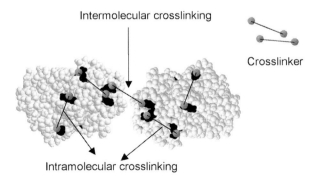

Scheme 6.22 Intra-molecular cross-linking–imprinting.

with dimethyl adipimidate, dimethyl suberimidate, or glutaraldehyde. The cross-linked enzymes were released by reduction of the disulphide linkage between the enzyme and the support. Interestingly, it was found that the cross-linked papain was significantly more stable against autolysis and thermal denaturation than the native enzyme, indicating that intramolecular cross-linking could make the enzyme conformation more rigid [180].

We believe that the presence of an enzyme conformation selector can also induce a conformation and that intramolecular cross-linking might be able to trap the enzyme conformation as it is.

6.5.7
Post-immobilization Imprinting

Until now, preparation of imprinted immobilized enzymes has been mainly based on soluble enzymes. It might, however, be beneficial to use immobilized enzymes instead of the free soluble enzymes for the enzyme imprinting.

Compared with the imprinting immobilization approach, post-immobilization imprinting is obviously advantageous in that the immobilized enzymes are often stabilized. Thus, the immobilized enzymes might be more stable than the native enzymes under the conditions used for the imprinting. Also, more radical conditions such as organic solvents can be used to alter the properties of the enzymes, as mentioned above. Chemistry that is not possible in aqueous medium can also be used.

Few examples are available of the use of immobilized enzymes for enzyme imprinting. It was recently exemplified that entrapment of lipase adsorbed on Celite in the presence of R-(–)-2-octanol was further entrapped in gel formed from n-butyltrimethoxysilane (n-BuTrMOS) and tetramethoxysilane (TMOS) [171].

6.5.8
Lyophilization Imprinting

In recent decades the use of enzyme powders in non-aqueous media has led to the discovery that the activity of enzymes in such media is usually several magnitudes lower than in aqueous media [181]. Although a fair comparison between the enzyme activity in aqueous and organic media is not possible (because they often catalyse different reactions in different media), the huge difference in the reaction rates is an indication that molecular flexibility or enzyme conformation might be quite different under these different conditions. Several methods have been used to improve enzyme activity in organic solvents:

- formation of organic-soluble surfactant–enzyme complexes (reduction of diffusion limitation),
- use of polymer-modified enzymes, for example pegylated enzymes,
- use of other enzyme–polymer complexes, and
- use of lyophilized enzyme additives such as salt, polymers, or other excipients [182].

Among these, lyophilization in the presence of an excipient is the most attractive, because of its simplicity and the avoidance of complicated chemical modification procedures and unexpected effects [183–186].

It has, for example, been found that penicillin amidase preparation prepared by lyophilization in the presence of potassium chloride was 750 times more active in hexane and 225 times more active in acetonitrile than the untreated enzyme, suggesting that the activation was strongly dependent on the solvent [187]. It was also found that the presence of amphiphilic polymers such as PEG or poly(vinyl pyrrolidone) in the lyophilized enzyme preparation could increase enzyme activity by several orders of magnitude. For instance, α-chymotrypsin–poly(vinylpyrrolidone) significantly enhanced enzyme activity in isooctane for transesterification of Ac-PheOEt and 1-propanol [188].

That the stability and activity of lyophilized enzymes often depend on the nature of the excipient used clearly suggests that the lyophilization process is indeed an efficient imprinting approach [182]. It has, however, been found that the conditions for maximum enzyme activity are very marginal. The factors that affect the activity of the final lyophilized enzyme depend not only on the pH, water activity, nature of buffers, nature of the salts or polymers, but also on the ratios of polymer or salts to the enzyme used.

6.6
Stabilization–Immobilization

Since the end of the 1970s, it has been widely recognized that proper immobilization of enzymes often enhances their stability against several denaturing factors such as heat, extreme pH, and organic solvents, owing to one or more of following effects which result from the immobilization techniques chosen:

- multiple point attachment on a complimentary surface – a concept developed in 1970s [164] – and stabilization by multiple-point attachment to solid surface [181];
- stabilization by the favourable microenvironment created in the proximity of the immobilized enzymes;
- stabilization by the modification effect, which results from the modification of enzyme surface functionality (the native form is stabilized);
- stabilization by confinement of the enzyme molecules in a limited space, thus the entropy of enzyme unfolding is reduced;
- stabilization by multiple non-covalent interactions, as for entrapment protein in the gel matrix or protein crystalline lattice.

Consequently many immobilization/stabilization strategies have been developed, in order to obtain stabilized enzymes [189].

In contrast to the immobilization–stabilization strategies developed in recent decades, it has been also recognized that the stability of the immobilized enzymes can be improved before enzyme immobilization [190]. Whereas the stabilization effects of the immobilization–stabilization process all result from the immobiliza-

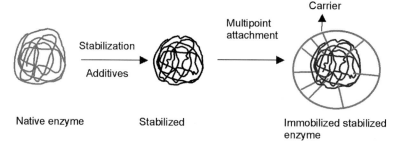

Scheme 6.23 Schematic illustration of stabilization and immobilization.

tion, the "stabilization–immobilization" concept proposed here implies that the enzyme can first be stabilized by use of suitable chemical or physical means before immobilization.

The subsequent immobilization technique chosen serves only to freeze the enzyme in a predetermined stabilized conformation. This technique therefore differs completely from what is called the "immobilization/stabilization" technique, in which stabilization is one result of the immobilization technique used [191, 192].

Although stable enzymes can be obtained from several sources, for example thermophilic strains or engineered strains, these techniques might offer a simple and complimentary method to other approaches for enzyme immobilization/stabilization, because it can provide a rational means of designing an immobilized enzyme with tailor-made properties, namely enhanced stability and selectivity. The principle is illustrated in Scheme 6.23

In fact, many modification–immobilization methods to be discussed below belong to this category. Enzymes stabilized by physical modification, for example use of binding ligands or addition of additives such as sugars and polyols can also be further immobilized [255].

6.6.1
Stabilization by Ligand Binding

It is has long been known that binding substrates or substrate analogues or ligands to a protein can stabilize a protein or enzyme [374]. This effect, the tightening of the enzyme conformation by binding ligands, is a widely used strategy for stabilizing proteins or enzymes during processing procedures including enzyme purification and enzyme immobilization [375], with the aim of avoiding conformation change or, in covalent enzyme immobilization, modification of amino acid residues close to the active site.

According to the model proposed by Schmid, however (Scheme 6.24), binding ligands to enzymes does not always stabilize the enzyme. If binding substrate to the enzyme leads to a conformation of higher internal energy, denaturation through conformation B is facilitated. In contrast, if a conformation of lower internal energy is obtained, the resulting complex is better protected against denaturation [255].

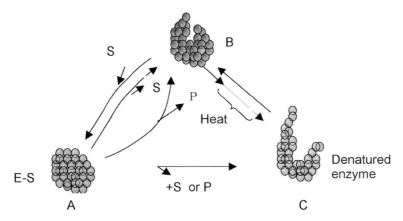

Scheme 6.24 Substrate (or product)-induced deactivation of the enzyme.

The fact that protein denaturant can change the protein structure by stabilizing the denatured form might validate this mechanism. Thus, it is essential to screen a suitable ligand, which is able to reduce the internal energy of the enzyme–substrate complex. In particular, for enzymes involving several substrates or products, selection of suitable ligands is crucial with respect to stabilization of the protein.

For example, penicillin G acylase, which can catalyse the formation of semi-synthetic β-lactam antibiotics, e.g. ampicillin, can also catalyse the hydrolysis of pen G to 6-APA and phenylacetic acid. It was, however, found that 6-APA (6-aminopenicillanic acid) and benzylpenicillin (BP) stabilize *E. coli* penicillin G acylase immobilized in PAA gel (polyacrylamide gel) against thermal inactivation, whereas phenylacetic acid (PAA) increased the rate of inactivation [375]. This result is actually consistent with the discovery by Guizan's groups that the inhibitor could induce an active but less stable conformation [376]. This phenomenon can be explained by the deactivation model proposed by Schmid [255].

As long as the bound ligands induce a stable conformation, the newly induced conformation might be trapped by immobilization or multipoint attachment to the carriers, leading to the formation of stabilized immobilized enzyme. Conversely, a destabilized immobilized enzyme might be obtained, if the enzyme is destabilized by binding a ligand.

It was recently found that the thermostability of immobilized penicillin G acylase prepared in the presence of some inhibitors was generally reduced compared with the control derivatives [376], suggesting that the presence of some inhibitors had induced a conformation that is very close to the deactivated form. Despite the decreased stability of the immobilized penicillin G acylase obtained in the presence of the corresponding inhibitor, this study further implied that stabilization–immobilization was practically possible.

6.6.2
Stabilization by Addition of Stabilizer as the Excipient of the Conformation

Apart from ligands, which can bind to cavities in the enzyme domains, other protein stabilizers, for example sugars, polyols, carboxylic acids, surfactants, some salts, amino acids, cryoprotectants, and metal ions can protect the protein by stabilizing it in a conformation that is close to the native form. Although the stabilizing mechanisms are not binding, but exclusion, these stabilizers can principally also be used for this purpose, i.e. as "conformation excipients". For example, deposition of lipase on the carrier surface in the presence of additives such as albumin, gelatine, casein, and PEG often protect the enzyme from deactivation [377]. Although this was interpreted as a result of preventing the enzyme from deformation on the carrier by occupying the surface-active adsorption site, the stabilizing effect of these additives before deposition cannot be excluded.

In entrapment of enzymes in sol–gel processes it is often found that the presence of additives is crucial with regard to activity retention [378, 379]. Although the mechanism of action of these additives is complex, it is possible to postulate that the pre-immobilized enzyme conformation and the structure of the gel can be affected by the additives.

The stability induction by the additives was also demonstrated by immobilizing epoxide hydrolase from Nocardia EH1 on DEAE-cellulose via ionic bonding. The fact that the thermal and operational stability of the immobilized epoxide hydrolase can be significantly improved by adding Triton X-100 during the immobilization undoubtedly suggested that even the enzyme stability can be imprinted [389].

6.6.3
Stabilization by Pre-immobilization Modification

In recent decades much attention has been paid to the immobilization–stabilization strategy [134, 380]. One important conclusion is that immobilization might not always be able to stabilize the enzyme. Consequently, much endeavour has also been devoted to the development of stabilization–immobilization strategies, based on the pre-modification of the native enzymes before binding to the carriers. The aim is often to strengthen binding of the enzyme to the carrier, control of the mode of binding (properties of linkage and number), or improve enzyme performance, for example activity, selectivity, and stability.

In general, stabilization by pre-immobilization modification can be classified into two types – stabilization by alteration of the microenvironment and stabilization by rigidification of the enzyme conformation by cross-linking.

6.6.3.1 Stabilization by Pre-immobilization Modification with Soluble Polymer
Covalently attaching soluble polymer–enzyme adducts was probably pioneered by Nakhapetyan and Akparov at the beginning of 1980s, when subtilisin was modified with dextran and subsequently immobilized on aminolated silica by several ap-

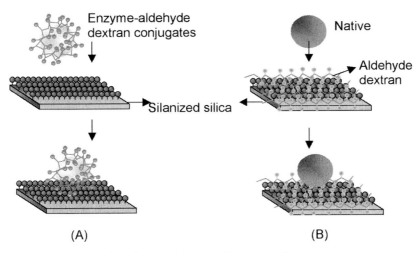

Scheme 6.25 Two methods for immobilization of enzyme on the amino silica mediated by aldehyde dextran.

proaches [318]. Following this work many other enzymes, for example invertase, trypsin, α-amylase [316], β-glucosidase [319, 320], and penicillin amidase [321], have been modified and immobilized on the carriers.

The α-amylase–dextran complex was prepared by attaching enzyme to periodate oxidized dextran by reductive alkylation in the presence of sodium cyanoborohydride. Subsequent binding of this enzyme–polymer conjugate to an aminalkylated silica led to the formation of immobilized enzyme with greater retention of activity (45 %, compared with 4 % for the native enzyme) [316]. The immobilized enzyme-polymer conjugate was, moreover, more thermostable than the enzyme immobilized on silica without dextran modification. This difference was obviously attributed to variation of the length of the spacer between the silica and the enzyme and the extra stabilization effect of dextran on the enzyme (alteration of the microenvironment) [147], as shown in Scheme 6.25.

Recently, the overwhelming potential of using the chemically modified enzyme for the further immobilization was exemplified by a comparative study of the immobilization of β-glycosidase by two different but related methods. In the first method, a conjugate of β-glucosidase with aldehyde dextran was immobilized further on aminolated silica whereas in the second approach the enzyme was immobilized on aldehyde dextran-coated aminolated silica. Surprisingly, it was found that not only high volume activity but also high thermostability can be obtained with the first approach [319].

Interestingly, it has been found that aldehyde dextran–penicillin G acylase immobilized on various aminoalkylated silica carriers led to approximately threefold enhancement of the thermal stability of the enzyme [321]. This could be ascribed to the alteration of the microenvironment (more polar) or confinement of the enzyme molecules (Scheme 6.26).

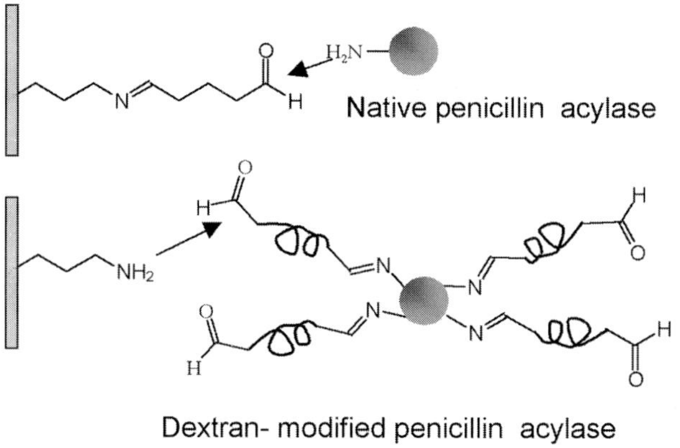

Scheme 6.26 Immobilization of penicillin G acylase and aldehyde dextran-penicillin G acylase on aminoalkylated silicas.

6.6.3.2 Stabilization by Pre-immobilization Chemical Modification

For chemical modification a variety of chemical modification techniques can be performed on the enzyme to be immobilized. Some of the chemical modification techniques are often able to stabilize the enzyme molecules. For example, it has been found that the stability and activity of the immobilized invertase on BrCN-activated Sepharose are very dependent on the modification and the properties of the modifying agents. The highest activity and stability were obtained by periodate treatment followed by reaction with ethylenediamine, as shown in Scheme 6.27.

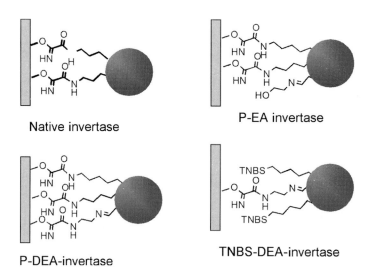

Scheme 6.27 Immobilization of chemically modified invertase on BrCN-activated Sepharose 2,4,6-trinitrobenzene sulphonic acid (TNBS).

From a stability test at 60 °C in 0.25 M sodium acetate buffer (pH 5.0), it was concluded that immobilization enhanced enzyme stability relative to the native enzyme. The stabilization effect on the immobilized enzymes obtained by different modifications is, however, mainly contributed by the first chemical modification [192]. The stability of the immobilized enzymes followed the order: SP-TNBS-invertase < SP-Invertase < SP-EA-Invertase < SP-DEA-invertase, i.e. stability was highest for SP-DEA-invertase.

It is highly probable that the stability of the immobilized invertase is related to the number of bonds and the microenvironment. Introduction of extra amino groups might enhance the multipoint attachment. This might explain the stability of SP-DEA-invertase. On the other hand, introduction of a polar hydroxyethyl group into invertase might also improve the stability of the enzyme, as shown in Scheme 6.27.

6.6.3.3 Chemical Cross-linking/Covalent Immobilization

Moreno and Fagain were probably the first to demonstrate that chemically cross-linked and thus stabilized enzymes can be further immobilized on carriers [320]. It was found that alanine aminotransferase modified with dimethysuberimidate and subsequently immobilized on a pre-activated agarose had high retention of activity (75 % instead of 50 %). More interestingly, it was found that the immobilized modified enzyme was approximately five times more thermally stable than the immobilized unmodified enzyme and 20 times more thermally stable than the free counterpart [320].

It was, nevertheless, found that the major stabilization effect originated from the first step – the modification step, suggesting that the enzyme structure was probably rigidified by chemical modification (intramolecular cross-linking).

6.7
Modification-based Enzyme Immobilization

Modification-based enzyme immobilization encompasses methods by which native enzymes are first chemically modified, using a chemical reagent or a chemical cross-linker, often leading to improvement of the enzyme function. Subsequently, the modified enzyme preparations (either soluble or insoluble) can be further immobilized with a suitable immobilization technique. The immobilized enzymes can, moreover, also be further modified with the aim of further improving enzyme stability, activity, and selectivity.

Although immobilization of modified enzymes on carriers was proposed by Glacer in 1962, with the aim of controlling the binding mode and number of bonds, the idea of improving the enzyme properties such as stability by pre-immobilization modification (PIM) was first proposed by Zaborsky at the beginning of 1970s [190]. In general, the modification-based enzyme immobilization techniques can be classified as modification and immobilization or immobilization and modification.

6.7.1
Immobilization then Modification

It has recently been found that the properties of immobilized enzymes such as thermostability, operational stability, activity and selectivity can be improved by so-called consecutive treatment, which can be one of two types – physical post-treatment and chemical post-treatment. With this method the immobilized enzyme can be subsequently modified physically or chemically, with the objective of further improving the performance, i.e. activity, stability, and selectivity, as illustrated in Scheme 6.28

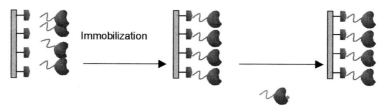

Scheme 6.28 Immobilization–modification.

It was recently found that physical post-immobilization treatment is crucial for some immobilized enzymes. For example, the pH imprinting technique is very crucial if the immobilized enzyme is to have optimum activity in anhydrous organic solvents [193]. Careful washing of immobilized enzyme such as CLEA or CLEC with organic solvents has also proven to be crucial to activity [193]. Co-lyophilization of the immobilized enzymes with additives such as sugars and PEG has long been known to be crucial for storage stability or activity in organic solvents [194].

Instead of physical treatment the immobilized enzyme can be further subjected to chemical modification. For instance, chemical modification of immobilized penicillin acylase from *E. coli* with formaldehyde led to a much more stable preparation, compared with the control [195]. More interestingly, changing the pH after immobilization was able to enhance the multiple attachments, leading to the formation of stabilized immobilized penicillin G acylase [196]. Consecutive modification of the matrix with compounds such as bovine serum albumin has also led to stabilization of immobilized penicillin V acylase [197].

A more detailed discussion of this type of enzyme immobilization technique can be found in Section 6.8.

6.7.2
Modification then Polymerization

Jaworek was an early pioneer who copolymerized an enzyme bearing an unsaturated bond to form an entrapped enzyme in a gel matrix with high specific activity, high retention of activity, and other advantages such as no swelling or shrinkage of the gel, no adsorption of charged substrates or reactants, unchanged kinetic properties, and high activity yield of the proteins subunits [162]. After this work, several publications describing the preparation of such entrapped enzymes appeared in the mid 1970s [197, 198].

It has been found that introduction of polymerizable functional groups into a protein enables the preparation of spherical granular enzymes [163], as shown in Scheme 6.29. Enzyme plastic has been formed by polymerizing AMP deaminase with cross-linking of glycidyl methacrylate in the presence of a cross-linker such as methylene bisacrylamide to form the plastic immobilized enzyme with ca 15% retention of activity. It has been found that the immobilized enzymes obtained are usually stable compared with the native enzymes or other forms of immobilization, probably because of rigidification of the conformation [199].

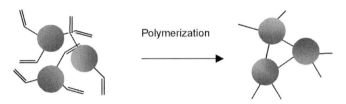

Modified enzyme with unsaturated bond   Chemically crosslinked enzyme

Scheme 6.29 Copolymerization of enzyme modified with unsaturated compounds.

Other enzymes, for example immobilized penicillin G acylase, which was first modified with maleic anhydride, were subsequently co-polymerized with acrylamide via a cross-linking agent. In this spherical granular form they are more resistant to heating, have lower affinity for benzylpenicillin, and are less inhibited by phenylacetate than the native enzyme. Substrate specificity and optimum pH are unchanged [201, 202].

This technique has also been used to increase the thermostability by so-called complimentary multipoint attachment [203]. The immobilized enzymes obtained, for example α-chymotrypsin, were also found to be more stable in organic–aqueous media [204].

This method has also recently been combined with bio-imprinting techniques for imprinting of several enzymes, for example protease [202, 166] and epoxide hydrolase [167], or pegylated enzyme entrapped in a PVA gel [207] with altered selectivity.

Apart from entrapment [208] or cross-linking in the presence of monomers, for example PAAm, modified enzymes bearing double bonds, for example glycidyl methacrylate-treated horseradish peroxidase, can also be grafted on to a carrier surface, leading to the formation of partially bound and partially entrapped immobilized enzymes with greater retention of activity and less diffusion limitation [209]. A method which takes advantage of this technique and enzyme–polymer conjugates was developed by Yang et al., who entrapped the enzyme-PEG conjugate (which bears an acrylate group at the end of PEG) in polyacrylates and found that the immobilized enzymes obtained were more stable toward heat and organic solvents [150].

6.7.3
Pre-immobilization Improvement Techniques (PIT)

For the purpose of this discussion, these techniques are called pre-immobilization improvement techniques (PIT) because they are used before enzyme immobilization. The use of PIT can be dated back to the early 1960s, when trypsin was modified with N-carboxyanhydrides of amino acids, leading to the formation of enzyme derivatives with polytyrosine residues, which can, consequently, be bound to the carrier [210]. Although the original aim was to avoid direct linkage of the enzyme and the carrier and, particularly, to control the mode and amount of bonding [382], new attempts to use this method for enzyme immobilization mainly focused on the pro-immobilization improvement of the enzyme, with regard to the activity, stability, and selectivity to be expected in the subsequent immobilization. The concept of stabilization and immobilization was probably proposed by Zaborsky in 1972 [190]. Since then, increasing attention has been paid to combination of enzyme modification with enzyme immobilization.

Current PIT can be placed in many categories, depending on the purpose and the methods used for immobilization. Those most frequently encountered in enzyme immobilization are:

- introduction of extra charges,
- interconversion of AAR (amino acid residues),
- alteration of enzyme hydrophobicity,
- introduction of unsaturated bonds,
- increasing enzyme solubility in organic solvents,
- mitigation of the toxic effect of the carrier surface.

Although originally the soluble modified enzyme was covalently immobilized on another insoluble activated polymeric carrier, several other techniques have been developed to make use of the improved enzyme derivatives; these include entrapment of the modified enzymes [211], or adsorption or covalent enzyme immobilization, encapsulation, and cross-linking, as shown in Scheme 6.30.

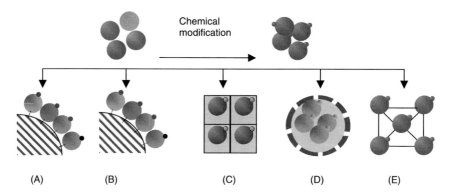

Scheme 6.30 Chemical modification and subsequent enzyme immobilization.

6.7.3.1 Introduction of Extra Charge

Since the first industrial biocatalytic process catalysed by amino acid acylase immobilized ionically on an ion exchanger, immobilization of enzyme by ionic adsorption has gained increasing attention, because of the simplicity of the technique and the mild conditions used for immobilization. One intrinsic drawback is, however, that the adsorptive force is not sufficiently strong to prevent enzyme wash off, compared with covalent enzyme immobilization.

To intensify the interaction of the enzyme with the charged carrier, an interesting method, based on the use of succinated enzymes instead of the native enzymes, was introduced in the early 1970s. Because of the introduction of extra negatively charged groups on the protein surface, adsorption of enzymes on the anionic exchanger was enhanced [212] as illustrated in Scheme 6.31.

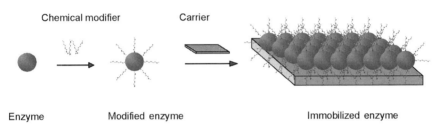

Scheme 6.31 Pre-immobilization Modification Immobilization (PIMI) Techniques.

It was also found that pyromellitic dianhydride-modified amyloglucosidase could be strongly adsorbed by DEAE-cellulose. The modified enzyme had stronger binding and could be eluted only with 0.25 M NaCl [213]; succinylated cyclodextrin glycosyltransferase of an alkalophilic Bacillus was also adsorbed on a vinylpyridine copolymer with 25 % retention of activity and enhanced optimum temperature (from 50 to 55 °C) [214].

Theoretically, positively charged groups can also be introduced to the proteins or enzymes. For the purpose of enzyme immobilization, however, this method has been not intensively exploited. One example is introduction of amino functionality to oxidized enzymes (aldehyde-bearing enzyme), which was subsequently immobilized on CNBr-activated Sepharose [53].

It is, however, worth mentioning that it is essential to control the degree of derivatization, because introduction of the extra charges will distort the protein structure, for example causing expansion of the structure as a result of repulsion between the newly formed charges, or conformation change, because of elimination of the original salt bridges. These effects will obviously disturb the forces maintaining the native structure, leading to deactivation or destabilization of the enzyme owing to the changes of enzyme conformation [215].

6.7.3.2 Alteration of Enzyme Hydrophobicity

By analogy with ionic adsorption, hydrophobilization of the enzyme surface by attachment of aromatic compounds such as methyl benzimidate hydrochloride enhanced the adsorption of the enzymes on hydrophobic carriers such as Amberlite XAD 7. It has been found that modification of enzymes such as trypsin, yeast alcohol dehydrogenase, and *Escherichia coli* asparaginase with hydrophobic imido esters led to almost quantitative adsorption of protein on XAD-7 polymer beads [216].

Other modifiers, for example monomethoxy PEG (1900) activated with *p*-nitrophenyl chloroformate or amino PEG activated with glutaraldehyde can also be used to modify enzymes and thus enhance their hydrophobic adsorption on polymeric carriers [217, 218]. Often, not only enhanced adsorption but also enhanced activity of the immobilized enzyme in organic solvents can be achieved [219].

Interestingly, the hydrophobic interaction of hen egg-white lysozyme on DEAE-cellulose was also enhanced by succination [220]. It seems like that the adsorptive force was determined by the pH used. Remarkably, succinated cyclodextrin glucanotransferase (CGTase) adsorbed on Amberlite IRA 900 has a lower optimum temperature than the native enzyme [221].

6.7.3.3 Formation of Polymer–Enzyme Conjugate

Among the different PIMI techniques, modification of native enzymes with water or organosoluble polymers to form different insoluble or soluble enzyme derivatives might be of widespread interest [222, 223], because these soluble enzyme–polymer complexes, for example glucose oxidase–polyamine (a primary amine derivative of poly(vinyl alcohol)) [223], pegylated enzymes [233], dextranylated enzymes [222], ethylene–maleic anhydride copolymer–glucoamylase, alginic acid–lysozyme [119], gelatine, starch [129], porcine pancreatic α-amylase alginate [224], and soluble oxidized cellulose-trypsin complex, are often stabilized. The first soluble enzyme-polymer was probably pioneered by Wykes and co-workers at the beginning of 1971 [225], with the aim of retaining the enzyme in a membrane reactor [120].

In general, water-soluble covalent enzyme–polymer conjugates can be formed by reacting enzyme with activated water-soluble polymers such as activated dextran [316], gelatine and a variety of activated PEG, etc. The advantages of the method might be the stabilization effect of the chemical modification step [226, 227], because conjugation of enzyme molecules with soluble polymers such as PEG and dextran often leads to thermo-stabilization [228]. The resulting water-soluble immobilized enzymes can be directly applied in membrane reactors, because of enlargement of the molecular size. They can, however, also be further immobilized by means of a variety of immobilization techniques, for example cross-linking, entrapment [229–231], adsorption, and covalent attachment techniques [316, 321].

Entrapment of Soluble Enzyme–Polymer Conjugates Entrapment of enzyme–polymer complexes was possibly first proposed in the 1970s by Hueper [129]. As noted above, this technology has the advantage that the enzymes are usually stabilized

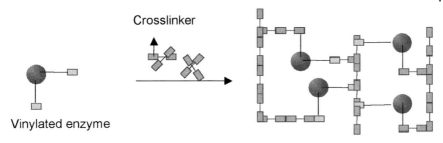

Scheme 6.32 In-situ entrapment and cross-linking.

by the first modification step. The soluble enzyme–polymer complex can also be directly used in membrane reactors – because of the size enlargement, the enzyme cannot pass through the membrane. Among the different modification–stabilization–immobilization strategies used, an interesting one is the cross-linking of pegylated enzymes with an acrylate group at the terminus of the PEG chain, as illustrated in Scheme 6.32.

It is known that not only PEGylation but also the cross-linking of enzyme often stabilizes the enzyme in organic solvents. Thus, this type of immobilized enzyme might have enhanced thermal stability and resistance against water-miscible organic solvents [150]. However, the enzyme–polymer complex can be also directly immobilized in the polymer matrix, as illustrated in Scheme 6.33.

For example, soluble covalent conjugates of carboxypeptidase B (CPB) and trypsin with poly(vinylpyrrolidone-co-acrolein) were further entrapped in poly-N-vinylcaprolactam, resulting in about 90% and 75% of original trypsin and CPB activity, respectively. Remarkably, the optimum temperature of the entrapped enzymes was approximately 25 °C higher than that of the soluble enzymes [232].

α-Chymotrypsin from bovine pancreas covalently modified with homo-bifunctional polyethylene glycol derivatives and entrapped in Ca-alginate gel particles had improved thermal and operational stability in the hydrolysis of N-acetyl-l-phenylalanine Me ester (APME) in both batch and continuous fixed bed reactors [231], suggesting that the first step undoubtedly stabilizes the enzyme. Similarly, native bovine serum amine oxidase (BSAO) and poly(ethylene glycol) (PEG)-treated ("PEGylated") BSAO were immobilized in a hydrogel during its synthesis, with improved operational stability in the presence of substrate [230, 231].

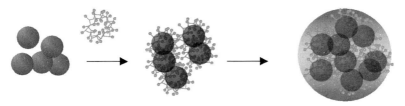

Scheme 6.33 Enzyme conjugation with soluble polymer before formation of gel beads [129].

Occasionally, however, when diffusion constraints dominate, activity retention is very low. For example, pegylated enzymes entrapped in a PVA cryogel can be used in organic solvent but the high diffusion constraints in the gel matrix reduce the activity of the enzyme to 10% of the pegylated enzymes [233].

Leakage of entrapped enzyme from the gel matrix is often ascribed to the ability of the enzyme to re-dissolve in an aqueous medium. In a recent report, enzyme was first cross-linked and then entrapped in a gel matrix. For example, β-galactosidase has been co-cross-linked with bovine serum albumin then entrapped in agarose beads to prevent enzyme leakage [121]. In line with this idea, enzyme–Con A complexes (polymeric) can be also entrapped in a gel matrix, for example entrapment of glycoenzyme–con A complex in alginate matrix [122].

Covalent Binding of Polymer–Enzyme Conjugates to Carrier   Covalently attachment of enzyme–polymer conjugates follows the principles developed for conventional covalent enzyme immobilization. Briefly, the soluble polymers are covalently bound to the enzyme, resulting in the formation of the soluble enzyme–polymer complex which can subsequently be immobilized on carriers by covalently binding. The principle is illustrated in Scheme 6.34.

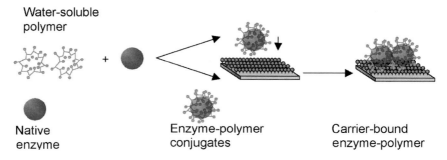

Scheme 6.34 Principle of immobilization of conjugated enzyme–polymer complexes.

Among the different polymers, for example dextran, soluble starch, gelatine or soluble cellulose, that most frequently used is dextran, which can be easily oxidized to the aldehyde form. Thus the enzyme can be easily bound to dextran aldehyde. The resulting enzyme–polymer can be bound to aminolated carriers such as silanized silica.

In general, the introduction of polar polymeric chains renders the microenvironment polar (hydrophilic), thus stabilizing the enzyme, as in the discussion of stabilization by pre-immobilization modification. For more detail the reader should return to Section 6.6.3.2.

Table 6.9 Conjugation–entrapment.

Polymer modifier/ entrapment matrix	Method	Remark	Enzyme	Ref.
Eudragit S-100	Covalent	The covalent enzyme-polymer conjugate retained 93% retention of proteolytic activity. Moreover, thermal stability of the immobilized enzyme was more than two fold enhanced relative to the free enzyme at 40 °C for 1 hour exposure	α-Chymotrypsin	120
BSA/Agarose or gellan gum	Conjugation–entrapment	Immobilization of β-galactosidases by co-cross-linking with bovine serum albumin, followed by entrapment in agarose beads is able to avoid enzyme leakaged	Cross-linked β-galactosidases-BSA	121
Con A/Alginate	Conjugation–entrapment	Non-covalent enzyme-Con A complex entrapped in alginate gel matrix	Glycoenzymes	122
Activated PEG/PVA cryogel	Conjugation–entrapment	The pegylation of the enzyme not only enhanced the enzyme stability but also prevented the enzyme from leakage	Pegylated glucose oxidase	123
Dextran and a copolymer of maleic acid anhydride and acrylic acid	Covalent	It has been revealed that the stability and activity of the enzyme-polymer conjugates are largely dictated by the nature of the polymer	α-Chymotrypsin	124
Dextran	Covalent	The conjugate retained complete activity but also high thermostability relative to the native enzyme	Catalase and trypsin	125
Agar, carrageenan, alginate, gelatine or low-Me pectin	Entrapment	Enhanced thermostability	Tannin-aminopeptidases	126
Alginate	Entrapment by gelation	Enhanced activity (75%) and good operational stability were obtained	Aminoacylase-alginate covalent complex	127
Cross-linked PGA	Entrapment	Mitigation of substrate inhibition at high concentration	The β-glucosidase-CAS complex	128
Gelatine gel spheres	Entrapment	Mitigation of substrate inhibition at high concentration	The β-glucosidase-CAS complex	128
Dextran or starch	Entrapment	Covalent bonding to water-soluble hydrophilic polymer, followed by embedding in a gel matrix enhanced enzyme stability and utility	Penicillin G acylase	129

Table 6.10 Classification of post-treatment of immobilized enzymes.

Nature	Methods	Comments	Ref.
Physical method	Solvent washing (rinsing)	Solvent washing can homogenize the immobilized enzyme by denaturing the adsorbed enzymes	246, 247
	pH raising	Increasing pH is able to enhance multipoint attachment, thus further improving enzyme stability	248, 254
	pH-imprinting	Optimization of enzyme activity in low water medium	388
	Addition of additives	Modulation of enzyme activity	179
	Lyophilization in the presence of additives	Modulation of enzyme activity and selectivity	187, 188
	Coating, entrapment or encagement	Immobilized enzymes (carrier-bound, covalent or other non-covalent) can be further physically entrapped or encaged, with the aim of improving activity, stability and selectivity	249–250
	Adjustment of water activity	Equilibrating the solid enzymes with selected salt with desired water activity in a closed vessel	279
Chemical method	Extra cross-linking and conformation imprinting	Enhancement of enzyme stability and conformation stability	134
	Consecutive chemical modification	Enhancement of enzyme activity	197
	Neutralization of excess carrier-active binding sites	Modulation of enzyme activity and stability by alteration of the microenvironment	253
	Alteration of catalytic activity	Synthetic enzymes	244
	Consecutive intramolecular cross-linking	Enhancement of conformational stability and thermostability	294

6.7.3.4 Introduction of Active Functionality for Covalent Binding

Chemically introduced enzyme-bound functionality (EBF) that takes part in the binding are functional groups introduced to the enzyme by chemical modification; these artificial EBF subsequently react with the carriers without further activation of the carriers. Because the aim of introducing artificial EBF is to avoid heterogeneous binding chemistry, controlled modification of native EBF, for example amino residues, with other mild reagents, for example mercaptopropyl imido ester, is often required to produce soluble enzyme derivatives with limited numbers of artificial EBF, with the aim of achieving higher orientation of the enzyme molecules, thus improving activity retention and enzyme stability. Further immobilization via these newly introduced EBF might produce an immobilized enzyme with a controlled number of bonds and minimized heterogeneity, for example the binding shown in Scheme 6.35.

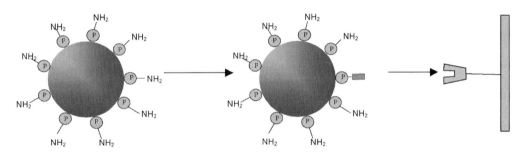

Scheme 6.35 Controlled modification of enzyme residues bearing amino groups.

Binding modified enzyme to the carrier was pioneered by Glazer in 1962, when modified enzymes with tyrosine residues were bound to a carrier bearing diazonium functionality [210]. Although this method did not receive enough attention, introduction of an artificial group into the protein molecules was regarded as the only method for controlling the mode of binding and number of bonds of the immobilized enzymes [234].

This idea has since been regarded as the cornerstone of modification–immobilization techniques and was intensively studied in 1980s and 1990s, with the aim of improving the activity, stability, and selectivity of the enzyme before it was immobilized [190]. For the purpose of enzyme orientation or minimizing the heterogeneity of bonding, isocyanate (–NCO), formyl (–CHO), sulphydryl (–SH), tyrosinyl (–PhOH), or phenylamine (–PhNH$_2$) functionality can be introduced, as shown in Scheme 6.36.

Among these enzyme modifications, thiolation, developed by Benesch and Benesch in 1956 [234], has been intensively studied for modification of protein function and immobilization of enzyme by use of artificial EBF. For example, sulphydryl functionality was introduced into several enzymes, for example α-chymotrypsin, RNase A, and trypsin, by thiolation with *N*-acetyl-*dl*-homocysteine thiolactone.

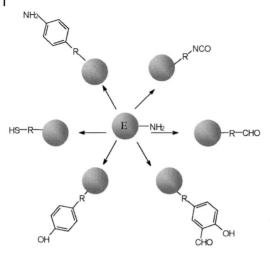

Scheme 6.36 Introduction of artificial enzyme binding functionality.

The modified enzymes were subsequently selectively immobilized on to polymers such as a copolymer of acrylonitrile and 1,4-divinylbenzene (with the latter as cross-linking reactant) and a gel cross-linked copolymer of acrylonitrile, acrylic acid, and 1,4-divinylbenzene bearing maleimide groups [235].

Modification of enzymes with difunctional cross-linkers such as di-isothiocyanate and difunctional aldehydes can also be used to activate enzymes bearing free amino groups, which can subsequently be bound to carriers bearing the corresponding functionality. For example, trypsin and pepsin modified by treatment with 3-isothiocyanato-1-propylisocyanate can be bound to carriers bearing NH_2 groups [234].

In a strict sense, activation of carboxyl residues with carbodiimides for immobilization of enzyme on carriers bearing alkyl or arylamino groups [236] also belongs to this category. Remarkably, it has been observed that most enzyme activity was retained by coupling the carbodiimide-activated enzyme to the carrier with alkyl or arylamino groups at the end of a longer spacer, implying that the activated carbonyl groups have some orientation effect [236]. Similarly, conversion of other non-amino acids residues, for example sugar moieties, into active amino acid residues by oxidation, followed by derivatization by use of a diamino compound or a non-protein amino acid such as glycyltyrosine, led to the formation of enzyme derivatives with new bonding functionality that could be immobilized on carriers bearing active groups [53]. Conversion of enzyme tyrosine residues into aminotyrosine is another efficient method of oriented enzyme immobilization, because the amino tyrosine residue is active at low pH at which other BEF might be not active.

Along with covalent bonding, a limited number (1 or 2 amino groups per enzyme molecule) of artificial chelatable functionality, for example N-succinimidyl 3-formyl-4-hydroxybenzoate, can also be introduced into an enzyme. Enzymes modified in this way can be immobilized on a suitable carriers by coordination bonding [237]. Because of the availability of vast number of chemical modifiers, this method might be a potentially powerful means of engineering a protein with desired orientation groups or improving performance such as stability and activity before enzyme immobilization. The plurality of some amino acid residues in the enzymes often makes it difficult to selectively introduce the artificial orientation functionality to the enzymes, however. Further research should be directed at development of site-specific modification techniques. In practice, this method of oriented immobilization is of less importance, because the chemical orientation groups can be introduced to the carriers instead of modifying the enzyme molecules, as will be discussed below.

Activation of enzymes with a difunctional cross-linker such as PEG–biooxirane can combine introduction of an active EBF and a spacer into the enzyme to be immobilized. Subsequent immobilization of the activated enzyme on a carrier bearing amino or sulphuryl groups will lead to the formation of an immobilized enzyme with a long hydrophobic spacer. With this method, α-amylase and β-galactosidase were immobilized on thioagarose with 10% and 75% retention of activity, respectively [238]. Interestingly it was found that increasing the PEG-to-enzyme ratio usually lead to reduced residual activity, suggesting that reducing the molecular mobility also reduces enzyme activity, irrespective of the presence of the spacer.

6.7.3.5 Introduction of Mediators
In addition from the pre-immobilization techniques discussed above, other pre-immobilization techniques developed with the objective of improving electron-transfer efficiency between redox enzymes, have also been developed. For example, phenothiazine mediators with poly(ethylene oxide) spacers attached to the surface of glucose oxidase significantly enhanced the electron-transfer rate [239].

6.7.3.6 Interconversion of Amino Acid Residues (AAR)
Occasionally the enzyme to be immobilized might contain few EBF and stabilization by multipoint attachment might be difficult. In such circumstances inter-conversion techniques (converting one type of abundant amino acid residue into another type) might extend the possibility of using multipoint attachment to stabilize the enzymes. One of the early examples was furnished by Hsiao and Royer, who converted the carbohydrate moieties of several enzymes such as glucosamylase, peroxidase, glucose oxidase, and carboxypeptidase Y into amino groups before immobilization [53].

Recently, the same concept was applied to penicillin G acylase and glutaryl acylase. The surface carboxyl groups of these two enzymes were partially concerted to amino groups. Thus, the number of bonds formed between the enzyme and the active carrier (glyoxyl agarose) can be enhanced, leading to the formation of immobilized enzymes with enhanced stability as compared with the conventionally immobilized enzymes on the same carrier [387].

6.7.3.7 Cross-linking/Immobilization

Before the binding to the carriers, enzymes can be cross-linked, with the objective of combining the stabilizing effect of the cross-linking with easy recovery of the enzyme by selecting an appropriate carrier [240].

6.8
Post-Immobilization Techniques

6.8.1
Introduction

Since the 1970s it has been realized that binding of an enzyme to a carrier is not the whole story of enzyme immobilization. Often, the immobilized enzyme must be subjected to a variety of physical and chemical treatment, with the aim of further improving its activity and stability, for instance improving storage stability by lyophilization, etc. These techniques, which are used to improve ready-made immobilized enzymes, are usually called post-immobilization techniques or post-treatment techniques [239]. In the past 50 years much knowledge has been accumulated on the various immobilization techniques. Post-treatment of immobilized enzymes – an important step of enzyme immobilization would still benefit from further exploration, however.

Although the term "post-treatment" was not mentioned until the beginning of 1990s, use of these techniques can be dated back to the mid of 1960s, when immobilized acylase was further treated with concentrated urea solution, resulting in significant enhancement of its activity [241]. In 1970s, different immobilized enzymes were occasionally lyophilized with sugars to improve the stability of the immobilized enzymes [242]. In the last decade, inspired by non-aqueous enzymology, the post-treatment of immobilized enzymes is gaining increasing importance in enzyme immobilization, because of its interesting potential, for example:

- improvement of storage stability by lyophilization with polyols [236],
- improvement of linkage stability,
- improvement of enzyme activity [243],
- improvement of enzyme stability,
- enhancement of conformation stability,
- conversion to a chemo enzyme [244],
- introduction of new functionality by use of solid-phase chemistry,
- homogenizing the immobilized enzyme by washing with organic solvent,
- prevention of enzyme leakage by post-immobilization cross-linking or other techniques,
- control of the water activity for use in organic solvents.

In recent decades many strategies have been developed for post-treatment of immobilized enzymes. These will be discussed below.

6.8.2
Classification of Post-treatments

As noted above, an immobilized enzyme contains two essential functions, its catalytic function, e.g. activity, stability, and selectivity, and its non-catalytic functions, for example the physical and chemical stability of the non-catalytic mass, especially geometric properties such as shape, length and size. Thus, the goal of the post treatment can be classified as follows:

- improvement of activity,
- improvement of stability,
- improvement of selectivity,
- improvement of geometric properties,
- improvement of chemical and physical stability.

Consequently, post-treatment techniques can be classified into two groups, physical post-treatment and chemical post-treatment. The former includes the pH-imprinting technique, solvent washing [247], and co-lyophilization with additives such as polyols [242]. The later refers to any chemical means subsequently used to modify the already prepared immobilized enzyme, which includes chemical modification of the enzyme, enhancement of multipoint attachment by increasing the pH of the immobilization medium [248], neutralization of excess active binding functionality (blocking or quenching treatment) [302, 303] or alteration of the microenvironment of the immobilized enzymes (Table 6.11).

6.8.3
Physical Methods

Although the phrase "physical methods" is used to denote the use of purely "physical" means to modify the immobilized enzyme, chemical reaction might be provoked by physical methods such increasing the pH of the medium containing the immobilized enzyme (Scheme 6.37). Currently used physical methods can be classified in the following groups:

- increasing the pH, with the objective of enhancing multipoint attachment;
- pH imprinting, with the objective of increasing the activity of the immobilized enzyme in organic solvent, in which it is not possible to adjust the ionic state of the enzyme molecules;
- solvent-washing, with the objective of denaturing non-immobilized enzymes or activating the enzymes;
- lyophilization, with the objective of improving storage stability or activity in organic solvents;
- thermal activation, with the objective of increasing the activity of the enzyme;
- activation, with the objective of increasing the conformation flexibility of the immobilized enzymes, thus enhancing activity;
- adjustment of water activity.

Table 6.11 Modification of enzyme and attachment to carriers.

Modifier	Carrier	Comments	Enzyme	Ref.
Oxidized dextran	An aminalkylated silica	Immobilized enzyme-polymer with higher activity retention (45%, compared with 4% for the native enzyme). Moreover, higher thermostability was obtained	α-Amylase	316
N-Acetylhomocysteine thiolactone, introducing 5–6 mol additional SH groups mol^{-1} FAD	Thiolated 6-aminohexyl-Sepharose	Immobilization of polythiolated PHLDH to a short spacer group, l-cysteiny-Sepharose, reduced specific activity but enhanced thermal stability and stability in aqueous dioxane by 800% and 770%, respectively, relative to free native I	Pig heart lipoamide dehydrogenase (PHLDH)	317
Dextrans of different molecular weights	Aminosilochrome	Modified subtilisin, immobilized on aminosilochrome via activation of carbohydrate with CNBr, had high stability	Subtilisins BPN and 72 (Carlsberg type)	318
Dextrans of different molecular weights	Aminosilochrome	Modified subtilisin, immobilized on aminosilochrome via activation of carbohydrate with periodate, was less stable	Subtilisins BPN and 72 (Carlsberg type)	318
Dextran aldehyde	Aminolated silica	Not only high volume activity but also high thermostability can be obtained with this method, compared with immobilization of β-glucosidase on dextran aldehyde-coated aminolated silica	β-Glucosidase	319
–	Dextran aldehyde-coated aminolated silica	This enzyme immobilized on dextran aldehyde-coated aminolated silica was less stable than the dextran-aldehyde modified enzyme immobilized on aminolated silica	β-Glucosidase	319
Con A	Sephadex G-50-GAH	Enhancement of thermostability and activity compared with the free enzyme	β-Galactosidase	320
Dextran	Amino-activated silica (Promaxon, Spherosil, and Aerosil)	Conjugated penicillin amidase immobilized by a classical method was more thermostable than native penicillin G acylase immobilized on the same carrier	Conjugated penicillin amidase	321

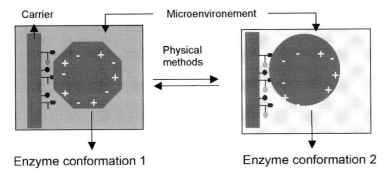

Scheme 6.37 Physical methods for post-treatment of immobilized enzymes.

Remarkably, some physical methods, for example pH, adjustment, solvent washing, lyophilization and thermal treatment might provoke a change in the flexibility of enzyme conformation or microenvironment, thus improving the enzyme activity and stability.

6.8.3.1 Increasing the pH

For covalent bonding with activated carriers bearing active functionality, the reactivity of the active functionality of the carriers with the amino acid residues of enzymes is highly dependent on the pH of the medium and the microenvironment of the carriers. Increasing the pH thus reinforces the multipoint attachment, leading to enhancement of enzyme stability [248]. The mechanism is illustrated in Scheme 6.38.

For instance, post-treatment by incubation at high pH (pH 10.5 instead of pH 7.0) led to 100-fold enhancement of the thermal stability of thermophilic catechol 2,3-dioxygenase immobilized on highly activated glyoxyl agarose beads [254]. Most probably, the increase of the pH enhanced multipoint attachment. Similarly, the stability of α-chymotrypsin and penicillin G acylase can be increased 5 and 18-fold compared with the conventional immobilized enzymes on Eupergit c carriers [248].

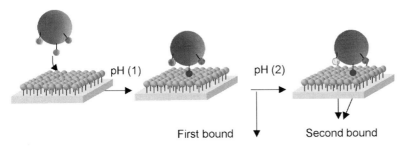

Scheme 6.38 Increasing the pH for enhancement of multipoint attachment.

6.8.3.2 Solvent Washing

Solvent washing was originally used to dehydrate the carrier-bound immobilized enzyme, to enable their use in non-aqueous media, especially non-polar organic media, in which the water activity must be controlled [245]. Thus, this technique is a result of the development of non-aqueous enzymology. In addition to the initial aim, dehydration of immobilized enzymes, solvent washing as a post-treatment technique has also been used to homogenize the immobilized enzymes. It is known that covalent immobilization of enzymes on carriers can lead to the formation of heterogeneous species, because some enzymes might be adsorbed on the carrier by non-specific adsorption or self affinity interaction [247]. On the basis of on this assumption, solvent washing techniques are used to homogenize the immobilized enzyme by deactivation of non-covalently immobilized enzymes – 20–30% of the non-immobilized enzyme that is probably adsorbed by self-affinity interaction can usually be selectively deactivated by solvent washing [247]. In addition to elimination of non-covalent immobilized enzymes, in the last decade this technique has been used to improve enzyme activity in organic solvents. Stepwise solvent washing is often needed to mitigate the speed of dehydration, thus keeping the immobilized enzyme more active in organic solvents [246]. Co-lyophilization of immobilized enzymes with additives such as sugars and PEG has long been known to be crucial for storage stability or activity in organic solvents [194]. This is largely attributed to the fact that the SCO_2 (supercritical CO_2) process gives the gel a different structure.

6.8.3.3 Lyophilization/Drying/Addition of Additives

Lyophilization/drying is one of the oldest methods used for post-treatment of immobilized enzymes with the aim of improving their storage stability. Lyophilization can be performed in the presence of additives [242, 255], for example stabilizers such as organic solvents and sugars, used to improve enzyme stability during storage in liquid media [255].

Early in the 1970s, co-lyophilization of the immobilized enzymes with sugars was found to increase the retention of activity after lyophilization and storage [257]. In the 1990s lyophilization of enzymes (mainly dissolved enzymes) in the presence of conformer selectors or other additives, for example sugars, polyols, and salts [183], was often used to imprint the enzymes for enhanced activity and selectivity [183] or to enhance enzyme activity in organic solvents [186].

For example, the activity and substrate specificity of subtilisin-catalysed acylation of nucleosides in organic solvents can be controlled by lyophilizing the enzyme from an aqueous solution in the presence of the substrate [185]; the presence of a ligand, N-Ac-l-Phe-NH_2 dramatically enhanced the activity of several enzymes (four proteases and three lipases) in organic solvents [258].

Often, the use of crown ethers such as 18-crown-6, 15-crown-5, and 12-crown-4 substantially (i.e. by at least one order of magnitude) improved the activity of enzymes such as subtilisin Carlsberg [258, 259] or lipase in polar organic solvents. The use of trapping methods in the presence of amphiphilic interfaces (TPI) was able to improve lipase activity at least 90-fold in the lipase-catalysed synthesis of flavour esters [260]. The presence of a transition-state analogue (biomolecular imprinting) can even induce new catalytic activity in native bovine serum albumin [261].

Remarkably, this technique is currently applied mainly to dissolved enzymes. It is, however, also conceivable that use of carrier-bound immobilized enzymes might have the same advantages as use of the dissolved enzymes, with regard to other post-immobilization techniques. For example, it has been found that the activity of enzymes in organic solvents is largely determined by hydration history and water activity. Thus, an enzyme might lose activity during continuous reaction recycling in organic solvents, because of the dehydrating effect of the solvent. In such circumstances it is necessary to restore the activity of the enzyme by rehydration in a suitable buffer. It is obviously an advantage to be able to use this process both on carrier-bound enzymes and on dissolved enzymes.

Addition of additives such as sugars, salts, and PEG to the immobilized enzyme, then lyophilization has been found to be effective in improvement of enzyme activity. It was recently found that the activity of cellulose-bound HRP bearing cellulose binding domains (CBD) was highly dependent on drying method. The activity of the immobilized HRP was in the order: wet enzyme (100%) > enzyme lyophilized in the presence of sucrose (65%) > lyophilized enzyme (14%) > enzyme kept dry in a desiccator (2.7%) [65].

Dabulis and Klibanov found that for some enzymes the activity remains higher even after removal of the excipients [258] whereas others have found that removing the sorbitol by washing reduced the activity of lipase and protease [262].

For gel-entrapped enzymes it is often found that the nature of the drying process has large influence on activity retention. Study has revealed that PCL entrapped in silica and aluminosilicate gels and dried by a supercritical CO_2 method was more active than that dried by conventional solvent evaporation [263]. This is largely attributed to the fact that the nature of the drying process affects both the texture of the gel and the properties of the enzyme [3].

6.8.3.4 pH Imprinting

The pH imprinting method is the process of adjusting the ionic state of dried immobilized enzymes by immersing them in an aqueous solution of defined pH then performing a suitable dehydration process, for example lyophilization. Thus, the ionic state of the immobilized dehydrated enzyme is dictated by the pH of the aqueous buffer from which the enzyme is lyophilized or dehydrated.

This technique is of crucial importance to the activity, selectivity, and stability of immobilized enzymes designed for use in anhydrous organic solvents [193], because enzyme powder cannot usually change its pH in anhydrous organic solvents. Many studies have revealed that enzyme activity at the optimum pH might be several times that at non-optimum pH. It has recently been shown that the activity of enzyme–PEG complexes in non-polar solvent was also very dependent on the pH of the solution from which the polymer-enzyme complex is lyophilized [188].

This technique is often used to produce enzyme powder to be used in non-aqueous media. Fewer studies have been reported for carrier-bound immobilized enzymes [388].

6.8.3.5 Physical Entrapment

Apart from these physical post-treatment processes, physical entrapment (or encagement) of the ready-made immobilized enzymes is also worth mentioning (Scheme 6.39). These methods have been developed to overcome some of the drawbacks of pre-immobilized enzymes by:

- enlargement of the geometric properties of enzymes immobilized on small particles,
- enhancement of the activity by avoidance of enzyme leakage,
- enhancement of selectivity,
- enhancement of stability.

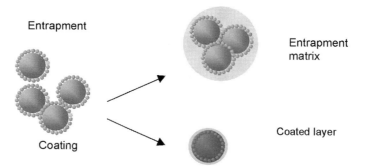

Scheme 6.39 Physical entrapment of pre-immobilized enzymes.

Entrapment of immobilized enzymes (adsorbed enzymes, covalently immobilized enzymes) in a polymeric gel is an example of a double-immobilization technique, which can be dated back to 1977, when Kumakura for the first time entrapped enzymes adsorbed on inorganic additives [262]. More recently the technique has been used by Singh et al. for preparation of double-entrapped penicillin G acylase for production of 6-APA [263]. Interested readers are advised to consult the previous chapter on enzyme entrapment.

6.8.3.6 Thermal Activation

It has been found that after thermal treatment at 70 °C of acylase immobilized on silanized silica (by the glutaraldehyde method) had twice the activity of the normal immobilized enzyme without thermal treatment [264]. Similarly, it has also been found that the residual enzyme activity of horseradish peroxidase (HRP) immobilized in a mesoporous material (folded sheets mesoporous materials, FSM-16) and then entrapped in a hybrid organic/inorganic gel increased from 73 to 99% after thermal treatment (70 °C, 60 min) [265]. These results suggest that thermal treatment might induce a new active conformation that is stabilized by the carrier used. It is, however, still not clear how long the enhanced activity can be maintained, compared with the immobilized enzymes without thermal treatment.

6.8.3.7 Activation by Denaturants

Studies of the effect of denaturants on the activity of carrier-bound immobilized enzymes was pioneered by Tosa et al. in 1969, when acylase immobilized on DEAE-cellulose or DEAE-Sephadex was subjected to treatment with concentrated (6 M) urea for 1 h. Surprisingly, the activity of the urea-treated immobilized enzyme was found to be much more active than the untreated immobilized acylase [241, 266]. This activation effect was interpreted as a result of the unfolding–refolding action of denaturants, which can unfold the native enzyme. For the immobilized enzyme unfolding is less possible than for the native enzyme, but the structure might become looser than that of the native enzyme, thus making it more active.

Similarly, it has recently been found that pig muscle triosephosphate isomerase covalently attached to a silica-based support activated with p-benzoquinone had high activity after treatment with urea, suggesting activation of the enzyme. The extent of activation depended on pH, and on buffer and salt concentrations. Increasing the ionic strength reduced or eliminated the activation. The phosphate ion also had a specific effect on the thermal inactivation [267].

Many denaturants, for example SDS [268, 269], bromoethanol [270], dimethyl sulphoxide [271], and the guanidinium ion [272] all have significant effect on protein structure. Thus, exploitation of a denaturant for activation of immobilized enzymes might be an interesting approach for acquisition of active immobilized enzymes.

6.8.3.8 Post-immobilization by Physical Coating

Post-immobilization by physical coating refers to the process by which immobilized enzymes are further coated with a layer of polymer with the aim of further enhancing the stability of enzyme function. For example: urease adsorbed on activated charcoal has been further coated with an ultrathin layer of cellulose nitrate [273]; enzyme coated on a solid surface in the presence of a polyelectrolyte can be coated with polyelectrolyte counter-ions, leading to the formation of a symplex layer [274]; fibrous carriers soaked with asparaginase has been further coated with poly(2-HEMA-co-EEA-co-HPA) [275]; urease adsorbed on Dowex 50W-X8 has been encapsulated further [276]; and choline oxidase immobilized on DEAE-Sepharose can be further coated with PVA-SbQ photo-cross-linked polymer [277]. In all these examples the coated immobilized enzymes were more stable than their untreated counterparts.

6.8.3.9 Rehydration/water Activity Adjustment

In the last two decades it has been widely observed that the enzyme activity in organic solvents is highly dependent on the water activity of the enzyme preparations and the reaction medium. For a specific enzyme preparation it is essential to control the water activity to achieve maximum activity of the enzyme in the selected medium. Control of water activity is easily achieved by equilibrating the enzyme preparation and the reaction medium with the salts of defined water activity [278]. This technique is, however, only applicable on the laboratory scale. For large-scale biotransformation it is difficult to equilibrate the enzyme and solvent to obtain maximum activity.

Although it is not known how to control water activity in large-scale biotransformation in organic solvents, it is conceivable that the water content of an immobilized enzyme preparation might decease during the repeated use, because of redistribution of the water between the enzyme and the solvent used, leading to a decrease in activity. This has been confirmed for enzyme-catalysed transesterification in toluene with immobilized *Candida Antarctica* Lipase B (water activity <0.1) [278].

Remarkably, it has been found that enzymes deactivated in this way can be reactivated by so-called rehydration processes, as shown in Scheme 6.40.

Briefly, lyophilization of the enzyme led to significant (~20%) recovery of the activity after fourfold recycling. This experiment has important implications:

- control of the water activity is essential to keep the enzyme maximally active in organic solvents;
- use of immobilized enzyme is advantageous over the lyophilized enzymes; although the latter can be also recovered from the reaction mixture, rehydration is much difficult for free enzyme powders;
- deactivation of enzymes in organic solvents might be reversible.

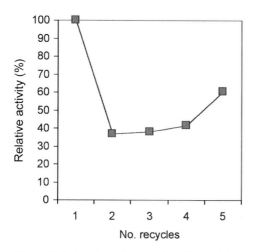

Scheme 6.40 Activity during recycling of immobilized CAL-B lipase and rehydration experiment.

A similar result was recently obtained with the CLEA of lipase CAL-B. It was found that the activity can be almost 100% regenerated by rehydration after several cycles in organic solvents, thus justifying the use of the immobilized enzymes instead of enzyme powder [279].

6.8.3.10 Sonication

In the 1980s it was found by several groups that ultrasonication of an enzyme–carrier complex could increase the activity of the enzyme or restore activity that had been lost during storage or under process conditions. For example, glucoamylase immobilized on an anionic polystyrene carrier (Wofatit Y 58), by a method which included activation of the carrier with 1,4-benzoquinone, lost 60% of its original activity after several weeks of use. More than 25% of the native activity could be recovered by treatment with ultrasound [280, 281].

6.8.3.11 Acid or Alkaline Treatment

Like other denaturants, strong acid or base can be also used to reactivate immobilized enzymes. A study has revealed that trypsin immobilized on a polyurethane carriers can be activated by treating the immobilized enzyme with 3 M NaOH at 90 °C. In this way, the activity of the immobilized enzyme can be increased eightfold [282]. This effect was explained by the author as the result of change in the structure of the carrier (less diffusion limitation) and the change of the bonds between enzyme and carrier (more flexible enzyme conformation) [282]. Similarly, acid treatment might also lead to enhancement of activity.

6.8.4
Chemical Methods

Chemical methods are methods which exploit active chemical reactants to provoke a chemical reaction on the enzyme or carrier, with the aim of improving the performance of the immobilized enzyme. As shown in Scheme 6.41, chemical meth-

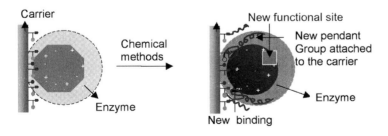

Scheme 6.41 Chemical post-treatment of immobilized enzyme.

ods can be simply classified into the two subgroups, namely the chemical modification of the immobilized enzyme molecules and the chemical modification of the carriers.

In the former group the enzyme molecules can be further chemically modified [192, 283] or intramolecularly cross-linked, or further embedded in a sandwich structure, whereas the carriers in the later group can be further modified by introduction of other so-called quenching agents (or blocking agents) with the aim of altering the microenvironment or avoiding the formation of the extra attachments, which often causes enzyme deactivation during storage.

6.8.4.1 Consecutive Cross-linking of the Immobilized Enzymes

Subsequent cross-linking of immobilized enzyme adsorbed on carriers has been pursued since the 1970s and 1980s as an efficient means of improving enzyme stability or avoiding the enzyme leakage [284, 285]. Consecutive cross-linking often significantly improved enzyme stability against denaturation induced by heat and tryptic digestion [285].

An interesting example was demonstrated by Woodward and Wohlpart, who found that the thermal stability of the β-glucosidase from *A. niger* immobilized on CNBr-activated Sepharose after aminoalkylation of the carbohydrate side chains of the enzyme was significantly improved by subsequent cross-linking with glutaraldehyde [286]. More interestingly, it was found that the thermal stability of the same enzyme immobilized on microcrystalline cellulose with glutaraldehyde was barely more than that of the soluble enzyme.

Cross-linking might not always be able to improve enzyme stability, however. For example, further formation of disulphide bridges in immobilized liver phenylalanine (Phe H) coupled to activated thiol-Sepharose 4B by oxidation did not increase the thermal stability [287].

Consecutive chemical cross-linking and chemical modification has also been used to improve enzyme stability. One intentional post-treatment of a covalently immobilized enzyme was first reported in 1989, when immobilized penicillin G acylase was chemically modified with formaldehyde followed by sodium borohydride reduction, leading to the formation of much more stable immobilized enzyme than the original unmodified one [283]. It was suggested that this solid-phase chemical modification of the immobilized enzyme is advantageous in that more drastic reaction conditions can be used to modulate and control the performance of the immobilized enzymes obtained, because the previously immobilized enzyme is usually stabilized and thus resistant to the deleterious effect of the chemical modification [283].

Apart from the chemical modification such as intramolecular cross-linking of the free enzyme [293], it is obvious that immobilized enzymes can also be further intermolecularly and intramolecularly cross-linked, with the aim of improving conformational stability and thermostability [294]. More interestingly, it has been found that conversion of the carboxyl groups of immobilized penicillin G acylase (on aldehyde agarose) into amino groups, then cross-linking with glutaraldehyde or formaldehyde led to at least one order of magnitude enhancement of the thermostability of the enzyme.

Remarkably, a lipase from a *Bacillus sp* immobilized on hydrophilic carrier HP-20 followed by glutaraldehyde cross-linking retained 97% activity after 25 cycles in aqueous medium, suggesting that cross-linking can not only improve enzyme stability but also improve the strength of bonding between the enzyme and the carrier [288].

Remarkably, although consecutive cross-linking can improve the stability of immobilized enzymes, it has been shown that the nature of the carrier is also of crucial importance. For example, pectinlyase has been covalently immobilized on XAD-7, Eupergit C, and Nylon 6 by different chemistry (XAD-TCT, Eupergit C-Oxirane, Nylon 6-GAH) followed by glutaraldehyde cross-linking. The maximal stabilization effect was obtained with Nylon 6. The different stability can (as shown in Scheme 6.42) possibly be ascribed to the nature of the carriers. In other word, the nature of the immobilized enzyme is mainly dictated by the first immobilization. Due to the difference in the enzyme conformation and orientation, pectinlyase immobilized on different carriers can not be stabilized to the same levels [289].

Instead of the chemical cross-linker, the stability of the covalently immobilized enzyme on the aminoalkylated silica can be further improved by chelation using a transition metal to form a metal–salt bridge adduct [295]. Subsequent glutaraldehyde-aided cross-linking of the non-covalently immobilized endo-polygalacturonase on trimalethyl-chtosan (TMC) is proved to be superior to other types of the same enzyme immobilized on a number of carriers of natural or synthetic origin, regarding the high activity and stability [296]. Similarly, cross-linking of immobilized porcine pancreas lipase on PAAm beads (by carbodiimide coupling) with glutaraldehyde or 3,5-difluoronitrobenzene also improved thermal stability [297].

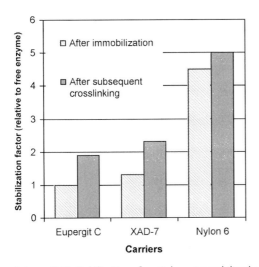

Scheme 6.42 Stabilization of pectinlyase immobilized on different carriers.

More recently it has been found that bovine liver catalase immobilized on highly activated glyoxyl agarose, followed by cross-linking with dextran-aldehyde, was able to significantly stabilize this labile tetrameric enzyme, which is usually dissociated in dilute solution and cannot be efficiently stabilized by normal covalent immobilization, because of limited coverage by multipoint attachment [298].

Consecutive cross-linking of enzymes immobilized on carriers by different types of adsorption has been well-established technology since the 1960s. In principle, the cross-linking step can be regarded as a post-treatment technique. For example, invertase has been immobilized on CC lectin-bound seralose, followed by cross-linking with glutaraldehyde, resulting in the formation of immobilized invertase with higher retention of activity. Higher thermal and operational stability was also obtained by use of an extra cross-linking step compared with the soluble enzyme [192]. This technology has not, however, yet been fully exploited. Combining this technique with imprinting technique might be an interesting means of engineering immobilized enzymes.

Entrapped enzymes can be also further cross-linked. For example, β-amylase from *Bacillus megaterium* B6 has been entrapped in a gel matrix then covalently cross-linked, resulting in an approximately 14-fold increase in catalytic half-life [322]. It is worthwhile pointing out that the consecutive cross-linking of the immobilized enzyme might lead to the formation of both internal and intra-cross-linkage as illustrated in Scheme 6.43.

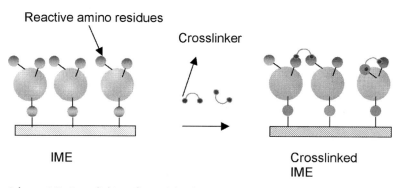

Scheme 6.43 Cross-linking of immobilized enzymes.

6.8.4.2 Consecutive Chemical Modification of Immobilized Enzymes

In addition to chemical cross-linking, other chemical functionality can be introduced to enzyme molecules immobilized on carriers. An early example was consecutive modification with proline of α-amylase immobilized on BrCN activated Sepharose [291]. Interestingly, it was found that the thermostability at 60 °C of α-amylase immobilized on BrCN-activated Sepharose can be increased at least 60 times compared with that of the free enzyme. It is most probable that introduction of proline by zero-coupling chemistry eliminates NH_2 groups around the active centre thus making the active centre more stable.

Because lyophilized enzyme powder is also an immobilized enzyme, modification of lyophilized enzymes in organic solvents can be also regarded as a post-immobilization technique [292].

6.8.4.3 Conversion

Immobilized enzymes are interesting precursors for creation of enzymes with new catalytic properties by chemical modification, because they are often more stabilized against hazardous modification procedures than the native enzymes [299]. One interesting example is the conversion of cross-linked microcrystalline protease to semi-synthetic peroxidase seleno-subtilisin [244, 300], as illustrated in Scheme 6.44.

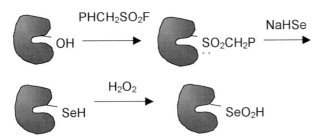

Scheme 6.44 Conversion of immobilized enzymes into a new catalyst by chemical modification.

6.8.4.4 Consecutive Modification of the Carrier

Another technology of post-treatment of the immobilized enzymes, which has received increasing attention since the 1990s, is consecutive modification of the carrier used for covalent bonding of the enzyme, with the aim of improving the microenvironment of the immobilized enzymes, as shown in Scheme 6.45.

The root of this technology stems from the observation that the reactive groups of the carrier are in large excess (often 1 : 300–600), compared with the amount of enzymes to be immobilized, so quenching of excess reactive groups is often required. Otherwise, the carrier-bound covalently immobilized enzyme often loses its activity during storage, because the excess reactive groups can react further with some amino acid residues [301]. In contrast, quenching of these active groups after immobilization can often effectively mitigate the loss of the activity during storage or use [302, 303] and can also enhance stability, for example the coupling of yeast mitochondrial alcohol dehydrogenase to glutaraldehyde-activated silane glass [301].

In addition to the effect on the activity retention, it was recently observed that blocking agents used to the block carrier-bound active groups (CAG) can also substantially improve the enzyme stability [304]. The stabilizing effect was, however, found to be related to the nature (charged or neutral) of the quenching agents. For example, among the quenching agents l-lysine, l-glycine, and ethanolamine, ethanolamine was the best quenching agent for immobilization of GL-7-ACA acylase on aldehyde silica, in terms of activity retention and the enzyme stability [305]. This suggests that the microenvironment of the carriers was significantly changed by use of these "deactivating agents".

Similarly, it has been found that the thermostability of immobilized β-galactosidase on to aldehyde-functionalized ceramic can be enhanced by addition of different amino acids, for example lysine, arginine, and histidine (lysine > arginine > histidine), compared with the immobilized enzyme without any treatment. The effect of these quenching agents depends on enzyme loading.

At higher enzyme loading, the quenching agents have no effect on the stabilization of these immobilized enzymes, suggesting that occupation of the surface is essential for enzyme stability. Thus, at high surface occupation enzymes are stabilized by the weak interaction with each other. In contrast, at lower enzyme loadings the enzymes are often destabilized at high temperature, because of deformation of the enzyme molecules and subsequent immobilization of the deformed enzyme on the carrier with excess active groups. Blocking these excess active groups should, however, prevent these problems.

Such work suggests that alteration of carrier-bound pendant groups can change the nature of the microenvironment of the immobilized enzyme. Consequently, enzyme stability, activity, and selectivity can be designed by selecting appropriate blocking agents.

Not only small organic compounds (charged or neutral, hydrophilic, or hydrophobic) but also synthetic polymers such as PEI [306, 311], natural polymers such as BSA [197], and derivatives such as aminodextran can also serve as quenching agents. It has, for example, been found that consecutive modification of penicillin V acylase from S. lavendulae immobilized on Eupergit C with bovine serum albumin led to a new biocatalyst (ECPVA) not only with activity enhanced 1.5-fold in the hydrolysis of penicillin V but also enhanced stability compared with its soluble counterpart. This biocatalyst could be recycled for at least 50 consecutive batch reactions without loss of catalytic activity [197]. It has been shown by the same group that thermostability of d-amino acid oxidase can be increased several times by a consecutive treatment of the immobilized enzyme with aldehyde dextran [307]. The principle is illustrated in Scheme 6.45.

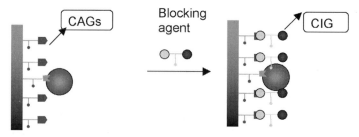

Scheme 6.45 Illustration of consecutive treatment of covalently immobilized enzymes with blocking agents.

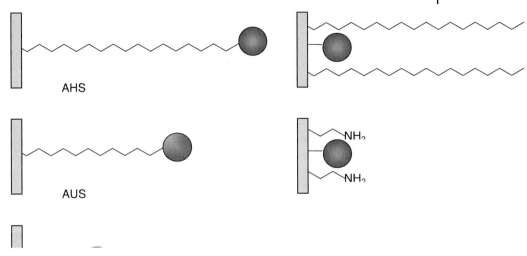

Scheme 6.46 Immobilization of chlorophyllase with different types of carriers and methods.

The role of the microenvironment was interestingly demonstrated by Sudina and his co-workers, who immobilized chlorophyllase (I) on aminohexadecyl-Sepharose (AHS), aminoundecyl-Sepharose (AUS), and heptyl-Sepharose (HS) by adsorption as a result of hydrophobic interactions, with retention of activity of 29.4%, 36 and 35.8% respectively. The storage stability of these immobilized enzymes was in the order 21%, 17.14%, and 7.5%, respectively, after storage for 1 month, suggesting that the enzyme was more stabilized by the hydrophobic carriers. In contrast, covalent immobilization of this enzyme on CNBr-activated Sepharose with subsequent fixation of either diethylamine or dodecylamine to the carrier led to the discovery that the immobilized enzyme was better stabilized by diethylamine than by dodecylamine [308], suggesting that enzyme stability was dictated mainly by its microenvironment, which can be altered by post-modification (Scheme 6.46).

6.8.4.5 Chemical Coating

Immobilized enzymes whether covalently immobilized, non-covalently adsorbed, or entrapped can be further coated with a layer by chemical methods, as illustrated in Scheme 6.47.

For example, α-amylase adsorbed on particles of copolymers of methyl methacrylate, trimethylolpropane trimethacrylate, and acrylonitrile was further encapsulated in a polymer layer formed by cross-linking xylene diamine and glutaraldehyde. Remarkably, encagement of the immobilized enzyme dramatically enhanced the stability of the immobilized enzymes [309]. Encagement of an enzyme in a sandwich structure can be achieved by reacting the immobilized enzyme with a polyaldehyde polymer, which can in turn react with polyamine, leading to the formation of bilayer-encaged immobilized enzyme [310].

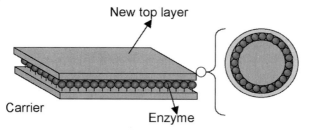

Scheme 6.47 Illustration of encagement of adsorbed enzyme/ Immobilization of chlorophyllase with different types of carriers and methods.

6.8.5
Outlook

Post-immobilization as an essential part of enzyme-immobilization techniques is attracting increasing attention, because of the unique potential promised by the techniques for improvement of the performance of the immobilized enzymes. Because most single immobilization techniques, for example adsorption, encapsulation, and covalent attachment often fail to provide a robust immobilized enzyme, consecutive improvement of the immobilized enzyme by use of a variety of double-immobilization and post-immobilization techniques is clearly justified. Because of stabilization of the enzymes by the preliminary immobilization techniques, their molecules can be further chemically and physically modified under harsher conditions that are fatal to the native enzymes. Thus, it is expected that more unprecedented results, for example alteration of activity, stability, and selectivity, can be achieved than are possible with conventional immobilization techniques.

6.9
Reversibly Soluble Immobilized Enzymes

Reversibly soluble immobilized enzymes are often needed to combine the advantages of immobilized enzymes, for example reusability, with the advantages of soluble enzymes, for example less, or no, diffusion limitation. Thus, reversibly soluble immobilized enzymes should be completely soluble under the reaction conditions but insoluble when the pH, temperature, or solvent is changed. In general, a special polymer is needed to form reversibly soluble polymer–enzyme conjugates. The polymer used can be synthetic or natural polymer or semi-synthetic polymers. The insolubilization mechanism can be pH-stimulated or temperature-stimulated.

6.9.1
pH-responsive Smart Polymer

Immobilization of enzymes on smart polymers can be dated back to the early and mid 70s [327], when several reversibly insoluble–soluble enzyme–polymer com-

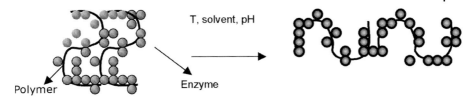

Insoluble state   Soluble state

Scheme 6.48 Illustration of enzyme immobilization on smart polymer and phase inversion in response to the pH, temperature or solvents change. The associated polymer-enzyme conjugates can collapse, when pH or temperature, solvent is changed.

plexes were prepared, for example lysozyme–alginic acid complex [119] and trypsin–acrolein–acrylic acid copolymer [327]. These soluble–insoluble polymers, which were mainly used in the membrane process [312], could be precipitated in a narrow pH range (Scheme 6.48). No practical application of this method was reported, however.

Enzymes can form complexes with polyelectrolytes. Insoluble enzyme–polymer complexes can be dated back to 1974, when Maeda and co-workers prepared invertase–collagen complex.

The effect of polysaccharides and other polyelectrolytes on the catalytic properties of enzymes such as chymotrypsin has been studied [313]. It was found that electrostatic interactions between the polyelectrolytes, the enzyme molecules, and the reaction components exerted an important effect on enzyme kinetics, protein interactions, and the reaction rate with glutaraldehyde. In the early 80s, penicillin G acylase was immobilized in water-soluble non-stoichiometric polyelectrolyte complexes (PEC) formed by poly(4-vinyl-N-ethylpyridinium bromide) (polycation) and polymethacrylic acid (polyanion). The structures of these PEC particles resulted in cooperative phase transitions of these systems when pH and ionic strength were varied slightly. It was found that reducing the pH from 6.1 to 5.7 led to a reversible phase transition followed by a by five to tenfold increase of K_m for immobilized penicillin amidase, depending on the substrate used. The phase transition induced by increasing the ionic strength to 0.27 m NaCl did not significantly change the K_m value of the enzymic reaction [372].

Interestingly, penicillin G acylase was also covalently linked to the polycation in PEC (formed from poly(4-vinyl-N-ethylpyridinium bromide and poly(methacrylic acid) in 1:3 ratio) previously modified with cyanuric chloride. Despite negligible diffusional difficulties, the catalytic efficiency of the enzyme action changes slightly in the homogeneous aqueous solution used for benzylpenicillin hydrolysis. Remarkably, it has been shown that the negatively charged shell of the polyelectrolyte complex has a large effect on the kinetic behaviour of the reaction–because the pH optimum of the reaction was broadened, $k$cat was increased in alkaline media and a sharp increase of the product-inhibition constant (from 2×10^{-5} to 1×10^{-3} m) was also observed.

A remarkable advantage of this kind of immobilized penicillin G acylase is the reversibility of the PEC–it could be reversibly converted to the insoluble state by a slight change in the pH or ionic strength of the solution. Transition to the insoluble state temporally inactivates the enzyme. The conditions for the phase separation of immobilized enzymes in solution of salts depend on the compound of the polyelectrolyte complex and on the nature of the low-molecular-weight electrolytes. Dissolution of the precipitate leads to quantitative recovery of the initial catalytic activity. In practice this intrinsic property enables a homogeneous catalysis and recovery of the catalyst in a heterogeneous phase [372].

Similarly, penicillin G acylase could be immobilized on a copolymer of two or more monomers from the group consisting of acrylic acid, or an alkyl ester thereof, methacrylic acid, or and alkyl ester or dialkylamino ester thereof, and a vinylpyridine derivative. The resulting immobilized enzyme usually has high activity and can be dissolved in an organic solvent or in a solution containing organic solvent [390] on changing the pH of the medium. The resulting immobilized enzyme can be precipitated at pH 5 or below and dissolves at pH 6 or higher.

It has been found that formation of a polymer–enzyme conjugate can potentially reduced the accessibility of the active centre of the enzyme [314].

A product-regulated immobilized enzyme has been prepared by immobilization of enzyme by entrapment. For example, glucose oxidase (GOD) entrapped within cross-linked copolymer networks consisting of N-isopropylacrylamide (NIPAAm) and acrylic acid (AA) can be used to oxidize glucose to gluconic acid, thus resulting in a decrease of the pH within and in the vicinity of the gel phase, leading to collapse of the gel because of hydrophobic interaction (attractive force) between NIPAAm residues [315].

In addition to the covalent complex, other reversibly soluble–insoluble immobilized enzymes can be formed by using an enteric coating polymer as a carrier. For example, proteases (e.g. papain) can be immobilized by methacrylic acid–methyl acrylate–methyl methacrylate copolymer (MPM-06) as the most active soluble–insoluble immobilized papain, which had an insoluble form below pH 4.8 and a soluble form above pH 5.8; it was also soluble in water-miscible organic solvents. It was reusable, more heat-stable, and more stable in water-miscible organic solvents than native proteases [326].

Modification of natural polymers, especially cellulose, often leads to the preparation of semi-synthetic soluble polymers which can be attached covalently to enzymes, resulting in the formation of reversibly soluble–insoluble enzyme–polymer complexes [333, 339]. This method makes it possible for soluble enzyme–polymer complexes to act on insoluble substrates and enables recovery of the enzyme from the reaction mixture, thus enabling reuse of the enzymes; one example is the use of insoluble carrier-bound immobilized enzymes [339]

Currently, the semi-synthetic polymer frequently used for this purpose is hydroxypropyl methylcellulose acetate succinate (AS). The covalently bound enzyme–polymer complex can be precipitated at ~pH 4 and dissolved at pH > 5–6 (Table 6.12).

6.9 Reversibly Soluble Immobilized Enzymes

Table 6.12 Reversibly soluble immobilized enzymes based pH sensitive polymers.

Polymer	Methods	Remarks	Enzyme	Ref.
Eudragit L	Covalent	pH-dependent reversibly soluble-insoluble behaviour	TV cellulase	323
Eudragit L100-55	Covalent?	pH-dependent reversibly soluble–in soluble; similar performance as NE and enhanced stability at 25–45 °C	Endo-pectinlyase	324
Eudragit S-100 Poly(MA-co-AAc)	Covalent	Activity retention depends on the pH of the medium	AP, GL, GA, Trypsin xylanase	325
pH-responsive Poly(MAAc-MA-MMA)	Covalent	High retention of activity; enhanced thermostability and stability in water-miscible organic solvent	Papain, chymotrypsin	326
Poly(ACr-co-AAc)	Covalent	Easily precipitated at pH 4–4.5 and re-dissolved at pH 7.0 with complete retention of activity	Trypsin	327
Eudragit S-100	Covalent	64% activity retention; 30% decreased K_m	Trypsin	328
PEI	Non-covalent	Enhanced stability and use in membrane reactor	PGA	329
PEI/Eudragit		96% activity retention on Eudragit and more than 100% activity was found with PEI; the payload is approximately 0.02–0.05	α-Amylase	330
NIPAAm	Covalent	52% activity retention; reduced K_m and thermostability	Isoamylase	331
NIPAAm	Covalent	37% coupling yield was obtained	Penicillin G acylase	332
Hydroxypropyl methylcellulose acetate succinate (AS)	Covalent	The immobilized enzyme–polymer complex was soluble above pH 5.2 and insoluble below 4.5	Chitinase	333
Hydroxypropyl methylcellulose acetate succinate (AS)	Covalent	41% activity retention; pH responsive solubility	Lysozyme	334
AS-1	Covalent	The immobilized enzyme (LY-AS) was soluble above pH 5.5 and insoluble below pH 4.5; stability was increased on immobilization	Lysozyme	335
Hydroxypropyl methylcellulose acetate succinate (AS)	Covalent	66% activity retention; the immobilized enzyme (LY-AS) was soluble above pH 6 and precipitated below pH 4.5. Half-life is nine cycles	Lysozyme	336
Hydroxypropyl methylcellulose acetate succinate (AS)	Covalent	Reversibly soluble–autoprecipitating at pH 4.0; a sharp response of solubility to slight changes of pH without decrease in enzymic activity	Amylase	337
Hydroxypropyl methylcellulose acetate succinate (AS)	Covalent	—	Human IgG	338
Hydroxypropyl methylcellulose acetate succinate) (AS-L)	Covalent	pH-dependent solubility; reversibly soluble–autoprecipitating at pH 3.8	Cellulase	339

6.9.2
Temperature-sensitive Smart Polymers

Temperature-sensitive smart polymers have been used for immobilization of enzymes. It has been found that penicillin amidase or chymotrypsin covalently attached to poly(N-vinylcaprolactam) could be precipitated from solution when the temperature was elevated [371]. In addition, increasing the NaCl concentration from 0.01 to 1.0 m shifted the half-precipitation temperature maximum from 34.5 to 24.5 °C.

β-Galactosidase has been immobilized within thermally reversible hydrogel beads and was studied in batch and packed bed reactor systems. The enzyme was entrapped in a copolymer hydrogel of N-isopropylacrylamide (NIPAAm) and acrylamide (AAm) as beads were formed in inverse suspension polymerization [349, 350].

Similarly, thermally reversible bioconjugates of α-chymotrypsin can also be prepared with other polymers, for example poly(N-isopropylacrylamide-co-acrylamido-2-deoxy-d-glucose) [351]. The solution–precipitation phase separation was rapid, entirely reversible, and recyclable, thus making this method more attractive for certain applications. Immobilization of α-chymotrypsin also significantly enhanced the thermal stability of the enzyme.

Thermo-responsive polymers bearing suitable functional groups can also be prepared by selection of one monomer with thermo-responsive properties and one monomer with a suitable functional group for subsequent coupling [131].

6.9.3
Solvent-sensitive Enzyme–Polymer Conjugates

The development of non-aqueous biocatalysts has spurred the development of organo-soluble–insoluble enzymes. To make the enzyme soluble in organic solvent, hybridisation of amphiphilic polymers with the enzyme molecules can often make the enzyme soluble. Among these polymers PEG is usually used to prepare soluble enzyme conjugates [205, 206].

In addition to the increased solubility, the stability, selectivity, and activity can often also be improved [354] and recovery of the enzymes can be also facilitated [370] when they are applied in the soluble states (Table 6.14). The recovery of the organo-soluble enzyme can be achieved by addition of other organic solvents, for example, diethyl ether or n-hexane. Thus, the enzyme can be used in the soluble state in chlorinated or aromatic organic solvents and recovered by addition of ether or n-hexane. The recovered enzyme can be used for another step of biocatalysts or dried out for later use.

PEGylated enzymes were developed in the 1970s. The original goal was use of the pegylated enzymes to reduce the immunogenicity and antigenicity of some protein drugs. In the 1980s the development of non-aqueous enzyme-aided biotransformation spurred the use of PEGylated enzymes, with the main objective of increasing their dispersion state in organic solvents or making them soluble in

6.9 Reversibly Soluble Immobilized Enzymes

Table 6.13 Reversibly soluble immobilized enzymes based temperature-sensitive polymers.

Polymer	Methods	Remarks	Enzyme	Ref.
Soluble poly(NIPA) chains	Covalent	–	Lipase	342
Thermally sensitive poly(NIPA)	Covalent	–	Trypsin	344
Thermally sensitive poly(NIPA)	Covalent	–	Trypsin	345
Thermally sensitive poly(NIPA)	Covalent	The conjugates increased in enzymatic activity with increasing oligomer conjugation to the native trypsin	Trypsin	346
A copolymer of N-isopropylacrylamide (NIPAAm) and N-acryloxysuccinimide (NAS)	Covalent	The immobilized enzyme had enhanced thermal stability compared with free enzyme and had similar activation energy. It could be recycled with repeated precipitation/dissolution cycles with high enzyme activity	α-Chymotrypsin	347
Poly(N-isopropylacrylamide-co-acrylamido-2-deoxy-d-glucose)	Covalent	The conjugates dissolved in aqueous solution had an array of lower critical solution temperatures depending on the amount of acrylamido-2-deoxy-d-glucose in the copolymer backbone	α-Chymotrypsin	351
Conjugates of poly(N-vinylcaprolactam) and poly(N-isopropylacrylamide), with soybean trypsin inhibitor, Cibacron blue F3G-A, Cu-iminodiacetic acid, and p-aminobenzamidine	Non-covalent	Affinity thermoprecipitation is restricted to those systems with ligands that can provide binding constants of 10^{-9} to 10^{-11} m or alternatively multipoint attachment of the target protein molecules	Trypsin and lactate dehydrogenase	351
Copolymers of N-isopropylacrylamide (NIPAM) and hydroxyethylmethacrylate (HEMA)	Covalent	Copolymers and trypsin-conjugates precipitated reversibly – for poly-NIPAM by increasing the temperature within a small range (approx. 2 °C). Addition of salt reduced the precipitation temperature of poly-NIPAM	Trypsin	352

Table 6.14 Solvent-sensitive immobilized enzymes.

Polymer	Methods	Remarks	Enzyme	Ref.
PEG	Covalent	The conjugate is soluble in benzene and has ca 20% of the activity of the free enzyme in aqueous buffer	Horseradish peroxidase	353
PEG	Covalent	Enhanced activity, stability	Lipase	354
PEG	Covalent	Conjugates are soluble in organic solvents and aqueous buffers	Catalase, dehydrogenase, lipases, protease, glycosidase	355
Poly(acryloylmorpholine)	Covalent	–	Lipase from *Pseudomonas* sp	357
PEG	Covalent	–	*Candida rugosa* lipase	358
PEG	Entrapment	One-tenth the activity of the enzyme-PEG in the free form	Lipase	360
Electrogenerated PLY(3,4-ethylenedioxythiophene) (PEDT) films	Entrapment	Better stability was obtained with modified PEG-GOD electrodes	Glucose oxidase	361
AA poly(ethyleneimine) (CEPEI)	Covalent	Redissolved in organic solvent with full restoration of catalytic activity and with remarkably high storage stability in the dry state	Chymotrypsin laccase	362
PEG-5000	Covalent	Conjugate of 2–3 molecules of PEG500 with one peroxidase molecule was soluble and active in organic solvent	Peroxidase	363
	Covalent	Modified enzyme is soluble and active in organic solvent	Peroxidase	364
Triazine-activated PEG	Covalent	Enzyme is organosoluble and highly active in organic solvent	Lipoprotein lipase	365
Triazine-activated PEG	Covalent	The enzyme–polymer conjugate has increased solubility and activity in organic solvent	*Candida rugosa* lipase	366
Methoxypoly(ethylene glycol)	Non-covalent	The non-covalent enzyme–mPGE can be dissolved in polar organic solvents	Lipase *Pseudomonas cepacia*	367

organic solvents. Enzymes modified with PEG are soluble in some non-polar organic solvents and can thus act on substrates which are sparingly soluble in organic solvents [357]. In addition, the enzymes are often stabilized on PEGylation [354].

The activity of enzymes completely solubilized in organic solvents is often higher than that of the un-modified counterpart. For example, many lipases have been pegylated to enhance their activity in organic solvents [365–367]. Other enzymes, for example peroxidases or catalases [355], have been also pegylated; all had higher stability and activity.

Besides PEG, many other amphiphilic polymers can also be used to endow enzymes with solubility in organic solvents. For instance, lipase hybridized with poly(N-acryloylmorpholine) had better catalytic activity in organic media [357]. Subtilisin chemically modified with poly(ethylene glycol) monomethacrylate is soluble at levels up to 5 g L^{-1} in a variety of organic solvents [146].

It might be worth pointing out that the activity of the lyophilized enzyme preparations is largely dictated by enzyme–polymer ratio, pH, and the nature of the buffer [188].

To make use of the improved enzymes as heterogeneous catalysts, the enzyme–polymer complexes can be entrapped or covalently immobilized further as carrier-bound immobilized enzymes which might have not only enhanced activity, stability, and selectivity relative to the unmodified enzyme but also ease of handling [361].

6.9.4
Reversibly Soluble Immobilized Enzyme Based on Ionic Strength-sensitive Polymers

Similarly, ionic strength-sensitive reversibly soluble enzyme–polymer covalent conjugates have been prepared by reaction of modified α-chymotrypsin with N-isopropylacrylamide or with N-vinylcaprolactam [370, 386]. The formation of enzyme–polymer conjugates by use of smart polymers also provides possibility of recovery of reversibly soluble–insoluble enzyme–polymer complexes from aqueous media merely by slightly altering the pH, temperature, or ionic strength [331, 332, 370]. Thus, the diffusion constraints characteristic of carrier-bound immobilized enzymes can be mitigated with this technique.

6.9.5
Reversibly Soluble Immobilized Enzyme Based on Light-sensitive Polymers

Subtilisin has been covalently attached to a methacrylate co-polymer containing spiropyran groups. The modified enzyme was soluble and active in toluene, but was precipitated by ultraviolet irradiation, which changed the spiropyran to merocyanine groups. Visible irradiation reversed the process [132].

Table 6.15 Reversibly soluble immobilized enzyme based on ionic strength-sensitive polymers.

Polymer	Methods	Remarks	Enzyme	Ref.
Poly(N-ethyl-4-vinylpyridinium bromide) (polycation) (PNEVPE) and excess poly(methylacrylic acid) (polyanion)(PMAAc)	Non-covalent	Phase transitions induced by slight change in pH or ionic strength; a change in the catalytic activity and thermal stability of N-PEC-bound I was fully reversible and reproducible	Penicillin G acylase	368
Poly(N-ethyl-4-vinylpyridinium bromide) (polycation) (PNEVPE) and excess poly(methylacrylic acid) (polyanion)(PMAAc)	Non-covalent	Phase transitions induced by slight change in pH or ionic strength	α-Chymotrypsin	368
Poly(N-ethyl-4-vinylpyridinium bromide) (polycation) (PNEVPE) and excess poly(methylacrylic acid) (polyanion)(PMAAc)	Non-covalent	Phase transitions induced by slight change in pH or ionic strength	Urease	368
A reversibly soluble–insoluble polymer, GA-AS (hydroxypropyl-methylcellulose acetate succinate)	Covalent	Its response ranges were between pH 4.0 and 5.0 at 50°C and between pH 3.5 and 4.5 at 4°C. Soluble GA-AS had higher specific activity toward phloridzin than the conventional enzyme	β-Glucosidase (GA)	369
PNIPAAm or PNVC	Covalent	The EPC were removed from reaction mixtures by precipitation with 5 m NaCl	α-Chymotrypsin	370
Poly(N-vinylcaprolactam)	Covalent	Increasing the NaCl concentration from 0.01 to 1.0 m shifted the half-precipitation temperature maximum from 34.5 to 24.5°C	Penicillin G acylase	371
Poly(N-vinylcaprolactam)	Covalent	Increasing the NaCl concentration from 0.01 to 1.0 m shifted the half-precipitation temperature maximum from 34.5 to 24.5°C	Chymotrypsin	371
PVNEPB-PAAc (1:3)	Covalent	Slight change in pH and Ionic strength can reverse the solubility	Penicillin G acylase	372

6.10
References

1. Tosa T, Mori T, Fuse N, Chibata I: Studies on continuous enzyme reactions Part V Kinetics and industrial application of aminoacylase column for continuous optical resolution of acyl-dl amino acids. *Biotechnol Bioeng* 1967, 9:603–615
2. Cao L: Immobilized enzymes: science or art? *Curr Opin Chem Biol* 2005, in press
3. Haynes R, Walsh KA: Insolubilized biologically active enzymes. US 1974, US 3796634
4. Broun G, Selegny E, Avrameas S, Thomas D: Enzymatically active membranes: some properties of cellophane membranes supporting crosslinked enzymes. *Biochim Biophys Acta* 1969, 185:260–262
5. Manecke G: Immobilization of enzymes by various synthetic polymers. *Biotechnol Bioeng Symp* 1972, 3:185–187
6. Gopinathan C, Balan TP: Immobilization of the enzyme urease in radiation crosslinked poly(vinyl alcohol) gel. *J Microb Biotechnol* 1989, 4:80–83
7. Lee DM, Nishioka GM, Swann WE, Nolf CA: Enzyme immobilization on non-porous glass fibres. U.S. 1988, US 85-789530
8. Favre-Bulle O, Le Thiesse JC: Preparation and use of a coated granular enzyme-containing catalyst. *Eur Pat App* 2002, WO 0200869 A2
9. Drevon GF, Danielmeier K, Federspiel W, Stolz DB, Wicks DA, Yu PC, Russell AJ: High-activity enzyme-polyurethane coatings. *Biotechnol Bioeng* 2002, 79:785–794
10. Kulik EA, Kato K, Ivanchenko MI, Ikada Y: Trypsin immobilization on to polymer surface through grafted layer and its reaction with inhibitors. *Biomaterials* 1993, 14:763–769
11. Matoba S, Tsuneda S, Saito K, Sugo T: Highly efficient enzyme recovery using a porous membrane with immobilized tentacle polymer chains. *Bio/Technology* 1995, 13:795–797
12. Albayrak N, Yang ST: Immobilization of β-galactosidase on fibrous matrix by polyethyleneimine for production of galacto-oligosaccharides from lactose. *Biotechnol Prog* 2002, 18:240–251
13. Sato K, Tamada M, Sugo T, Kawamoto H: *J Membr Sci* 2002, 205:175–182
14. Miura S, Kubota N, Kawakita H, Saito K, Sugita K, Watanabe K, Takanobu S: High-throughput hydrolysis of starch during permeation across α-amylase-immobilized porous hollow-fibre membranes. *Radiat Phys Chem* 2002, 63:143–149
15. Godjevargova T, Dimov A: Grafting of acrylonitrile copolymer membranes with hydrophilic monomers for immobilization of glucose oxidase. *J Appl Polym Sci* 1995, 57:487–491
16. Choi SH, Lee KP, Lee JG: Adsorption behavior of urokinase by polypropylene film modified with amino acids as affinity groups. *Microchem J* 2001, 68:205–213
17. Pessela BCC, Fernandez-Lafuente R, Fuentes M, Vian A, Garcia JL, Carrascosa AV, Mateo C, Guizan JM: Reversible immobilization of a thermophilic β-galactosidase via ionic adsorption on PEI-coated Sepabeads. *Enzyme Microb Technol* 2003, 32:369–374
18. Cui AI, Tang J, Li WL, Wang ZC, Sun CQ, Zhao MY: Preparation of catalytically active enzyme thin film by alternate deposition of horseradish peroxidase and bipolar quaternary ammonium on solid surface. *Mater Chem Phys* 2001, 71:23–27
19. Caruso F, Fiedler H, Haage K: Assembly of β-glucosidase multilayers on spherical colloidal particles and their use as active catalysts. *Colloid Surf A: Physicochem Eng Aspects* 2000, 169:287–293
20. Li WJ, Xian M, Wang ZC, Sun CQ, Zhao MY: Alternate deposition of horseradish peroxidase and bipolar pyridine salt on the solid surface to prepare electrocatalytically active enzyme thin film. *Thin Solid Films* 2001, 386:121–126
21. Lee YW, Stanish I, Rastogi V, Cheng TC, Singh A: Sustained enzyme activity of organophosphorus hydrolase in polymer

encased multilayer assemblies. *Langmuir* 2003, 19:1330–1336

22 Ho GH, Liao CC: Multi-layer immobilized enzyme compositions. US 1985, US 4506015 A

23 Rao SV, Anderson KW, Bachas LG: Controlled layer-by-layer immobilization of horseradish peroxidase. *Biotechnol Bioeng* 1999, 65:389–396

24 Kobayashi Y, Anzai JI: Preparation and optimization of bienzyme multilayer films using lectin and glyco-enzymes for biosensor applications. *J Electroanal Chem* 2001, 507:250–255

25 Gemeiner P, Docolomansky P, Vikartovska A, Stefuca V: Amplification of flow-microcalorimetry signal by means of multiple bioaffinity layering of lectin and glycoenzyme. *Biotechnol Appl Biochem* 1998, 28:155–161

26 Goudd FE, Ronel SH: Formation of a stabilized enzyme by inclusion in a water-soluble, hydrophilic polymer. Ger Offen 1973, DE 2305320

27 Wei X, Zhang MG, Gorski W: Coupling the lactate oxidase to electrodes by ionotropic gelation of biopolymer. *Anal Chem* 2003, 75:2060–2064

28 Horvath C: Pellicular immobilized enzymes. *Biochim Biophys Acta* 1974, 358:164–177

29 Ho GH, Liao CC: Multi-layer immobilized enzyme compositions. US 1985, US 4506015 A

30 Kawakita H, Sugita K, Saito K, Tamada M, Sugo T, Kawamoto H: Optimization of reaction conditions in production of cycloisomaltooligosaccharides using enzyme immobilized in multilayers on to pore surface of porous hollow-fibre membranes. *J Membr Sci* 2002, 205: 175–182

31 Kobayashi S, Yonezu S, Kawakita H, Saito K, Sugita K, Tamada, M, Sugo T, Lee W: Highly multilayered urease decomposes highly concentrated urea. *Biotechnol Prog* 2003, 19:396–399

32 Kamata Y, Ishikawa E, Motoki M: Enzyme immobilization on ion exchangers by forming an enzyme coating with transglutaminase as a crosslinker. *Biosci Biotechnol Biochem* 1992, 56:1323–1324

33 Pandey PC, Upadhyay S, Pathak HC: A new glucose sensor based on encapsulated glucose oxidase within organically modified sol-gel glass. *Sens Actuators B* 1999, 60:83–89

34 Trojanowicz M, Geschke O, Krawczynski vel Krawczyk T, Cammann K: Biosensors based on oxidases immobilized in various conducting polymers. *Sens Actuators B* 1995, 28:191–199

35 Hidaka M, Aizawa M: Electrochemically synthesized polyaniline/enzyme membrane for a choline biosensor. *Denki Kagaku Oyobi Kogyo Butsuri Kagaku* 1995, 63:1113–1120

36 Ege H: A method for immobilizing a polypeptide in a polymer, a membrane produced thereby, and an electrochemical sensor containing the membrane. PCT Int Appl 1989, WO 8907139 A1

37 Sung WJ, Bae YH: A glucose oxidase electrode based on electropolymerized conducting polymer with polyanion-enzyme conjugated dopant. *Anal Chem* 2000, 72:2177–2181

38 Mihaela B, Antonella C, Giuseppe P: Oxidase enzyme immobilization through electropolymerized films to assemble biosensors for batch and flow injection analysis. *Biosens Bioelectron* 2003, 18:689–698

39 Kurota A, Kamata Y, Yamauchi F: Enzyme immobilization by the formation of enzyme coating on small pore-size ion-exchangers. *Agric Biol Chem* 1990, 54:1557–1568

40 Dekker RF: Immobilization of a lactase on to a magnetic support by covalent attachment to polyethyleneimine-glutaraldehyde-activated magnetite. *Appl Biochem Biotechnol* 1989, 22:289–310

41 Butterfield DA, Bhattacharyya D, Daunert S, Bachas L: Catalytic biofunctional membranes containing site-specifically immobilized enzyme arrays: a review. *J Membr Sci* 2001, 181:29–37

42 Shao WH, Zhang XE, Liu H, Zhang ZP, Cass AEG: Anchor-chain molecular system for orientation control in enzyme immobilization. *Bioconj Chem* 2000, 11:822–826

43 Clark DS, Bailey JE: Characterization of heterogeneous immobilized enzyme

subpopulations using EPR spectroscopy. *Biotech Bioeng* 1984, 26:231–240

44 Bernath FR, Vieth WR: Lysozyme activity in the presence of nonionic detergent micelles. *Biotechnol Bioeng* 1972, 14:737–752

45 Wallace EF, Lovenberg W: Carbohydrate moiety of dopamine β-hydroxylase. Interaction of the enzyme with concanavalin A. *Proc Natl Acad Sci USA* 1974, 71:3217–3220

46 Froehner SC, Eriksson KE: Properties of the glycoprotein laccase immobilized by two methods. *Acta Chem Scand B* 1975, 29:691–694

47 Viswanath S, Wang J, Bachas LG, Butterfield DA, Bhattacharyya D: Site-directed and random immobilization of subtilisin on functionalized membranes: activity determination in aqueous and organic media. *Biotechnol Bioeng* 1998, 60:608–616

48 Persson M, Bulow L, Mosbach K: Purification and site-specific immobilization of genetically engineered glucose dehydrogenase on thiopropyl-Sepharose. *FEBS Lett* 1990, 270:41–44

49 Saleemuddin M: Bioaffinity based immobilization of enzymes. *Adv Biochem Eng/Biotechnol* 1999, 64:203–226

50 Viswanath S, Wang J, Bachas LG, Butterfield DA, Bhattacharyya D: Site-directed and random immobilization of subtilisin on functionalized membranes: activity determination in aqueous and organic media. *Biotechnol Bioeng* 1998, 60:608–616

51 Bilkova Z, Mazurova J, Churacek J, Hora D, Turkova J: Oriented immobilization of chymotrypsin by use of suitable antibodies coupled to a nonporous solid support. *J Chromatogr A* 1999, 852:141–149

52 Huang XL, Walsh MK, Swaisgood HE: Simultaneous isolation and immobilization of streptavidin-β-galactosidase: some kinetic characteristics of the immobilized enzyme and regeneration of bioreactors. *Enzyme Microb Technol* 1996, 19:378–383

53 Hsiao HY, Royer GP: Immobilization of glycoenzymes through carbohydrate side chains. *Arch Biochem Biophys* 1979, 198:379–385

54 Bilkova Z, Churacek J, Kucerova Z, Turkova J: Purification of anti-chymotrypsin antibodies for the preparation of a bioaffinity matrix with oriented chymotrypsin as immobilized ligand. *J Chromatogr B* 1997, 689:273–279

55 Jan U, Husain Q, Saleemuddin M: Preparation of stable, highly active and immobilized glucose oxidase using the anti-enzyme antibodies and F (ab) 2′. *Biotechnol Appl Biochem* 2001, 34:13–17

56 Solomon B, Koppel R, Pines G, Katchalski-Katzir E: Enzyme immobilization via monoclonal antibodies. I. Preparation of a highly active immobilized carboxypeptidase A. *Biotechnol Bioeng* 1986, 28:1213–1221

57 Fusek M, Turkova J, Stovickova J, Franek F: Polyclonal anti-chymotrypsin antibodies suitable for oriented immobilization of chymotrypsin. *Biotechnol Lett* 1988, 10:85–90

58 Vishwanath SK, Watson CR, Huang W, Bachas LG, Bhattacharyya D: Kinetic studies of site-specifically and randomly immobilized alkaline phosphatase on functionalized membranes. *J Chem Technol Biotechnol* 1997, 68:294–302

59 Huang XL, Catignani GL, Swaisgood HE: Immobilization of biotinylated transglutaminase by bioselective adsorption to immobilized avidin and characterization of the immobilized activity. *J Agric Food Chem* 1995, 43:895–901

60 Latyshko NV, Gudkova LV, Degtiar RG, Gulyi MF: Immobilization of *Penicillium vitale* catalase on aminoethyl cellulose and properties of the obtained preparations. *Ukrainskii Biokhimicheskii Zhurnal* 1981, 53:48–52

61 Andersson L, Jornvall H, Akeson A, Mosbach K: Separation of isoenzymes of horse liver alcohol dehydrogenase and purification of the enzyme by affinity chromatography on an immobilized AMP analog. *Biochim Biophys Acta* 1974, 364:1–8

62 Siegbahn N, Maansson MO, Mosbach K: Immobilized and soluble site-to-site directed enzyme complexes composed of alcohol dehydrogenase and lactate dehydrogenase. *Methods Enzymol* 1987, 136:103–113

63 Fukui S, Ikeda S: Immobilized enzyme requiring pyridoxal-5′-phosphate as coenzyme. *Jpn Kokai Tokkyo Koho* 1975, JP 50042086

64 Turkova J: Oriented immobilization of biologically active proteins as a tool for revealing protein interactions and function. *J Chromatogr B* 1999, 722:11–31

65 Bhardwaj A, Lee J, Glauner K, Ganapathi S, Bhattacharyya D, Butterfield DA: Biofunctional membranes: an EPR study of active site structure and stability of papain non-covalently immobilized on the surface of modified poly(ether) sulfone membranes through the avidin-biotin linkage. *J Membr Sci* 1996, 119:241–252

66 Fishman A, Levy I, Cogan U, Shoseyov O: Stabilization of horseradish peroxidase in aqueous-organic media by immobilization on to cellulose using a cellulose-binding domain. *J Mol Catal B* 2002, 18:121–131

67 Royer GP: Immobilization of glycolenzymes through carbohydrate chains. *Methods Enzymol* 1987, 135:141–146

68 Braun B, Klein E: Immobilization of *Candida rugosa* lipase to nylon fibres using its carbohydrate groups as the chemical link. *Biotechnol Bioeng* 1996, 51:327–341

69 Zuzana B, Marcela S, Antonin L, Daniel H, Jiri L, Jaroslava T, Jaroslav C: Oriented immobilization of galactose oxidase to bead and magnetic bead cellulose and poly(HEMA-co-EDMA) and magnetic poly(HEMA-co-EDMA) microspheres. *J Chromatogr B* 2002, 770:25–34

70 Pugniere M, San JC, Coletti-Previero MA, Previero A: Immobilization of enzymes on alumina by means of pyridoxal 5′-phosphate. *Biosci Rep* 1988, 8:263–269

71 Voivodov K, Chan WH, Scouten W: Chemical approaches to oriented protein immobilization. *Makromol Chem, Macromol Symp* 1993, 70/71:275–283

72 Lu ZZ, Yang JJ, Leng LC, Feng XD, Li DC: Studies on biologically active p-(methacrylamido) benzoic acid esters. *Kexue Tongbao (Chinese Edition)* 1981, 23:1433–1435

73 Ahmad A, Bishayee S, Bachhawat BK: Novel method for immobilization of chicken brain arylsulfatase A using concanavalin A. *Biochem Biophys Res Commun* 1973, 53:730–736

74 Kovba GV, Rubtsova MY, Egorov AM: Chemiluminescent biosensors with immobilized peroxidase. *J Biolumin Chemilumin* 1997, 12:33–36

75 Mansson MO, Siegbahn N, Mosbach K: Site-to-site directed immobilization of enzymes with bis-NAD analogues. *Proc Natl Acad Sci USA* 1983, 80:1487–1491

76 Schlereth DD: Preparation of gold surfaces with biospecific affinity for NAD (H)-dependent lactate dehydrogenase. *Sens Actuators B* 1997, 43:78–86

77 Schlereth DD, Kooyman RPH: Self-assembled monolayers with biospecific affinity for NAD(H)-dependent dehydrogenases: characterization by surface plasmon resonance combined with electrochemistry 'in situ'. *J Electroanal Chem* 1998, 444:231–240

78 Liu XC, Scouten WH: Studies on oriented and reversible immobilization of glycoprotein using novel boronate affinity gel. *J Mol Recognit* 1996, 9:462–467

79 Iborra JL, Cortes E, Manjon A, Ferragut J, Llorca F: Affinity chromatography of frog epidermis dopa-oxidase. *J Solid-Phase Biochem* 1976, 1:91–100

80 Ikeda S, Fukui S: Preparation of pyridoxal 5′-phosphate-bound sepharose and its use for immobilization of tryptophanase. *Biochem Biophys Res Commun* 1973, 52:482–488

81 Kasche V, Löffler F, Scholzen T, Krämer DM, Boller T: Rapid protein purification using phenylbutylamine-Eupergit: a novel method for large-scale procedures. *J Chromatogr* 1990, 510:149–154

82 Fonseca LP, Cabral JMS: Optimization of a pseudo-affinity process for penicillin acylase purification. *Bioprocess Eng* 1999, 20:513–524

83 Santarelli X, Fitton V, Verdoni N, Cassagne C: Preparation, evaluation and application of new pseudo-affinity chromatographic supports for penicillin acylase purification. *J Chromatogr B* 2000, 739:163–172

84 Stella AM, Batlle DC, Alcira M: Porphyrin biosynthesis - immobilized enzymes and ligands. V. Purification of aminolevulinate dehydratase from bovine liver by affinity chromatography. *Int J Biochem* 1977, 8:353–358

85 Ikeda S, Hara H, Sugimoto S, Fukui S: Immobilized derivative of pyridoxal 5′-phosphate. Application to affinity chromatography of tryptophanase and tyrosine phenol-lyase. *FEBS Lett* 1975, 56:307–311

86 Woychik JH, Wondolowski MV: Covalent bonding of fungal β-galactosidase to glass. *Biochim Biophys Acta* 1972, 289:347–351

87 Burton NP, Lowe CR: Design of novel affinity adsorbents for the purification of trypsin-like proteases. *J Mol Recognit* 1992, 5:55–68

88 Takahara H, Okamoto H, Sugawara K: Affinity chromatography of peptidyl-arginine deiminase from rabbit skeletal muscle on a column of soybean trypsin inhibitor (Kunitz)-Sepharose. *J Biochem* (Tokyo, Japan) 1986, 99:1417–1424

89 Anspach B, Unger KK, Davies J, Hearn MTW: Affinity chromatography with triazine dyes immobilized on to activated non-porous monodisperse silicas. *J Chromatogr* 1988, 457:195–204

90 Filippusson H, Erlendsson LS, Lowe CR: Design, synthesis and evaluation of biomimetic affinity ligands for elastases *J Mol Recognit* 2000, 13:370–381

91 Rosemeyer H, Seela F: Quantitative affinity chromatography of adenosine deaminase on polymer-bound inosine: the assessment of binding constants by biospecific elution. *Anal Biochem* 1981, 115:339–346

92 Turkova J, Fusek M, Stovickova J, Kralova Z: Biospecific complex formation as a tool for oriented immobilization. *Makromol Chem Macromol Symp* 1988, 17:241–256

93 Horowitz PM, Purification of thiosulfate sulfurtransferase by selective immobilization on blue agarose. *Anal Biochem* 1978, 86:751–753

94 Joshi VK, Shahani KM, Kilara A, Wagner FW: Regenerable affinity chromatography support. *J Chromatogr* 1979, 176:11–18

95 Friend, BA, Shahani KM: Characterization and evaluation of *Aspergillus oryzae* lactase coupled to a regenerable support. *Biotechnol Bioeng* 1982, 24:329–345

96 Hughes P, Lowe CR, Sherwood RF: Metal ion-promoted binding of proteins to immobilized triazine dye affinity adsorbents. *Biochim Biophys Acta* 1982, 700:90–100

97 Small DAP, Atkinson T, Lowe CR: High-performance liquid affinity chromatography of enzymes on silica-immobilized triazine dyes. *J Chromatogr* 1981, 216:175–190

98 Bouriotis V, Galpin IJ, Dean PDG: Applications of immobilized phenyl-boronic acids as supports for group-specific ligands in the affinity chromatography of enzymes. *J Chromatogr* 1981, 210:267–278

99 Ni WC, Klee CB: Selective affinity chromatography with calmodulin fragments coupled to Sepharose. *J Biol Chem* 1985, 260:6974–6981

100 Ngo, TT: Novel protein binding capacity and selectivity of immobilized pyridinium pseudo-affinity gels: applications in immunoglobulin G purification and quantitative determination. *Anal Lett* 1993, 26:1477–1491

101 Huang XL, Catignani GL, Swaisgood HE: Immobilization of biotinylated transglutaminase by bioselective adsorption to immobilized avidin and characterization of the immobilized activity. *Agric Biol Chem* 1995, 43:895–901

102 Palanizamy UD, Winzor DJ, Lowe CR: Synthesis and evaluation of affinity adsorbents for glycoproteins: an artificial lectin *J Chromatogr B* 2000, 746:265–281

103 Persson M, Bulow L, Mosbach K: Purification and site-specific immobilization of genetically engineered glucose dehydrogenase on thiopropyl-Sepharose. *FEBS Lett* 1990, 270:41–44

104 Madoz-Gurpide J, Abad JM, Fernandez-Recio J, Velez M, Vazquez L, Gomez-Moreno C, Fernandez VM: Modulation of electroenzymatic NADPH oxidation through oriented immobilization of ferredoxin:$NADP^+$ reductase on to

modified gold electrodes. *J Am Chem Soc* 2000, 122:9808–9817

105 Baneyx F, Schmidt C, Georgiou G: Affinity immobilization of a genetically engineered bifunctional hybrid protein. *Enzyme Microb Technol* 1990, 12:337–342

106 Kondo A, Teshima T: Preparation of immobilized enzyme with high activity using affinity tag based on proteins A and G. *Biotechnol Bioeng* 1995, 46:421–428

107 Min DJ, Andrade JD, Stewart RJ: Specific immobilization of in vivo biotinylated bacterial luciferase and FMN:NAD(P)H oxidoreductase. *Anal Biochem* 1999, 270:133–139

108 Wang JQ, Bhattacharyya D, Bachas LG: Orientation specific immobilization of organophosphorus hydrolase on magnetic particles through gene fusion. *Biomacromolecules* 2001, 2:700–705

109 Zhang JK, Cass AEG: A study of His-tagged alkaline phosphatase immobilization on a nanoporous nickel- titanium dioxide film. *Anal Biochem* 2001, 292: 307–310

110 Stempfer G, Hoell-Neugebauer B, Kopetzki E, Rudolph R: A fusion protein designed for noncovalent immobilization: stability, enzymic activity, and use in an enzyme reactor. *Nat Biotechnol* 1996, 14:481–484

111 Bircakova M, Truksa M, Scouten WH: Oriented immobilization of restriction endonuclease EcoRI. *J Mol Recognit* 1996, 9:683–690

112 Huang W, Wang JQ, Bhattacharyya D, Bachas LG: Improving the activity of immobilized subtilisin by site-specific attachment to surfaces. *Anal Chem* 1997, 69:4601–4607

113 Armizen P, Mateo C, Cortes E, Barredo JL, Salto F, Diez B, Rodes L, Garcia JL, Fernandez-Lafuente R, Guizan JM: Selective adsorption of poly-His tagged glutaryl acylase on tailor-made metal chelate supports. *J Chromatogr A* 1999, 848:61–70

114 Clare DA, Valentine VW, Catignani GL, Swaisgood HE: Molecular design, expression, and affinity immobilization of a trypsin-streptavidin fusion protein. *Enzyme Microb Technol* 2001, 28:483–491

115 Mrabet NT: One-step purification of *Actinoplanes missouriensis* d-xylose isomerase by high-performance immobilized copper-affinity chromatography: functional analysis of surface histidine residues by site-directed. *Biochemistry* 1992, 31:2690–2702

116 Allard L, Cheynet V, Oriol G, Mandrand B, Delair T, Mallet F: Versatile method for production and controlled polymer-immobilization of biologically active recombinant proteins. *Biotechnol Bioeng* 2002, 80:341–348

117 Mateo C, Fernandez-Lorente G, Abian O, Fernandez-Lafuente R, Guizan JM: Multifunctional epoxy supports: A new tool to improve the covalent immobilization of proteins. the promotion of physical adsorptions of proteins on the supports before their covalent linkage. *Biomacromolecules* 2000, 1:739–745

118 Brown E, Racois A, Gueniffey H: Preparation and properties of urease derivatives insoluble in water. *Tetrahedron Lett* 1970, 25:2139–2142

119 Charles M, Coughlin RW, Hasselberger FX: Soluble-insoluble enzyme catalysts. *Biotechnol Bioeng* 1974, 16:1553–1556

120 Sharma S, Kaur P, Jain A, Rajeswari MR, Gupta MN: A Smart Bioconjugate of Chymotrypsin. *Biomacromol* 2003, 4:330–336

121 Berger JL, Lee BH, Lacroix C: Immobilization of β-galactosidases from *Thermus aquaticus* YT-1 for oligosaccharides synthesis. *Biotechnol Tech* 1995, 9:601–606

122 Husain S, Iqbal J, Saleemuddin M: Entrapment of concavalin-glycoenzyme complexes in calcium alginate gels. *Biotechnol Bioeng* 1985, 27:1102–1109

123 Doretti L, Ferrara D, Gattolin P, S Lora, Schiavonc F, Veronesec FM: PEG-modified glucose oxidase immobilized on a PVA cryogel membrane for amperometric biosensor applications, *Talanta* 1998, 45:891–898

124 Lasch J, Bessmertnaya L, Kozlov LV, Antonov VK: Thermal stability of immobilized enzymes: Circular dichroism, fluorescence and kinetic measurements of α-Chymotrypsin attached to soluble carriers. *Eur J Biochem* 1976, 63:591–598

125 Marshall JJ, Rabinowitz ML: Stabilization of catalase by covalent attachment to dextran. *Biotechnol Bioeng* 1976, 18:1325–1329

126 Kikkoman Corporation: Immobilization of enzymes. *Jpn Kokai Tokkyo Koho* 1984, JP 59220187 A2

127 Lee PM, Lee KH, Siaw YS: Covalent immobilization of aminoacylase to alginate for l-phenylalanine production. *J Chem Technol Bioeng* 1993, 58:65–70

128 Woodward J, Clarke KM: Hydrolysis of cellobiose by immobilized β-glucosidase entrapped in maintenance-free gel spheres. *Appl Biochem Biotechnol* 1991, 28:277–283

129 Hueper F: Water soluble, polymeric substrate covalently bound penicillin acylase for preparing 6-aminopenicillanic acid. Ger Offen 1974, DE 2312824

130 Steinke K, Vorlop KD: Precipitable enzyme-polymer conjugates with high activity. *DECHEMA Biotechnol Conf* 1990, 4:889–492

131 Hoshino K, Akakabe S, Morohashi S, Sasakura T: Immobilization of enzymes on thermo-responsive polymers. *Methods Biotechnol* 1997, 1:101–108

132 Ito Y, Sugimura N, Kwon OH, Imanishi Y: Enzyme modification by polymers with solubilities that change in response to photoirradiation in organic media. *Nat Biotechnol* 1999, 17:73–75

133 Wykes JR, Dunnill P, Lilly D: Immobilization of α-amylase by attachment to soluble support materials. *Biochem Biophys Acta* 1971, 250:522–529

134 Fernandez-Lafuente R, Rosell CM, Rodriguez V, Guizan JM: Strategies for enzyme stabilization by intramolecular crosslinking with bifunctional reagents. *Enzyme Microb Technol* 1995, 17:517–523

135 Monsan et al.: Nouvelle methode de preparation d'enzymes fixes sur des supports mineraux, *CR Acad Sci Paris*, T. 1971, 273:33–36

136 Bartling GJ, Brown HD, Chattopadhyay S K: Synthesis of matrix-supported enzyme in non-aqueous conditions. *Nature* 1973, 243:342–344

137 Bartling GJ, Brown HD, Chattopadhyay SK: Protein modification in nonaqueous media. preparation and properties of matrix-supported lysozyme. *Biotechnol Bioeng* 1974, 14:361–369

138 Stark MB, Kolmberg K: Covalent immobilization of lipase in organic solvents. *Biotechnol Bioeng* 1989, 34:942–950

139 Brzozowski AM, Thim L: A model for interfacial activation in lipases from the structure of a fungal lipase–inhibitor complex. *Nature* 1991, 351:491–494

140 Van Tilbeurgh H, Egloff MP, Martinez C, Rugani N, Verger R, Cambillau C: Interfacial activation of the lipase–procolipase complex by mixed micelles revealed by X-ray crystallography. *Nature* 1993, 362:814–818

141 Linko Y, Pohjola LA: Simple entrapment method for immobilizing enzymes within cellulose fibres. *FEBS Lett* 1976, 62:77–80

142 Nakamura T, Ono Y: Immobilized enzyme. Japan Kokai 1977, JP 76-19374

143 Kaputskii FN, Bil'dyukevich AV, Khlyustov SV, Rytik PG, Zhavrid SV, Yurkshtovich TL, Protsenko VE, Alinovskaya VA, Shavrova EV et al.: Enzyme immobilization on cellulose. SU 1567625 A1

144 Norvick SJ, Dordick JS: Protein-containing hydrophobic coatings and films, *Biomaterials* 2002, 23:441–448

145 Fukunaga K, Minamijima N, Sugimura Y, Zhang ZZ, Nakao K: Immobilization of organic solvent soluble lipase in non-aqueous conditions and properties of the immobilized enzymes. *J Biotechnol* 1996, 52:81–88

146 Yang Z, Williams D, Russell AJ: Synthesis of protein-containing polymers in organic solvents. *Biotechnol Bioeng* 1995, 45:10–17

147 DeSantis G, Jones JB: Chemical modification of enzymes for enhanced functionality. *Curr Opin Biotechnol* 1999, 10:324–330

148 Lee S, Takahashi T, Anzal J, Suzuki Y, Osa T: Preparation of enzyme membranes by use of organic solvent. A composite membrane of poly(vinyl stearate) and poly(ethylene glycol)-modified glucose oxidase for enzyme sensors. *Pharmazie* 1994, 49:620–621

149 Blinkovsky AM, Khmelmitsky YL, Dordick JS: Organosoluble enzyme-polymer complex, a novel type of biocatalysts for non-aqueous media, *Biotechnol Tech* 1994, 8:33–38

150 Yang Z, Mesiano AJ, Venkatasubramanian S, Gross SH, Harris JM, Russell AJ: Activity and stability of enzymes incorporated into acrylic polymers. *J Am Chem Soc* 1995, 117:4843–4850

151 Wang P, Sergeeva MV, Lim L, Dordick JS: Biocatalytic plastics as active and stable materials for biotransformations. *Nat Biotechnol* 1997, 15:789–793

152 Castro GR: Enzymatic activities of proteases dissolved in organic solvents *Enzyme Microb Technol* 1999, 25:689–694

153 Knubovets T, Osterhout JJ, Connolly PJ, Klibanov AM: Structure, thermostability, and conformational flexibility of hen egg-white lysozyme dissolved in glycerol. *Proc Natl Acad Sci USA* 1999, 96:1262–1267

154 De Oliveira PC, Alves GM, de Castro HF: Immobilization studies and catalytic properties of microbial lipase on to styrene-divinylbenzene copolymer. *Biochem Eng J* 2000, 5:63–71

155 Castro HF, Silva MLCP, Silva GLJP: Evaluation of inorganic matrixes as supports for immobilization of microbial lipase. *Brazilian J Chem Eng* 2000, 17:849–857

156 Brown E, Racois A: Immobilization of trypsin previously protected by a synthetic macromolecular inhibitor. In: Thomas D, Kerneyez JP (Eds) *Anal Control Immobilized Enzyme Syst, Proc Int Symp* 1976:111–114

157 Miwa N, Ohtomo K: Enzyme immobilization in the presence of substrates and inhibitors. *Jpn Kokai Tokkyo Koho* 1975, JP 56045591

158 Brown E, Racois A: Immobilized enzymes. 13. Immobilization of trypsin previously protected by a synthetic water-soluble macromolecular inhibitor. *Makromol Chem* 1981, 182:1605–1616

159 Van Leeputen E, Horisberger M: Immobilization of enzymes on magnetic particles. *Biotechnol Bioeng* 1974, 16:385–396

160 Alvaro G, Fernandez-Lafuente R, Blanco RM, Guizan JM: Stabilizing effect of penicillin G sulfoxide, a competitive inhibitor of penicillin G acylase: its practical applications. *Enzyme Microb Technol* 1991, 13:210–214

161 Rosell CM, Fernandez-Lafuente R, Guizan JM: Modification of enzyme properties by the use of inhibitors during their stabilization by multipoint covalent attachment. *Biocatal Biotransform* 1995, 12:67–76

162 Jaworek D: New immobilization techniques and supports. *Enzyme Eng* 1974, 105–114

163 Suzuki S, Hirano K, Takagi Y: Plastic immobilized enzyme. *Japan Kokai* 1976, JP 51070872

164 Martinek K, Klibanov AM, Coldmacher VS, Berezin LV: The principles of enzyme stabilization, *Biochim Biophys Acta* 1977, 485:1–12

165 Staahl M, Mansson MO, Mosbach K: The synthesis of a d-amino acid ester in an organic media with α-chymotrypsin modified by a bio-imprinting procedure. *Biotechnol Lett* 1990, 12:161–166

166 Peissker F, Fischer L: Crosslinking of imprinted proteases to maintain tailor-made substrate selectivity in aqueous solutions. *Bioorg Med Chem* 1999, 7:2231–2237

167 Kronenburg NAE, de Bont JAM, Fischer L: Improvement of enantioselectivity by immobilized imprinting of epoxide hydrolase from *Rhodotorula glutinis*. *J Mol Catal B* 2001, 16:121–129

168 Tischer W, Kasche V: Immobilized enzymes: crystals or carriers? *Trends Biotechnol* 1999, 17:326–335

169 Kumakura M, Yoshida M, Asano M, Kaetsu I: Immobilization of enzymes by radiation-induced polymerization of glass-forming monomers. Double entrapping of enzymes in the presence of various additives. *J Solid-Phase Biochem* 1977, 2:279–288

170 Glad M, Norrloew O, Sellergren B, Siegbahn N, Mosbach K: Use of silane monomers for molecular imprinting and enzyme entrapment in polysiloxane-coated porous silica. *J Chromatogr* 1985, 347:11–23

171 Furukawa S, Ono T, Ijima H, Kawakami K: Effect of imprinting sol-gel immobilized lipase with chiral template

substrates in esterification of (R)-(+)- and (S)-(–)-glycidol. *J Mol Catal B* 2002, 17:23–28

172 McIninch JK, Kantrowitz ER: Use of silicate sol-gels to trap the R and T quaternary conformational states of pig kidney fructose-1,6-bisphosphatase. *Biochim Biophys Acta* 2001 547:320–328

173 Huang DB, Clint F, Ainsworth FJS, Marianne S: Three quaternary structures for a single protein. *Proc Natl Acad Sci USA* 1996, 93:7017–7012

174 Cao Y, Musah RA, Wilcox SK, Goodin DB, Mcree DE: Protein conformer selection by ligand binding observed with crystallography. *Prot Sci* 1998, 7:72–78

175 Schrag JD, Li Y, Cygler M, Lang D, Burgdorf T, Hecht HJ, Schmid RD, Schomburg D, Rydel TJ, Oliver JD, Strickland LC Dunaway CM, Larson S, Day J, McPherson A:The open conformation of a Pseudomonas lipase. *Structure* 1997, 5:187–202

176 Margolin AL: Novel crystalline catalysts. *Trends Biotechnol* 1996, 14:223–230

177 Lalonde J: Practical catalysis with enzyme crystals. *Chemtech* 1997, 27:38–45

178 Cao L, van Rantwijk F, Sheldon RA: Cross-linked enzyme aggregates: A simple and effective method for the immobilization of penicillin acylase. *Org Lett* 2000, 2:1361–1364

179 Lopez-Serrano P, Cao L, van Rantwijk F, Sheldon RA: Cross-linked enzyme aggregates with enhanced activity: application to lipases. *Biotechnol Lett* 2002, 24:379–1383

180 Royer GP, Ikeda S, Aso K: Cross-linking of reversibly immobilized enzymes. *FEBS Lett* 1977, 80:89–94

181 Klibanov AM: Why are enzymes less active in organic solvents than in water? *Trends Biotechnol* 1997, 15:97–101

182 Triantafyllou AO, Wehtje E, Adlercreutz P, Mattiasson B. How do additives affect enzyme activity and stability in nonaqueous media? *Biotechnol Bioeng* 1997, 54:67–76

183 Ru MT, Wu KC, Lindsay JP, Dordick JS, Reimer JA, Clark DS: Towards more active biocatalysts in organic media: Increasing the activity of salt-activated enzymes. *Biotechnol Bioeng* 2001, 75: 187–196

184 Rich JO, Dordick JS: Imprinting enzymes for use in organic media. *Methods Biotechnol* 2001, 15:13–17

185 Rich JO, Dordick JS: Controlling subtilisin activity and selectivity in organic media by imprinting with nucleophilic substrates. *J Am Chem Soc* 1997, 119:3245–3252

186 Rich JO, Mozhaev VV, Dordick JS, Clark DS, Khmelnitsky YL: Molecular imprinting of enzymes with water-insoluble ligands for nonaqueous biocatalysts. *J Am Chem Soc* 2002, 124:5254–5255

187 Lindsay JP, Clark DS, Dordick JS: Penicillin amidase is activated for use in nonaqueous media by lyophilizing in the presence of potassium chloride. *Enzyme Microb Technol* 2002, 120:1–5

188 Murakami Y, Hoshib R, Hiratab A: Characterization of polymer–enzyme complex as a novel biocatalyst for nonaqueous enzymology. *J Mol Catal B* 2003, 22:79–88

189 Monsan P, Combes D: Enzyme stabilization by immobilization. *Methods Enzymol* 1988, 137:584–598

190 Zaborsky OR: Alteration of enzymic properties prior to immobilization. *Biotechnol Bioeng Symp* 1972, 3:211–217

191 Hernaiz MJ, Crout DHG: Immobilization/stabilization on Eupergit C of the β-galactosidase from *B. circulans* and an β-galactosidase from *Aspergillus oryzae. Enzyme Microb Technol* 2000, 27:26–32

192 Husain S, Jafri F, Saleemuddin M: Effects of chemical modification on the stability of invertase before and after immobilization. *Enzyme Microb Technol* 1996, 18:275–280

193 Costantino HR, Griebenow K, Langer R, Klibanov AM: On the pH memory of lyophilized compounds containing protein functional groups. *Biotechnol Bioeng* 1997, 53:345–348

194 Rocha JMS, Gil MH, Garcia FAP: Effects of additives on the activity of a covalently immobilized lipase in organic media. *J Biotechnol* 1998, 66:61–67

195 Blanco RM, Guizan JM: Additional stabilization of PGA-agarose derivatives by chemical modification with aldehydes. *Enzyme Microb Technol* 1989, 11:360–366

196 Guizan JM, Alvaro G, Fernandez-Lafuente R, Rosell CM, Garcia JL, Tagliani A: Stabilization of heterodimeric enzyme by multipoint covalent immobilization: penicillin G acylase from *Kluyvera citrophila*. *Biotechnol Bioeng* 1993, 42:455–464

197 Torres-Bacete J, Arroyo M, Torres-Guzman R, De la Mata I, Castillon MP, Acebal C: Stabilization of penicillin V acylase from *Streptomyces lavendulae* by covalent immobilization. *J Chem Technol Biotechnol* 2001, 76:525–528

198 Batkai L, Horvath I, Horvath-Feher E, Boross L, Li VP: Insoluble enzymes. *Fr Demande* 1974, FR 2219172

199 Kestner A, Kreen M: Enzymes immobilized in polymerized gels. *USSR* 1974, SU 443902

200 Hamsher JJ: Immobilized enzymes and intermediates for their preparation. *Fr Demande* 1974, FR 2212340

201 Szewczuk A, Ziomek E, Mordarski M, Siewinski M, Wieczorek J: Properties of penicillin amidase immobilized by copolymerization with acrylamide. *Biotechnol Bioeng* 1979, 21:1543–1552

202 Robak M, Szewczuk A: Penicillin amidase from *Proteus rettgeri*. *J Acta Biochim Pol* 1981, 28:275–284

203 Martinek K, Klibanov AM, Goldmacher VS, Berezin IV: The principles of enzyme stabilization, I. Increase in thermostability of enzymes covalently bound to a complimentary surface of a polymer support in a multipoint fashion. *Biochim Biophys Acta* 1977, 485:1–12

204 Mozhaev VV: Multipoint attachment to a support protects enzyme from inactivation by organic solvents: α-chymotrypsin in aqueous solutions of alcohols and diols. *Biotechnol Bioeng* 1990, 35:653–659

205 Takahashi K, Saito Y, Inada Y: Lipase made active in hydrophobic media. *J Am Oil Chem Soc* 1988, 65:911–916

206 Yang Z, Domach M, Auger R, Yang FX, Russell A:Polyethylene glycol-induced stabilization of subtilisin. *Enzyme Microb Technol* 1996, 1882–1889

207 Doretti L, Ferrara D, Lora S, Schiavon F, Veronese FM: Acetylcholine biosensor involving entrapment of acetylcholinesterase and poly(ethylene glycol)-modified choline oxidase in a poly(vinyl alcohol) cryogel membrane. *Enzyme Microb Technol* 2000, 27:279–285

208 Podual K, Doyle FJ 3rd, Peppas NA: Glucose-sensitivity of glucose oxidase-containing cationic copolymer hydrogels having poly(ethylene glycol) grafts. *J Controlled Release* 2000, 67:9–17

209 Cremonesi P, Mazzola G, Focher B, Vecchio G: Peroxidase immobilized on acrylic copolymer beads. *Angew Makromol Chem* 1975, 48:17–27

210 Glazer AN, Bar-Eli A, Katchalski E: Preparation and characterization of polytyrosyl trypsin, *J Biol Chem* 1962, 237:1832–1838

211 Broun GB: Chemically aggregated enzymes. *Methods Enzymol* 1976, 44:263–280

212 Kamogashira T, Mihara S, Tamaoka H, Doi T: 6-Aminopenicillanic acid by penicillin hydrolysis with immobilized enzyme. Jpn Kokai Tokkyo Koho 1972, JP 47028187

213 Tyagi R, Gupta MN: Noncovalent and reversible immobilization of chemically modified amyloglucosidase and β-glucosidase on DEAE-cellulose. *Proc Biochem* 1994, 29:443–448

214 Nakamura N, Horikoshi K: Production of Schardinger β-dextrin by soluble and immobilized cyclodextrin glycosyl-transferase of an alkalophilic Bacillus sp. *Biotechnol Bioeng* 1977, 19:87–99

215 Hollecker M, Creighton TE: Effect on protein stability of reversing the charge on amino groups. *Biochim Biophys Acta* 1982, 701:395–404

216 Ampon K, Means GE: Immobilization of proteins on organic polymer beads. *Biotechnol Bioeng* 1988, 32:689–697

217 Basri M, Ampon K, Yunus WMZW, Razak CNA, Salleh AB: Immobilization of hydrophobic lipase derivatives on to organic polymer beads. *J Chem Technol Biotechnol* 1994, 59:37–44

218 Cao LQ, Bornscheuer UT, Schmid RD: Lipase-catalysed solid-phase synthesis of sugar esters. Influence of immobilization on productivity and stability of the enzyme. *J Mol Catal B* 1999, 6:279–285

219 Koops BC, Papadimou E, Verheij HM, Slotboom AJ, Egmond MR: Activity and stability of chemically modified *Candida antarctica* lipase B adsorbed on solid supports. *Appl Microb Biotechnol* 1999, 52:791–796

220 Barroug A, Fastrez J, Lemaitre J, Rouxhet P: Adsorption of succinylated lysozyme on hydroxyapatite. *J Colloid Interface Sci* 1997, 189:37–42

221 Lee SH, Shin HD, Lee YH: Evaluation of immobilization methods for cyclodextrin glucanotransferase and characterization of its enzymic properties. *J Microbiol Biotechnol* 1991, 1:54–62

222 Marshall JJ, Rabinowitz ML: Stabilization of catalase by covalent attachment to dextran. *Biotechnol Bioeng* 1976, 18:1325–1329

223 Hixson HF: Water-soluble enzyme-polymer grafts. Thermal stabilization of glucose oxidase. *Biotechnol Bioeng* 1973, 15:1011–1016

224 Gomez L, Ramirez HL, Villalonga R: Modification of α-amylase by sodium alginate. *Acta Biotechnol* 2001, 21: 265–273

225 Wykes JR, Dunnill P, Lilly D: Immobilization of α-amylase by attachment to soluble support materials. *Biochem Biophys Acta* 1971, 250:522–529

226 Blomhoff HK, Chrittensen TB: Effect of dextran and dextran modifications on the thermal and proteolytic stability of conjugated bovine testis-galactosidase and human serum albumin. *Biochem Biophys Acta* 1983, 743:401–407

227 Sundaram P, Venkatesh R: Retardation of thermal and urea induced inactivation of α-chymotrypsin by modification with carbohydrate polymers. *Protein Eng* 1998, 11:699–705

228 Kazan D, Ertan H, Erarslan A: Stabilization of *Escherichia coli* penicillin G acylase against thermal inactivation by crosslinking with dextran dialdehyde polymers. *Appl Microbiol Biotechnol* 1997, 48:191–197

229 Markvicheva EA, Bronin AS, Kudryavtseva NE, Rumsh LD, Kirsh YE, Zubov VP. Immobilization of proteases in composite hydrogel based on poly(N-vinyl caprolactam). *Biotechnol Tech* 1994, 8:143–148

230 Demers N, Agostinelli E, Averill-Bates DA, Fortier G: Immobilization of native and poly(ethylene glycol)-treated ('PEGylated') bovine serum amine oxidase into a biocompatible hydrogel. *Biotechnol Appl Biochem* 2001, 33:201–207

231 Mohapatra SC, Hsu JT: Immobilization of α-chymotrypsin for use in batch and continuous reactors. *J Chem Technol Biotechnol* 2000, 75:519–525

232 Markvicheva EA, Bronin AS, Kudriavtseva NE, Kuz'kina IF, Pashkin II, Kirsh YE, Rumsh LD, Zubov VP: A new method of immobilizing proteolytic enzymes in polymeric hydrogels. *IOORGANICHESKAIA KHIMIIA* 1994, 20:257-62

233 Veronese FM, Mammucari C, Schiavon F, Schiavon O, Lora S, Secundo F, Chilin A, Guiotto A: Pegylated enzyme entrapped in poly(vinyl alcohol) hydrogel for biocatalytic application. *Farmaco* 2001, 56:541–547

234 Gemeiner P, Halak P, Polakova K: Two-step covalent immobilization of enzymes as a way for study of effects influencing catalytic activity. *J Solid-Phase Biochem* 1980, 5:197–209

235 Manecke G, Middeke HJ: Reactive carriers with maleimide groups for the immobilization of enzymes, 2. *Angew Makromol Chem* 1984, 121:27–39

236 Drobnik J, Saudek V, Svec F, Kalal J, Vojtizek V, Barta M: Enzyme immobilization techniques on poly(glycidyl methacrylate-co-ethylene dimethacrylate) carrier with penicillin amidase as model. *Biotechnol Bioeng* 1979, 21:1317–1332

237 Moriya K, Tanizawa K, Kanaoka Y: Immobilized chemotrypsin by means of Schiff base copper (II) chelate. *Biochem Biophys Res Commun* 1989, 162:408–414

238 Manta C, Ferraz N, Betancor L, Antunes G, Batista-viera F, Carlsson J, Caldwell K: Polyethyleneglycol as a spacer for solid phase enzyme immobilization. *Enzyme Microb Technol* 2003, 33:890–898

239 Ban K, Ueki T, Tamada Y, Saito T, Imabayashi SI, Watanabe M: Fast electron transfer between glucose oxidase and electrodes via phenothiazine mediators with poly(ethylene oxide)

240 Lenders JP, Crichton RR: Chemical stabilization of glucoamylase from *Aspergillus niger* against thermal inactivation. *Biotechnol Bioeng* 1988, 31:267–277

241 Tosa T, Mori T, Chibata I: Studies on continuous enzyme reactions Part VI. Enzymatic properties of DEAE-Sephadex-aminoacylase complex. *Agric Biol Chem* 1969, 33:1503–1509

242 Kennedy JF, Barker SA, Rosevear A: Use of a poly(allyl carbonate) for the preparation of active, water insoluble derivatives of enzymes. *J Chem Soc, Perkin Trans 1* 1972, 20:2568–2573

243 Van Unen DJ, Sakodinskaya IK, Engbersen JFJ, Reinhoudt DN: Crown ether activation of crosslinked subtilisin Carlsberg crystals in organic solvents. *J Chem Soc Perkin Trans 1* 1989, 20:3341–3343

244 Haering D, Schreier P: Chemical engineering of enzymes: altered catalytic activity, predictable selectivity and exceptional stability of the semisynthetic peroxidase seleno-subtilisin. *Naturwissenschaften* 1999, 86:307–312

245 Ivanova AE, Schneider MP: Methods for the immobilization of lipases and their use for ester synthesis. *J Mol Catal B: Enzymatic* 1997, 3:303–309

246 Partridge J, Halling PJ, Moore BD: Practical route to high activity enzyme preparations for synthesis in organic media. *Chem Commun* 1998, 7:841–842

247 Yang YG, Chase HA: Immobilization of α-amylase on poly(vinyl alcohol)-coated perfluoropolymer supports for use in enzyme reactors, *Biotechnol Appl Biochem* 1998, 2:145–154

248 Mateo C, Abian O, Fernandez-Lafuente R, Guizan JM: Increase in conformational stability of enzymes immobilized on epoxy-activated supports by favoring additional multipoint covalent attachment. *Enzyme Microb Technol* 2000, 26:509–515

249 Muecke I: Enhancement of the stability of immobilized enzymes with bifunctional crosslinkers and polyamine. DE 87-3719324

250 Furukawa SY, Ono T, Ijima H, Kawakami K: Activation of protease by sol-gel entrapment into organically modified hybrid silicates. *Biotechnol Lett* 2002, 24:13–16

251 Fernandez-Lafuente R, Rodriguez V, Mateo C, Fernandez-Lorente G, Arminsen P, Sabuquillo P, Guizan JM: Stabilization of enzymes d-amino acid oxidase against hydrogen peroxide via immobilization and post-immobilization techniques. *J Mol Catal B: Enzymatic* 1999, 7:173–179

252 Rogalski J, Dawidowicz A, Leonowicz A: Lactose hydrolysis in milk by immobilized β-galactosidase. *J Mol Catal B: Enzymatic* 1994, 93:233–245

253 Ovsejevi K, Brena B, Batista-Viera F, Carlsson J: Immobilization of β-galactosidase on thiolsulfonate-agarose. *Enzyme Microb Technol* 1995, 17:151–156

254 Fernandex-Lafuente R, Cowan DA, Wood ANP: Hyperstabilization of a thermophilic esterase by multipoint covalent attachment. *Enzyme Microb Technol* 1995, 17:366–372

255 DePaz RA, Dale DA, Barnett CC, Carpenter JF, Gaertner AL, Randolph TW: Effects of drying methods and additives on the structure, function, and storage stability of subtilisin: role of protein conformation and molecular mobility. *Enzyme Microb Technol* 2002, 31:765–774

256 Schmid RD: Stabilized soluble enzymes. *Adv Biochem Eng* 1979, 12:41–118

257 Gray CJ, Livingstone CM: Properties of enzymes immobilized by the diazotized m-diaminobenzene method. *Biotechnol Bioeng* 1977, 19:349–364

258 Dabulis K, Klibanov AM: Drastic enhancement of enzymatic activity in organic solvents by lyoprotectants. *Biotechnol Bioeng* 1993, 41:566–571

259 Santos AM, Vidal M, Pacheco Y, Frontera J, Baez C, Ornellas O, Barletta G, Griebenow K: Effect of crown ethers on structure, stability, activity, and enantioselectivity of Subtilisin Carlsberg in organic solvents. *Biotechnol Bioeng* 2001, 74:295–308

260 Gonzalez-Navarro H, Braco L: Lipase-enhanced activity in flavour ester

reactions by trapping enzyme conformers in the presence of interfaces. *Biotechnol Bioeng* 1998, 59:122–127
261 Slade CJ, Vulfson EN: Induction of catalytic activity in proteins by lyophilization in the presence of a transition state analog. *Biotechnol Bioeng* 1998, 57:211–215
262 Triantafyllou AO, Wehtje E, Adlercreutz P, Mattiasson B: Effects of sorbitol addition on the action of free and immobilized hydrolytic enzymes in organic media. *Biotechnol Bioeng.* 1995, 54:67–76
263 Buisson P, Hernandez C, Pierre M, Pierre AC: Encapsulation of lipases in aerogels. *J Non-Crystal Solids* 2001, 285:295–302
264 He F, Zhuo RX, Liu LJ, Xu MY: Immobilization of acylase on porous silica beads: preparation and thermal activation studies. *React Funct Polym* 2000, 45:29–33
265 Li B, Takahashi H: New immobilization method for enzyme stabilization involving a mesoporous material and an organic/inorganic hybrid gel. *Biotechnol Lett* 2000, 22:1953–1958
266 Tosa T, Mori T and Chibata I: Activation of water-insoluble aminoacylase by protein denaturing agents. *Enzymologia* 1971, 40:49–53
267 Abraham M, Alexin A, Szajani B: Immobilized triosephosphate isomerases. A comparative study. *Appl Biochem Biotechnol* 1992, 36:1–12
268 Yonath A, Podjarny A, Honig B, Sielecki A, Traub W: Crystallographic studies of protein denaturation and renaturations. 2. Sodium dodecyl sulfate induced structural changes in triclinic lysozyme. *Biochemistry* 1977, 16:1418–1424
269 Yonath A, Podjarny A, Honig B, Traub W, Sielecki A, Herzberg O, Moult J: Structural analysis of denaturant-protein interactions: comparison between the effects of bromoethanol and SDS on denaturation and renaturation of triclinic lysozyme. *Biophys Struct Mech* 1978, 4:27–36
270 Lehmann MS, Stansfield RFD: Binding of dimethyl sulfoxide to lysozyme in crystals, studied with neutron diffraction. *Biochemistry* 1989, 28:7028–7033

271 Snape KW, Tijan R, Blake CCF, Koshland DE. Crystallographic study of the interaction of urea with lysozyme. *Nature* 1974, 250:295–298
272 Hibbard LS, Tulinski A: Expression of functionality of α-chymotrypsin. Effects of guanidine hydrochloride and urea in the onset of denaturation. *Biochem* 1978, 17:5460–5468
273 Piskin K, Chang TMS: A new combined enzyme-charcoal system formed by enzyme adsorption on charcoal followed by polymer coating. *Int J Artif Organs* 1980, 3:344–346
274 Mizutani F, Yabuki S, Hirata Y: Amperometric biosensors using poly-l-lysine/poly(styrenesulfonate) membranes with immobilized enzymes. *Denki Kagaku Oyobi Kogyo Butsuri Kagaku* 1995, 63:1100–1105
275 Sakurada Y, Yamane T: Fibrous substance entrapping an enzyme. *Japan Kokai* 1975, JP 50029728
276 Miyawaki O, Nakamura K, Yano T: Mass transfer and reaction with microcapsules containing enzyme and adsorbent. *Enzyme Eng* 1978, 3:79–84
277 Leca B, Blum LJ: Luminol electrochemiluminescence with screen-printed electrodes for low-cost disposable oxidase-based optical sensors. *Analyst* 2000, 125:789–791
278 Secundo F, Carrea G, Soregaroli C, Varinelli D, Morrone R: Activity of different *Candida Antarctica* lipase B formulations in organic solvents, *Biotechnol Bioeng* 2001, 73:157–163
279 Cao LQ: Regeneration of activity of CLEAs of CAL-B by rehydration, unpublished results, 2002
280 Schmidt P, Fischer J, Hettwer W, Mansfeld HW, Wahl G, Schellenberger A, Millner R, Rosenfeld E: Reactivation of immobilized enzymes contaminated by adsorbed materials. *Ger (East)* 1981, DD 150628 Z
281 Schmidt P, Rosenfeld E, Millner R, Schellenberger A: Effects of ultrasound on the catalytic activity of matrix-bound glucoamylase. *Ultrasonics* 1987, 25:295–299
282 Chuprina LM, Tsirkel VA, Pkhakadze GA, Lipatova TE: Increased activity of immobilized trypsin in the presence of

alkali. *Dopovidi Akademii Nauk Ukrains'koi RSR, Seriya B: Geologichni, Khimichni ta Biologichni Nauki* 1984, 10:81–83

283 Fernandez-Lafuente R, Rosell CM, Alvaro G, Guizan JM: Additional stabilization of penicillin G acylase-agarose derivatives by controlled chemical modification with formaldehyde. *Enzyme Microb Technol* 1992, 14:489–495

284 Svensson B, Ottesen M: Immobilization of β-glucanase and studies on its degradation of barley β-glucan Carlsberg. *Carlsberg Res Commun* 1978, 43:5–14

285 Husain Q, Saleemuddin M: An inexpensive procedure for the immobilization of glycoenzymes on Sephadex G-50 using crude concanavalin A. *Biotechnol Appl Biochem* 1989, 11:508–512

286 Woodward J, Wohlpart DL. Properties of native and immobilized preparations of β-d-glucosidase from *Aspergillus niger*. *J Chem Technol Biotechnol* 1980, 32:547–552

287 Weiss B, Hui M, Lajtha AV: Efforts at stabilization of rat liver phenylalanine hydroxylase. *Res Commun Chem Pathol Pharm* 1979, 25:153–164

288 Dosanjh NS, Kaur J: Immobilization, stability and esterification studies of a lipase from a *Bacillus Sp. Biotechnol Appl Biochem* 2002, 36:7–12

289 Spagna G, Pifferi PG, Tramontini M: Immobilization and stabilization of pectinlyase on synthetic polymers for application in the beverage industry. *J Mol Catal A* 1995, 101:99–105

290 Tatsumoto K, Oh KK, Baker JO, Himmel ME: Enhanced stability of glucoamylase through chemical crosslinking. *Appl Biochem Biotechnol* 1989, 20:293–308

291 Arasaratnam V, Balasubramaniam K: Thermal stabilization of immobilized α-amylase by coupling with proline. *Proc Biochem* 1995, 30:299–303

292 Tarlap A, Kaplan H: Chemical modification of lyophilized proteins in nonaqueous environments. *J Protein Chem* 1997, 16:183–193

293 Martinek K, Torchilin VP: Stabilization of enzymes by intramolecular crosslinking using bifunctional reagents. *Methods Enzymol* 1988, 137:615–626

294 Fernandez-lafuente R, Rosell CM, Rodriguez V, Guizan JM: Strategies for enzyme stabilization by intramolecular crosslinking with bifunctional reagents. *Enzyme Microb Technol* 1995, 17:517–523

295 Maladkar NK: Immobilization of penicillin G acylase on silica gel based supports. *Ind Biotechnol* 1992, 485–491

296 Spagna G, Pifferi PG, Tramontini M: Immobilization and stabilization of pectinlyase on synthetic polymers for application in the beverage industry. *J Mol Catal A* 1993, 101:99–105

297 Bagi K, Simon LM, Szajani B: Immobilization and characterization of porcine pancreas lipase. *Enzyme Microb Technol* 1997, 20:531–535

298 Betancor L, Hidalgo A, Fernandez-Lorente G, Mateo C, Fernandez-Lafuente R, Guizan JM: Preparation of a stable biocatalyst of bovine liver catalase using immobilization and postimmobilization techniques. *Biotechnol Prog* 2003, 33:8756–7938

299 Kaiser ET, Lawrence DS: Chemical mutation of enzyme active sites. *Science* 1984, 226:505–511

300 Häring D, Schreier P: Novel biocatalysts by chemical modification of known enzymes: cross-linked microcrystals of the semi-synthetic peroxidase selenosubtilisin. *Angew Chem Int Ed Engl* 1998, 37:2471–2473

301 Brougham MJ, Johnson DB: Studies on the stability of soluble and immobilized alcohol dehydrogenase from yeast mitochondria. *Enzyme Eng* 1980, 5:431–434

302 Chae HJ, In M-J, Kim RY: Optimization of protease immobilization by covalent binding using glutaraldehyde. *Appl Biochem Biotechnol* 1998, 73:195–204

303 Piller K, Daniel RM, Petach HH: Properties and stabilization of an extracellular β-glucosidase from the extremely thermophilic archaebacteria Thermococcus strain AN1: enzyme activity at 130°C. *Biochim Biophys Acta* 1996, 1292:197–205

304 Saito T, Yoshida Y, Kawashima K, Lin KH, Inagaki H, Maeda S, Kobayashi T:

Influence of aldehyde groups on the thermostability of an immobilized enzyme on an inorganic support. *Mater Sci Eng C* 1997, 5:149–152

305 Park SW, Lee JW, Hong SI, Kim SW: Enhancement of stability of gL-7-ACA acylase immobilized on silica gel modified by epoxide silanization. *Proc Biochem* 2003, 39:359–366

306 Guizan JM, Sabuquillo P, Fernandez-Lafuente R, Fernandez-Lorente G, Mateo C, Halling PJ, Kennedy D, Miyata E, Re D: Preparation of new lipases derivatives with high activity-stability in anhydrous media: adsorption on hydrophobic supports plus hydrophilization with polyethylenimine. *J Mol Catal B* 2001, 11:817–824

307 Fernandez-Lafuente R, Rodriguez V, Mateo C, Fernandez-Lorente G, Arminsen P, Sabuquillo P, Guizan JM: Stabilization of enzymes (d-amino acid oxidase) against hydrogen peroxide via immobilization and post-immobilization techniques. *J Mol Catal B* 1999, 7:173–179

308 Sudina EG, Samartsev MA, Golod MG, Dovbysh E: Study of the role of the molecular environment in chlorophyllase functioning during its immobilization on organic carriers. *Ukrainskii Biokhimicheskii Zhurnal* 1979, 51:404–408

309 Nitto Electric Industrial Co Ltd: Preparation of immobilized enzymes. Jpn Kokai Tokkyo Koho 1984, JP 82-154856

310 Tor R, Dror Y, Freeman A: Enzyme stabilization by bilayer "encagement". *Enzyme Microb Technol* 1989, 11:306–312

311 Abian O, Wilson L, Mateo C, Fernandez-Lorente G, Palomo JM, Fernandez-Lafuente R, Guizan JM, Re D, Tam A, Daminatti M: Preparation of artificial hyper-hydrophilic micro-environments (polymeric salts) surrounding enzyme molecules. New enzyme derivatives to be used in any reaction medium. *J Mol Catal B* 2002, 20:295–303

312 O'Neill SP, Wykes JR, Dunnill P, Lilly MD: Ultrafiltration-reactor system using a soluble immobilized enzyme. *Biotechnol Bioeng* 1971, 13:319–321

313 Strelzowa SA, Tolstogusow WB: Properties of proteins in complexes with acid polysaccharides and other polyelectrolytes. *Colloid Polym Sci* 1977, 255:1054–1066

314 Arasaratnam V, Galaev IY, Mattiasson B: Reversibly soluble biocatalyst: optimization of trypsin coupling to Eudragit S-100 and biocatalyst activity in soluble and precipitated forms. *Enzyme Microb Technol* 2000, 27:254–263

315 Matsuda K, Orii H, Hirata M, Kokufuta E: Construction of a biochemo-mechanical system using enzyme-loaded polyelectrolyte gels. *Polym Gel Network* 1994, 2:299–305

316 Germain P, Crichton, Robert R: Characterization of a chemically modified-amylase immobilized on porous silica. *J Chem Technol Biotechnol* 1988, 41:297–315

317 Lowe CR: Immobilized lipoamide dehydrogenase. 3. Preparation and properties of an immobilized polythiolated enzyme. *Eur J Biochem* 1977, 76:411–417

318 Nakhapetyan LA, Akparov VKh: Thermostability of soluble and immobilized subtilizins after their modification by dextrans and dextrins. *Enzyme Eng* 1980, 5:423–426

319 Matthijs G, Schacht E: Comparative study of methodologies for obtaining β-glucosidase immobilized on dextran-modified silica. *Enzyme Microb Technol* 1996, 19:601–605

320 Khare SK, Gupta MN: Preparation of concanavalin A-β-galactosidase conjugate and its application in lactose hydrolysis. *J Biosci* 1988, 13:47–54

321 Burteau N, Burton S, Crichton RR: Stabilization and immobilization of penicillin amidase. *FEBS Lett* 1989, 258:185–189

322 Ray RR, Jana SC, Nanda G: Biochemical approaches of increasing thermostability of β-amylase from *Bacillus megaterium* B6. *FEBS Lett* 1994, 356:30–32

323 Taniguchi M, Kobayashi M, Fujii M: Properties of a reversible soluble-insoluble cellulase and its application to repeated hydrolysis of crystalline cellulose. *Biotechnol Bioeng* 1989, 34:1092–1097

324 Dinnella C, Lanzarini G, Ercolessi P: Preparation and properties of an

immobilized soluble-insoluble pectin-lyase. *Proc Biochem* 1995, 30:151–157

325 Tyagi R, Roy I, Agarwal R, Gupta MN: Carbodiimide coupling of enzymes to the reversibly soluble insoluble polymer Eudragit S-100. *Biotechnol Appl Biochem* 1998, 28:201–206

326 Fujimura M, Mori T, Tosa T: Preparation and properties of soluble-insoluble immobilized proteases. *Biotechnol Bioeng* 1987, 29:747–752

327 Van Leemputten E, Horisberger M: Soluble-insoluble complex of trypsin, immobilized on acrolein–acrylic acid copolymer, *Biotechnol Bioeng* 1976, 18:587–590

328 Kumar A, Gupta MN: Immobilization of trypsin on an enteric polymer Eudragit S-100 for the biocatalysts of macromolecular substrate. *J Mol Catal B* 1998, 5:289–294

329 Bryjak J, Noworyta A: Kinetic behavior of penicillin acylase stabilized by poly(ethyleneimine). *Bioprocess Eng* 1995, 13:183–187

330 Cong L, Kaul R, Dissing U, Mattiasson B: A model study on Eudragit and polyethyleneimine as soluble carriers of α-amylase for repeated hydrolysis of starch. *J Biotechnol* 1995, 42:75–84

331 Chen JP, Lee JJ, Liu HS: Comparison of isoamylase immobilization to insoluble and temperature-sensitive reversibly soluble carriers, *Biotechnol Tech* 1997, 11:109–112

332 Ivanov AE, Edink E, Kumar, Galaer IY, Arendsen AF, Bruggink A, Mattiasson B, *Biotechnol Prog* 2003, 19:1167–1175

333 Chen JP, Chang KC: Immobilization of chitinase on a reversibly soluble-insoluble polymer for chitin hydrolysis. *J Chem Technol Biotechnol* 1994, 60: 133–140

334 Chen JP, Chen YC: Improvement of cell lysis activity of immobilized lysozyme with reversibly soluble-insoluble polymer as carrier. *Biotechnol Tech* 1996, 10:749–754

335 Chen SH, Yen YH, Wang CL, Wang SL: Reversible immobilization of lysozyme via coupling to reversibly soluble polymer. *Enzyme Microb Technol* 2003, 33:643–649

336 Chen JP, Chen YC: Preparations of immobilized lysozyme with reversibly soluble polymer for hydrolysis of microbial cells. *Bioresour Technol* 1997, 60:231–237

337 Hoshino K, Taniguchi M, Marumoto H, Fujii M: Repeated batch conversion of raw starch to ethanol using amylase immobilized on a reversible soluble-autoprecipitating carrier and flocculating yeast cells. *Agric Biol Chem* 1989, 53:1961–1967

338 Taniguchi M, Kobayashi M, Natsui K, Fujii M: Purification of staphylococcal protein A by affinity precipitation using a reversibly soluble-insoluble polymer with human IgG as a ligand. *J Ferment Bioeng* 1989, 68:32–36

339 Taniguchi M, Hoshino K, Watanabe K, Sugai K, Fujii M: Production of soluble sugar from cellulosic materials by repeated use of a reversibly soluble-autoprecipitating cellulase. *Biotechnol Bioeng* 1992, 39:287–292

340 Saito T, Yoshida Y, Kawashima K, Lin KH, Inagaki H, Maeda S, Kobayashi T: Influence of aldehyde groups on the thermostability of an immobilized enzyme on an inorganic support. *Mater Sci Eng* 1997, 5:149–152

341 Weetall HH, Vann WP, Pitcher WH Jr, Lee DD, Lee YY, Tsao G-T: Scale-up studies on immobilized, purified glucoamylase, covalently coupled to porous ceramic support. In: Weetall HH, Suzuki A (Eds) *Immobilized Enzyme Technology*, Plenum, New York, 1975, p 269

342 Takeuchi S, Omodaka I: Temperature responsive graft copolymers for immobilization of enzymes. *Macromol Chem* 1993, 194:1991–1999

343 Chen G, Hoffman AS: Synthesis of carboxylated poly(NIPAAm) oligomers and their application to form a thermoreversible polymer-enzyme conjugates. *J Biomater Sci, Polym Ed* 1994, 5: 371–382

344 Ding Z, Chen G, Hoffman AS: Synthesis and purification of thermally sensitive oligomer–enzyme conjugates of poly(N-isopropylacrylamide)–trypsin. *Bioconj Chem* 1996, 7:121–126

345 Matsukata M, Takei Y, Aoki T, Sanui K, Ogata N, Kikuchi A, Sakurai Y, Okano T: Effect of molecular architecture of poly(N-isopropylacrylamide)–trypsin conjugates on their solution and enzymatic properties. *Bioconj Chem* 1996, 7:96–101

346 Ding Z, Chen G, Hoffman AS: Unusual properties of thermally sensitive oligomer-enzyme conjugates of poly(N-isopropylacrylamide)-trypsin. *J Biomed Mater Res* 1998, 39:498–505

347 Chen JP, Hsu MS: Preparations and properties of temperature-sensitive poly(N-isopropylacrylamide)–chymotrypsin conjugates. *J Mol Catal B* 1997, 2:233–241

348 Kukhtin AV, Eremeev NL, Belyaeva EA, Kazanskaya NF: Relationship between state of a thermosensitive matrix and the activity of urease immobilized in it. *Biochemistry* 1997, 62:371–376

349 Park TG, Hoffman AS: Effect of temperature cycling on the activity and productivity of immobilized β-galactosidase in a thermally reversible hydrogel bead reactor. *Appl Biochem Biotechnol* 1988, 19:1–9

350 Park TG, Hoffman AS: Immobilization and characterization of β-galactosidase in thermally reversible hydrogel beads. *J Biomed Mater Res* 1990, 24:21–38

351 Kim HK, Park TG: Synthesis and characterization of thermally reversible bioconjugates composed of γ-chymotrypsin and poly(N-isopropylacrylamide-co-acrylamido-2-deoxy-d-glucose). *Enzyme Microb Technol* 1999, 25:31–37

352 Galaev IY, Mattiasson B: Affinity thermoprecipitation: contribution of the efficiency of ligand-protein interaction and access of the ligand. *Biotechnol Bioeng* 1993, 41:1101–1106

353 Takahashi K, Nishimura H, Yoshimoto T, Saito Y, Inada Y: A chemical modification to make horseradish peroxidase soluble and active in Benzene. *Biochim Biophys Res Commun* 1984, 121:261–265

354 Takahashi K, Yoshimoto T, Ajima A Matsushima A, Inada Y: Ester-Exchange catalysed by lipase modified with polyethylene glycol. *Biochem Biophys Res Commun* 1985, 131:532–536

355 Takahashi K, Yoshimoto T, Ajima A, Inada Y: Poly(ethylene glycol)-modified catalase exhibits unexpectedly high activity in benzene. *Biochem Biophys Res Commun* 1984, 125:761–766

356 Matsushima Y, Kodera M, Hiroto H, Nishimura H, Inada Y: Bioconjugates of proteins and polyethylene glycol: potent tools in biotechnological processes. *J Mol Catal B* 1996, 2:21–17

357 Bovara R, Ottolina G, Carrea G, Ferruti P, Veronese FM: Modification of lipase from *Pseudomonas* sp., with poly(acryloyl morpholine) and study of its catalytic properties in organic solvents. *Biotechnol Lett* 1994, 16:1069–1074

358 Baillergon MW, Sonnet PE: Polyethylene glycol modification of *Candida rugosa* lipase. *J Am Oil Chem Soc* 1988, 65:1812–1815

359 Inada Y, Takahashi K, Yoshimoto T, Ajima A, Matsushima A, Saito Y: Application of poly(ethylene glycol)-modified enzymes in biotechnological processes: organic solvent soluble enzymes, *Tibtech* 1986, 4: 190–194

360 Veronese FM, Mammucari C, Schiavon F, Schiavon O, Lora S, Secundo F, Chilin A, Guiotto A: Pegylated enzyme entrapped in poly(vinyl alcohol) hydrogel for biocatalytic application. *Farmaco* 2001, 56:541–547

361 Piro B, Dang LA, Pham MC, Fabiano S, Tran-Minh C: A glucose biosensor based on modified-enzyme incorporated within electropolymerized poly(3,4-ethylenedioxythiophene) (PEDT) films. *J Electroanal Chem* 2001, 512:101–109

362 Vakurov AV, Gladilin AK, Levashov AV, Khmelnitsky YL: Dry enzyme-polymer complexes: stable organosoluble biocatalysts for nonaqueous enzymology. *Biotechnol Lett* 1994, 16:175–178

363 Wirth P, Souppe J, Tritsch D, Biellmann JF: Chemical modification of horseradish peroxidase with ethanal-methoxypolyethylene glycol: solubility in organic solvents, activity and properties. *Bioorg Chem* 1991, 19:133–142

364 Souppe J, Urrutigoity M, Levesque G: Chemical modification of horseradish peroxidase and papain with α,ω-diamino-poly(oxyethylene)/carboxymethyl bis(dithiocarboxylate) poly-

365 Inada Y, Nishimura H, Takahashi K, Yoshimoto T, Saha AR, Saito Y: Ester synthesis catalysed by polyethyleneglycol modified lipase in benzene. *Biochem Biophys Res Commun* 1984, 122:845–850

366 Baillargeon MW, Sonnet PE: Lipase modified for solubility in organic solvents. *Ann N.Y. Acad Sci* 1988; 542:244–249

367 Secundo F, Carrea G, Vecchio G, Zambianchi F: Spectroscopic investigation of lipase from *Pseudomonas cepacia* solubilized in 1,4-dioxane by non-covalent complexation with methoxy-poly(ethylene glycol). *Biotechnol Bioeng* 1999, 64:624–629

368 Margolin AL, Sherstyuk SF, Izumrudov VA, Zezin AB, Kabanov VA: Enzymes in polyelectrolyte complexes. The effect of phase transition on thermal stability. *Eur J Biochem* 1985, 146:625–632

369 Hoshino K, Taniguchi M, Ueoka H, Ohkuwa M, Chida C, Morohashi S, Sasakura T: Repeated utilization of β-glucosidase immobilized on a reversibly soluble-insoluble polymer for hydrolysis of phloridzin as a model reaction producing a water-insoluble product. *J Ferm Bioeng* 1996, 82:253–258

370 Gololobov MY, Ilyashenko VM: Reversibly soluble enzyme-polymer conjugates. *PCT Int Appl* 2002, WO 0234902 A2

371 Kirsh YE, Galaev IY, Karaputadze TM, Margolin AL, Svedas V: Thermoprecipitating polyvinylcaprolactame–enzyme conjugate. *Bioteknologina* (Russian) 1987, 3:184–189

372 Margolin AL, Izumrudov VA, Svedas V, Zezin AB, Kabanov VA, Berezin IV: Preparation and properties of penicillin amidase immobilized in polyelectrolyte complexes. *Biochim Biophys Acta* 1981, 660:359–365

373 Ahmad S, Anwar A, Seleemunddin M: Immobilization and stabilization of invertase on Cajanus cajan lectin support, *Bioresour Technol* 2001, 79:121–127

374 Stellwagen E, Cronlund MN, Barnes LD: A thermostable enolase from the extreme thermophile thermos aquaticus YT-1. *Biochemistry* 1973, 8:1552–1559

375 Illanes A, Altamirano C, Zuniga ME: Thermal inactivation of immobilized penicillin acylase in the presence of substrate and products. *Biotechnol Bioeng* 1996, 50:609–616

376 Rosell CM, Fernandez-Lafuente R, Guizan JM: Modification of enzyme properties by the use of inhibitors during their stabilization by multipoint covalent attachment. *Biocatal Biotransform* 1995, 12:67–76

377 Wehjite E, Adlercreutz P, Mattiasion B: Improved activity retention of enzyme deposited on solid supports. *Biotechnol Bioeng* 1992, 41:171–180

378 Reetz MT, Zonta A, Simpelkamp J: Efficient immobilization of lipases by entrapment in hydrophobic sol–gel materials. *Biotechnol Bioeng* 1996, 49:527–534

379 Gill I, Ballesteros A: Encapsulation of biologicals within silicate, siloxane and hybrid sol–gel polymers: an efficient and generic approach. *J Am Chem Soc* 1998, 120:8587–8598

380 Blanco RM, Calvete JJ, Guizan JM: Immobilization-stabilization of enzymes: variables that control the intensity of the trypsin (amine) agarose (aldehyde) multi-point covalent attachment. *Enzyme Microb Technol* 1988, 11:353–359

381 Moreno JM, Fagain CO: Stabilization of alanine aminotransferase by consecutive modification and immobilization. *Biotechnol Lett* 1996, 18:51–56

382 Katchalski E, Goldstein L, Levin Y, Blumberg S: Water-insoluble enzyme derivatives. US 1972, US 3706633

383 Fernandez-Lafuente R, Guizan JM: Enzyme and protein engineering via immobilization and post-immobilization techniques. *Recent Res Devel Biotech Bioeng* 1998, 1:299–309

384 Mitchell DJ, Baker JO, Oh KK, Grohmann K, Himmel ME: Enhanced utility of polysaccharidases through chemical cross-linking and immobilization. Application to fungal β-d-glucosidase. *ACS Symposium Series* 1991, 460:137–151

385 Singh D, Goel R, Johri BN: Production of 6-aminopenicillanic acid through double entrapped *E. coli* NCIM 2563. *Curr Sci* 1988, 57:1229–1231

386 Hao Y, Andersson Maria V, Carmen G, Igor Y, Mattiasson B, Hatti-Kaul R: Stability properties of thermoresponsive poly(N-isopropylacrylamide)-trypsin conjugates. *Biocatal Biotransform* 2001, 19:341–359

387 López-Gallego F, Montes T, Fuentes M, Alonso N, Grazu V, Betancor L, Guisán JM, Fernández-Lafuente R: Improved stabilization of chemically aminated enzymes via multipoint covalent attachment on glyoxyl supports. *J Biotechnol* 2005, 116: 1–10

388 Kwon DY, Hong YJ, Yoon SH: Enantiomeric synthesis of (S)-2-methylbutanoic acid methyl ester, apple flavor, using lipases in organic solvent. *J Agri Food Chem* 2000, 48:524–530

389 Kroutil W, Orru RVA, Faber Kurt: Stabilization of Nocardia EH1 epoxide hydrolase by immobilization. *Biotechnol Lett* 1998, 20:373–377

390 Tosa T, Mori T, Fujimura M: Immobilized enzyme having reversible solubility. US-patent 4, 783, 409, 1988

Subject Index

A

Accurel EP 82, 103
Acetaldehyde 10, 30, 31, 89, 92
Acetate kinase 125
N-acetyl-l-tyrosine ethyl ester 64
Acrylic acid 142–144, 206, 240–244, 249–254, 264, 334, 342, 503, 522
Acryloxysuccinimide 248, 353, 527
Acrylonitrile 63, 238, 255–257, 286, 436, 437, 504, 521
Activation
– carbonyldiimidazole 272
– chloroformate 272
– cyanogen bromide 253
– epoxidation 269
– imidation 286
– interfacial 69, 73
– mesylation 269
– tosylation 269
– water-soluble carbodiimide 274
– Woodward reagent 274
Action pattern 94
Active site 27, 78, 95, 190, 197, 200–212, 219, 229, 277, 420
– accessibility of 462, 524
– amino acid residues near 199, 478
– availability of 12
– blockage of 462
– enlargement of 478
– entrance to 97
– essential amino acid residues close to 210, 467, 473
– geometry of, 69, 92, 97
– involvement of 188, 248
– modification of 28, 29, 288, 462, 488
– number of 34
– occupation of 113, 468
– presence of substrate in 482
– protection of the active centre 277
– size of 69–77
– titration of 9, 64, 68, 195

Activator 75, 91, 111, 112, 114, 202, 254, 274, 454
Activity
– additive-dependent 75, 328
– apparent 62
– binding density-controlled 80, 81, 210, 211
– binding nature-controlled 78
– carrier nature-dependent 74
– carrier-bound active group-controlled 196
– carrier-bound inert group-controlled 200
– carrier-dependent 137
– conformation-controlled 27, 68, 327
– conformational flexibility-controlled 213
– diffusion-controlled 67, 68, 211, 212, 326
– enzyme nature-dependent 75
– enzyme orientation-controlled 210
– expression of 60, 175, 181, 182
– hydrolytic 66, 68
– hydrophilicity-dependent 75
– loading-controlled activity 70, 72, 213, 325
– matrix-dependent 342
– medium-dependent 72
– microenvironment-dependent 72
– molecular orientation-controlled 28
– native 68
– orientation-determined 77, 78
– pore-size-dependent 81, 93, 194
– protection of 482
– reactive amino acid residue-controlled 211
– reactor-dependent 81
– retention of 177
– spacer-controlled 205
– specific 57, 62, 67, 82, 136, 171, 213
– substrate-controlled 69
– substrate-dependent 69
– synthetic 66, 68, 75, 82, 269
– volume 34, 62, 80, 171
– water-activity-dependent 82

Acyl amide   142, 200, 238–241, 248, 249, 257–259, 272–283, 321, 353, 409
Acyl azide   6, 12, 265, 277, 291
– conversion of amide to   283
– conversion of carboxylic acid to   273
– polymeric   186, 237
Acyl donor   32, 90
Acylase   9, 128
Additive
– as conformation inducer   19
– enhancement of carrier's coverage by   56
– in immobilization medium   26, 36
– modulation of enzyme activity by   72, 75, 137, 361
– modulation of enzyme selectivity   95–99
– modulation of enzyme stability   91
Adsorbent
– affinity   78, 86
– backbone of   119
– bio-affinity   113
– biospecific   134
– hydrophobic   21, 108
– immobilized metal   119
– ionic   5
– non-specific   99
– polysaccharide-based   78
– selective   1
– specific   99
Adsorption   5–9, 90, 116–119
– affinity   54
– behaviour   75, 81
– binding type of   97-145
– bio-specific   54, 113
– capacity   65, 80
– classification of   54
– conventional   97
– covalent   144
– direct   66
– hydrophobic   72, 108
– ionic   5, 78, 100
– mediated   124
– non-covalent   55–97, 188, 231
– non-specific physical   54, 99
– of enzyme on carriers   34
– physical   5, 6, 21, 75
– specific   6, 21, 72
– type of   65
– unconventional   120
Adsorption–deposition   56
Aerosol   279
Affinity
– bio-specific   90
– dye-based   90, 117, 118
– group-specific   113
– metal   116, 119–121
Agarose
– aldehyde   219, 284, 290–292, 517
– butyl   75
– cyanogen bromide-activated   23, 210
– epoxy-activated   23
– glutaraldehyde-activated   28, 218
– glyoxal   30
– octyl   93–97
– oxidized   287
– thiolsulfinated   28, 218
– p-toluenesulphonyl chloride-activated   269
– tosylated   33
Aldehyde   12, 285, 292
– conversion to amino group   292
– conversion to carboxyl group   292
– conversion to diazonium salt   293
Aldehyde dextran   208
Aldehyde oxidase   82
Alginate   33
Alkaline phosphatase   84
Alkylation   192, 217, 477, 481
Alkylsulphatase   30, 94
Alumina   5, 124
Amberlite   6, 28, 40, 101
Amberlite CG-50   65, 101, 278
Amberlite DP-1   67, 108
Amberlite IRA 900   130, 498
Amberlite IRA-904   101
Amberlite IRA 938   71, 272
Amberlite IRC 50   101, 137
Amberlite XAD 4   22, 82, 129, 130, 498
Amberlite XAD-7   82
Amberlite XE-97   5
Amidation   192
Amidohydrolase   134, 351
Amino acid
– as ligand   122
– chiral   21
– essential   210, 289
– N-terminal   190
Amino acid ester   21, 29
Amino acid oxidase   7
Amino acid residues
– active   169–192
– buried   193
– charged   100
– covalent modification of   77
– essential   77
– exposed   192
– for covalent binding   190
– hydrophilic   192, 193
– hydrophobic   192, 193

– reactive 190, 191
– reactivity of 191
Amino acylase 7, 20, 21, 53, 66, 78, 128, 144, 223
6-Aminopenicillanic acid 14, 22
Amoxicillin 244
Ampicillin 244
AMP deaminase 5
Amylase
– α-amylase 5, 8, 9, 20, 30, 64, 80
– Taka-amylase 225
Amyloglucosidse
Aspergillus niger 71, 84
Analogue
– of a covalently bound immobilized enzyme 323
– substrate 98, 113, 332
– transition state 19
Anhydride, polymeric 186, 238–240
Antibiotic
– enzymatic synthesis of semi-synthetic antibiotics 170
– kinetically controlled synthesis of 31, 32, 232–234
– nucleus of semi-synthetic b-lactam 22
– semi-synthetic b-lactam antibiotics 64, 241, 488
– synthesis of semi-synthetic b-lactam 4, 14, 19
Antibody 113, 465
– carrier-bound 462
– immobilized 116
– monoclonal antibody 114, 115
Aphron 421
Aquaphilicity
– activity influenced by 35, 75
– definition of 76, 77
– enzyme binding influenced by 143
– introduction of 15, 16
– selectivity influenced by 228–235
Arm
– PEG 352
– polypeptide 352
Artificial cell 398
Arylation 186, 192
Aspartase 110
Assembly, layer-by-layer 405, 429
Attachment
– adsorptive covalent 142
– carrier-bound multipoint 214–217
– complementary multipoint 214
– covalent 5, 14, 68, 270, 333
– multipoint 169, 170, 214, 218, 219, 224, 232, 278

– number of 233
– "over-dosed" multipoint 189, 196–200, 211, 218
– single-point 218, 222, 232
Azalactone 15, 23, 237, 252

B
Backbone 61, 98, 187, 197
– hydrophobic 199
– nature of 66
– of carrier 183
Barrier
– membrane-like physical 397
– physical 319
Bead
– agarose 369
– calcium alginate gel 131
– chitosan 206
– DEAE-cellulose 64
– phenyl-sepharose 84
– polyacrylamide 86
– synthetic polymeric 364
– titania 362
Bentonite, enzyme adsorbed on 86, 131, 142
Benzaldehyde 80
– *p*-Benzoquinone 197, 256, 265, 271, 282
Binding
– affinity 9
– chemistry of 7, 23
– coordination 9, 256
– covalent 7, 169 f
– mode of 9, 12, 26, 28, 169, 192
– non-covalent 63, 318
– non-specific 54
– position of 25, 26
Binding chemistry 170, 233
Binding density 59–63, 73, 74, 80, 215, 216
Binding functionality
– active 176, 186, 235
– concentration of 73, 176
– density of 63, 64
– enzyme activity affected by 71, 78
– enzyme orientation affected by 77
– enzyme selectivity affected by 94
– enzyme stability affected by 82, 91
– nature of 65, 78
– specific 55
Binding mode
– definition of 189
– enzyme stability affected by 145, 217
– heterogeneity of 191
– multipoint 260
– number of binding 230

– orientation   230
Binding sites
– non-specific   54
– number of   63
– specific   81
Bioaffinity
– antibody-antigen based   113
– group-specific   113
Biocatalyst   1
Biodegradability   236
Biogel CM-100   274
Bio-immobilization   171, 188
Bio-separation   188, 236
Biosensor   322
Blocking agent   222, 223, 227
Bond
– nature of   55
– number of   169
Bovine serum albumin   190, 275, 285
Bromoperoxidase   110

C
Capsule
– liposome based   416
– macroporous   402
– microporous   402
– pre-designed   405, 429
N-Carbamyl-D-amino acid amido hydrolase   137
Carbodiimide, water-soluble   274, 275
Carbonic anhydrase, bovine   123 128, 326
Carbonyldiimidazole   272
Carrier
– aminoalkylated   289
– aminolated   186, 279, 291
– beaded   322
– benzoquinone-activated aminolated   282
– cellulose-based   264
– chemical nature of   65
– chlorotriazine-activated   280
– cyanogen bromide-activated   198, 199, 207
– epoxy synthetic   197
– glycophase-coated   187
– hydrophilic   68, 76
– hydrophobic   68, 84, 232, 324
– inorganic   85
– internal structure of   63
– macroporous   243, 254, 321
– mesoporous   243
– non-porous   180, 211, 212
– non-specific   98
– PEI-coated   136
– polyacrylic acid ester   251

– polyamide   284
– polysaccharide-based   232
– polyvinylalcohol-based   89
– porous   180
– predesigned   184, 235
– prefabricated   364
– ready-made   184, 200, 227, 236, 250, 333
– silanized   187
Casein   124
Catalase   118
CD-spectrum   57
Celite   66, 74, 82
Celite Hyflo-supercel   68, 76
Cephalosporin acetylesterase   86
Cetyl trimethylammonium bromide   94
Charcoal   53
– granule, activated   436
Chelating support   476
Chelating ligand   119
Chelation   90
China clay   134
Chitopearl   137
Chlorofomate   272
Chloroperoxidase   30
Cholesterol oxidase   325
Choline oxidase   131
Chromatography
– affinity   188
– immobilized metal affinity   116
α-Chymotrypsin   60, 64, 78, 85, 124, 142
Cibacron blue   90, 116
Cluster
– hydrophilic   222
– hydrophobic   123, 196, 198, 203, 210
CM-cellulose   204, 277
Conformation selector   327
Coacervation   423
– interfacial   424
– phase inversion   423
Coating   184, 402
– combination with encapsulation   402
– enzymatically gelified   456
– enzyme immobilization by   450–458
– enzyme-polymer complex   456
– ligand-mediated   455
– monolayer   451
– multiple layer   452
– non-covalent   123
– phase-inversion   451
– sol–gel supported   371, 457
Confinement   330
– imprinting by   484
– molecular   225
– stabilization by   486

Configuration, reactor 171, 180
Combi-methods 14, 334
Complexation 400
– interfacial 400
Composite, 237
Compressibility, adiabatic 421
Condensation 409
– interfacial 409
Conformation
– denatured 60
– freezing of 170
– open 70
Conformation change 221
Conjugate, enzyme-polymer 240
Coordination 54
Co-polymerisation 244, 352
Core decomposition 405
Criteria for robust immobilized enzyme 3
Cross-linking
– by light 336, 339
– chemical 429
– glutaraldehyde aided 85
– interfacial 400
– stabilization by 490, 493
Cross-linker, bifunctional 190, 269, 277
Cryogel 344
Crystal, cross-linked enzyme 18, 97, 135
Crystalline enzyme 7
Crown ether 95, 332
Cut-off 403
Cutinase 60, 103
Cyanogen bromide 7, 263
Cyanuric chloride
– activation of amino group by 258
– activation of hydroxyl group by 258, 270
Cyclohexane 398
Cystamine 356
Cytochrome C 105

D

Deactivation, protein 363
DEAE-cellulose 90, 205
DEAE-Sephadex 90
Dehydration 510
Dehydration history 26
Density of the binding functionality 111, 171
Denaturation
– heat-induced 123
– pH-induced 123
– reversible 123
Deposition 371, 399
– alternating 452
– interfacial 400

Dextran
– aldehyde 190
– amino 190
– oxdized 288
Dextransucrase 57
Diaion CR 20, 279
Diazotization 192, 288
Diazonium salt 5,196, 199, 248,265–268
Di-cross-linker 504
Dicyclohexylcarbodiimide 291
3,5-difluoronitrobenzene 86
Diffusion constraints 57, 178, 326, 364, 406
Diffusion, inter-particle 65
Diffusion limitation 176, 213, 320, 331
Diisopropylflupro phosphate 64
Bis-dimethyladipimidate 86
Dioxane 240
Disperant 257
Distribution
– amino acid residues in enzyme 192
– of enzyme in carrier 60
Divinylsulphone 270
Duolite resin
– Duolite 568 82
– Duolite 761 66, 68, 82
– Duolite A-147 213
– Duolite A7 89
– Duolite S-30 248
– Duolite S562 82
– Duolite XAD 761 66, 82
Dowex 1X4 101
Drug delivery 363, 403, 416

E

Effect
– chemical modification 26
– confinement 329
– crowding 329
– hydration 329
– interfacial activation 73
– microenvironment 27, 73, 80
– molecular orientation 77
– multipoint attachment 211
– negative partitioning 204
– of binding functionality 78
– positive partitioning 26, 204
– shielding 131
– tentacle 136
Effector 19
Electrode 7, 322, 374
Emulsion, double 423–429
Enantioselectivity
– conformation-controlled 230, 331–333

- diffusion-controlled 230
Encagement 377, 434–436
Encapsulation 397, 434
- conventional 399
- definition of 397
- interfacial 400
- non-conventional 401, 430
- post-loading 429
Enzyme
- carrier-bound 6 ff, 77, 510 ff
- cell-free 16
- covalently bound 36, 113
- coverage 71
- cross-linked 7, 18, 181
- crystalline 7
- glycosylated 190, 290
- hard 57
- insoluble 1
- lipid coated 369
- plastic 10
- pre-immobilized 181
- soft 57
- sol–gel entrapped 333
- sphere granule 351
- supported so–gel entrapped 376
- whole cell associated 401
Enzacryl AH 283
Enzyme aggregate 181, 485
- cross-linked 17
Enzyme conformation 184
- destabilized 59
- native 59
- selective 232
Enzyme crystal 8, 181, 485
Enzyme deformation 85–87
Enzyme distribution 60
Enzyme envelope 7
Enzyme immobilization, site-specific 193
Enzyme loading 71, 242
- minimum 56
- relationship with diffusion limitation 71
Enzyme orientation 170
Engineering
- carrier 183
- genetic 23
- medium 92, 93
- of microenvironment 25
- process 91
- substrate 91
Entrapment 317
- classic 334
- concomitant 334
- conventional 334
- covalent 351

- definition 319
- diffusion 362
- diffusion-controlled 363
- impregnation 364
- in situ 376
- of cross-linked enzyme 371
- of polymer–enzyme covalent conjugates 368
- of polymer–enzyme non-covalent complex 371
- post-loading 362
- selective protein partition 363
- supported 371, 372
Entrapment method 319
Environment 225
- compatible 231
- hydrophilic 225
Epichlorohydrin 254
Epoxide hydrolase 110
Erythrocyte haemolysate 398
ESR spectroscopy 69, 189
Eupergit C 63, 85, 175, 186, 195, 210, 290
Eupergit C 250 L 175, 195

F
Fiber 369
- cellulose 263
- hollow 290, 405
- non-woven 372
Filament 436
Film 369
- hydrophobic polymer 369
- silicon rubber 64
- supported enzyme 371
Flexibility, conformational 189, 205, 206, 216–218, 226, 333
Foam, polyurethane 371
Fractogel 128
Fructose-1,6-bisphosphatase 30
Function
- catalytic 1–3, 320, 333
- non-catalytic 1–3, 320, 333
Functionality
- active 169, 170, 183, 248
- aldehyde 275, 288
- amino 264
- built-in intrinsic 237
- hydroxyl 253
- inert 170, 236, 324
- non-active 170, 183
- oxirane (epoxy) 265, 288
- pendant 172, 183
- photo-crosslinkable terminal 369
- polycarbonate 245–247

Subject Index | 557

G

β-Galactosidase 68, 217
Endo-d-galacturonase 206
Gel
– Ca alginate 365
– CM-agarose 81
– cold-set 344
– glycidyl methylacrylate-derived dextran 363
– hybrid 343
– hydrophobic 323
– polyacrylamide 329
– synthetic 321
Gelation
– by action of temperature 344
– by solvent removal 351
– chemical 320
– interfacial 400, 403
– physical 320, 372
Geometry
– of carrier 182
– of the active centre 94–97
– pore 171
Glass
– controlled pore 178
– cyanogen bromide activated 281
– dextran-coated porous 262
– glycophase porous 75
– hydrophobic controlled pore 81
– plate 372
– polyethyleneimine-coated non-porous 86, 137
– polymethylglutamate (PMG)-coated 124
Glucoamylase 30, 86
Glucose isomerase 64, 81, 83
Glucose oxidase 225, 249, 259, 267, 275, 280, 284, 319
β-Glucosidase 67, 207, 208
Glutamate dehydrogenase 83
Glutaraldehyde
– as activating agent for inert carrier 277
– as cross-linker 7
Glycerol carbonate 200
Glycoenzyme 235, 284
Glycosylation 114
Graft copolymer 284
Grafting 184, 236
Granule 438
Group
– aldehyde 249, 290
– amino 277
– carboxyl 281
– carrier-bound active 183–185
– carrier-bound inert 183–187

– epoxy 186, 241–244
– inert pendant 203
– isocyanate 199
– orientation 330

H

Hemoglobin 78
Heterogeneity of binding nature 192
Heteropolymer 250
1,6-Hexanediamine 398
Homopolymer 247, 377
Horse liver alcohol dehydrogenase 69, 93
Horseradish peroxidase 57, 85
Host 376
Hydration 405
Hydrogel 80, 259
– degradable 259
Hydrolysis
– mandelic acid methyl ester 95
– of penicillin G 134
– of sec-alkyl sulphates 94
Hydrophilicity 7, 20, 65, 75
– enzyme activity influenced by 75, 76
– enzyme loading influenced by 65
– of the carrier 7, 324, 325
– tailor-made 12
Hydrophilization
– of the carrier 233
– of the enzyme 34, 227
Hydrophobicity 238, 326
Hydrophobic core 192
Hydroxyl group
– separate 268
– vicinal 190

I

Imidoester 282
Imprinting 19, 332
Immobilization
– adsorption-based 53–145
– biospecific 75, 80
– double 128–145, 330
– heterogeneous 193
– of enzyme, covalent 169
– of enzyme in organic medium 170
– pseudo-affinity 113
– pseudo-covalent 123
– reversible denaturation 123
– site-specific 113, 193
Immobilized enzyme
– carrier-bound 173
– industrial 244
– plastic 352
– ready-made 227

- robust 3
- spacer-mediated 233
Impregnation 333
Indion 48-R 278
Indion 850 74
Interaction
- biospecific 78
- charge-charge 73, 87
- cooperative 91
- hydrophobic 83
- incompatible 204
- multivalent 91
- non-specific 80
- polymer-metal oxide 346
- polymer–polymer 346
- polymer–salt 346
- specific non-covalent 90
- synergetic interaction 67
Interconversion 202
- of active carrier 288–293
- of inert carrier 260–287
Invertase 5
Ionic strength 72
Irradiation 336
Isocyanate 283
- polymeric 186
Isoelectric point 57
Isothiourea 279
Isourea 279

L

Laccase 205, 405
β-Lactamase 199
Lactase 113, 114, 124, 345, 372
Lactose 370
Langmuir film 76
Layer
- dense 364
- partition 178
Layer-by-layer 15, 405
Leakage 86, 103, 121 f
Leaving group 201, 224, 247
Lewait 258, 278
Ligand
- biospecific 137
- metal chelating ligand 119
Linkage, guanidine 281
Lipase 204
- *Bacillus stearothermophilus* 227, 228
- *Candida antarctica* B 30, 70, 94, 326
- *Candida rugosa* 30, 70, 73, 89, 92, 95, 200, 202, 319, 326, 328, 466, 480, 485
- *Chromobacterium viscosum* 366
- *Mucor miehei* 30, 74, 81, 228

- *Nigellaa sativa* seed 72
- porcine pancreas 76
- *Pseudomonas cepacia* 31, 365
- *Pseudomonas fluorescens* 68, 93, 206, 230, 332, 365
- *Rhizopus niveus* 66, 74, 76, 82
- *Rhizopus oryzae* 68, 92
- *Staphylococcus carnosus* 226
Lipid
- complex 26
- lipid-coated enzyme 368, 408
- pairs 14
Liposome
- capsules based on 416
- enzyme immobilized on 124–128
- polymeric 416
Loading 211–213
Luciferase 76, 127
Lysozyme 79, 118

M

2-Mannosidase 30
Mass transfer 194
Matrix
- bovine serum albumin 365
- charged 240
- film-like polymer 371
- hydrophobic 324
- hydrophobic sol-gel 326
- peo 328
- predesigned 405
- pva 326, 331
- ready-made gel 365
- rigid sol-gel 327
- soft polymeric 328
- sol-gel 325, 329
- synthetic gel 330
Mediator 128
Medium aqueous 11
Medium enzyme immobilization in organic 11, 477
Membrane
- hollow fiber 136
- hollow sphere 182
- PAN 287
- ready-made hollow sphere 429
- semi-permeable 401
- spherical 397, 409, 412
Mesylate 269
Method
- benzoquinone 12, 271
- carbodiimide 274
- combined 20
- divinylsulphone 270

– generic 21
– glutaraldehyde 276
– hybrid 134
– s-triazine 270, 277
– Ugi 287
Micelle
– encapsulation of enzyme 16, 404
– polymeric 416
– preparation 416
Microbead 136, 137
Microemulsion
– encapsulation of enzyme in 412
– oil-in-water 421
Microdevice 405
Microenvironment
– favourable 228
– hydrophilic 21, 227
– non-compatible 189
– of the carrier 8–10, 63, 64, 73, 187, 199–201, 211, 260
– of the enzyme 29, 55, 183
– of the medium 331
Microencapsulation 398
Microcapsule
– magnetic 281
– non-swellable 405
– polyelectrolyte-coated 432
– soft 405
– swellable 405
Microsphere
– affinity 98
– hydrogel 320
– liquid hollow 412
– monosized 208
– plain 207
– polyacrolein microspheres 250
– polystyrene 7, 76
– polyvinylalcohol 272
– soft hollow microsphere 19
Milk xanthine oxidase 68
Mitochondria 325
Mobility
– conformational 233
– molecular 189, 224, 228, 343
– of immobilized enzyme molecule 175
Modification
– chemical 184
– controlled 202, 228
– no chemical 53
– of protein surface 128
– post-immobilization 184, 227, 506–522
– pre-immobilization 430, 490–493
Modifier 129
– chemical 223

– physical 223
Monomer 200
– hydrophilic 200
– hydrophobic 239
Monolayer
– coverage 57
– hydrophobic 344
– macromolecular 338
– occupation 174
– organic silane 377
Montmorillonite 78
Morphology 194
Moving front theory 60
Multilayer 70, 136

N

Nanocapsules 421
Nylon
– fiber 469
– hydrolyzed 86, 134, 138, 278, 284
– modified 212
Nitrophenol 251
p-Nitrophenyl ester 251
1,2-Naphthoquionone 282

O

Occupation
– active centre 113
– carrier's surface 173, 519
– monolayer 173
Organic phase 399
Organophosphorus hydrolase 79
Orientation
– mismatched 189
– random 458
– site-specific 91, 113, 458
Osmolyte 223
Oxidase
– amino acid 204
– glucose 75, 90, 131, 180, 227, 259, 330
– glycolate 330
– lactate 330
Oxirane ring 12, 203, 242, 289
Oxygen 78
Ovalbumin 425

P

Papain 105, 173
Particle size 64, 65, 81
Payload 34, 173, 247
Pectin 342
Pectin esterase 103
Pectinase 85

PEG   131, 227
PEGylation   499, 528
Penicillin G acylase   4, 7, 9, 14, 20, 26, 31, 74, 79, 105, 112, 114, 124, 127, 135, 143, 144, 172–176, 179, 194–197, 200–205, 211, 213, 221, 232–235, 240–245, 248, 254–257, 371
Penicillin V acylase   131, 132, 227
Penetration of enzyme molecules in the carrier   67–70, 179, 180
Peptide formation   192
Permeability, substrate   402
Phase inversion   401
Phosgene   279
Phospholipid, polymerizable   416
Polarization phenomena   328
Polyacid   346
Polyacrylamide   254, 255
Polyacylazide   185
Polyaldehyde   185, 249–251
Polyaminostyrene   278
Polyanhydride   185
Polyaphrons   421
Polyazlactone   185
Polybase   346
Polycarbonate   185, 245–247
Polycarboxylic acid phenyl ester   185
Polyepoxide   185
Polyelectrolyte, non-biodegradable   429
Polyethyleneimine   364
Polyethylene terephthalate   84
Polyisocyanate   5, 185
Polyisothiocyanate   7
Polymer
– natural   376, 520
– semi-synthetic   346
– synthetic   376, 520
Polynitrile   257
Polyphenolic polymer   249
Polysaccharide
– cyanogen bromide activated   264, 288
– linear   258
– natural   342
– oxidized   186
– semi-synthetic   258
Polystyrene   85
Polyurethane
– foam   372
– porous   141
– prepolymer   336
– sheet   141
Polyvinylalcohol   257, 261, 319
Pore
– closed   321

– dead-ended pore   173
– open   321
– permanent   321
– wall pore   430
Pore diameter   179, 201
Pore distribution   171, 175–178
Pore radius   171
Pore size   61–63, 81, 173, 174
– distribution   177
– mean   81
– minimum   178
Pore volume   171–177, 201
Porosity   177, 194, 328, 333
Polymerisation
– dispersion   246
– interfacial   400, 401
– precipitation   236, 255
– suspension   200, 240, 255
Post-synthesis   74, 247
Post-treatment   245, 368, 506–522
Pre-carrier   253–268
Precursor
– gel   344
– gel-forming   320
Pregel   334
Pre-immobilization   430, 490–493
Prepolymer   325, 336
– cross-linking of   336
Pressure drop   180
Pricon brown   116
Pricon green   116
Pricon red   116
Pricon yellow   116
Principle
– monolayer   56, 87
– stability   56
Product map   31, 32, 234
Productivity   1, 22
Prostaglandin synthetase   105
Protease   30, 31, 85, 131
Protein drug   1
Pseudo-immobilization   171
Porogen   257
Pseudo-affinity
Pullulan   274

Q
QAE-Sephadex   101
Quenching agent   227, 243, 520

R
Reaction medium   2, 172
Reaction system   19, 22, 181
Reaction type   172

Subject Index

Reactor
- column 2
- membrane 359
- monolithic 252, 253
- plug-flow 2
- shallow-packed bed 81
- sieve-plate 16, 323
- stir-tank 2
- stirred batch 81

Reactor configuration 2, 180, 183
Recyclability 3
Regioselectivity 29, 92, 230
Rehydration 511, 514
Release, controlled 320
Resin
- anion-exchange 68
- epoxy-amine 338
- phenol-formaldehyde 248
- polyethyleneimine-coated 97

Resin plate 364
Reversibility
- of adsorption 113
- of binding 11, 114, 471

Rigidification
- by crosslinking 34, 490
- by multipoint attachment 26, 72
- lower activity caused by 18, 210–215, 232, 458
- stabilized by 343, 352

Rigidity
- carrier 125, 362
- enzyme conformation 211, 345

S

Scaffold
- enzyme 15, 34
- insoluble carrier 9, 34, 235
- ligand 115, 470

Schiff base formation 192, 217–219, 235, 250, 268
SDS 422
Sebacoyl chloride 398
Selectivity
- additive-controlled 95
- aquaphilicity-controlled 233
- binding functionality-controlled 94
- carrier nature-dependent 333
- carrier-bound active group-controlled 228
- carrier-bound inert group-controlled 230
- conformation-controlled 92, 97, 233
- definition of 29
- diffusion-controlled 93, 233
- functional group 230
- medium-controlled 95

- microenvironment-dependent 331
- orientation-controlled 95
- reaction 92, 230, 235
- spacer-controlled 232, 233
- substrate 92, 230
- water activity-dependent 332

Sepabead 95, 186, 377
Sephadex 59, 67, 86, 95
Sephadex G-25 67
Sepharose
- cyanogen bromide-activated 69, 71
- n-hexyl-substituted 83, 84
- n-octylamine-substituted 68, 83, 90
- octyl 69, 93, 124, 128
- palmityl-substituted 83, 123
- phenyl 84

Shape
- irregular 182
- regular 182

Shrinking core theory 60
Size
- pore 179
- protein globule 179

Silica
- aldehyde 234
- aminopropyl 234
- colloidal 174
- controlled pore 19, 266
- macroporous 74
- mesoporous 73
- octyl 70
- polyethyleneimine-coated 66, 86
- silanized 79
- $TiCl_4$-activated 271
- tosylated 269

Softanol 30, 422
Sol–gel
- classification of sol-gel precursor 360
- conventional process 355
- hydrophobic 358
- precursor 361

Solvent
- organic
- pore-generating 177

Spacer
- activity influenced by 205, 206
- albumin 208
- amino dextran 208
- definition of 188–190
- hydrophilic 189, 225, 227, 231
- hydrophilic macromolecular 225
- hydrophobic 189, 210, 221
- length of 170, 208
- role of 66

Spheron   206, 300
Spherosil   66, 81
Spherosil XOA 200   68
Stability
– additive-dependent   91
– binding density-controlled   91
– carrier nature-controlled   86
– carrier-bound active group-controlled   217
– carrier-bound inert group-controlled   221
– confinement-controlled   85, 225, 329
– conformation-controlled   83
– cross-linking-dependent   85
– diffusion-controlled   85
– enzyme loading-dependent   85, 87
– enzyme orientation-dependent   91
– enzyme structure-dependent   331
– enzyme-dependent   91, 330
– matrix-nature-dependent   329
– mechanical   364, 377
– medium-controlled   87
– microenvironment-controlled   89, 277
– operational   377
– temperature-dependent   87
Strategy
– adsorption–crosslinking   136
– adsorption–deposition   56
– double immobilization   215
– immobilization–stabilisation   215
– post-immobilization   214
– pre-immobilization stabilization   20
– stabilization–immobilization   17, 490
Structure
– hyper-activated   97
– open   97
– sandwich   436
Stereoselectivity   92, 230
Steric hindrance   69, 188, 196, 205–210, 224, 264, 481
Subtilisin   30, 58
Subtilisin BPN'   91
Subtilisin Carlsberg   196, 291
Subtilisin Novo   291
Sugar   223, 327, 356, 423
Sulphurtransferase   118
Support
– chitosan-based   135
– polyacrylamide-based   225
– polystyrene-based   225
– silica-based   513
Surface
– accessible   174, 179
– available   61
– external   172–174, 195
– hydrophobic   228
– internal   172–174
– outer surface   195
– solid   58
– specific   194, 201, 244
Surface area   61, 171, 174
Surface occupation   176
Surface overage   57
Surface utility   175
Surfactant   336, 415
– anionic   415
– cationic   415
– non-ionic   415
Swelling factor   239, 247, 258
Symplex   349–351
Synperonic A   20, 422
Synthesis
– asymmetric   29
– solid-to-solid   181

T

Tag
– chemically introduced   117, 503
– genetically engineered   16, 117, 473
Technique
– activation   185
– derivatization   228
– entrapment technique   319
– grafting   243
– interconversion   185, 202, 270
– modification-immobilization   9
– post-immobilization   222
– template-leaching   327
– water-soluble carbodiimide   275
Template
– activator   359
– as conformer selector   31, 481–486
– for microcapsule   405
– pore generating   325, 328, 357
Thermostability
– cross-linking dependent   21
– matrix-dependent   329
Thermolysin   206
Thiolphosgene   279, 280
Toluene   238, 249
Tortuosity   63, 123, 171, 194
Tosylation   15, 230, 268, 271
Toyonite   28, 82
Transglutaminase   134, 336, 456
Tresyl chloride   220, 369
Triazine   220, 265, 280
Trisacyl gel   272
Triton X-100   19

α-Trypsin 56, 57, 65, 72–77, 123, 173–178, 183
Tween 80 96
Tyrosinase 83

U
Uigi reaction 192, 284, 287
Ultrasonication 515
Urease 104, 118
Urokinase 30

V
Vesicle 416
Void of the matrix 368
Volume activity 34, 61, 67, 74, 78, 171, 323

W
Water activity 3, 82
Water content 241
Water partition 27
Water regain capacity 3
Water retention capacity 216
Wofatit Y 56, 515
Woodward reagent K 7, 202, 274

X
Xanthine oxidase 112
Xerogel 285
Xylene diamine 249, 438, 522
D-Xylose isomerase 110

Y
Yeast alcohol dehydrogenase 84, 87, 129, 320, 336, 498
Yeast cell 324, 325, 366
Yeast hexokinase 474
Yeast microsome 345
Yeast mitochondrial alcohol 519
Yeast RNA 283
Yield, space-time 22

Z
Zero cross-linking 277
Zeolite
– FSM-16 105, 133, 226
– hydrophobic type Y 104, 105
– MCM-41 85, 104, 376
– MCM-48 105
– NaY 102